Meteorology of Tropical West Africa

# Meteorology of Tropical West Africa

## The Forecasters' Handbook

*Edited by*
*Douglas J. Parker*
*Mariane Diop-Kane*

WILEY Blackwell

This edition first published 2017 © 2017 by John Wiley & Sons Ltd

*Registered Office*
John Wiley & Sons Ltd, The Atrium, Southern Gate, Chichester, West Sussex, PO19 8SQ, UK

*Editorial Offices*
9600 Garsington Road, Oxford, OX4 2DQ, UK
The Atrium, Southern Gate, Chichester, West Sussex, PO19 8SQ, UK
111 River Street, Hoboken, NJ 07030-5774, USA

For details of our global editorial offices, for customer services and for information about how to apply for permission to reuse the copyright material in this book please see our website at www.wiley.com/wiley-blackwell.



*Library of Congress Cataloging-in-Publication data applied for*

ISBN: 9781118391303

A catalogue record for this book is available from the British Library.

Wiley also publishes its books in a variety of electronic formats. Some content that appears in print may not be available in electronic books.

Cover image: Françoise Guichard
Cover design: Wiley

Set in 10/12pt Warnock by SPi Global, Pondicherry, India

10 9 8 7 6 5 4 3 2 1

# Contents

# Contributors

*Nicolas Chapelon*
CISMF, Météo-France
Toulouse
France

*Andrew Colman*
Met Office Hadley Centre
Exeter
UK

*Rosalind Cornforth*
University of Reading
Reading
UK

*Andreas H. Fink*
Karlsruhe Institute of Technology
Karlsruhe
Germany

*Peter Knippertz*
Karlsruhe Institute of Technology
Karlsruhe
Germany

*Jean Philippe Lafore*
CNRM, Météo-France and CNRS
Toulouse
France

*Arlene Laing*
Cooperative Institute for Research in the Atmosphere (CIRA)
Colorado State University
Fort Collins, Colorado
USA

*Sean Milton*
Met Office
Exeter
UK

*Zilore Mumba*
University of Zambia
Lusaka
Zambia

*Douglas J. Parker*
University of Leeds
Leeds
UK

*Rita D. Roberts*
National Center for Atmospheric Research
Boulder, CO
USA

*Wassila M. Thiawa*
Climate Prediction Center, NCEP, NOAA
College Park, MD
USA

*James W. Wilson*
National Center for Atmospheric Research
Boulder, CO
USA

## Foreword

This book represents the first effort to document modern weather forecasting science for a specific region of the world. As the Secretary-General of the World Meteorological Organization (WMO), I am particularly pleased to have seen this book evolve as a collaboration between operational forecasters, numerical modelling agencies and the academic community. This partnership, which has its roots in the African Monsoon Multidisciplinary Analyses (AMMA) international programme, has led to a fusion of material and ideas in the book, bringing together theory and practice.

The book documents scientific knowledge about the West African weather systems and summarizes the principles and practice of operational prediction. The material has been organized in such a way to take the reader seamlessly from the regional climate, through a study of the synoptic and mesoscale weather systems, to practical forecasting methods on timescales of hours, days, weeks and a whole season. Modern weather forecasters need to take into account all these timescales when delivering their prediction services. New tools including kilometre-scale prediction models and new satellite products have the potential to assist forecasters in their delivery of forecast information to address user requirements.

WMO was one of the partners in the preparation of this book, through its Education and Training Office, the Data-processing and Forecasting System Division of the Weather and Disaster Risk Reduction Services Department and the World Weather Research Programme within the Atmospheric Research and Environment Branch. In addressing the needs of the meteorological community in West Africa, the book makes a significant contribution to the WMO Capacity Development Programme.

Moreover, natural disasters have been well documented in West Africa, and it is expected that this book will improve the ability of the National Meteorological and Hydrological Services (NMHSs) to deliver accurate hazard information for risk assessments, prevention, response and recovery. The publication of this book is also timely, as in 2017 WMO initiates the implementation of the well-known Severe Weather Forecasting Demonstration Project (SWFDP) in the West African region. I expect this book to be the primary scientific reference for training in severe weather forecasting associated with this project and a key source for academic training of students and professionals in meteorology in West Africa, a critical region for weather forecasting. This is a zone in which the population is particularly vulnerable to weather and climate events, where the science of meteorology has a strong track record, including the Global Atmospheric Research Programme (GARP) Atlantic Tropical Experiment (GATE) experiment of the 1970s, and where there has been a recently renewed research impetus through the AMMA programme.

I look forward to this book making a significant contribution to the quality of weather forecasts and warnings in West Africa in the coming years. I also hope that it will stimulate new research and operational activities, in particular, in the use of numerical weather prediction (NWP) systems and, in return, through feedback, increase the quality of NWP evaluation activities. Furthermore, the emphasis on the use of in situ and remotely sensed measurements in forecasts should provide additional motivation for the implementation and exploitation of measurement systems in West African countries.

*Petteri Taalas*
Secretary-General,
World Meteorological Organization

# Preface

This book aims to synthesise the latest knowledge of West African meteorology with operational tools and methods for weather forecasting in the region. We hope that it will be valuable for the training of weather forecasters in West Africa, for weather forecasters around the world having an interest in tropical predictions, and for students of meteorology. It is our intention that the material should appeal to students who have an interest both in tropical dynamics and operational forecasting, and the book is targeted at both the undergraduate and postgraduate levels. A basic meteorological knowledge is assumed.

The book is particularly demanded by national meteorological services, whose general forecasting services are still in their early stages. For many years, focus has been on forecasting for aviation, but with climate change and its associated hazards and impacts, there is increasing demand for specific meteorological forecasts at a wider range of time and space scales. The main users need predictions for agriculture and food security, water resources, civil protection and disaster risk reduction, and health. Although the book does not address these issues directly, the phenomena needed to issue such forecasts at the required timescales have been covered throughout the book, with the aim of giving the forecaster enough materials to extrapolate and deliver such weather and climate services. The book also gives some insight for further research needed.

The book emerged from discussions between the two editors and Chris Thorncroft about West African forecasting for the JET2000 experiment in 2000. For that campaign, we were flight-planning for the UK research aircraft and trying to use modern numerical weather prediction (NWP) products alongside the conceptual models known to operational forecasters in the region. It was clear that there has been a lack of documentation of the basic meteorology of West Africa, a lack of documentation of forecasting methods (a more widespread problem for the tropics), and a lack of communication between the research and forecasting communities (again, not a problem unique to West Africa).

The African Monsoon Multidisciplinary Analysis (AMMA, 2002 onwards) offered an opportunity to deal with some of these problems, and a forecasting programme was integrated with the intensive AMMA field campaigns of 2006. The AMMA Operational Centre (AOC) included a team of forecasters selected from different national meteorological services in West Africa, and hosted by the African Centre of Meteorological Applications for Development (ACMAD) in Niamey. This group worked on the production of systematic, standardised daily forecasts and analyses for presentation in the daily briefings at the AOC. The AOC forecasting programme was led by Jean Philippe Lafore and Zilore Mumba, and achieved some significant successes: a network of forecasters and researchers was created, important dialogues were initiated, and the standardisation of diagnostic plots, in the form of the West African Synthetic Analysis/Forecast (WASA/F), was put in place. However, the AOC forecasting also highlighted (at least for the academics involved) the massive challenge of developing reliable daily forecasts in the region, especially of convective systems.

Following the AOC forecasting, further development of WASA/F was conducted by ACMAD and the Centre National de Recherches Météorologiques, Météo-France (CNRM) over the next couple of years. We regard this framework for consistent plotting of weather features to be an essential part of the scientific development of forecasting in this region. There are a number of important conceptual models of weather systems in West Africa (African easterly waves (AEWs) and so on), and it is necessary to analyse, monitor, track and predict the behaviour of such features. By demarking relatively simple structures, such as an AEW trough, other aspects of the same phenomenon can be implied, if the user knows the typical configuration of such features: in the case of an AEW, the user can infer likely areas of rainfall, and rainfall type, relative to the trough location. In order for this conceptual framework to be useful it is necessary that there is scientific rigour to the process, and a great deal of the effort undertaken in creating this book has gone into the consideration of these principles. Features to be drawn schematically should correspond to agreed measures and thresholds in observational or model data, so that in principle those measures could be inferred within quantitative limits from the schematic features.

The plots should also be 'repeatable', in the sense that another forecaster making the same chart would produce the same result. Considerable effort has been put into the achievement of these goals; a summary of the rules for plotting of weather features on the WASA/F maps is provided in Chapter 11.

Preparation and writing of the book began formally in October 2009 at a workshop hosted by the International Centre for Theoretical Physics (ICTP) and co-funded by the World Meteorological Organization's (WMO's) World Weather Research Programme (WWRP) in association with THORPEX-Africa. A one-day workshop was used to explore some processes of interaction between researchers and forecasters, and the outcome was a structured plan of chapters and their content (including, importantly, the list of lead authors). This workshop also formalised the Editorial Committee for the project, and led to the agreement of the basic modus operandi of the preparation of the book.

Given that the important results to be documented are universal, and mainly are not country dependent, it was agreed that we could work with a small core group of forecasters alongside the various academics who were participating. Some other principles were at this point agreed:

- The Handbook will always describe the state of the art, assuming the availability of a full set of tools. This will give forecasters at the main centres the best information on the current state of knowledge, and will serve to drive demand for such tools.
- The Handbook will also describe pragmatic approaches to forecasting, including for example the plotting of synoptic charts from regional observations and the computation of stability indices from upper air data.
- Methods that are untested may be included in the Handbook, if there is good reason to do so, such as demonstrated success of the method in another part of the world. This approach will allow forecasters to perform the necessary testing for their region.
- Most chapters are separated into two parts: background scientific knowledge and operational methods.

Each chapter has one or two lead authors, plus a group of contributing authors. All chapters were peer reviewed by one or more people. In one or two cases the reviewer made substantial and specific contributions to the material and was invited to join the list of contributing authors.

An extended workshop was held in Leeds in the period October to December 2012. A forecaster core group worked together for 8 weeks with various European and African academics and forecasters visiting for shorter periods. This activity was the main opportunity to make progress on intensive collaboration over scientific details.

A final workshop co-funded by WMO, Met Office and ACMAD was hosted by Mariane Diop-Kane at Agence Nationale de l'Aviation Civile et de la Météorologie du Sénégal (ANACIM) in Dakar in March 2013, with the aim of allowing the whole West African forecasting community (with representatives from eight national forecasting agencies, Agence pour la Securite de la Navigation Aerienne en Afrique et a Madagascar (ASECNA) and West African Universities) to scrutinise and review the handbook material in order that we could move to completion of the book within a short time after this. In fact, completion of the book took another 2 years, due to iterations needed in resolving some of the ideas raised at that workshop.

Given that West African meteorology involves many interacting processes and spatio-temporal scales, while prior knowledge has hitherto been dispersed among many sources, organisation of the material has been quite complicated, and we have had to make a few choices. In particular, we have made deliberate decisions regarding the terminology of some of the main climatological features. We decided against using the term ITCZ (intertropical convergence zone) over land, because the convergence structure varies significantly with height and this causes confusion in the literature. While we retain 'ITCZ' over the ocean, over the continent we refer, when needed, to specific observable features such as the zone of maximum precipitation. At the same time, we have chosen to retain both of the terms ITD and ITF (intertropical discontinuity and intertropical front). These terms are interchangeable and are each used very widely in the literature and in operational centres. Similarly, we use both tephigrams and skew-*T* diagrams in the presentation of thermodynamic profiles, since both are in common use and the student should be prepared for this.

The main challenges in the preparation of this work have been the limited contributor time available, the need for scientific discourse in order to agree and approve the material to be presented, and the effort needed in ensuring cross-consistency between chapters. We are very grateful to the funding agencies that have supported us, and the time contributed by many scientists.

Chapter 1 presents an overview of the West African climate, without a specific consideration of forecasting methods. The following chapters build from synoptic meteorology through specific weather phenomena to nowcasting. Chapters 7 and 8 move to longer, subseasonal and seasonal timescales. The book then covers methods of remote sensing, numerical weather prediction and finally plotting conventions and standards in Chapter 11.

*Douglas J. Parker*
*Mariane Diop-Kane*
February 2017

## Acknowledgements

This book is the product of a cooperative programme of work between the ACMAD, the AMMA programme and the WMO WWRP, all of whom have generously provided financial support and the support of their staff.

Based on a French initiative, AMMA was built by an international scientific group and funded by a large number of agencies, especially from Africa, France, the UK and USA. It has been the beneficiary of a major financial contribution from the European Community's Sixth Framework Research Programme. Financial and other support for this book has also been gratefully received from ANACIM, ICTP, Météo-France/CNRM, Met Office, NERC (NE/G018499/1), the University of Leeds, RIPIECSA, THORPEX, the Walker Institute, the WMO Training Activities Division and the WMO Development and Regional Activities Department, Offices for Africa and Least Developed Countries. Doug Parker has been supported by a Royal Society Wolfson Research Merit Award.

Paul Hardaker, as Chief Executive of the Royal Meteorological Society (RMetS), lent great support for publication of this work as a complete book by our publishers, Wiley-Blackwell. Both the RMetS and the American Meteorological Society have kindly allowed us to reproduce many figures from their academic journals.

The Steering Committee for the project represents the major stakeholders in the book and its impacts, and comprises Aida Diongue-Niang, Mariane Diop-Kane, Ibrahima Hamza, Abdoulaye Harou, Jean Philippe Lafore, Benjamin Lamptey, Ifeanyi Nnodu, Douglas Parker, Patrick Parrish and Leon Guy Razafindrakoto. The Editorial Committee has been responsible for the selection of chapter authors, and has provided strategic guidance on the project throughout. Its members have been Douglas Parker, Ernest Afiesimama, Jim Caughey, Rosalind Cornforth, Mariane Diop-Kane, Aida Diongue-Niang, Andreas Fink, Ibrahima Hamza, Jean Philippe Lafore, Arlene Laing, Peter Lamb, Benjamin Lamptey, Zilore Mumba, Ifeanyi Nnodu, Jerome Omotosho, Steve Palmer, Wassila Thiaw, Chris Thorncroft and Adrian Tompkins.

In addition to those already named, and the chapter lead authors, we are particularly grateful for contributions made by a very wide group of operational forecasters, and other specialists in West African forecasting, namely Abdou Kassimou, Abébé Abdoulaziz, Garba Adamou, Raphael Agbabu, Leonard Amekudzi, Aboubacar Angoua, Nicole Asencio, Alpha Boubacar Barry, Florent Beucher, Cathryn Birch, Nicolas Chapelon, Fleur Couvreux, Okoloye Cyprian, Diakaria Kone, Abdou Aziz Diop, Marie-Christine Dufresne, Florence Favot, Jim Galvin, Adamou Garba, Amadou Tidiane Gaye, Françoise Guichard, Hama Hamidou, Alioune Kamara, André Kamga, Cheikh Kane, Yassin Khan, Diakaria Kone, Touré Koure, Gnatoulouma Kpabeba, Laurent Labbe, Yerima Ladan, Sidi Ould Mohamed Lemine, Serge Lepape, Ousmane Ndiaye, Justin Noukpozounkou, Mathieu Nuret, Bernard Orji, Deve Orkuma, David Perry Osika, Dazangwendé Emmanuel Poan, Felix Randrianavalona, Romain Roehrig, Diori Saley, Maré Dominique Sangare, Abdoulaye Sarr, Massamba Seck, Tinni Seydou, Muller Tsatsu Siameh, George Emmanuel Stafford, Abdoulaye Taye, Augustin Tchouanwo, Moussa Yoro Thiam, Madiagne Thiaw, Mamadou Bassando Traore, Henk Verschuur, Patrick Vidal, Jean-Christoph Vincendon, Charles Yorke and Jingmay Yu.

We are especially grateful to Abdou Malam Kassimou for his efforts in coordinating forecaster contributions to the chapters, and to Rosaleen McDonnell who acted as Editorial Manager for the project in its final stages, and was instrumental in bringing the whole work to its conclusion.

## Acronyms

| | |
|---|---|
| AAI | absorbing aerosol index |
| ACARS | Aircraft Communications Addressing and Reporting System |
| ACMAD | African Center of Meteorological Applications for Development |
| AEJ | African easterly jet |
| AERONET | AErosol RObotic NETwork |
| AEW | African easterly wave |
| AI | aerosol index |
| AIRS | Atmospheric Infrared Sounder |
| ALMIP | AMMA Land Surface Model Intercomparison Project |
| AMDARS | Aircraft Meteorological Data Relay |
| AMJO | African Madden–Julian oscillation |
| AMMA | African Monsoon Multidisciplinary Analysis |
| AMMA | Analyse Multidisciplinaire de la Mousson Africaine |
| AMSU | Advanced Microwave Sounding Unit |
| ANACIM | Agence Nationale de l'Aviation Civile et de la Météorologie du Sénégal |
| ANASYG | Graphic SYnoptic ANAlysis |
| AOT | Aerosol Optical Thickness |
| ARM | Atmospheric Radiation Measurement |
| ARPEGE | Operational forecasting global model of Météo-France |
| ASCAT | Advance SCATterometer |
| ASDAR | Aircraft to Satellite Data Relay |
| ASECNA | Agence pour la Sécurité de la Navigation Aérienne en Afrique et à Madagascar |
| ATOVS | Advanced TIROS Television and Infrared Observational Satellite Operational Vertical Sounder |
| AVHRR | Advanced Very High Resolution Radiometer |
| BoDEx | Bodélé Dust Experiment |
| BRN | bulk Richardson number |
| BTD | brightness temperature difference |
| CALIPSO | Cloud-Aerosol Lidar and Infrared Pathfinder Satellite Observations |
| CAPE | convective available potential energy |
| CAT | clear air turbulence |
| CATCH | Couplage de l'Atmosphère Tropicale et du Cycle Hydrologique |
| CBL | convective boundary layer |
| CCA | canonical correlation analysis |
| CCEW | convectively coupled equatorial wave |
| CDAS | Climate Data Assimilation System |
| CFS | Climate Forecast System |
| CIN | convection/convective inhibition |
| CLAUS | Cloud Archive User Service |
| CLIVAR | Climate and Ocean: Variability, Predictability and Change |
| CMC | Canadian Meteorological Centre |
| CNRM | Centre National de Recherches Météorologiques, Météo-France |
| CPC | Climate Prediction Center |

| | |
|---|---|
| CRM | Cloud Resolving Model |
| DCAPE | downdraught CAPE |
| DMCPAs | Disaster Management and Civil Protection Authorities |
| ECMWF | European Centre for Medium-Range Weather Forecasts |
| EJ | easterly jet |
| ENSO | El Niño Southern Oscillation |
| EOF | empirical orthogonal function |
| EPS | ensemble prediction system |
| ER | equatorial Rossby |
| ERA-Interim | European Centre for Mid-Range Weather Forecast ReAnalysis Interim |
| ET | actual evapotranspiration |
| EUMETSAT | EUropean organisation for the exploitation of METeorological SATellites |
| GARP | Global Atmospheric Research Programme |
| GATE | GARP Atlantic Tropical Experiment |
| GCM | general circulation model |
| GEFS | Global Ensemble Forecast System |
| GEO | geostationary or geosynchronous Earth orbit |
| GEWEX | Global Energy and Water Exchanges Project |
| GFS | Global Forecast System |
| GIMMS | Global Inventory Modeling and Mapping Studies |
| GMDSS | Global Maritime Distress and Safety System |
| GNSS | Global Navigation Satellite System |
| GOME-2 | Global Ozone Monitoring Experiment-2 |
| GPCP | Global Prediction Climatology Project |
| GPM | Global Precipitation Mission |
| GPS | Global Positioning System |
| GRAS | GNSS Receiver for Atmospheric Sounding |
| GTS | Global Telecommunications System |
| GWD | gravity wave drag |
| HAPEX-Sahel | Hydrology-Atmosphere Pilot Experiment in the Sahel, 1990–1992 |
| HDH | Harmattan dust haze |
| HI | heat index |
| HIRS | High-Resolution Infrared Radiation Sounder |
| HL | heat low |
| HOAPS | Hamburg Ocean Atmosphere Parameters and Fluxes from Satellite Data |
| HRV | high-resolution visible |
| HSS | Heidke skill score |
| IASI | Infrared Atmospheric Sounding Interferometer |
| ICRISAT | International Crops Research Institute for the Semi-Arid Tropics |
| IGRA | Integrated Global Radiosonde Archive |
| IHC | interhemispheric thermal contrast |
| IR | infrared |
| IRD | Institute de Recherche pour le Développement |
| IRI | International Research Institute for Climate and Society |
| ISCCP | International Satellite Cloud Climatology Project |
| ISD | integrated surface daily data set |
| ITB | intertropical boundary |
| ITCZ | intertropical convergence zone |
| ITD | intertropical discontinuity |
| ITF | intertropical front |
| JAS | July–August–September |
| JMA | Japan Meteorological Agency |
| LAM | limited-area model |
| LCL | lifted/lifting condensation level |
| LEO | low Earth orbit |

| | |
|---|---|
| LFC | level of free convection |
| LLJ | low-level jet |
| LLWAS | Low-Level Wind-Shear Alert System |
| LLWJ | low-level westerly jet |
| LST | local solar time |
| MCA | maximum covariance analysis |
| MCGE | multi-centre grand ensemble |
| MCS | mesoscale convective system |
| MCV | mesoscale convective vortex |
| MD | monsoon depth |
| METAR | meteorological aviation routine weather report |
| METAR | aerodrome routine meteorological report |
| METAR | meteorological aviation report |
| MetUM | Met Office Unified Model |
| MFC | moisture flux convergence |
| MFD | moisture flux divergence |
| MIDAS | Met Office Integrated Data Archive System |
| MISR | Multi-angle Imaging Spectroradiometer |
| MISVA | Monitoring and Forecasting of Intraseasonal Variability over Africa |
| MJO | Madden–Julian oscillation |
| ML | monsoon layer |
| MODIS | Moderate Resolution Imaging Spectroradiometer |
| MOS | model output statistics |
| MPE | Multisensor Precipitation Estimate |
| MRG | mixed Rossby–gravity |
| MSE | moist static energy |
| MSG | Meteosat Second Generation |
| MSL | mean sea level |
| MSLP | mean sea-level pressure |
| MT | monsoon trough |
| MTG | Meteosat Third Generation |
| MVIRI | Meteosat Visible and Infra-Red Imager radiometer |
| NBL | night-time, or nocturnal, planetary boundary layer |
| NCEP | National Centers for Environmental Prediction |
| NDVI | normalized differential vegetation index |
| NLLJ | night-time low-level jet |
| NMHSs | national meteorological and hydrological services |
| NMS | National Met Services |
| NOAA | National Oceanic and Atmospheric Administration |
| NOAA/WPC | National Oceanographic and Atmospheric Administration Weather Prediction Center |
| NPP | National Polar-orbiting Partnership |
| NWP | numerical weather prediction |
| OI | optimum interpolation |
| OLR | outgoing longwave radiation |
| OMI | Ozone Monitoring Instrument |
| OMP | Observatoire Midi-Pyrénées |
| $P$ | Precipitation |
| PBL | planetary boundary layer |
| PC | principal component |
| PJ | polar jet |
| POES | Polar-orbiting Environmental Satellites |
| PPI | plan position indicator |
| PRESAO | PREvision Saisonairre des pluis en Afrique del l'Ouest |
| PRESYG | Graphic Synoptic Prediction |
| PSANAF | Potential dust source areas in northern hemispheric Africa |

| | |
|---|---|
| PV | potential vorticity |
| PVU | potential vorticity unit |
| PW | precipitable water |
| PW* | precipitable water anomaly |
| PWAT | precipitable water |
| PWV | precipitable water vapour |
| QBZD | quasi-biweekly zonal dipole |
| RCOF | Regional Climate Outlook Forum |
| RDT | Rapid Development Thunderstorm |
| RFE2 | Rainfall Estimate Version 2 |
| RGB | red, green, blue |
| RGB | red, green and blue (primary colours combined for image visualisation) |
| RHI | range–height indicator |
| RIPIECSA | Recherches Interdisciplinaires et Participatives sur les Interactions entre les Ecosystèmes, le Climat et les Sociétés en Afrique de l'ouest |
| RMDCN | Regional Meteorological Data Communications Network in Europe |
| RMM | Real-time Multivariate MJO series |
| ROC | relative operating characteristic |
| RTTOV | Radiative Transfer for TOVS |
| S4CAST | Sea Surface Temperature-based Statistical Seasonal Forecast Model |
| SAFNWC | Satellite Application Facility on Support to Nowcasting and Very Short Range Forecasting |
| SAL | Saharan air layer |
| SAMUM | Saharan Mineral Dust Experiment |
| SAWS | South Africa Weather Service |
| SDS-WAS | Sand and Dust Storm Warning Advisory and Assessment System |
| SeaWiFS | Sea-viewing Wide Field-of-view Sensor |
| SEDOO | Service de DOnnée |
| SEVIRI | Spinning Enhanced Visible and InfraRed Imager |
| SHL | Saharan heat low |
| SMAP | NASA's Soil Moisture Active Passive mission, January 2015 |
| SMOS | ESA's current Soil Moisture and Ocean Salinity mission launched on 2 November 2009 |
| SOLAS 74 | International Convention for the Safety of Life at Sea 1974 |
| SOP | Specific Observation Period |
| SOP-1, SOP-2 | Special Observing Periods of AMMA |
| SPECI | aerodrome special meteorological report |
| SSM/I | Special Sensor Microwave/Imager |
| SST | sea surface temperature |
| STJ | subtropical jet |
| SVD | singular value decomposition |
| SWFDP | WMO Severe Weather Forecast Demonstration Project |
| SYNOP | surface synoptic observation |
| SYNOP | synoptic observation |
| TAF | terminal aerodrome forecast |
| TB | brightness temperature |
| TCW | total column water |
| TD | tropical disturbance |
| TDWR | terminal Doppler weather radars |
| TEJ | tropical easterly jet |
| THORPEX | The Observing-System Research and Predictability Experiment |
| TIGGE | THORPEX Interactive Grand Global Ensemble |
| TIROS | Television and Infrared Observational Satellite |
| TOA | top-of-atmosphere |
| TOMS | Total Ozone Mapping Spectrometer |
| TRMM | Tropical Rainfall Measuring Mission |
| UTC | Universal Time Coordinated |

| | |
|---|---|
| UTT | upper tropospheric trough |
| UV | ultraviolet |
| VIIRS | Visible Infrared Imaging Radiometer Suite |
| VIS | visible |
| VP | velocity potential |
| WAHL | West African heat low |
| WAM | West African monsoon |
| WASA/F | West African Synthetic Analysis/and Forecast |
| WAWJ | West Africa westerly jet |
| WCRP | World Climate Research Programme |
| WMO | World Meteorological Organization |
| WV | water vapour |
| WWRP | World Weather Research Programme |
| YOTC | Year of Tropical Convection |
| ZDR | differential reflectivity |

# 1

# Mean Climate and Seasonal Cycle

**Lead author:** *Andreas H. Fink*[a],

**Contributing authors:** *Thomas Engel*[b], *Volker Ermert*[b], *Roderick van der Linden*[b], *Malvin Schneidewind*[b], *Robert Redl*[b,g], *Ernest Afiesimama*[c], *Wassila M. Thiaw*[d], *Charles Yorke*[e] *and Matthew Evans*[f]

**Reviewer:** *Serge Janicot (Institut Pierre-Simon Laplace, Paris, France)*

[a] Institute of Meteorology and Climate Research – Troposphere, Karlsruhe Institute of Technology, Karlsruhe, Germany
[b] Institute of Geophysics and Meteorology, University of Cologne, Cologne, Germany
[c] Nigerian Meteorological Agency, Abuja, Nigeria
[d] NOAA/National Centers for Environmental Prediction/Climate Prediction Center, Camp Springs, MD, USA
[e] Ghana Meteorological Agency, Accra, Ghana
[f] Department of Chemistry, University of York, York, UK
[g] Meteorological Institute, Ludwig-Maximilians-University Munich, Munich, Germany

## 1.1 Introduction

Tropical West Africa is one of the most densely populated regions of Africa, and the annual climate is of critical importance to the livelihoods of people in the region. The weather and climate of West Africa have a major impact on food production, water resources and health, as well as other economic sectors, such as power generation, transportation and fisheries. The West African monsoon (WAM), one of the three major monsoon systems in the tropics, dominates the climate of the region. The WAM represents a seasonal reversal of the prevailing surface winds over the greater part of the West African region, with the wintertime northeasterly Harmattan winds being replaced in the summer months by southwesterly monsoon winds. The summer monsoon brings moisture from over the Atlantic Ocean into the continent, and therefore feeds the annual rainfall, which is so critical to the local populations. Variations in the WAM from year to year determine the difference between good and bad rainfall; or on a regional basis, between sufficient rain and drought.

The travel reports of famous 19th century explorers of North Africa, like René Caillié, the first European who reached Tombouctou, and Heinrich Barth, who crossed the Sahara to travel in the Sahel for years, already described impressive dust- and rainstorms. However, it was not until the 1920s to 1940s that these phenomena, today known as being characteristic to the region, were described in the meteorological literature (e.g. Sutton, 1925). The German meteorologists Regula and Piersig were likely the first that simultaneously published results on what are now known as 'African easterly waves' (AEWs; Fink, 2012; Section 2.1.3). Hubert (1939) linked an AEW to the New England 1938 hurricane that destroyed Long Island in the USA, a connection that is now common knowledge in meteorology. The typical morning maximum in surface wind speed in semi-arid northern tropical Africa has been explained by Farquharson (1939) to be associated with the morning breakdown of the night-time inversion by turbulent mixing (see Section 4.1.3.2). In terms of the synoptic structure of the monsoon system, Hamilton *et al.* (1945) published a seminal survey on weather in Nigeria and were among the first who introduced the concept of four weather zones that migrate north–south with the seasonal cycle.

While our understanding of the other aforementioned phenomena has advanced over the intervening decades, especially after the advent of weather satellites in the 1970s, the 'four weather zone concept' is, with alterations and amendments, still used as a basic conceptual approach to the WAM system. Thus, it seems reasonable to introduce this historic concept first. Figure 1.1 shows the four traditional zones in a south–north schematic for the wet monsoon month of July, based on a modern atmospheric three-dimensional analysis and satellite observations of clouds and precipitation.

The northernmost zone, zone A, is located to the north of the intertropical discontinuity (ITD; also known as intertropical front, ITF) that separates, at about 20°N, the hot and dry northerly surface flow to the north from the moist and cool southwesterly monsoon flow to the south.

*Meteorology of Tropical West Africa: The Forecasters' Handbook*, First Edition. Edited by Douglas J. Parker and Mariane Diop-Kane.

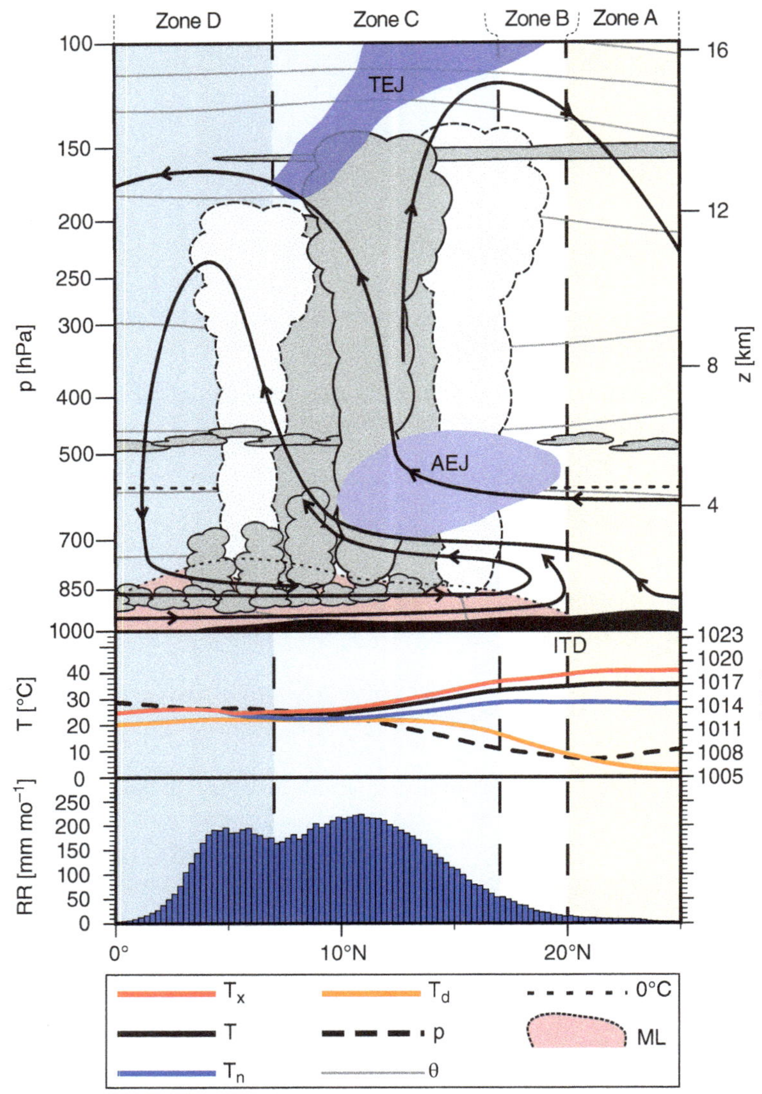


**Figure 1.1** Schematic cross-section of the atmosphere between 10°W and 10°E in July and illustration of the weather zones A–D of the West African monsoon. Shown are the positions of the ITD, upper-level jet streams (African easterly jet (AEJ), tropical easterly jet (TEJ)), the monsoon layer (ML) (as defined by westerly, i.e. positive, zonal winds), streamlines, clouds, the freezing level (0°C isotherm), isentropes (θ), minimum ($T_n$), maximum ($T_x$) and mean ($T$) and dew point temperatures ($T_d$), atmospheric pressures ($p$), and mean monthly rainfall totals (RR).

Precipitation is rare in this zone (Figure 1.1, lower panel), and daytime maximum temperatures in excess of 40°C are frequent (Figure 1.1, middle panel). The dryness is expressed by low dew point temperatures (<15°C), and this zone also hosts the West African surface heat low; that is, the zone of minimum surface pressure (Figure 1.1, middle panel).

Zone B is the narrowest zone in terms of latitude, where the moist monsoon layer, defined here as the lower tropospheric layer with a westerly wind component, is shallow, and only short-lived outbreaks of thunderstorms occur (Section 3.2.3.1.1).

Zone C is the zone of maximum thunderstorm and shower activity and associated precipitation, as indicated by the grey-shaded deep convective cloud (Figure 1.1, upper panel). Here, low-level southwesterlies change to mid-level easterlies associated with the AEJ that forms a waveguide to the AEWs. Zone C is also overlaid by the upper level TEJ (Figure 1.1, upper panel).

In zone D, deep convective clouds are infrequent and shallow stratus or mid-level altostratus clouds dominate, except for some rainy parts along the Guinea Coast, as for example the Niger delta and the region to the southwest of Cape Palmas.[1] In the other zone D regions the 'little dry season' commences in July and sometimes in August. The moist monsoon layer is deepest at the Guinea Coast (~4–6°N) and shoals further inland. It meets the surface at around 20°N to define the ITD.

The dotted line in Figure 1.1 (top panel) demarcates the boundary between the moist monsoon layer (red shading) and the drier air above, and intersects the 850 hPa level near the boundary of zones B and C. In terms of maps of geopotential height, this line represents a trough of low pressure, and may be termed the 'monsoon trough.' However, as will be seen in later chapters (e.g. Sections 2.1.2, 3.1.4.1.4 and 11.10), forecasters prefer to indicate the monsoon trough in terms of the line of higher 850 hPa vorticity, typically linking closed vortices at this level, and this feature lies to the south of the geopotential height trough, typically to the south of the AEJ.

The four weather zones retreat southward from the continent after the peak of the monsoon season to leave, in the classical view, only zones B and A on the continent in January, with the ITD being located just 200–300 km north of the Guinea coastline. As we will see at the end of this chapter, some changes to the monsoon system are noteworthy in boreal winter, and sketches of both seasons will be discussed in some more detail at the end of this chapter.

Operational weather forecasting begins with an appreciation of the climatology of a region, and therefore this chapter introduces the physical and meteorological conditions that define the climate of West Africa on the continental scale. The seasonal evolution of the WAM is known to be driven by the thermodynamic contrasts between the hot and dry Sahara in the north, and the cool, moist Atlantic Ocean in the south and west. The continental-scale thermodynamic gradient from the ocean to the desert is modulated by the availability of moisture over the continent, due to vegetation patterns and soil moisture availability. Since the vegetation and soil moisture patterns are themselves controlled by the seasonal availability of rainfall, the monsoon is a climatic system in which the land, ocean and atmosphere are closely coupled. The spatial contrasts in the land, ocean and atmospheric systems are therefore explained and documented here, beginning with the patterns of rainfall, showing how these influence the water and thermodynamic energy cycles, and in turn explaining how these drive the three-dimensional monsoon circulation. Since the monsoon winds deliver the moisture needed for the seasonal rains, this completes the cycle of precipitation, thermodynamics and circulation that define the coupled monsoon system.

Tropical West Africa, considered here as the region between 0° and 30°N, the West Coast and 20°E (Figure 1.2a) comprises several climate zones, from the hyper-arid desert climates of the southern Sahara, the semi-arid Sahel zone, the wet and dry climates of the Soudanian and parts of the Guinea coastal zones, to the wet climates along the Pepper, the Niger delta and the Bight of Bonny coasts. Compared with other locations, West Africa is a data-sparse region. This is reflected in the reporting frequency of synoptic stations to the Global Telecommunication System (GTS) of the World Meteorological Organisation (WMO) (Figure 1.2b). For this chapter, several national weather services and individuals have provided additional data. Thus, if data availability permits, the climate normal period 1981–2010 is taken as the reference period. The surface and upper-air climates of West Africa will be described by showing annual mean and seasonal climatologies of key climate variables. In order to sample the seasonal cycle, horizontal maps for January, April, July and October will be discussed. In terms of the peak of the rainy season in the Sahel, August would have been more appropriate. However, differences to July are generally small, and the climatology of the cold dry season is better depicted in January.

It is intended that this chapter, and the many climatological details it explains, will serve as both a valuable educational resource and as an important 'handbook' reference for the West African climate. This chapter will reference more detailed discussions of particular climatic features, in which the climate of a local region has its own special conditions that can be found in later chapters.

## 1.2 Rainfall and Evaporation Climatologies

Two of the most important climatic surface variables in West Africa are rainfall $P$ and actual evapotranspiration $ET$. The evapotranspiration is the sum of evaporation from the bare soil, transpiration from plants and evaporation from plant tissues. The difference between $ET$ and $P$ links the atmospheric and terrestrial water budgets and, if infiltration below the root zone of plants is subtracted, determines the water availability in the topsoil layer. Therefore, $P$ and $ET$ determine the strength of the coupling between the land surface and the atmosphere, and are strongly linked to the northward penetration of the monsoon system every year. As the southwesterly monsoon air advances across the continent, it loses water through precipitation to the surface, but recovers water from evapotranspiration. The degree of water vapour recycling controls the northward extent of the water vapour available for rainfall. Food and livelihood security of the vast majority of the West African population presently depend on rain-fed agriculture, thus underpinning the importance of $ET$ and $P$.

Rainfall climatologies from surface rain gauges for 1981 to 2010 suffer from the degradation of the network for various reasons, or unavailability of existing data. Figure 1.3 shows a gauge-based climatology for the 39-year period 1951 to 1989. The following three salient features are noteworthy: (a) the strong rainfall decrease from the northern Soudanian zone to the Sahara by about 145 mm per 100 km; (b) particular wet regions at the southwest coast and the Guinea Mountains, to the west of the Cape of The Three Points, at the mouth of the Niger River, and in the Cameroon Line Mountains; and (c) a dry zone that stretches from the Coast of Ghana across southern Togo and Central Benin into northwestern Nigeria. This region is called the Ghana–Benin dry zone[2] (Vollmert *et al.*, 2003). The wet regions occur where the moisture-laden monsoonal southwesterlies blow perpendicular to the coastline or are lifted due to coastal orography. The dry zone results from a

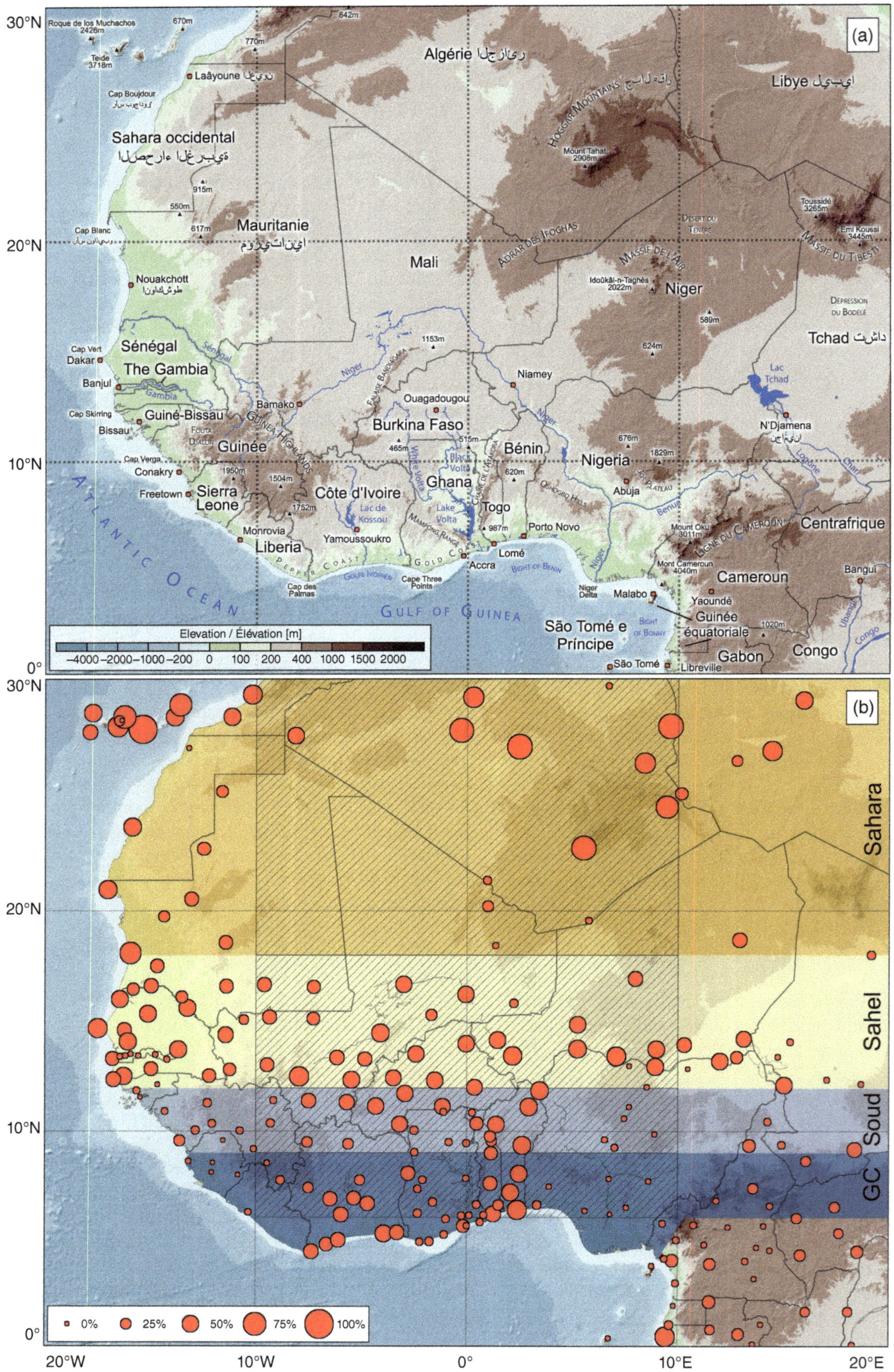

(a)
30°N
20°N
10°N
0°
Roque de los Muchachos 2426m
670m
842m
770m
Teide 3718m
Laâyoune
Cap Boujdour
Algérie
Libye
Hoggar Mountains
Mount Tahat 2908m
Sahara occidental
915m
550m
617m
Cap Blanc
Mauritanie
Mali
Adrar des Ifoghas
Désert du Ténéré
Massif de l'Aïr
Tousside 3265m
Emi Koussi 3445m
Massif du Tibesti
Nouakchott
Idoûkâl-n-Taghès 2022m
Niger
589m
Dépression du Bodélé
Cap Vert
Dakar
Sénégal
1153m
624m
Tchad
Lac Tchad
The Gambia
Banjul
Gambia
Niamey
Bamako
Ouagadougou
Cap Skirring
Guiné-Bissau
Bissau
Fouta Djallon
Guinée
Guinea Highlands
Burkina Faso
465m
515m
Bénin
676m
1829m
N'Djamena
Cap Verga
Conakry
1950m
1504m
Ghana
620m
Nigeria
Freetown
Sierra Leone
1752m
Côte d'Ivoire
Lac de Kossou
Lake Volta
Togo
Abuja
Jos Plateau
Benue
Logone
Chari
Monrovia
Liberia
Yamoussoukro
987m
Porto Novo
Mount Oku 3011m
Ligne du Cameroun
Centrafrique
Lomé
Accra
Gold Coast
Pepper Coast
Golfe Ivoirien
Cape Three Points
Bight of Benin
Cap des Palmas
Mont Cameroun 4040m
Cameroun
Bangui
Atlantic Ocean
Gulf of Guinea
Niger Delta
Malabo
Yaoundé
Guinée équatoriale
São Tomé e Príncipe
Bight of Bonny
1020m
Gabon
Congo
Ubangi
São Tomé
Libreville
Elevation / Élévation [m]
−4000 −2000 −1000 −200 0 100 200 400 1000 1500 2000
(b)
Sahara
Sahel
Soud
GC
0% 25% 50% 75% 100%
20°W
10°W
0°
10°E
20°E

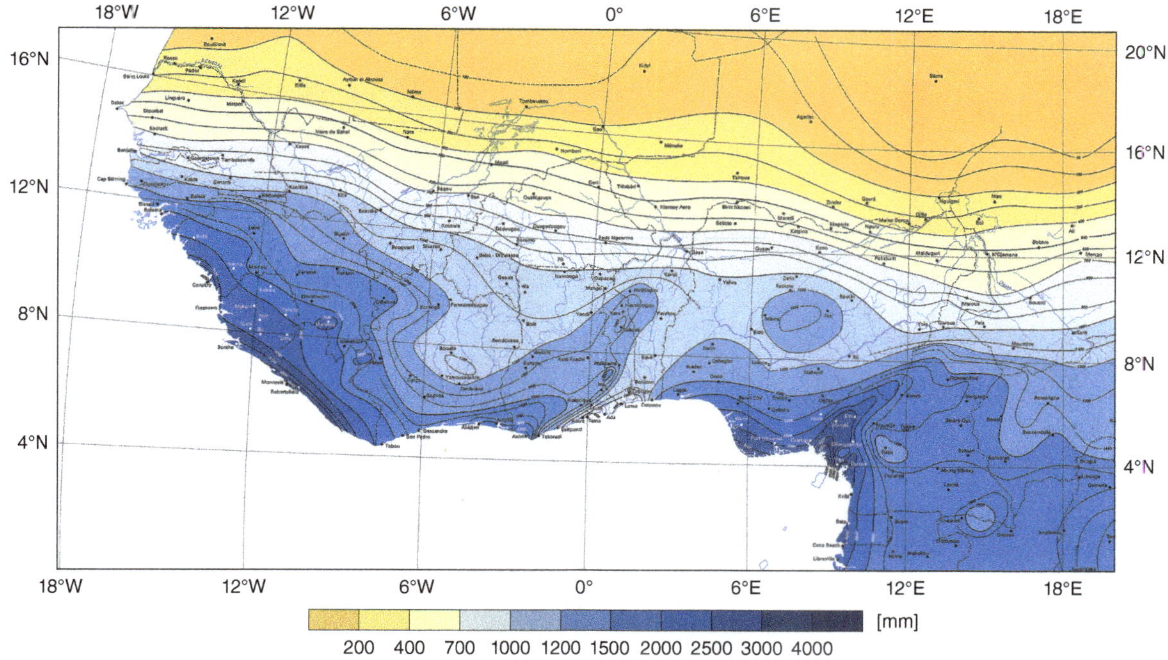


**Figure 1.3** Mean annual precipitation for the period 1951–1989. The isohyets are based on 890 rainfall stations. *Source*: Institut de Recherche pour le Développement (IRD).

combination of coastal upwelling of cold waters and frictional divergence along SW–NE-oriented coastlines (Vollmert *et al.*, 2003; Nicholson, 2011).

Figure 1.4 shows annual rainfall totals and the seasonal march of rainfall using the blended satellite estimates from the Tropical Rainfall Measuring Mission (TRMM) 3b42 Version 7 three-hourly rainfall product for the 15-year period 1998–2012. According to the assessment of rainfall products for Africa by Parker *et al.* (2011), the TRMM product is one of the best rainfall estimates available. However, over land, substantial differences to gauge-based estimates must be taken into account, especially over highlands and where rainfall from warm clouds is significant. Figure 1.4a shows a similar annual rainfall pattern when compared with Figure 1.3. The TRMM rainfall product discloses rainfall amounts in excess of 3000 mm over the oceanic intertropical convergence zone (ITCZ) off the southwest coast and over the Bight of Bonny. The product also reveals that the Ghana–Benin dry zone extends into the western Gulf of Benin waters.

The Hovmöller (time–space) diagram in Figure 1.4b elucidates the seasonal south–north migration of rainfall in West Africa. Worthy of mention are (a) a slower inland penetration of rains until August when compared with the withdrawal; (b) a stronger first rainy season in May/June along the Guinea Coast; and (c) the 'monsoon jump' around 24 June (Sultan and Janicot, 2003) and the subsequent 'little dry season' at the Guinea Coast. Figure 1.5 shows the rainfall maps for the months of January, April, July and October 1998 to 2012. In January, the maximum rainfall zone[3] is located over the equatorial Gulf of Guinea and infrequent rains amounting to 25–75 mm occur in the Cape Palmas and Niger delta areas (Figure 1.5a). The majority of the continent records less than 25 mm at the peak of the dry season. However, typically up to two major rainfall events do occur during an average dry season between November and February in the region between 7.5° and 15°N in association with tropical–extratropical interactions (Knippertz and Fink, 2009; see also Section 2.1.4). The 75 mm isohyet (contour of mean rainfall) reaches up to 10°N in April (Figure 1.5b), and, based on rainfall thresholds, by this time the monsoon has already commenced in the Niger delta and southern Guinea Coast (e.g. Bello, 1996). In this month, convective rainfall events are still often related to disturbances

**Figure 1.2** Maps of West Africa. (a) West Africa is defined here as the region from the western coast to 20°E. *Source*: Open Street Maps. (b) Location of synoptic stations and their reporting frequency to the GTS. *Source*: Met Office Integrated Data Archive System (MIDAS), 1981–2010. The coloured areas indicate the positions of the climate zones: Guinea Coast (GC), Soudanian (Soud), Sahel and Sahara. Hatched areas indicate the part of climates zones used in the Hovmöller diagrams of Figures 1.4, 1.10, 1.11 and 1.23 and the cross-sections in Figure 1.30.

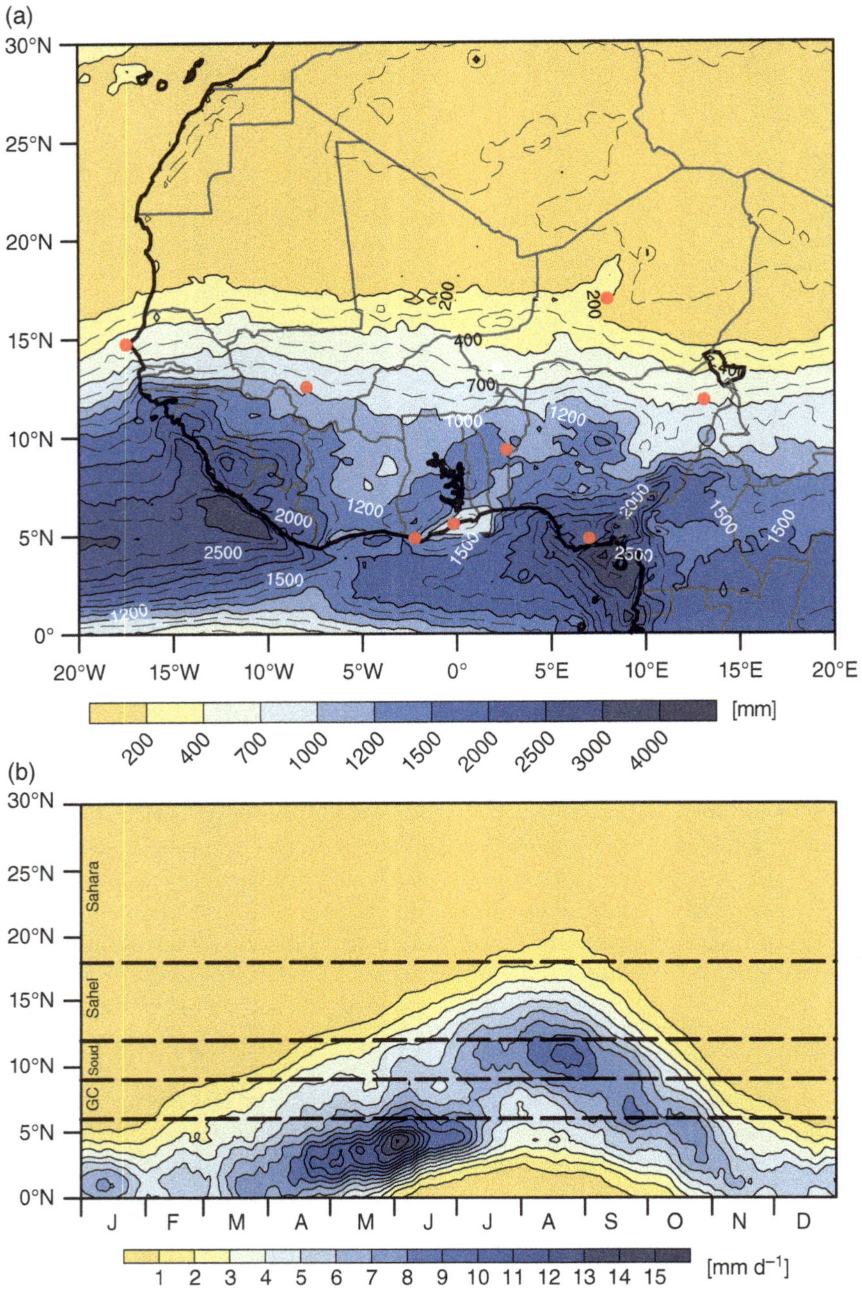


**Figure 1.4** (a) Mean annual precipitation. Red dots denote the positions of the stations shown in Figures 1.6 and 1.7. (b) Hovmöller diagram of the 15-day running mean of daily precipitation, averaged between 10°W and 10°E. Regions are shown in Figure 1.2b. *Source*: TRMM 3B42 Version 7, 1998–2012.

in the upper-level westerlies that mostly originate from extratropical latitudes. In July, after the dynamical onset of the monsoon, rainfall in the Soudano-Sahelian zone is dominated by organised convective systems that are often weakly coupled to synoptic features and only weakly influenced by extratropical systems. On the continent, maximum monthly rainfall on the order of 200 mm is recorded in the Soudanian zone (9–12.5°N), but rainfall in excess of 75 mm reaches as far north as 16–17°N into the central Sahel (Figure 1.5c). The rainfall map discloses the presence of the little dry season at the Coast of Ghana, a weaker little dry season in the western Ivorian Gulf and the persistence of stronger rains in most parts of coastal Nigeria. In October (Figure 1.5d), the rainfall belt retracts equatorward, with substantial rains persisting on the Jos Plateau and Guinea Mountains.

Figure 1.6 shows the monthly and 15-day running mean seasonal cycle of rainfall at nine selected stations (for locations see Figure 1.4a). Stations at the Guinea coast to the east of Cape Palmas have a bimodal rainfall distribution with more abundant rains during the first rainy season in May/June, less rain during the second rainy season in October, and a little dry season in July–August (see Axim and Accra in Figure 1.6g and h). The little dry season weakens toward the Niger delta, and Port Harcourt (Figure 1.6i) shows only a weak bimodality, if any. Accra, in the centre of the Ghanaian part of the Ghana–Benin dry zone, only receives 661 mm of annual rainfall, compared with

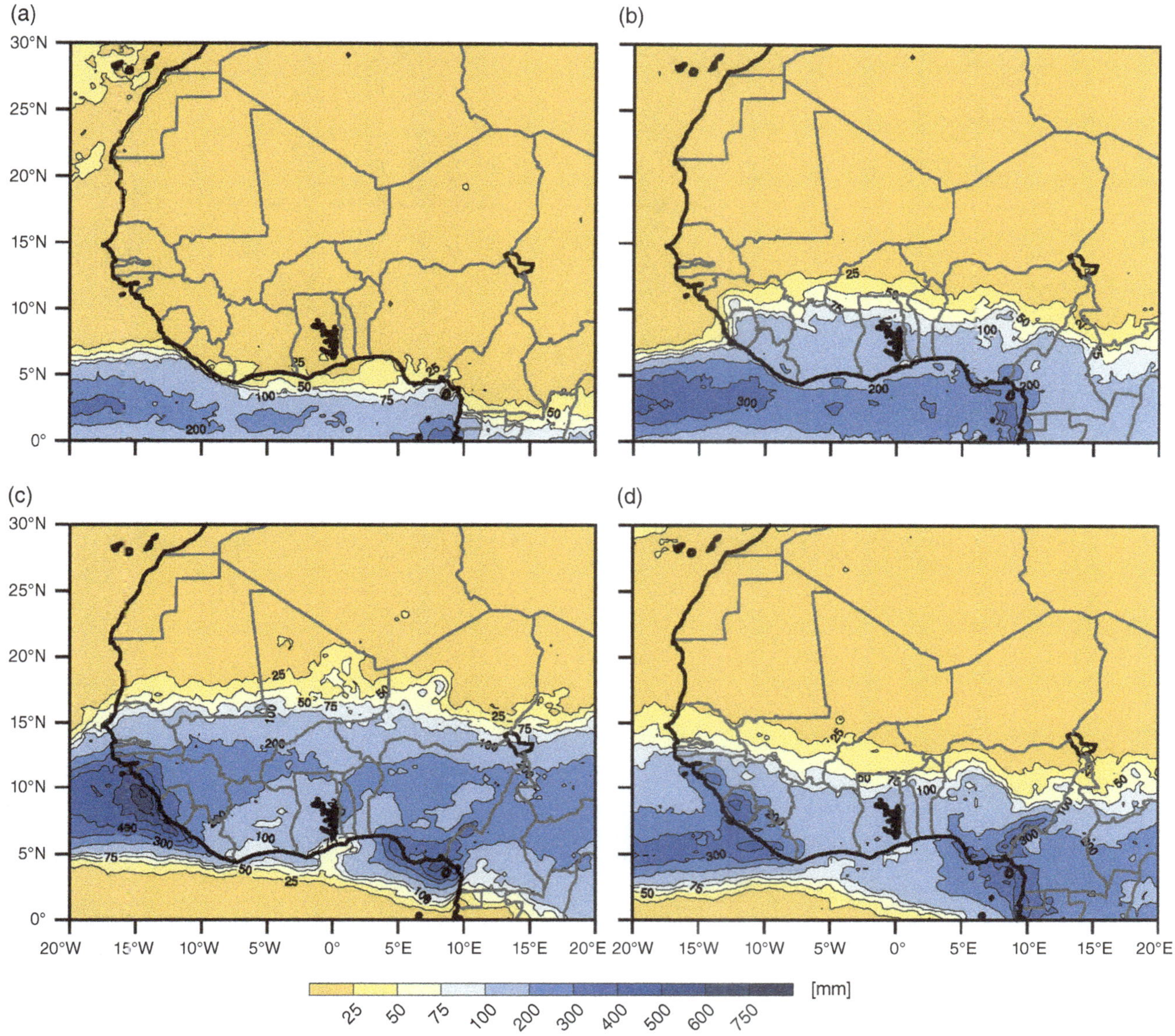


**Figure 1.5** Mean monthly rainfall for (a) January, (b) April, (c) July and (d) October. *Source*: TRMM 3B42 Version 7, 1998–2012.

1859 mm in Axim further west. All stations further north have a unimodal rainfall distribution. Parakou, located at around 9°N in the transition zone between the Guinea coastal and Soudanian zones, has a single, but broad rainfall peak culminating in August (Figure 1.6e). The peak gets narrower and shallower the more northward the stations are located (Bamako, Maiduguri, Dakar, and Niamey; Figure 1.6d, f, a and b respectively). Agadez is located at the southern fringes of the Sahara and receives only 136 mm in an average year (Figure 1.6c). Figure 1.7 exemplifies empirical rainfall probability density functions both in terms of daily rainfall amount and of frequency in bins of 5 mm for the same station sample as in Figure 1.6. It is generally seen that daily rainfall totals between 0 and 5 mm are most frequent (green bars in Figure 1.7). This is especially true for the wettest stations Port Harcourt and Axim (Figure 1.7i and g respectively). For these stations, daily rainfall totals between 5 and 10 mm contribute the largest fraction to the annual rainfall. Bamako and especially Parakou are stations at which daily rainfall of 15–20 mm contributes the most to the annual rainfall total (Figure 1.7d and e). Several other stations, like Dakar, Agadez, Maiduguri and Accra, have a rather equal contribution of rainfall intensities of up to 25 mm to the annual total (Figure 1.7a, c, f and h respectively).

The main contribution of daily rainfall amounts of about 10–25 mm to the annual total hints at the dominance of rainfall from strong mesoscale convective systems (MCSs). An MCS is the generic term for an organised convective system that has an extensive and cold cloud area in infrared satellite imagery and produces a contiguous precipitation area on the order of 100 km or more in at least one direction (Glickmann, 2000). Figure 1.8, taken from the 20-year (1986 to 2005) climatology presented in Bennartz and Schröder (2012), shows that the coldest (i.e. highest cloud tops), fastest

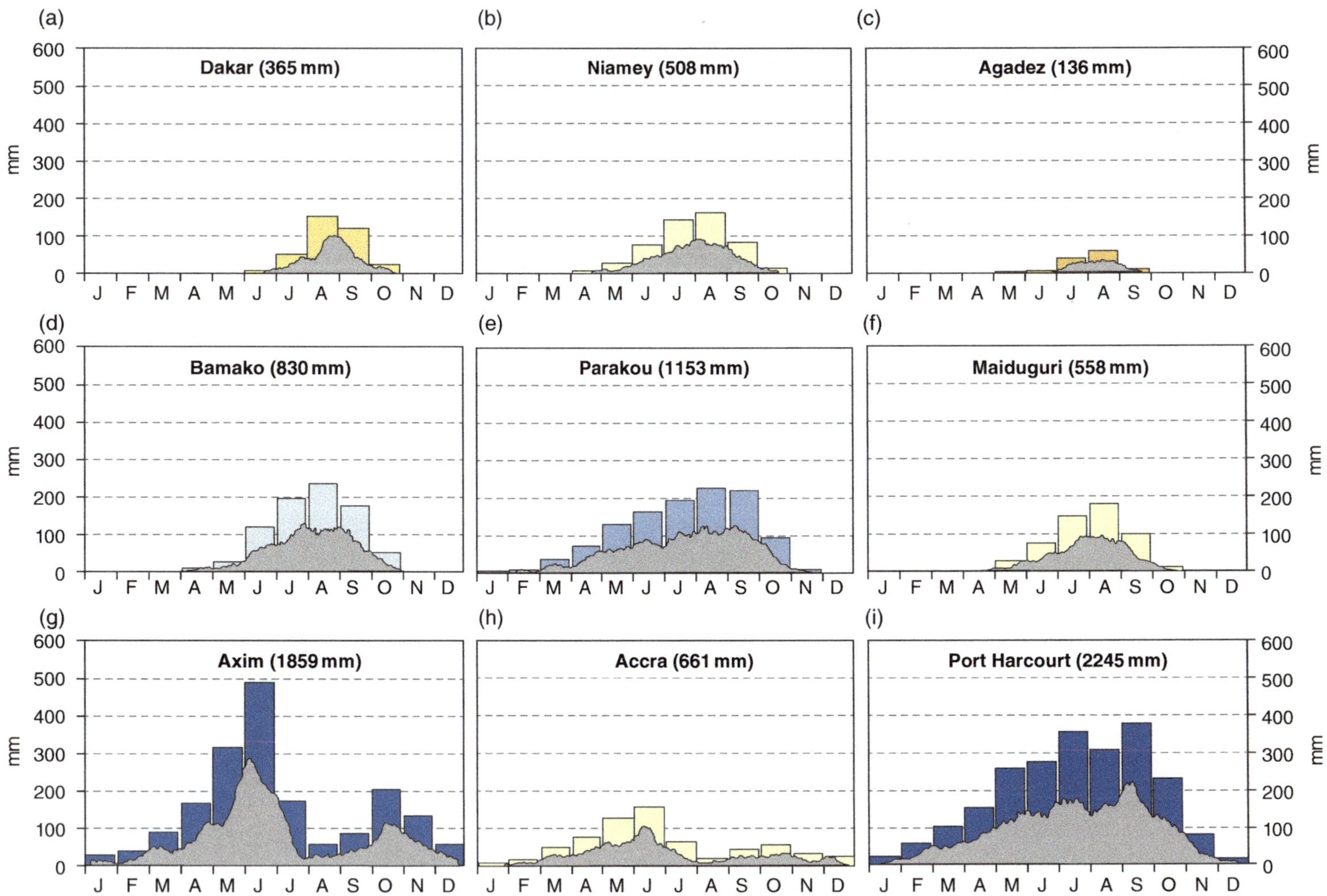


**Figure 1.6** *Colour bars:* mean monthly precipitation between 1981 and 2010 in mm (left and right axes) for (a) Dakar, 14.73°N, 17.50°W, (b) Niamey, 13.48°N, 2.17°E, (c) Agadez, 16.97°N, 7.98°E, (d) Bamako, 12.53°N, 7.95°W, (e) Parakou, 9.35°N, 2.62°E, (f) Maiduguri, 11.85°N, 13.08°E, (g) Axim, 4.87°N, 2.23°W, (h) Accra, 5.60°N, 0.17°W, and (i) Port Harcourt, 4.85°N, 7.02°E. *Grey areas:* 15-day running mean precipitation amount between 1981 and 2010 in millimetres. Mean annual total is given at the top of the panel. Station locations are shown in Figure 1.4a. The colour filling of the bars corresponds to Figure 1.4a.

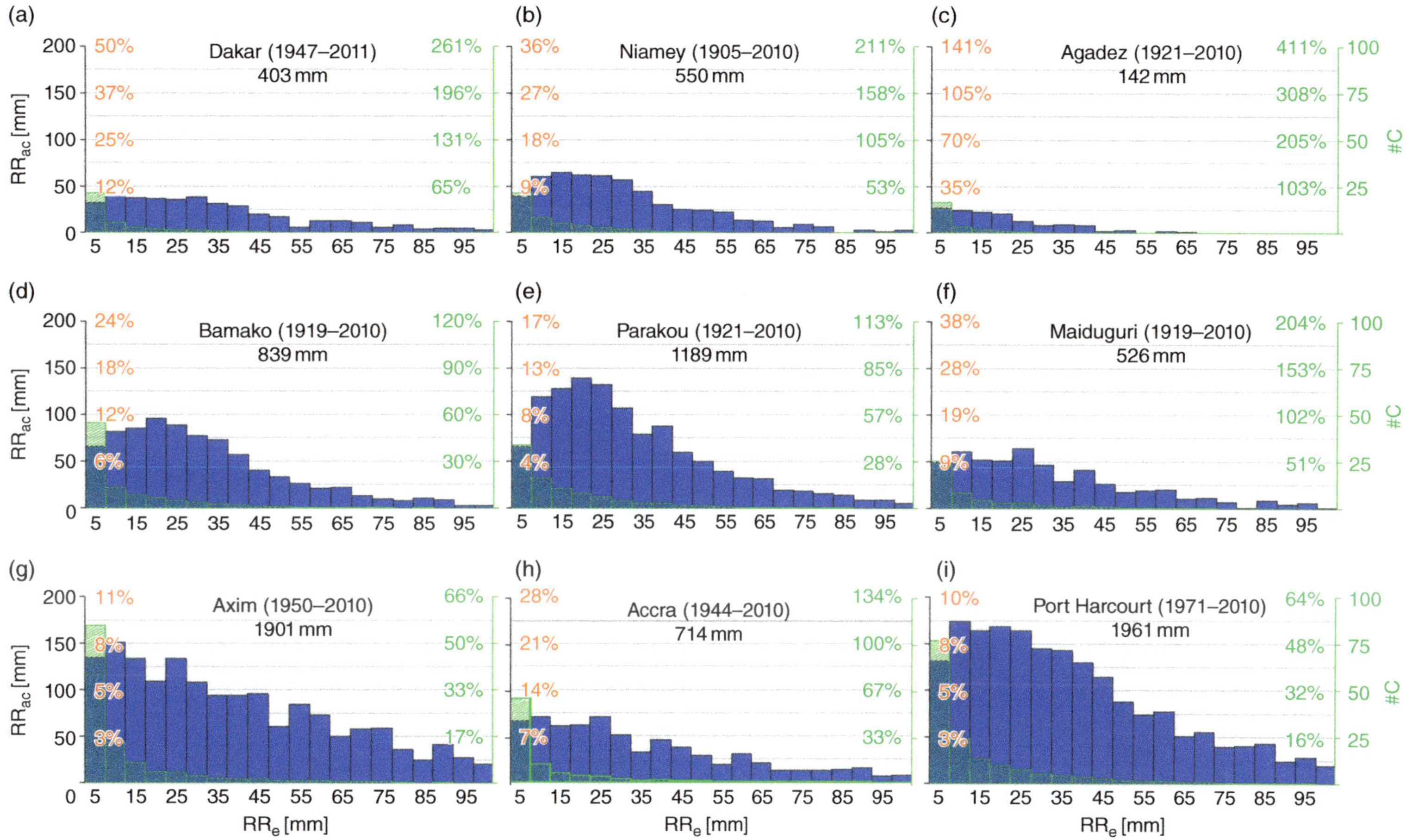


**Figure 1.7** Accumulated rainfall per 5 mm interval in mm (blue bars, scale on left axis) and percentage (red numbers) of the annual total. Number (green bars, scale on right axis) of rainfall events per 5 mm interval in % of total (green numbers). Stations in panels (a)–(i) are the same as in Figure 1.6. Reference period (as long as possible) and mean annual total is given at the top of the panel.

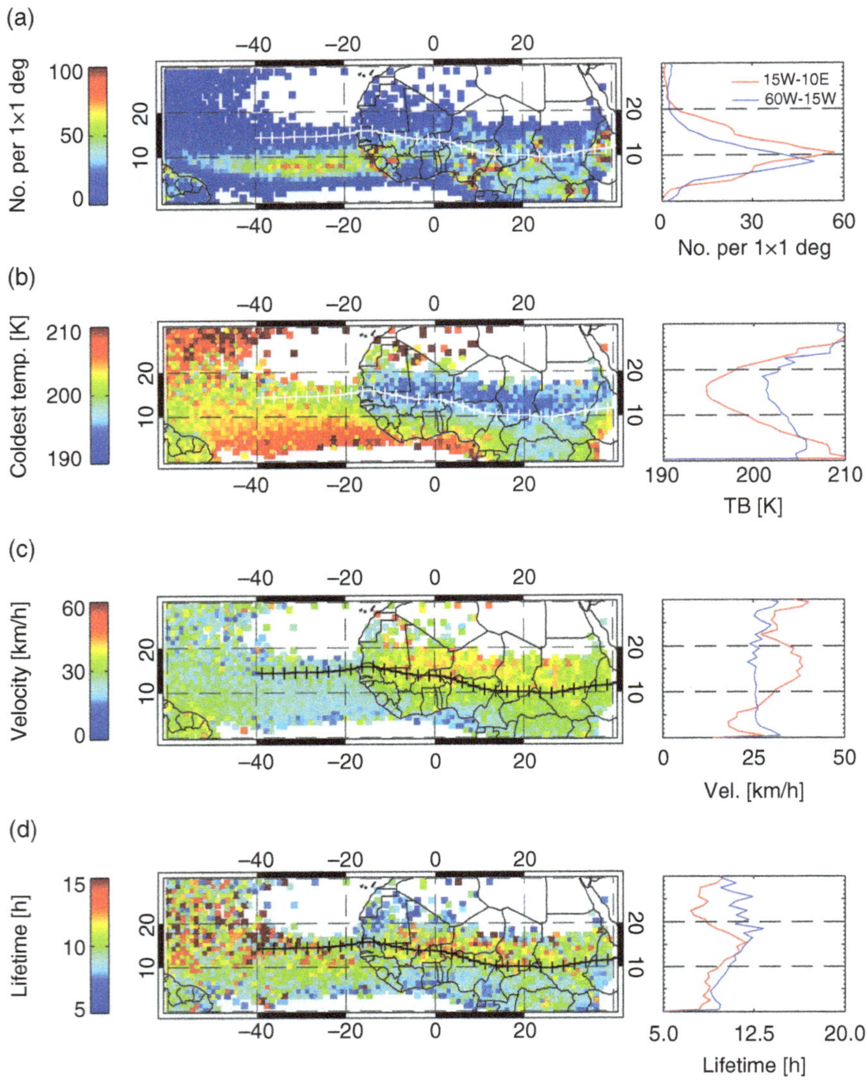


**Figure 1.8** Mean features of convective systems derived from 20 years (July–September 1986–2005) of Meteosat Visible and Infra-Red Imager radiometer observations. Data are binned in 1°×1° boxes according to their origin. Values are provided for: (a) the total number of convective events identified; (b) the coldest temperature reached during system lifetime (mean over all systems in box); (c) mean system velocity; (d) the mean lifetime of the systems. The right panels provide corresponding zonal mean plots for two longitude bands (60°W–15°W and 15°W–10°E). The line between 10° and 20°N is the average position of the 600 hPa zonal wind maximum for the 20-year period (from NCEP reanalysis). *Source*: figure taken from Bennartz and Schröder (2012). ©American Meteorological Society. Used with permission.

moving and longest living MCSs are observed over West Africa in the 10–20°N latitude belt between July and September. In fact, MCSs contribute between 75% and 90% to the annual rainfall in the Soudanian and Sahelian climate zones respectively (Mathon *et al.*, 2002; Fink *et al.*, 2006; see also Section 3.1.3).

Measuring actual *ET* is difficult. It involves the deployment of lysimeters that are costly and pose a challenge in terms of installation and maintenance. Furthermore, *ET* varies very strongly over short distances of even a few metres, due to the heterogeneity of most land surfaces, so it is a challenge to relate point measurements to regional averages. In general, such measurements are not available for West Africa. WMO 'class A' evaporation pans are maintained at synoptic stations, but they yield potential evaporation over a water body and cannot directly be linked to the *ET* minus *P* (*ET* – *P*) differences relevant for soil moisture. A common approach to infer reasonable patterns of *ET* is to force an ensemble of land surface models with observed rainfall, radiation and other reanalysed surface meteorological variables. For West Africa, this was recently carried out under the AMMA (African Monsoon Multidisciplinary Analyses) Land Surface Model Intercomparison Project (ALMIP; Boone *et al.*, 2009). Results from ALMIP are discussed in Meynadier *et al.* (2010). Figure 1.9 shows annual patterns of *ET*, *P* (taken from TRMM) and *ET* – *P* taken from the Meynadier *et al.* (2010) study over land and from the Hamburg Ocean Atmosphere Parameters and Fluxes from Satellite Data (HOAPS) over the ocean. *ET* reaches up to 2–3 mm day$^{-1}$ in the vegetated zones, but decreases over the Sahel to values of about 1 mm day$^{-1}$ due to the lack of soil moisture and vegetation (Figure 1.9a). The TRMM rainfall in Figure 1.9b is similar to Figure 1.4a, but for a shorter reference period 2002 to 2007. *ET* – *P* is negative over the Guineo-Soudanian zone, with the highest values of –3 to –5 mm day$^{-1}$ over the Niger delta and the southern part of the Cameroon Line Mountains

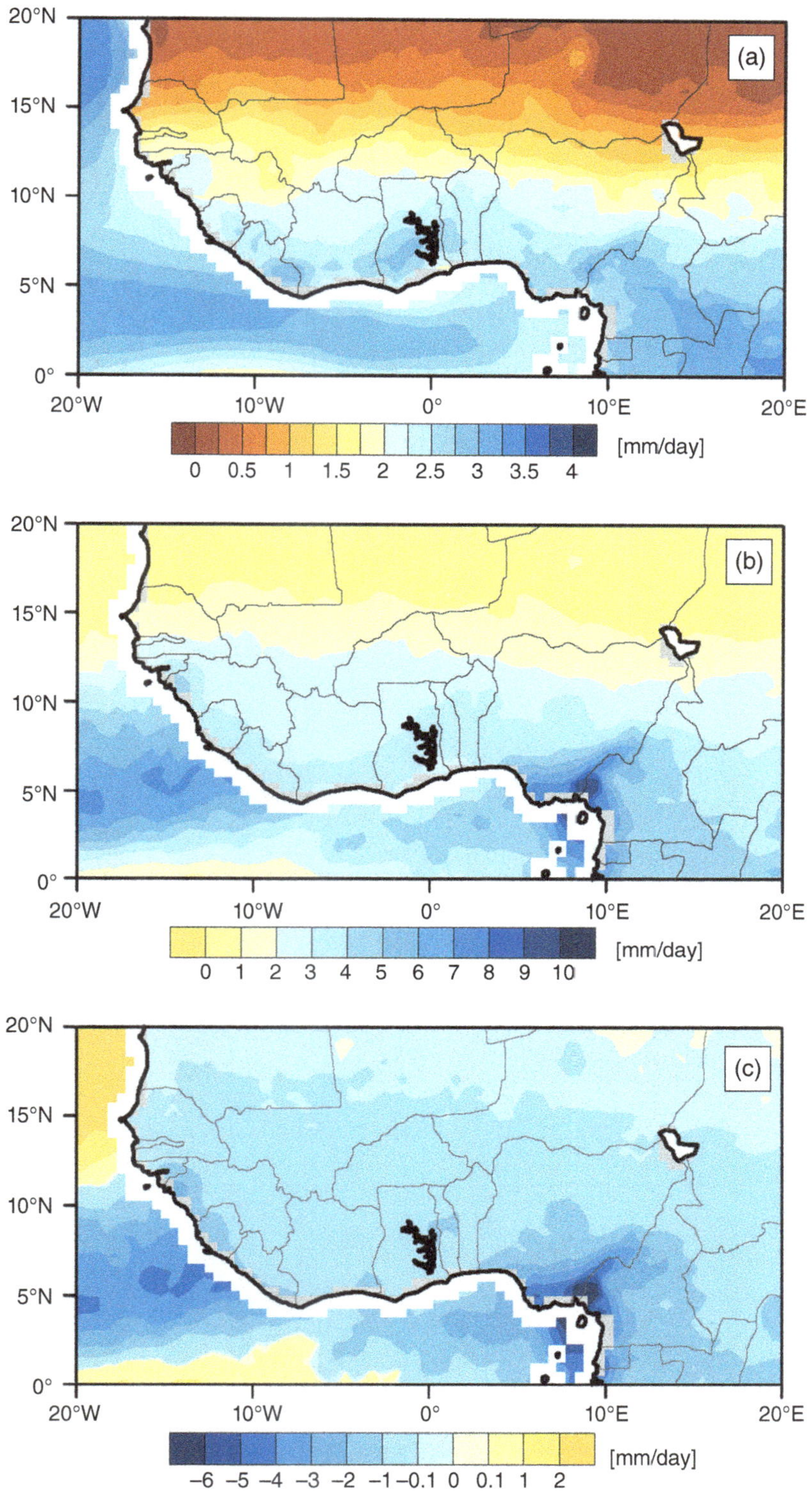


**Figure 1.9** (a) Annual (2002–2007) actual evapotranspiration (*ET*), (b) annual rainfall (*P*), and (c) Annual *ET* − *P*. Units are millimetres per day. *Source*: *ET* from HOAPS V 3.2 over ocean and after Meynadier *et al.* (2010) over land and P from TRMM 3B42 version 7.

(Figure 1.9c). A negative *ET* − *P* is indicative of surface runoff in rivers and/or soil infiltration. The Atlantic Ocean waters adjacent to West Africa also reveal an excess of rainfall over evaporation, except at the west coast north of Guinea and over the equatorial Gulf of Guinea. In the Sahel and Sahara, *ET* − *P* is near zero, indicating arid regions without permanent river flow. It shall be stressed that – owing to errors in the TRMM rainfall estimation, radiation data, surface winds, vegetation and soil cover inventories, as well as deficiencies in the physics of land surface model – large uncertainties remain with the *ET* − *P* patterns depicted in Figure 1.9c.

## 1.3 Water Vapour and Clouds

The inland penetration of the WAM in the summer months is determined by the recycling of water at the surface, through *P–ET*, and by the horizontal transport and mixing of water vapour and clouds in the atmosphere. Therefore, in this section, climatologies of variables related to the atmospheric branch of the hydrological cycle of the WAM are discussed. Vertically integrated water vapour amounts, alternatively known as precipitable water (PW), delineate the moist tropical belt and can be used as a measure of the inland penetration of the WAM, or in moist cases for AEW detection (see Section 2.1.3.2). In the annual mean, PW is highest over the northern Gulf of Guinea and Bight of Bonny, with values between 40 and 50 mm (Figure 1.10a). The highest value of 52 mm occurs in the eastern Bight of Bonny southeast of Malabo. PW contours are zonally oriented and decrease toward the Sahara to values between 10 and 20 mm (Figure 1.10a). The lowest value, of 11 mm, is observed over the Hoggar Mountains due to the desert climate and high terrain. The Hovmöller diagram reflects the annual march of higher tropospheric moisture onto the continent (Figure 1.10b). Isohyets and the ITD, inferred from the 14°C surface isodrosotherm (dew point contour), are also plotted (see Section 11.2 for discussion of the ITD definition).

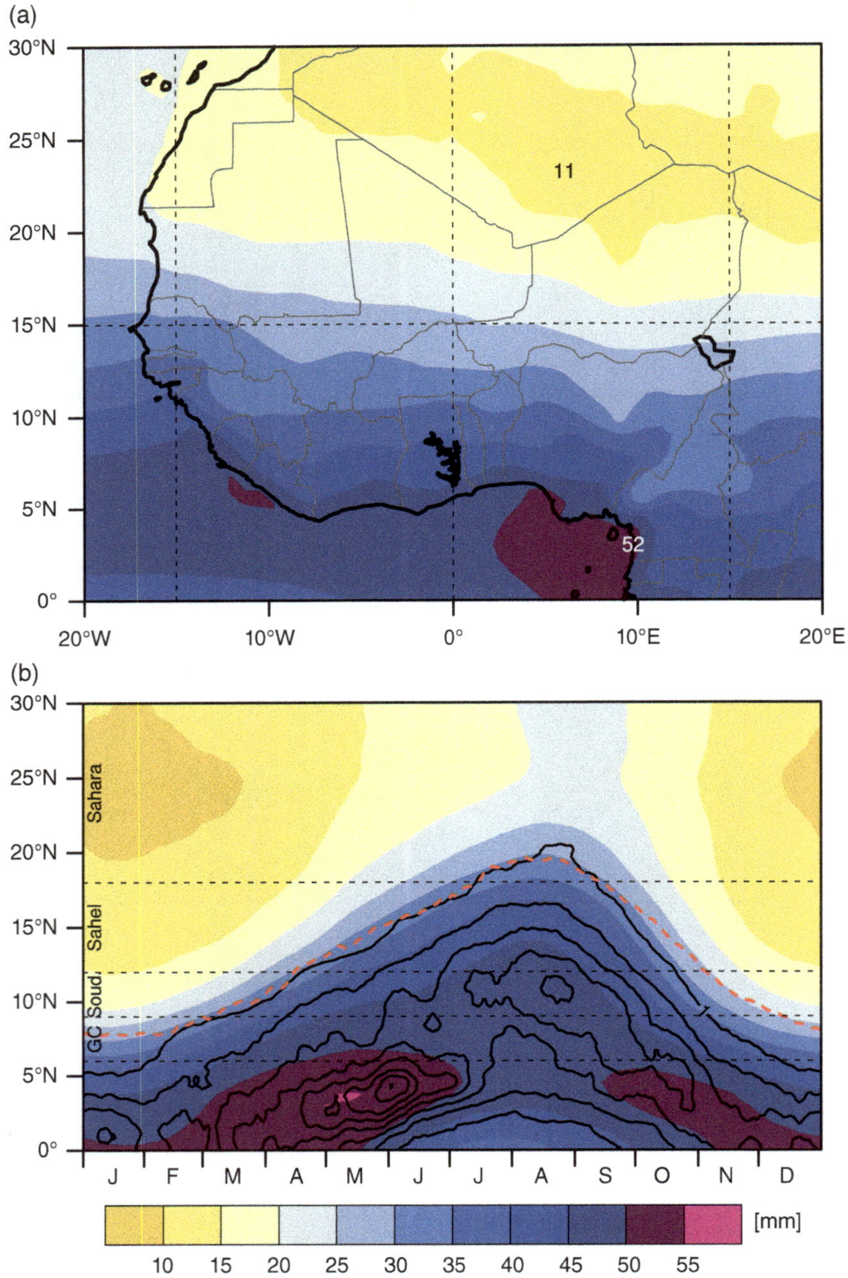


**Figure 1.10** Integrated (1000–100 hPa) PW (mm). (a) Long-term annual mean, (b) Hovmöller diagram of the 15-day running mean of PW and daily precipitation (contours as in Figure 1.4b, source TRMM 3B42 Version 7, 1998–2012), averaged between 10°W and 10°E. Dashed red line indicates the location of the ITD, inferred from the 14°C 2 m dew point isodrosotherm. Regions are shown in Figure 1.2b. *Source*: ERA-Interim, 1981–2010.

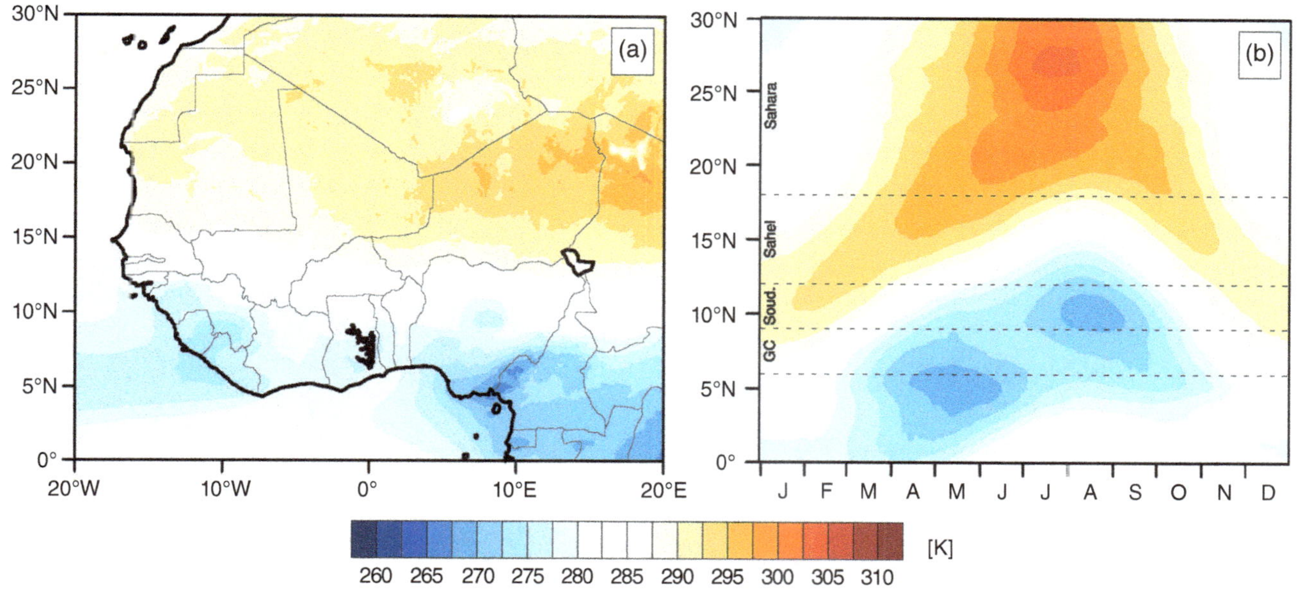


**Figure 1.11** Infrared brightness temperatures (K): (a) annual mean; (b) Hovmöller diagram of the 15-day running mean of daily brightness temperatures. Regions are shown in Figure 1.2b. *Source*: GridSat, 1983–2009, cf. Knapp *et al.* (2011).

The ITD location indicates the most poleward penetration of the moist monsoonal southwesterlies onto the continent (see Section 1.5). Between February and August, the 1 mm day$^{-1}$ (i.e. about 30 mm month$^{-1}$) isohyet aligns well with the ITD and both are located between the 30 and 35 mm PW contours. Overall, the continental penetration of moist tropospheric air and its withdrawal are much smoother when compared with rainfall; the 'monsoon jump', seen in the latter variable between June and July in the sudden displacement of maximum rainfall from about 4°N to 11°N, is not clearly distinguishable in PW (compare Figures 1.4b and 1.10b). This may be understood through the role that PW plays in the rainfall generation process: PW provides the atmospheric moisture reservoir, but rainfall generation and maintenance processes, which may occur on small scales, determine the actual precipitation that reaches the ground. In other words, PW values higher than 30–35 mm are a necessary but not sufficient condition for significant rainfall at the ground.

Minima in maps of satellite infrared (at wavelengths between 10 and 12 μm) brightness temperature are caused by cold cloud tops that are often related to deep cumulonimbus clouds (deep convection) and associated rainfall. Thus, not surprisingly, the monsoon jump is clearly evident in the Hovmöller diagram of brightness temperature (Figure 1.11b), and the annual brightness temperature distribution shows the maximum precipitation zones over the southwest coast and the Cameroon Line Mountains (Figure 1.11a). This brightness temperature signal over the mountains is, however, amplified when compared with hotter lowlands by the fact that during clear conditions the satellite senses a colder land surface in the mountains. Other deficiencies of using brightness temperature as an indicator of mean precipitation are evident over parts of the Gulf of Guinea; brightness temperature would not indicate as much precipitation as is being observed in TRMM (cf. Figure 1.4a). This is likely to be due to rain from warmer clouds over the oceanic ITCZ. Note that the 37 GHz channel of passive microwave sensors aboard TRMM and other satellites receive a direct signal from precipitation over oceans.

The seasonal distribution of brightness temperature in Figure 1.12 not only mimics the northward migration of cold clouds, but also follows hot surface temperatures in the often cloud-free Saharan heat low region (see Section 1.5). Again, brightness temperature and TRMM rainfall are not consistent over the warm waters of the Gulf of Guinea. Compare Figures 1.5b and 1.12b: TRMM points to larger April rains over the Gulf of Guinea, whereas brightness temperatures are coldest over land. One explanation is the presence of deeper and colder convective clouds over land as a result of higher instability that does not necessarily translate into higher rainfall; continental cumulonimbus clouds possess stronger updrafts and produce larger cirrus anvils. As a result, extensive high (about 15 km) and therefore cold cirrus decks – as remnants from deep cumulonimbus convection – especially 'contaminate' the brightness temperature signal in terms of a proxy for rainfall over land.

Figure 1.13 shows mean daytime and night-time cloud and precipitation[4] fractions from CloudSat and Cloud–Aerosol Lidar and Infrared Pathfinder Satellite observations (CALIPSO) for the period June to September between 2006 and 2010. At night, the cloud fraction

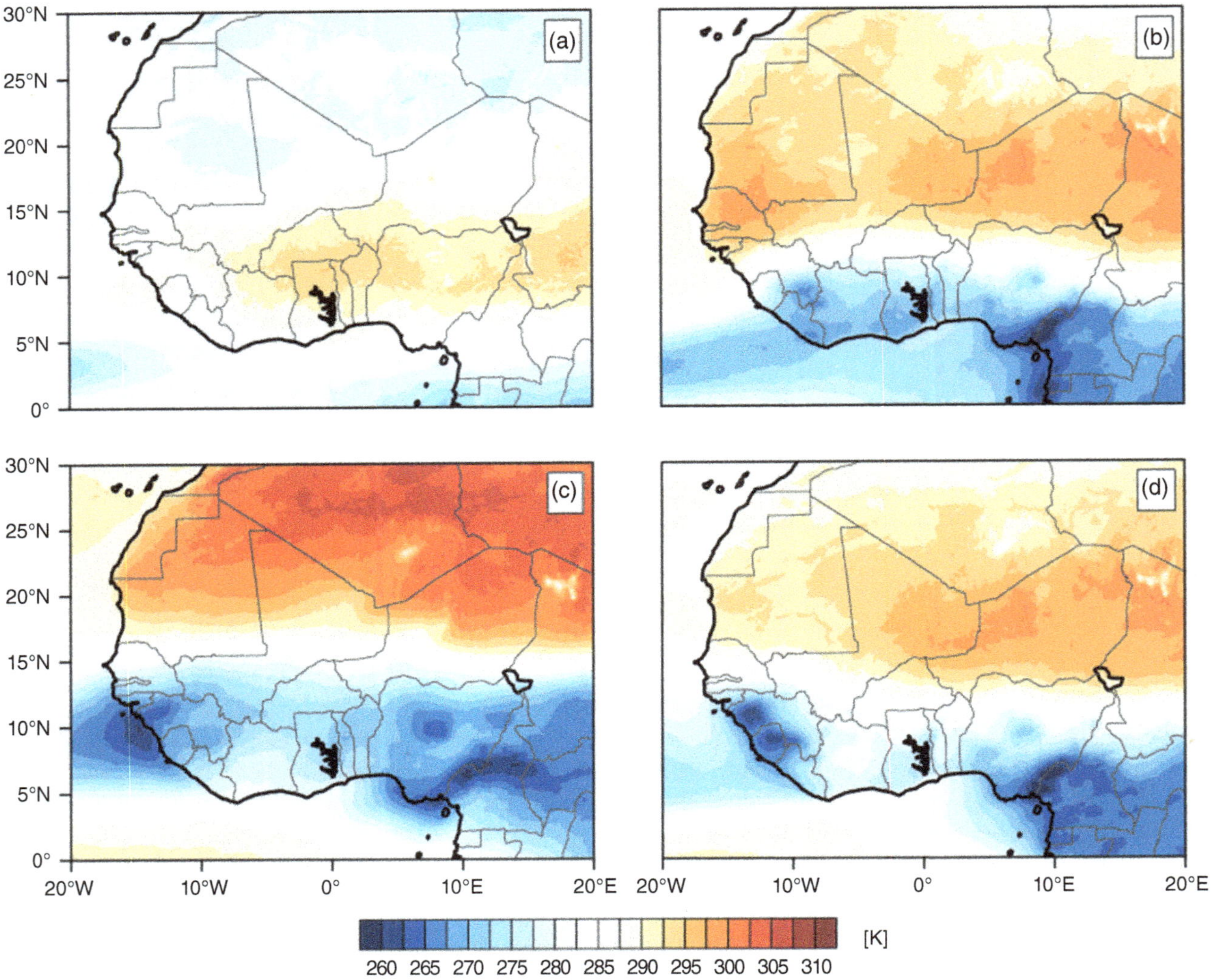


**Figure 1.12** Infrared brightness temperatures (K) for (a) January, (b) April, (c) July and (d) October. *Source*: GridSat, 1983–2009, cf. Knapp *et al.* (2011).

exceeds 50% near the tropopause in the rainy zone around 10°N. This is related to an increase of high-level cloudiness due to anvil outflow from afternoon and early night convection (Figure 1.13a). During the daytime, cloud fractions in excess of 30% occur with the land–sea breeze front at the Guinea Coast around 5–6°N (Figure 1.13b). In this zone, night-time low-level stratus, as described in Schrage and Fink (2012), is seen with fractions of about 20% (Figure 1.13a). This fraction seems to be an underestimation due to the presence of higher clouds and problems with cloud detection in CloudSat in the lowest kilometre above ground. Finally, note the comparably high cloud fraction of 10–15% at 500 hPa over the Sahara (Figure 1.13); this indicates the occurrence of clouds at the top of the Saharan boundary layer as described, for example, by Parker *et al.* (2005a), and by Marsham *et al.* (2013) for Bordj Badji Mokhtar (21°20′N; 0°57′E) in June 2011.

## 1.4 Radiation, Dust, Sunshine Duration, Vegetation and Soil Water

This chapter continues with observables closely tied to the energy and bio-terrestrial water cycle of the WAM. These parameters determine the link between the water cycle of the region and the thermodynamic energy budget, and therefore contribute to driving the dynamics of the WAM. For readers unfamiliar with the basic physics and terminology of the thermodynamic budget, these are outlined in Section 4.1.2.1. Figure 1.14a shows annual mean solar irradiation at the surface as contours. The highest values, in excess of 260 W m$^{-2}$, occur over the central northern Sahel and south central Sahara, as a result of the rather cloud-free conditions throughout the year in these zones. Further west, the occurrence of mid- to upper-level clouds associated with tropical plumes (Fröhlich *et al.*, 2013) reduces the surface irradiation

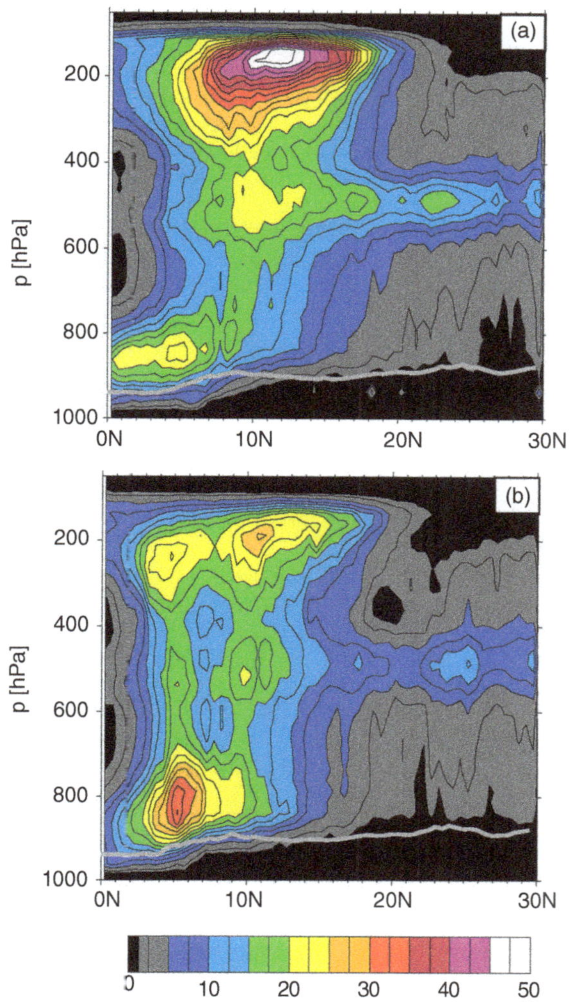


**Figure 1.13** Cloud and precipitation fraction for June to September (2006–2010) observed by CloudSat and CALIPSO for (a) night-time overpasses and (b) daytime overpasses. Fractions are shown at 2.5% intervals at a resolution of 0.5° in latitude and 25 hPa in pressure, averaged over all observations throughout the period between 10°W and 10°E. Values below 1.25% are shown in black with a thick grey line indicating the mean pressure at 600 m above the surface. *Source*: amended from Stein *et al.* (2011).

somewhat. Minimum values, below 220 W m$^{-2}$, occur in a zonal band just north of the Guinea Coast. Here, deep convection, land–sea-breeze-related cloudiness and the presence of morning stratus (Knippertz *et al.*, 2011) reduce solar irradiation.

The mean annual net surface radiation is at maximum over the waters of the Atlantic Ocean, where a low surface albedo increases absorption and where, compared with hot land surfaces, outgoing longwave radiation (OLR) is lower. Values peak west of Cape Verde, where a 180 W m$^{-2}$ radiative gain at the surface is observed (Figure 1.14a). Over land the largest radiative energy surplus at the surface occurs over the Guinea Mountain region and parts of southern Ghana and Togo. A high surface albedo of over 0.5 combined with high values of OLR yield only small positive values of net surface radiation of 30 W m$^{-2}$ in the Ténéré sand desert in eastern Niger (Figure 1.14).

In general, the high surface albedo in large parts of the Sahara desert (Figure 1.14b), the high radiative loss to space due to hot desert surfaces and a dry troposphere (Figure 1.10a) cause the entire Sahara to be a region of annual-mean net radiation deficit at the TOA (top-of-atmosphere, Figure 1.14b). In the absence of precipitation, significant warm air advection or soil heat storage (it can be assumed that these processes are small or negligible), adiabatic warming due to sinking motion must compensate the radiative cooling. Thus, the TOA radiation deficit reinforces the dryness of the region by augmenting the subsidence, which reduces relative humidity. Under the impression of an ongoing strong West African drought in the early 1970s, this radiative mechanism was put forward by Charney (1975) to postulate that land degradation in the Sahel will reduce cloud cover and therefore cause desertification. This process has, however, subsequently been shown to be much more complex, with evidence of both land recovery and degradation in the Sahel, and so the Charney mechanism is no longer linked to desertification at timescales smaller than centuries. The 15°N parallel divides the region of annual-mean TOA radiation deficit in the north from radiation surplus in the south. In other tropical regions this dividing line is located further poleward between 30 and 40°N latitude. Thus, it is prudent to state that the huge and hot Sahara desert cools the Earth's atmosphere. The net TOA radiation surplus peaks in values exceeding 60 W m$^{-2}$ over the tropical Atlantic Ocean away from the ITCZ, and over the northern Congo basin region where dark rainforests are present.

The seasonal cycle of solar incoming radiation is a combination of the Sun's zenith angle and cloudiness. Generally, incoming solar radiation peaks before the overhead position of the Sun occurs, due to simultaneously increasing cloudiness in all West African regions (Figure 1.15a). The Soudanian zone (9–12.5°N) is the only region that exhibits two clear minima in surface solar irradiation in August and January. Here, the first peak in March is higher than in November, most likely due to the presence of more clouds just after the rainy season. The net surface radiation in Figure 1.15b clearly shows the differences among the regions. In the Sahel, the net radiative gain at the surface increases until August despite more cloudiness. This is a result of the surface greening and associated albedo lowering, as well as increased atmospheric humidity after the poleward passage of the ITD that reduces the night-time loss of energy through OLR (Guichard *et al.*, 2009). Figure 1.15c clearly shows that the radiation balance at the TOA is negative over the Sahara throughout most of the year. This is also true for the Sahel between November and March.

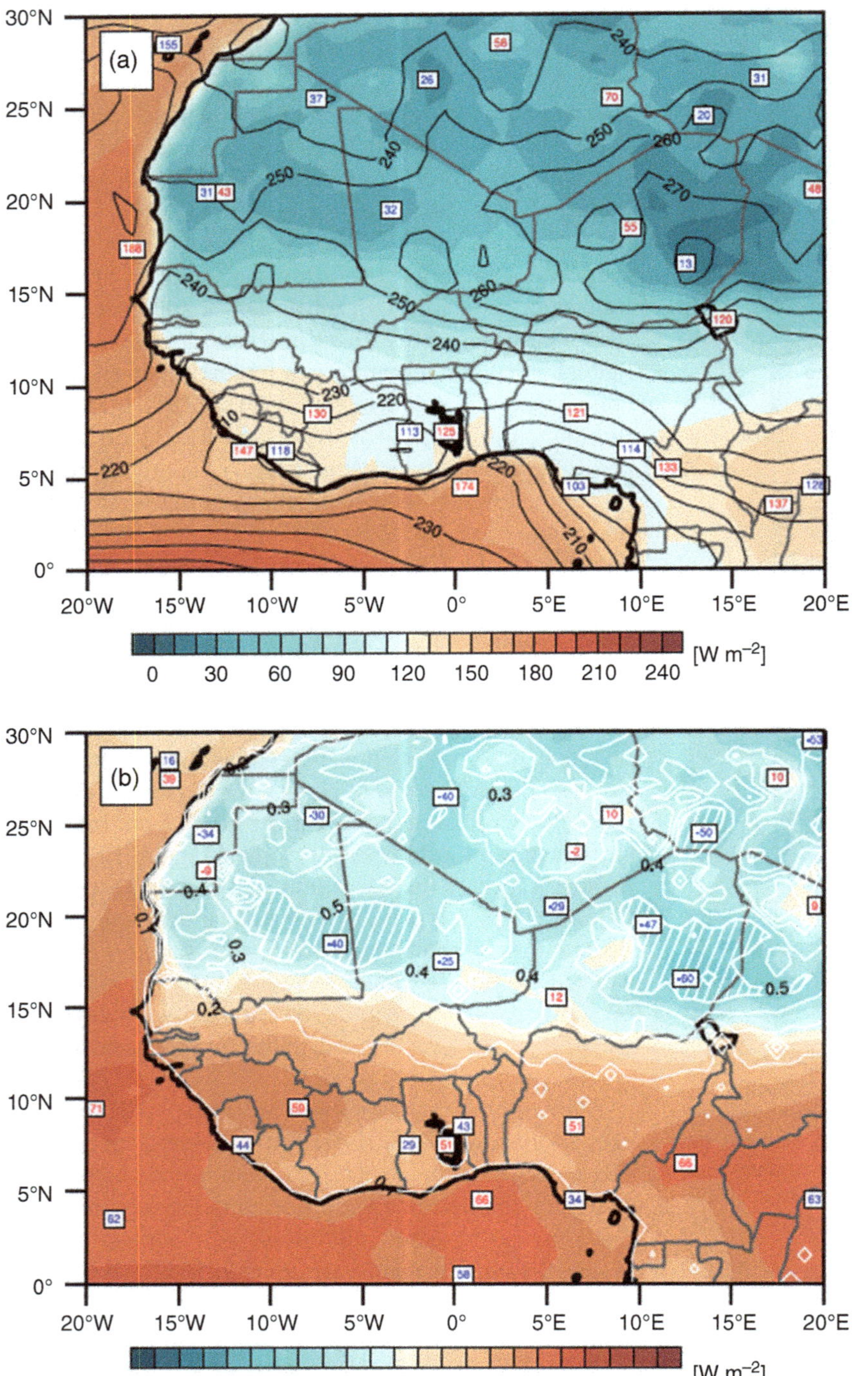


**Figure 1.14** Mean annual net radiation at (a) the surface and (b) the top of the atmosphere (TOA). Contours in (a) and (b) denote the incoming surface solar radiation and surface albedo, respectively. The hatched areas in (b) denote areas with albedo larger than 0.5. Boxed red (blue) numbers denote local maxima (minima) of net radiation at the surface in (a) and at the TOA in (b). *Source*: Global Energy and Water Exchanges Project (GEWEX), 1984–2007 and Global Inventory Modeling and Mapping Studies (GIMMS), 1982–2005.

The Sahara is the largest source of Aeolian mineral dust (see Chapter 5). Figure 1.16 shows the frequency of dust weather reports at synoptic stations between 1983 and 2010 for January, April, July and October. In winter, many stations north of the ITD (see Section 1.5) observe dusty weather between 20 and 50% of all available observations (Figure 1.16a). The TOMS/OMI (Total Ozone Mapping Spectrometer/Ozone Monitoring Instrument) AI (aerosol index) indicates the largest aerosol contents in winter, south of the ITD along the Guinea Coast. The TOMS/OMI AI is a semi-quantitative measure of the column aerosol content, is sensitive to both dust aerosols and soot from biomass burning, and is more sensitive to elevated aerosol plumes than plumes nearer the ground (Section 5.2.2.3). Thus, the combination of biomass burning in vegetated southern West Africa and the equatorward advection of Saharan dust aloft causes the TOMS/OMI AI peak in the Guinea Coast region. In April, the zonal band of high TOMS/OMI AI values is located over the Sahel, while the dust frequencies at

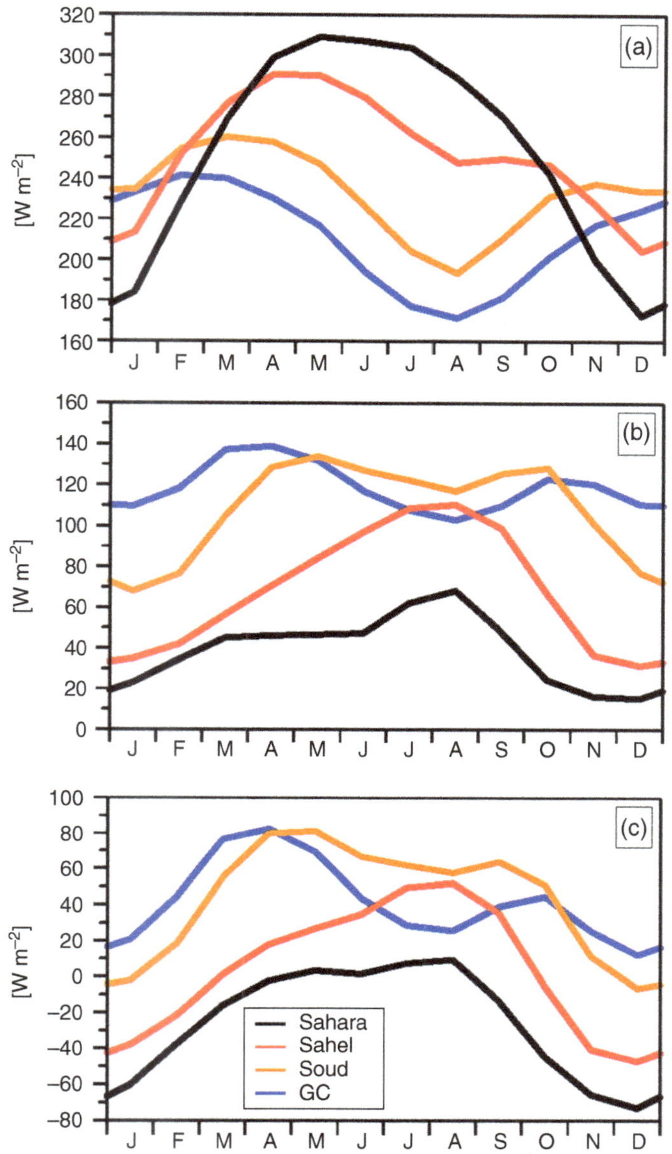


**Figure 1.15** Monthly means of (a) surface solar irradiation, (b) surface net radiation and (c) TOA net radiation (W $m^{-2}$), averaged between 10°W and 10°E. GC: Guinea Coast zone; Soud: Soudanian zone. Zones are depicted in Figure 1.2b. *Source*: GEWEX, 1984–2007.

weather stations do not change much (Figure 1.16b). In July, the western part of the Sahara is the dust hot spot (Knippertz and Todd, 2010). In the summer season, haboobs (see Chapter 5) also lead to dust lifting south of the ITD (Figure 1.16c). Overall, autumn seems to be the season with the lowest dust activity (Figure 1.16d), related to both increased vegetation cover (Figure 1.18d) and less active dust lifting processes (Section 5.1.1.1). In the annual mean, stations in the Sahel have the highest frequencies of dusty weather, and Klose *et al.* (2010) coined the term 'Sahel dust zone'.

Soil moisture is important for agricultural practices. However, spatially distributed information on soil moisture in the plant root zone is lacking. With the help of satellite sensors, soil moisture in the topmost few centimetres can be estimated by, for example, the diurnal variation of the skin brightness temperature, at least in regions with a not too dense vegetation. The volumetric soil water content in the topsoil layer for January 1983 to 2010 is shown in Figure 1.17a. Values below 0.15 $m^3\ m^{-3}$ (i.e. 0.15 $m^3$ water in 1 $m^3$ soil) prevail to the north of the region with dense vegetation and are indicative of the dry soils during the dry season. April already shows an increase in the volumetric soil water content in the Guinea Coast and Soudanian zones as a consequence of the first rains (Figure 1.17b). Here, values of 0.3 $m^3\ m^{-3}$ are reached in July (Figure 1.17c). The wetting of the soils in the Sahel is also evident. In October, some drying is evident in the northern Sahel, while in the remaining area the soil moisture stays near the July levels (Figure 1.17d).

Typically, areas of rainforest and savannah forests can be found to the south of around 12°N, while trees are increasingly sparse to the north of this parallel. However, the vegetation cover also has a strong seasonal variation due to changes in the leaf area index: increased soil moisture goes along with a greening of the vegetation.

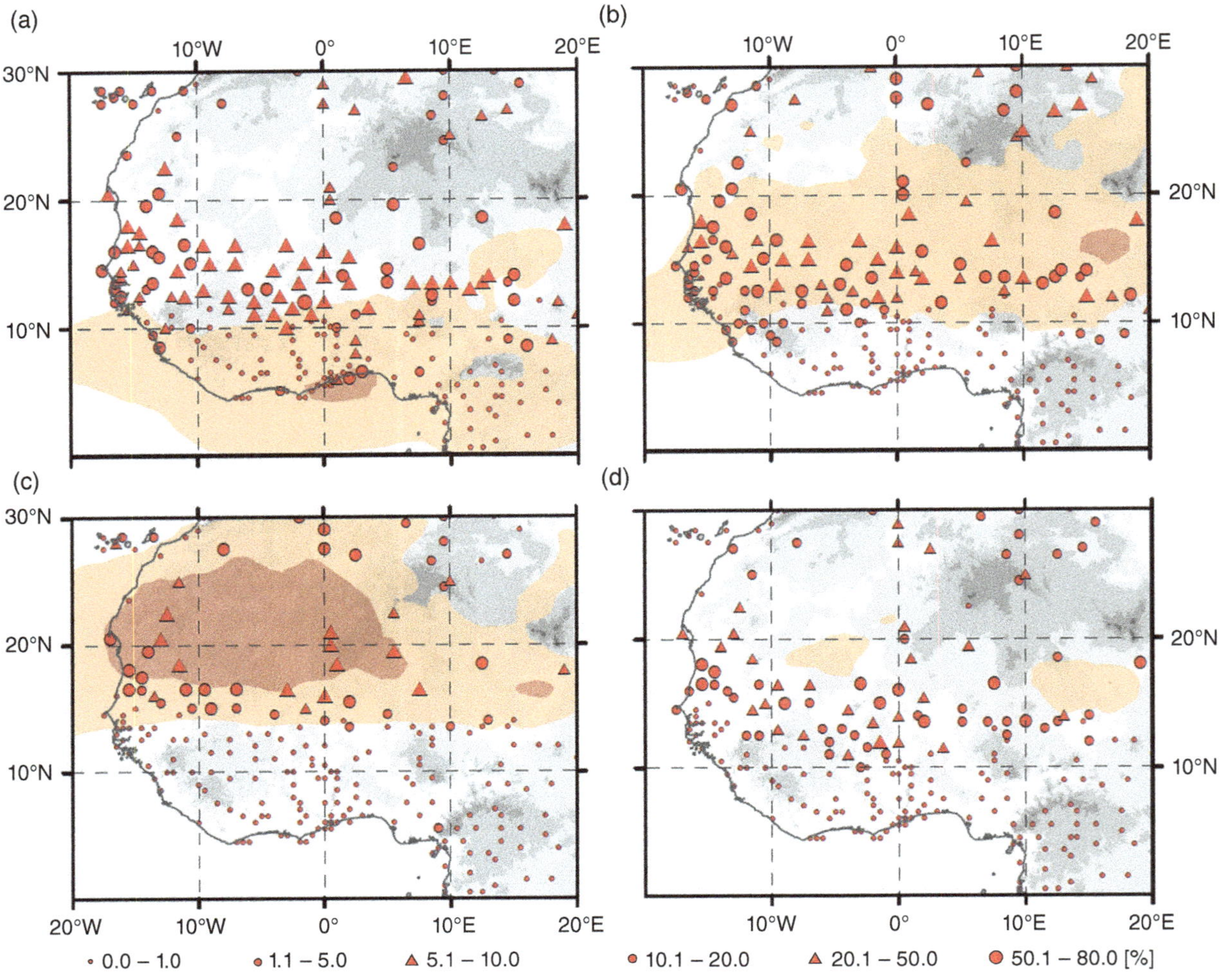


**Figure 1.16** Frequency of dust weather in SYNOP reports for (a) January, (b) April, (c) July and (d) October. *Source*: MIDAS, 1983–2010. Light brown and brown shaded areas indicate regions of monthly averaged TOMS/OMI AI for the analysis period exceeding an index value of 1.8 and 3 respectively (updated from Klose *et al.* (2010)).

The seasonal evolution of vegetation can be diagnosed with the help of the normalised differential vegetation index (NDVI). The climatological greening of the vegetation from January to April is evident, as is a further wilting of the vegetation in the Sahel, causing a strong gradient in 'vegetation greenness' at around 10°N (Figure 1.18a and b). This gradient is even stronger, but displaced northward by about 2° latitude, in July (Figure 1.18c). Not much change is seen between July and October, though even some weak greening occurs in the northern Sahel as a late response to summer rains (Figure 1.18d).

## 1.5 Pressure, Temperature, Humidity and Wind at the Surface

This section deals with climatologies of some classical surface observables recorded at weather stations. Over the Sahara, surface net radiation is primarily responsible for air temperature fluctuations due to the low cloudiness throughout all seasons, with temperature highest in areas or months with maximum net surface radiation surplus. Increasing humidity, cloudiness and precipitation in the summer solstice season make this relation more complex in the Sahel (Guichard *et al.*, 2009). This also holds for the mean surface pressure response to the net surface solar radiation. However, surface pressure is influenced by temperature, such that low-pressure systems tend to develop in areas of high temperature, and particularly in the Saharan heat low in summer. In addition, thermal gradients near the equator develop airflow that tends to move from high to low pressure, resulting in the low-level moist monsoonal southwesterlies. This chapter will elucidate that surface temperature and humidity, mean sea-level pressure (MSLP), and surface winds fields can be understood as this kind of response to the seasonal variations in the net solar radiation (see Section 1.4 and Figure 1.15).

Figure 1.19 shows MSLP and 10 m wind vectors for January, April, July and October. The MSLP threshold for changing from red to blue colours was chosen to be 1008 hPa, so that the blue colours denote the area of the

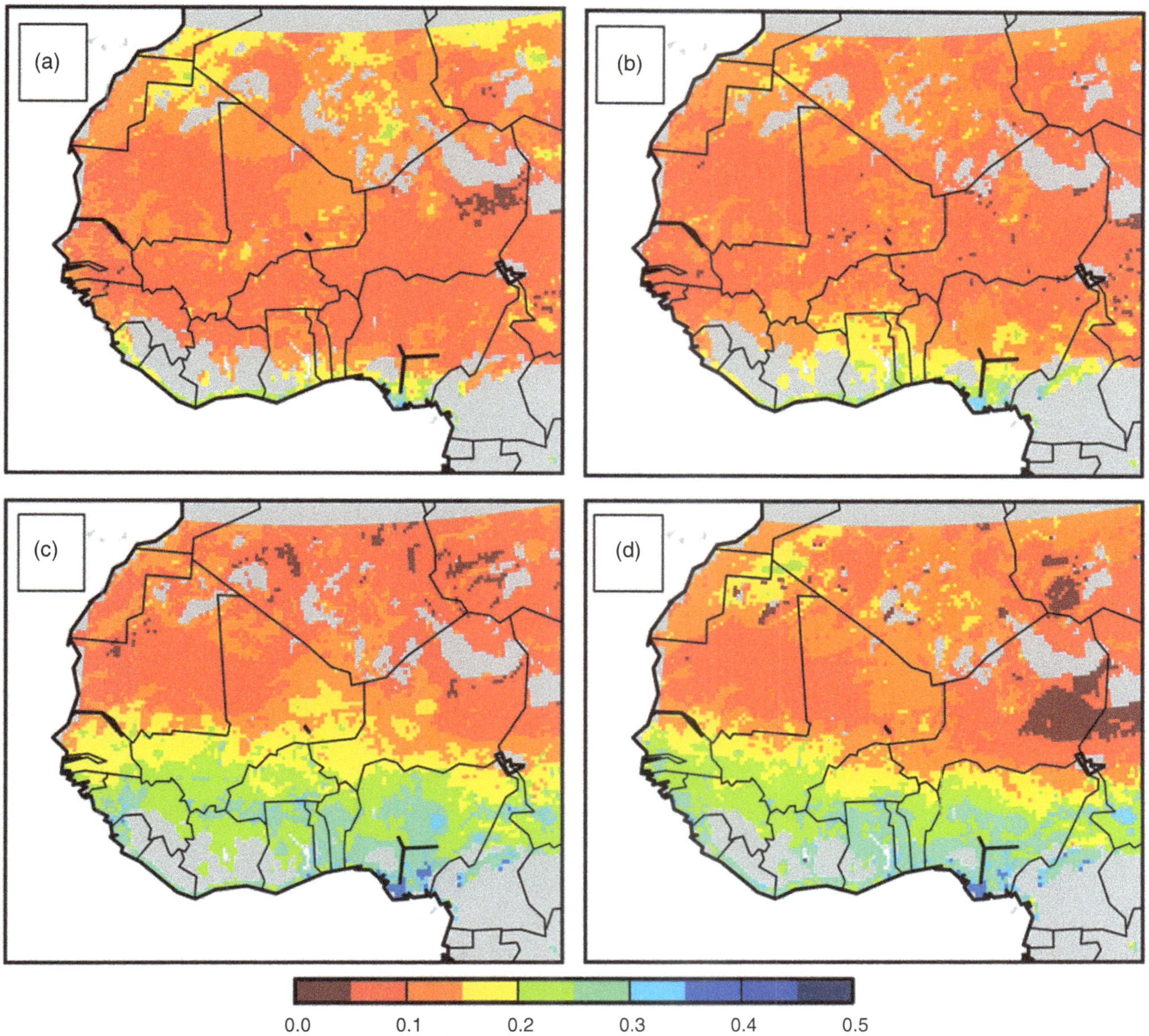


**Figure 1.17** Climatology of volumetric topsoil moisture ($m^3\ m^{-3}$) for (a) January, (b) April, (c) July and (d) October. Grey areas depict regions where the vegetation density did not allow for a satellite assessment of topsoil moisture. *Source*: Meteosat, 1983–2010.

heat low, as defined by Lavaysse *et al.* (2009). In winter, an east–west zone of low pressure between 1009 and 1011 hPa is located just north of the Guinea Coast at about 7–8°N. The Guinea Coast thus experiences a weak onshore monsoonal flow (Figure 1.19a). To the north of this weak winter-time heat low, the Harmattan winds blowing from the northeast are stronger, reaching about 4–5 m $s^{-1}$. In April, the heat low establishes itself in the central Sahel, with lowest values below 1007 hPa in the Lake Chad area (Figure 1.19b). Monsoon winds intensify over the Gulf of Guinea and penetrate the continent as far as 12°N. The Harmattan winds weaken due to the weakening of the high pressure in the northern Sahara. In July, the heat low shifts northwestward and is situated over the western part of the Sahara, over northern Mali. Therefore, it is termed 'West African heat low' (WAHL) to distinguish it from its stronger counterpart over the eastern Sahara. The low-level southwesterly monsoon winds reach the 20°N parallel in July (Figure 1.19c). Note the strong trade winds off the northwest African coast that also ventilate the WAHL at its northwestern corner. In October, the heat low weakens and the monsoonal southwesterlies start their retreat from the continent (Figure 1.19d). Finally, it shall be stressed that although the mean monsoonal southwesterlies are very weak, hurricane-strength gales can occur in association with gusts from thunderstorms and squall lines (a linear type of MCS; Section 3.1.3.2). Fink *et al.* (2010) show the observed gusts at Gaya (Niger, 11°53′N; 3°27′E) in the 2005 rainy season with a peak gust of 153 km $h^{-1}$

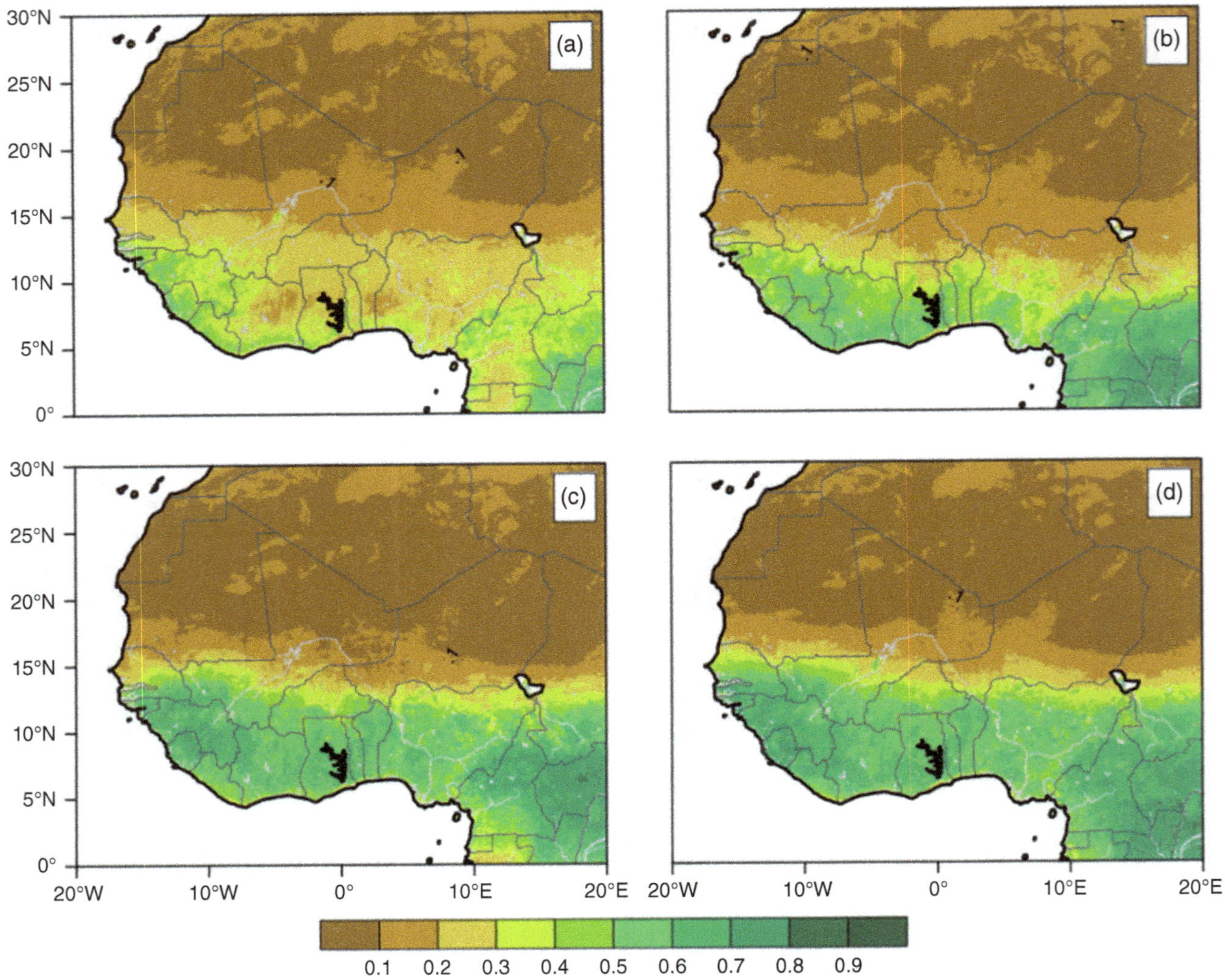


**Figure 1.18** Mean NDVI for (a) January, (b) April, (c) July and (d) October. *Source*: GIMMS, 1982–2005.

(or 83 kt) on 14 June 2005. Gales from cold microbursts and gust fronts flowing out of thunderstorms can cause localised severe damage (see Box 3.1).

Figure 1.20 shows 'wind roses' from weather stations across the continent. The coastal stations Cotonou and Abidjan observe wind from the southwestern sector at most times throughout the year. At the Guinea Coast, Harmattan winds occur only during a few days when the humidity drops, dust content increases and 'heat waves' are possible due to the reversal of winds from onshore to offshore (Knippertz and Fink, 2006). In the Sahel, winds occur from the southwest and northeast at about the same frequency per year (e.g. Niamey and Bamako). Further north, Harmattan winds prevail (Tomboctou) or are dominant (Bilma). Tamanrasset in the Hoggar Mountains shows prevailing easterly winds that occur in northern summer, but also a large fraction of WNW winds that are due to westerlies in winter. Dakar is an exception to the Sahelian stations since northerlies dominate; the ITD dips equatorward toward the West Coast and the station is only rarely on its equatorward side. Thus, Dakar is influenced by the trade wind system of the North Atlantic throughout the year. Further inland, the frequency distribution of wind directions at Tambacounda is more similar to, for example, Bamako, though a peak in westerly wind indicates the penetration of land–sea breeze circulation (or 'Atlantic inflow'; Grams *et al.*, 2010).

The seasonal cycles of the 2 m temperature (coloured) and dew point temperature (contours, with the 14°C dew point temperature contour highlighted as a bold dashed line to indicate the ITD location) are shown in Figure 1.21. Consistent with the area of confluent surface winds, the ITD is located just north of the Guinea Coast in January (Figure 1.21a). Over land this is also the area of the highest 2 m temperature (26–28°C). The Guinea Coast remains in the humid air with dew point temperatures in excess of 22°C. In April, the continent heats up, with highest daily mean temperatures of 32–34°C at about 15°N (Figure 1.21b). Here, daytime maximum temperature can well exceed 40°C, and even 45°C in

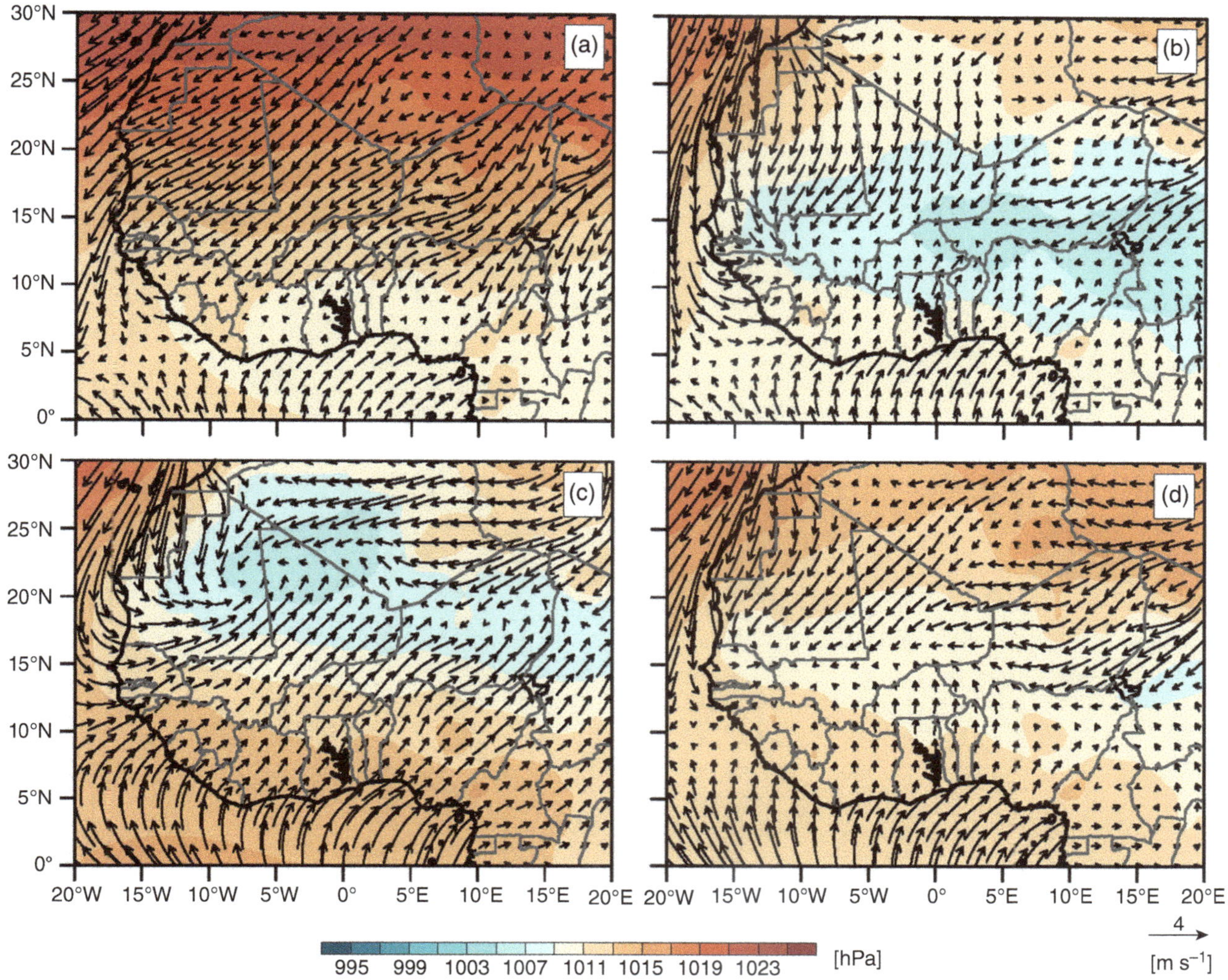


**Figure 1.19** Monthly mean of MSLP (contours) and 10 m wind vectors for (a) January, (b) April, (c) July and (d) October. *Source*: ERA-Interim, 1981–2010.

places, putting considerable thermal stress on the population (see Section 4.2.1). The cold oceanic Canary Current off the West Coast considerably cools the coastal areas from Guinea-Bissau northward in January and April and causes a pleasant climate (Figure 1.21a and b). In July, the cold coastal waters retract northward and the Cape Verde Peninsula area experiences warm waters exceeding 26°C. In the same month, hottest daily mean temperatures of 36–38°C occur in the WAHL region over the western part of the Sahara. The ITD is now located at about 20°N. Note the relatively cool temperatures in the southern zone. At the Guinea Coast the little dry season (see Section 1.1) is rather cool (24–26°C, Figure 1.21c) due to the cold air advection from the Gulf of Guinea and frequent low-level cloud cover (Schuster *et al.*, 2013; van der Linden *et al.*, 2015). The cold air advection with the monsoonal southwesterlies is strongest along the Gold Coast due to coastal upwelling of cool waters (Vollmert *et al.*, 2003). In July, coastal SSTs regularly drop well below 25°C at Ghanaian harbours and can be as chilly as 20°C (not shown). The area of very cold waters along the Coast of Ghana does not reach far into the open waters of the Gulf of Guinea, and thus it is not very well represented even in the NOAA SSTs with full 0.25° latitude–longitude resolution (not shown). Rather, parts of the larger scale equatorial cold tongue are evident in Figure 1.21c. In October, the surface temperature over the continent decreases and the ITD retreats equatorward (Figure 1.21d), though it is still located further north than in April.

A more detailed analysis of the ITD in the central Sahel is presented in Lélé and Lamb (2010) and shown in Figure 1.22 taken from their study. These authors delineated the ITD[5] based on the 15°C isodrosotherm calculated from surface station observations of temperature and humidity. Figure 1.22a shows that, in this part of West Africa, the ITD moves northeastward rather than straight north in the course of the season between the

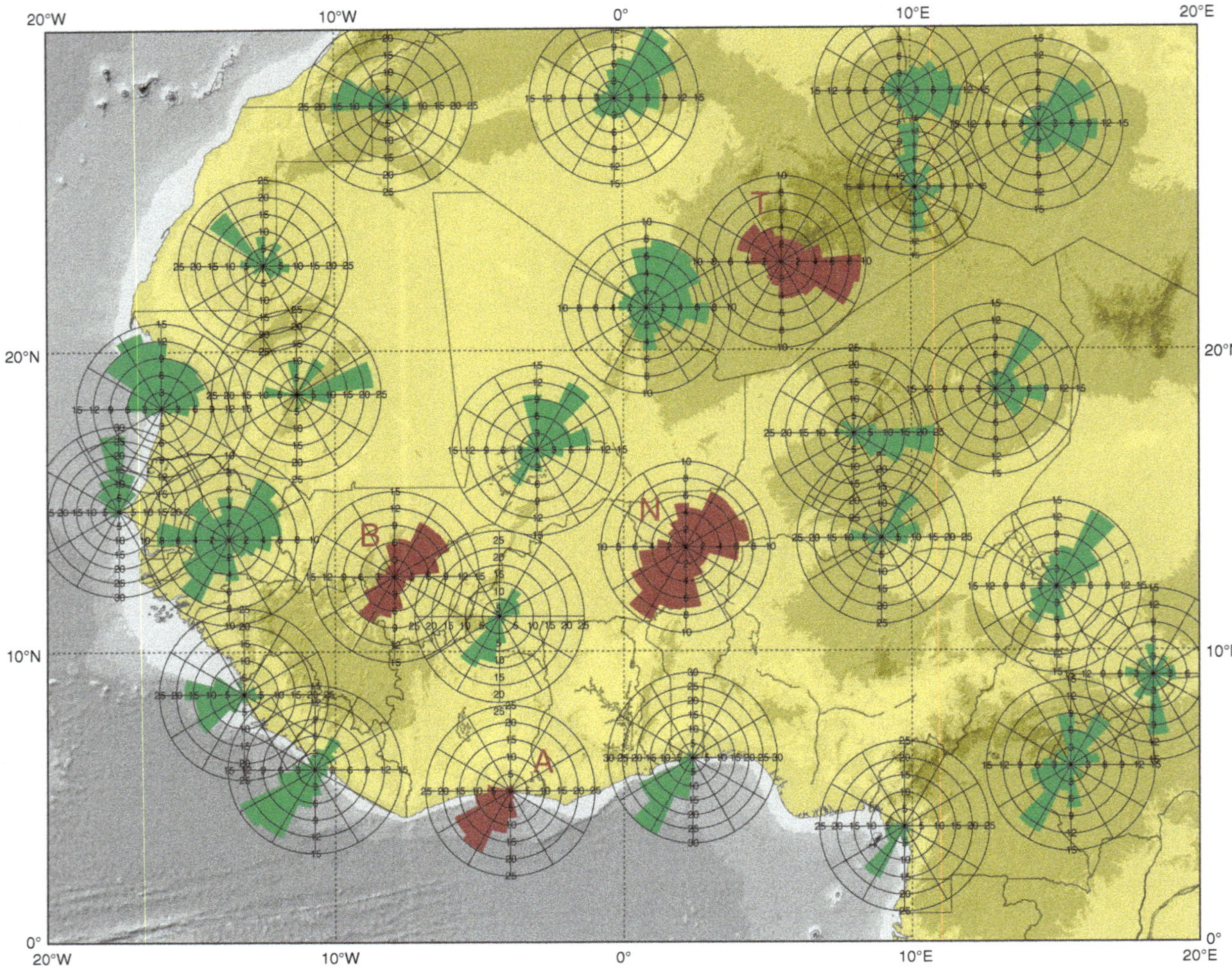


**Figure 1.20** Wind roses from 27 West African synoptic weather stations (Abidjan (A): 5.25°N, 3.93°W; Adrar: 27.88°N, 0.28°W; Agadez: 16.97°N, 7.98°E; Bamako (B): 12.53°N, 7.95°W; Bilma: 18.68°N, 12.92°E; Bobo-Dioulasso: 11.17°N, 4.32°W; Bordji Badji Mokhtar: 21.33°N, 0.95°E; Bouar: 5.97°N, 15.63°E; Cotonou: 6.35°N, 2.38°E; Dakar: 14.73°N, 17.50°W; Douala: 4.00°N, 9.73°E; Ghat: 25.13°N, 10.15°E; Grand Bassa/Roberts Field: 06.23°N, 10.23°W; In Amenas: 28.05°N, 9.63°E; Lungi: 8.62°N, 13.20°W; Ndjamena: 12.13°N, 15.03°E; Niamey (N): 13.48°N, 2.17°E; Nouakchott: 18.10°N, 15.95°W; Sarh: 9.15°N, 18.38°E; Sebha: 27.02°N, 14.45°E; Tamanrasset (T): 22.78°N, 5.52°E; Tambacounda: 13.77°N, 13.68°W; Tidjikja: 18.57°N, 11.43°W; Tindouf: 27.67°N, 8.13°W; Tombouctou: 16.72°N, 3.00°W; Zinder: 13.78°N, 8.98°E; Zouerate: 22.75°N, 12.48°W). The wind direction was categorised in bins of 20°. Four radiosonde stations (red wind roses) are used in Figure 1.31. *Source*: MIDAS and Integrated Surface Daily data sets, 1981–2010.

last dekad of April and the first August dekad, when it reaches its northernmost position. The retreat of the ITD starts in the second dekad of August, is faster than its progression on the continent, as evident in the larger spatial distance of the dekadal isochrones, and reaches southern Burkina Faso/northern Ghana in the second dekad of November (Figure 1.22b).

The Hovmöller diagram in Figure 1.23 illustrates the seasonal cycle of 2 m temperatures and its association with rainfall from TRMM for the West African continent. It demonstrates nicely (a) the asymmetry with a slow onset and swift withdrawal in terms of temperature and rainfall, (b) how the rains penetrate the continent in the wake of the heat low and (c) the cooler and drier little dry season at the Guinea Coast around August.

## 1.6 Upper Air Wind, Mass and Humidity Fields

### 1.6.1 An Eulerian Perspective on the Wind, Circulation and Dynamics

The upper-air climatologies of variables related to the wind and mass fields as well as to the vertical stratification of the troposphere above West Africa are now discussed. In terms of wind, the 925 hPa wind vectors shown in Figure 1.24 share many similarities with the near-surface wind in Figure 1.19. However, the 925 hPa wind systems are more pronounced, and the ITD is usually displaced somewhat equatorward. The features worthy of note are: (a) the strong Harmattan winds in Chad and Niger that

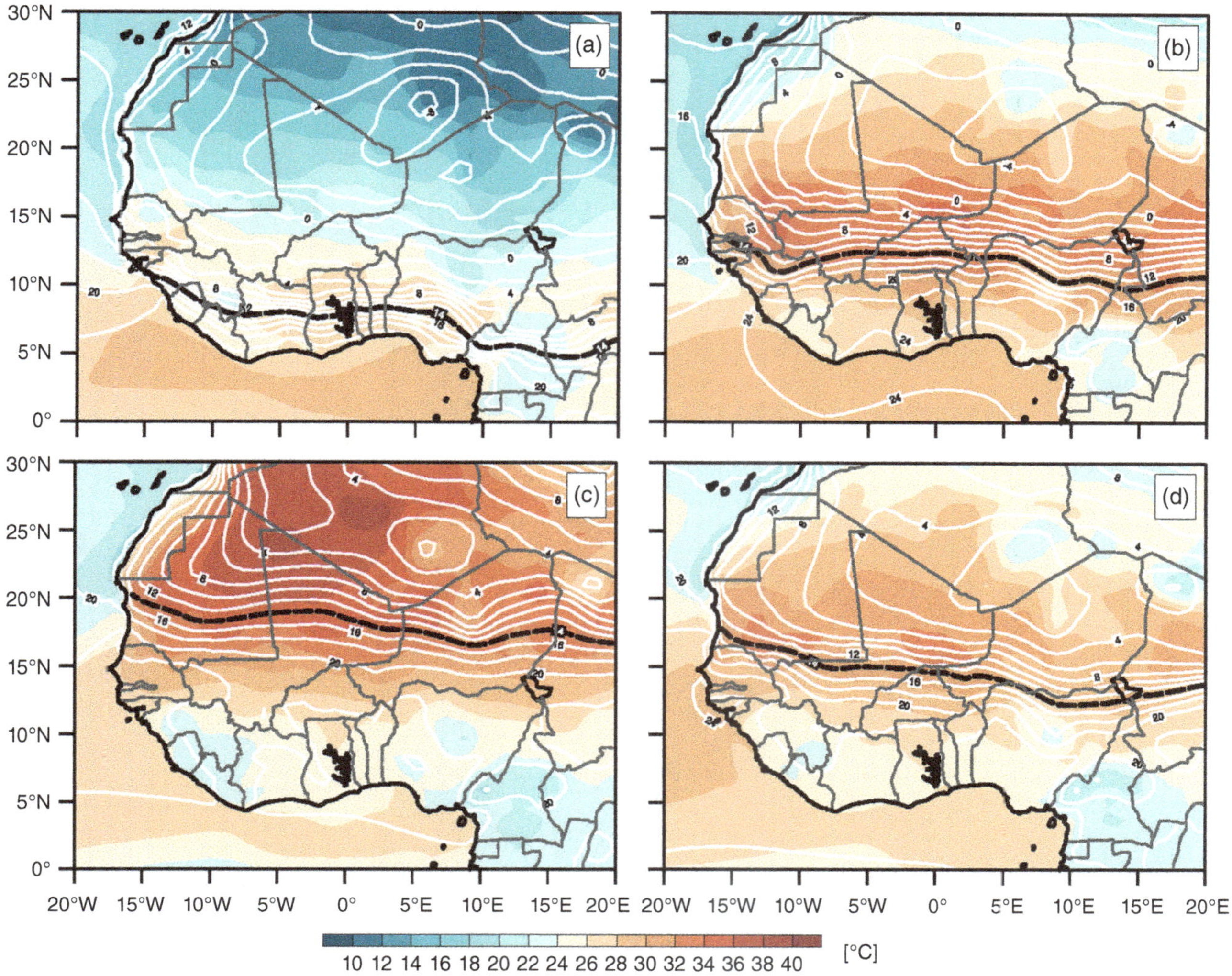


**Figure 1.21** Mean monthly 2 m temperature and dew point temperature for (a) January, (b) April, (c) July and (d) October. The ITD is indicated by the black, dashed 14°C 2 m dew point temperature contour. *Source*: ERA-Interim, 1981–2010. Over ocean the National Oceanographic and Atmospheric Agency (NOAA) sea surface temperatures (SSTs), converted from the original 0.25° × 0.25° latitude–longitude resolution to the ERA-Interim resolution of 0.75°, are displayed for the period 1982–2010.

occur in combination with strong geopotential height gradients and a channelling effect between the Tibesti and Ennedi Mountains over the Bodelé depression in Chad in January (Figure 1.24a); (b) the appearance of stronger (5–7 m s$^{-1}$) monsoonal winds south of the monsoon trough (i.e. the axis of minimum geopotential height) with a channelling effect in the Benue valley between the Jos Plateau and Cameroon Line Mountains in April (Figure 1.24b); (c) continental monsoon winds up to 20°N in July with maxima along the coast and in the northern Sahel (Figure 1.24c); and (d) a considerable weakening of the monsoon winds due to weakened geopotential height gradients south of the monsoon trough, with a reappearance of strong ENE winds south of the Tibesti Mountains in October (Figure 1.24d). At 850 hPa, the trough in geopotential height is much less pronounced, and a trough axis can be delineated only in April and July (Figure 1.25b and c). It can also be seen that some features, like the strong winds in the Bodelé depression mentioned earlier, are confined to the lowest levels. In October, the channelling of the easterlies between the Hoggar and Tibesti Mountains is better seen at 850 hPa, due to the higher terrain and low surface pressure in the area separating these two major Saharan mountain massifs (Figure 1.25d).

The 700 hPa level is generally near the altitude of the maximum speed of the AEJ. In July, however, the AEJ is lifted to somewhat higher altitudes over the West African continent. The higher altitude can be understood from the physical origin of the AEJ: this jet stream is the consequence of the thermal wind equation (Holton, 2004), with higher boundary-layer temperatures to the north of the AEJ over the Sahara. This meridional temperature gradient reverses at higher altitudes of about 4–5 km in

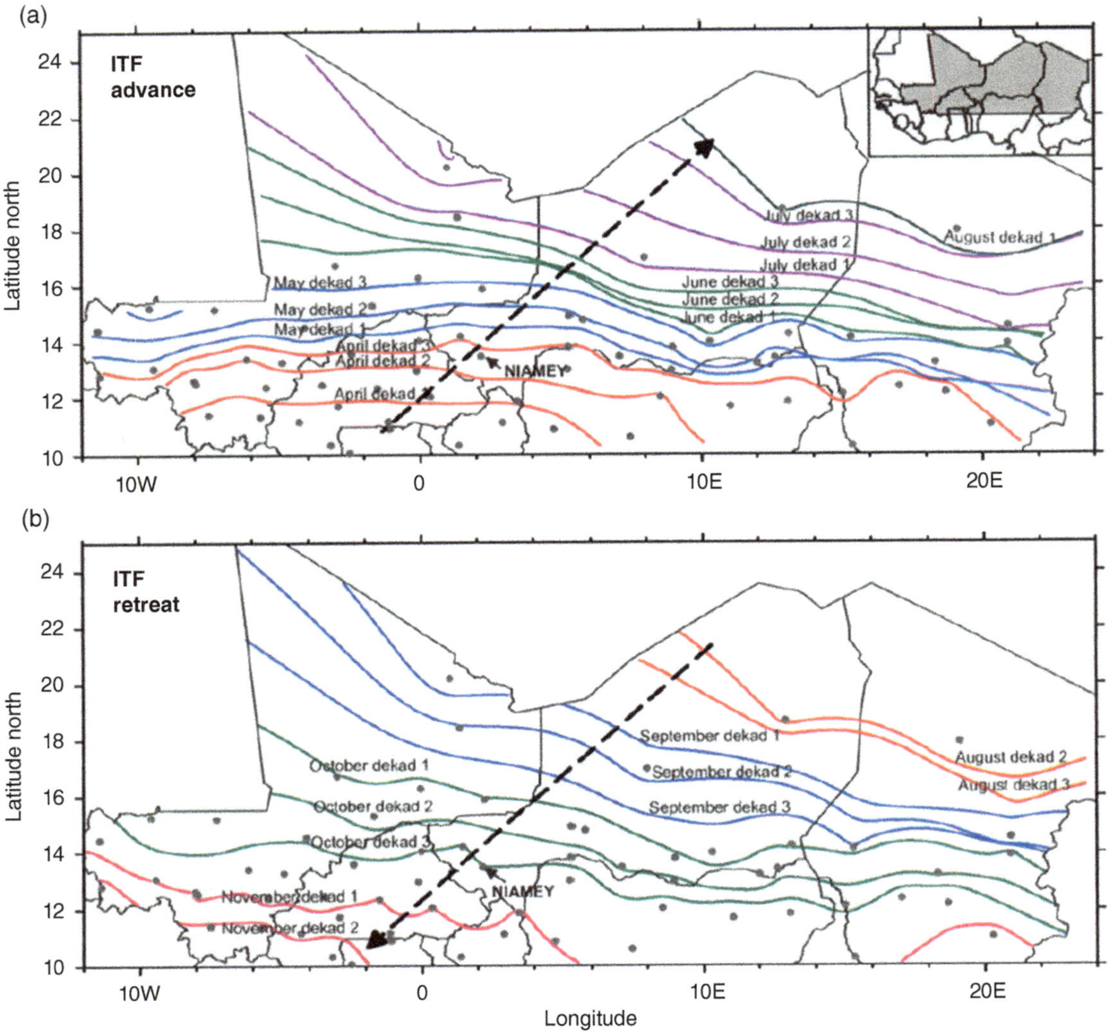


Figure 1.22 Long-term dekadal-average position of the West African ITD for 1974–2003 for (a) the northward advance from early April until early August and (b) the southward retreat from mid-August to mid-November. Boldface broken-line arrow indicates direction of advance and retreat for the central part of the study region. Dots locate synoptic stations from which daily minimum temperature and daily maximum relative humidity were employed to obtain daily dew point temperature and delineate the ITD using a 15°C dew point isodrosotherm. Location of Niamey station is indicated. *Source*: from Lélé and Lamb (2010). ©American Meteorological Society. Used with permission.

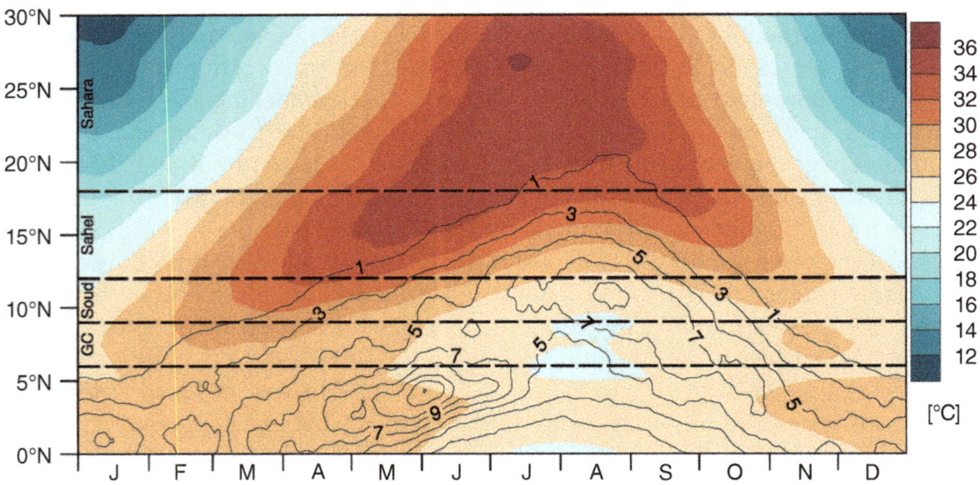


Figure 1.23 Hovmöller diagram of the 15-day running mean of the daily mean temperature. *Source*: ERA-Interim, 1981–2010. Contours denote the 15-day running mean of daily precipitation. *Source*: TRMM 3B42 Version 7, 1998–2012. Variables were zonally averaged between 10°W and 10°E. Zones are displayed in Figure 1.2b.

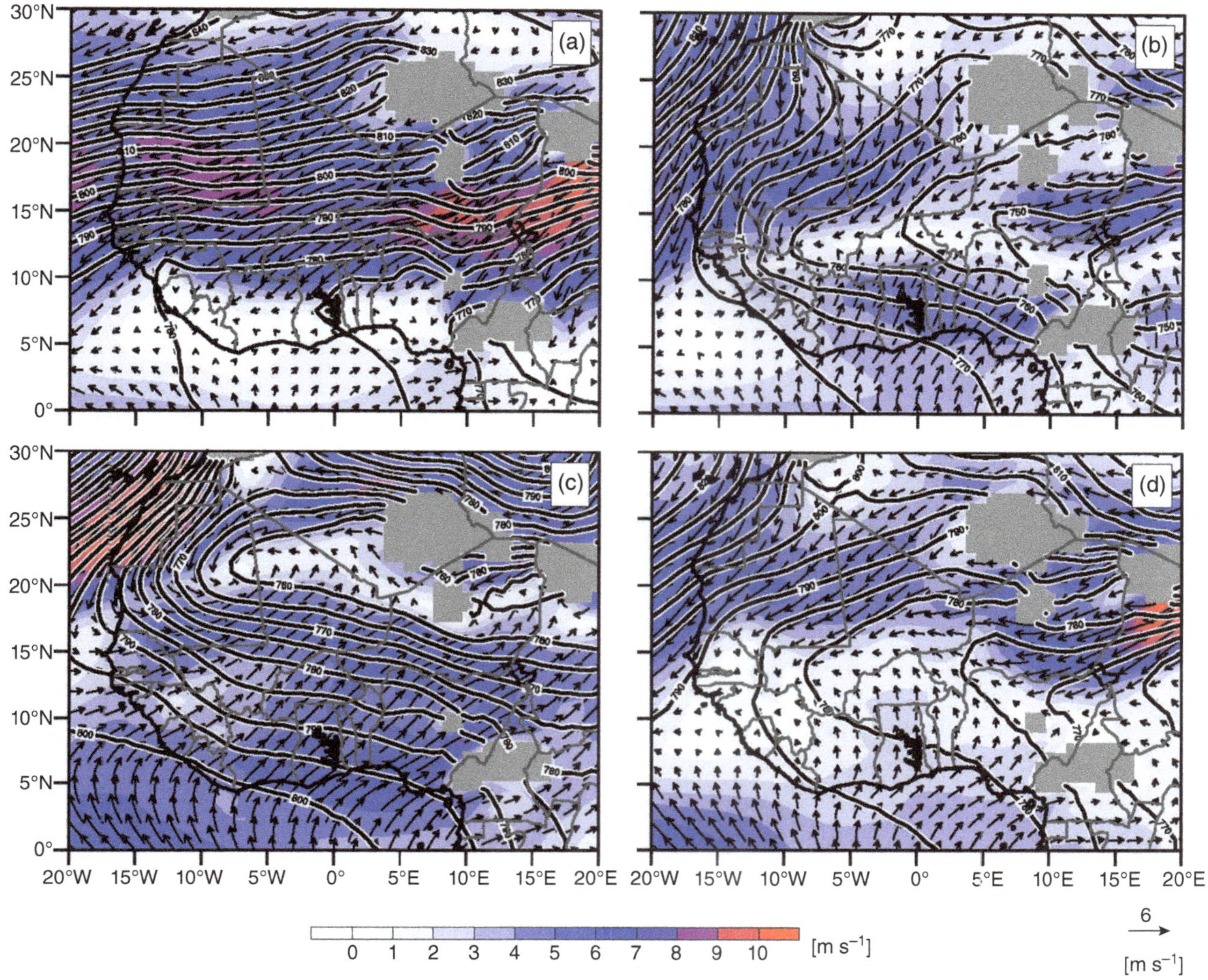


**Figure 1.24** Mean monthly 925 hPa geopotential height (contours), wind vectors and wind speed (colours) for (a) January, (b) April, (c) July and (d) October. Grey areas show regions below ground level. *Source*: ERA-Interim, 1981–2010.

July, where the air over the Sahara is cooler than the air to the south due to the steeper, less stable vertical temperature gradient within the Saharan boundary layer, when compared with the equatorial flank of the AEJ. The temperature gradient reversal determines the altitude of the AEJ core by thermal wind balance. Consistently, Fink *et al.* (2004) note that the AEJ core was at higher altitudes over land when compared with the East Atlantic Ocean during all three phases of the Global Atmospheric Research Programme (GARP) Atlantic Tropical Experiment (GATE) between June and August 1974. Owing to the importance of the location of the AEJ for the genesis of MCSs and as a waveguide for AEWs, we show here the 600 hPa wind and geopotential height analyses for July, rather than at the 700 hPa level.

In January, there is a comparatively weak AEJ wind speed maximum of 7–8 m $s^{-1}$ at the western Guinea Coast and over equatorial Africa (Figure 1.26a). In April, the mean AEJ is situated over the Guinea Coastal zone and supports the genesis of MCSs in this region (Figure 1.26b; see Chapter 3). In July, the AEJ is at its farthest northward position at about 15°N near the West Coast (Figure 1.26c), while it has begun its southward displacement in October (Figure 1.26d). In April, July and October the AEJ exhibits mean peak wind speeds of more than 13 m $s^{-1}$ (25 kt). On individual days, the AEJ wind speed can exceed 20 m $s^{-1}$ (or 40 kt; Thorncroft *et al.*, 2003).

Between May and October, AEWs develop along the AEJ. During this period, a strong horizontal wind shear at the anticyclonic northern flank of the AEJ leads to a reversal in the poleward potential vorticity gradient at the AEJ level and is thus instrumental in satisfying the barotropic instability criterion of a jet (Thorncroft and Hoskins, 1994). More recent work (e.g. Berry and Thorncroft, 2005; Mekonnen *et al.*, 2006) suggests that

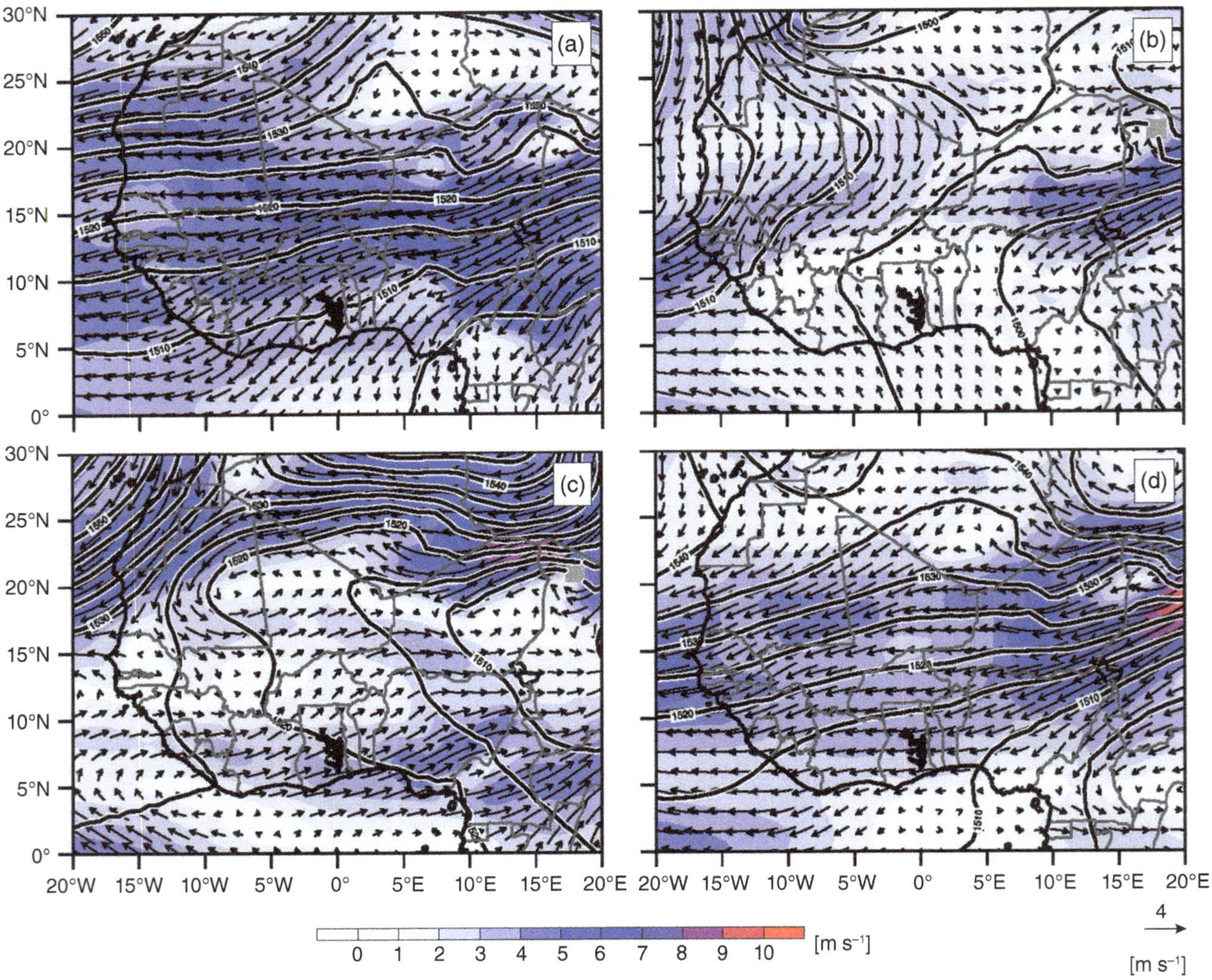


**Figure 1.25** As in Figure 1.24, but for 850 hPa.

MCSs play a role both in triggering wave-like disturbances in the AEJ entrance region over the eastern Sahel, and in promoting the growth of the AEW trough during its westward propagation across the subcontinent (see Section 2.1.3). AEWs are associated with two tracks of cyclonic vortices: a northerly, near-surface track north of the AEJ in the ITD region and a southerly track south of the AEJ (e.g. Fink *et al.*, 2004). The southerly vortices in particular are physically related to the occurrence of squall lines, and this relation gets stronger toward the West African coast (Fink and Reiner, 2003). Figure 1.27 shows the track density of AEWs for 1981 to 2010 for the northerly (850 hPa) and southerly (600 hPa) vortices. For the northerly AEWs, a maximum of occurrence is observed west of the Greenwich Meridian along the ITD region at 20°N (Figure 1.27a) and just south of the WAHL (Figure 1.19c). The southerly AEWs travel between 8 and 12°N (Figure 1.27b), just south of the zone of maximum precipitation (Figure 1.5c). Off the West African coast, the AEW density increases further and the tracks of the northerly and southerly vortices apparently merge. This does not necessarily imply that the vortices physically merge (e.g. Fink *et al.*, 2004). However, such a physical merger might be one favourable mechanism for an AEW-related tropical cyclogenesis near the Cape Verde Islands provided that the environmental conditions are favourable (e.g. Ross and Krishnamurti, 2007).

The 200 hPa maps show the presence of the subtropical jet (STJ) in all months except for July, when it is north of 30°N (Figure 1.28). The TEJ is seen in July, with its zonal axis at about 8°N (Figure 1.28c). It can be disputed whether this TEJ is simply the exit region of the TEJ that stretches from northern India to the eastern Sahel – this TEJ emanating from the Indian Monsoon region is located at higher altitudes and much further north (15°N). Rather, the southern TEJ at 8°N is likely to be the result of the geopotential height gradient south of the upper-level Saharan anticyclone and upper-level outflow from convection over the Sahel and central equatorial Africa. The distinction between the two TEJs has recently been

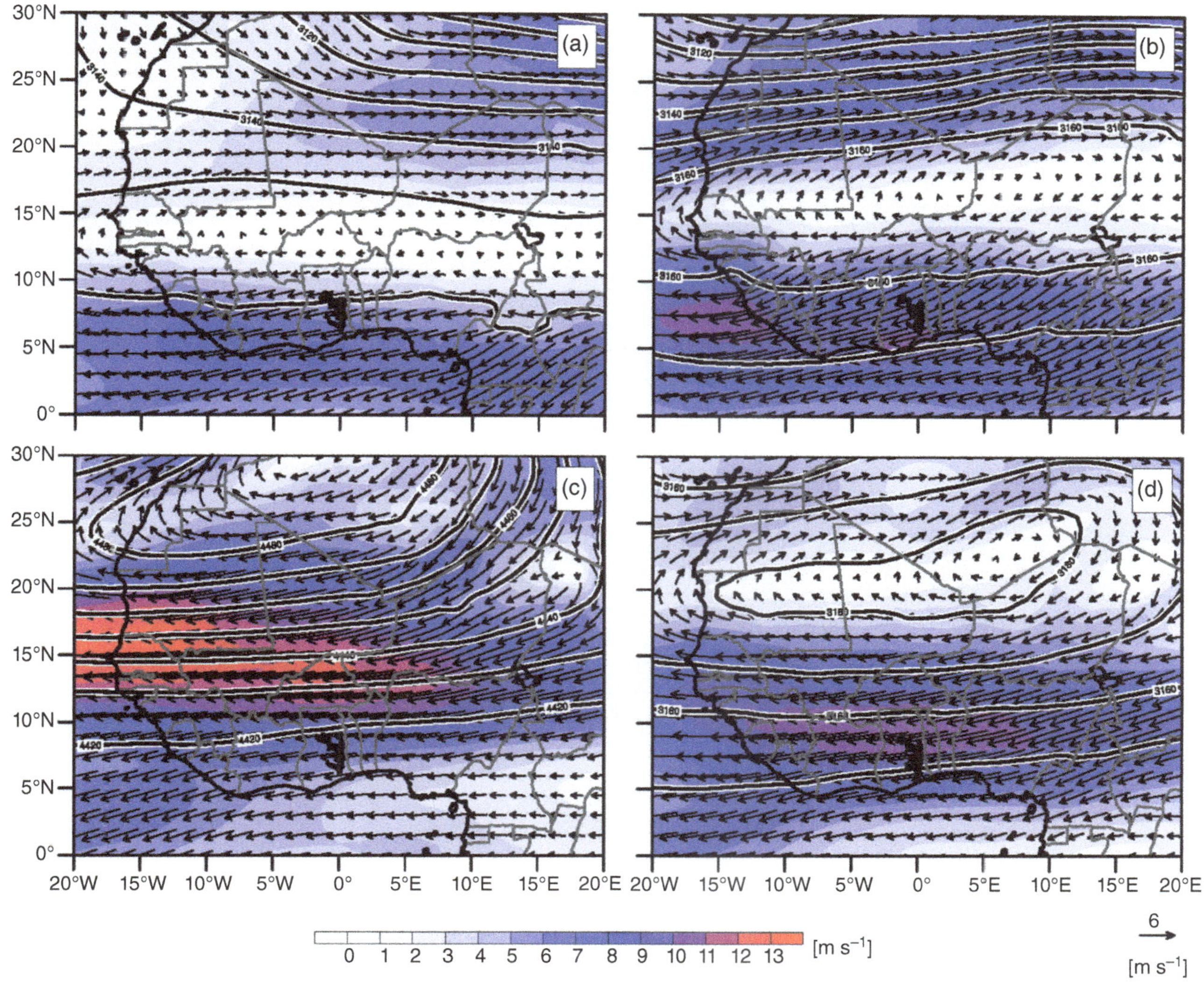


**Figure 1.26** As in Figure 1.24, but for 700 hPa. For July the 600 hPa level is shown in (c).

noted in the West African Synthetic Analyses and Forecasting (WASA/F) guide (Section 11.7). Nicholson *et al.* (2007) note wave activity on the TEJ and relate it to an intense monsoon rainfall event in 1950. The climatology and origin of waves at the TEJ, however, are not well known.

Several studies have highlighted the importance of the diurnal cycle of the monsoon and pointed to the existence of a night-time low-level jet (NLLJ) – for example, Parker *et al.* (2005b) and Lothon *et al.* (2008) for the Sahelian zone, and Schrage and Fink (2012) and Schuster *et al.* (2013) for the Guineo-Soudanian zone. The NLLJ is important for the northward transport of moisture (Parker *et al.*, 2005b) and the night-time genesis of an extensive stratus cloud deck in southern West Africa during the rainfall season (Schuster *et al.*, 2013, van der Linden *et al.*, 2015). At 925 hPa, the early morning wind maxima and late afternoon wind minima are clearly seen in the July climatology (Figure 1.29a and b). Note also that a westerly NLLJ occurs at 0600 UTC in Sénégal and Mauritania (Figure 1.29a); this NLLJ has been recently described by Grams *et al.* (2010) and Todd *et al.* (2013) in association with a cool, humid 'Atlantic inflow' that serves to ventilate the Saharan heat low on its western edge in summer.

A north–south cross-section of the seasonal wind fields represented by streamlines and isotachs, as well as isentropes in the $y$–$z$ plane, is shown in Figure 1.30. Figure 1.30a shows a strong STJ between 150 and 200 hPa with strong subsidence underneath, over the Sahara, between 25° and 30°N in January. The near-surface Harmattan winds are stronger than the monsoonal southerlies, and their confluence is located at around 5–7°N (Figure 1.30a). At this time of the year, the ascending branch of the Hadley cell is represented by an oceanic ITCZ over the equatorial Gulf of Guinea. In April, the STJ maintains its strength over the Sahara, but is displaced 5° poleward with reduced subsidence underneath

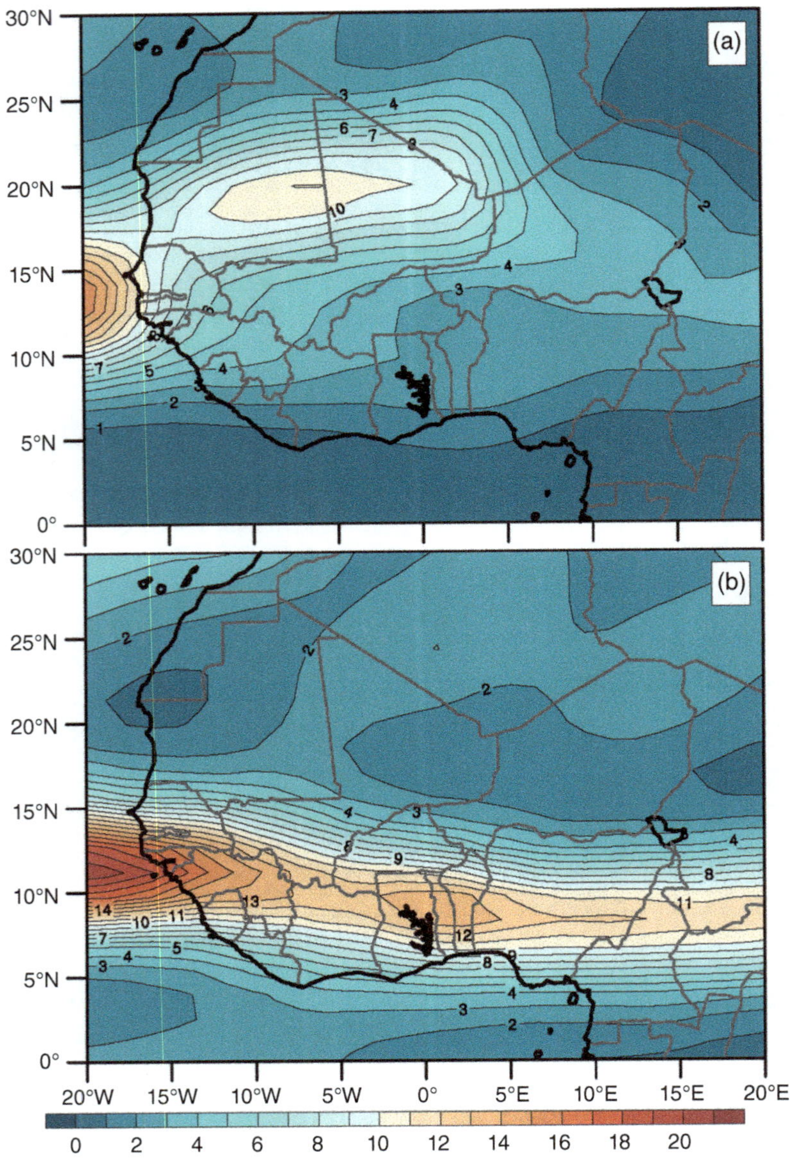


**Figure 1.27** Track density climatology for AEWs for the June to October period between 1981 and 2010 based on the relative vorticity at (a) 850 hPa and (b) 600 hPa, from the ERA-Interim reanalysis. Units are number density per month per unit area, where the unit area is equivalent to a 5° spherical cap (~$10^6$ km$^2$).

(Figure 1.30b). A similar northward shift is seen in the ascending branch of the oceanic ITCZ that approaches the Guinea coastline while strengthening. Here, the first rainy season is about to start. A continental heat low is present in April between 10 and 20°N as evidenced by the vertical slope and surfacing of the 310 K isentrope, indicating near-neutral static stability. The westerly monsoon flow has crossed 10°N and forms a wedge underneath the hot, northerly Harmattan winds. This near-surface layer in which there is a continuous westerly wind component has been referred to as the monsoon layer in Section 1.1. By July, the heat low has moved to 20–30°N with a concomitant northward shift in the low-level monsoonal westerlies to about 20°N (Figure 1.30c). The ascending branch of the Hadley cell is less clearly seen when compared with April, and is located around 10°N. The low-level upward motion over the heat-low area between 20 and 25°N reaches the 700 hPa level, where it diverges with the equatorward branch feeding into the deep upward motion at 10°N. At upper levels, the TEJ system, peaking at 100 hPa around 15°N and between 150 and 200 hPa at 8°N, is flanked by STJs in either hemisphere. Mid- to upper-level subsidence dominates the southern hemisphere and the region north of 20°N. In October, the monsoon system retreats equatorward with a strengthening of the Northern Hemispheric STJ (Figure 1.30d). Maximum vertical motion is again located near the Guinea coastline and reflects the time of the second rainy season in this area (Section 1.1).

Climatological skewed $T$–log $p$ diagrams of radio-soundings at Abidjan, Niamey, Bamako and Tamanrasset are displayed in Figure 1.31 for January, April, July and October. In January, the troposphere over the latter three stations is very dry (Figure 1.31a). Tamanrasset

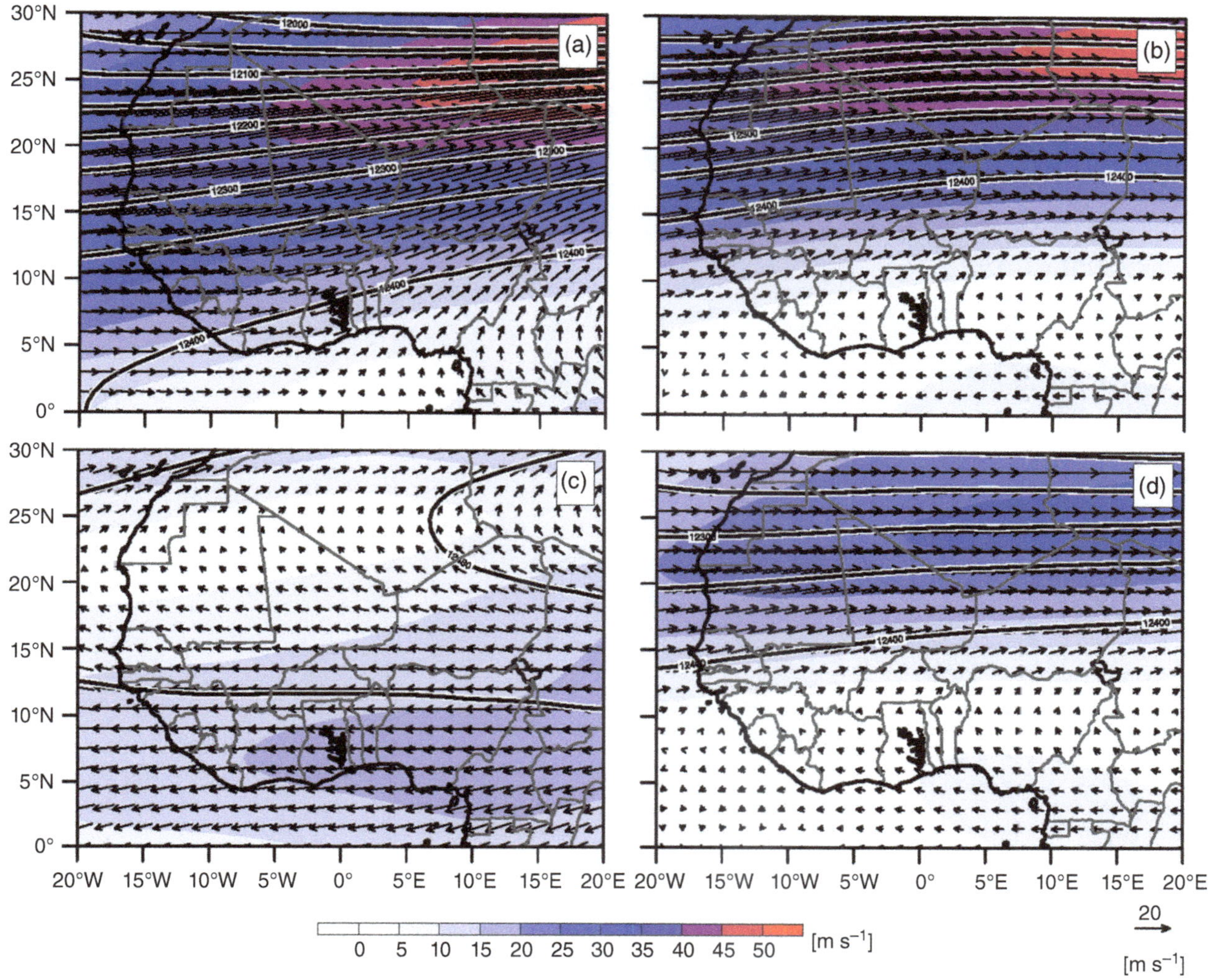


Figure 1.28 As in Figure 1.24, but for 200 hPa.

(22°47′N; 5°31′E) also shows a colder troposphere with a lower tropopause at 200 hPa instead of near 100 hPa; this clearly reflects that Tamanrasset is in subtropical air masses; this is also observed in April (Figure 1.31b). Consistent with the earlier noted location of the heat low between 10 and 20°N in April, lower-tropospheric temperatures at Bamako (12°32′N; 7°57′W) and Niamey (13°29′N; 2°10′E) almost follow a dry adiabatic lapse rate up to 700 hPa. In July, the monsoon progresses inland and the moisture profiles at these two Sahelian stations are similar to the Abidjan (5°15′N; 3°56′W) soundings (Figure 1.31c). The Tamanrasset soundings now shows an almost dry static stability up to the 500–600 hPa layer. This is the top of the Saharan air layer (SAL), where the spread (i.e. the difference between the temperature and the dew point temperature) attains its minimum value, reflecting the occasional occurrence of clouds at the top of the SAL (see Section 1.3). Tamanrasset now also observes a tropical tropopause near 100 hPa. The October soundings reflect the retreat of the monsoon (Figure 1.31d). Three final observations are noteworthy. First, the Niamey soundings are drier than the Bamako soundings in all seasons, reflecting the somewhat wetter and cloudier condition in the western WAM region. Second, except for perhaps January, near-surface temperature and static instability at Abidjan are considerably lower than at Bamako and Niamey, reflecting the fact that Abidjan is situated in the cool and stable monsoon layer with southwesterly wind throughout the year. Third, the low-level shear from southwesterly monsoon to mid-level AEJ easterlies can clearly be seen at Bamako and Niamey in July. Here, the lower troposphere is still sufficiently dry to promote downdrafts from falling raindrops that are instrumental in the organisation of MCSs (see Section 3.1.2.2)

### 1.6.2 A Lagrangian Perspective on the Circulation

The streamlines shown in Figure 1.30b only hint at the origins of air masses, and do not allow for an assessment of mixing processes in the rainfall zone or at the ITD. Long-term assessments are not available, but aircraft

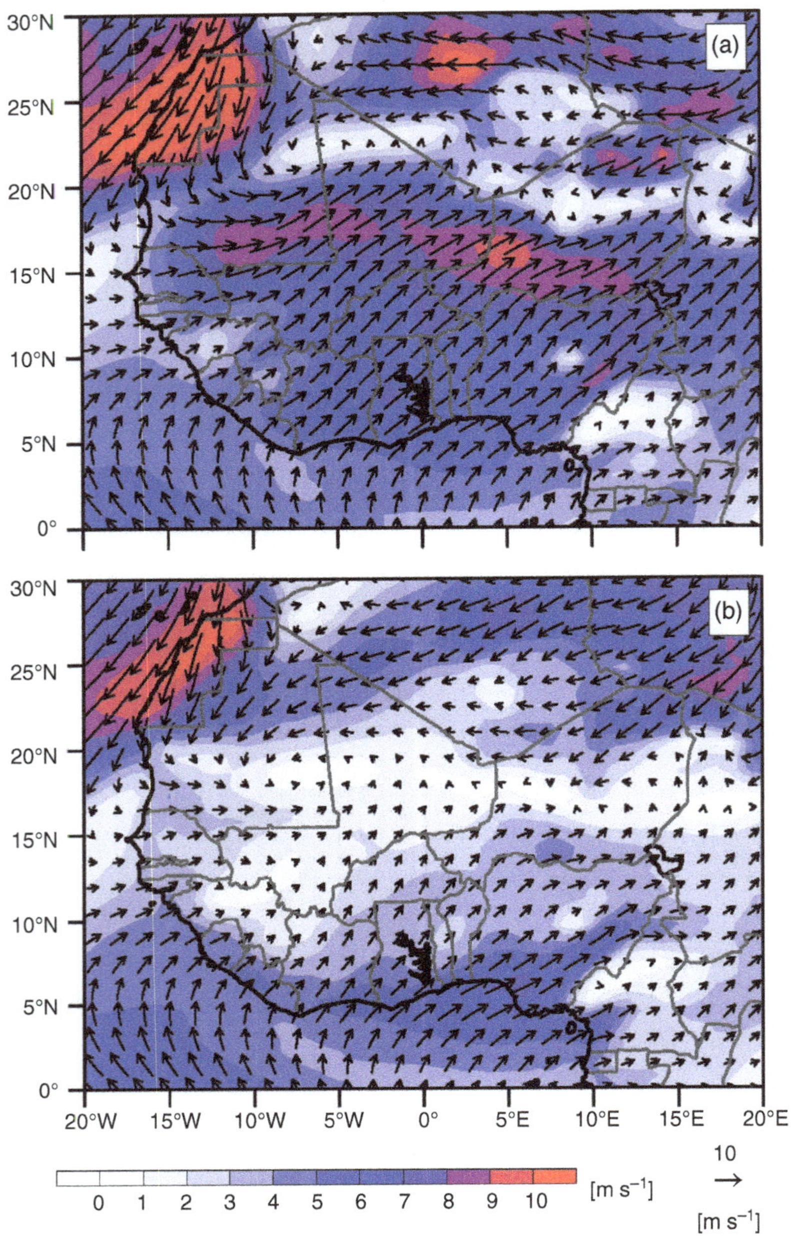


**Figure 1.29** Mean 925 hPa winds for July at (a) 0600 and (b) 1800 UTC. *Source:* ERA-Interim, 1981–2010.

measurements of atmospheric composition conducted during AMMA in early August 2006 provided an indication of air mass origin, vertical transports and mixing. Figure 1.32 shows the location (latitude versus altitude) of measurements made by the British BAe146 aircraft. Different chemical species are emitted by different processes, acting in different geographical regions: high concentrations of one compound can indicate a particular source or process. Measurements of a large range of chemicals were made from the aircraft, some in situ and some made by subsequent analysis of canisters of air collected during the flight. A statistical clustering technique can link air masses with similar chemical composition. A clusters algorithm (*k*-mean) was applied to the AMMA dataset using the observations of ozone, carbon monoxide, isoprene, ethene and relative humidity. In Figure 1.32, each filled box represents the location of one canister, and the colours represent the different clusters. Although this was a relatively short (1 month) sampling period from a climatological perspective, the measurements represent the Lagrangian history of air movement over a longer period, and nicely illustrate the air mass characteristics of the monsoon system.

Five clear clusters emerge in Figure 1.32, which reflect the differing sources and processes of compounds in the region, which can be attributed to anthropogenic (human) activity, biomass (forest) burning, biogenic (forest) emissions, the SAL and the monsoon layer. Human activity (petrochemicals and high-temperature combustion) leads to the emission of hydrocarbons and

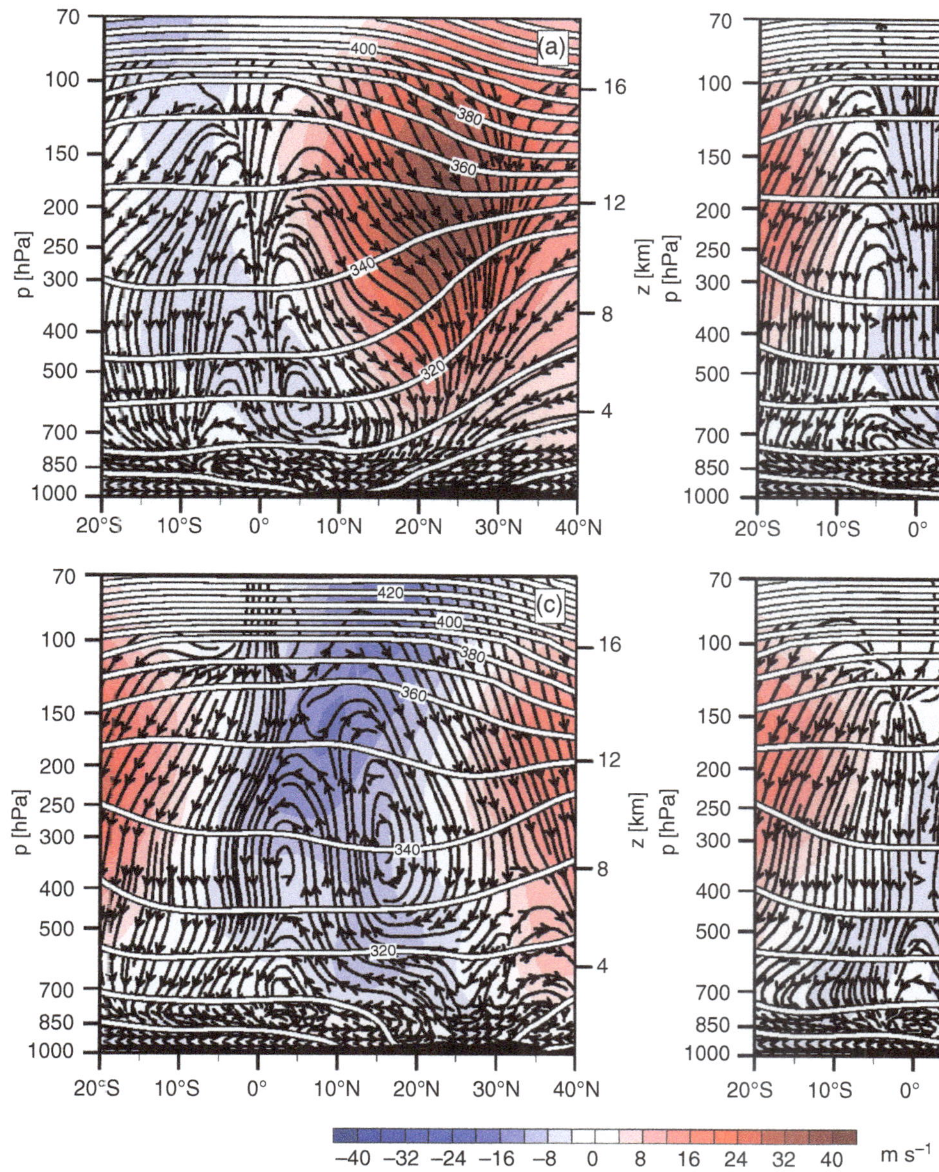

Figure 1.30 Mean monthly isentropes (contours), zonal wind speed (colours), and streamlines (v-wind, w-wind × 1000) for (a) January, (b) April, (c) July and (d) October, averaged between 10°W and 10°E. *Source*: ERA-Interim, 1981–2010.

carbon monoxide from the surface. These kinds of air masses were observed when the aircraft was close to its campaign base in Niamey (Niger) and around the mega-city Lagos (Figure 1.32a). Biomass burning air is high in carbon monoxide and ozone from the large-scale fires in southern hemisphere Africa (where August is the middle of the dry season) and is transported around 4000 km northward to West Africa (Figure 1.32b), advected at mid-tropospheric levels and entering the region of monsoonal convection (zone C) from the south. There are also occasional observations of the products of similar but smaller local burning around Niamey. Biogenic air masses are high in isoprene. This is a molecule uniquely emitted by forests, and its short lifetime (~1 h) means that it is not transported far from its emission source, staying close to the ground in the forested regions of West Africa (Figure 1.32c). The SAL represents air that is characteristic of that found over the Sahara desert, and which is transported southward in the mid-troposphere (Figure 1.32d). SAL air has high ozone, low carbon monoxide and low relative humidity and is apparently lifted over the monsoon layer by northerly winds. This flow is seen in the streamlines of Figure 1.34b. Figure 1.32d also shows that some of the SAL air makes it to the south of the region at mid-levels, even as high as 8 km. The monsoon layer (Figure 1.32e) represents the boundary-layer air over southern West Africa, with high relative humidity, lower ozone and lower carbon monoxide; this air is also found intermittently at higher altitudes in zone C, where it has been lifted by moist convection.

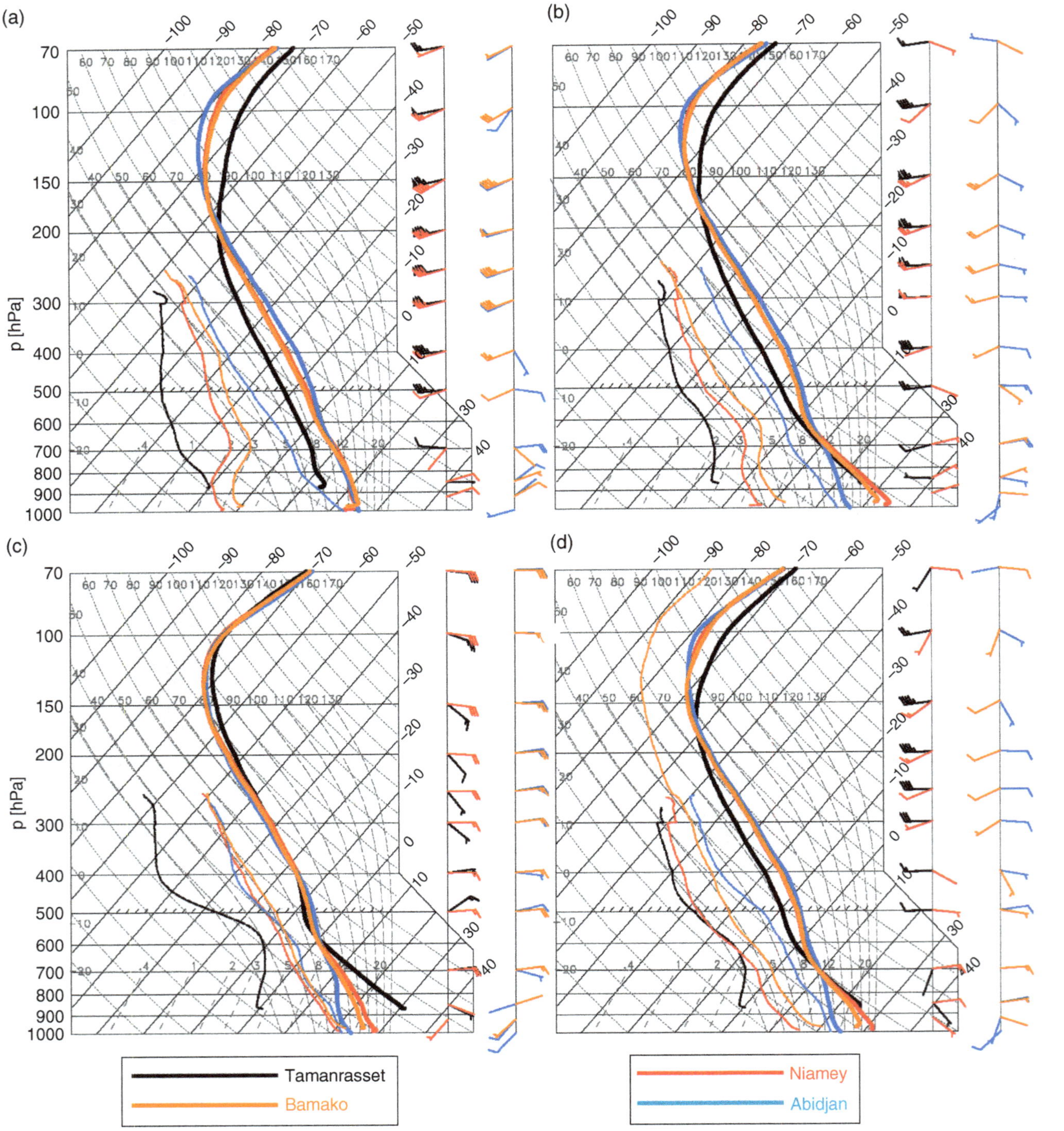


**Figure 1.31** Composite radiosonde profiles between 1981 and 2010 from Tamanrasset (black lines and wind barbs), Niamey (red), Bamako (orange) and Abidjan (blue) for (a) January, (b) April, (c) July and (d) October. Both the 0000 and 1200 UTC soundings were used for the climatology. The upper-air observations are displayed into a skew *T*–log *p* diagram. Included are the measured air temperature (solid lines), dew point temperature (thin lines) and wind observations (wind barbs). The data were extracted from the Integrated Global Radiosonde Archive (http://www.ncdc.noaa.gov/oa/climate/igra), were interpolated onto a vertical grid with a 1 hPa resolution and finally averaged. Stations are in the map of Figure 1.20.

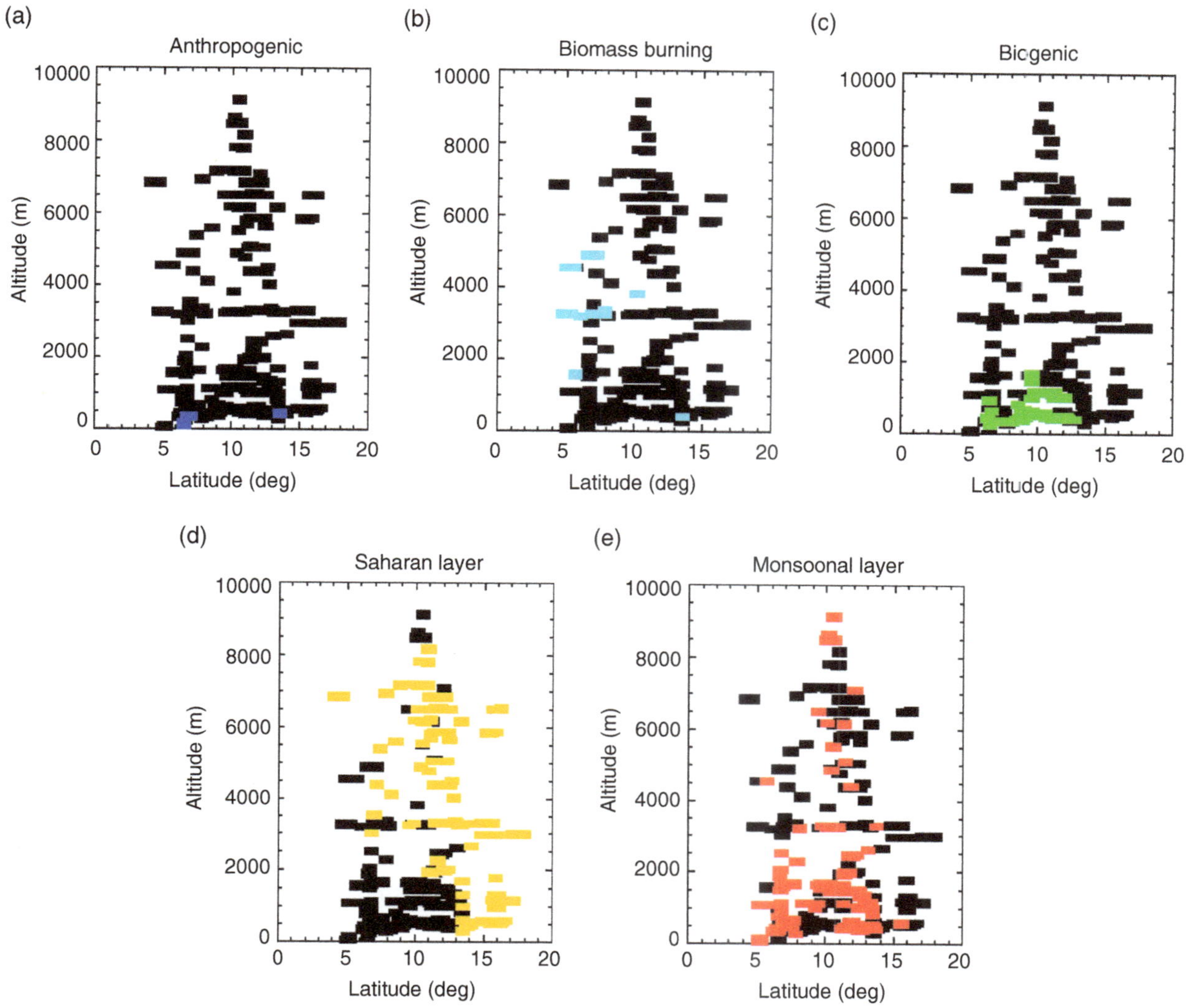


Figure 1.32 Cluster analysis of composition observations collected by the BAe146 aircraft during the AMMA 2006 campaign. Each rectangle represents a canister measurement, plotted by latitude and altitude. Colours represent samples belonging to clusters that are representative of (a) anthropogenic, (b) biomass-burning, (c) biogenic, (d) Saharan and (e) monsoon-layer air masses.

The panels of Figure 1.32 illuminate the physical behaviour of the climatological features described in this chapter. The monsoon layer intrudes northward into the continent at low levels, and its composition is dominated by exchanges with the land surface (biogenic and monsoon-layer air in Figure 1.32c and e). SAL air moves southward from the Sahara and meets the monsoon-layer air at the ITD, which can be seen to be a very sharp air mass boundary in Figure 1.32d and e. South of the ITD, the SAL air is lifted above the monsoon layer in northerly winds; monsoon-layer air is also found in the mid-troposphere (above 3 km) as a result of moist convection lifting air from the monsoon layer out of the boundary layer (Figure 1.32e). In these levels, in zone D and the southern part of zone C, air masses are also seen to be intruding from the southern hemisphere, rich in biomass-burning products. Therefore, the mid-tropospheric air in zone C of the monsoon system is a zone of active mixing and exchange between monsoon layer air, SAL air and air masses moving from the south. These processes of convective exchange have been discussed in more detail by Parker *et al.* (2005a).

## 1.7 Summary Schematics of the January and July West African Monsoon System

The climatologies presented in this chapter and the conceptual understanding of the WAM that has been put forward have been condensed into two types of schematics for January and July: one map and one meridional cross-section. Similar schematics have been produced in the scientific literature over the years (e.g. Hamilton *et al.*, 1945; Lamb, 1980; Buckle, 1996; Parker *et al.*, 2005a), but here the diagrams are updated to make use of the most recent reanalysis data and new satellite resources. Note that (a) the geographical sections were expanded somewhat for the schematics and (b) the drawing of the sketches was backed up by additional maps not shown here. The January map and meridional cross-section are displayed in Figures 1.33a and 1.34a respectively. The most salient features are as follows.

1) The continental heat low is still situated in the northern hemisphere, but strongest over eastern Africa with a weaker zonal branch over southernmost West Africa (Figure 1.33a).
2) Correspondingly, the ITD is located just north of the Guinea Coast at around 7–8°N. The shallow, moist monsoon layer and land–sea breeze circulation give rise to some cloudiness and infrequent showers directly at the coast (Figures 1.33a and 1.34a).
3) Whereas the Guinea Coast is climatologically under the influence of a low-level moist southwesterly monsoon flow throughout the year, the vast majority of West Africa is, in January, influenced by the dry northeasterly Harmattan flow. The flow emanates from the anticyclonic recurvature of westerlies around the Libyan High, which is connected to the Azores High (Figure 1.33a).
4) Features 1–3 are nicely reflected in the meridional profiles of pressure, temperature and dew point temperature depicted in the middle panel in Figure 1.34a.
5) The cold waters of the Canary Current extend south to the Cape Verde area, causing cold temperatures at the West Coast (Figure 1.33a).
6) A strong, SSW- to NNE-oriented STJ is found over the Sahara at 12 km altitude. A weak, though identifiable, AEJ is located just south of the Guinea Coast. An upper-level easterly jet (EJ) is located in the southern hemisphere (Figure 1.33a).
7) Maximum rainfall is located just south of the Guinea Coast over the equatorial Gulf of Guinea, with a maximum axis still in the northern hemisphere (Figure 1.33a; see also rainfall cross-section in the lowermost panel of Figure 1.34a). This maximum rainfall is associated with the ascending branch of the Hadley cell.

The corresponding July map and meridional cross-section are shown in Figures 1.33b and 1.34b respectively. Here, features worthy of note are as follows.

1) The WAHL is situated over northern Mali (Figure 1.33b).
2) Correspondingly, the ITD is located within the centre of the WAHL near the 20°N parallel (Figure 1.33b).
3) The monsoon layer, characterised by the moist southwesterly monsoon flow, forms a wedge deep into the continent, extending to 20°N, with its largest depth up to almost 700 hPa at the Guinea Coast (Figure 1.34b). Nicholson and Grist (2003) found the largest depth of the moist layer to lie in between the latitudinal positions of the AEJ and TEJ in the maximum rainfall zone. Aside from data problems, this discrepancy may relate to various, partly non-wind-based definitions of the moist monsoon layer. In and north of the WAHL region, the Harmattan flow is now more easterly (Figure 1.33b). Note that on a day-to-day basis, monsoon-layer definitions based on the wind field have high variability. For this reason, a PW-based definition of the monsoon layer depth is presented in Section 11.10.
4) Features 1–3 are again nicely reflected in the meridional profiles of pressure, temperature and dew point temperature depicted under the cross-section in Figure 1.34b. In the WAHL region, maximum daytime temperatures exceed 40°C.
5) The chilliness at the Guinea Coast is augmented where coastal upwelling occurs in the parts of the Ivorian Gulf and along the Coast of Ghana. At this time of the year, the equatorial SST cold tongue is most pronounced (Figure 1.33b).
6) The STJ axis is found just north of the Mediterranean Coast, while the AEJ is aligned along the 15°N parallel. As discussed in the text, the STJ is distinguished between a higher (100 hPa), more northerly TEJ over eastern Africa and a lower (200 hPa) TEJ at about 8°N over the WAM region (Figure 1.33b). In the meridional cross-section these TEJs merge to show a tilted region of enhanced easterlies (Figure 1.34b). Nicholson and Webster (2007) found jet-like near-surface westerlies in the monsoon layer, especially in the wet year of 1950. They relate the inertial instability of this westerly jet to a strong poleward flow of moist air and a corresponding wet Sahel rainy season. In multidecadal climatologies from ERA-Interim and other reanalyses, such a jet-like feature is not discernible. Therefore, it was decided not to draw a low-level westerly jet, though it may be present in some (wet WAM) years.
7) AEW activity is split into a lower (850 hPa) northerly track emanating just south of the WAHL region and a higher (600 hPa) southerly track that extends upstream to the Ethiopian Highlands. Both tracks apparently merge over the Eastern Atlantic Ocean, but this does not necessarily reflect a merger of the AEW vortices (Figure 1.33b). Note that the regions of AEW activity have been chosen by an arbitrary contour of their track density, and therefore extend further to the north and south than the cores of the AEW vortices.
8) Maximum rainfall is found between 10–12°N; that is, south of the AEJ and within the southerly AEW track region (see bar graph in the lowermost panel of Figure 1.34b). The maximum rainfall zone is sandwiched between the 200 hPa TEJ and the AEJ (Nicholson, 2008). From the Guinea Coast to the maximum rainfall region, a small diurnal temperature

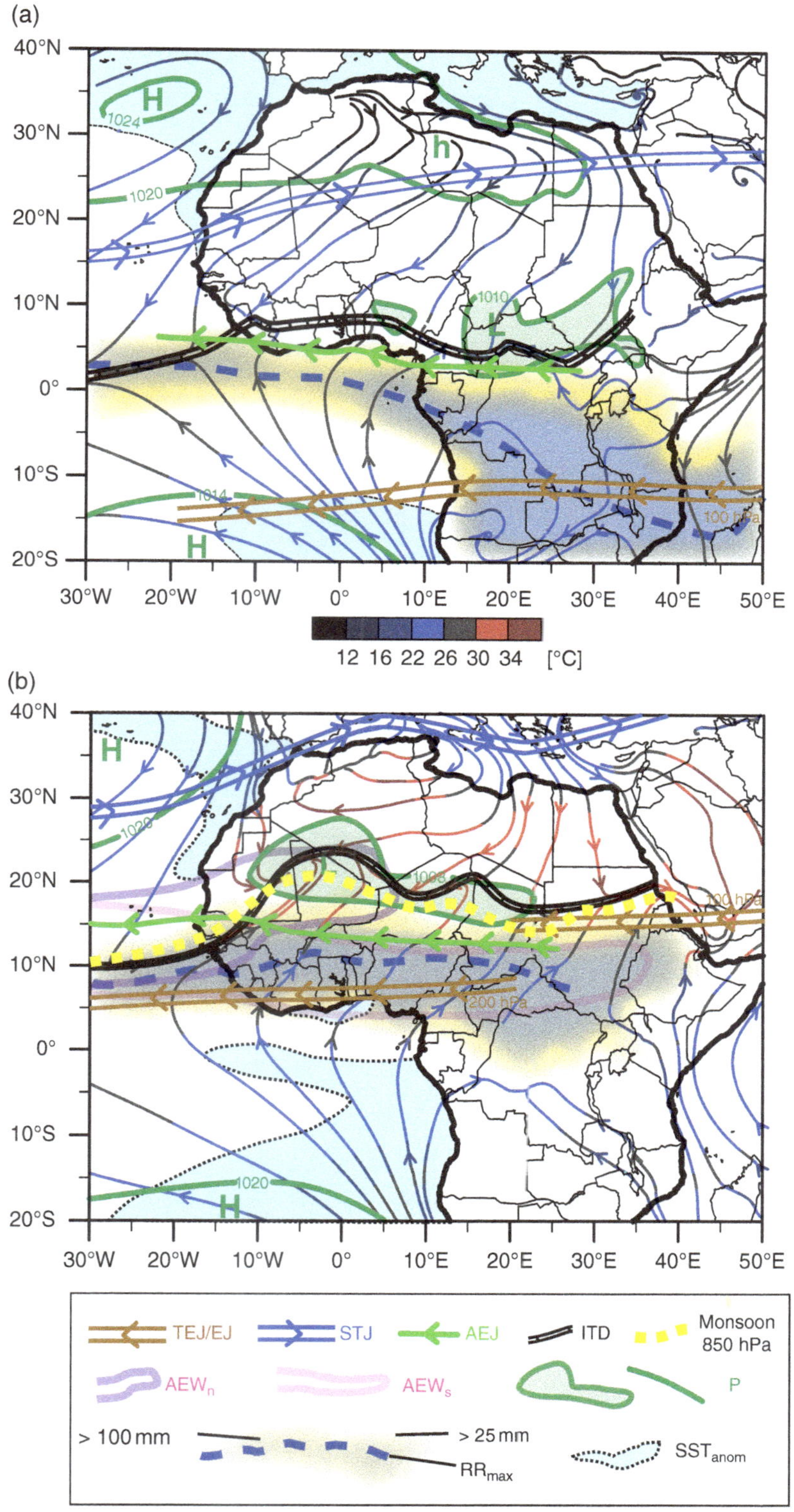


Figure 1.33 Schematic illustration of atmospheric and oceanic features determining the weather and climate of West Africa in (a) January and (b) July. Shown are the positions of the ITD, the monsoon trough, upper-level air streams (AEJ, TEJ/EJ (Easterly Jet) and STJ), surface winds coloured according to the 2 m air temperature (see colour bar), the tropical rain belt with the maximum axes of rainfall ($RR_{max}$), northerly and southerly AEW vortices propagation zones ($AEW_n$ and $AEW_s$ respectively), areas with relatively cold SSTs ($SST_{anom}$) and example pressure lines marking the Azores ('H'), Libyan ('h') and Saint Helena ('H') highs and the heat low ('L').

range is observed, with relatively cool temperatures at the coast (Figure 1.34b).

9) The deepest cumulonimbus clouds occur in the AEJ region, with those in the more southerly rainy zone being less deep (Figure 1.34b).
10) Low-level convergence between the monsoon and Harmattan flows, as well as vertical motion, occurs in the dry heat-low ITD region. An equatorward return flow at about 700 hPa feeds the deep ascent in the rainy zone (Figure 1.34b). The July and January situations reflect neither a maritime trade wind nor monsoon ITCZ. The term ITCZ over West Africa is flawed and should be avoided – see Ramage (1995) and Fink *et al.* (2010).

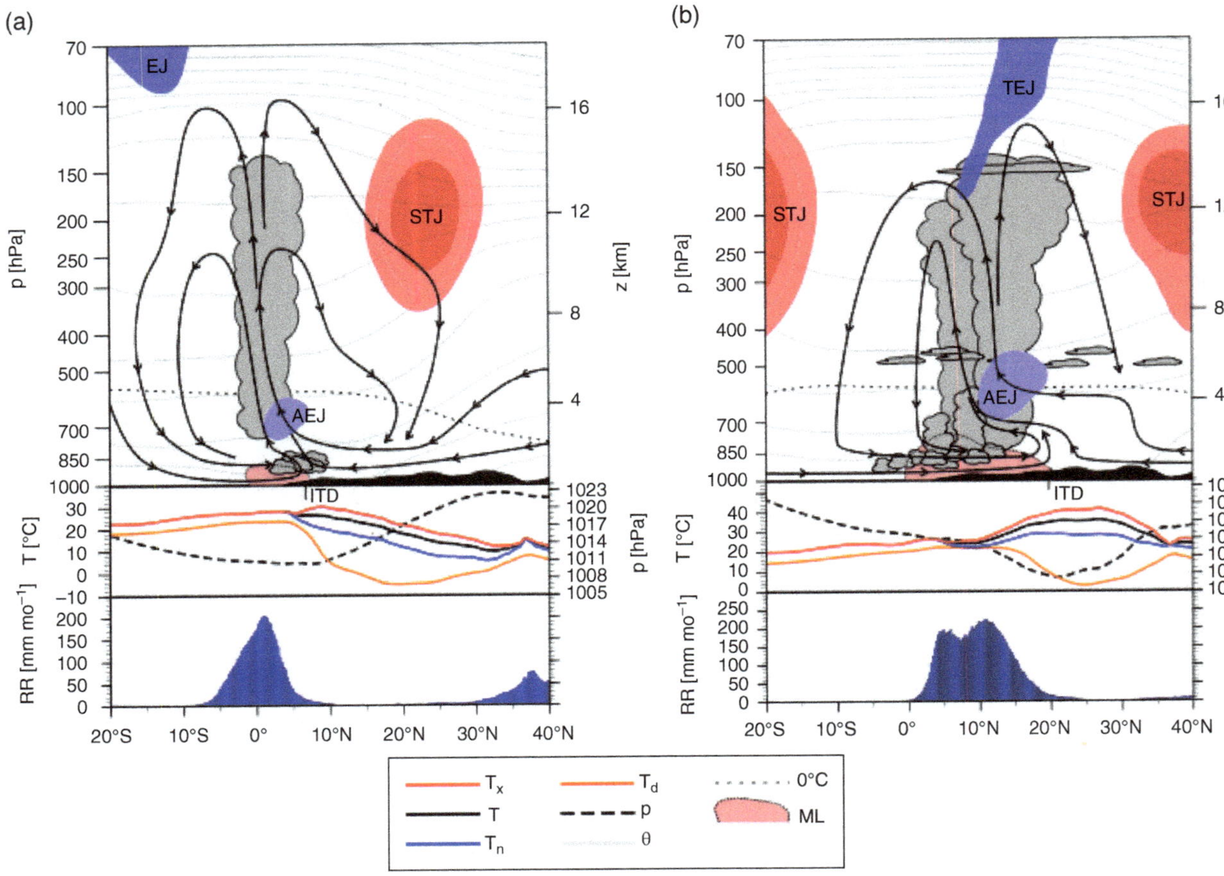


**Figure 1.34** Schematic cross-section of the atmosphere between 10°W and 10°E in (a) January and (b) July. Shown are the positions of the ITD, upper-level jet streams (AEJ, TEJ/EJ and STJ), the monsoon layer (ML) (as defined by westerly, i.e. positive zonal winds), streamlines, clouds, the freezing level (0°C isotherm), isentropes, minimum ($T_n$), maximum ($T_x$) and mean temperatures ($T$), dew point temperatures ($T_d$), atmospheric pressures ($p$) and mean monthly rainfall totals (RR).

While the notion of seasonally migrating 'four weather zones' is still a very useful concept (see Section 1.1 and Figure 1.1), some amendments are worthy of note. For example, the monsoon trough line in Buckle (1996: figure 8.9) descends to the north of the AEJ, but really the top of the monsoon layer descends towards the ITD. If defined by monsoonal southwesterly winds, the monsoon layer also barely extends above 700 hPa, as indicated by Buckle's and other schematics. Moreover, the reversal from a southerly to a northerly wind component lies below the AEJ. Satellite observations of clouds also suggest that zone D on the continent is more characterised by low-level stratiform clouds than by mid-level clouds. While it is true that in January zone B type weather is observed at the Guinea Coast, the land–sea breeze circulation (Section 4.1.4.2) is a major component of this zone in winter at the coast.

In this chapter, a climatological description of many observables describing the WAM has been presented. The latest observational platforms and reanalyses were used as the basis for presenting the key features of the regional climate, and for explaining the physical linkages between these systems. The chapter will serve as reference for the remaining chapters of the book, and we hope that it will also serve as a standard reference document for forecasters in the region. It should be stressed, however, that the paucity of data and difficulties in observing certain variables, like soil moisture or *ET*, still preclude an even deeper insight into the WAM system. This problem will not solely be solved by new satellite platforms. Rather, enhanced in-situ ground and upper-air observations are mandatory to better understand the WAM system that is important for the livelihood security of a growing population.

## Acknowledgements

We would like to acknowledge the help of several colleagues without which this chapter could not have been written. Remi Meynadier (LATMOS, Université Pierre et Marie Curie, CNRS, Paris, France) provided the data to plot Figure 1.9. Alexander Löw (Max-Planck Institute for Meteorology (MPIfMet) Hamburg, Germany) kindly provided the data to plot the albedo distribution in

Figure 1.14b and the NDVI data to plot Figure 1.18. He also provided Figure 1.17. Martina Klose (USDA-ARS Jornada Experimental Range, Las Cruces, USA) gratefully furnished Figure 1.16. Thorwald Stein and Kevin Hodges (both University of Reading, Reading, UK) kindly provided Figures 1.13 and 1.27 respectively. We are indebted to those colleagues who provided additional rainfall data: Issa Lélé, Pete Lamb (both University of Oklahoma, Norman, OK, USA), Mariane Diop Kane (Agence Nationale de l'Aviation Civile et de la Météorologie du Sénégal (ANACIM), Dakar, Sénégal), Abdou Kassimou (African Centre of Meteorological Applications for Development (ACMAD), Niamey, Niger), Guillaume Quantin (IRD, France), Bayo Omotosho (Federal University of Technology – Akure (FUTA), Akure, Nigeria), Ben Orji (Nigerian Meteorological Agency (NIMET), Lagos, Nigeria), Sidi Mohammed (National Meteorological Office, Nouakchott, Mauretania), Francis Didé (Agence pour la Sécurité de la Navigation Aérienne en Afrique et à Madagascar (ASECNA) Bénin, Cotonou, Benin) and Gregor Leckebusch and Auwal Abdussallam (both University of Birmingham, Birmingham, UK).

## Appendix: Acronyms

| | |
|---|---|
| AEJ | African easterly jet |
| AEW | African easterly wave |
| AI | aerosol index |
| ALMIP | AMMA Land Surface Model Intercomparison Project |
| AMMA | African Monsoon Multidisciplinary Analyses |
| CALIPSO | Cloud–Aerosol Lidar and Infrared Pathfinder Satellite Observations |
| EJ | easterly jet |
| ERA-Interim | European Centre for Medium-Range Weather Forecasts ReAnalysis Interim |
| *ET* | actual evapotranspiration |
| GATE | Global Atmospheric Research Programme (GARP) Atlantic Tropical Experiment |
| GARP | Global Atmospheric Research Programme |
| GEWEX | Global Energy and Water Exchanges Project |
| GIMMS | Global Inventory Modeling and Mapping Studies |
| GTS | Global Telecommunication System |
| HOAPS | Hamburg Ocean Atmosphere Parameters and Fluxes from Satellite Data |
| IRD | Institut de Recherche pour le Développement, France |
| ITCZ | intertropical convergence zone |
| ITD | intertropical discontinuity |
| ITF | intertropical front |
| MCS | mesoscale convective system |
| MIDAS | Met Office Integrated Data Archive System |
| MSLP | mean sea-level pressure |
| NCEP | National Centers of Environmental Prediction |
| NDVI | normalised differential vegetation index |
| NLLJ | night-time low-level jet |
| NOAA | National Oceanographic and Atmospheric Agency |
| OLR | outgoing longwave radiation |
| OMI | Ozone Monitoring Instrument |
| *P* | precipitation |
| PW | integrated precipitable water |
| SAL | Saharan air layer |
| SST | sea surface temperature |
| STJ | subtropical jet |
| TEJ | Tropical Easterly Jet |
| TOA | top-of-atmosphere |
| TOMS | Total Ozone Mapping Spectrometer |
| TRMM | Tropical Rainfall Measuring Mission |
| UTC | Universal Time Coordinated |
| WAHL | West African heat low |
| WAM | West African monsoon |
| WMO | World Meteorological Organisation |

## Notes

1 All geographic place names mentioned in the text are shown in Figure 1.2a. The climatic zones are indicated in Figure 1.2b.

2 Also termed in older literature as 'Ghana–Dahomey dry zone'.

3 Note that the use of the term 'ITCZ' over the continent is ambiguous and misleading, so it is avoided here; see Ramage (1995) and Fink *et al.* (2010).

4 Stein et al. (2011) did not distinguish falling hydrometeors (rain and snow) from those suspended in the cloud. Thus, strictly speaking, Figure 1.13 shows clouds and precipitation fractions.

5 In common with a number of authors, Lélé and Lamb (2010) used the term 'intertropical front' (ITF), an alternative terminology used in the literature to indicate the baroclinicity associated with the ITD.

## References


Bello NJ. 1996. The late onset of the rains in the coastal area of Nigeria. *Weather* **51**: 219–223. doi: 10.1002/j.1477-8696.1996.tb06216.x.

Bennartz R, Schroeder M. 2012. Convective activity over Africa and the tropical Atlantic inferred from 20 years of geostationary satellite observations. *J. Climate* **25**: 156–169. doi: 10.1175/2011JCLI3984.1.

Berry GJ, Thorncroft C. 2005. Case study of an intense African easterly wave. *Mon. Weather Rev.* **133**: 752–766. doi: 10.1175/MWR2884.1.

Boone A, Decharme, B, Guichard F, *et al.* 2009. The AMMA Land Surface Model Intercomparison Project (ALMIP). *Bull. Am. Meteorol. Soc.* **90**: 1865–1880. doi: 10.1175/2009BAMS2786.1.

Buckle C. 1996. *Weather and Climate in Africa*. Addison Wesley Longman Ltd; 312 pp.

Charney JG. 1975. Dynamics of deserts and drought in Sahel. *Q. J. R. Meteorol. Soc.* **101**: 193–202. doi: 10.1002/qj.49710142802.

Farquharson JS. 1939. The diurnal variation of wind over tropical Africa. *Q. J. R. Meteorol. Soc.* **65**: 165–183. doi: 10.1002/qj.49706528004.

Fink AH. 2012. To the 75th anniversary of the discovery of African easterly waves. 30th Conference on Hurricanes and Tropical Meteorology, 15–20 April 2012, Ponte Vedra Beach, FL, USA, S3.1. http://ams.confex.com/ams/30Hurricane/webprogram/Manuscript/Paper206073/extended_abstrext_30stamshurr_2012_fink_AEWhistory.pdf.

Fink AH, Reiner A. 2003. Spatio-temporal variability of the relation between African easterly waves and West African squall lines in 1998 and 1999. *J. Geophys. Res.* **108**(D11): 4332, ACL 5-1–5-17. doi: 10.1029/2002JD002816.

Fink AH, Vincent DG, Reiner P, Speth P. 2004. Mean state and wave disturbances during phases I, II, and III of GATE based on ERA-40. *Mon. Weather Rev.* **132**: 1661–1683. doi: 10.1175/1520-0493(2004)132<1661:MSAWDD>2.0.CO;2.

Fink AH, Vincent DG, Ermert V. 2006. Rainfall types in the West African Soudanian zone during the summer monsoon 2002. *Mon. Weather Rev.* **134**: 2143–2164. doi: 10.1175/MWR3182.1.

Fink AH, Paeth H, Ermert V, *et al.* 2010. Meteorological processes influencing the weather and climate of Benin. In: *Impacts of Global Change on the Hydrological Cycle in West and Northwest Africa*, P Speth, M Christoph, B Diekkrüger, *et al.* (eds). Springer, pp. 135 – 149.

Fröhlich LP, Knippertz P, Fink AH, Hohberger E. 2013. An objective climatology of tropical plumes. *J. Climate* **26**: 5044–5060. doi: 10.1175/JCLI-D-12-00351.1.

Glickmann TS. 2000. *Glossary of Meteorology*. American Meteorological Society: Boston, MA, USA; 855 pp.

Grams CM, Jones SC, Marsham JH, *et al.* 2010. The Atlantic inflow to the Saharan heat low: observations and modelling. *Q. J. R. Meteorol. Soc.* **136**: 125–140. doi: 10.1002/qj.429.

Guichard F, Kergoat L, Mougin E, *et al.* 2009. Surface thermodynamics and radiative budget in the Sahelian Gourma: seasonal and diurnal cycles. *J. Hydrol.* **375**: 161–177. doi: 10.1016/j.jhydrol.2008.09.007.

Hamilton RA, Archbold JW, Douglas CKM. 1945. Meteorology of Nigeria and adjacent territory. *Q. J. R. Meteorol. Soc.* **71**: 231–264. doi: 10.1002/qj.49707130905.

Holton JR. 2004. *An Introduction to Dynamic Meteorology*, 4th edn. Academic Press: Oxford, 549 pp.

Hubert H. 1939. Origine africaine d'un cyclone tropical atlantique. *Ann. Phys. France d'Outre-Mer* **6**: 97–115.

Klose M, Shao Y, Karremann MK, Fink AH. 2010. Sahel dust zone and synoptic background. *Geophys. Res. Lett.* **37**: L09802. doi: 10.1029/2010GL042816.

Knapp KR, Ansari A, Bain CL, *et al.* 2011. Globally gridded satellite observations for climate studies. *Bull. Am. Meteorol. Soc.* **92**: 893–907. doi: 10.1175/2011BAMS3039.1.

Knippertz P, Fink AH. 2006. Synoptic and dynamic aspects of an extreme springtime Saharan dust outbreak. *Q. J. R. Meteorol. Soc.* **132**: 1153–1177. doi: 10.1256/qj.05.109.

Knippertz P, Fink AH. 2009. Prediction of dry-season precipitation in tropical West Africa and its relation to forcing from the extratropics. *Weather Forecast.* **24**: 1064–1084. doi: 10.1175/2009WAF2222221.1.

Knippertz P, Todd MC. 2010. The central west Saharan dust hot spot and its relation to African easterly waves and extratropical disturbances. *J. Geophys. Res.* **115**: D12117. doi: 10.1029/2009JD012819.

Knippertz P, Fink AH, Schuster R, *et al.* 2011. Ultra-low clouds over the southern West African monsoon region. *Geophys. Res. Lett.* **38**: L21808. doi: 10.1029/2011GL049278.

Lamb PJ. 1980. Sahelian drought. *New Zeal. J. Geogr.* **68**: 12–16. doi: 10.1111/j.0028-8292.1980.tb00657.x.

Lavaysse C, Flamant C, Janicot S, *et al.* 2009. Seasonal evolution of the West African heat low, a climatological perspective. *Clim. Dynam.* **33**: 313–330. doi: 10.1007/s00382-009-0553-4.

Lélé IM, Lamb PJ. 2010. Variability of the intertropical front (ITF) and rainfall over the West African Sudan–Sahel zone. *J. Climate* **23**: 3984–4004. doi: 10.1175/2010JCLI3277.1.

Lothon M, Saïd F, Lohou F, Campistron B. 2008. Observation of the diurnal cycle in the low troposphere of West Africa. *Mon. Weather Rev.* **136**: 3477–3500. doi: 10.1175/2008MWR2427.1.

Marsham JH, Hobby M, Allen CJT, *et al.* 2013. Meteorology and dust in the central Sahara:

Observations from Fennec supersite-1 during the June 2011 Intensive Observation Period. *J. Geophys. Res.* **118**: 4069–4089. doi: 10.1002/jgrd.50211.

Mathon V, Laurent H, Lebel T. 2002. Mesoscale convective system rainfall in the Sahel. *J. Appl. Meteorol.* **41**: 1081–1092. doi: 10.1175/1520-0450(2002)041 < 1081:MCSRIT > 2.0.CO;2.

Mekonnen A, Thorncroft CD, Aiyyer AR. 2006. Analysis of convection and its association with African easterly waves. *J. Climate* **19**: 5405–5421. doi: 10.1175/JCLI3920.1.

Meynadier R, Bock O, Guichard F, *et al.* 2010. West African monsoon water cycle: 1. A hybrid water budget data set. *J. Geophys. Res.* **115**: D19106. doi: 10.1029/2010JD013917.

Nicholson SE. 2008. The intensity, location and structure of the tropical rainbelt over West Africa as factors in interannual variability. *Int. J. Climatol.* **28**: 1775–1785. doi: 10.1002/joc.1507.

Nicholson SE. 2011. Sub-Saharan Africa. In: *Dryland Climatology*, SE Nicholson (ed.). Cambridge University Press: Cambridge; pp. 293–322.

Nicholson SE, Grist JE. 2003. The seasonal evolution of the atmospheric circulation over West Africa and Equatorial Africa. *J. Climate* **16**: 1013–1030. doi: 10.1175/1520-0442(2003)016 < 1013:TSEOTA > 2.0.CO;2.

Nicholson SE, Webster PJ. 2007. A physical basis for the interannual variability of rainfall in the Sahel. *Int. J. Climatol.* **28**: 1775–1785. doi: 10.1002/qj.104.

Nicholson SE, Barcilon AI, Challa M, Baum J. 2007. Wave activity on the tropical easterly jet. *J. Atmos. Sci.* **64**: 2756–2763. doi: 10.1175/JAS3946.1.

Parker DJ, Thorncroft CD, Burton RR, Diongue-Niang A. 2005a. Analysis of the African easterly jet, using aircraft observations from the JET2000 experiment. *Q. J. R. Meteorol. Soc.* **131**: 1461–1482. doi: 10.1256/qj.03.189.

Parker DJ, Burton RR, Diongue-Niang A, *et al.* 2005b. The diurnal cycle of the West African monsoon circulation. *Q. J. R. Meteorol. Soc.* **131**: 2839–2860. doi: 10.1256/qj.04 52.

Parker D, Good E, Chadwick R. 2011. Reviews of observational data available over Africa for monitoring, attribution and forecast evaluation. Hadley Centre Technical Note No. 86. http://www.metoffice.gov.uk/media/pdf/b/o/HCTN_86_revised.pdf (accessed 23 July 2016).

Ramage CS. 1995. *Forecaster's Guide to Tropical Meteorology*, AWS/TR-95/001, Air Weather Service, Scott Airforce Base, IL, 392 pp.

Ross RS, Krishnamurti TN. 2007. Low-level African easterly wave activity and its relation to Atlantic tropical cyglogenesis in 2001. *Mon. Weather Rev.* **135**: 3950–3964. doi: 10.1175/2007MWR1996.1.

Schrage JM, Fink AH. 2012. Nocturnal continental low-level stratus over tropical West Africa: observations and possible mechanisms controlling its onset. *Mon. Weather. Rev.* **140**: 1794–1809. doi: 10.1175/MWR-D-11-00172.1.

Schuster R, Fink AH, Knippertz P. 2013. Formation and maintenance of nocturnal low-level stratus over the southern West African monsoon region during AMMA 2006. *J. Atmos. Sci.* **70**: 2337–2355. doi: 10.1175/JAS-D-12-0241.1.

Stein THM, Parker DJ, Delanoë J, *et al.* 2011. The vertical cloud structure of the West African monsoon: a four-year climatology using CloudSat and CALIPSO. *J. Geophys. Res.* **116**: D22205. doi: 10.1029/2011JD016029.

Sultan B, Janicot S. 2003. The West African monsoon dynamics. Part II: the 'preonset' and 'onset' of the summer monsoon. *J. Climate* **16**: 3407–3427. doi: 10.1175/1520-0442(2003)016 < 3407:TWAMDP > 2.0.CO;2.

Sutton LJ. 1925. Haboobs. *Q. J. R. Meteorol. Soc.* **51**: 25–30. doi: 10.1002/qj.49705121305.

Thorncroft CD, Hoskins BJ. 1994. An idealized study of African easterly waves. I: a linear view. *Q. J. R. Meteorol. Soc.* **120**: 953–982. doi: 10.1002/qj.49712051809.

Thorncroft CD, Parker DJ, Burton RR, *et al.* 2003. The JET2000 Project: aircraft observations of the African easterly jet and African easterly waves. *Bull. Am. Meterol. Soc.* **84**: 337–351. doi: 10.1175/BAMS-84-3-337.

Todd MC, Allen CJT, Bart M, *et al.* 2013. Meteorological and dust aerosol conditions over the Western Saharan region observed at Fennec supersite-2 during the Intensive Observation Period in June 2011. *J. Geophys. Res.* **118**: 8426–8447. doi: 10.1002/jgrd.50470.

Van der Linden R, Fink AH, Redl R. 2015. Climatology of low-level continental stratus in southern West Africa for 2006–2011. *J. Geophys. Res.* **120**: 1186–1201. doi: 10.1002/2014JD022614.

Vollmert P, Fink AH, Besler H. 2003. 'Ghana Dry Zone' und 'Dahomey Gap': Ursachen für eine Niederschlagsanomalie im tropischen Westafrika. *Erde* **134**: 375–393 (English abstract and figure captions).

# 2

# Synoptic Systems

**Lead authors:** *Rosalind Cornforth[a], Zilore Mumba[b] and Douglas J. Parker[c]*

**Contributing authors:** *Gareth Berry[d], Nicolas Chapelon[e], Kone Diakaria[f], Mariane Diop-Kane[g], Volker Ermert[h], Andreas H. Fink[i], Peter Knippertz[i], Jean Philippe Lafore[j], Arlene Laing[k], Serge Lepape[e], Ross Maidment[m], John Methven[n], Ben Orji[o], David Osika[p], Emmanuel Poan[j], Rémy Roca[r], Simon Rowell[n], Roger Smith[t], Thomas Spengler[u], Christopher M. Taylor[v], Chris Thorncroft[w], Jean-Christoph Vincendon[e] and Charles Yorke[p]*

**Reviewer:** *Chris Thorncroft (University at Albany, State University of New York, USA)*

[a] Walker Institute, University of Reading, Reading, UK
[b] Department of Mathematics and Statistics, University of Zambia, Lusaka, Zambia
[c] School of Earth and Environment, University of Leeds, Leeds, UK
[d] School of Mathematical Sciences, Monash University, Clayton, VIC, Australia
[e] Commandement Interarmées de l'Espace (CISMF), Institut National Polytechnique de Toulouse (INP-Toulouse), Toulouse, France
[f] Agence pour la Securite de la Navigation Aerienne en Afrique et a Madagascar (ASECNA), Abidjan, Côte d'Ivoire
[g] Agence Nationale de l'Aviation Civile et de la Météorologie du Sénégal (ANACIM), Dakar, Sénégal
[h] Institute of Geophysics and Meteorology, University of Cologne, Cologne, Germany
[i] Institute of Meteorology and Climate Research, Karlsruhe Institute for Technology, Karlsruhe, Germany
[j] Centre National de Recherches Météorologiques (CNRM-UMR3589), Météo-France and CNRS, Toulouse, France
[k] Cooperative Institute for Research in the Atmosphere (CIRA), Colorado State University, Fort Collins, Colorado, USA
[m] TAMSAT, Department of Meteorology, University of Reading, Reading, UK
[n] Department of Meteorology, University of Reading, Reading, UK
[o] Regional Training Centre, Nigerian Meteorological Agency, Oshodi-Lagos, Nigeria
[p] Ghana Meteorological Agency, Accra, Ghana
[r] LMD/IPSL, CNRS, Université Pierre et Marie Curie, Paris, France
[t] Meteorological Institute, University of Munich, Munich, Germany
[u] Geophysical Institute, University of Bergen and Bjerknes Centre for Climate Research, Bergen, Norway
[v] Centre for Ecology and Hydrology, Wallingford, UK
[w] Atmospheric and Environmental Sciences, University at Albany, SUNY, Albany, NY, USA

## 2.1 Scientific Background

### 2.1.1 Introduction

The synoptic meteorology of West Africa has been studied for many decades. Since the 1930s at least, it has been known that coherent synoptic weather systems exist in the region, in the form of African easterly waves (AEWs; e.g. Piersig, 1944[1936]). More recently we have developed synoptic models of the Saharan heat low (SHL; e.g. Rácz and Smith, 1999; Parker *et al.*, 2005b; Lavaysse *et al.*, 2010, 2011; Smith and Spengler, 2011), the intertropical discontinuity (ITD; Flamant *et al.*, 2007), and various kinds of waves (e.g. Thorncroft and Hoskins, 1994; Mounier *et al.*, 2007; Mekonnen and Rossow, 2011; Ventrice and Thorncroft, 2013; Wu *et al.*, 2013) and vortices in the region (e.g. Tyner and Aiyyer, 2012). Understanding such coherent weather systems, which evolve in a familiar pattern over periods of a couple of days, helps the analysis and forecasting of the weather (Davis *et al.*, 2012). However, unlike mid-latitude regions, in West Africa the rainfall, dominated by cumulonimbus rainstorms, tends to be much less closely controlled by the prevailing synoptic conditions, being more strongly conditioned by the diurnal cycle of stability (e.g. Laing *et al.*, 2008; Bouniol *et al.*, 2012) and the land-surface forcing. Because rainfall is controlled by a combination of synoptic state, time of day and surface forcing, the relationships between synoptic state and rainfall can be difficult to evaluate (e.g. Hsieh and Cook, 2005; Cornforth *et al.*, 2009; Poan *et al.*, 2015). Probably for these reasons, the description of some aspects of the synoptic meteorology remains incomplete, in the literature and in teaching. Recent years have seen substantial progress, not least through the enhanced observational efforts and related modelling and forecasting improvements achieved during the African Monsoon Multidisciplinary Analysis

*Meteorology of Tropical West Africa: The Forecasters' Handbook*, First Edition. Edited by Douglas J. Parker and Mariane Diop-Kane.

(AMMA) field campaigns (e.g. Janicot *et al.*, 2008; Lebel *et al.*, 2010; Lafore *et al.*, 2011). Here, we present a consensus of current thinking on synoptic weather systems in West Africa.

Synoptic meteorology is an important area of weather forecasting in West Africa, where the role of the human forecaster is particularly important (Fink *et al.*, 2011), relative to numerical model forecasts. From the enhanced observations of the AMMA period, we know that numerical weather prediction (NWP) models have some useful skill in representing the synoptic state over West Africa, but we also know that there remain some gross errors, notably in the representation of rainfall (e.g. Diallo *et al.*, 2014). Through a knowledge of synoptic dynamics and conceptual models, the forecaster can use NWP products alongside observations, to improve the quality of short-range forecasts in their region.

On any given day, the synoptic conditions are determined by a combination of remote atmospheric dynamics, subseasonal patterns such as Kelvin waves, regional circulations, and local responses of the weather to the synoptic state (e.g. Mekonnen and Rossow, 2011). In this chapter we focus on discussing the synoptic circulations that occur within the West Africa region, encompassing scales of a few thousand kilometres, and on timescales of 1–10 days. The larger scale circulations controlling the synoptic environment, and their seasonal variations, have already been introduced, in Chapter 1. Convectively coupled Kelvin waves also fall within this 1–10-day timescale, but are discussed in Section 7.1.4. The specific weather patterns developed within a given synoptic state are discussed in Chapters 3, 4 and 5, which address convective rainfall, local weather and atmospheric dust respectively. Here, in this chapter, we begin by presenting synoptic features on the continental scale, which drive variability. In particular, we introduce those features that need to be analysed by a forecaster on a day-to-day basis; rules for plotting the features on a synoptic chart are presented in Chapter 11. The SHL is just such a feature. The variability of the SHL influences synoptic dynamics in West Africa, as well as the occurrence of wind-blown dust; we also discuss the structure of the African easterly jet (AEJ) and the monsoon layer. In Section 2.1.3 we then addresses the most important synoptic patterns in this region – AEWs, extratropical wave intrusions and Guinea Coast weather systems.

### 2.1.2 Continental-scale Synoptic Features

The continental-scale features that have been introduced in Chapter 1 show significant synoptic variability. As explained in Chapter 11, these features are vitally important to plot on a synoptic chart. It is important for forecasters to be aware of how these features vary on intraseasonal and diurnal timescales. Subseasonal variability in these features is particularly important because of the associated change in environment that supports or interacts with known synoptic disturbances. Diurnal variability in the continental-scale features is important as this can significantly impact the nature and impact of the known synoptic disturbances on local weather.

#### 2.1.2.1 Saharan Heat Low

Heat lows and heat troughs are prominent climatological features in many arid land areas of the world during the warmer months, especially in low latitudes where insolation is at its peak. Such regions include the Saudi Arabian Peninsula, northwestern and northeastern Australia, northern and southwestern Africa, West Pakistan and northern India, the Qinghai-Xizang plateau in China, southwestern North America and the Iberian Peninsula. Some authors refer to these systems as thermal lows or thermal troughs. One of the most prominent examples is the summer heat low over the Sahara Desert. Although the SHL lies in the extreme north of the West African region in summer, its dynamics and intraseasonal variability are very important in their influence on the region's diurnal and synoptic circulations.

Heat lows are warm-cored, surface-based low pressure systems situated over land, and are generally contained within the lower troposphere below about 600 hPa. The theoretical study by Rácz and Smith (1999) used an idealised model to understand the essential dynamical features of a continental-scale heat low that forms over an idealised land mass, when that region is subjected to a diurnal cycle of heating and cooling. It was shown that many of the basic features of observed heat-low circulations, such as the effects associated with the coastal sea breezes and the nocturnal low-level jet, are explained by the diurnal cycle of boundary layer stability (Section 4.1.3). These concepts have been found subsequently to explain important aspects of the West African monsoon (WAM) system and the SHL (Parker *et al.*, 2005b; Lothon *et al.*, 2008; Grams *et al.*, 2010; Schuster *et al.*, 2013).

Some of the key features of the SHL can be gleaned from Figure 2.1 (see also Figure 1.33), which shows selected time-mean fields for the June–July–August season during the period from 1989 until 2009. The SHL is evident as an area of low mean sea-level pressure (MSLP) (Figure 2.1a) centred at about 22°N, 3°W. The heat low is discernible in the 850 hPa and 2 m temperature fields also (Figure 2.1b and c respectively), with maxima coinciding approximately with the local pressure minimum. The low is bounded by mountain ranges to the north (Atlas) and east (Ahaggar), and by the Atlantic Ocean to the west (see Figure 2.1d). It is reasonable to surmise that the occurrence of the heat low in that area is a result of the interplay between the orography as well as distribution of sea surface temperature (SST) and the prevailing broad-scale flow conditions.

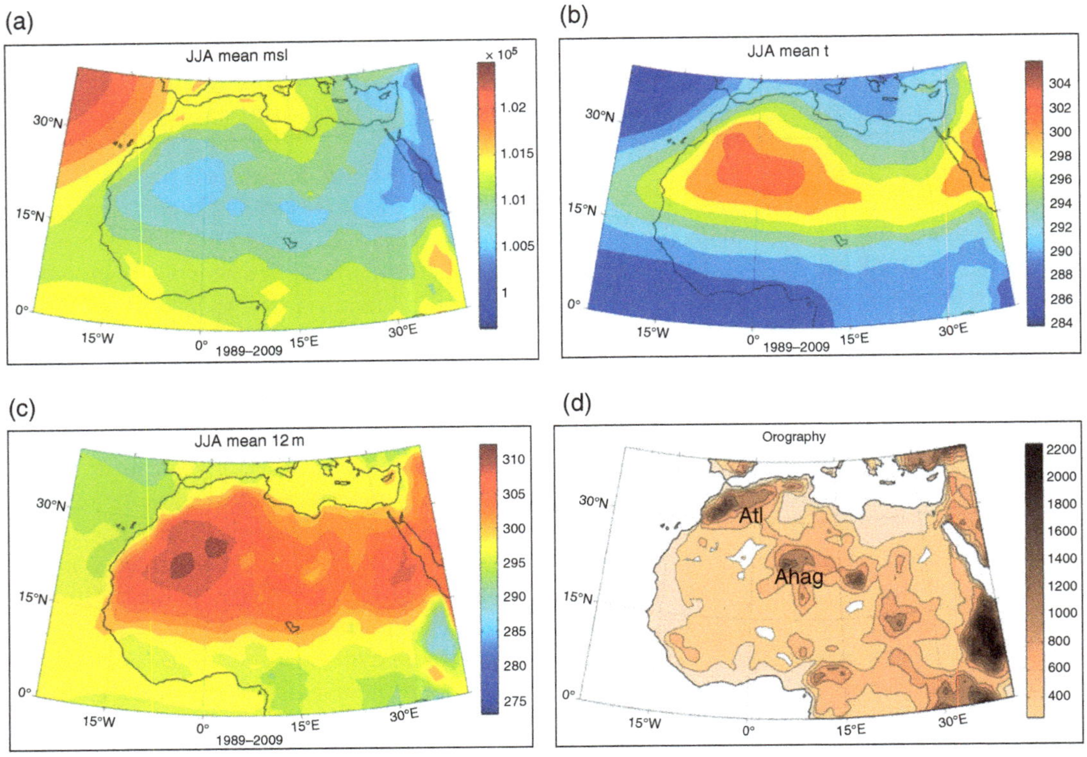


**Figure 2.1** Characteristics of the SHL seen in ECMWF ERA Interim climatological mean (1989–2009) fields for June–July–August for (a) mean sea-level pressure (MSLP), (b) temperature at 850 hPa, (c) 2 m temperature and (d) orographic height.

On subseasonal and synoptic timescales, the SHL migrates around the location shown in Figure 2.1 (Lavaysse *et al.*, 2009) and pulsates in strength (Lavaysse *et al.*, 2010) as it interacts with mid-latitude weather systems (Thorncroft and Flocas, 1997; Chauvin *et al.*, 2010; Roehrig *et al.*, 2011; see Section 7.1.5.1).

A prominent feature of idealised models of the heat low is the large diurnal variation in the low-level cyclonic circulation, as found by Rácz and Smith (1999) and Spengler *et al.* (2005). The mean winds are relatively calm throughout the day, while the growth of a nocturnal low-level jet leads to strong convergence and the formation of a super-geostrophic low-level cyclonic circulation during the night. Under the influence of large-scale background flows the nocturnal convergent flow can also be strongly frontogenetic (Spengler *et al.*, 2005) and, therefore, may contribute to the nocturnal intensification of the intertropical front (ITF). The formation of the cyclonic circulation can be understood through the inertial turning of the unbalanced low-level flow toward the low-pressure centre (e.g. see Blackadar (1957); Section 4.1.3.2). Therefore, while the minimum surface pressure in the idealised model occurs at 1700h, the maximum tangential wind speed at low levels is attained around 0530h and the minimum at 1600h. Although the surface low is most intense in late afternoon, the cyclonic winds weaken during the daytime and restrengthen again at night. The behaviour at low levels is in stark contrast to that aloft, where there is almost no diurnal variation in the strength of the anticyclone (less than 2 m $s^{-1}$).

The nocturnal low-level jet can yield high values of vertical low-level shear that might be hazardous for air traffic (Sections 4.1.3.2 and 6.2.3). Over northern Africa, the effect of a nocturnal increase of the low-level convergence is also important for the advection of moisture into the Sahel region (Parker *et al.*, 2005b; Lothon *et al.*, 2008), particularly during the monsoon onset period in June.

The diurnal variability at low levels is in contrast to the quasi-stationary flow behaviour at the mid to upper tropospheric levels (Spengler and Smith, 2008). The low-level cyclone is surmounted by an upper-level anticyclone in the upper levels of the Saharan boundary layer, around 700–500 hPa. The anticyclone is associated with the divergent branch of the overturning circulation due to the warm Saharan boundary layer, in conjunction with the approximate conservation of absolute angular momentum. The anticyclone extends through much of the troposphere, but has its maximum strength in the lower troposphere in the upper levels of the Saharan boundary layer. To the south of the SHL, the easterly winds associated with this anticyclonic circulation correspond to the AEJ. Thorncroft and Blackburn (1999) stressed the importance of the mean state of the SHL on the AEJ.

Forecasters in the Sahel and Guinea Coast zone often look to the synoptic station temperatures in Mauritania, Mali and Niger as indicators of a likely strengthening of the monsoon circulation in subsequent days, due to stronger south–north temperature and pressure gradients. Practice has been confirmed by theory – Parker *et al.* (2005b) confirmed that the strength of the thermal anomalies in the heat low region is correlated with strengthening nocturnal monsoon winds further south.

Further impacts of the variability of the SHL have been shown by Chauvin *et al.* (2010) and Roehrig *et al.* (2011). On timescales of 10–25 days, variability in the SHL (with some extratropical influences) can be strongly related to subsequent anomalies of rainfall in the Sahel about 5 days later. 'Ventilation' of the SHL by cool advection from the Mediterranean can weaken the Sahelian pressure gradient and reduce subsequent rainfall over the Sahel.

Ventilation (see Chou *et al.* (2001)) of the heat low can occur from all of its flanks, in association with different kinds of weather systems. For instance, Grams *et al.* (2010) showed how a synoptically varying 'Atlantic inflow' around Mauritania and Sénégal interacts with the strength of the SHL in this western region, while convective cold pools emerging from storms over the Atlas Mountains of Morocco commonly bring cool air into the desert from the northwest. The interaction between the SHL and the cool ventilation from the south, in the monsoon flow and in embedded convective cold pools within the monsoon is of particular dynamical interest, as described in Section 2.1.2.2. Observations and data-assimilation results (Garcia-Carreras *et al.*, 2013) and model sensitivity studies (Marsham *et al.*, 2013) are consistent in showing that the convective cold pools from within the monsoon make a substantial contribution to the thermodynamic budget of the SHL. This raises an important possibility of feedback between the SHL and the monsoon. An intense SHL leads, for example, to a stronger monsoon circulation and increased convection in the Sahel in the subsequent days. The result of this convection is to drive stronger cold pools, which then act to ventilate the Saharan boundary layer. The relationship between these processes, which are not well captured in general circulation models (GCMs), remains an interesting area of research. Some of these mechanisms are discussed in more detail in Section 7.1.6.

#### 2.1.2.2 The Intertropical Discontinuity or Intertropical Front

The ITD or ITF (henceforward ITD) is a zone of confluence of the winds, and an air-mass boundary, which separates the hot, dry Saharan air from the cooler, moist monsoonal air in the spring and summer months. Despite its name, the ITD is neither a true discontinuity nor a simple 'front', because it has a very mobile and dynamic structure, related to the diurnal variability of 'zone B' in the mean climatic section (Section 1.1 and Figure 1.1). It is one of the key features used by forecasters to analyse the state of the monsoon on a synoptic chart, since it maps the northward extent of the moist monsoon air (Figure 1.1, and guidance in Section 11.2 for plotting this feature).

On a daily basis the ITD is governed by the same physical processes that control the heat low. At night, when boundary-layer mixing is very weak, the cool monsoonal air to the south of the ITD accelerates northward, into the low pressure of the heat low. This means that the ITD can surge northward by several hundred kilometres overnight. Stations in the northern Sahel/southern Sahara that have experienced a hot, dry afternoon with light winds can experience cooler, moister winds with a southerly component arriving overnight. The nocturnal surge of the ITD also tends to involve a sharpening of its front, into a form of gravity current (see Section 4.1.4), which can be strong enough to lift dust and reduce visibility (Bou Karam *et al.*, 2008). The nocturnal surge of the ITD is followed by a rather shallow monsoon layer, perhaps 1000 m deep, resembling a gravity current.

During the following day, the air in the shallow monsoon layer is mixed by dry convection in the boundary layer, and eventually through the morning and into the afternoon will be mixed into the residual Saharan boundary layer (or Saharan air layer (SAL)). This cycle brings moisture into the southern margins of the Sahara. Temperature increases and humidity decreases at stations in the southern Sahara, as the cool moist monsoonal air becomes mixed with the hot and dry SAL, and conversely the SAL is moistened by this process, so that altocumulus clouds are common at the top of the SAL, around 550 hPa (Parker *et al.*, 2005a; Stein *et al.*, 2011). The dry convective mixing weakens the daytime winds, so that they become relatively light and typically gain a northeasterly component through mixing with the SAL winds.

As the day proceeds, the ITD diagnosed in synoptic station data retreats southward – the cool, moist air is mixed up with Saharan air from above, and the confluence zone also moves southward as the winds in the northern region become more northerly. It is important to realise that the southward retreat of the ITD during the day is not a material process – the air itself is not moving southward at this time – but a process due to the change in the thermodynamic characteristics of the air to the south of the ITD as it is mixed up, and changes from 'monsoonal' to more 'Saharan'. This is in contrast to the nocturnal ITD surge, when air moves materially northward at low levels.

The dynamical nature of the ITD is further complicated by the role of convective cold pools, emerging from convective storms in the monsoon air to the south of the ITD. These cold pools are generated by evaporative cooling of precipitation within the storms (Section 3.1.2.2) and also propagate as gravity currents. The convective cold pools commonly surge northward, in the late afternoon and overnight, travelling several hundred kilometres. As the cold pools enter the ITD region, they tend to merge with the ITD itself (which may be forming a sharp front around this time anyway) and then be indistinguishable from the ITD in routine observations (Flamant *et al.*, 2007, 2009; Marsham *et al.*, 2008). Thus, we see that when deep convective storms are active just to the south of the ITD, a northward surge of the ITD through the arrival of convective cold pools can be expected in that region in the coming hours.

#### 2.1.2.3 African Easterly Jet

The AEJ is a distinct and robust feature of the tropical North African climate which has a significant influence on synoptic variability, and in particular AEWs. The AEJ is characterised by a peak in the zonal easterly wind of about 15 m $s^{-1}$ around 600–700 hPa that is in approximate thermal wind balance with the lower tropospheric temperature gradient from south to north (or baroclinicity; Cook, 1999; see Figure 2.2). Consistent with thermal wind balance, the baroclinicity (south–north temperature gradient) reduces to zero between the surface and the level of the AEJ. Thorncroft and Blackburn (1999) showed that the reduction in baroclinicity with height is consistent with north–south variations in convection; with moist convection predominant equatorward of the AEJ and dry convection predominant poleward of the AEJ. Parker *et al.* (2005a) analysed observations from the JET2000 experiment (Thorncroft *et al.*, 2003) to highlight, in some detail, the meridional variations in convection and associated thermodynamic profiles in the AEJ region, and the schematic of the AEJ shown in Figure 2.2 is based on this work. While deep moist convection and a peak mean rainfall are located just south of the AEJ peak and dry convection and the associated SAL are located north of the AEJ, there is an important transition zone between. In this transition zone the shallow monsoon layer undercuts the SAL that sits above it (see Figure 2.2). The relatively warm SAL air sitting above the moist monsoon air is associated with significant convective inhibition (CIN) and the build-up of convective available potential energy (CAPE) in the boundary layer in these regions. Combined with the vertical shear of the AEJ and the dryness of the SAL, which favours intense convective downdraughts (Section 3.1.2.2), the region just poleward of the AEJ is favourable for the establishment of tropical squall lines. While these convective systems are included in Figure 2.2 we do not expect convection to occur to the north of the AEJ as frequently as the region south of the AEJ (which has lower CIN).

Consistent with the meridional contrasts in moist and dry convection highlighted earlier, the AEJ is characterised by two distinctive potential vorticity (PV) anomalies. A negative PV anomaly is established in the vicinity of the SAL, poleward of the AEJ (consistent with the low stability in the SAL) and a positive PV anomaly is established equatorward of the AEJ in association with moist convection (e.g. Thorncroft and Blackburn, 1999). The positive PV anomaly is predominantly at the level of the AEJ, which is consistent with a dominant contribution from stratiform rainfall and the frequent generation of mesoscale convective vortices (MCVs) there. The so-called monsoon trough at 850 hPa (see Sections 1.1 and 2.1.3.3) is viewed here as a downward reflection of the mid-level PV peak and associated vorticity, and may represent a focal zone for MCVs.

The distinctive PV anomalies linked to the AEJ are associated with different-signed meridional gradients in PV. Combined with the meridional gradients of surface temperature, the AEJ satisfies the necessary conditions for both barotropic and baroclinic instability

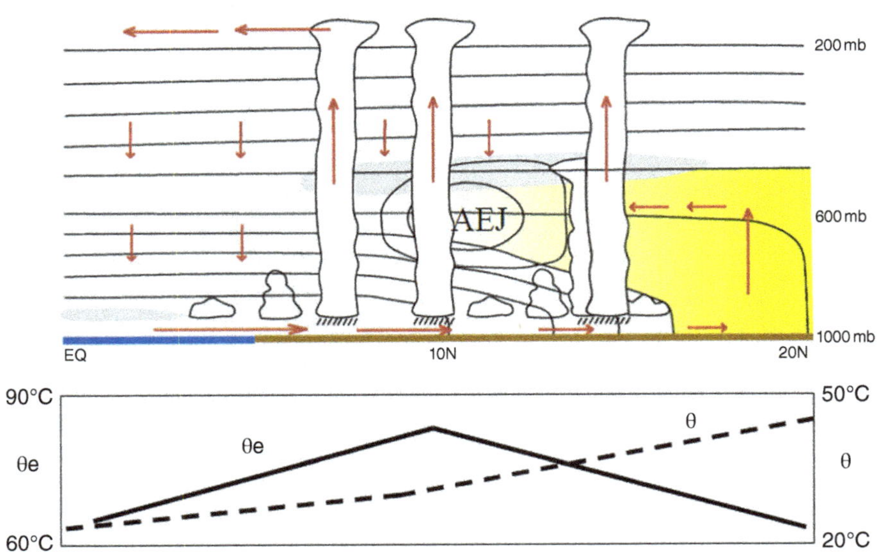


**Figure 2.2** Schematic south–north vertical cross-section along the Greenwich Meridian, highlighting the SHL–AEJ–ITCZ system and SAL (shaded yellow), with meridional variations in atmospheric boundary-layer potential temperature $\theta$ (contoured in upper panel) and equivalent potential temperature $\theta_e$. The monsoon trough is located just to the south of the AEJ. *Source*: adapted from Figure 10 of Parker *et al.* (2005a) for the AMMA International Science Plan (http://amma-international.org/library/docs/AMMA_ISP_May2005.pdf), and to be compared with Figures 1.1 and 1.34. Used with permission.

(e.g. Pytharoulis and Thorncroft, 1999). While recent studies have questioned the extent to which observed AEWs owe their existence to a linear instability mechanism (Hall *et al.*, 2006; Thorncroft *et al.*, 2008); the presence of these PV gradients is still essential for the development of AEWs and should be monitored.

### 2.1.3 African Easterly Waves

#### 2.1.3.1 Background

AEWs develop and propagate on the AEJ over tropical North West Africa during the northern hemisphere summer. Their presence was recognised during the first half of the twentieth century from African rain gauge data that showed a distinct periodicity at around a 3-day timescale (Piersig, 1944[1936]). It was then noted by forecasters (e.g. Simpson *et al.*, 1968) that the same disturbances that brought about this periodicity were moving westward into the Atlantic basin and, in some cases, developing into tropical cyclones. Further studies have revealed that AEWs are the dominant synoptic-scale weather phenomena within the WAM system and tropical Atlantic during boreal summer. They can be observed between May and October, and peak in intensity during August and September (e.g. Carlson, 1969a,b; Burpee, 1974; Karyampudi and Carlson, 1988; Ross, 1991) and are one of the best known examples of a coherent, synoptic weather pattern in the tropics.

In the scientific literature there has been some inconsistency in the way in which AEW structures are described, and the theoretical models do not always agree with the experience of operational forecasters. It has become clear that AEWs are variable, from case to case, according to geographical location, and varying through the lifecycle of a single wave. For these reasons, we describe here some key observations and analyses of AEWs in Section 2.1.3.2, followed by a consensus description of the elements making up an AEW, and by which AEW structure may be objectively defined (in Section 2.1.3.3).

#### 2.1.3.2 Observed Synoptic Structure of African Easterly Waves, Analyses and Statistical Composites

AEWs are most prominent in the lower troposphere, peaked at the AEJ level at around 650 hPa (approximately 3.5 km) over tropical North Africa and are characterised by periodicities of 2–6 days and wavelengths of 2000–5000 km (Reed *et al.*, 1977). Carlson (1969a,b) conducted synoptic analyses of several AEWs from the summer of 1967 and noted the presence of a wave-like perturbation to the wind field at 10,000 ft (approximately 3 km, or 700 hPa), along with two cyclonic vortices at 2000 ft (approximately 0.75 km, or 925 hPa), either side of the AEJ. Most AEWs tend to form somewhere between 15°E and 30°E over central or eastern Africa, and propagate westward through the WAM and into the tropical Atlantic, where they are a common precursor of tropical cyclones (e.g. Carlson, 1969a; Avila and Pasch, 1992; Mekonnen *et al.*, 2006). In some instances AEWs have been observed to propagate coherently as far west as the eastern Pacific Ocean (e.g. Avila and Pasch, 1992). The AEW structure evolves as it moves westward. Before reaching the coast of West Africa, the AEWs that develop into tropical cyclones have a distinctive cold-core structure below the level of the AEJ, consistent with a vorticity maximum at the level of the AEJ (e.g. Reed *et al.*, 1977; Kwon and Mak, 1990; Hopsch *et al.*, 2007). They often begin to transform toward more warm-core structures as they move toward the Guinea Highlands region (5–13°N, 8–15°W), with regions of deep convection becoming more confined to the trough, as shown by Hopsch *et al.* (2010). The wave trough often slopes eastward with height below the AEJ level (Reed *et al.*, 1977), as expected for baroclinic growth on an easterly jet.

Kiladis *et al.* (2006), based on a regression analysis, highlighted the typical coherent synoptic evolution of AEWs as they grow and propagate across tropical North Africa and into the East Atlantic. The regression is keyed on a space–time, filtered time-series of outgoing longwave radiation (OLR) at 10°N, 10°W in the so-called 'tropical depression' band. The day-0 analysis in Figure 2.3 shows the regressed OLR together with streamfunction and wind vectors at 850 hPa and indicates a very coherent structure. The AEW trough–ridge structures are tilted against the horizontal shear north and south of the jet, consistent with AEWs that are undergoing barotropic growth (Section 2.1.3.5). It is also apparent in this figure that AEWs typically have a dynamical influence that crosses the Equator (Section 1.7; Figure 1.33) and also reaches as far north as 25–30°N. Given the meridional extent of the AEWs, we should expect them to interact with known convectively coupled equatorial waves to the south (Section 7.1.4) and mid-latitude weather systems to the north (Section 2.1.4). At day-0 the peak in convection, depicted by a minimum in OLR, is located in the AEW trough.

Figure 2.3 depicts the evolution of the AEWs through considering negative and positive time lags. The first indication of the AEW that reaches 10°N, 10°W at day-0 is a region of anomalous convection in the Darfur region at day-4. This is consistent with the hypothesis that AEWs are triggered by finite-amplitude disturbances such as deep convection (see Hall *et al.* (2006) and Thorncroft *et al.* (2008)). As the AEWs develop in the following days, the peak convection is located ahead of the AEW trough, a feature seen in many earlier studies. Interestingly, as the AEWs move westward the peak in convection shifts into the trough and eventually shifts into the southerlies behind the AEW trough when they are over the ocean.

Figure 2.4 shows the vertical structure of the meridional wind and corresponding OLR for an easterly wave passage over Bamako (continental), and then Dakar (west coast).

While both patterns are tilted eastward with height up to the AEJ level (~700 hPa), at Dakar there is a westward tilt above 500 hPa as expected on a shear flow, while Bamako still has an eastward tilt. Radiosonde data have shown that at Bamako (12.5°N, 8.0°W) most AEWs had maximum amplitudes at the level of the AEJ, while at Dakar (14.7°N, 17.5°W) the waves had maximum amplitude between 850 and 950 hPa (Pytharoulis and Thorncroft, 1999). Although AEWs occur as waves with maximum amplitude at low levels north of the AEJ and waves with maximum amplitude close to the AEJ level further south, both represent a single dynamic mode and propagate simultaneously across Africa (Pytharoulis and Thorncroft, 1999; Cornforth *et al.*, 2009).

Figure 2.5 displays the composite horizontal wind field and relative vorticity on several pressure levels associated with AEWs from the European Centre for Medium-Range Weather Forecasts (ECMWF) ERA-Interim reanalysis dataset, which shows the prominent structures also noted by observation-based studies. Below the level of the AEJ (1000 hPa and 850 hPa), twin vortices are noted either side of the AEJ axis (near 15°N, denoted by the highest

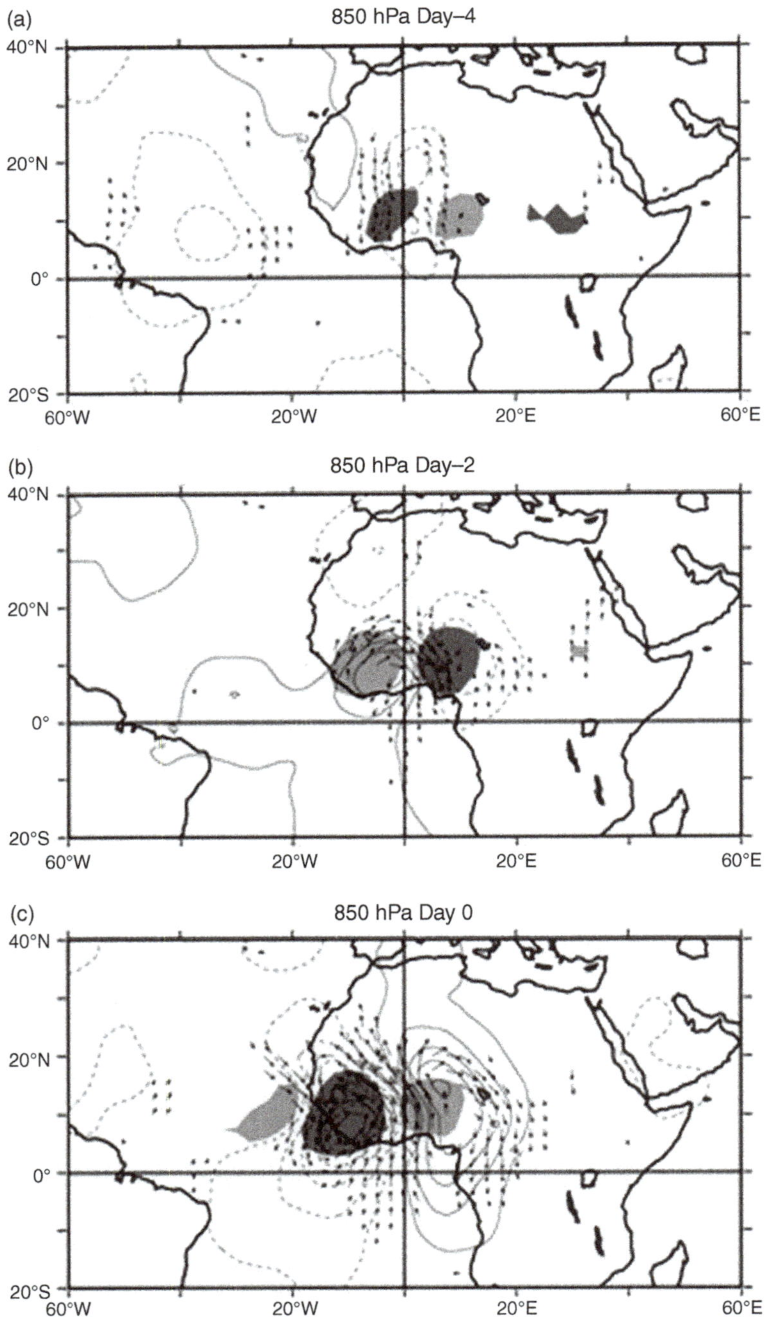


**Figure 2.3** Composite AEW structure of 850 hPa streamfunction with anomalous convection. The plots are derived from a regression based on a space–time filtered time-series of OLR at 10°N, 10°W in the so-called 'tropical depression' band. Dark shading denotes OLR anomalies less than $-10$ W m$^{-2}$ (active convection), while light shading denotes OLR anomalies greater than $+10$ W m$^{-2}$ (suppressed convection). *Source*: Figure 3 from Kiladis *et al.* (2006). ©American Meteorological Society. Used with permission.

Figure 2.3 (Cont'd)

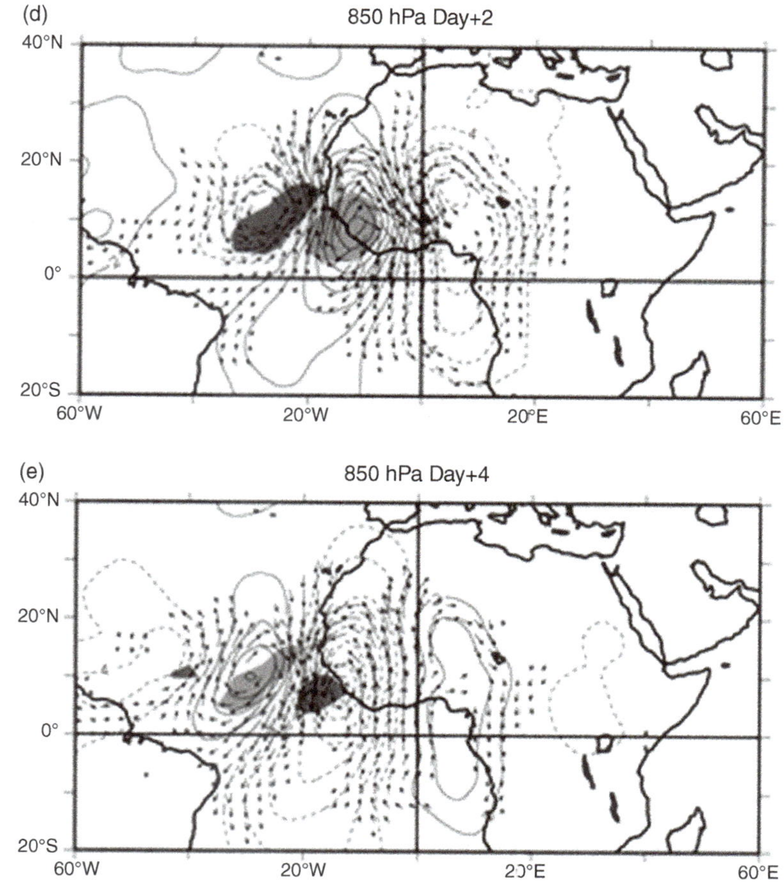


wind speeds at the 700 hPa level). The northernmost of these vortices is collocated with the maximum surface temperature gradient (i.e. along the ITD), and the southern vortex is located near the peak rainfall. Peak relative vorticity amplitudes are observed at the 700 hPa level, where only a single maximum is seen close to, and equatorward of, the AEJ axis. Consistent with AEWs being a structure of the lower troposphere, at the 200 hPa level there is little discernible AEW structure, with the wind field being dominated by the upper-level tropical easterly jet (TEJ).

While most climatological studies of AEWs have stressed their dynamical structure, focusing on the dynamical trough at the AEJ level, Poan *et al.* (2013) have highlighted a moist-thermodynamic perspective of AEW structures, by basing their statistical composite AEW on the precipitable water (PW) anomaly (PW*). Although both perspectives are really analysing the same atmospheric structures, the choice of compositing field gives a different perspective on the same systems. In particular, using PW* to identify the AEWs emphasises structures in the northern Sahel, to the north of the AEJ, where the background PW gradient is particularly strong. From the PW* perspective, the mean period is 5–6 days, slightly longer than found with dynamical criteria. Also, the wavelength is longer, at about 3800 km. AEW structures detected with PW* are emphasised on the northern flank of the AEJ where the meridional gradient of PW is maximum, and are consistent with earlier studies for this northern zone (Reed *et al.*, 1977; Kiladis *et al.*, 2006). Using the PW* perspective, some relationships between the thermodynamic structure and the winds are better identified, and the spatial and temporal footprint is larger so that AEWs can be detected earlier and further to the east. However, this approach to analysing AEWs remains an open area of research; for example, how the gradient in the PW* might influence AEW development remains unclear.

AEWs are very effectively analysed using 'Hovmöller' diagrams, contouring a signature field of AEW structure such as meridional wind, vorticity, OLR or PW*, on axes of longitude against time; these diagrams can even be a useful forecasting tool in real time. Figure 2.6 shows such Hovmöller diagrams for the 2011 monsoon season, comparing PW* and relative vorticity patterns. In Figure 2.6, the westward-moving AEW structures can be seen in both fields, and the speed of each AEW can be computed by the gradient of a linear feature in

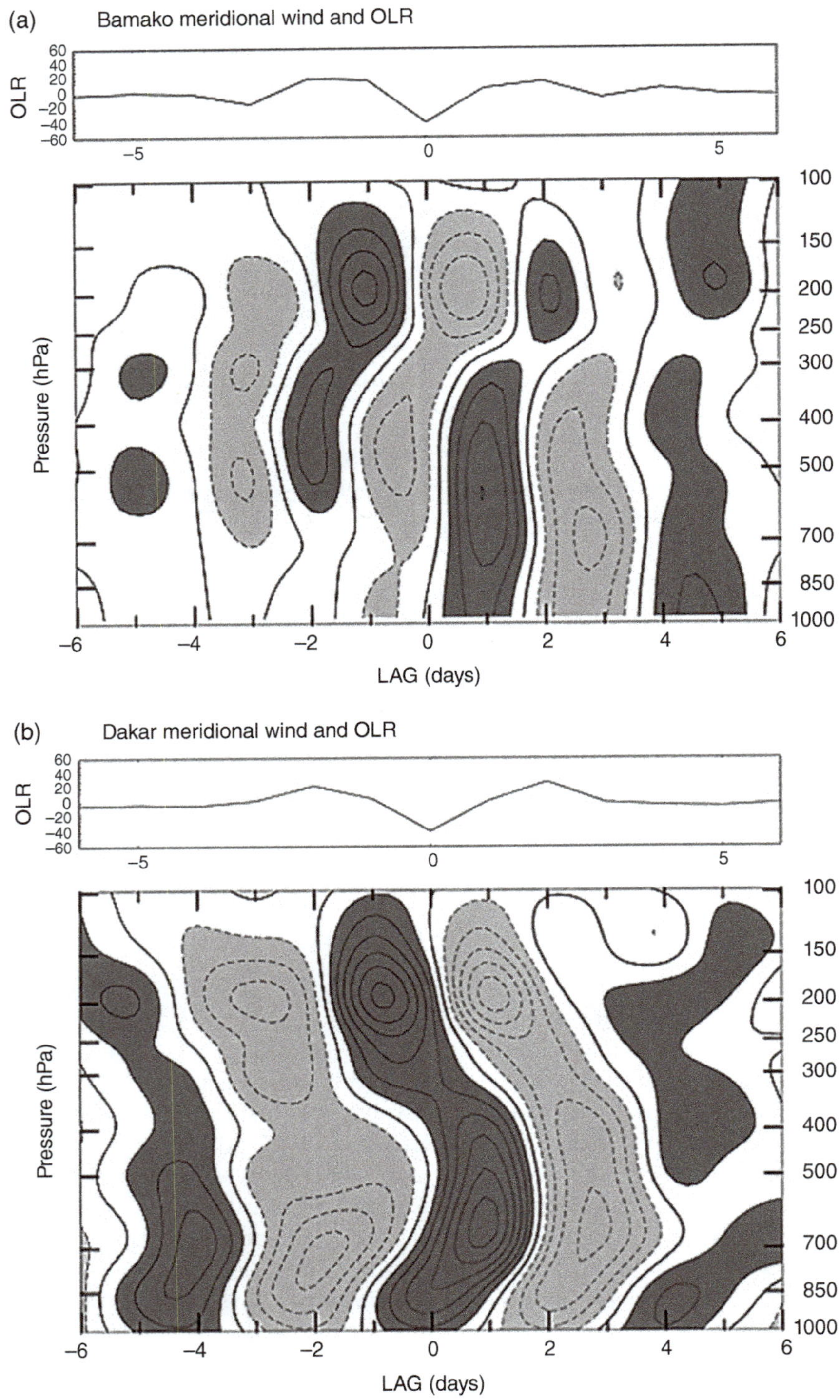


**Figure 2.4** Time–height cross-section of the meridional wind anomaly at (a) Bamako and (b) Dakar, scaled to a $-40\ \mathrm{W\ m^{-2}}$ perturbation in tropical depression-filtered OLR at the nearest grid point to each station. Contour interval is $0.5\ \mathrm{m\ s^{-1}}$, negative contours dashed. Dark (light) shading denotes anomalies greater than (less than) $0.5\ \mathrm{m\ s^{-1}}$. The associated OLR anomaly ($\mathrm{W\ m^{-2}}$) is shown at the top. *Source:* Figure 10 from Kiladis *et al.* (2006) ©American Meteorological Society. Used with permission.

this diagram. Typically, the mean PW* lies to the east of the wave trough, coincident with, and to the east of, the low-level southerly winds. These southerly winds cause an increase in PW* over time, and therefore the PW* builds on its western side, while decaying to its east where there are relatively dry northerlies. Figure 2.14 is another example of the use of a Hovmöller to analyse AEWs.

Figure 2.7 shows the horizontal structure of AEWs composited on PW* by Poan *et al.* (2013). Maximum PW* is located to the east of the wave trough, while maximum low-level southerly winds are in and to the east of the trough. This combination of winds and PW* leads to an amplification and easterly propagation of the AEW (with southerly advection of moist air maximised in and to the west of the PW*). Note that, as in Figure 2.3, the

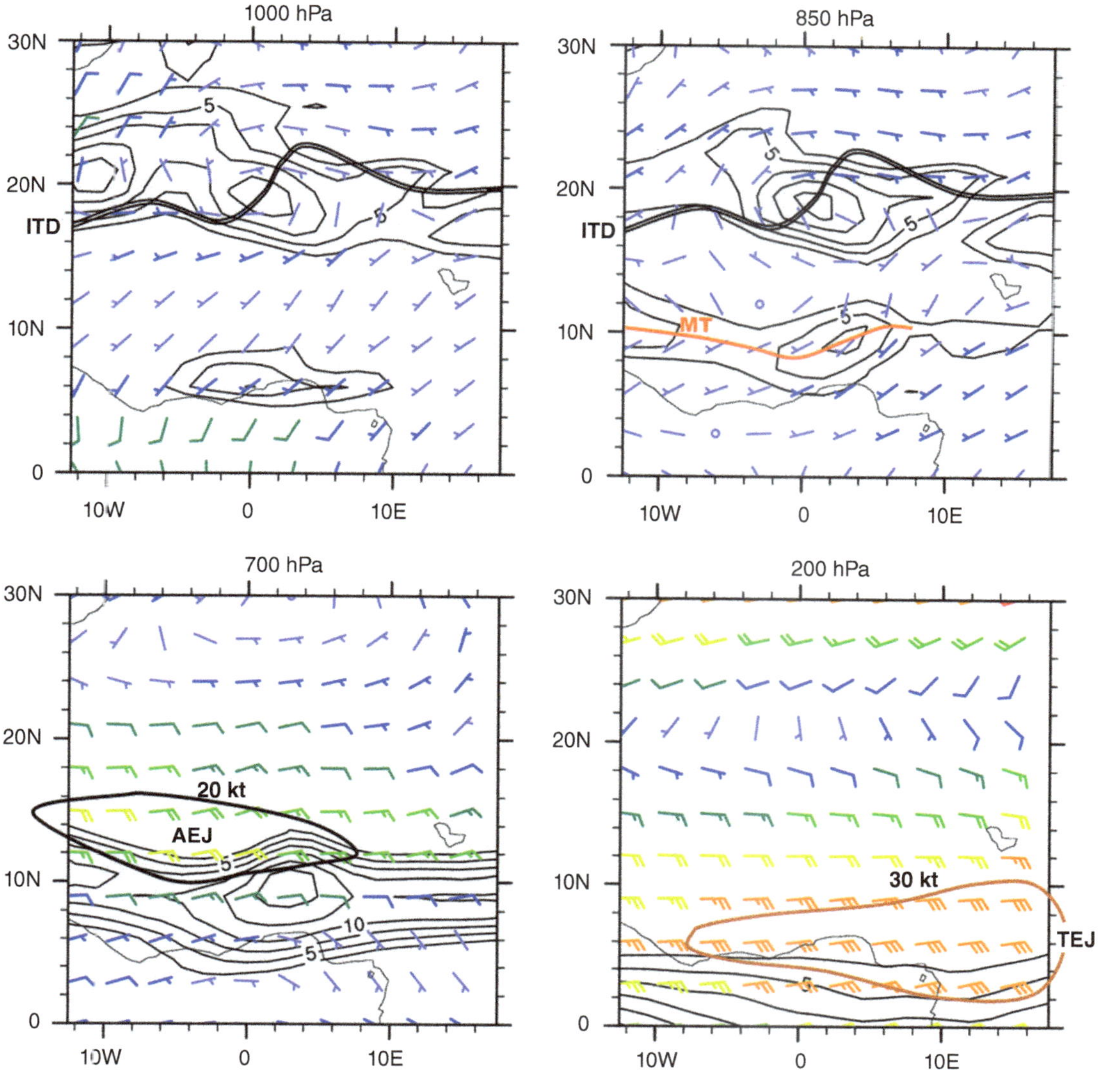


**Figure 2.5** Vertical structure of AEW vorticity and winds, from a composite of AEW cases, based upon the 700 hPa trough axis passing the longitude of Niamey (2.5°E), and comprising 803 cases between 1989 and 2008, identified by the objective identification method of Berry *et al.* (2007). Composite winds (barbs, coloured in knots) and relative vorticity (positive values greater than $5 \times 10^{-6}\ s^{-1}$ contoured every $5 \times 10^{-6}\ s^{-1}$) are shown on different pressure levels from the ERA-Interim reanalysis dataset. The ITD, AEJ and TEJ have been overplotted consistently with the methods of Chapter 11.

trough in Figure 2.7 is tilted 'upshear' in a configuration that is consistent with barotropic growth of the AEWs (feeding on the kinetic energy of the AEJ). However, considerable variability in the trough orientation is seen from case to case.

The OLR patterns of Figure 2.7 indicate the peak in deep convection within and just ahead of the PW*, which means a peak in rainfall within and slightly to the east of the trough. However, this OLR pattern must be treated with considerable caution: these composites are based on the PW* and, therefore, will not emphasise the important and intense squall lines that develop and propagate into the areas of lower PW* to the west of the AEW trough. Figure 2.3 is an example of the link of observed OLR with the dynamical trough.

Figure 2.8 confirms the upshear tilt of the wave trough with height (for instance, the zero-line of $v^*$ in Figure 2.8a tilts to the east with height at low levels); comparable patterns have been derived in other composite studies. The humidity and temperature fields, in contrast, are almost vertically aligned, and therefore the relationship between the trough and the thermodynamic (temperature and humidity) fields changes with height. Near the ground, the cool, moist air is mainly within the ridge (anticyclonic circulation), and in the southerly flow to the west of that.[1] At the AEJ level of 600–700 hPa the cool, moist air lies to the west of the southerly winds and to the east of the AEW trough. This eastward tilt is consistent with the composite waves gaining energy from baroclinic instability mechanisms (Section 2.1.3.5).

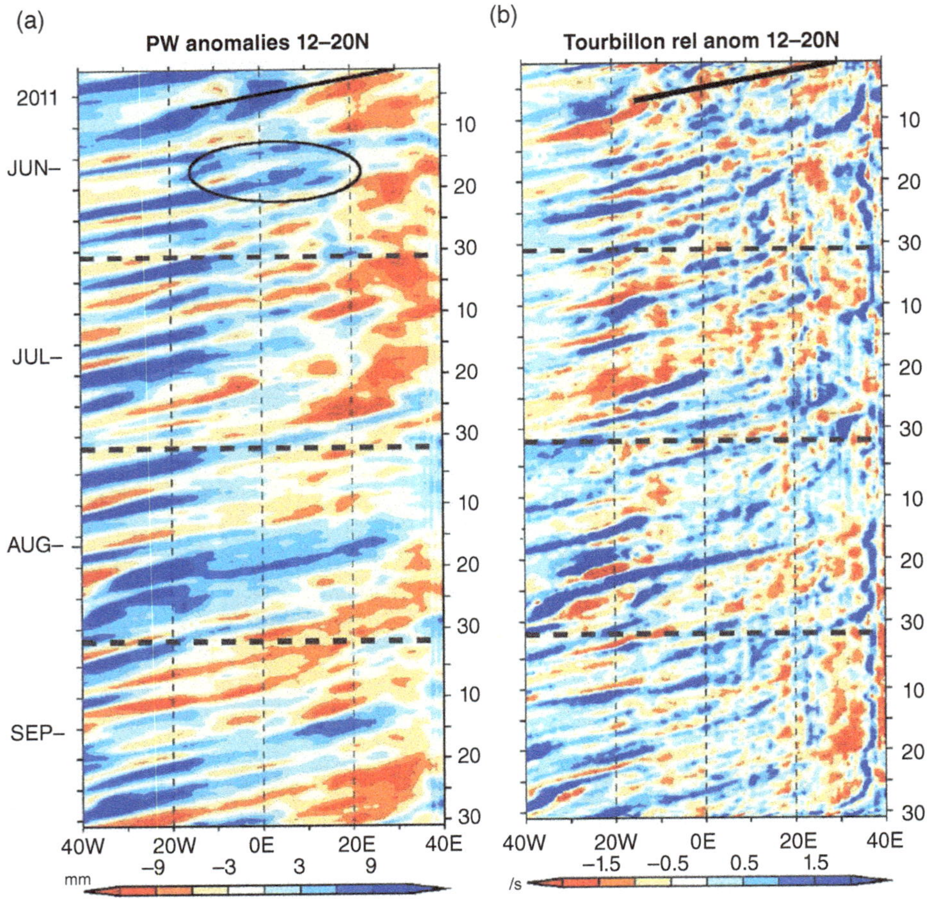


Figure 2.6 Time–longitude diagram of the 2011 June to September intraseasonal anomalies averaged over the 12–20°N band: (a) PW (mm) and (b) 850-hPa relative vorticity ($\times 10^{-5}$ $s^{-1}$). The black solid lines materialise a slope of a $-9$ m $s^{-1}$ velocity. *Source*: from Poan (2013).

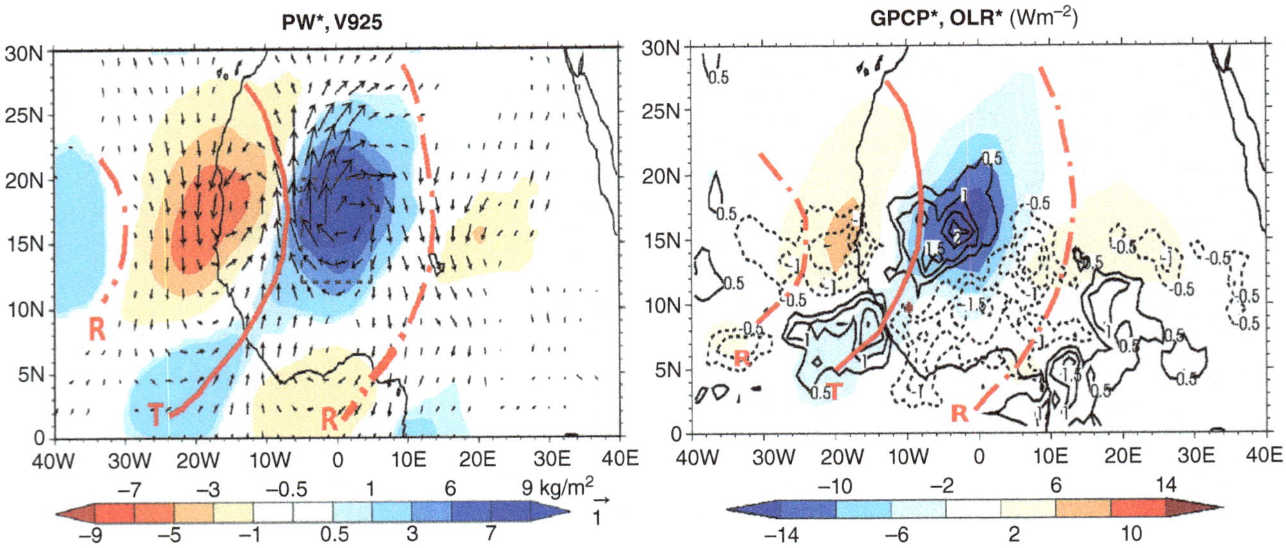


Figure 2.7 PW*, 925 hPa winds (left) and GPCP and OLR anomaly fields (right), based on a composite on the PW* field of AEWs (Poan *et al.*, 2013). The (red) trough and ridge lines are based on the wind fields at 600 hPa. Note that the relationship of OLR with the wave structure must be treated with caution because these structures do not effectively capture the very intense squall lines that propagate to the west of the AEW trough into areas of lower PW*. *Source*: ©American Meteorological Society. Used with permission.

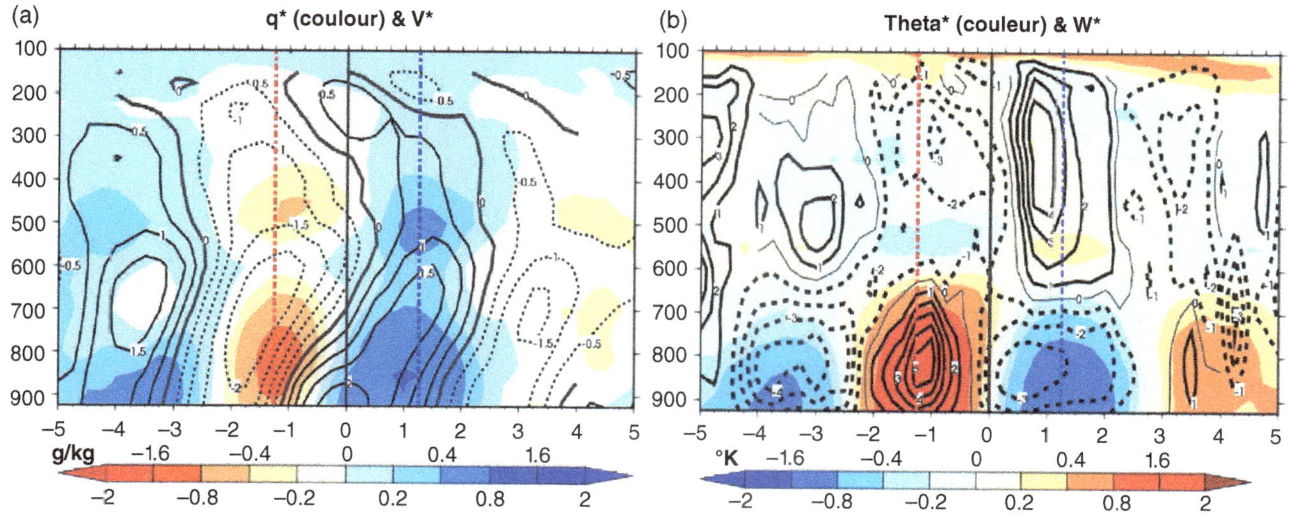


Figure 2.8 Vertical–lag composite of (a) specific humidity (g kg$^{-1}$, shaded) and meridional wind (m s$^{-1}$, contours) anomalies, (b) potential temperature (K, shaded) and vertical wind (mm s$^{-1}$, contours) anomalies, based on a composite on the PW* field of AEWs. The vertical black line corresponds to the trough location at $t_0$, whereas vertical dashed red and blue lines correspond to the passage of the dry/warm and moist/cold stages respectively (at $t_0 - 1.25$ and $t_0 + 1.25$). *Source*: from Poan (2013).

Although AEWs occur at low latitudes, there is some degree of thermal wind balance apparent in these composite plots. For instance, at zero longitude in Figure 2.8a there is northerly shear of the wind with height, with the cooler air to the east, consistent with thermal wind balance.

The composite maps of streamlines at low, medium and high levels, from the PW* perspective (Figure 2.9), confirm these relationships. 850 hPa is close to the steering level of the AEWs, and therefore this is a level where strong, closed vortices exist. The winds at low levels act both to reinforce the high PW* through cool, moist advection, and to cause it to propagate to the west.

The statistical and climatological perspectives on AEWs can hide the fact that there appears to be very great variability in AEW structure from case to case. Furthermore, the strong diurnal cycle of the boundary layer and of deep convection over West Africa means that there is a diurnal cycle of AEW structure, which is still not very well understood or documented. Bain *et al.* (2011), analysing an AEW case from 27 July 2006, showed how nocturnal acceleration of the low-level winds led to a surge of moisture northward into the wave trough, while daytime dry convective mixing in the north of the AEW led to a deceleration of the AEJ in that region.

#### 2.1.3.3 African Easterly Wave Structure

In Section 2.1.3.2 we introduced evidence of the basic structure of AEWs, as it has been defined in various research studies. However, given that there is considerable variability in AEW structure from case to case, according to the maturity of an AEW, its geographic location and the time of day, it is difficult to generate a single consensus for a single AEW structure. Furthermore, the scientific literature does not always accord with forecasters' experiences of daily synoptic analysis. The scientific papers have been dominated by description of composite structures, while synoptic case studies of AEWs are surprisingly rare. For instance, there appears to be a difference between the synoptic analysis of AEWs, very often showing 'downshear' tilt of the trough (e.g. Figure 2.16), and statistical composites showing 'upshear' tilt.[2]

For these reasons, presenting an AEW as a single structure would likely be misleading, because it would not reflect the wide range of observed structures. Instead, in describing AEW structures here, we emphasise their diversity. After lengthy discussion between researchers and forecasters, a modular approach is used here to describe AEW structures. This takes into account the key features, or elements of an AEW. The composite AEW structures, as well as any individual case study, may then be explained and discussed in terms of these elements. In listing the elements, we follow common forecasting practice and begin with the surface features.

1) *Surface or low-level structure.* The very first analyses of AEWs in the 1930s were based on the study of surface pressure tendencies. Forecasters still use these changes to infer the wave passage and its influence on weather conditions. Owing to the strong diurnal cycle of the surface pressure, the tendencies must be looked over 24 h or after removing the climatological diurnal cycle.

   The wave structure has a very strong control on the boundary-layer structure, and the associated

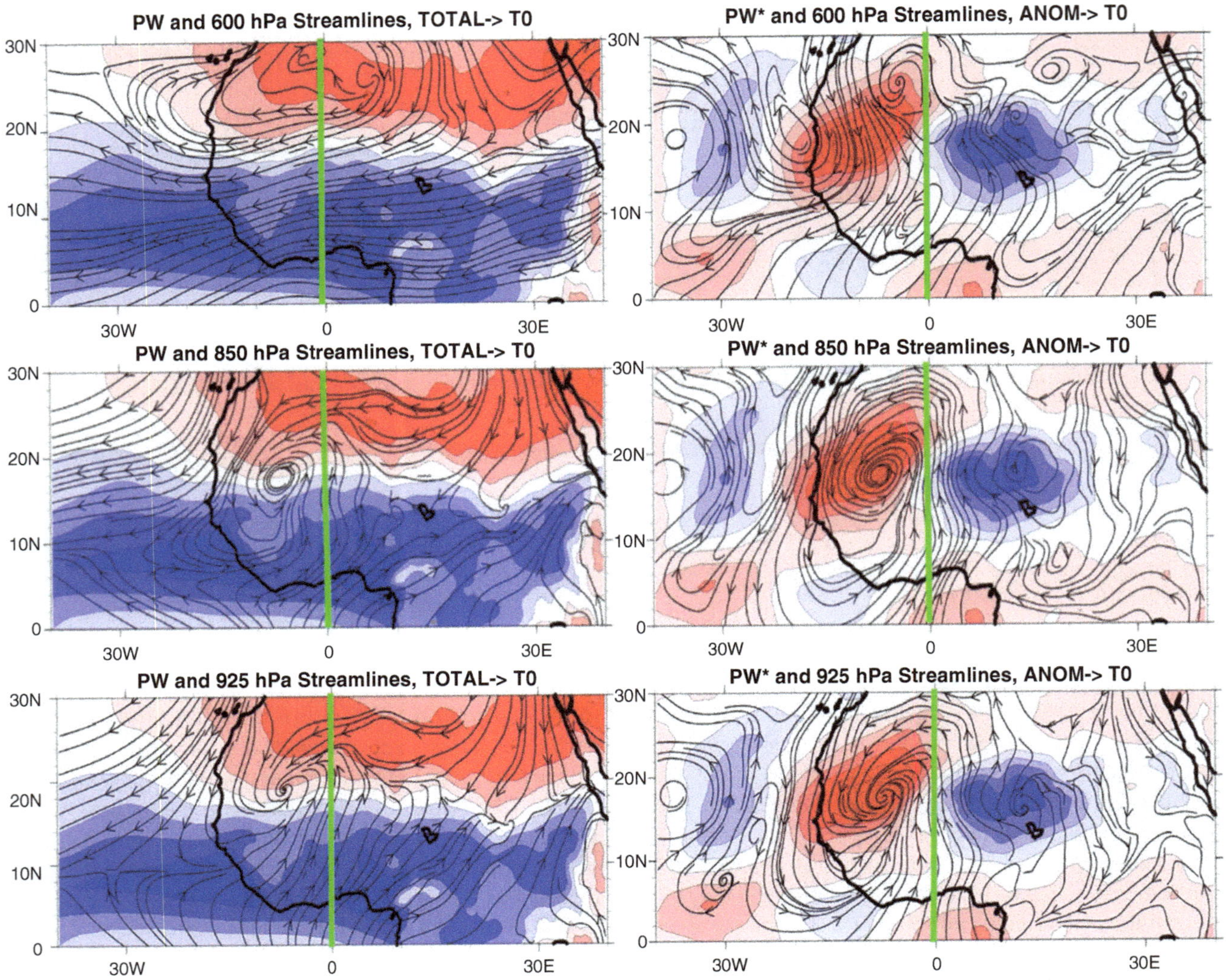


**Figure 2.9** Composites of streamlines and PW (mm, shaded) fields at 925, 850 and 600 hPa at the passage ($t_0$) of the AEW at 0°E. The left and right columns correspond to the full and anomaly fields respectively. *Source*: from Poan (2013).

cloud fields. Air that is displaced northward typically has a cooler, shallower and more humid boundary layer, with more extensive shallow cumulus (collectively, a deeper monsoon layer) than air that has been displaced southward.

2) *ITD displacement.* The ITD typically has a strong northward incursion in association with the AEW, normally in and to the east of the AEW trough. The ITD has a very strong diurnal cycle, which tends to mask its synoptic evolution during the day, as the low-level air in zone B is mixed strongly with Saharan air above it. At night, the ITD regenerates strongly, pushing the ridge of cold air at low levels northward (Bain *et al.*, 2011).

3) *The AEW trough at AEJ level (e.g. 600 or 700 hPa).* The AEW trough is generally strongest at the AEJ level, and corresponds to the region of cyclonic curvature of the winds. In composite studies, the axis of the trough tilts upshear; that is, westward to the north and to the south of the AEJ axis (e.g. Figures 2.3 and 2.7). In some synoptic events, the opposite shear is observed, with the trough tilting eastward to the north and south of the AEJ (e.g. Figure 2.16b).

4) *Waves and vortices on the monsoon trough (at 850 hPa).* AEWs propagate slower than the AEJ (~7m $s^{-1}$ as opposed to 12–15 m $s^{-1}$), and therefore the wave *steering level* lies below the jet, often close to 850 hPa. The vorticity associated with the wave trough extends vertically below the AEJ level, and the vorticity at 850 hPa very often represents closed circulations, or vortices, which can be diagnosed from 850 hPa wind vectors or streamline maps for instance. These vortices may be enhanced by convective systems and are very important in the regeneration of convection on a subsequent day. Vortices may also on some occasions extend to higher and lower altitudes, reflecting a greater depth of the system.

5) *Precipitable water.* The PW is an integrated measure of the northward penetration of the monsoon system, and it shows a strong variability associated with AEW structure. AEW patterns are strongly highlighted in the PW* field (Poan *et al.*, 2013), and in some cases[3] PW* may present a more robust diagnosis of AEW structure, and its precipitation patterns in particular, than the wind fields, since it integrates the greater variability of the winds and advection. In composite analyses, the peak PW* lies to the east of the AEW trough, where there are southerly winds at the AEJ level.

   The patterns of deep convective stability (CAPE and CIN) are more complex, because the low-level equivalent potential temperature $\theta_e$, which controls CAPE, has a maximum just to the south of the AEJ (e.g. see Figure 2.2 and Thorncroft *et al.* (2003)), while CIN is also controlled by the temperatures above the boundary layer. Typically, this leads to a southwest-to-northeast tilt in the pattern of high CAPE associated with an AEW, because low-level $\theta_e$ is high in the northward surge of moist air to the northeast of the trough (Figure 2.7), but is also high where warm (and humid) air is carried in the northerlies to the southwest of the trough (see Figure 2.7 and Figure 11.17b).
6) *Monsoon depth.* Various measures of the monsoon depth have been proposed, based on changes in the wind direction or on depth of the thermodynamic monsoon layer (see Section 11.10). Where the monsoon layer is deep, there tends to be a deep layer of cool, humid air with a general southerly advection. This intrusion of cool monsoon air is essentially collocated with the high PW*. The coolness of the air, through the thermal wind relation, tends to lead to relative anticyclonic flow at low levels (or a weakening of the cyclonic flow). For instance, in composite wave structures (Reed *et al.*, 1977) the waves are cold-core, with the deep, cool monsoon layer close to the AEW trough axis, and the vorticity weakening with height below the trough.

Analysis and drawing rules for each of these elements are provided in Chapter 11.

We can discuss the typical AEW structures in terms of these features (e.g. the downstream changes in the vertical structure). A schematic interpretation of the composite AEW structures in these terms is now shown in Figure 2.10.

Figure 2.10 has been drafted for this book as a result of meetings and discussions among operational forecasters, academics and modellers, and represents a new consensus on AEW structure. A number of important points must be made in relation to the fields and their relationships.

1) *Panels 1 and 2 at 600–700 hPa represent the most common levels for diagnosing the presence of AEWs,* close to the AEJ level, with trough and ridge lines related to the curvature of the streamlines (here defined where $v = 0$). The typical wavelength is 3000 km or 30° of longitude from trough to trough. Open waves are more typical at this level than strong, closed circulation centres. However, closed circulation centres are possible in some cases, especially in the south (as shown in panel 1).

   *The horizontal tilt of the troughs, as diagnosed in composites from model analyses, is as shown in panel 1*; that is, upshear tilt (extending to the west, both north and south of the AEJ core). However, individual cases show variations in this tilt, or the tilt evolving as a wave moves toward the west.

   *A westward tilt of the wave trough leads to faster AEJ winds to the east of the trough,* as shown in panel 2 (because in this region the streamlines are closer together). Eastward tilt is also possible, as in the example in Figure 2.16. Breaks in the jet core are commonly possible, when an AEW is intense (see online case study CS01[4], 1–10 August 2012, or the extreme flood case CS04 of Ouagadougou 1 September 2009).
2) *Initiation of deep convection has a wide spread of possible locations, but a consensus for the peak and distribution are shown here.* Fast-moving mesoscale convective systems (MCSs) typically occur along the location of the AEJ core ahead of the trough and are favoured by mid-level dry air and strong shear (Sections 3.1.2.2 and 3.1.2.3). Slower-moving convective systems are found to the south, where wind shear is weaker and the monsoon depth is deeper.
3) *Panels 3 and 4 show the winds, vortices and PW* at 850 hPa,* close to the steering level of typical AEWs. The thermodynamic signal of the AEWs is generally maximised in these levels below the AEJ. The PW* are approximately collocated with thermal anomalies, as cool, moist air is advancing northward to the east of the trough and warm, dry air moves southward to the west of the AEJ-level trough.
4) *A closed vortex is commonly observed at 850 hPa, and usually (though not always) to the west of the AEJ-level trough.* This is a typical signature of a baroclinic wave tilt. The 850 hPa vortices are connected with the line of maximum 850 hPa vorticity, to the south of the AEJ, to delineate the 'monsoon trough' (see Section 11.10).

   **It is important to note that forecasters associate this 850 hPa vortex with a likely area for the development of thunderstorms.**

   *It should be recognised that, in general, the monsoon trough may not be a continuous line.* The example of Figure 11.14 shows a rather continuous configuration of the monsoon trough, but on other occasions the

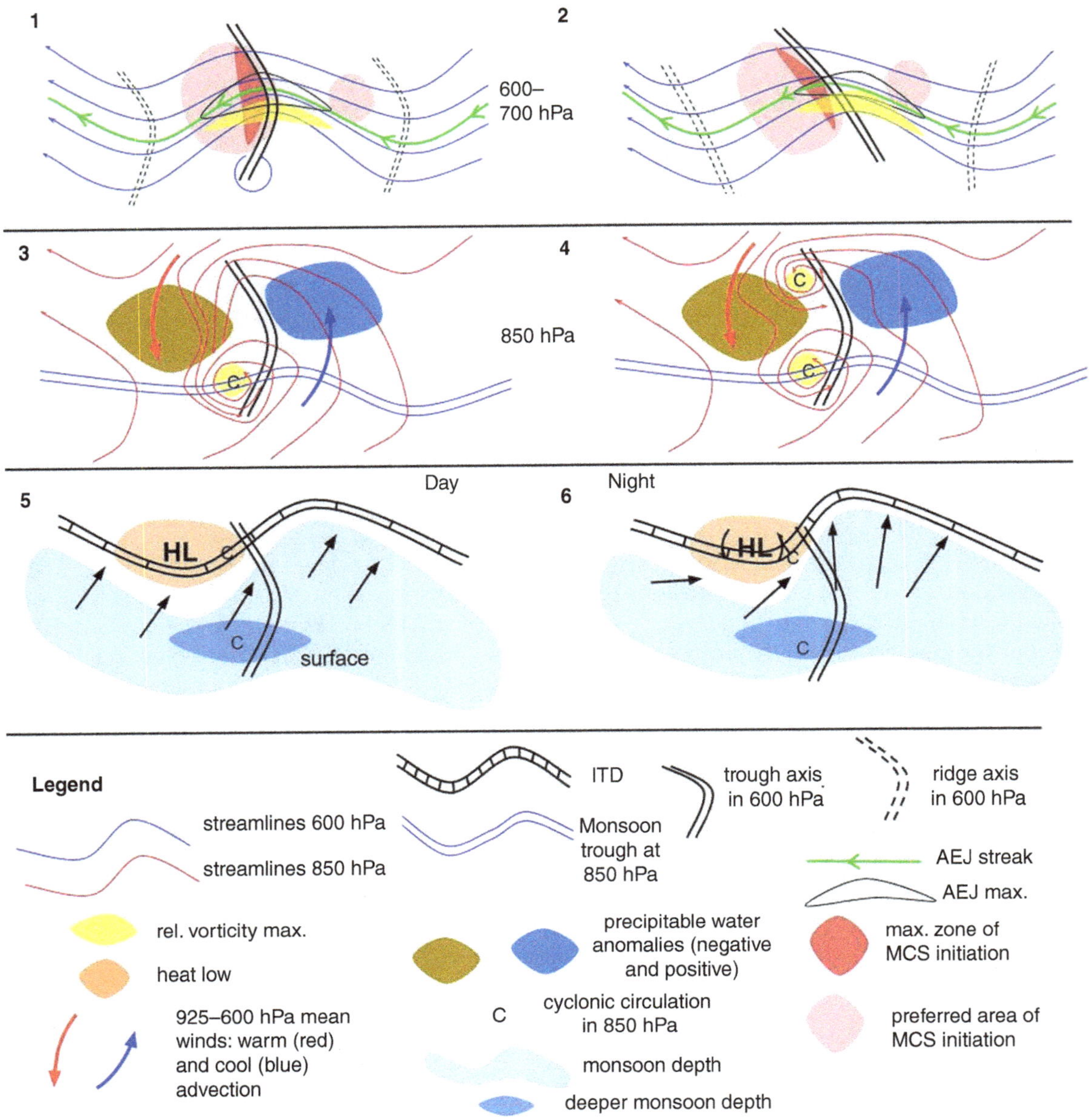


**Figure 2.10** Schematic of the various observable elements of an AEW, and likely relationships between these. Left-hand panels show a 'normal' situation, as far as this exists, while right-hand panels show common alternatives. For example, the structure in panel 2 would be expected in an environment with additional barotropic shear, with a stronger easterly wind to the north (i.e. $U_y < 0$).

monsoon flow may be so deep and strong that the monsoon trough disappears in the southerly flow at the rear of the trough (in which case the monsoon trough may effectively exist at a lower pressure level).

*On many occasions, a northern vortex is also seen on the ITD at the 850 hPa level* (panel 4). If it is absent, the vortex may simply be shallower, and seen on the 925 hPa level instead. When plotting the monsoon trough, it is important not to connect the ITD vortices with the monsoon trough.

5) *Panels 5 and 6 show the surface signature of the AEW.* A significant north–south deviation in the ITD follows the AEW (also seen at 925 hPa). Monsoon depth follows the wave of ITD extensions to the north (e.g. Figure 11.16), and is commonly observed to be deepest in the area of the cyclonic circulation south of the AEJ, as marked here. Forecasters use the northward extension of the ITD to describe associated shifts in zones B–D (Figure 1.1) of the monsoon system; for instance, where the ITD is pushed to the north, we expect zones B, C and D similarly to be pushed northward.

*Increased monsoon depth generally increases cloudiness in the northern region.* Associated with the movement of the ITD, there is a southward extension of the SHL to the west of the trough with lower cloud cover and high insolation.

6) *Surface fields have a substantial diurnal cycle, especially in the north (zone B).* A surge of low-level winds northward into the heat low, plus a cyclonic circulation around the heat low, which both intensify overnight, push the ITD northward by some 300–500 km, especially to the east of the heat low anomaly and AEW trough (panel 6).

Figure 2.10 relates to the structure of AEWs over the continent. As AEWs move westward over the coast and the Atlantic Ocean, three other elements play a role in the AEW structure:

- *A semi-permanent vorticity maximum over the Guinea Highlands, or Guinea/Sénégal coast.* This feature appears to be sustained by the regular deep convection occurring over and downstream of the Highlands.
- *Equatorward cold advection off the west coast of the continent.* As the wave trough moves over the coast, the northerly winds ahead of the trough no longer advect warm air, but instead advect cooler air at low levels, triggering baroclinic decay immediately. This explains why 'marine AEWs' are more barotropic in nature and have a weaker vertical tilt.
- *The intertropical convergence zone (ITCZ) over the Atlantic Ocean.* The reduced low-level thermal contrasts and lack of heat low over the ocean mean that the maximum rainy zone reverts to a more classical ITCZ structure, with low-level convergence roughly collocated with the maximum deep convection and rainfall. Effectively the ITD and monsoon trough merge as the wave moves over the ocean, to form a more vertically aligned trough.

It has repeatedly been emphasised here that there is considerable variation in AEW structure from case to case. Section 2.2 describes typical combinations of the dynamical elements that make up typical AEW patterns in different contexts.

#### 2.1.3.4 African Easterly Wave Genesis

##### 2.1.3.4.1 Upstream Triggers

Observational studies have consistently demonstrated that AEWs are predominantly generated over the African continent to the east of 10°E. Commonly, upstream convection, in an East African region including the Darfur Mountains (15°N, 23°E) and the Ethiopian Highlands (10°N, 35°E), initiates AEWs near the entrance of the AEJ (Hall *et al.*, 2006; Kiladis *et al.*, 2006; Mekonnen *et al.*, 2006; Thorncroft *et al.*, 2008; Leroux and Hall, 2009). Hall *et al.* (2006) showed in idealised modelling studies that intense localised heating (a proxy for mesoscale convective systems) in the entrance region of the AEJ could excite synoptic structures resembling AEWs downstream. The hypothesis that finite-amplitude impulses of heating result in the genesis of AEWs is supported by case studies (e.g. Berry and Thorncroft, 2005), showing AEWs generated following a large burst of deep convection over central Africa. Thorncroft *et al.* (2008) demonstrated that the genesis of AEWs is sensitive to the location of the initial convective perturbation; they are more efficiently triggered by heating close to the entrance of the AEJ. The implication is that the intermittency of observed easterly waves may be in part explained by the variability of convective activity toward the east, instead of purely by considerations of the jet structure. More generally, Leroux *et al.* (2010) found that subseasonal variability of AEWs tends to be preceded by anomalous convection in the Darfur region. Leroux and Hall (2009) suggested that the AEW response also depends on the AEJ structure, with wider and stronger AEJ (strong shear and PV reversals over an extended region near the AEJ) producing more intense AEWs – entirely consistent with baroclinic growth (see also Hsieh and Cook (2005)). In idealised simulations with moist processes, Cornforth *et al.* (2009) showed that moist AEWs were characterised by intermittent periods of growth and decay that were correlated with increases and decreases in the mean rainfall respectively but anticorrelated with jet strength. The strong interdependence between the AEJ, the AEWs, the moist convection and the upper levels establishes an *internal variability* on a timescale of 8–10 days in the idealised model consistent with the observed average 9-day cycle of wet (dry) sequences (e.g. Sultan *et al.*, 2003).

Influences from mid-latitudes (e.g. Leroux *et al.*, 2011) may also help modulate the amplitude of AEW activity on intraseasonal timescales. A long integration of a simple GCM in an artificially dry context simulating the northern summertime global circulation produces a summertime climatology that includes a realistic AEJ and westward-propagating 3–5-day disturbances over West Africa. These simulated waves display intraseasonal intermittency, as do the observed AEWs. Further experiments designed to discern the source of this intermittency in the model show that the simulated waves were mainly triggered by dynamical precursors coming from the North Atlantic storm track.

##### 2.1.3.4.2 Large-scale Dynamics

In order to explain the full structure of the AEWs, it is essential to consider not only the role of convection but also the large-scale dynamical 'mode' of wave dynamics. The vertical and horizontal shears associated with the mean AEJ satisfy the Charney and Stern (1962) criteria for baroclinic and barotropic instability (e.g. Burpee, 1974). Idealised modelling studies (e.g. Thorncroft and Hoskins, 1994; Cornforth *et al.*, 2009) have shown that the most unstable dynamic mode growing on the AEJ has similar spatial structures to observed AEWs and also demonstrated that AEWs can amplify by barotropic and baroclinic energy exchanges. In light of these results it has been contended that the genesis of AEWs is associated with initially small perturbations to the flow over central and eastern Africa that increase in amplitude owing to the unstable nature of the background flow. However, analogous to the genesis of baroclinic waves in the mid-latitudes, it is recognised that the dynamic instability of the background flow is too slow (relative to observed wave lifecycles and observed wave precursors) to explain the genesis of AEWs, although such dynamical instability does appear to influence the growth and structure of waves once they have been initiated. Although a coherent instability mode may not be essential to wave initiation, nevertheless a 'growing configuration' of wave modes on the basic state AEJ will emerge. Recent work has also highlighted an alternative mechanism for AEW genesis over Africa that, instead of downstream development from upstream triggers, involves upstream energy dispersion from the west (Diaz and Aiyyer, 2013).

#### 2.1.3.5 African Easterly Wave Dynamics and Evolution

The AEJ satisfies the Charney and Stern (1962) barotropic and baroclinic instability criterion, meaning that the sign reversal of the meridional PV gradient (Figure 2.11) satisfies a necessary condition for instability of the mean flow (Burpee, 1972; Thorncroft and Blackburn, 1999; Dickinson and Molinari, 2000). The sign reversal of PV at the AEJ level is consistent with the presence of a strip of high PV on the southern flank of the AEJ, which is in turn augmented by deep convection during the active monsoon.

The genesis of AEWs through small-amplitude barotropic or baroclinic instability is highly relevant to the structure of AEWs over West Africa even if time dependence

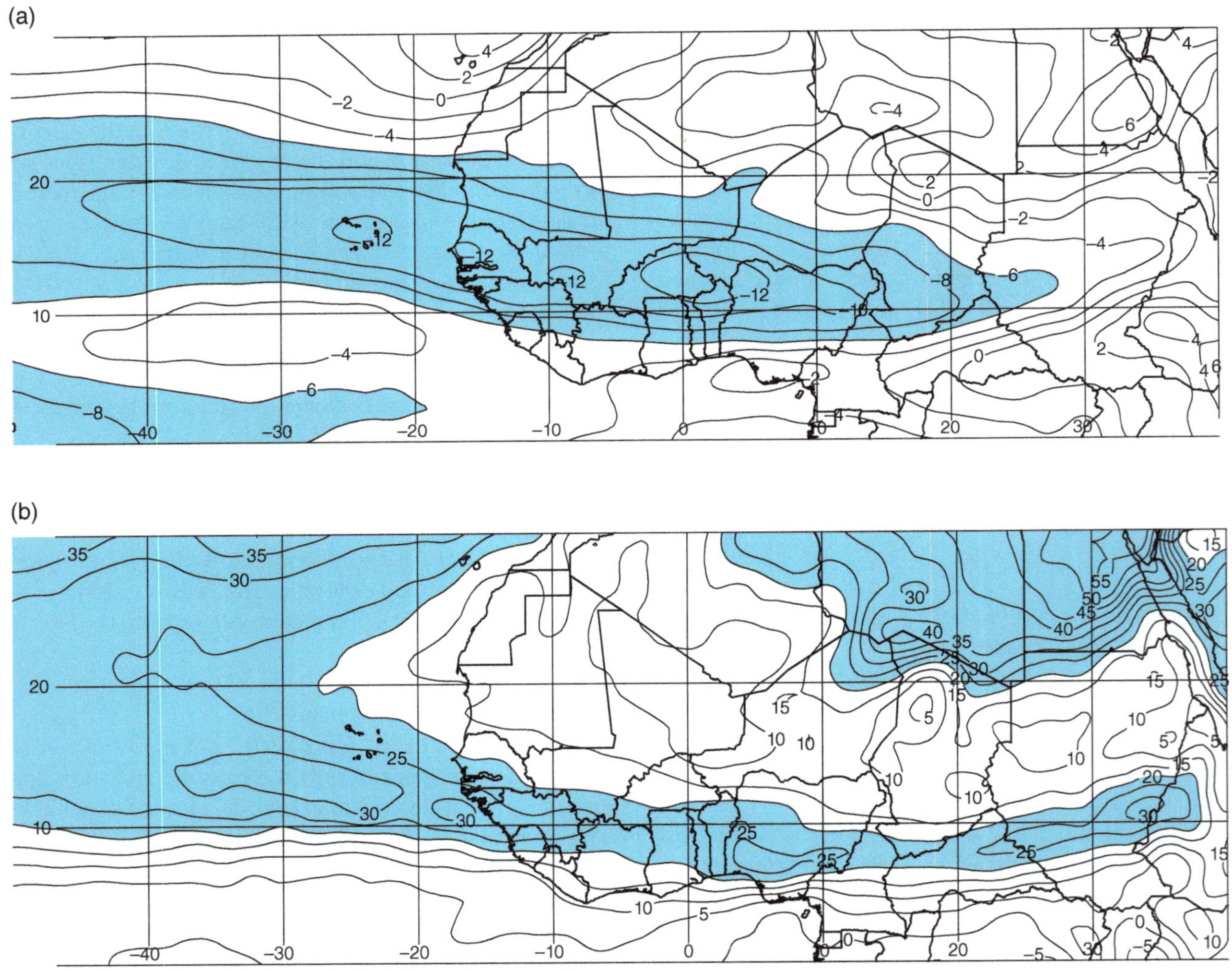


**Figure 2.11** Illustration of the meridional PV reversal. Panels (a) and (b) (based on operational ECMWF analyses and adapted from Figure 1 in Thorncroft *et al.* (2003)) show (a) zonal wind and (b) PV, on the 600 hPa surface. The maximum PV on this level is to the south of the AEJ maximum (each shaded blue), with low PV to the north of the AEJ. *Source*: ©American Meteorological Society. Used with permission.

differs in the initial-value problem with finite-value precursors (Section 2.1.3.4). AEWs are large-amplitude waves whose evolution is strongly influenced by these barotropic and baroclinic processes, and Rossby wave dynamics in particular. For example, the typically slow growth of an AEW once its structure has emerged and it then continues to propagate westward through the WAM system is entirely consistent with a slow baroclinic growth rate. As such, understanding the behaviour of these waves and instabilities can help us to understand the evolution of the mature AEWs.

Basic theory of synoptic-scale instabilities shows that the waves can gain energy from barotropic processes if the trough tilts 'upshear' in the horizontal – that is, if the trough extends westward with latitude, both north and south of the AEJ core latitude (as in Figure 2.10, panel 1). In this case, the wave will gain energy from the basic kinetic energy of the AEJ itself. The composite AEW patterns seen in Figures 2.3 and 2.7 confirm that, on average, AEWs have a significant element of barotropic instability (although individual cases may differ from this structure). Similarly, waves can gain energy through baroclinic mechanisms if the wave trough tilts upshear (eastward; see Figures 2.8 and 2.12) with height, below the AEJ core, in which case the wave gains energy through thermal advection, with warm northerlies intensifying the warm sector of the AEW. Again, the structures seen in Figure 2.8 agree with the baroclinic instability model, where there is cool, moist southerly advection at low levels within and to the west of the low-level cool anomaly; this pattern acts both to strengthen the cool anomaly and to cause it to propagate to the west (as a surface Rossby wave – see Parker (2008)). In turn, the westward propagation at low levels enables the low-level structure of the wave to keep up with the easterly motion of the wave at the AEJ level. The whole system has a steering level some way between the surface and the AEJ level, typically around 850 hPa.

The relationships between the trough, the winds and the vorticity structures at different levels are demonstrated using fields taken from an AEW case during August 2000. In Figure 2.13a on the large scale, equatorward of the 700 hPa AEJ axis (denoted by the thick red line), there is a zonally orientated (east–west) strip of positive relative vorticity. Similarly, at 925 hPa the temperature field exhibits a well-defined, zonally orientated zone of temperature gradient, reflecting the zonal nature of the AEJ and the monsoon. The AEW structure in this case is evident as a wavelike disturbance to this zonally orientated background; the middle of the relative vorticity perturbation is marked by the trough axis (black line), and the warm temperature perturbation is located slightly ahead (west) of the trough. Baroclinic growth will result if the flows due to the vorticity (and PV) at the different levels interact, as their configuration means that the flow around each would increase the amplitude of the other. It is also expected that barotropic growth would result from the flow around the 700 hPa vorticity maximum distorting the large-scale background 'strip' of vorticity. These dynamic growth concepts are identical to those applied in mid-latitude systems and consistent with the observed structures.

Although the AEW structures observed have a configuration that is consistent with growth via dry dynamics, calculations from idealised simulations of AEWs have shown that the dry dynamics act too slowly to fully describe the observed growth rate. Based upon a case study of the event shown in Figure 2.13, Berry and

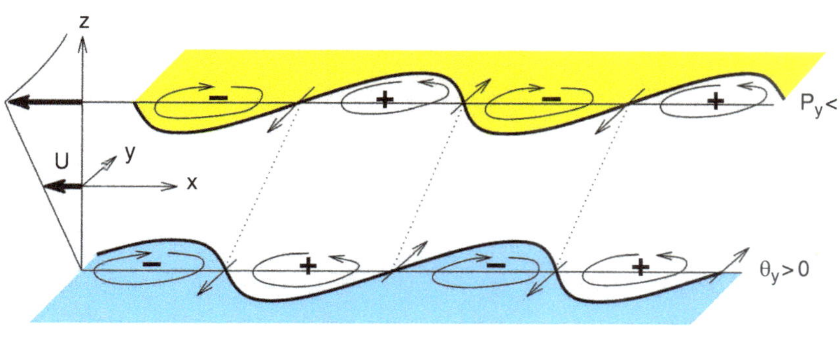


Figure 2.12 A schematic depicting the meridional displacements of air and associated circulations within an AEW that are predicted for a 'baroclinic' growth configuration. An 'upper Rossby wave component' propagates along the African easterly jet and interacts with a 'lower wave' propagating westward. The upper wave exists in the jet core where the meridional PV gradient ($P_y$) is negative. At this level, *southward displacement* of air from above the Sahara (shaded yellow) must create a negative PV anomaly and anticyclonic circulation, enabling westward *propagation* relative to the surface flow. The lower wave exists on the meridional potential temperature gradient, which is positive, so that *northward displacement* brings cooler, moist air (shaded blue) from the south which is associated with anti-cyclonic circulation, enabling westward *propagation* relative to the surface flow. When the circulations induced by the upper wave are positioned east of the circulations induced by the lower wave (velocity components tilt eastward as indicated by dotted lines), their interaction by meridional advection results in mutual growth.

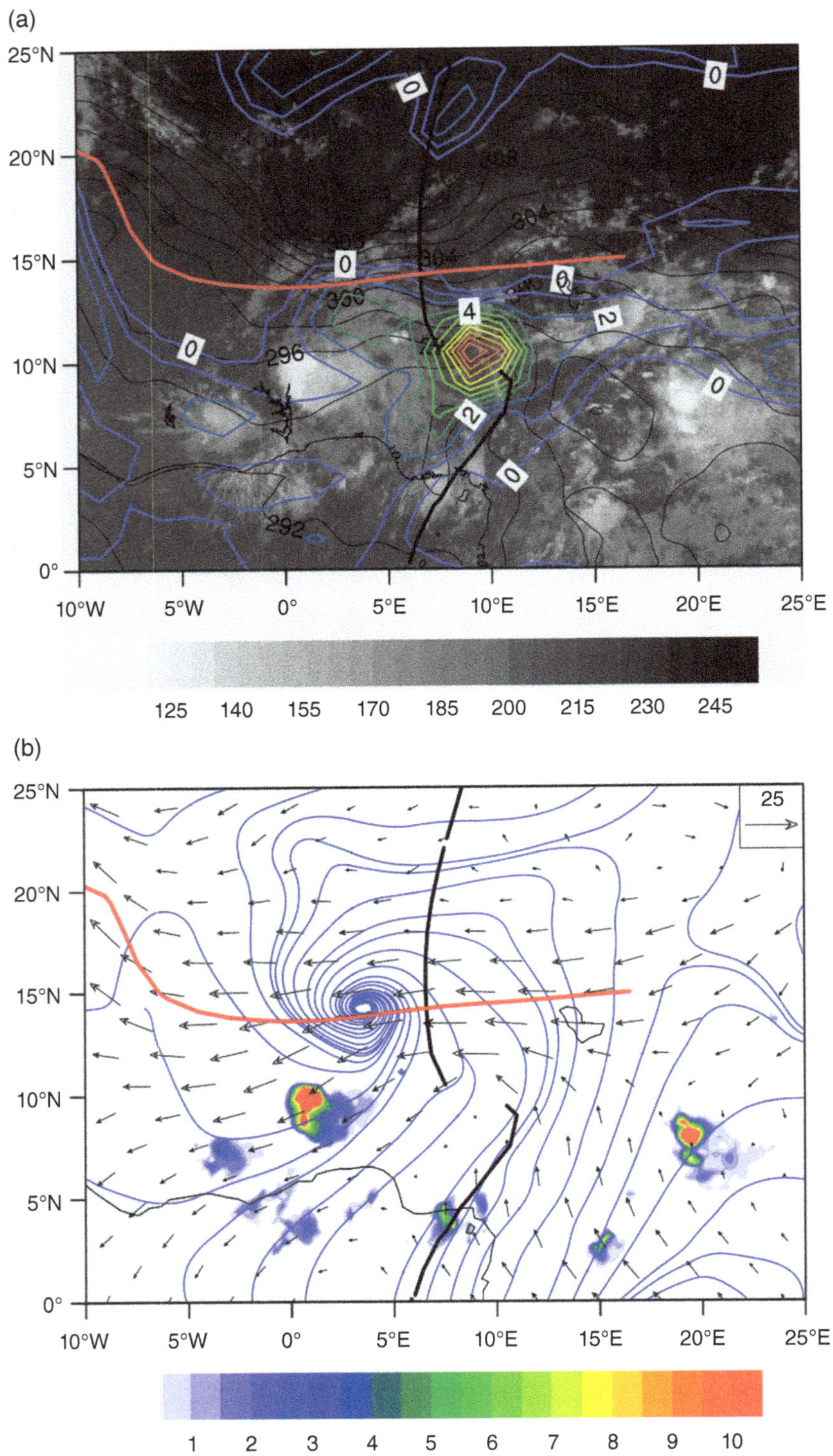


**Figure 2.13** Synoptic maps of ECMWF ERA-Interim reanalysis fields on 1 August 2000, 1200 UTC: (a) 700 hPa relative vorticity (coloured lines), 925 hPa temperature (thin black lines), 700 hPa objective trough (thick black line) and objective jet axes (thick red line, see Berry *et al.* (2007) for definition). (b) Mean TRMM 3B42 estimated precipitation rate (mm $h^{-1}$, shaded according to legend), Streamlines at 925 hPa (blue streamlines) and wind vectors at 700 hPa. The same objective trough and jet axes as in (a) are overlaid.

Thorncroft (2005) suggested that the presence of organised convective systems increases the growth rate of the synoptic AEW (Cornforth *et al.*, 2009). It was hypothesised that the very intense mesoscale vorticity centres generated by the convection within the AEW are a key catalyst to AEW growth. These authors suggested that, as convection is preferred near to and ahead of the trough axis, the summation of the flow associated with the small-scale vorticity centres will project upscale and enhance the mid-level flow around the AEW, increasing both the AEW amplitude and its growth rate via dry dynamics. In a numerical modelling study, Berry and Thorncroft (2012) revealed that an AEW will weaken on a timescale of a few days if convective systems are removed, underlining the importance of convection and its coupling to AEWs.

#### 2.1.3.6 Relationship with Deep Convection

In the WAM, AEWs have a modulating effect on rainfall, with some estimates suggesting that in excess of 60% of organised squall lines during summer are associated with AEWs (e.g. Fink and Reiner, 2003).[5] Composite analyses (e.g. Reed *et al.*, 1977) indicate that rainfall and cloudiness are primarily peaked ahead (west) of the mid-level trough axis and south of the AEJ core. Some studies have also noted a secondary peak of convective rainfall behind (east of) the mid-level trough axis, poleward of the AEJ axis (as illustrated by Poan's composite of Figure 2.7b).

More detailed analysis of individual convective systems associated with AEWs (e.g. Payne and McGarry, 1977) have shown that MCSs tend to propagate twice as fast as AEWs, consistent with being driven by downdrafts originating near the mid-level AEJ (e.g. see Rowell and Milford (1993), Chong and Hauser (1989) and Lafore and Moncrieff (1989)). This suggests that the preferred regions of rainfall consist of many convective systems travelling through a particular phase of the AEW, rather than having a single persistent area of rainfall moving at the same speed as the AEW. In this sense, the genesis, maturity and decay locations for MCSs within AEWs may be distinct. The movement of convective systems through AEWs, the variety of possible AEW configurations and the distinct diurnal cycle of the convection all add complexity to the question of how rainfall is related to mean AEW structure.

The observed relationship between AEWs and rainfall is illustrated by Figure 2.14, which shows a Hovmöller space–time diagram of 700 hPa curvature vorticity from the National Centers for Environmental Prediction (NCEP) Global Forecast System (GFS) operational analysis (a diagnostic of AEW trough location used by Berry *et al.* (2007)) overlaid with rainfall rate estimates from the CMORPH dateset (Joyce *et al.*, 2004); both fields are averaged over 5–15°N for the month of August 2004. In this figure the passage of AEW troughs can be distinguished by the westward-moving curvature vorticity maxima, while convective systems can be noted by the smaller scale bands of rainfall that also propagate westward. Note that the rainfall does not show individual convective systems, as a consequence of the averaging in latitude; instead, it signifies groups or 'envelopes' of convective systems.

To show more details of a particular case, Figure 2.15 shows MCSs associated with easterly waves during August 2006, including the precursors to Hurricane Helene. In the figure, the trough of the westernmost AEW and the inverted 'V' cloud pattern are marked by dashed lines at 1800 UTC on 6 September (upper left panel). At that time MCSs are on both sides, but on subsequent days convection is enhanced ahead of the trough.

There have been many observational studies explaining the location and type of convective clouds relative to the AEW trough. Several studies have concluded that deep convective clouds tend to occur at or ahead of the AEW trough (Carlson 1969a,b; Reed *et al.*, 1977; Payne and McGarry, 1977; Diedhiou *et al.*, 1999; Kiladis *et al.*, 2006). Some of these studies included waves over the Atlantic and along the West African coast and focused on synoptic-scale convection. Other studies of mesoscale convection over West Africa found that organised convective systems are also initiated east of the AEW cyclonic centres, particularly to the north of the AEJ (Fink *et al.*, 2006; Laing *et al.*, 2008; Cifelli *et al.*, 2010). Those studies also observed MCSs that developed behind the AEW trough and propagated into the region ahead of the AEW trough (e.g. Figure 2.16).

Earlier studies by Burpee (1974), Duvel (1990), Mathon *et al.* (2002) and Fink and Reiner (2003) observed enhanced convective precipitation ahead of the AEW ridge for systems north of 12.5°N (e.g. Figure 2.17). The enhancement of convection and precipitation at or behind the AEW trough, observed mainly in the northern Sahel, is primarily influenced by moisture availability – the northward transport of moisture from the ocean and equatorial forests – and therefore matches the location of high PW* seen in Figure 2.7, relative to the wave trough. Note also in Figure 2.17 the large spatial spread in initiation locations relative to the wave trough. In contrast with mid-latitude

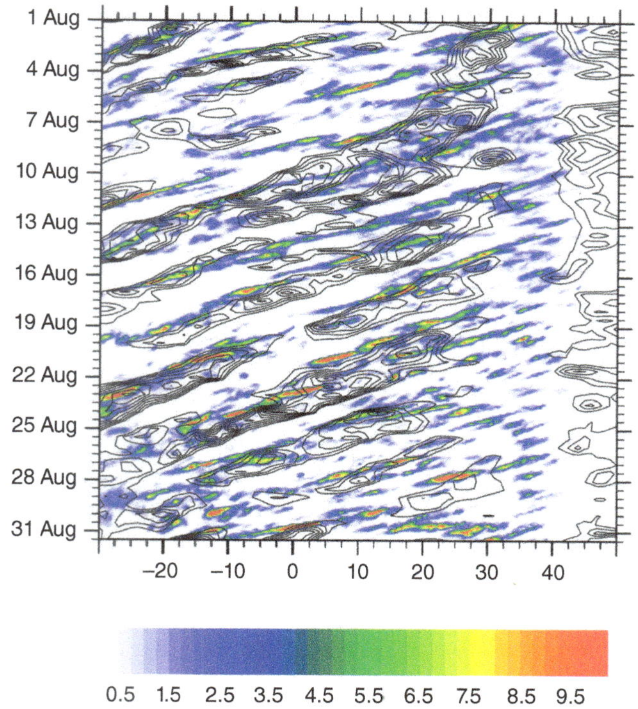


**Figure 2.14** Hovmöller space–time diagram of 700 hPa curvature vorticity (positive values greater than $2 \times 10^{-6}$ s$^{-1}$ contoured every $2 \times 10^{-6}$ s$^{-1}$) from GFS operational analyses and CMORPH precipitation estimate (shaded in mm h$^{-1}$, according to colour bar), both averaged 5–15°N for the month of August 2004.

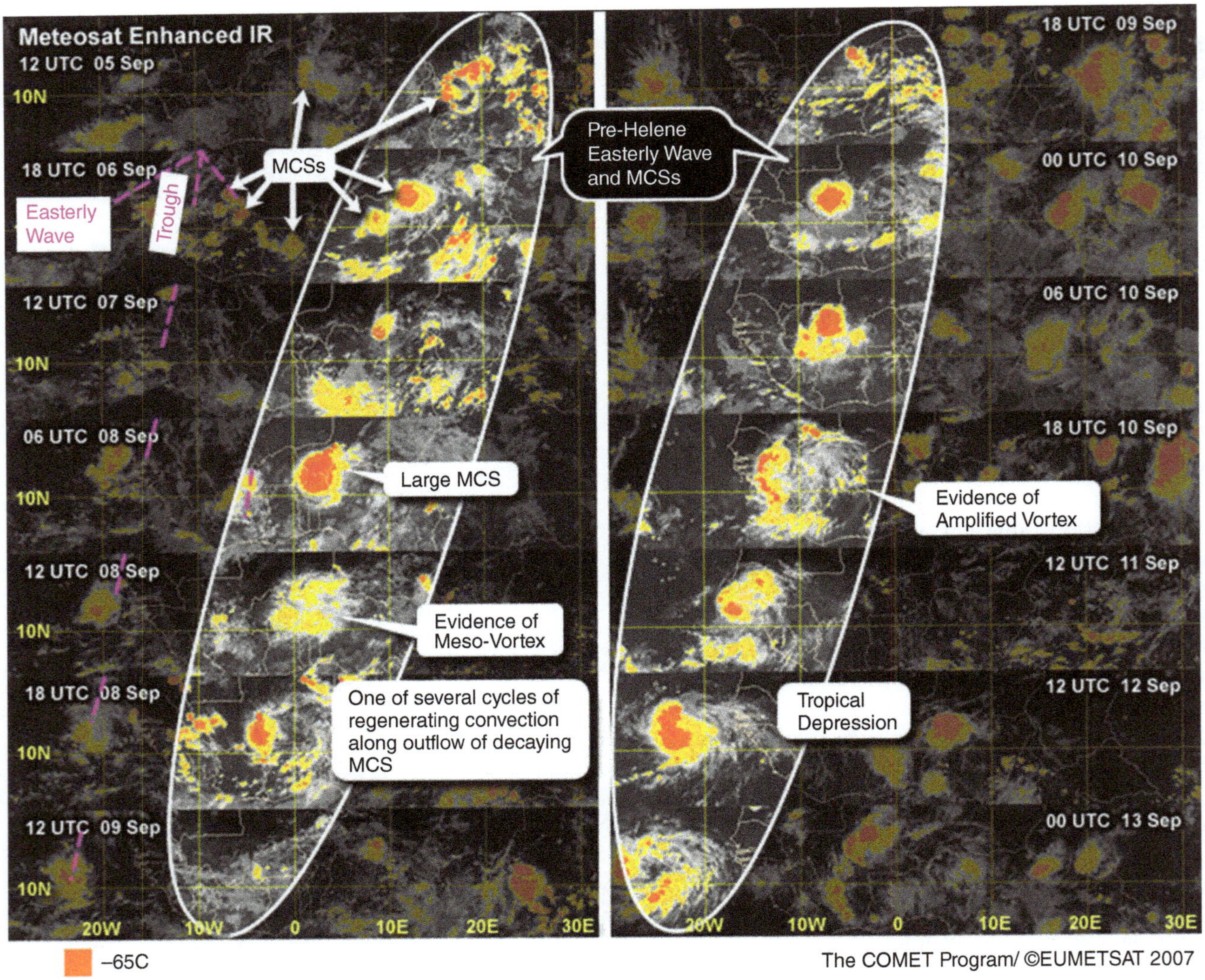


Figure 2.15 Enhanced infrared images of AEWs and MCSs between 1200 UTC 5 September and 0000 UTC 13 September 2006, precursors to Helene (2006). *Source*: from Laing and Evans (2011).

weather systems, the convective initiation is only relatively loosely dictated by the AEW structure, and other factors such as topographic patterns have an important role also in determining the location of convective initiation.

Note that the tilt of the trough axis to the north and south of the AEJ core means that the location of convection ahead of and behind the trough can have some ambiguity: convection to the south of the AEJ, which is collocated with the trough at this latitude, may be significantly to the west of the trough at the AEJ latitude, as in Figure 2.7.

The interaction of the vorticity associated with AEWs and the mesoscale vorticity associated with MCSs is an area of active research. Some tropical cyclones, such as Alberto (in 2000), begin as AEW–MCS systems that initiate near the mountains of East Africa (the Darfur Mountains and Ethiopian Highlands) and undergo cycles of decay and regeneration while moving westward. The swirling cloud pattern at 1200 UTC 8 September and 1800 UTC 10 September (Figure 2.15) indicates the presence of an MCV, a feature that sometimes forms in the stratiform region of decaying MCSs. The pre-Helene vortices appear to amplify over the continent within the AEW structure, a strong vortex emerges from the continent and tropical cyclogenesis occurs soon after. These observations support the notion that mesoscale convection gives rise to positive PV anomalies in the lower troposphere for tropical cyclogenesis.

#### 2.1.3.7 Downstream Transformation

AEWs are frequently observed to traverse the Atlantic Ocean and move into the Caribbean region. During their transition across the ocean they tend to retain regions of deep convection (that can be tracked using satellite imagery) and thus remain a hazard to shipping and aviation. In the Caribbean region, AEWs can bring disturbed weather to the islands and remain an important feature

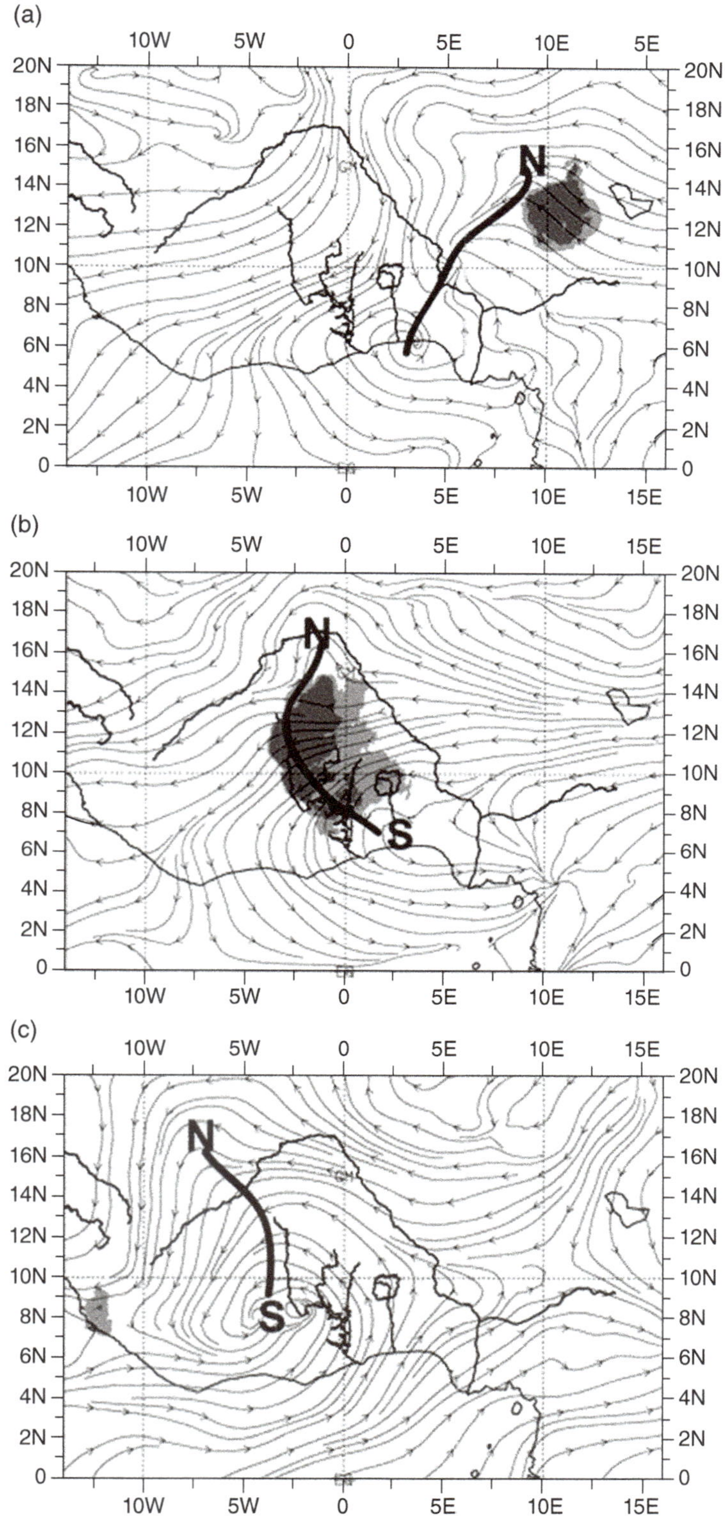


**Figure 2.16** (a) Streamlines on the 850 hPa surface at 0600 UTC 3 September, (b) 0000 UTC 4 September and (c) 1800 UTC 4 September 2002. Also shown are the location of the Upper Ouémé Valley and the areas enclosed by the 233 K (light grey) and 213 K (dark grey) brightness temperatures used to track this and all other sub-type Ia (advective organised convective systems) cloud systems. The analysed cyclonic centres of northerly and southerly AEWs are labelled 'N' and 'S' respectively. The bold AEW trough lines delineate the subjectively analysed maximum cyclonic curvature between the two AEW vortices. *Source:* Figure 11 from Fink *et al.* (2006). ©American Meteorological Society. Used with permission.

for weather forecasters to consider. In addition to the rainfall component, synoptic ridges associated with AEWs can transport large amounts of Saharan dust across the Atlantic Ocean (e.g. see Zipser *et al.* (2009) and Dunion and Velden (2004)). On some occasions, AEWs have been observed to interact with mid-latitude disturbances and act as the moisture source for rainfall in northwestern Africa (Knippertz *et al.*, 2003).

AEWs are an important precursor to Atlantic and eastern Pacific tropical cyclones. Although the tropical cyclones occur outside the domain served by West African forecasters, it remains important for forecasters

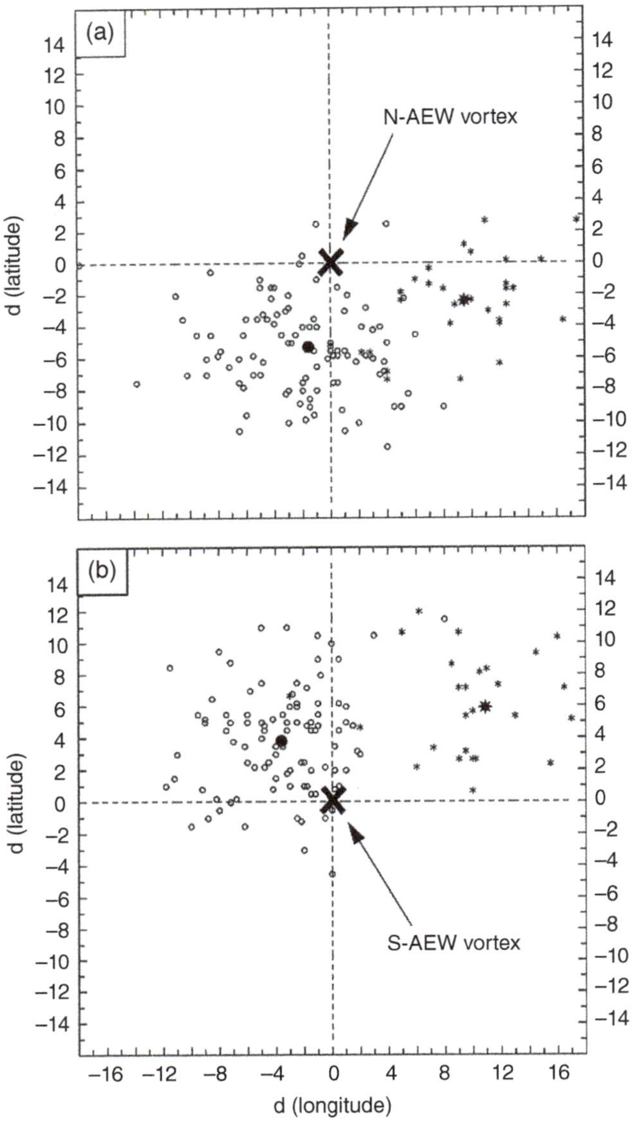


Figure 2.17 Scatterplots of the positions of origin points of squall lines relative to the simultaneous locations of the accompanying (a) northerly and (b) southerly AEW vortices. The abscissa (ordinate) denotes the longitudinal (latitudinal) distance, with negative values indicating that the squall line origin is west (south) of the respective vortices. The open circles represent squall lines that formed west of the AEW trough, while the stars denote squall lines assigned to AEW phase intervals in the southerly flow east of the trough, north of 12.5°N. The mean positions of either class of squall lines are indicated by the large bold circles and stars respectively. In both diagrams, *d* is difference, and both cover the periods May–October 1998 and 1999. *Source*: Figure 11 from Fink and Reiner (2003). Copyright John Wiley and Sons. Reproduced with permission.

to have some understanding of the process of transformation from an AEW into a tropical cyclone. While the locations at which the cyclone forms vary from case to case depending on the environment and the nature of the AEW, there are well-documented cases of tropical cyclogenesis occurring close to the West African coastline – for example, Hurricane Alberto in 2000 (e.g. Lin *et al.*, 2005) and Tropical Storm Debby in 2006 (e.g. Jenkins *et al.*, 2008; Ventrice *et al.*, 2012; Lin *et al.*, 2013). In such cases, intensification of convection within the AEW toward a tropical cyclone may begin partially over land, posing a potential flooding risk (e.g. Jenkins *et al.*, 2010). Furthermore, tropical cyclones forming close to the coastline can have a negative impact on shipping, fishing, and so on in the countries of West Africa.

A high fraction of North Atlantic hurricanes have AEWs as their trigger. For example, 2005 was a particularly active hurricane season, with 23 out of 31 named storms originating from AEWs. In comparison, 2006 was a quiet season, with 7 out of only 10 named storms originating from AEWs (National Hurricane Center, 2015). Over Africa, AEWs are often characterised by a low-level vortex north of the AEJ and an AEJ-level vortex south of the AEJ (see Section 2.1.3.2). However, as the AEWs move westward across the West African coast and into the East Atlantic the southern vortex tends to develop a low-level circulation consistent with increased frequency of deep convection in the coastal region (Janiga and Thorncroft, 2013). Hopsch *et al.* (2007) compared the mean structure of AEWs that were associated with East Atlantic tropical cyclogenesis and those that were not. They noted that the mean structure of AEWs that developed tended to have stronger low-level vorticity, stronger vertical motion and higher humidities throughout the troposphere in the trough. Several recent studies have extended this work (e.g. Agudelo *et al.*, 2011; Leppert *et al.*, 2013a,b; Cecelski and Zhang, 2013). Rowell *et al.* (2016) indicate that developing AEWs usually have a stronger AEJ on the poleward side of the relative vorticity maximum. Consistent with the AEJ strength, developing AEWs generally have a stronger relative vorticity maximum over the lifetime of the AEW and developers tend to develop faster, with a statistically significant increase over non-developers in the first 48 h once west of 20°W. Developing AEWs tend to have a mean SST greater than 27°C, and this is particularly significant over the first 48 h, where the great majority of developers have an SST greater than 27°C. Also, the developing AEWs pass through a more humid troposphere, particularly at the AEJ level and above. The consensus from these studies is that AEWs that have stronger low-level circulations, moister columns and have more ubiquitous convection as they track into the East Atlantic are more likely to be associated with tropical cyclogenesis.

### 2.1.4 Mid-latitude Troughs and Upper-level Troughs

Upper-level troughs from the mid-latitudes significantly affect tropical West Africa during the boreal winter half year (Knippertz, 2007). A widely used method for identifying such systems is to detect significant anomalies in

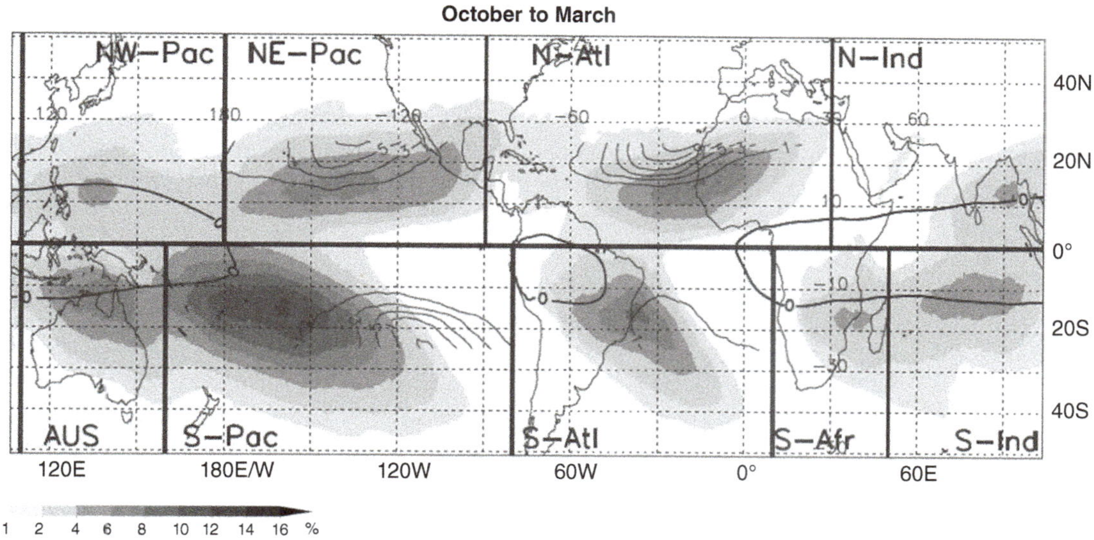


**Figure 2.18** Mean tropical plume frequency (%; shaded) for October–March between 1983/84 and 2005/06. The contours display the mean frequency of troughs at low latitudes (%) detected by Fröhlich and Knippertz (2008) in the ERA-40 data between 1980 and 2001. The bold grey contour presents the zero line of the mean zonal wind at 200 hPa indicating equatorial easterlies from the western Pacific to the eastern Atlantic and over South America, with westerlies elsewhere. *Source*: Figure 4a from Fröhlich *et al.* (2013). ©American Meteorological Society. Used with permission.

PV, either on an isentropic surface that intersects the subtropical tropopause, or averaged over the upper troposphere/lower stratosphere. Climatologies of these so-called PV intrusions or cut-offs show a frequency maximum over the Atlantic Ocean near northwestern Africa (Figure 2.18). For lower isentropic levels, the maximum shifts northeastward into the Mediterranean Sea – see Wernli and Sprenger (2007: figure 4). The frequency of PV intrusions is largely reduced during the summer half year (Fröhlich and Knippertz, 2008).

Linear Rossby wave theory provides a first-order explanation for the geographical, vertical, seasonal and even interannual behaviour of mid-latitude trough penetration to low latitudes. Mean zonal winds near the tropopause are strongly westerly over the tropical East Pacific and Atlantic Oceans (see Figure 2.18). This allows a propagation of Rossby wave energy into these regions, which have been referred to as 'westerly ducts' for this reason (Webster and Holton, 1982; Hoskins and Ambrizzi, 1993; Kiladis and Weickmann, 1997). The westerly winds weaken during the summer half of the year and toward lower levels, associated with a reduction in trough occurrence (Tomas and Webster, 1994; Fröhlich and Knippertz 2008). The El Niño/Southern Oscillation modifies the equatorial winds on interannual timescales, causing a tendency of reduced PV intrusion activity over the Pacific Ocean and enhanced PV intrusion activity over the Atlantic (Waugh and Polvani, 2000).

A clear indicator of the penetration of an upper-level trough into low latitudes in satellite imagery is the occurrence of a tropical plume on its eastern flank (Iskenderian, 1995; Knippertz, 2005). Tropical plumes are elongated continuous bands of mainly upper- and middle-level clouds that stretch for thousands of kilometres poleward and eastward from the ITCZ into the subtropics (Figure 2.19). Usually, an anomalously strong subtropical jet (STJ) streak is aligned along the often-sharp poleward edge of the tropical plume (Figure 2.19). Climatological studies show a tropical plume frequency maximum over the eastern Atlantic/western West Africa, consistent with the PV intrusion climatology (Figure 2.18). Amounts of precipitation associated with tropical plumes differ largely between systems and along the axis of the tropical plume. Knippertz (2005), for example, describes a tropical plume that is rooted in ITCZ convection over South America, stretches across the Atlantic as an upper-tropospheric cloud band and is finally associated with heavy precipitation over northwestern Africa.

The precipitation events associated with upper-level troughs over tropical West Africa during the dry season November to March are sometimes referred to as 'Heug' or 'Mango' rains. They are irregular in time, space and magnitude. Their climatological contribution to annual rainfalls is estimated to be on the order of a few per cent only, which is hardly relevant for agriculture (Seck, 1962; Gaye *et al.*, 1994). There are, however, examples of much more significant dry-season rainfalls associated with upper-level troughs. Such events can have substantial impacts on local hydrology and human activities, from the greening of pastures to the flooding and rotting of harvests (Knippertz and Martin, 2005; Fall *et al.*, 2007; Knippertz and Fink, 2008).

The climatology and past case studies suggest a distinction of two different stereotypical cases and

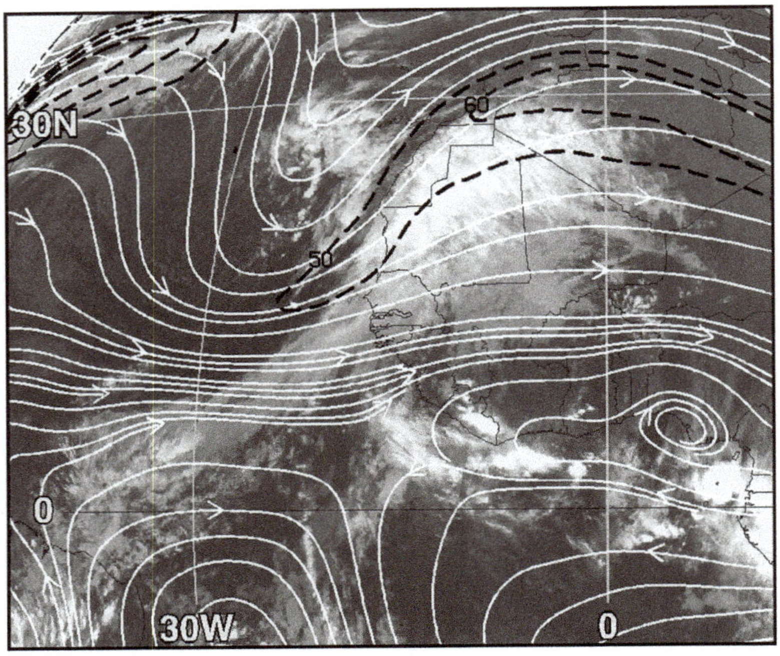


**Figure 2.19** Meteosat infrared satellite image for 0000 UTC 10 January 2002 with superimposed isotachs (dashed black in m $s^{-1}$) and streamlines (in white), both at the 345 K isentropic level. *Source*: slightly modified version of Figure 2c from Knippertz and Martin (2005). Used with the permission of the Royal Meteorological Society.

mechanisms. Although in each of these mechanisms the flow pumps the ITD northward together with deep moisture, the trough and STJ positions differ.

1) Sahelian/Saharan countries close to the west coast of West Africa, such as Sénégal, Gambia, Mauritania and Mali, can be directly affected by upper-level troughs penetrating southward over the eastern Atlantic Ocean, as the ones shown in the climatology and case example in Figures 2.18 and 2.19. Significant events have been described by a number of authors, including Seck (1962), Borgne (1979), Gaye *et al.* (1994) and Knippertz and Martin (2005). An interesting case worth mentioning is the record-breaking, high-impact case during 9–11 January 2002 (Figure 2.19). Storm totals of more than 50 mm fell over a widespread area covering northwestern Sénégal and southwestern Mauritania. New January precipitation records were broken at several observing stations, led by Podor in northern Sénégal with 115.8 mm. Several people were killed in collapsing houses or through hypothermia, buildings were damaged or destroyed by flooding, and the loss of crop and livestock was substantial (Fall *et al.*, 2007). Several dynamical ingredients are necessary to create such an extreme event (Knippertz and Martin, 2005), as depicted in Figure 2.20a: (a) A previous low-latitude upper-trough instigates a transport of moist air from the Atlantic and the inner tropics toward the western Sahel without generating any significant precipitation; (b) the moistening of lower levels causes a significant destabilisation in the region of interest; (c) a subsequent trough releases the convective instability through weak forcing for mid-level ascent; (d) there are indications that inertial instability in the right entrance region of the STJ streak might further enhance uplift. Knippertz and Martin (2005) show that a succession of two troughs is not necessary toward the end of the rainy season, when the spatial distance to the main moisture reservoir is smaller, so that sufficient transport can be achieved by one persistent system.
2) The entire Guineo-Soudanian zone (~7–12°N) can be affected by occasional northward shifts of the main rain zone, which usually end just north of the Guinea Coast during the dry season. Knippertz and Fink (2008) document that these shifts can be triggered by the penetration of an upper-level disturbance into northern Africa (Figure 2.20b). This is more indirectly communicated through falling pressures over the Sahara and Sahel in the right entrance region of the anomalous STJ streak to the southeast of the positively tilted trough, where warm advection prevails at upper levels (this corresponds to ascending poleward flow ahead of the trough). The associated northward shift and intensification of the usually weak continental heat low allows moist southerlies to penetrate further than normal into the continent. Daytime heating of the (at this time of the year) rather dry land surface can then trigger deep moist convection and dry-season precipitation. As for the western Sahel cases, dry-season precipitation events are often preceded by very persistent or repeated troughing over northern Africa, which allows the slowly responding tropical circulation to attune to the forcing (Knippertz and Fink, 2009).

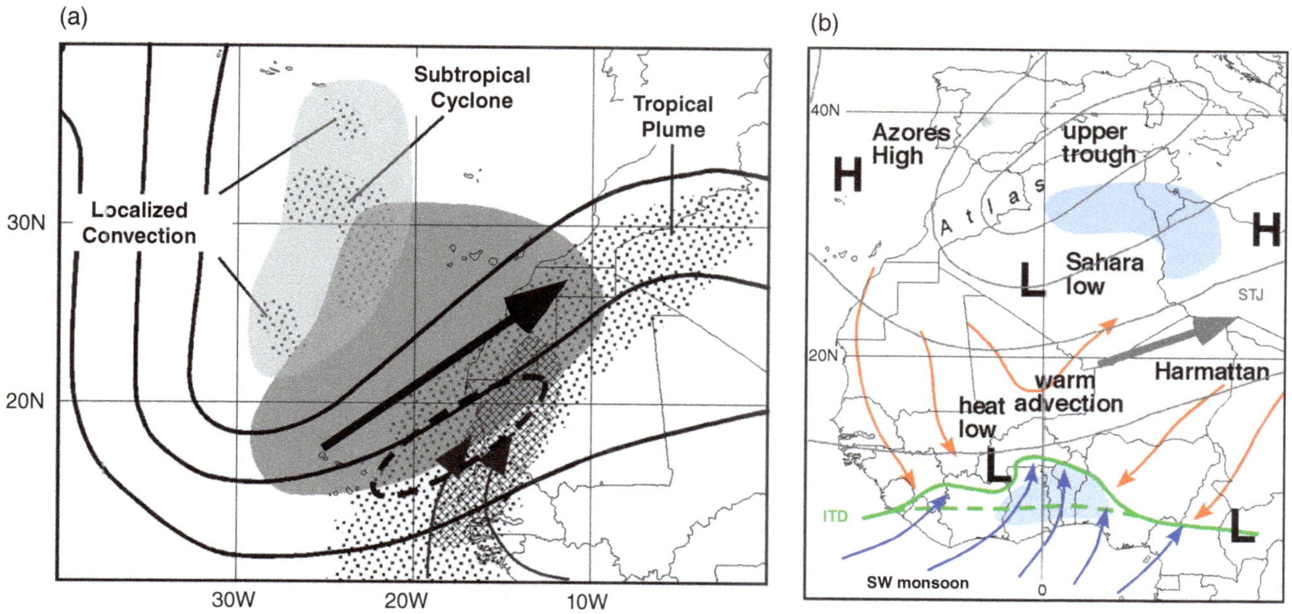


Figure 2.20 Schematic depictions of the large-scale circulation associated with dry-season precipitation over West Africa associated with low-latitude upper-level disturbances from the extratropics. (a) Cases with a direct influence form the upper trough, mostly affecting the western Sahel. Thick black lines delineate the low-latitude upper trough and the thick arrow indicates the associated STJ streak. Thin grey arrows show mid-level moisture transports from the deep tropics. Stippled regions indicate high clouds, and hatching delineates the major precipitation zone. Light (dark) grey shading depicts the region of convective instability under the coldest air at upper levels (positive quasi-geostrophic forcing for mid-level ascent). The dashed lines bound a region of upper-level inertial instability along the anticyclonic shear-side of the jet. *Source*: Figure 11 from Knippertz (2007). Used with permission. (b) Cases with an indirect influence from the upper trough, mostly affecting the Soudano-Sahelian zone. Grey lines depict upper-level geopotential height. The green dashed and solid lines show the climatological and the actual position of the intertropical front that separates dry continental air and northerly flow from moister air and southerly flow. Regions with precipitation are shaded in blue. For more details, see text. *Source*: modified version of Figure 14 from Knippertz and Fink (2008). ©American Meteorological Society. Used with permission.

A scientific result with great relevance for operational weather forecasting in tropical West Africa during the dry season is the enhanced predictability of rainfall directly or indirectly associated with extratropical troughs. The detailed modelling case study by Meier and Knippertz (2009) shows fairly robust predictions of the extraordinary January 2002 case over Sénégal and Mauritania up to lead times of 7 days, consistent with the successful reproduction of the case with regional models by Knippertz and Martin (2005) and Fall *et al.* (2007). Statistical evaluations of 5-day forecasts of unusual dry-season rainfall events in the Guineo-Soudanian zone with the model of the ECMWF indicate that extratropical influence enhances predictive skill (Knippertz and Fink, 2009). These results should encourage forecasters to take advantage of available dry-season precipitation forecasts in terms of the dissemination of early warnings.

### 2.1.5 Extratropical Dry Air Intrusions

Layers of very dry air (relative humidity (RH) <10%) are observed over many tropical regions in the free troposphere. In a generally rather moist immediate environment, such dryness can only be fed by air originating from the upper level of the extratropics, where similar potential temperatures and water vapour mixing ratios exist. As a result, these dry layers are known as dry 'intrusions' to highlight their non-tropical origins. These layers have been shown to modulate convective activity strongly over the tropical oceans at various scales (Brown and Zhang, 1997; Redelsperger *et al.*, 2002).

Such extratropical dry air intrusions have also been identified in the WAM mid-troposphere (Roca *et al.*, 2005). The analysis revealed that, at 500 hPa over the Sahel, dry air originating in the upper levels (200–250 hPa) on the anticyclonic side of the polar jet stream around 50°N is frequently observed at various scales. The dry intrusions start subsiding over Europe and travel south on the 330 K isentropic level reaching, a few days later, the Sahelian mid-troposphere (Figure 2.21).

Mid-tropospheric dry air influences convective activity and rainfall variability in various ways. Dry air generally has an inhibiting effect on the first initiation of convection, as a result of the temperature inversion induced by long-wave radiation at the bottom of the dry layer (Mapes and Zuidema, 1996) or by entrainment of dry air into ascending parcels that decreases their buoyancy and limits the vertical development of the convective clouds (e.g. Parsons *et al.*, 2000). In contrast, mid-level dry air favours the organisation

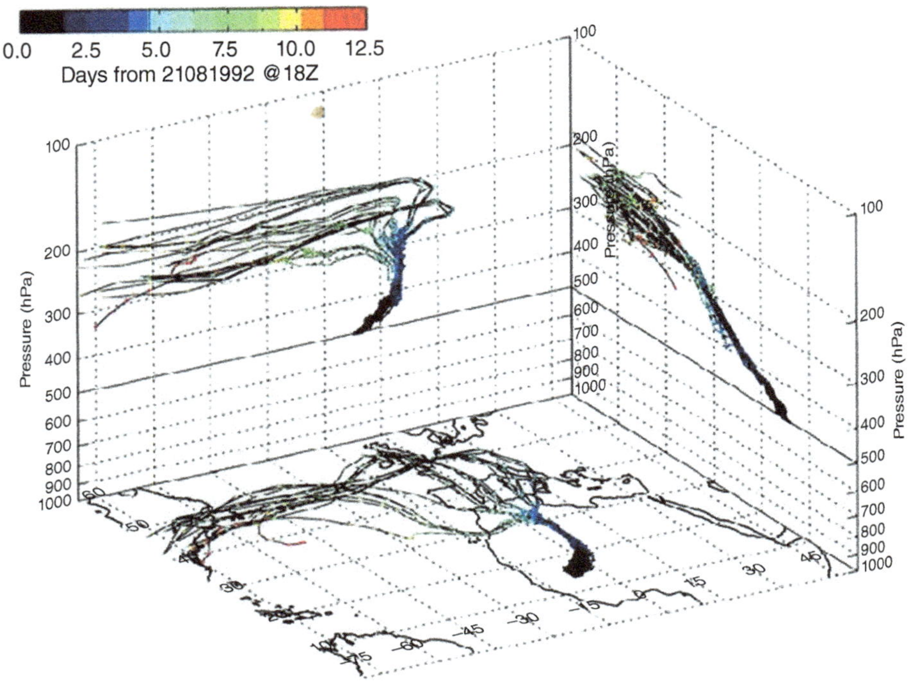


**Figure 2.21** Three-dimensional view of an ensemble of back trajectories initially located at 500 hPa centred at (a) 15°N, 6°W. The colour scale indicates the duration of the air mass along the trajectory in days since 1800 UTC 21 August 1992. *Source*: Figure 9(a) from Roca *et al.* (2005). ©American Meteorological Society. Used with permission.

of convection into squall lines, by feeding downdraughts and mid-level rear-inflow currents, known to be key elements of organised convective systems (Section 3.1.3).

Figure 2.22 shows the Hovmöller diagram of the RH at 500 hPa averaged over the 12.5°N–17.5°N latitude belt for the 2006 season, along with the latitude of last saturation of the air masses, and reveals some features of dry intrusion events. Three substantial events, labelled 1 to 3, correspond to pan-West African dry intrusions. These plumes are composed of very dry air with RH below 10% and even below 5% at some points, and last from 3 to 5 days. These dry structures travel from around 25°E to the Atlantic coast. The three events are separated by around 40–50 days and are thought to be associated with Madden–Julian oscillation (MJO) events that were clearly depicted in the equatorial wind anomaly during the season (Janicot *et al.*, 2008), although the link between the pan-African dry events and the MJO is not direct.

A second type of event with smaller-scale dry structures with RH below 10% is also observed. This latter mode is more restricted to the Sahel region and exhibits synoptic timescales (see also Roca *et al.* (2005)).

The three dry intrusions highlighted in Figure 2.22 each led to a substantial reduction in West African rainfall. South of the ITD, down to the coastal regions, the average rainfall anomaly associated with these three events was negative, around 5 mm day$^{-1}$; a significant portion of the seasonal rain. The cloudiness and MCS anomalies concur with the rainfall anomaly, leading to a large, well-spread, negative anomaly associated with the dry intrusions. Note that during the three events, AEWs were active over the region but did not yield significant rainfall or deep convective events (Deme and Roca, 2008).

### 2.1.6 Guinea Coast Systems

#### 2.1.6.1 Guinea Coast Cloud and Rainfall

The Guinea Coast is defined here as the region between Cape Palmas and the Niger delta in the south and the ninth parallel in the north. Between 8°N and 9°N, the annual cycle of rainfall changes from bimodal in the south to unimodal in the Soudano-Sahelian zone. The Guinea Coast region has annual precipitation in excess of 1000 mm, with more than 2000 mm in some places (Section 1.2). Consequently, the natural vegetation ranges from rainforest to a semi-deciduous tree savannah.

The Guinea Coast experiences different types of rainfall. A categorisation into different types of rainfall has been made by, for example, Acheampong (1982), Omotosho (1985), Kamara (1986), and Fink *et al.* (2006, 2010). In principle, these authors distinguish between squally MCSs or disturbance lines (Section 3.1.3.2), monsoon or vortex rainfall, and isolated thunderstorms. Other types of rainfall are afternoon thunderstorms that

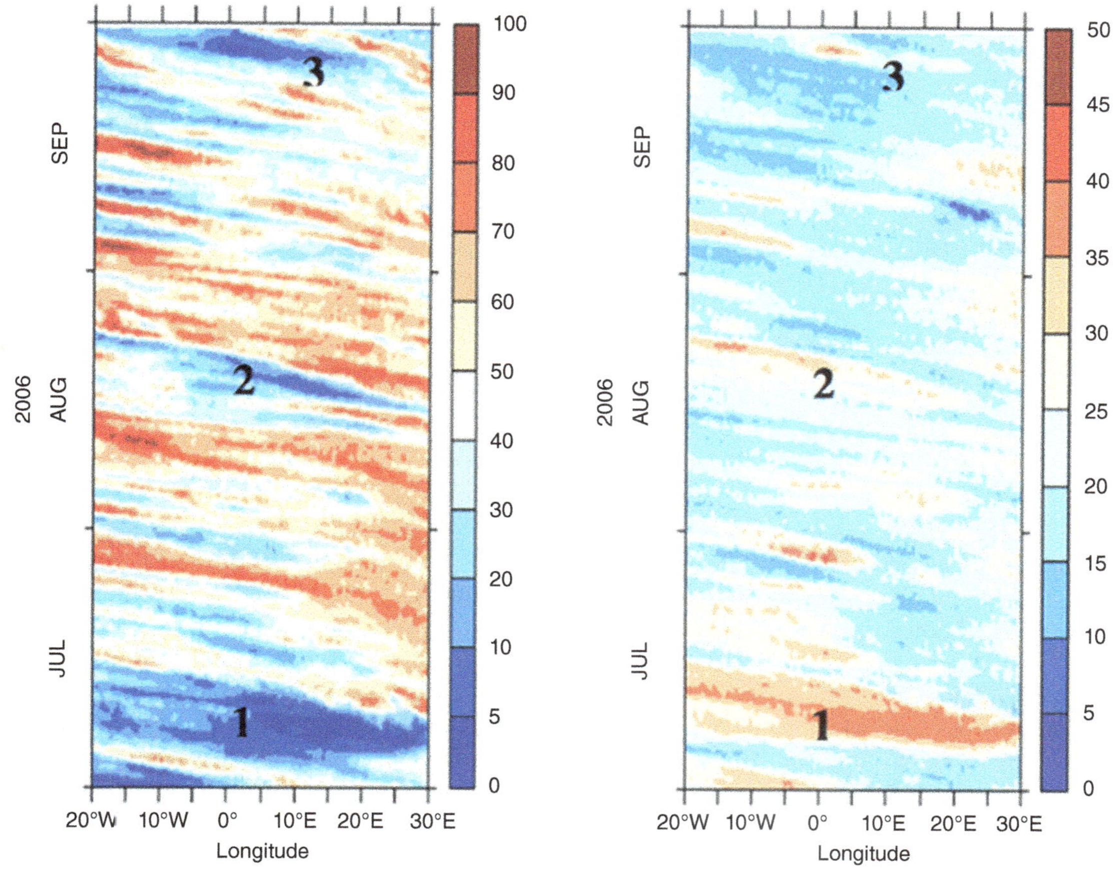


Figure 2.22 Time–longitude diagram averaged over the 12.5–17.5°N latitude belt for the 2006 season. Left: RH at 500 hPa. Right: latitude of last saturation (degrees north). The bold numbers correspond to the three identified large events. Both the humidity and the last saturation information come from a Lagrangian modelling effort (Pierrehumbert, 1998; Pierrehumbert and Roca, 1998) fed by NCEP/GFS analysis.

are aligned parallel to the coast and are associated with the land–sea breeze (Section 4.1.4.2), and dry-season rainfall events in relation to the tropical–extratropical interactions described in Section 2.1.4. In the following, the synoptic conditions associated with MCSs and vortex rainfall are briefly described.

MCSs at the Guinea Coast can occur throughout the year, but they are rare between December and February and confined to the immediate coastline. They are more frequent between March and November, and peak in April–May before the first rainy season, and October after the second rainy season (Acheampong, 1982; Omotosho, 1985). This is because, at this time of the year, a mid-level northeasterly wind component (Figure 1.25) provides the low-level vertical shear and mid-level dryness necessary for the organisation of the convection into MCSs (Section 3.1.3.2). Contrary to the conditions in the Sahel during its rainy season, the Guinea Coast zone is characterised by moister and less sheared vertical profiles throughout most of the year (Parker *et al.*, 2005a). Therefore, MCSs in the coastal zone tend to have a slower westward propagation speed and reveal warmer cloud-top temperatures in infrared satellite imagery than systems that occur further north (Schrage and Fink, 2007). Owing to the moister troposphere and slower propagation speed, rainfall amounts are not necessarily lower than for Sahelian MCSs and can cause severe local flooding. However, peak gusts, temperature and specific humidity drops are less pronounced than in the Sahel even though hurricane-force winds have been reported along the Guinea Coast in association with the arrival of a convective cold pool (Section 3.1.2.2).

Another type of rainfall at the Guinea Coast has been termed 'monsoonal rains'. These are prolonged periods (longer than 4–6 h) of steady or intermittent, light to moderate rains from convection with low electrical activity (Eldridge, 1957; Kamara, 1986; Buckle, 1996). Fink *et al.* (2006) speculated that some monsoon rainfall events might be dynamically linked to low- to mid-tropospheric vortices and used the term 'vortex-type rainfall'[6] for a weather pattern when deep westerlies were observed as a consequence of a nearly stationary mid-level vortex to the north of the Guinea Coast. This rainfall type was associated with a very moist troposphere with low CAPE, small CIN and low shear in the lower troposphere, but strong TEJ winds aloft. While this environment is not conducive

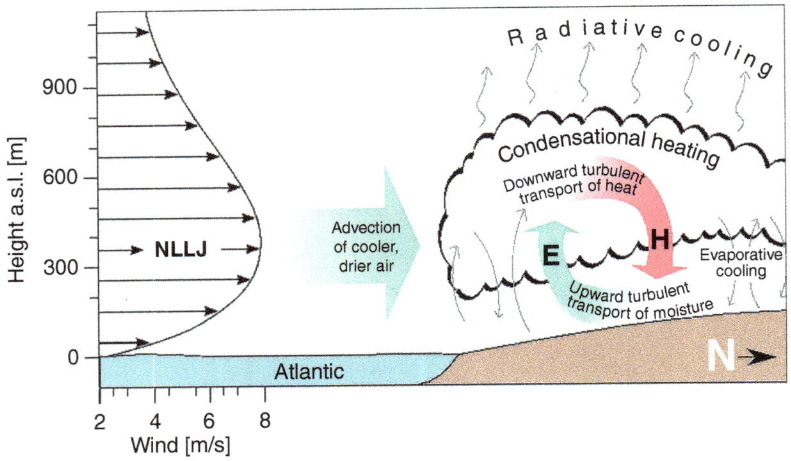


**Figure 2.23** Meteorological processes associated with the formation of low-level stratiform cloud decks at the Guinea Coast. The NLLJ is driven by the onshore pressure gradient, enhanced by the sea breeze, and causes cold air advection from the Atlantic Ocean, the turbulent upward mixing of moisture (E) and downward mixing of warmer air (H). These processes dominate over detrimental effects like dry air advection by the NLLJ and early night near-surface inversions (not shown) during most nights of the wet season, and thus the night-time cloud formation is widespread. *Source:* Figure 12(a) from Schuster *et al.* (2013). ©American Meteorological Society. Used with permission.

to the formation of MCSs, the vortex in the lower troposphere may be associated with low-level synoptic-scale convergence that triggers the convection, especially at daytime. However, not much is known about the dynamics of this rainfall type, nor the origin of the vortices. Achaempong (1982) speculates the low-level vortices are related to orographic effects. It may not be excluded, however, that some of the slow-moving vortices are decelerated or stalling AEWs – an AEW behaviour that has been noted for the 1 September 2009 extreme rainfall event at Ouagadougou (see Section 2.2.2.2).

In 2002, a vortex-type event occurred at the end of the summer monsoon. The presence of a mid- and upper-level trough from the mid-latitudes blocked the usual easterly air current of the AEJ. As a result, deep westerly winds were present up to a level of about 500 hPa. This period was characterised by frequent rain events, indicating that the troposphere was supplied by sufficient water vapour from the Gulf of Guinea. Over various parts of Togo, Benin and Nigeria, deep convection was initiated during the morning hours of 1 October 2002. The cloud cells merged to one single MCS that moved toward the southwest. This cloud system finally collapsed east of Lake Volta during the night hours on 2 October 2002. The radiosonde observations from Parakou from 1 October 2010 at 1200 UTC show an unstable layering of the troposphere, up to about the 200 hPa level, leading to the widespread occurrence of deep convection. In Benin's synoptic stations this deep convection resulted in rainfall amounts of 21 mm, 13 mm, 2 mm and 0.5 mm at Savè, Bohicon, Natitingou and Cotonou respectively.

A rather recurrent weather feature at the Guinea Coast during the wet season is the southwesterly night-time low-level jet (NLLJ; see Section 4.1.3.2) and associated, extensive low-level stratiform cloudiness. Clearly, for aviation weather, the NLLJ and low-level clouds are important, especially for airports inland of the Guinea Coast. The cloud base is so low that it sometimes rests at the surface and causes fog, especially over higher terrain (Section 4.1.3.3). The low-level cloudiness was first mentioned by Schrage and Fink (2007) and it is the typical situation for nights that lack deep convection, though some clear nights occur as well. Van der Linden *et al.* (2015) created a low-level cloud climatology and showed that, after sunset, the zone of abundant low-level stratiform clouds is confined to latitudes south of 6–7°N but starts to spread inland during early night hours. Its maximum northward extent is reached between 0900 and 1000 UTC when its northern edge is located between 10°N and 11°N. The area affected by a large fraction of nocturnal and morning low-level clouds between 10°W and 10°E is in the order of 800 000 km$^2$. Schuster *et al.* (2013) summarised the mechanisms that favour or disfavour the genesis of the low-level cloudiness (Figure 2.23). A strong night-time low-level jet promotes upward mixing of moisture and cooler air as well as cold air advection from the cool waters of the Gulf of Guinea. Detrimental to the formation of the low-level clouds is a strong surface inversion early in the night and dry air advection from the Atlantic Ocean. Once formed, the OLR at the top of the cloud layer seems to stabilise the cloud layer. It is important to note that the night-time inland spreading of the low-level clouds, as described in van der Linden *et al.* (2015), is not an advection feature, but a northward propagation of turbulence underneath the evolving NLLJ. Apparently, there are subtle balances during non-precipitative nights that determine whether clouds form or not. As a consequence, about one-third of the nights remained clear at Djougou (central Benin) during June–October 2006 (Schrage and Fink, 2012), and models struggle to have a realistic climatological representation or to forecast the cloud layer (Knippertz *et al.*, 2011; Schuster *et al.*, 2013).

### 2.1.6.2 The Little Dry Season

The areas southwest of Nigeria, and coastal parts of Benin, Togo, Ghana and Cote d'Ivoire, experience the influence of the "little dry season" (which corresponds to 'zone D' of Figure 1.1). This phenomenon occurs between July and August when negative anomalies are observed in the rainfall pattern. A drop in static energy is associated with its occurrence. Cold winds, low surface temperatures, morning drizzle, sunny afternoons and partly cloudy conditions characterise this period.

The effect of a prolonged little dry season can be positive or negative. During this period, some crops that have suffered from excess water normally get some relief. Also, this period presents a favourable environment for cold-related diseases.

The mechanism of formation of the little dry season has been linked to some of the meteorological parameters – rainfall, pressure, equivalent potential temperature, surface temperature, surface wind and SST. Most simply, it is linked to the northward progression of the ITD and the whole monsoon system around the time of monsoon onset. Though the little dry season occurs annually, its onset, duration and cessation vary from year to year.

Over the years it has been observed that the peak surface pressures are, by July, linked to the little dry season. During this period, the high pressure builds in the Gulf of Guinea and the St. Helena high pressure cell exhibits significant anticyclogenesis. This effect spreads to the coastal areas of West Africa, where sinking motion is established and convection is reduced to a minimum. The classical flow over the Gulf of Guinea during this period is anticyclonic. Owing to upwelling off the Gulf of Guinea, there is a cooling in the SST during this period. It shall be stressed, however, that the little dry season is most pronounced where coastal upwelling and cold SSTs are strongest; that is, along the coasts of Ghana east of the Cape of the Three Points as well as along the coasts of Togo and Benin. The little dry season is weak or absent where warm onshore waters persist and the coastline is perpendicular to the southwest monsoon winds; for example, the Niger delta in Nigeria and the Pepper Coast of Liberia (see Figures 1.5 and 1.25).

### 2.1.7 Synoptic-scale Land-surface Feedbacks

Moist convection is strongly modulated by synoptic variability, as discussed in Section 2.1, and the diabatic heating associated with convection is considered to play an important role in the growth of AEWs. Here, we consider an additional feedback between rainy systems and the synoptic dynamics, which operates via the impact of rainfall on soil moisture and the surface energy balance. Surface fluxes of sensible and latent heat in sparsely vegetated regions of Africa are very sensitive to rainfall in previous days. For 2–3 days after rainfall, the daytime turbulent fluxes are dominated by latent heat, with only weak exchange of sensible heat with the planetary boundary layer (PBL; Kohler *et al.*, 2010).

An example of synoptic-scale variability in surface fluxes is shown in Figure 2.24. Rain occurred across much of the southern part of the domain in the preceding 36 h,

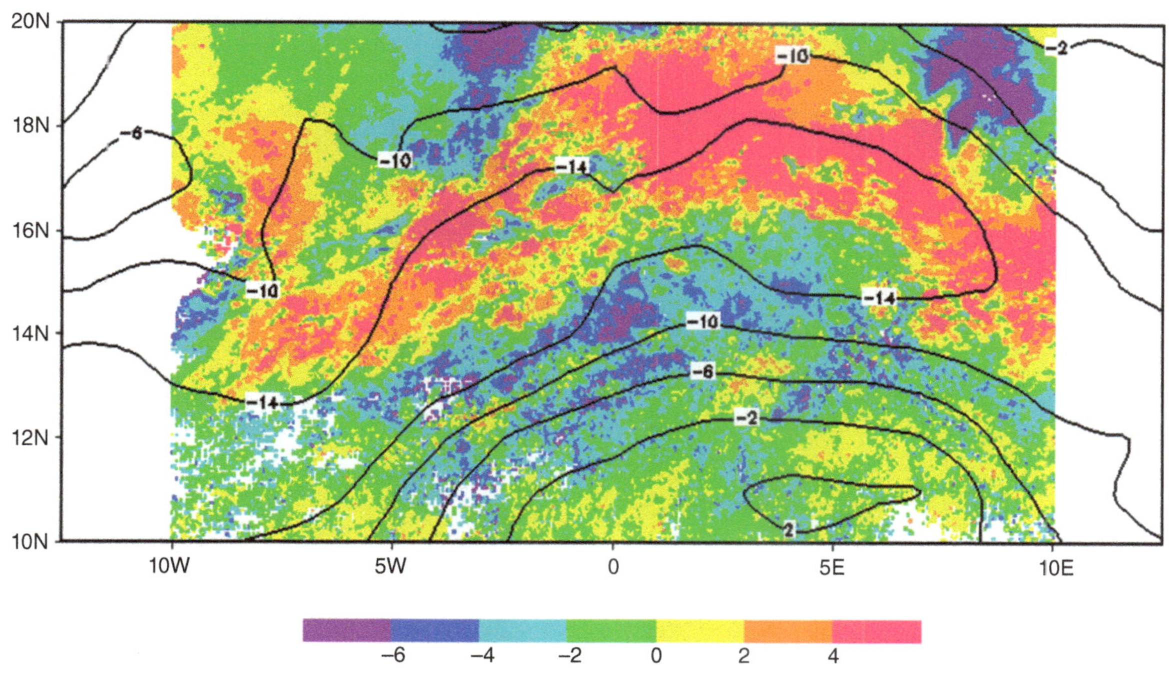


Figure 2.24 Daytime land surface temperatures (shading; K) on 26 July 2006 derived from Meteosat Second Generation (http://landsaf.meteo.pt) anomalies from the 3-week mean. The (pink) band of positive temperature anomalies denotes a dry region, whilst the relatively cool (blue) conditions to the south indicate high soil moisture from storms in the previous 36 h. The zonal wind (m s$^{-1}$) at 700 hPa from the ECMWF operational analysis for 1200 UTC on that day is overlain, indicating a significant kink in the AEJ.

extending as far north as 16°N in the central region. This rain occurred in association with a developing wave, depicted by the 700 hPa zonal wind (Bain *et al.*, 2010). In this case, the atmosphere induced a wave-like structure in soil moisture and surface fluxes. According to estimates from an ensemble of land models forced by observations as part of the AMMA Land Model Intercomparison Project (ALMIP; Boone *et al.*, 2009), the antecedent rainfall pattern resulted in very marked spatial variability of surface fluxes; 24 h mean values of sensible and latent heat fluxes were of the order of 70 W $m^{-2}$ and 20 W $m^{-2}$ respectively in the dry (pink) areas, compared with sensible and latent heat fluxes of 0 W $m^{-2}$ and 120 W $m^{-2}$ respectively in the recently wetted (blue/purple) region at 14–16°N. Such strong contrasts in surface fluxes have been shown to have a substantial impact on the lower atmosphere (Kohler *et al.*, 2010).

Figure 2.25 illustrates the geographical distribution of variability in surface fluxes on synoptic timescales. The data, again derived from the ALMIP dataset, depict a band of strong variability in sensible heat (and also latent heat, not shown) across the Sahel belt during the monsoon season. The location of this band is determined largely by day-to-day variability in near-surface soil moisture and its impact on evapotranspiration. South of the Sahel, the vegetation is dense, and latent heat flux is dominated by transpiration of soil water stored in the root zone. In the Sahel, evaporation direct from the soil becomes an important term in the energy balance, driven by rainfall typically every few days. Finally, in the north, though the land is sensitive to soil water, lack of rainfall restricts variations in surface fluxes. On top of this hydrological driver one can also add radiative forcing, where cloud cover restricts the total surface energy flux.

There have been several modelling studies that have examined in detail how soil moisture feedbacks influence West Africa at the synoptic scale. In an early study, Walker and Rowntree (1977) found a strong sensitivity of a developing AEW to soil moisture in their simulations. Cook (1999) highlighted the importance of the meridional gradient in soil moisture for the maintenance of the AEJ, but did not consider the temporal fluctuations in the soil moisture field. A subsequent GCM study (Taylor and Clark, 2001) examined the sensitivity of AEWs to the underlying land surface. They found increased synoptic variability in a simulation where a description of dense vegetation in the Sahel was replaced by a more realistic parameterisation of sparse vegetation. The latter produced increased day-to-day variability in surface fluxes, which, it was argued, enhanced synoptic variability in the atmosphere.

Taylor *et al.* (2005) examined the temporal variability in the surface energy balance using cloud-free thermal infrared imagery. They found strong synoptic-scale fluctuations in surface temperature across the Sahel driven by rainfall, with minima in regions of wet soil. The surface temperatures (and hence surface fluxes) were found to be well-organised at the synoptic scale, linked to the westward propagation of MCS and AEWs. They developed a conceptual model of the land–atmosphere interaction within an AEW, which is depicted schematically in Figure 2.26.

Considering an extensive, synoptic-scale dry zone such as depicted on the left (west) of Figure 2.26, strong horizontal gradients in temperature will develop in response

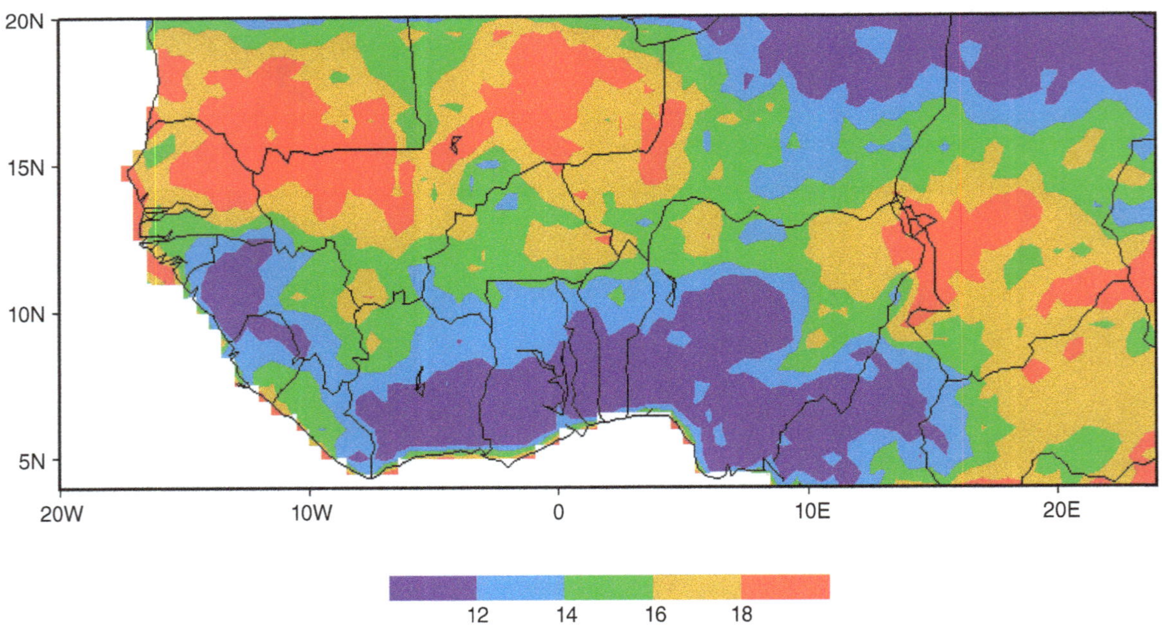


**Figure 2.25** Standard deviation of sensible heat flux *H* (W $m^{-2}$) averaged over July–September 2004–2007 from the ALMIP model ensemble (Boone *et al.*, 2009). The daily data have been filtered to remove variability with periods exceeding 10 days.

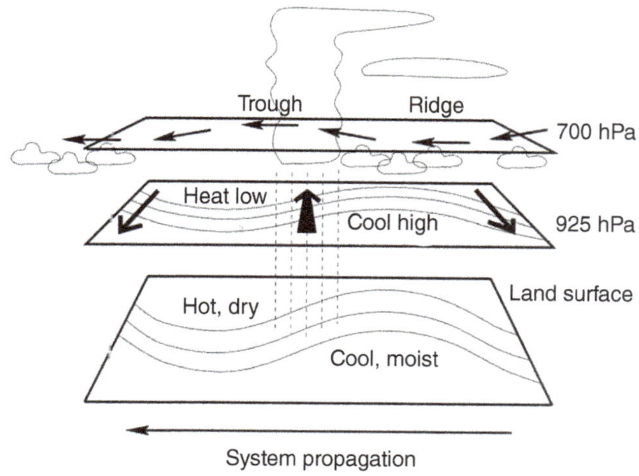


**Figure 2.26** Schematic depicting land–atmosphere coupling within an AEW. The modulation of convection by a wave at 700 hPa creates synoptic variability in soil moisture, with wet soils favoured behind the trough. Zonal contrasts in availability of moisture for evapotranspiration drive sensible heat flux patterns, which perturb the isentropes at 925 hPa. In turn, this configuration favours maximum southerly flow between the dry and wet soils, enhancing new convection there. *Source*: Figure 7 from Taylor *et al.* (2005). Used with the permission of the Royal Meteorological Society.

to the daytime surface sensible heat flux contrasts. These perturb the climatological meridional gradient of temperature. Following Rácz and Smith (1999), during the day, when horizontal low-level flow is weak, a heat low anomaly tends to develop over the dry soil. Once the sun sets and coupling with the land surface effectively switches off (Section 4.1.3.2), the atmosphere responds to the thermal gradient generated during the day. This induces the development overnight of positive vorticity above dry soil, superimposed on the existing atmospheric dynamics. In the West African context, this flow favours enhanced southerly advection of moisture to the east of the dry soil, providing a favourable environment for new a MCS there (recalling from Section 2.1.3.6, that in the northern Sahel, where rainfall is limited by PW, rain occurs in the positive PW*, to the east of the low-level trough). As these systems propagate westward, they will tend to erode the dry anomaly from the east, in effect shifting the surface features westward. Similarly, on a timescale dictated by the drying of the top soil after rain (~2–3 days), the pre-existing wet soil to the east (right of Figure 2.26) will weaken in the absence of further rain. On top of this diabatic coupling, Parker (2008) explored the impact of adiabatic Rossby wave dynamics. Surface-induced zonal temperature anomalies embedded within the strong meridional gradient propagate westward through this mechanism. Analysis of a linear model showed that the fastest growing mode occurs when the lower atmosphere has a thermal anomaly to the east of a land surface temperature anomaly.

Modelling and observational evidence have been presented to illustrate how this form of coupling influences monsoon dynamics on timescales of the order of 15 days (see Chapter 7 discussion of the Sahelian mode of intraseasonal variability). Wet sequences within the rainy season produce large-scale soil moisture features, which, via sensible heating, can feed back on the atmosphere. Taylor (2008) used passive microwave soil moisture data to estimate the modulations in surface heating across West Africa driven by soil moisture. By compositing wet and dry features, and examining the atmospheric state associated with large-scale soil moisture variability, that study was able to demonstrate that surface variability led (i.e. induced) variability in daytime PBL temperatures. These in turn amplified circulation anomalies considered important for this mode of variability (Janicot and Sultan, 2001), specifically low-level vorticity in the Sahel and southern Sahara. The data in Figure 2.27 illustrate temperature and wind anomalies at 925 hPa (derived from ECMWF analysis) associated with a wet phase of the mode. Negative anomalies in sensible heat flux drive negative PBL temperatures, which in turn produce anticyclonic vorticity.

This mechanism was subsequently explored in a GCM study (Lavender *et al.*, 2010). In simulations where soil moisture was imposed, heat low anomalies developed above dry soils and cool high anomalies developed above wet soils, and the associated modulation of moisture advection induced enhanced rainfall to the west of pre-existing wet soil regions, consistent with Taylor (2008). On the other hand, similar intraseasonal rainfall features were found in a simulation where soil moisture was held constant, though westward propagation of precipitation was less coherent. These results imply the atmosphere can support the Sahelian mode in the absence of surface hydrological feedback. However, analysis of the runs also showed how the atmospheric mode became phase locked with the imposed soil moisture, suggesting an important (though not primary) role for land feedbacks.

## 2.2 Operational Methods and Canonical Synoptic Patterns

### 2.2.1 Synoptic Fields and Main Inferences

In developing a weather forecast for a given location (or area) and time range, the forecaster needs to get an understanding of the three-dimensional structure and evolution of a given synoptic situation. This can be achieved through the use of various combinations of charts/fields that can best capture the situation.

Table 2.1 summarises the major parameters that may assist in analysis of the synoptic situation. How to use these parameters, as well as some proposed diagnostics to better detect and analyse the synoptic structures, is explained in Chapter 11.

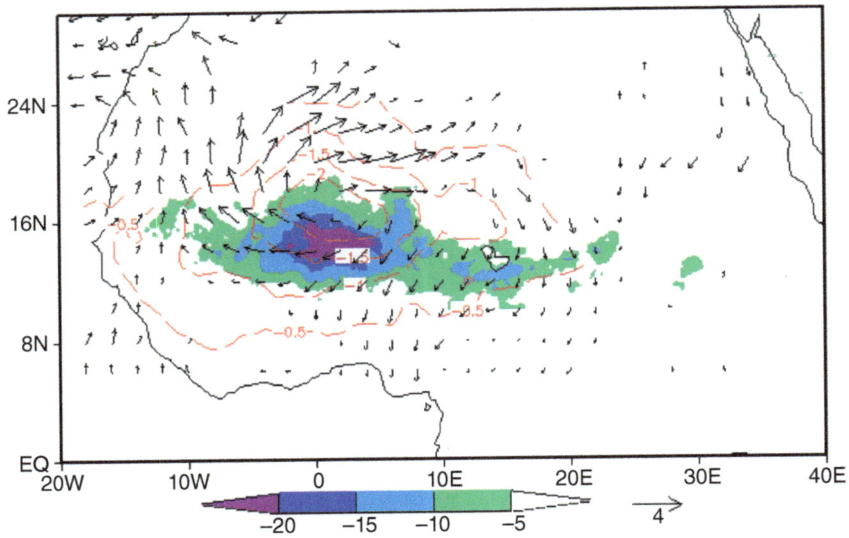


**Figure 2.27** Differences in surface sensible heat flux (shaded; W m$^{-2}$), 925 hPa temperature differences (at 1800 UTC; dashed contour; K), and wind speed (at 0600 UTC the following day; vectors; m s$^{-1}$) between wet and dry surface intraseasonal composites. The wet (dry) composite is defined at day 0 by the minimum (maximum) in filtered daytime mean surface heat flux over a rectangle centred on 0°E, 15°N. Data are only plotted where they differ significantly from zero at the 95% level.

### 2.2.2 African Easterly Wave Case Studies and Canonical Structures

Training of forecasters on real-world case studies is a key step to assimilate the basic knowledge presented in this book. In particular, it is necessary to realise the high diversity of situations encountered in reality, which can sometimes be quite different from the conceptual models presented in Section 2.1. Therefore, we have developed a set of online case studies (see online link[4]). Some particular illustrations from these case studies, as well as some additional examples, are presented here to show canonical configurations of AEW structure and behaviour as a real-world complement to the idealised AEW structures shown in Figure 2.10.

#### 2.2.2.1 An Archetypal African Easterly Wave Event: Case Study CS02[4], 12–16 August 2012

This case shows many features that may be regarded as archetypes for AEW behaviour. The wave was first detected at 0000 UTC on 13 August (see online material). At this time, the detection of this new perturbation is not easy based on the 600 hPa meridional wind, but the major visible features are: the AEJ acceleration; wet anomalies over northern Chad as detected by the PW* field, associated with a large area of convective activity; and a strong northerly mean transport in the 925-600 hPa layer occurring in the 10–25°E band from the Mediterranean Sea to the Equator. By 1200 UTC (Figure 2.28 at $D_0$ − 1.5) the perturbation is well organised with all key features in place; that is, the trough at 19°E, the AEJ core at its rear, the dry/wet dipole, the monsoon trough along 10°N, deep northerlies ahead and monsoon flux intensifying behind over south Sudan.

On 15 August at 0000 UTC (Figure 2.28 at $D_0$), the wave and the associated features, including deep convection, reach their maximum magnitude (Figure 2.29). The trough is tilted northeastward and shows the peak AEJ speeds to its east (as in Figure 2.10). The ITD shows a northward penetration corresponding to the moist phase of the wave behind (to the east of) the trough, and a southward ITD penetration in the dry phase, ahead of (west of) the trough. At 850 hPa (Figure 2.29b) there is a closed vortex at 0°E, 10°N, which is bringing strong southerly advection (Figure 2.29f) on its eastern side. Note also the well-defined SHL enhancement located at around 5°W, 23°N, ahead of the trough to the northwest, whose pressure gradient drives the monsoon burst (strong southerly winds in and ahead of the trough).

The following day the perturbation continues its westward progression toward the coast but slows down when arriving over the Fouta Djallon mountains on 16 August. At 1200 UTC (Figure 2.28 at $D_0$ + 1.5) the monsoon inflow from Guinea to the Liberian coast is deep and intense with a cyclonic centre C600_850 around Conakry. During this stage, the AEJ core for the first time precedes the trough. The monsoon burst reaches its maximum northward penetration as indicated by the ITD (in the numerical model data), reaching 28°N over southern Algeria. This moist region is cloudy, but the convection is less vigorous.

**Table 2.1** Summary of the major parameters that may assist in analysis of the synoptic situation, and which are usually readily available to forecasters. Other chapters, particularly Chapters 3, 7, 9 and 11, have proposed additional diagnostics for synoptic analysis, including CAPE, CIN, wind shear, PW, monsoon depth and mean meridional transport, plus their anomalies from the mean seasonal cycle. The difficulty can be that NWP centres often provide only the basic variables as listed here. Forecasters need convenient and interactive tools to handle all these fields and superpose them.

| Parameter | Useful levels | Application |
|---|---|---|
| MSLP | | The 1015 hPa line is a commonly used reference for the extent of southward push or retreat of the northerly high pressure influence. During the dry season, the inland ridge of the Azores High usually manifests itself as a closed 'Libyan High'. A tightening of the pressure gradient on the southern flank of the Libyan High is usually a prelude to an outbreak of a widespread dust episode (see Chapter 5). This is also a crucial field to monitor the SHL activity (see Section 11.3) characterised by a large diurnal oscillation.<br>Although the winds are influenced by the Coriolis force, especially in the north of the region, the geostrophic wind is not a good estimate of the low-level winds. Instead, winds typically flow toward low pressure, turned to the right by the Coriolis force. During the day, low-level wind speeds are much reduced by convective turbulent drag. |
| Wind charts | Surface/10 m, 925 hPa, 950 hPa, 850 hPa, 700 hPa, 500 hPa, 300 hPa, 200 hPa, 150 hPa | The low-level charts are of interest in determining the location of the ITD and hence the northward penetration of the monsoon flow (10 m to 925 hPa) and the location of vortices (10 m to 850 hPa). The middle levels (700 hPa) are examined for the strength and location of the AEJ, and for waves (AEWs), which may also be associated with low-level vortices at 850 hPa and the surface. At 300 hPa and above, extratropical troughs and the STJ are of interest.<br>The 850 hPa level, lying close to the steering level of AEWs, can show distinct closed vortices and allows drawing of the monsoon trough.<br>Streamline plots will suffice for most purposes, but commonly these do not show wind strength, which is useful for locating jet axis and for determining whether dust may be raised locally or at the main source regions. |
| Temperature | Surface, 850 hPa and 700 hPa | These charts are used in conjunction with humidity and winds to assess stability conditions and where the enhancement or suppression of convection may be favoured. They also help to identify the structure of AEWs below the AEJ level. The 850 hPa level is convenient to monitor the SHL (see Section 11.3). |
| RH | Surface, 925hPa, 850 hPa, 700 hPa | High values of low-level RH increase the CAPE of the column. Mid-level dryness (e.g. 700–500 hPa) can suppress convective initiation, but also can favour the organisation and maintenance of squall lines. |
| Dew point temperature. | Surface, 925 hPa, 850 hPa, 700 hPa, 500 hPa | The surface dew point temperature gradient (around 15 °C) is used among other criteria to locate the ITD (see Chapter 11).<br>The humidity profile over West Africa from the surface through 850 hPa, 700 hPa and up to 500 hPa is used in forecasting the intensity and growth of thunderstorms. At 500 hPa, areas with dew point temperatures less than a threshold value are marked as areas of mid-level 'dry air intrusion'. Such dry air intrusion zones, when located very close to very humid areas, can be associated with very violent thunderstorm activity. |
| Vorticity | 850 hPa, 700 hPa, 600 hPa, 500 hPa, 200 hPa | From the surface to 700 hPa (even to 500 hPa), closed vortices can be associated with significant weather, in particular thunderstorm initiation.<br>Vorticity is used at 850 hPa to define the monsoon trough, which also connects closed vortices.<br>At upper levels, vorticity is an indicator of mid-latitude troughs. |

On 16 August at 1200 UTC (Figure 2.28 + 1.5 at $D_0$ + 1.5), while the AEW exits Africa over the Atlantic, a new one is growing at the Sudan–Chad frontier. As for the previous perturbation, the precursor detected a day before is a deep northerly flow organised from the Mediterranean Sea to 5°N, followed to the east by northward monsoon penetration. Again, we notice a reinforcement of the AEJ that can be detected before the appearance of the trough.

#### 2.2.2.2 African Easterly Wave Breaking Events

AEW 'breaking' is here defined to occur when the curvature of the wave, seen in the 600 hPa streamlines for instance, becomes very steep, oriented north–south (or even in the northwest to southeast orientation), to the west of the wave trough. At the same time, the breaking wave becomes quite slow moving. Such events are important as they are favourable to high-impact weather systems with the possibility of flooding, and

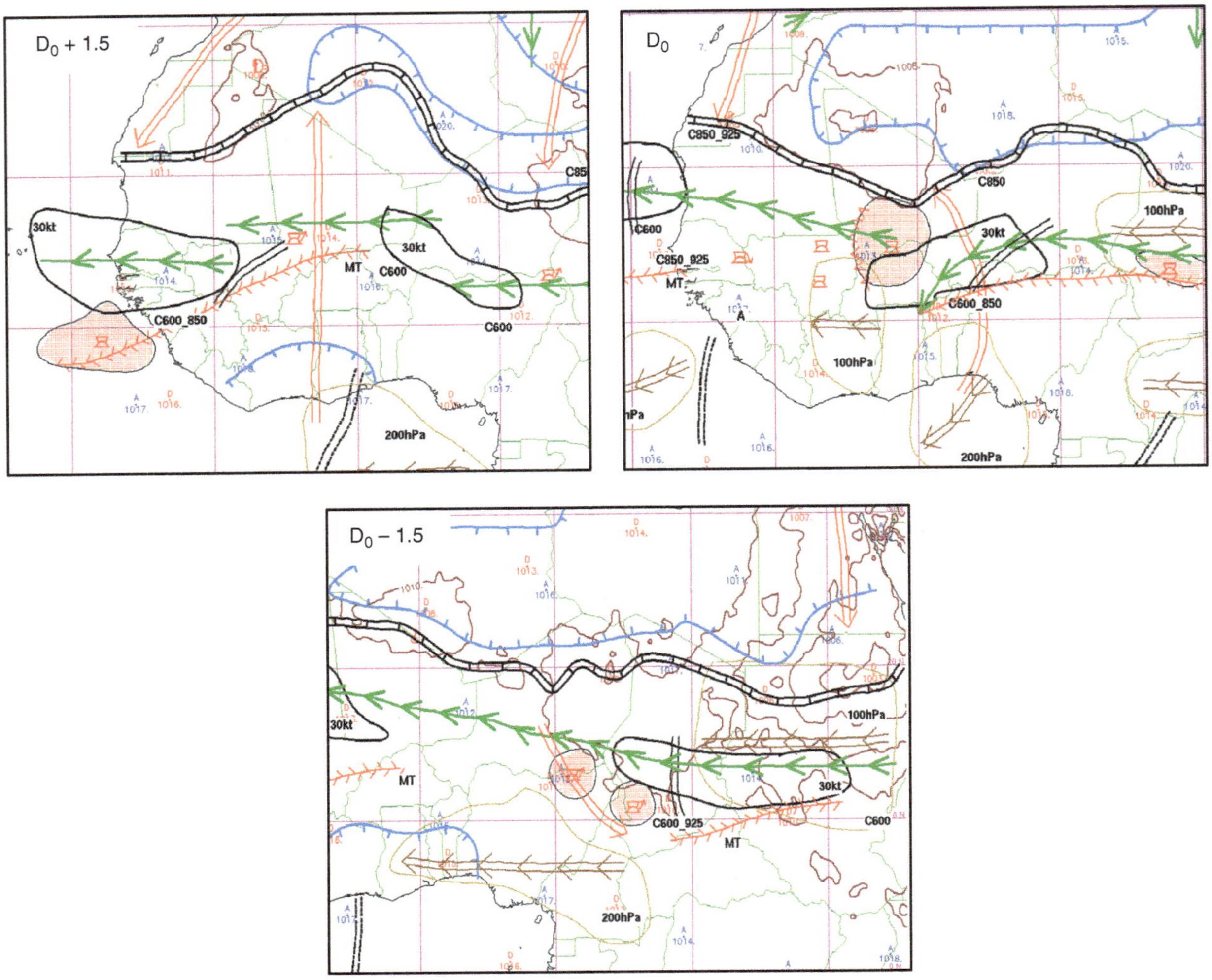


Figure 2.28 Sequence of West African Synthetic Analysis (WASA) charts showing the structure of the AEW of 12–16 August 2012. Day $D_0$ is 0000 UTC on 15 August 2012 corresponding to the mature and growing AEW stage. One and half day before ($D_0$ – 1.5) and later ($D_0$ + 1.5) correspond to the initiation stage and to its arrival at the coast. These charts may be compared with the conceptual model developed in Section 2.1.3.3 and Figure 2.10.

favour cyclogenesis when arriving at the coast. Owing to their strong vorticity signature, the predictability of breaking AEWs is relatively strong, but such events are not frequent.

Favourable factors for AEW breaking appear to be:

1) *A pre-existing significant monsoon penetration over the eastern Sahel, resulting in a larger wet spell* (positive PW*) over the Sahel, that propagates westward.
2) *A train of AEWs.* Indeed, it seems that a trough T2 can be reinforced by the previous trough T1 (with T1 to the west of T2) if the monsoon burst associated with this first trough T1 is intense and deep. A strong ridge develops between T1 and T2, with drier conditions, which slows down the propagation of the second trough T2, allowing T2 to intensify, with the development of deep convection. This is consistent with the upstream propagation of the group velocity discussed in Diaz and Aiyver (2013).
3) *An intense AEJ* core associated with a growing trough.
4) *Intense deep convection* to the south of the AEJ core.
5) *Interaction with a deep and intense monsoon flow from the equatorial band.* This monsoon burst can be detected by the mean wind intensity in the surface to 600 hPa layer, reaching more than 7.5 m $s^{-1}$ from the southwest sector.
6) *Forcing from mid-latitudes* can help the wave breaking, although the forcing from the equatorial band (the deep monsoon burst) appears as a more important ingredient for this scenario.

Two examples of AEW breaking are shown in Figures 2.30 and 2.31, both exhibiting these six favourable factors. Figure 2.30 illustrates the AEW 'breaking case' that occurred on 7 August 2012, well defined by the north–south orientation of the blue streamlines at 600 hPa to the west of the trough (around 0°E, 10°N). The AEJ core is stronger than 30 kt at the rear of an intense MCS around 2°W, 13°N, moving toward the southwest. Note the development of an intense ridge between the previous trough over the Atlantic Ocean and the breaking wave. This case illustrates factors (2) (train of AEWs) and (5) (forcing by the equatorial band), from the list of favourable factors above. For a detailed analysis of this case refer to case study CS01[4].

The second case corresponds to the Ouagadougou flood event that occurred on 1 September 2009 with a maximum of precipitation of 263 mm accumulated in less than 12 h. Figure 2.31 illustrates the formation of the breaking, in terms of the curvature of streamlines at 600 hPa (around 0°E, 10°N on 1 September). A first strong trough $T_1$ passes over Benin and west Niger on 29 August. As this moves to the west there is strong southerly flow from the equatorial band behind it, bringing dry air (see weak $T_d$ (green and blue) areas around 5°W on 31 August) into the region of the ridge at 600 hPa. In the lee of the first trough $T_1$, a second trough $T_2$ is detected on 29 August in the Lake Chad region. Owing to the previous ridge intensification, $T_2$ slows its westward propagation and intensities to form a closed circulation over northeast Nigeria on 31 August. Then $T_2$ moves faster and intensifies until on 1 September it is overturning, with the winds northwesterly on its western side. Owing to the pre-existing large wet spell (positive PW* – favourable factor (1)), intense precipitation occurs over Burkina Faso a few degrees to the west of the vortex in a wet environment where northerlies are maximum. It is worth noticing that the rainfall in this case is not a squall line system, as no cold pool was generated due to a lack of rain evaporation in this wet environment.

### 2.2.2.3 Coastal Development of African Easterly Waves

There are two AEW tracks over West Africa: one to the south and the other to the north of the AEJ (Section 2.1.3.2). The interactions between these waves at the coast and eventually their merging can have important impacts on the weather and give rise to tropical cyclogenesis (Chen *et al.*, 2008; Hankes *et al.*, 2015; Section 2.1.3.7). When arriving at the coast, often the AEWs amplify and change their structure with two possible configurations, corresponding to a reinforcement of either (configuration A) the southern vortex, or (configuration B) the northern vortex at 850 hPa. In general, the AEWs reach their maximum amplitude at the coast for all variables associated with their signature, such as oscillation of the streamfunction, meridional wind perturbation, vorticity and PW*. Also, the eastward vertical tilt of the trough below the AEJ level (600–700 hPa) is reduced near the coast.

#### 2.2.2.3.1 *Summary of Configuration A: Reinforcement of the Southern Vortex*

Favourable conditions for this configuration are:

1) A strong AEJ core (~45 kt, or more than 20 m $s^{-1}$) coupled with a strong southern vortex.
2) Very humid air in the column associated with a very strong and deep monsoon surge and strong onshore flow south to west of the vortex.

In addition, the interaction and merging with a northern vortex can contribute to reinforce the southern vortex. In the case of strong northerlies from mid-latitudes along the west coast, the northern vortex is blocked; it is forced to move southward, so as to meet the southern vortex and merge with it.

When passing over the Fouta Djalon the southern vortex is reinforced and deepened by the enhancement of convection. The effect is maximum if the southern vortex arrives over Fouta Djalou around 1800 UTC. The vortex then moves northwest toward Dakar and leaves the coast south of Dakar.

This type of situation is more typical around the end of July, into August. It is favourable to intense precipitation in the vicinity of the trough and to the north of the vortex. The slower propagation of the vortex than previously, over the continent, can explain large accumulations of precipitation. If other factors are favourable, cyclogenesis can occur in such situations. The 1–10 September 2014 period provides two nice examples of configuration A, driven by the 'southern vortex'. This period is characterised by the passage of an 'AEW train' triggered east of 20°E in mid-August and moving westward over the whole domain at about 9 m $s^{-1}$, with the AEWs amplifying over West Africa to reach their maximum over the ocean both in terms of PW* and vorticity anomalies. Figure 2.32 summarises the evolution of the synoptic situation during the 2–4 September period. On 2 September two trough axes $T_{N1}$ and $T_{S1}$ are well identified, north and south respectively, of the AEJ core located over Burkina. In agreement with the schematic (Figure 2.10), the trough axis $T_{N1}$ is well defined between the dipole of PW* over Mauritania, whereas the moist anomaly is in phase and ahead of $T_{S1}$, associated with a strong vortex over northern Ghana. Also, two ridge axes

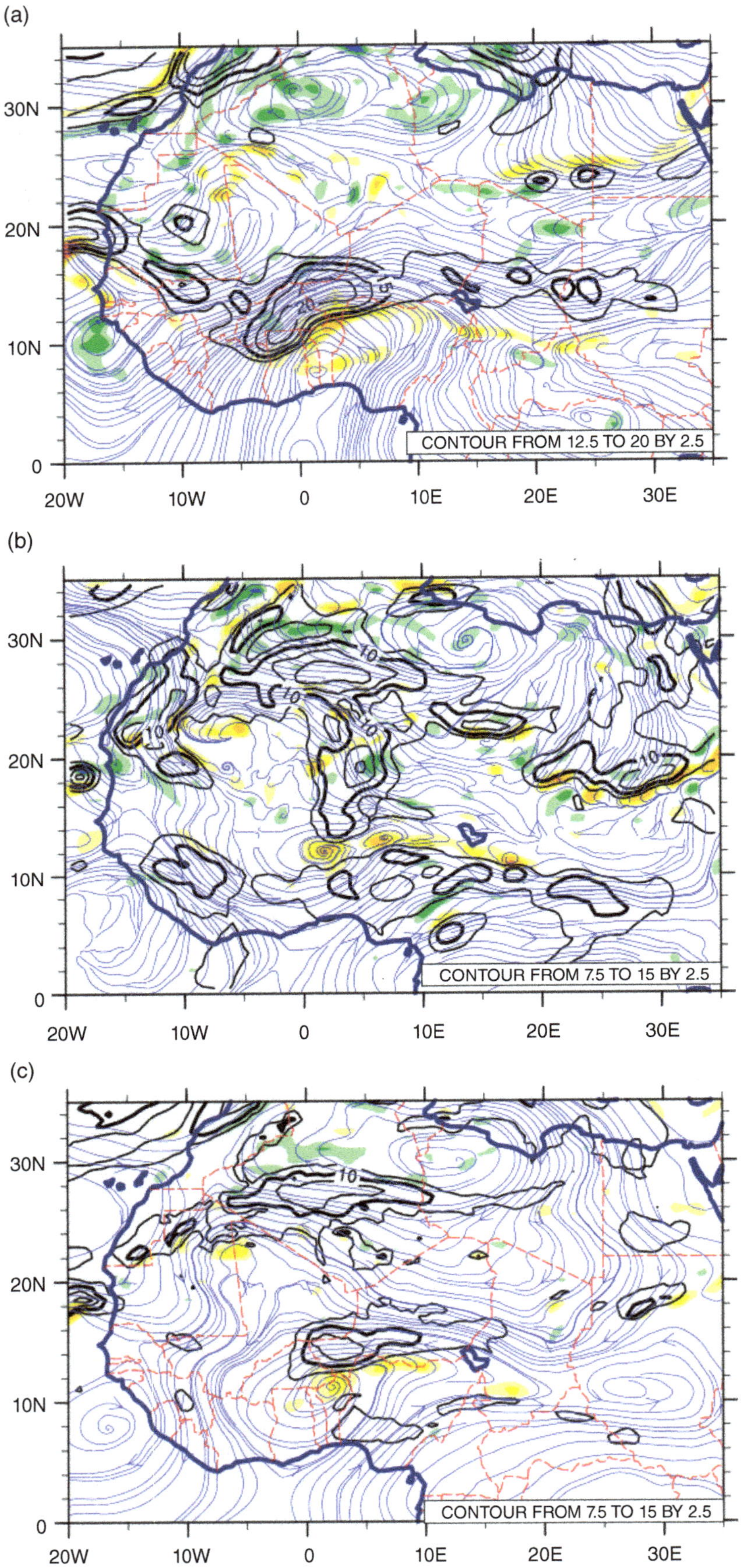


**Figure 2.29** Example of an archetypal AEW from 15 August 2012, 0000 UTC. Streamlines (blue) are shown at (a) 600 hPa, (b) 850 hPa and (c) for the mean flow in the 950–600 hPa layer, with mean horizontal wind intensity (isolines, m $s^{-1}$) and vorticity (colour, $10^{-5}$ $s^{-1}$); (d) shows the infrared Meteosat image with yellow and red colours for coldest clouds (less than −40 °C and −65 °C respectively); (e) shows PW* (colour, mm) and 925 hPa horizontal wind anomalies (m $s^{-1}$); (f) shows mean meridional wind in the 925–600 hPa layer (colour, m $s^{-1}$) and mean horizontal wind vector in the 925–850 hPa layer. The WASA map for this case is in Figure 2.28 (time $D_0$).

Figure 2.29 (Cont'd)

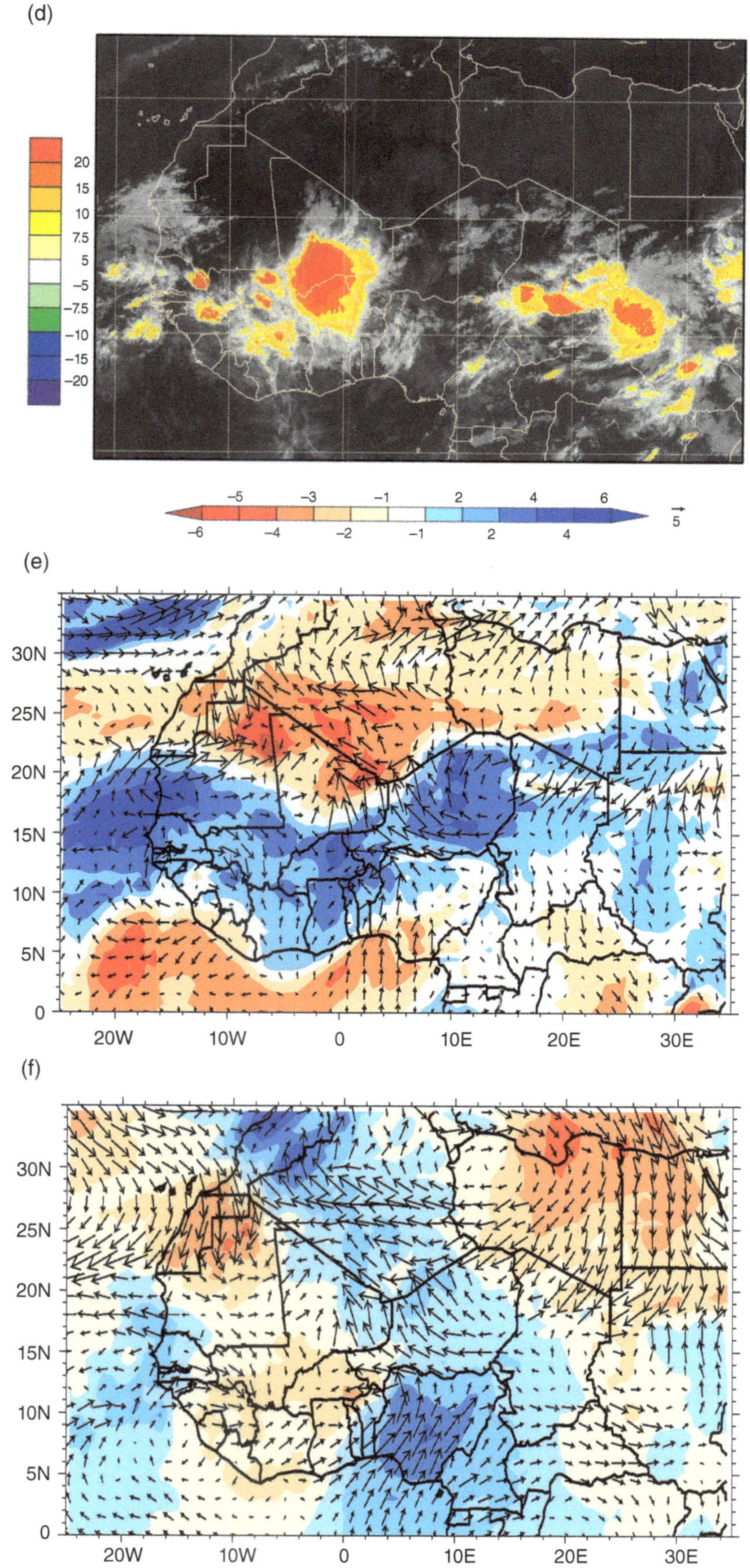

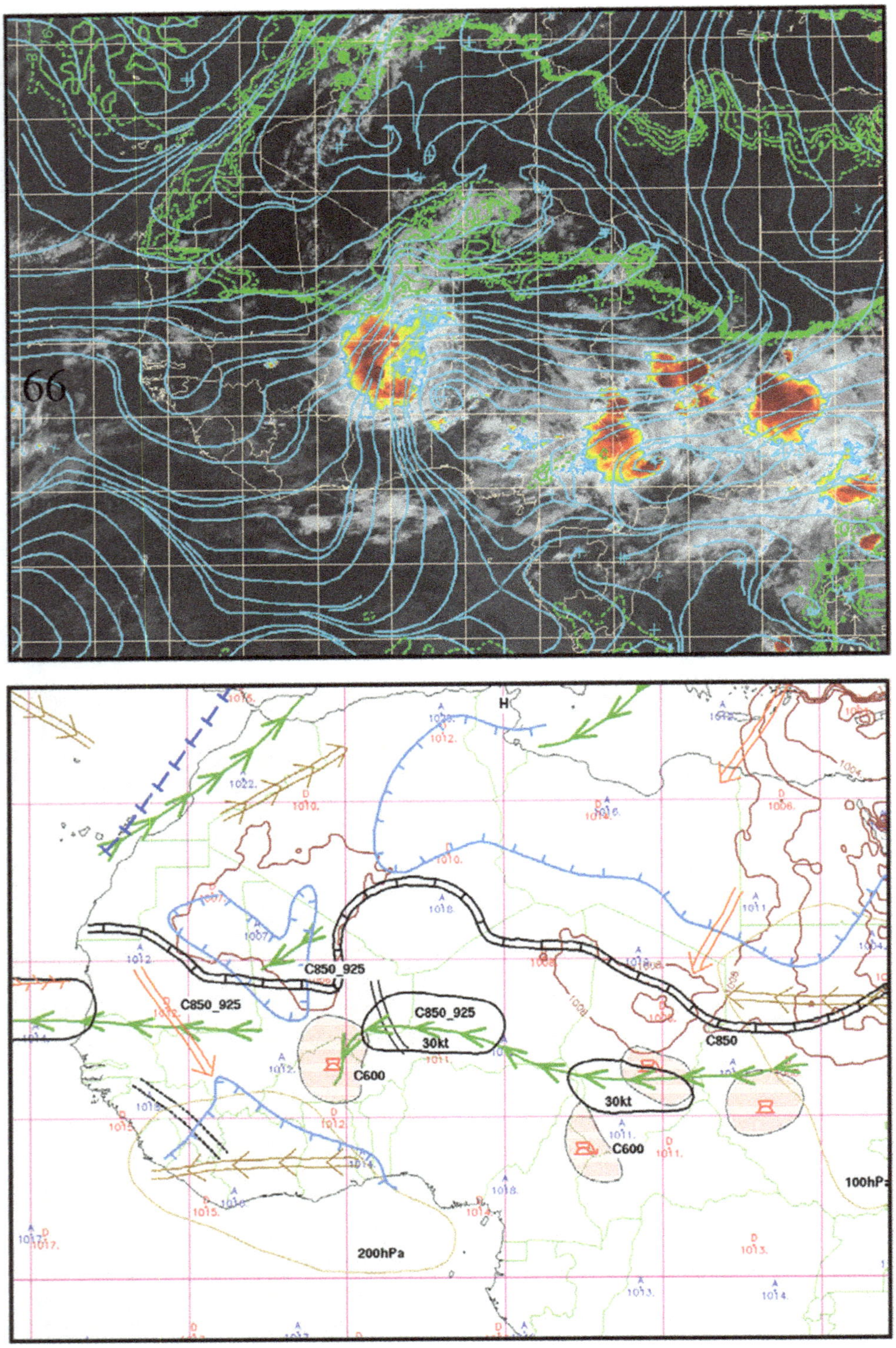


Figure 2.30 Example of an AEW 'breaking case' on 7 August 2012 at 0000 UTC (case study CS01[4]). Upper panel shows the infrared image with superposition of streamlines at 600 hPa and $T_d$ at the surface (green isolines to visualise the ITD); lower panel shows the WASA. Note that the breaking of the AEW is identified by the north–south orientation of the streamlines to the west of the trough.

are identified (dashed black lines), one over the ocean and the other crossing Niger and Nigeria, indicating the strength of the AEW train south of the AEJ. One day later (Figure 2.32, second row), the previous structure has moved westward and amplified. However, owing to strong northerlies (trade winds) over the eastern Atlantic, the propagation of the northern vortex associated with $T_{N1}$ is blocked over Mauritania, whereas the $T_{S1}$ axis has moved toward the coast and caught up with $T_{N1}$ to form a continuous trough axis. In addition to this merging process, the occurrence of intense convection over the Guinea Highlands during the previous night and day has

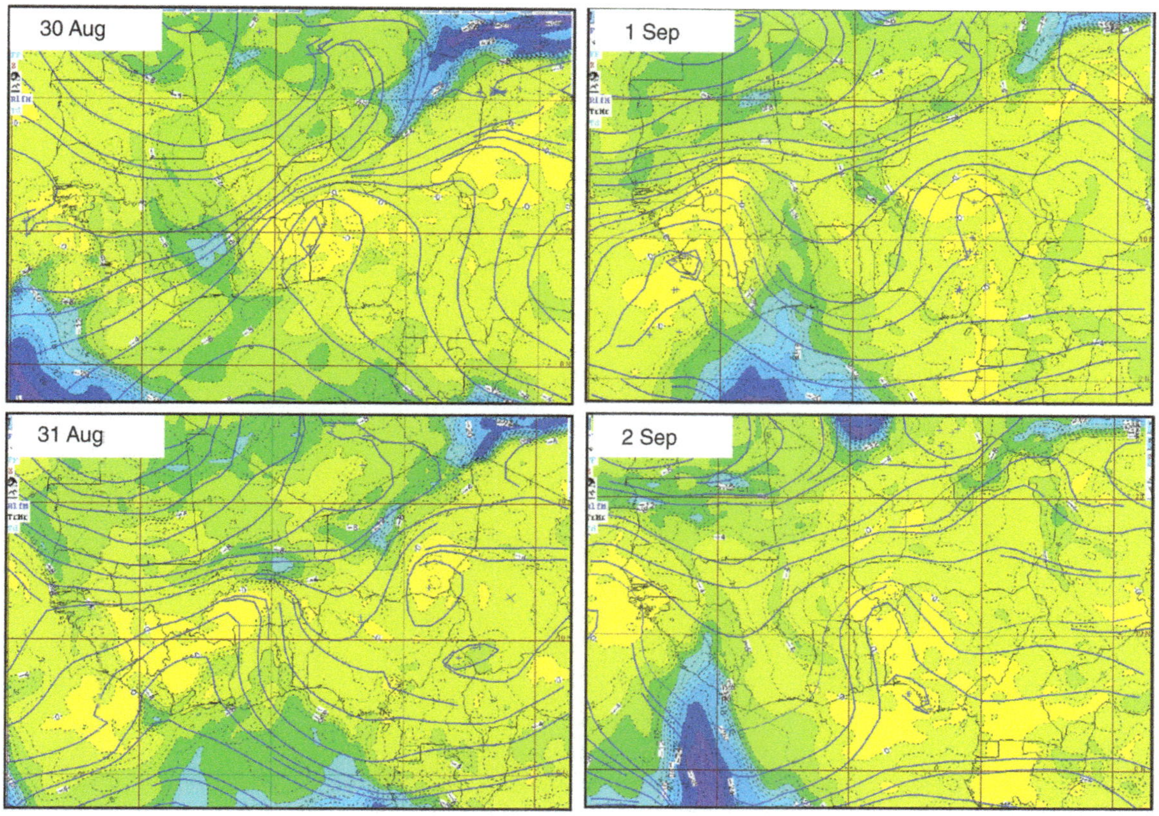


Figure 2.31 Illustration of the Ouagadougou flood case (case study CS04[4]) showing streamlines and dew point, $T_d$ (colour) at 600 hPa on 30 August–2 September 2009 at 0000 UTC.

contributed also to the intensification of the southern vortex. The following day the merging and convection processes continue to result in an intense, moist vortex reaching the west coast, first moving northward along the coast and then leaving the continent westward (Figure 2.32, last row). Behind this, we have the merging and amplification of the north and south ridges, as observed for the troughs, helping the development of a second trough to the east over Nigeria. The following evolution scenario was very similar to that shown in Figure 2.32, but the resulting southern vortex was stronger and moister due to its earlier development and a trajectory a little more to the south. This second AEW reached a stronger magnitude, allowing the triggering of tropical cyclone Edouard (11–19 September 2014).

#### 2.2.2.3.2 *Summary of Configuration B: Reinforcement of the Northern Vortex*

This situation is very common, representing perhaps 50% of cases in July. Favourable conditions are:

1) A northern vortex at 850 hPa is localised in the vicinity of the ITD and connected with the heat low reinforcement.
2) The northern vortex amplification could result from an interaction with mid-latitudes, in particular strong northerlies along the west coast blocking the propagation of the northern vortex. As with configuration A, northerlies along the west coast seem to support the development, but in this case they have a better connection and continuity with northerlies at mid-latitudes.
3) A weaker southern vortex can favour reinforcement of the northern vortex.

This type of situation is favourable for significant dust lifting over Mauritania, and export of the dust over the Atlantic Ocean. Convection can occur close to the ITD that penetrates far to the north, but over the ocean precipitation is shifted to the south along the ITCZ. This situation is not favourable to tropical cyclogenesis.

### 2.2.3 Dry-season Thunderstorms in West Africa

The dry season in West Africa is dominated by dry north-easterly winds that transport Harmattan dust haze across West Africa. However, thunderstorms do sometimes occur during the dry season. When this happens, it usually sparks off interest amongst the public, with impacts ranging from agriculture to ground water conservation.

A common configuration for such dry-season storms is the interaction of a low-pressure system or trough over West Africa, with a low over the North African region. The favourable interaction of these systems can lead to

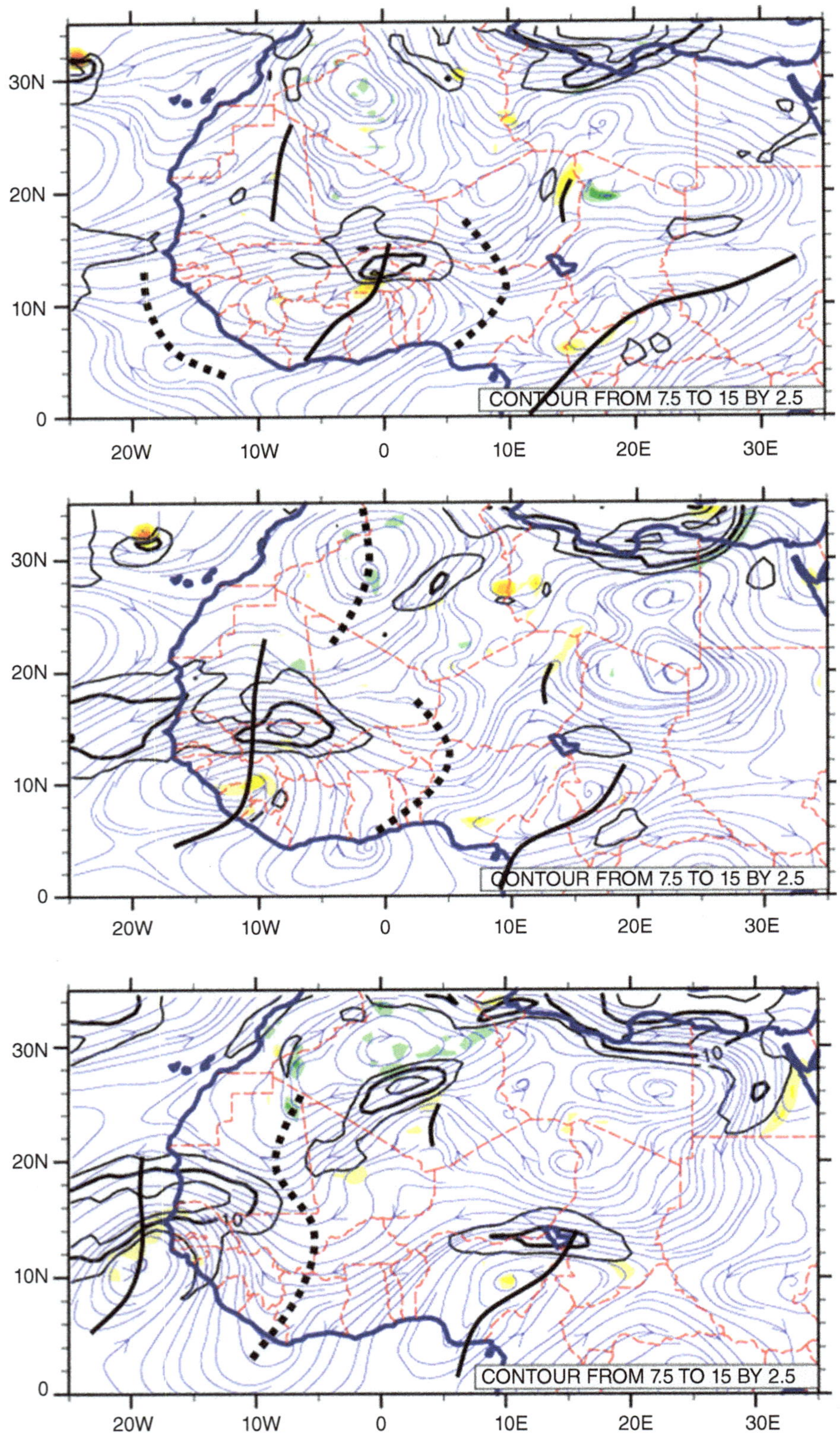


**Figure 2.32** Evolution of the synoptic situation on 2, 3 and 4 September 2014 (first, second and third rows respectively) as depicted by the ECMWF analysis at 1200 UTC. Left column: streamlines, horizontal wind intensity (2.5 m $s^{-1}$ isoline contours above 7.5 m $s^{-1}$) and vorticity (colour, $10^{-5}$ $s^{-1}$) for the mean flow averaged in the 950–600 hPa layer. Right column: PW* (colour, mm) and 925 hPa horizontal wind anomalies. Trough and ridge axes as detected by the above diagnostics are superposed by black solid and dashed lines respectively.

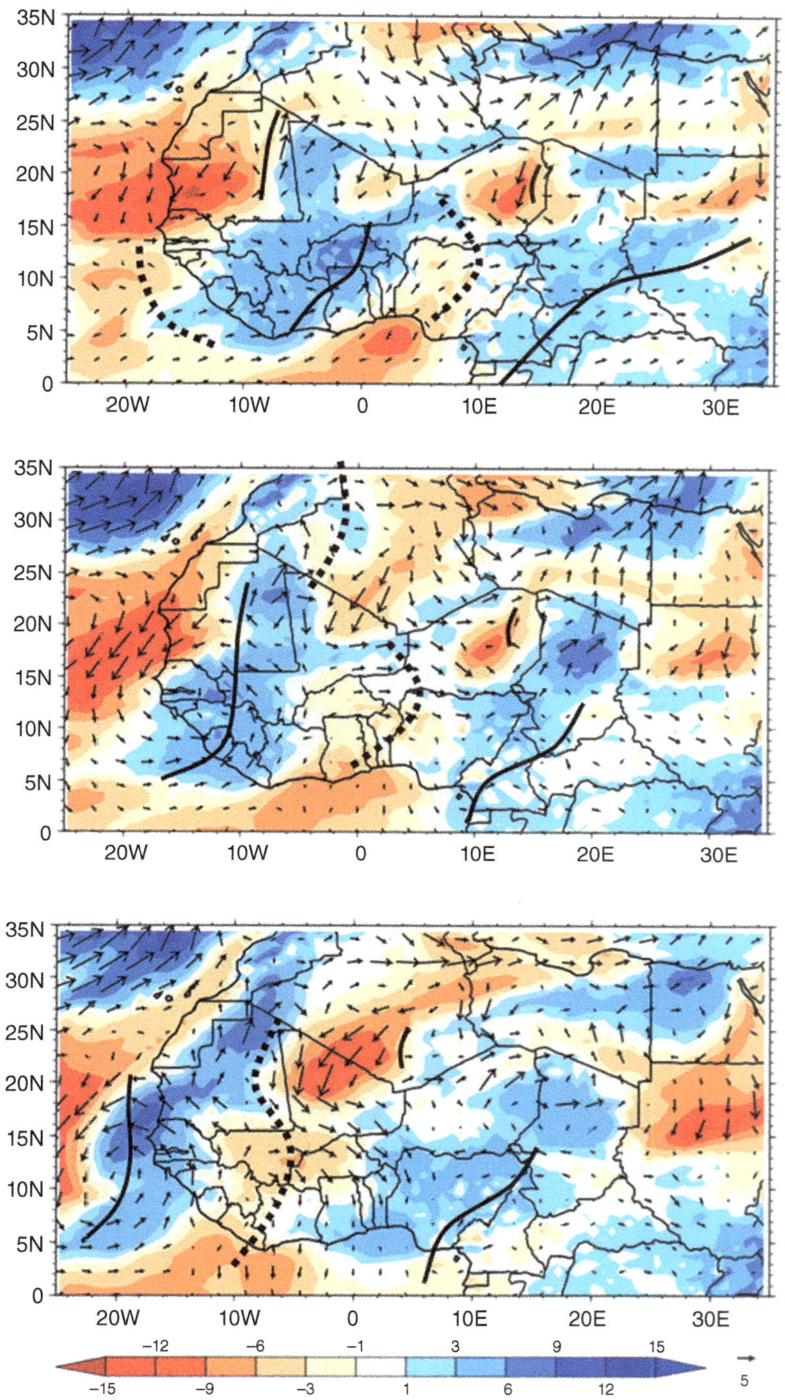


Figure 2.32 (Cont'd)

strong inland moisture transport from the west and Guinea coasts, supporting rainfall. Figure 2.33 illustrates an example, in the case of the thunderstorms that affected many coastal areas of West Africa during the period of 13–15 February 2010.

A mid-latitude trough that had been advancing eastward across the North Atlantic was penetrating deep into West Africa by the 12 February. There was significant interaction between the tropical and extratropical systems (Figure 2.33a).

This situation created a favourable environment for penetration of moisture-laden southwesterlies far inland, and the MSLP and humidity (not shown) charts show evidence of a poleward shift of the ITD: on the 12 February, the ITD was at about 9°N, while by 1200 UTC on the 13 February it had shifted to about

(a)

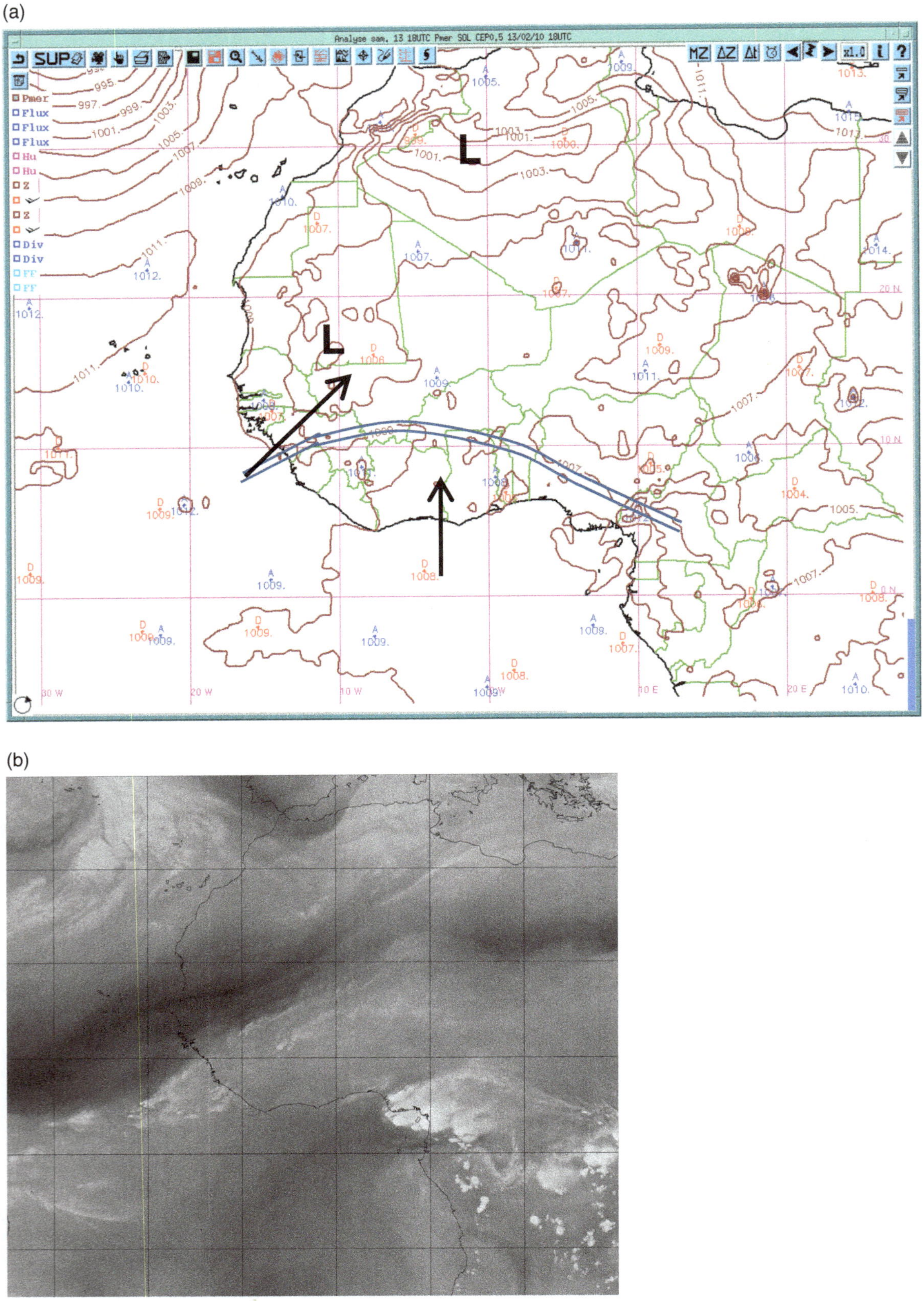


(b)

**Figure 2.33** (a) Synoptic situation at 1200 UTC on 13 February 2010, indicating ITD, low-level winds (schematic vectors) and trough connecting low-pressure systems over Algeria and southern Mauritania, and Meteosat SEVIRI water vapour imagery (b) at 1200 UTC 13 February and (c) at 0000 UTC 15 February 2010. Panel (a) original to authors. *Source*: panels (b) and (c) courtesy of NEODAAS/University of Dundee.

(c)

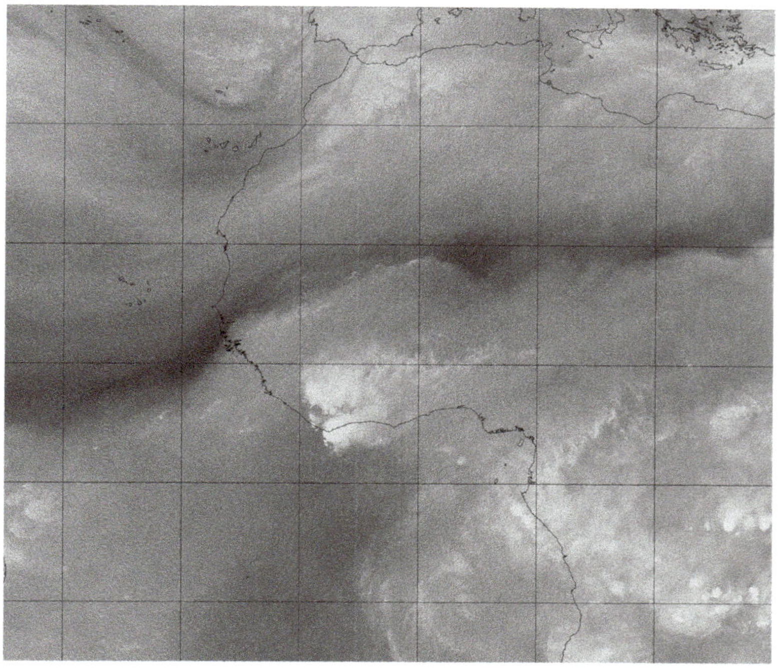

Figure 2.33 (Cont'd)

10.5°N (Figure 2.33a). A moist environment is an important component that is conducive for thunderstorm development. The trough is also dynamically favourable for thunderstorm initialisation since it generates vertical motion.

By 0200 UTC on 13 February, convective cloud clusters were observed over southeastern coastal areas of Nigeria as well as coastal areas of southwest Cameroon. The clusters became organised, with a distinct leading edge, and by 0900 UTC the system had covered much of southeastern parts of Nigeria. Westward propagation of the system ensued, and by around 1800 UTC (Figure 2.33b), the system had metamorphosed into a massive mesoscale convective complex. Around this time, the system was contiguous with deep convection all along the coast, south from Nigeria and Cameroon, and as far as Angola (Figure 2.33b). It is possible that some dynamical effects of this line of deep convection along the coast (presumably supported by local sea breezes) helped to intensify the system over Nigeria.

The storm continued to propagate westward and affected southern parts of Ghana, Togo and Benin. By 0000 UTC on 15 February (Figure 2.33c), the massive MCS had covered most parts of Ivory Coast. Finally, by 1100 UTC on 15 February the system showed consistent decay over Liberia. The organisation and westward propagation of the MCS suggests that the easterlies observed at 850 hPa over the coastal areas of West Africa could have played a significant role in their kinematics. Indeed, it is likely that the northward push of cool oceanic air, moving with the ITD, is associated with easterly thermal-wind shear (as it is in the rainy season), which can help to organise storms.

Interaction between tropical and extratropical systems is fundamental to thunderstorm development during the dry season. Forecasters in the countries of the Guinea Coast look for these patterns, where a West African trough couples with a North African low, as a sign of likely thunderstorm activity in the coming days.

## Acknowledgements

We are grateful to the ECMWF for providing the ERA Interim data used in this study and to Meteosat for providing the Meteosat data used in the satellite images shown in Sections 2.1 and 2.2. We are also grateful to the COMET MetEd programme for Figure 2.15 (from Laing and Evans (2011)).

## Appendix: Acronyms

| | |
|---|---|
| AEJ | African easterly jet |
| AEW | African easterly wave |
| ALMIP | AMMA Land Model Intercomparison Project |

| | |
|---|---|
| AMMA | African Monsoon Multidisciplinary Analysis |
| CAPE | convective available potential energy |
| CIN | convection/convective inhibition |
| ECMWF | European Centre for Medium-Range Weather Forecasts |
| GCM | general circulation model |
| GFS | Global Forecast System |
| ITCZ | intertropical convergence zone |
| ITD | intertropical discontinuity |
| ITF | intertropical front |
| MCS | mesoscale convective system |
| MCV | mesoscale convective vortex |
| MJO | Madden–Julian oscillation |
| MSLP | mean sea-level pressure |
| NCEP | National Centers for Environmental Prediction |
| NLLJ | night-time low-level jet |
| NWP | numerical weather prediction |
| OLR | outgoing longwave radiation |
| PBL | planetary boundary layer |
| PV | potential vorticity |
| PW | precipitable water |
| PW* | precipitable water anomaly |
| SAL | Saharan air layer |
| SHL | Saharan heat low |
| SST | sea surface temperature |
| STJ | subtropical jet |
| TEJ | tropical easterly jet |
| UTC | Universal Time Coordinated |
| WAM | West African monsoon |

## Notes

1 Figure 2.8 is a temporal composite, so, as the AEW is westward propagating, times before the trough passage correspond to locations west of the trough.

2 In fact, theoretical analysis has shown that both 'downshear' and 'upshear' tilts can be associated with 'baroclinic' wave growth: upshear – when the AEJ is narrow compared with the scale of the AEW; downshear – when the AEJ is broad compared with the scale of the AEW (Methven *et al.*, 2005).

3 Note that the classical definition of AEW structure is in terms of winds, and in some cases AEWs may be dry.

4 http://www.cnrm.meteo.fr/waf_handbook_casestudies/.

5 By the same token, it should be recognised that perhaps 40% of squall lines are *not* associated with AEWs, and the presence of a large area of organised convection does not automatically indicate the presence of an AEW.

6 Forecasters have also used this term for some time.

## References

Acheampong PK. 1982. Rainfall anomaly along the coast of Ghana – its nature and causes. *Geogr. Ann. A* **64**: 199–211.

Agudelo P, Hoyos CD, Curry J, Webster PJ. 2011. Probabilistic discrimination between large-scale environments of intensifying and decaying African easterly waves. *Clim. Dyn.* **36**: 1379–1401.

Avila LA, Pasch RJ. 1992. Atlantic tropical systems of 1991. *Mon. Weather Rev.* **120**: 2688–2696.

Bain CL, Parker DJ, Taylor CM, *et al.* 2010. Observations of the nocturnal boundary layer associated with the West African monsoon. *Mon. Weather Rev.* **138**(8): 3142–3156.

Bain CL, Parker DJ, Dixon N, *et al.* 2011. Anatomy of an African easterly wave in July 2006. *Q. J. R. Meteorol. Soc.* **137**: 923–933.

Berry GJ, Thorncroft CD. 2005. Case study of an intense African easterly wave. *Mon. Weather Rev.* **133**: 752–766.

Berry GJ, Thorncroft CD. 2012. African easterly wave dynamics in a mesoscale numerical model: the upscale role of convection. *J. Atmos. Sci.* **69**: 1267–1283.

Berry GJ, Thorncroft CD, Hewson T. 2007. African easterly waves during 2004 – analysis using objective techniques. *Mon. Weather Rev.* **135**: 1251–1267.

Blackadar AK. 1957. Boundary layer wind maxima and their significance for the growth of nocturnal inversions. *Bull. Am. Meteorol. Soc.* **38**: 283–290.

Boone A, de Rosnay P, Balsamo G, *et al.*. 2009. The AMMA Land Surface Model Intercomparison Project (ALMIP). *Bull. Am. Meteorol. Soc.* **90**: 1865–1880. doi: 10.1175/2009bams2786.1.

Borgne JL. 1979. Un exemple d'invasion polaire sur la région mauritano--sénégalaise (An example of a polar intrusion into the Mauritanian–Sengalese region). *Ann. Géograph.* **489**: 521–548.

Bou Karam D, Flamant C, Knippertz P, *et al.* 2008. Dust emissions over the Sahel associated with the West African monsoon intertropical discontinuity region: a representative case-study. *Q. J. R. Meteorol. Soc.* **134**(632): 621–634. doi: 10.1002/qj.244.

Bouniol, D, Couvreux F, Kamsu-Tamo P-H, *et al.* 2012. Diurnal and seasonal cycles of cloud occurrences, types, and radiative impact over West Africa. *J. Appl. Meteorol. Climatol.* **51**(3): 534–553. doi: 10.1175/JAMC-D-11-051.1.

Brown RG, Zhang C. 1997. Variability of midtropospheric moisture and its effect on cloud-top height distribution during TOGA COARE. *J. Atmos. Sci.* **54**: 2760–2774.

Buckle C. 1996. *Weather and Climate in Africa*. Longman; 312 pp.

Burpee RW. 1972. The origin and structure of easterly waves in the lower troposphere of North Africa. *J. Atmos. Sci.* **29**(1): 77–90.

Burpee RW. 1974. Characteristics of North African easterly waves during the summers of 1968 and 1969. *J. Atmos. Sci.* **31**: 1556–1570.

Carlson TN. 1969a. Some remarks on African disturbances and their progress over the tropical Atlantic. *Mon. Weather Rev.* **97**: 716–726.

Carlson TN. 1969b. Synoptic histories of three African disturbances that developed into Atlantic hurricanes. *Mon. Weather Rev.* **97**: 256–276.

Cecelski SF, Zhang D-L. 2013. Genesis of Hurricane Julia (2010) within an African easterly wave: low-level vortices and upper-level warming. *J. Atmos. Sci.* **70**(12): 3799–3817.

Charney JG, Stern ME. 1962. On the stability of internal baroclinic jets in a rotating atmosphere. *J. Atmos. Sci.* **19**: 159–172.

Chauvin F, Roehrig R, Lafore J-P. 2010. Intraseasonal variability of the Saharan heat low and its link with midlatitudes. *J. Climate* **23**: 2544–2561.

Chen TS, Wang SY, Clark AJ. 2008. North Atlantic hurricanes contributed by African easterly waves north and south of the African easterly jet. *J. Clim.* **21**: 6767–6776.

Chong M, Hauser D. 1989. A tropical squall line observed during the COPT 81 experiment in West Africa. Part II: water budget. *Mon. Weather Rev.* **117**: 728–744.

Chou C, Neelin JD, Su H. 2001. Ocean–atmosphere–land feedbacks in an idealized monsoon. *Q. J. R. Meteorol. Soc.* **127**: 1869–1891.

Cifelli R, Lang T, Rutledge SA, Guy N, *et al.* 2010. Characteristics of an African easterly wave observed during NAMMA. *J. Atmos. Sci.* **67**: 3–25.

Cook KH. 1999. Generation of the African easterly jet and its role in determining West African precipitation. *J. Climate* **12**: 1165–1184.

Cornforth RJ, Hoskins BJ, Thorncroft CD. 2009. The impact of moist processes on the African easterly jet–African easterly wave system. *Q. J. R. Meteorol. Soc.* **135**: 894–913.

Davis J, Knippertz P, Fink AH. 2012. The predictability of precipitation episodes during the West African dry season. *Q. J. R. Meteor. Soc.* **139**: 1047–1058. doi: 10.1002/qj.2014.

Deme A, Roca R. 2008. Subjective analysis of African easterly waves during the 2006 monsoon: comparison to other methods. AMMA-EU Deliverable 2.1.C.g. Available from the AMMA Bureau.

Diallo I, Bain CL, Gaye AT, *et al.* 2014. Simulation of the West African monsoon onset using the HadGEM3-RA regional climate model. *Clim. Dyn.* **43**, 575–594. doi: 10.1007/s00382-014-2219-0.

Diaz ML, Aiyyer A. 2013. Energy dispersion in African easterly waves. *J. Atmos. Sci.* **70**: 130–145. doi: 10.1175/JAS-D-12-019.1.

Dickinson M, Molinari J. 2000. Climatology of sign reversals of the meridional potential vorticity gradient over Africa and Australia. *Mon. Weather Rev.* **128**: 3890–3900.

Diedhiou A, Janicot S, Viltard A, *et al.* 1999. Easterly wave regimes and associated convection over West Africa and tropical Atlantic: results From the NCEP/NCAR and ECMWF reanalyses. *Clim. Dyn.* **15**(11): 795–822. doi: 10.1007/s003820050316.

Dunion JP, Velden CS. 2004. The impact of the Saharan air layer on the Atlantic tropical cyclone activity. *Bull. Amer. Meteorol. Soc.* **85**: 353–365.

Duvel J-P. 1990. Convection over tropical Africa and the Atlantic Ocean during northern summer. Part II: modulation by easterly waves. *Mon. Weather Rev.* **118**(9): 1855–1868.

Eldridge RH. 1957. A synoptic study of West African disturbance lines. *Q. J. R. Meteorol. Soc.* **83**: 303–314.

Fall S, Niyogi D, Mohanty UC, Kumar A. 2007. Application of weather prediction models for hazard mitigation planning: a case study of heavy off-season rains in Sénégal. *Nat. Hazards* **41**: 227–243. doi: 10.1007/s11069-006-9033-x.

Fink AH, Reiner A. 2003. Spatio-temporal variability of the relation between African easterly waves and West African squall lines in 1998 and 1999. *J. Geophys. Res.* **108**(D11): 4332. doi: 10.1029/2002JD002816.

Fink AH, Vincent DG, Ermert V. 2006. Rainfall types in the West African Soudanian zone during the summer monsoon 2002. *Mon. Weather Rev.* **134**: 2143–2164.

Fink AH, Paeth H, Ermert V, *et al.* 2010. Meteorological processes influencing the weather and climate of Benin. In: *Impacts of Global Change on the Hydrological Cycle in West and Northwest Africa*, P Speth, M Christoph, B Diekkrüger, *et al.* (eds). Springer; pp. 135–149.

Fink AH, Agustí-Panareda A, Parker DJ, *et al.* 2011. Operational meteorology in West Africa: observational networks, weather analysis and forecasting. *Atmos. Sci. Lett.* **12**: 135–141. doi: 10.1002/asl.324.

Flamant C, Chaboureau J-P, Parker DJ, *et al.* 2007. Airborne observations of the impact of a convective system on the planetary boundary layer thermodynamics and aerosol distribution in the inter-tropical discontinuity region of the West African monsoon. *Q. J. R. Meteorol. Soc.* **133**: 1175–1189. doi: 10.1002/qj.97.

Flamant C, Knippertz P, Parker DJ, *et al.* 2009. The impact of a mesoscale convective system cold pool on the northward propagation of the intertropical discontinuity over West Africa. *Q. J. R. Meteorol. Soc.* **135**(638): 139–159. doi: 10.1002/qj.357.

Fröhlich L, Knippertz P. 2008. Identification and global climatology of upper-level troughs at low latitudes.

*Meteorol. Z.* **17**(5): 565–573. doi: 10.1127/0941-2948/2008/0320.

Fröhlich L, Knippertz P, Fink AH, Hohberger E. 2013. An objective climatology of tropical plumes. *J. Climate* **26**: 5044–5060. doi: 10.1175/JCLI-D-12-00351.1.

Garcia-Carreras L, Marsham JH, Parker DJ, *et al.* 2013 The impact of convective cold pool outflows on model biases in the Sahara. *Geophys. Res. Lett.* **40**: 1647–1652. doi: 10.1002/grl.50239.

Gaye AT, Fongang S, Garba A, Badiane D. 1994. Study of Heug rainfall in Sénégal using conventional data and Meteosat imagery. *Veille Climatique Satellitaire* **49**: 61–71 (available from Ministre de la Cooperation–ORSTOM–Meteo-France, Lannion, France).

Grams CM, Jones SC, Marsham JH, *et al.* 2010. The Atlantic inflow to the Saharan Heat low: observations and modelling. *Q. J. R. Meteor. Soc.* **136**: 125–140. doi: 10.1002/qj.429.

Hall NMJ, Kiladis GN, Thorncroft CD. 2006. Three-dimensional structure and dynamics of African Easterly Waves. Part II: dynamical modes. *J. Atmos. Sci.* **63**: 2231–2245.

Hankes I, Wang Z, Zhang G, Fritz C. 2015. Merger of African easterly waves and formation of Cape Verde storms. *Q. J. R. Meteorol. Soc.* **141**(689): 1306–1319. doi: 10.1002/qj.2439

Hopsch SB, Thorncroft CD, Hodges KI, Aiyyer A. 2007. West African storm tracks and their relationship to Atlantic tropical cyclones. *J. Climate* **20**: 2468–2483.

Hopsch SB, Thorncroft CD, Tyle KR. 2010. Analysis of African easterly wave structures and their role in influencing tropical cyclogenesis. *Mon. Weather Rev.* **138**: 1399–1419.

Hoskins BJ, Ambrizzi T. 1993. Rossby wave propagation on a realistic longitudinally varying flow. *J. Atmos. Sci.* **50**: 1661–1671. doi: 10.1175/1520-0469(1993)050<1661:RWPOAR>2.0.CO;2.

Hsieh JS, Cook KH. 2005. Generation of African easterly wave disturbances: relationship to the African easterly jet. *Mon. Weather Rev.* **133**: 1311–1327. doi: 10.1175/MWR2916.1.

Iskenderian H. 1995. A 10-year climatology of Northern Hemisphere tropical cloud plumes and their composite flow patterns. *J. Climate* **8**: 1630–1637. doi: 10.1175/1520-0442(1995)008<1630:AYCONH>2.0.CO;2.

Janicot S, Sultan B. 2001. Intra-seasonal modulation of convection in the West African monsoon. *Geophys. Res. Lett.* **28**: 523–526.

Janicot S, Thorncroft CD, Ali A, *et al.* 2008. Large-scale overview of the summer monsoon over West Africa during the AMMA field experiment in 2006. *Ann. Geophys.* **26**: 2569–2595.

Janiga A, Thorncroft CD. 2013. Regional differences in the kinematic and thermodynamic structure of African easterly waves. *Q. J. R. Meteorol. Soc.* **139**: 1598–1614. doi: 10.1002/qj.2047.

Jenkins G, Kucera P, Joseph E, *et al.* 2010. Coastal observations of weather features in Sénégal during the African Monsoon Multidisciplinary Analysis Special Observing Period 3. *J. Geophys. Res.* **115**: D18108. doi: 10.1029/2009JD013022.

Jenkins GS, Pratt AS, Heymsfield A. 2008. Possible linkage between Saharan dust and tropical cyclone rain band invigoration in the eastern Atlantic during NAMMA-06. *Geophys. Res. Lett.* **35**: L08815. doi: 10.1029/2008GL034 072.

Joyce RJ, Janowiak JE, Arkin PA, Xie P. 2004. CMORPH: a method that produces global precipitation estimates from passive microwave and infrared data at high spatial and temporal resolution. *J. Hydrometeorol.* **5**: 487–503.

Kamara SI. 1986. The origins and types of rainfall in West Africa. *Weather* **41**: 48–56.

Karyampudi VM, Carlson TN. 1988. Analysis and numerical simulations of the Saharan air layer and its effect on easterly wave disturbances. *J. Atmos. Sci.* **45**: 3102–3136.

Kiladis GN, Weickmann KM. 1997. Horizontal structure and seasonality of large-scale circulations associated with submonthly tropical convection. *Mon. Weather Rev.* **125**: 1997–2013. doi: 10.1175/1520-0493(1997)125<1997:HSASOL>2.0.CO;2.

Kiladis GN, Thorncroft CD, Hall NMJ. 2006. Three-dimensional structure and dynamics of African easterly waves. Part I: observations. *J. Atmos. Sci.* **63**: 2212–2230.

Kohler M, Kalthoff N, Kottmeier C. 2010. The impact of soil moisture modifications on CBL characteristics in West Africa: a case study from the AMMA campaign. *Q. J. R. Meteorol. Soc.* **136**: 442–455. doi: 10.1002/qj.430.

Knippertz P. 2005. Tropical–extratropical interactions associated with an Atlantic tropical plume and subtropical jet streak. *Mon. Weather Rev.* **133**: 2759–2776. doi: 10.1175/MWR2999.1.

Knippertz P. 2007. Tropical–extratropical interactions related to upper-level troughs at low latitudes. *Dyn. Atmos. Oceans* **43**: 36–62. doi: 10.1016/j.dynatmoce.2006.06.003.

Knippertz P, Fink AH. 2008. Dry-season precipitation in tropical West Africa and its relation to forcing from the extratropics. *Mon. Weather Rev.* **136**: 3579–3596. doi: 10.1175/2008MWR2295.1.

Knippertz P, Fink AH. 2009. Prediction of dry-season precipitation in tropical West Africa and its relation to forcing from the extratropics. *Weather Forecast.* **24**: 1064–1084. doi: 10.1175/2009WAF2222221.1.

Knippertz P, Martin JE. 2005. Tropical plumes and extreme precipitation in subtropical and tropical West Africa. *Q. J. R. Meteorol. Soc.* **131**: 2337–2365. doi: 10.1256/qj.04.148.

Knippertz P, Fink AH, Reiner A, Speth P. 2003. Three late summer/early autumn cases of tropical–extratropical interactions causing precipitation in northwest Africa. *Mon. Weather Rev.* **131**: 116–135.

Knippertz P, Fink AH, Schuster R, *et al.* 2011. Ultra-low clouds over the southern West African monsoon region. *Geophys. Res. Lett.* **38**: L21808. doi: 10.1029/2011GL049278.

Kwon HJ, Mak M. 1990. A study of the structural transformation of the African Easterly Waves. *J. Atmos. Sci.* **47**: 277–292.

Lafore J-P, Moncrieff MW. 1989. A numerical investigation of the organization and interaction of the convective and stratiform regions of tropical squall lines. *J. Atmos. Sci.* **46**: 521–544. DOI:10.1175/1520-0469(1989)046<0521:ANIOTO>2.0.CO;2.

Lafore J-P, Flamant C, Guichard F, *et al.* 2011. Progress in understanding of weather systems in West Africa. *Atmos. Sci. Lett.* **12**: 7–12. doi: 10.1002/asl.335.

Laing A, Evans J-L. 2011. *Introduction to Tropical Meteorology*. UCAR/COMET. http://www.meted.ucar.edu/tropical/textbook_2nd_edition/(accessed 16 October 2015).

Laing AG, Carbone R, Levizzani V, Tuttle J. 2008. The propagation and diurnal cycles of deep convection in northern tropical Africa. *Q. J. R. Meteorol. Soc.* **134**: 93–109. doi: 10.1002/qj.194.

Lavaysse C, Flamant C, Janicot S, *et al.* 2009. Seasonal evolution of the West African heat low: a climatological perspective. *Clim. Dyn.* **33**: 313–330.

Lavaysse C, Flamant C, Janicot S, Knippertz P. 2010. Links between African easterly waves, midlatitude circulation and intraseasonal pulsations of the West African heat low. *Q. J. R. Meteorol. Soc.* **136**(S1): 144–158. doi: 10.1002/qj.555

Lavender SL, Taylor CM, Matthews AJ. 2010. Coupled land–atmosphere intraseasonal variability of the West African monsoon in a GCM. *J. Climate* **23**: 5557–5571.

Lebel T, Parker D, Flamant C, *et al.* 2010. The AMMA field campaigns: multiscale and multidisciplinary observations in the West African region. *Q. J. R. Meteorol. Soc.* **136**: 8–33. doi: 10.1002/qj.486

Leppert KD II, Cecil DJ, Petersen WA. 2013a. Relation between tropical easterly waves, convection, and tropical cyclogenesis: a Lagrangian perspective. *Mon. Weather Rev.* **141**: 2649–2668.

Leppert KD II, Petersen WA, Cecil DJ. 2013b. Electrically active convection in tropical easterly waves and implications for tropical cyclogenesis in the Atlantic and East Pacific. *Mon. Weather Rev.* **141**: 542–556.

Leroux S, Hall NMJ. 2009. On the relationship between African easterly waves and the African easterly jet. *J. Atmos. Sci.* **66**: 2303–2316.

Leroux S, Hall NMJ, Kiladis GN. 2010. A climatological study of transient–mean-flow interactions over West Africa. *Q. J. R. Meteorol. Soc.* **136**: 397–410. doi: 10.1002/qj.474.

Leroux S, Hall NMJ, Kiladis GN. 2011. Intermittent African easterly wave activity in a dry atmospheric model: influence of the extratropics. *J. Climate* **24**: 5378–5396. doi: 10.1175/jcli-d-11-00049.1.

Lin Y-L, Robertson KE, Hill CM. 2005. Origin and propagation of a disturbance associated with an African easterly wave as a precursor of Hurricane Alberto (2000). *Mon. Weather Rev.* **133**: 3276–3298.

Lin Y-L, Liu L, Tang G, *et al.* 2013. Origin of the pre-Tropical Storm Debby (2006) African easterly wave–mesoscale convective system. *Meteorol. Atmos. Phys.* **120**: 123–144. doi: 10.1007/s00703-013-0248-6.

Lothon M, Saïd F, Lohou F, Campistron B. 2008. Observation of the diurnal cycle in the low troposphere of West Africa. *Mon. Weather Rev.* **136**: 3477–3500.

Mapes BE, Zuidema P. 1996. Radiative–dynamical consequences of dry tongues in the tropical troposphere. *J. Atmos. Sci.* **53**: 620–638.

Marsham JH, Parker DJ, Grams CM, *et al.* 2008. Uplift of Saharan dust south of the intertropical discontinuity. *J. Geophys. Res. Atmos.* **113**(D21): D21102. doi: 10.1029/2008JD009844.

Marsham JH, Dixon N, Garcia-Carreras L, *et al.* 2013. The role of moist convection in the West African monsoon system – insights from continental-scale convection-permitting simulations. *Geophys. Res. Lett.* **40**: 1843–1849. doi: 10.1002/grl.50347.

Mathon V, Laurent H, Lebel T. 2002. Mesoscale convective system rainfall in the Sahel. *J. Appl. Meteorol.* **41**: 1081–1092.

Mekonnen A, Rossow WB. 2011. The interaction between deep convection and easterly waves over tropical North Africa: a weather state perspective. *J. Climate* **24**: 4276–4294. doi: 10.1175/2011jcli3900.1.

Mekonnen A, Thorncroft CD, Aiyyer AR. 2006. Analysis of convection and its association with African Easterly Waves. *J. Climate* **19**: 5405–5421.

Meier F, Knippertz P. 2009. Dynamics and predictability of a heavy dry-season precipitation event over western Africa – sensitivity experiments with a global model. *Mon. Weather Rev.* **137**: 189–206. doi: 10.1175/2008MWR2622.1.

Methven J, Hoskins BJ, Heifetz E, Bishop CH. 2005. The counter-propagating Rossby wave perspective on baroclinic instability. Part IV: nonlinear life cycles. *Q. J. R. Meteorol. Soc.* **131**: 1425–1440.

Mounier F, Kiladis GN, Janicot S. 2007. Analysis of the dominant mode of convectively coupled Kelvin waves in the West African monsoon. *J. Climate* **20**(8): 1487–1503.

National Hurricane Center (2015). Best Track Data (HURDAT2). http://www.nhc.noaa.gov/data/#hurdat (accessed 23 July 2016).

Omotosho JB. 1985. The separate contribution of line squalls, thunderstorms and the monsoon to the total rainfall in Nigeria. *J. Climatol.* **5**: 543–552.

Parker DJ. 2008. A simple model of coupled synoptic waves in the land surface and atmosphere of the northern Sahel. *Q. J. R. Meteorol. Soc.* **134**: 2173–2184. doi: 10.1002/qj.343.

Parker DJ, Thorncroft CD, Burton RR, Diongue-Niang A. 2005a. Analysis of the African easterly jet, using aircraft observations from the JET2000 experiment. *Q. J. R. Meteorol. Soc.* **131**: 1461–1482. doi: 10.1256/qj.03.189.

Parker DJ, Burton RR, Diongue-Niang A, *et al.* 2005b. The diurnal cycle of the West African monsoon circulation. *Q. J. R. Meteorol. Soc.* **131**: 2839–2860.

Parsons DB, Yoneyama K, Redelsperger JL. 2000. The evolution of the tropical western Pacific atmosphere–ocean system following the arrival of a dry intrusion. *Q. J. R. Meteorol. Soc.* **126**: 517–548.

Payne SW, McGarry MM. 1977. The relationship of satellite inferred convective activity to easterly waves over West Africa and the adjacent ocean during phase III of GATE. *Mon. Weather Rev.* **105**: 413–420.

Pierrehumbert RT. 1998. Lateral mixing as a source of subtropical water vapor. *Geophys. Res. Lett.* **25**: 151–154.

Pierrehumbert RT, Roca R. 1998. Evidence for control of Atlantic subtropical humidity by large scale advection. *Geophys. Res. Lett.* **25**: 4537–4540.

Piersig W. 1944[1936]. Variations of pressure and winds, a treatise on the weather of trade wind region of eastern North Atlantic. *Bull. Am. Meteorol. Soc.* **25**: 2–16. (English translation of parts 2 and 3 of 'Schwankungen von Luftdruck und Luftbewegung, sowie ein Beitrag zum Wettergeschehen im Passatgebiet des östlichen Nordatlantischen Ozeans', Archiv D, Deutschen Seewarte, Bd 54, No. 6, 1936.)

Poan DE. 2013. Documentation and physical interpretation of the African monsoon intra-seasonal variability for improved weather forecasts. PhD thesis, Université Paul Sabatier, Toulouse, France.

Poan DE, Roehrig R, Couvreux F, Lafore J-P. 2013. West African monsoon intraseasonal variability: a precipitable water perspective. *J. Atmos. Sci.* **70**: 1035–1052. doi: 10.1175/JAS-D-12-087.1.

Poan DE, Lafore J-P, Roehrig R, Couvreux F. 2015. Internal processes within African easterly wave system. *Q. J. R. Meteorol. Soc.* **141**: 1121–1136. doi: 10.1002/qj.2420.

Pytharoulis I, Thorncroft CD. 1999. The low-level structure of African easterly waves in 1995. *Mon. Weather Rev.* **127**: 2266–2280.

Rácz Z, Smith RK. 1999. The dynamics of heat lows. *Q. J. R. Meteorol. Soc.* **125**: 225–252.

Redelsperger JL, Parsons DB, Guichard F. 2002. Recovery processes and factors limiting cloud-top height following the arrival of a dry intrusion observed during TOGA COARE. *J. Atmos. Sci.* **59**: 2438–2457.

Reed RJ, Norquist DC, Recker EE. 1977. The structure and properties of African wave disturbances as observed during phase III of GATE. *Mon. Weather Rev.* **105**: 17–33.

Roca R, Lafore JP, Piriou C, Redelsperger JL. 2005. Extra-tropical dry air intrusions into the West African monsoon mid-troposphere: an important factor for the convective activity over Sahel. *J. Atmos. Sci.* **62**: 390–407.

Roehrig R, Chauvin F, Lafore JP. 2011. 10–25-day intraseasonal variability of convection over the Sahel: a role of the Saharan heat low and midlatitudes. *J. Climate* **24**: 5863–5878. DOI:10.1175/2011JCLI3960.1.

Ross RS. 1991. Energetics of African wave disturbances derived from the FSU Global Spectral Model. *Meteorol. Atmos. Phys.* **45**(3): 139–158. doi: 10.1007/BF01029651.

Rowell DP, Milford JR. 1993. On the generation of African squall lines. *J. Climate* **6**: 1181–1193.

Rowell S, Hodges KI, Cornforth RJ. 2016. Hurricane Katrina, August 2005 – an African easterly wave study. *Mon. Weather Rev.* in press.

Schrage JM, Fink AH. 2007. Use of a rain gauge network to infer the influence of environmental factors on the propagation of quasi-linear convective systems in West Africa. *Weather Forecast.* **22**: 1016–1030.

Schrage JM, Fink AH. 2012. Nocturnal continental low-level stratus over tropical West Africa: observations and possible mechanisms controlling its onset. *Mon. Weather Rev.* **140**: 1794–1809. doi: 10.1175/MWR-D-11-00172.1.

Schuster R, Fink AH, Knippertz P. 2013. Formation and maintenance of nocturnal low-level stratus over the southern West African monsoon region during AMMA 2006. *J. Atmos. Sci.* **70**: 2337–2355. doi: 10.1175/JAS-D-12-0241.1.

Seck A. 1962. Le Heug ou pluies de saison sèche au Sénégal (The Heug or dry season rainfall in Sénégal). *Ann. Géograph.* **385**: 225–246.

Simpson RH, Frank N, Shideler D, Johnson HM. 1968. Atlantic tropical disturbances, 1967. *Mon. Weather Rev.* **96**: 251–259.

Smith RK, Spengler T. 2011. The dynamics of heat lows over elevated terrain. *Q. J. R. Meteorol. Soc.* **137**: 250–263. doi: 10.1002/qj.737.

Spengler T, Smith RK. 2008. The dynamics of heat lows over flat terrain. *Q. J. R. Meteorol. Soc.* **134**: 2157–2172.

Spengler T, Smith RK, Reeder MJ. 2005. The dynamics of heat low in simple background flows. *Q. J. R. Meteorol. Soc.* **131**: 3147–3165.

Stein THM, Parker DJ, Delanoe J, *et al.* 2011. The vertical cloud structure of the West African monsoon: a 4 year climatology using CloudSat and CALIPSO. *J. Geophys. Res. Atmos.* **116**(D22): D22205. doi: 10.1029/2011JD016029.

Sultan B, Janicot S, Diedhiou A. 2003. The West African monsoon dynamics. Part I: documentation of intra-seasonal variability. *J. Climate* **16**: 3389–3406. doi: 10.1175/1520-0442(2003)016<3389:TWAMDP>2.0.CO;2.

Taylor CM. 2008. Intraseasonal land–atmosphere coupling in the West African monsoon. *J. Climate* **21**: 6636–6648.

Taylor CM, Clark DB. 2001. The diurnal cycle and African easterly waves: a land surface perspective. *Q. J. R. Meteorol. Soc.* **127**: 845–867.

Taylor CM, Parker DJ, Lloyd CR, Thorncroft CD. 2005. Observations of synoptic scale land surface variability and its coupling with the atmosphere. *Q. J. R. Meteorol. Soc.* **131**: 913–938.

Thorncroft CD, Blackburn M. 1999. Maintenance of the African easterly jet. *Q. J. R. Meteorol. Soc.* **125**: 763–786. doi: 10.1002/qj.49712555502

Thorncroft CD, Flocas HA. 1997. A case study of Saharan cyclogenesis. *Mon. Weather Rev.* **125**: 1147–1165.

Thorncroft, CD and Hoskins, BJ. 1994. An idealized study of African easterly waves. *Part I: a linear view. Q. J. R. Meteorol. Soc.* **120**: 953–982.

Thorncroft CD, Parker DJ, Burton RR, *et al.* 2003. The JET2000 Project – aircraft observations of the African easterly jet and African easterly waves. *Bull. Am. Meteorol. Soc.* **84**: 337–351. doi: 10.1175/BAMS-84-3-337.

Thorncroft CD, Hall NMJ, Kiladis GN. 2008. Three-dimensional structure and dynamics of African easterly waves. Part III: genesis. *J. Atmos. Sci.* **65**: 3596–3607.

Tomas RA, Webster PJ. 1994. Horizontal and vertical structure of cross-equatorial wave propagation. *J. Atmos. Sci.* **51**: 1417–1430. doi: 10.1175/1520-0469(1994)051<1417:HAVSOC>2.0.CO;2.

Tyner B, Aiyyer A. 2012. Evolution of African easterly waves in potential vorticity fields. *Mon. Weather Rev.* **140**: 3634–3652. doi: 10.1175/MWR-D-11-00170.1.

Van der Linden R, Fink AH, Redl R. 2015. Satellite-based climatology of low-level continental clouds in southern West Africa during the summer monsoon season. *J. Geophys. Res. Atmos.* **120**: 1186–1201. doi: 10.1002/2014JD022614.

Ventrice MJ, Thorncroft CD. 2013. The role of convectively coupled atmospheric Kelvin waves on African easterly wave activity. *Mon. Weather Rev.* **141**: 1910–1924. doi: 10.1175/MWR-D-12-00147.1.

Ventrice MJ, Thorncroft CD, Janiga MA. 2012. Atlantic tropical cyclogenesis: a three-way interaction between an African easterly wave, diurnally varying convection, and a convectively coupled atmospheric Kelvin wave. *Mon. Weather Rev.* **140**: 1108–1124.

Walker J, Rowntree PR. 1977. The effect of soil moisture on circulation and rainfall in a tropical model. *Q. J. R. Meteorol. Soc.* **103**: 29–46.

Waugh DW, Polvani LM. 2000. Climatology of intrusions into the tropical upper troposphere. *Geophys. Res. Lett.* **27**: 3857–3860. doi: 10.1029/2000GL012250.

Webster PJ, Holton JR. 1982. Cross-equatorial response to middle-latitude forcing in a zonally varying basic state. *J. Atmos. Sci.* **39**: 722–733. doi: 10.1175/1520-0469(1982)039<0722:CERTML>2.0.CO;2.

Wernli H, Sprenger M. 2007. Identification and ERA-15 climatology of potential vorticity streamers and cutoffs near the extratropical tropopause. *J. Atmos. Sci.* **64**: 1569–1586. doi: 10.1175/JAS3912.1.

Wu M-LC, Reale O, Schubert SD. 2013. A characterization of African easterly waves on 2.5–6-day and 6–9-day time scales. *J. Climate* **26**: 6750–6774. doi: 10.1175/JCLI-D-12-00336.1.

Zipser EJ, Twohy CH, Tsay S-C, *et al.* 2009. The Saharan air layer and the fate of African easterly waves – NASA's AMMA field study of tropical cyclogenesis. *Bull. Am. Meteorol. Soc.* **90**: 1137–1156.

# 3

# Deep Convection

**Lead author:** *Jean Philippe Lafore*[a]

**Contributing authors:** *Nicolas Chapelon*[b], *Mariane Diop*[c], *Balla Gueye*[d], *Yann Largeron*[a], *Serge Lepape*[b], *Ousmane Ndiaye*[c], *Douglas J. Parker*[e], *Emmanuel Poan*[a], *Rémy Roca*[f], *Romain Roehrig*[a] *and Chris Taylor*[g]

**Reviewer:** *Mitch Moncrieff (NCAR, Boulder, CO, USA)*

[a] *CNRM-UMR3589, Météo-France and CNRS, 42 avenue G. Coriolis, F-31057, Toulouse, France*
[b] *DF/CISMF, Météo-France, Toulouse, France*
[c] *Agence Nationale de l'Aviation Civile et de la Météorologie du Sénégal (ANACIM), Dakar, Sénégal*
[d] *Agence pour la Sécurite de la Navigation Aérienne en Afrique et à Madagascar (ASECNA), Dakar, Sénégal*
[e] *School of Earth and Environment, University of Leeds, Leeds, UK*
[f] *OMP-LEGOS, Toulouse, France*
[g] *Centre for Ecology and Hydrology (CEH), Wallingford, UK*

## 3.1 Scientific Background

### 3.1.1 Introduction

Cumulonimbus clouds deliver the majority of the seasonal rainfall to West Africa, and therefore forecasting these clouds is one of the primary challenges for weather and climate prediction in the region. These clouds represent a form of *atmospheric convection*, which is an essential part of the global tropical circulation. The cumulonimbus clouds are also controlled locally by the surface and atmospheric conditions that drive them. Cumulonimbus convection, therefore, links small-scale climate processes with the large-scale circulation, and so forecasting the rainfall demands consideration of both large-scale and small-scale processes. Unfortunately, cumulonimbus convection is very poorly handled by numerical weather prediction (NWP) models, and their performance has been shown to be particularly poor over West Africa. It is fair to say that forecasting convective rainfall is the most difficult task for a forecaster in the region, but there are various physical and climatological guidelines that will assist performance in this task.

Convection is a fundamental and general process in geophysics resulting from an unstable vertical distribution of thermodynamic energy, and tending to lead to a recovery to stable conditions. Owing to the excess of moist energy gained from the heated surface and the loss of energy by radiative processes in the free atmosphere, unstable conditions are generated universally over the tropics. The large-scale circulations such as the Hadley or Walker circulations are insufficient to reach an equilibrium without the key process of convection. As deduced from simple budget considerations by the pioneering work of Riehl and Malkus (1958), who introduced the 'hot towers' concept, convection occurs at very small scales (1 km and less) compared with the continental-scale circulations in which it is embedded; this raises tremendous difficulties in both understanding and forecasting the processes involved.

On our blue planet, convection involves both 'dry' and 'humid' regimes in the planetary boundary layer (PBL) and in clouds respectively, resulting in conditional instabilities depending on the state of saturation reached by the air mass (Section 3.1.2.1). This nonlinear process increases the complexity of the convection, which already involves numerous coupled processes (surface, PBL, turbulence, radiation, microphysics, dynamics, etc.) operating over a very large range of temporal scales (from seconds, for the microphysics, to days for organised storms), and of spatial scales (from the micrometre for drops, the metre for turbulent eddies in clouds, up to a thousand kilometres for the largest tropical convective systems).

Convection is essential and beneficial for humanity, as it allows the formation of most of the precipitation, especially in the tropics. However, convection is also a source of vulnerability as convective storms are associated with high-impact weather events (such as floods), destructive wind bursts and showers, erosion, dust formation, and dry spells when convection is suppressed for a long time.

*Meteorology of Tropical West Africa: The Forecasters' Handbook*, First Edition. Edited by Douglas J. Parker and Mariane Diop-Kane.

The role played by convection is also key for the atmospheric general circulation and for the climate. Notably, the associated latent heat release, with powerful vertical transports of mass, heat, humidity and momentum, partly drive the large-scale atmospheric circulation, and the resulting cloud and water vapour distributions strongly impact the Earth's radiative budget and its equilibrium.

#### 3.1.1.1 Is Convection Different over Africa?

Figure 3.1 shows the distribution of precipitation over the tropics during the boreal summer (JJAS). Africa is not the region where precipitation is the most intense: the Indian and Asian monsoons are much more copious rain producers. Also, the Pacific, Atlantic and Indian Ocean intertropical convergence zones (ITCZs) are efficient in producing rain, according to this Global Prediction Climatology Project (GPCP) estimate (Adler *et al.*, 2003). Owing to the Tropical Rainfall Measuring Mission (TRMM) satellite launched in late 1997, we have learned a lot about convection distribution in the tropics during the past decade. We have found that storms over land produce similar rainfall amounts to maritime ones, but that lightning is much more prevalent over land (Figure 3.2). Storms over land are also much taller, and have a greater number of large ice particles (Liu *et al.*, 2008), especially over Central Africa. The analysis by Zipser *et al.* (2006) indicates that the most intense thunderstorms on Earth occur over the tropical continents, and foremost over Africa, with the deepest storms and maximum amount of lightning over Central Africa (Figure 3.3). During the North African monsoon these extreme storms occur over the Sahel band. Their occurrence is much less during the Indian and Australian monsoons. For mid-latitudes, the storms over the Great Plains of the USA are extreme, as are the storms over South America.

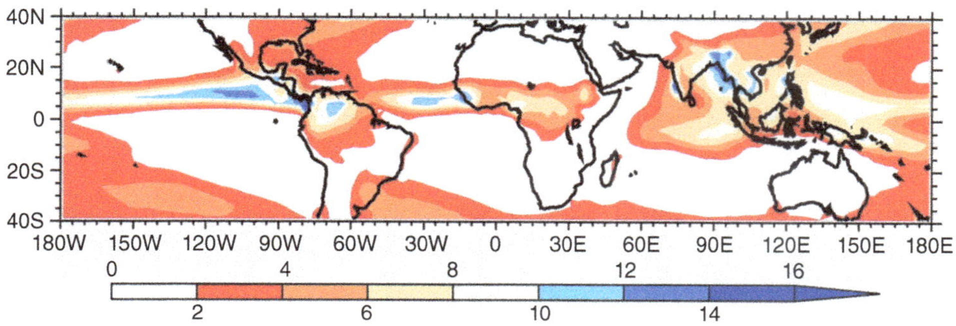


Figure 3.1 Boreal summer (JJAS) mean distribution of rainfall (mm day$^{-1}$) over the tropics for the 1979–2012 period, based on GPCP (http://www.esrl.noaa.gov/psd) version 2.2 monthly precipitation dataset (Adler *et al.*, 2003).

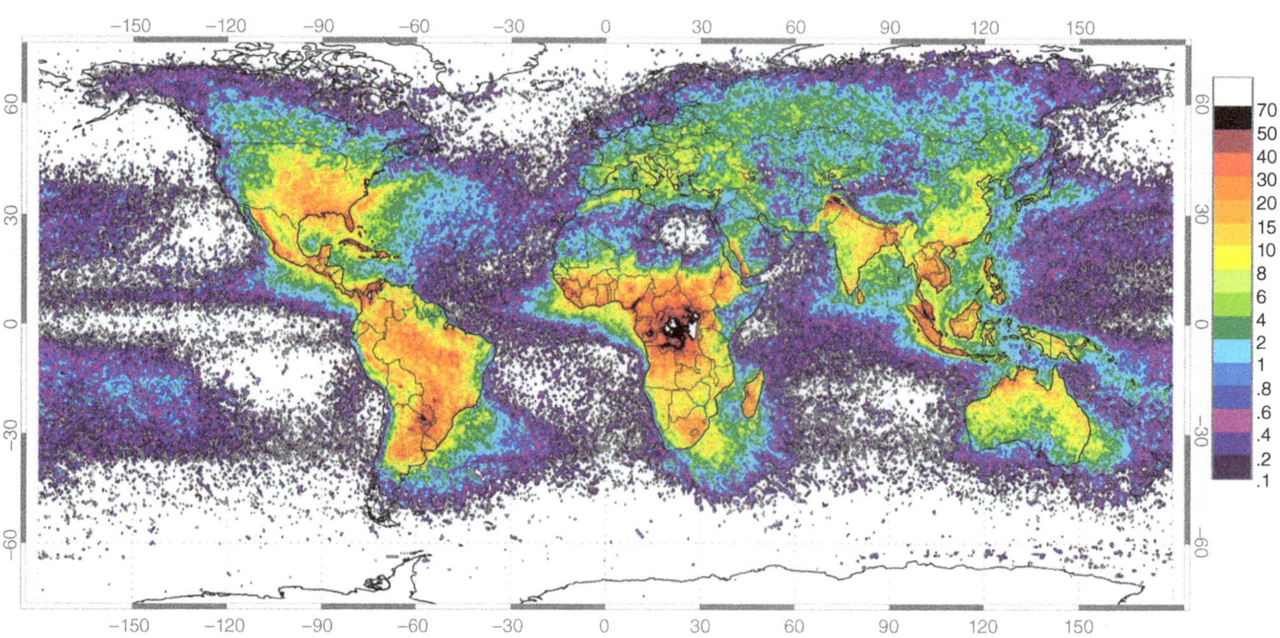


Figure 3.2 Annual number of lightning flashes by pixel over the 1995–2003 period. *Source*: Global Hydrology Resources Center.

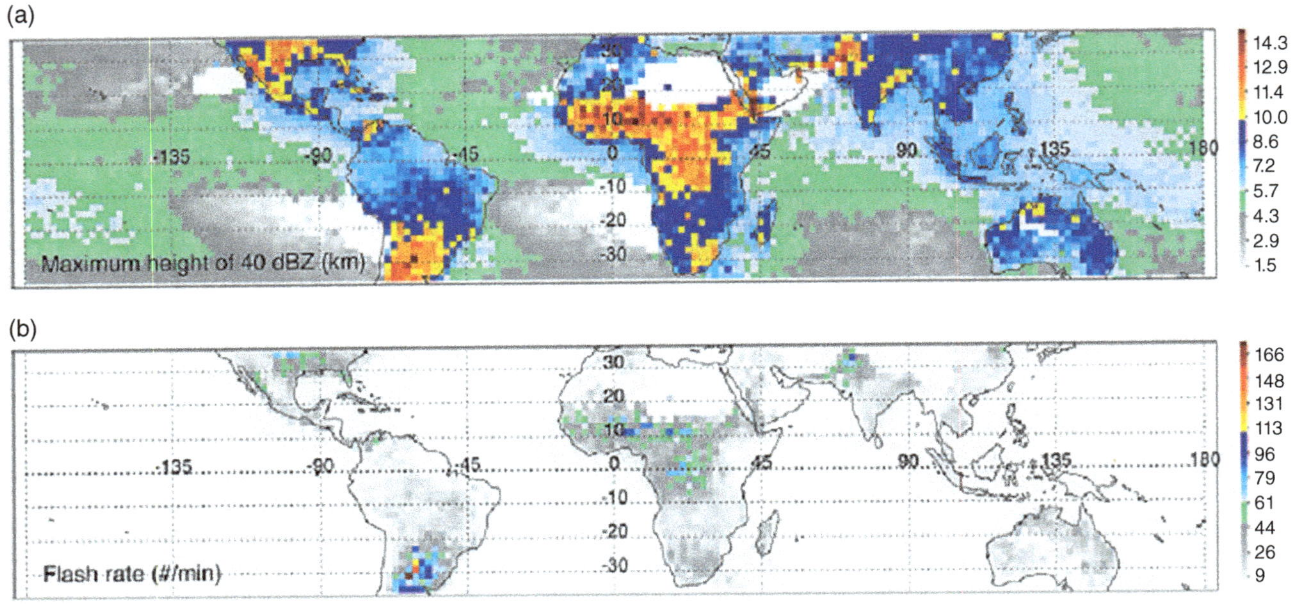


**Figure 3.3** Distribution of the most extreme storms (1% precipitation features – PF) for each 2° latitude × 2° longitude box, for: (a) the maximum height (km) and (b) the flash rate (min$^{-1}$). Boxes with less than 150 PFs are left blank. *Source*: Zipser *et al.* (2006). ©American Meteorological Society. Used with permission.

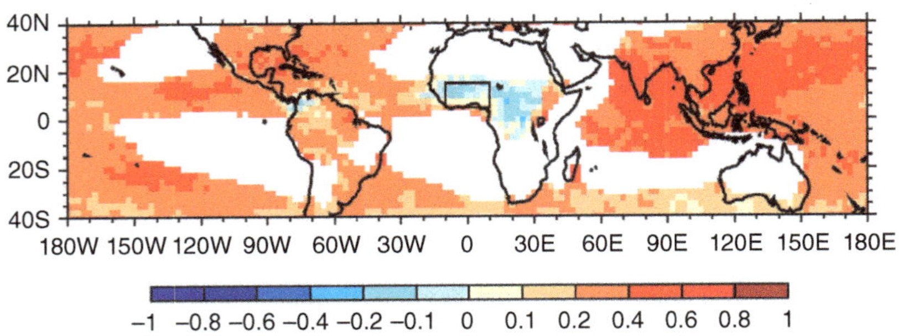


**Figure 3.4** Autocorrelation of GPCP precipitation anomalies at a 1 day lag in JAS, for the 1997–2006 period. *Source*: Roehrig *et al.* (2013). ©American Meteorological Society. Used with permission.

Compared with other continental regions, convection over Africa has some unique characteristics. For instance, Roehrig *et al.* (2013) examined the autocorrelation of GPCP precipitation anomalies at a 1 day lag to investigate the coherence of rainfall from one day to the next. These results demonstrate that Africa is a unique region where precipitation probabilities resemble white noise locally, with a lag correlation that is even negative in some places (blue areas on Figure 3.4). This is in contrast to other regions (red areas). There is no persistence of rainfall over Africa. CMIP5 climate and global forecast models are unable to reproduce this high variability of rainfall over Africa.

Neelin *et al.* (2009) documented a robust relationship observed over oceanic tropical regions between the rainfall rate ($P$) and the precipitable water (PW). As illustrated by Figure 3.5, a power law occurs above a critical value, increasing with the vertically averaged tropospheric temperature $\hat{T}$. We superposed on this figure red points corresponding to a similar study performed over Sénégal with the daily rainfall observed by the operational rain gauge network over a 12-year period (2000–2011), and the PW provided by the ERA-I reanalysis. First, as expected, the Sahel appears much warmer and drier ($\hat{T} \geq 273\text{K}$, $\text{PW} \leq 60$ mm) than the eastern Pacific. More surprisingly, convection appears to be more efficient over the Sahel, as the rainfall threshold is only 36 mm, compared with about 55 mm over the Pacific for similar temperatures $\hat{T}$. However, the analysis of rainfall probability density functions over the Sahel (not shown) indicates non-Gaussian distributions with strong variability. For a PW of 50 mm the rain rate is on average 5 mm day$^{-1}$, but most of this precipitation is concentrated on less than 20% of the area at a much greater intensity, raising a difficult forecasting issue. TRMM represents this $P$–PW relationship and its strong variability, unlike models that cannot simulate extreme and localised rain events over Sahel.

#### 3.1.1.2 Interaction Between Convection and the Larger Scale Flow

Cumulonimbus clouds are intense heating sources localised on small scales ('hot convective tower' concept of Riehl and Malkus 1958). It is important to understand how such localised heating gets transmitted to the larger

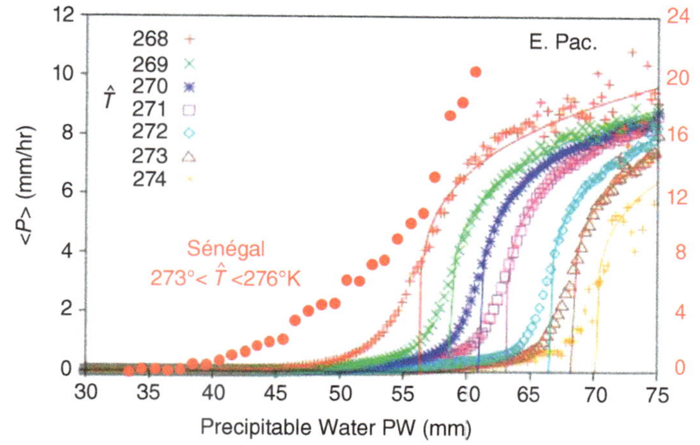


**Figure 3.5** Observed average precipitation rate $\langle P \rangle$ (in mm h$^{-1}$) as a function of precipitable water (PW, mm), for the eastern Pacific for 1 K bins of the vertically averaged tropospheric temperature $\hat{T}$. *Source*: Neelin *et al.* (2009). ©American Meteorological Society. Used with permission. Daily precipitation rates $\langle P \rangle$ (mm day$^{-1}$) observed by the Sénégal rain gauge network for the 2000–2011 period have been superposed with red points.

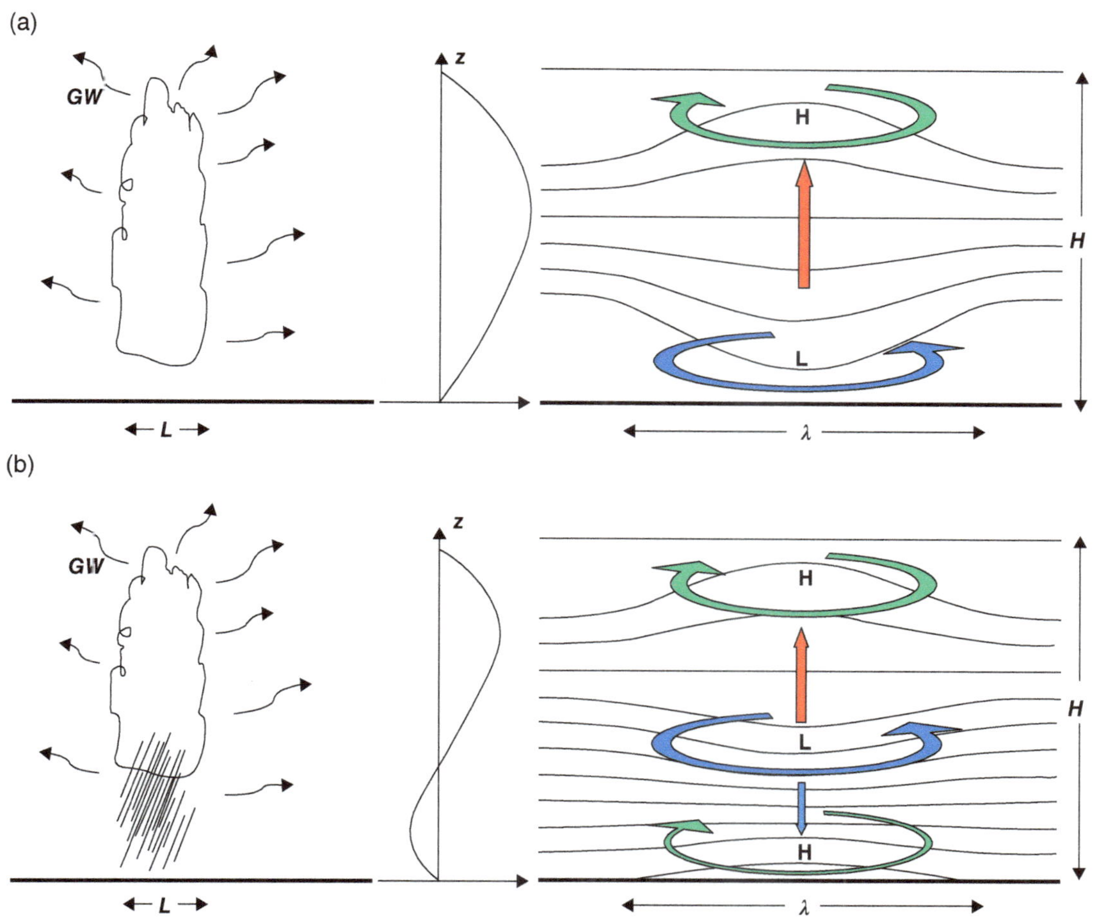


**Figure 3.6** Schematic representation of the 'adjustment problem' for (a) a simple convective heating profile and (b) a convective cooling–heating vertical dipole, as typically observed for semi-arid conditions characterized by an elevated cloud base. Left side is the vertical structure of convective cells with characteristic horizontal *L* and vertical *H* scales, emitting gravity waves (GW) in the stable environment. The corresponding vertical cooling–heating profiles are provided in the middle. Right side represents the resulting geostrophic equilibrium reached after an ~1/*f* timescale for the geopotential height field, and the associated mean vertical motion, cyclonic and anticyclonic circulations at the scale of the Rossby radius $\lambda$.

scales: the 'adjustment problem' illustrated by Figure 3.6. The highly localised perturbations (which can be idealised mathematically as Dirac delta-functions) generate gravity waves propagating rapidly (typically ~50 m s$^{-1}$ for the deepest and fastest) from the convective source, evacuating the energy (see Section 3.1.4.1.5). At the same time, the action of the rotation of the Earth (Coriolis term *f*) tends to balance the pressure fluctuations induced by the convective heating. After a typical timescale of 1/*f*, the heating is redistributed by the gravity waves over a horizontal scale called the Rossby radius[1] $\lambda = NH/f$, where $N$ is the stability of the atmosphere and $H$ its depth. The resulting dilatation of the atmosphere generates negative and positive pressure perturbations at low and upper levels respectively. The circulation reaches geostrophic equilibrium after a typical timescale ~$1/f$ (about 9 h for a latitude of 12.5°N) associated with cyclonic and anticyclonic circulations at low and upper levels respectively (Figure 3.6a). On scales smaller than the Rossby radius the circulation induced by convection assumes different dynamical characteristics. It is mainly convergent below the heated convective layer and divergent above. This circulation feeds the vertical motion in convective areas.

Owing to dry conditions prevailing over West Africa the cloud bases are elevated, resulting in strong rain evaporation (see Section 3.1.2.2) and cooling beneath hot convective towers (Figure 3.6b). This results in a contraction of the lower atmosphere, which, combined with its dilatation aloft, generates a vertical pressure triplet $H$–$L$–$H$ at low, mid and upper levels respectively. For scales larger than the Rossby radius, the geostrophic equilibrium is characterised by anticyclonic circulations at lower and upper levels, and cyclonic circulation at mid-levels corresponding to the African easterly jet (AEJ) level. At smaller scales, the maximum convergence occurs at mid-levels (around 700–500 hPa), feeding mean ascent above and mean subsidence underneath. So, in West Africa the flow tends to be divergent and anticyclonic close to the surface in convective areas. This is characteristic of semi-arid continental regions such as West Africa.

In the tropics, (except for tropical cyclones), the Rossby radius is large (greater than ~1000–5000 km) due to a weak Coriolis term and weak flow rotation. So convection in the tropics generates vorticity at a slow rate, and thus has difficulty in feeding the dynamics of the synoptic waves that can feed back on the convective activity. Over West Africa, African easterly waves (AEWs; see Section 2.1.3) are a major synoptic feature that strongly interact with convection, partly through gravity waves. Chapter 2 discusses this complex interplay between convection and AEWs that can increase the predictability in such situations.

Gravity waves generated by convection may also help to trigger convection at remote locations, as developed in Section 3.1.4.1.5.

#### 3.1.1.3 Aims of this Chapter

The aforementioned description of convective characteristics over Africa raises several questions: Why is convection different over Africa? Why is there such poor skill of models over this region? What is the intrinsic predictability of convection over Africa? Can forecasters add values to the NWP products available to them?

This chapter addresses these questions, based on a physical understanding of convection gained by research conducted over the last few decades. The chapter is structured around the four types of processes and factors that govern the convective activity, as summarised schematically in Figure 3.7; that is, (1) convective updraughts, (2) downdraughts and density currents, (3) vertical shear of the wind and (4) the external forcings both by small-scale processes (surface, boundary layer, orography, etc.) and by the larger scale flow, developed in Chapter 2. In this chapter we show that the unique characteristics of convection over West Africa can be explained in terms of the semi-arid climatic conditions that prevail in the region. This understanding provides a guideline to forecasting of the convective activity by analysing each of the four types of processes and the associated relevant indices or diagnostics as developed in Chapter 11.

### 3.1.2 Processes and Factors Governing Convection

This section analyses the first three factors that govern the convection as summarised by Figure 3.7.

#### 3.1.2.1 Convective Updraughts

Moist convection is characterised by upward vertical motion within clouds due to the conditional convective instability. The parcel theory, which neglects the pressure force and the small-scale mixing with the environment, approximates the parcel vertical acceleration by the buoyancy force

$$B = g\left(\theta_v' / \theta_{v0} - q_l - q_s\right) \text{ with } \theta_v = \theta\left(1 + \varepsilon q_v\right)$$

where $g$ is the acceleration due to gravity and $\theta_v'$ is the deviation of virtual potential temperature from the environment $\theta_{v0}$ accounting for the mass of vapour $q_v$. Parcels with a warm $\theta_v'$ anomaly will be more buoyant than the environment and will accelerate upward. The weight of liquid and solid hydrometeor ($q_l$ and $q_s$ mixing ratios respectively) will decelerate the ascent of an air parcel (hydrometeor drag). As illustrated by Figure 3.8 (see also Section 6.1.2.3 and Figure 6.10), during its lifting the parcel follows the dry adiabat up to the cloud base level $z_{lcl}$ and becomes less buoyant than the environment for stable conditions. Then it continues along the moist adiabat. Owing to the condensation (i.e. latent heat release) in the cloud, the parcel warms and its negative buoyancy decreases, to become null at the level of free convection $z_{lfc}$ where the parcel trajectory meets the ambient profile on the tephigram. Above $z_{lfc}$ the parcel's buoyancy becomes positive up to the level of equilibrium $z_{top}$, where it vanishes again. The surface between the moist adiabat and the environment in the layer of free convection ($z_{lfc}$ to $z_{top}$) is positive (Figure 3.8).

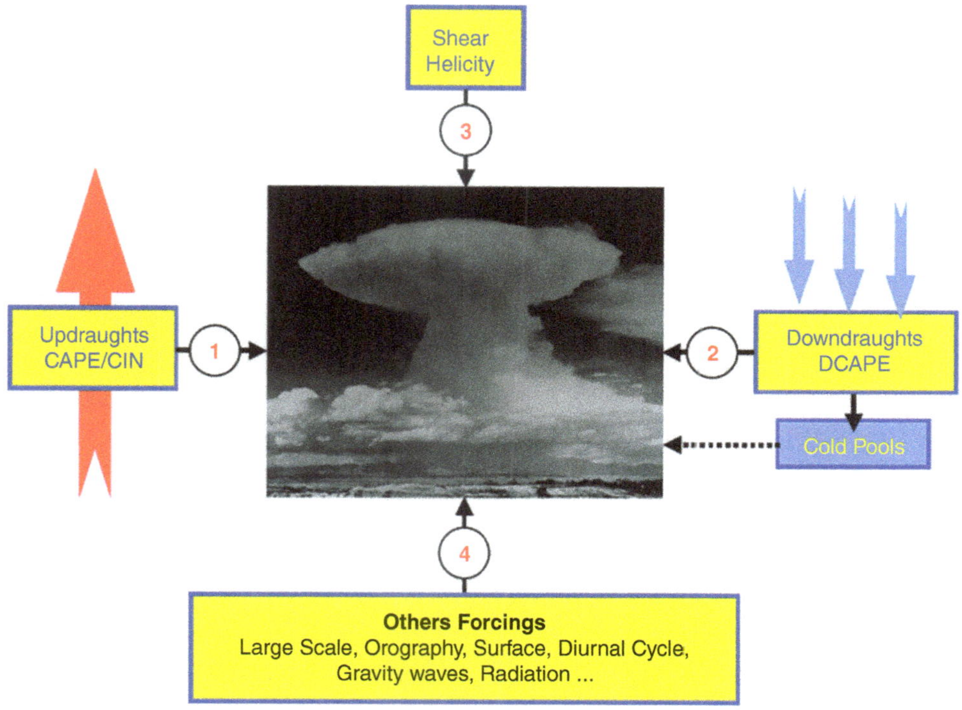


**Figure 3.7** Schematic representation of the four types of processes and related indices that govern the convective activity. CAPE: convective available potential energy; CIN: convective inhibition energy; DCAPE: downdraught convective available potential energy.

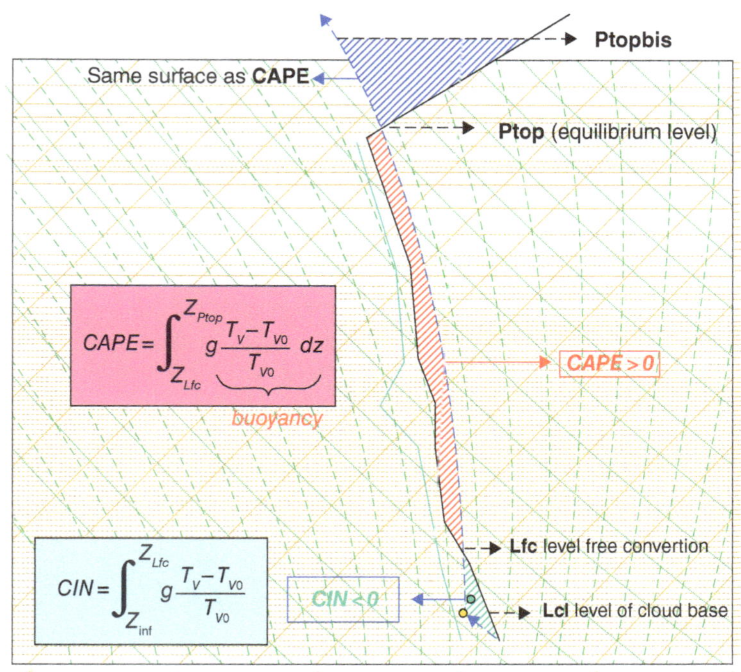


**Figure 3.8** Illustration of the parcel theory for an air mass lifted from the surface.

The vertical integral of the buoyancy above $z_{lfc}$ is called the convective available potential energy (CAPE) and provides an upward acceleration for the parcel. When completely converted into kinetic energy, CAPE provides the maximum vertical velocity for convective updraughts as $w_{max} = 1/2\sqrt{\text{CAPE}}$. Using this formula, the value of CAPE is typically a few thousand joules per kilogram, so $w_{max}$ reaches unrealistic values (such as 50 m s$^{-1}$ even for a moderate CAPE of 1250 J kg$^{-1}$). In reality the non-hydrostatic pressure force, the mixing of a rising parcel with the environment and the hydrometeor loading (drag) collectively reduce the updraught extrema to a few tens of metres per second for African storms. Above the $z_{top}$ level the parcel is decelerated as $B$ becomes negative (Figure 3.8). Owing to their upward momentum, the parcels can penetrate the stable lower stratosphere, injecting large amounts of vapour and ice crystals, which are important for the atmosphere chemistry of those levels. This penetration is called 'overshooting' and corresponds to the coldest temperatures observed on infrared (IR) imagery. Subsequently, the overshooting parcels descend back to their level of equilibrium around $z_{top}$ and feed the large anvils surrounding deep convective areas.

Conversely, the integral of the buoyancy from the surface up to the level of free convection $z_{lfc}$ is negative. This integral, the convective inhibition energy (CIN) constitutes a barrier of energy to be overcome to enable a parcel to reach its level of free convection, giving rise to the concept of *conditional* convective instability. The challenge of convection forecasting is rooted mainly in the difficulty of breaking this barrier, and hence triggering and maintaining convection; during the West African monsoon, sufficient CAPE is commonly available for storms, but the exact location of the storms depends on the detailed conditions that lead to triggering. External forcings needed to overcome the CIN are described in category (4) of Figure 3.7, and include, among many others, the large scale, the surface, the PBL and gravity wave processes. Over West Africa, owing to subsiding mean motion (see Chapter 1), the CIN increases northward to reach values up to 200 J kg$^{-1}$ over the Sahel, whereas CAPE is significantly large almost everywhere. So, over Africa, the triggering is a more crucial issue than the instability for understanding, simulating and forecasting convection.

Water vapour fuels the convection rain engine: latent heating provided by the surface evaporation and stored in the PBL and the sub-cloud layer is released in convective updraughts driven by high CAPE. The column water vapour, called PW, is a simple parameter to look at. As illustrated by Figure 3.5, rain increases with PW, and rain is not possible without sufficient PW. The PW threshold depends on the region (36 mm and 55 mm for Sénégal and the East Pacific respectively). Nevertheless, it is possible to get deep convection without rainfall at the surface due to rain evaporation, when the sub-cloud layers are very dry, as detailed in the following section.

In short, it is necessary to consider three parameters in the analysis of the convective updraughts: CAPE and PW are favourable and necessary but not sufficient, while CIN tends to suppress the development of convection. Notice that the CIN barrier delays the triggering of convection and allows the storage of more energy in the sub-cloud layer. This favours a build-up of CAPE and PW and the onset of more intense storms later in the day, or the following day. Although these three parameters are necessary in order to understand the occurrence of convective updraughts, taken alone they are not sufficient to forecast the convective activity; this also depends on other aspects of the storm environment that are developed in the following section.

#### 3.1.2.2 Downdraughts, Downdraught Available Convective Potential Energy and Density Currents

Owing to mass conservation, the development of updraughts requires the occurrence of compensating subsiding motion somewhere. The subsidence can occur at different scales.

1) *Cloud scale.* Within the convective areas, three types of downdraughts are considered:
   i) Intense microscale downdraughts may occur (with downward speeds >15 m s$^{-1}$, on horizontal scales <1 km), generating intense 'microbursts' when impacting the surface (Box 3.1) as illustrated in Chapter 6 (Sections 6.2.3 and 6.2.4). These events are highly dangerous for aviation, and may be generated from very small-scale clouds.
   ii) Convective downdraughts (1–15 m s$^{-1}$, over a few kilometres) are associated with convective precipitation and updraughts.
   iii) Mesoscale subsidence (a few tens of centimetres per second, over 10–100 km) associated with trailing precipitation underneath stratiform anvils. The subsidence results in characteristic 'onion or diamond' shape soundings in post squall-line regions as proposed by Zipser (1977) for a Caribbean system, and frequently observed over West Africa (Figure 3.9).
2) *Mesoscale.* In the vicinity of the convection, compensating subsidence is forced by gravity waves in the environment, allowing the atmosphere to redistribute heat around the convective source.
3) *Synoptic and planetary scale.* Over extensive areas, and far from the convection area, large-scale subsidence of a few centimetres per second occurs, such as within the subsiding branch of the Hadley cell. Radiation cooling is then a key process to equilibrate the adiabatic warming occurring within the large-scale subsidence.

Direct observation of vertical motion is possible only in research mode during field experiments. Nevertheless, the different types of downdraughts in convective areas may generate wind bursts near the surface at different scales that can be operationally observed, as illustrated by the wind distributions observed for two surface stations over the Sahel and displayed in Box 3.1.

When environmental conditions are favourable, both for updraughts (Section 3.1.2.1) and for downdraughts,

## Box 3.1 Observed distribution of 10 m wind over Sahel in 2006

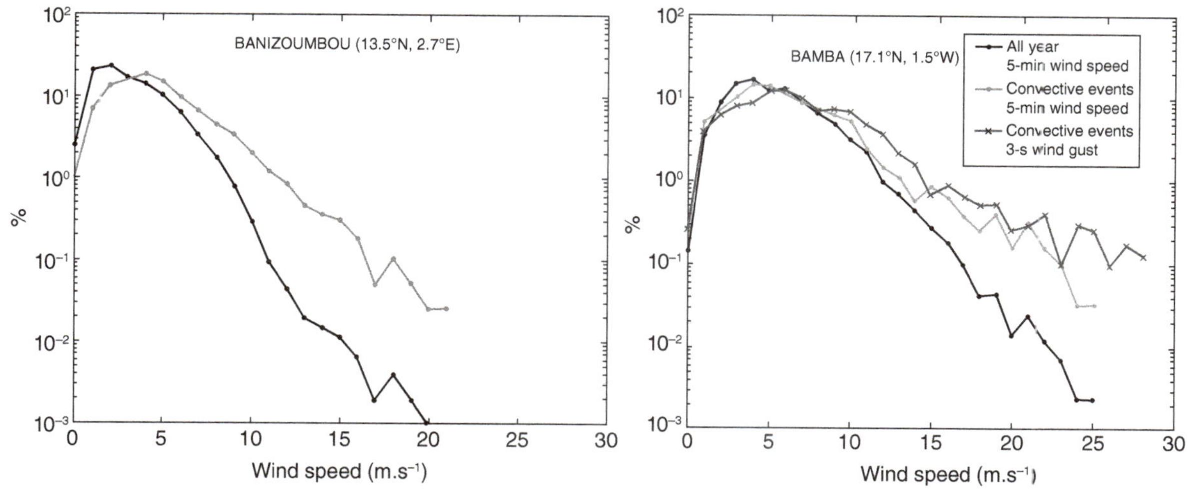


Probabilities are in percent per wind bin of 1 m $s^{-1}$ on a log-scale as measured by sonic anemometers at Banizoubou (13.5°N – provided by the AMMA Sahelian Dust Transect; Marticorena *et al.*, 2010) and Bamba (17.1°N – provided by AMMA-CATCH; Mougin *et al.*, 2009) sites, and treated as in Largeron *et al.* (2015).

Two sampling periods are considered: (i) the entire 2006 year (black curves) and (ii) the convective periods (light grey) at the considered station defined as the 1 h period surrounding an observation of rain or of a convective event detected by an IR tracking at this location. About 80 convective periods are detected in 2006 at both locations, with a mean duration of 3–4 h, corresponding to 3.5% of the year and 8.5% of the monsoon season. The total sampling size on this year is 105 120 and 3700 respectively, so corresponding for a 0.1% probability to about 100 and 4 cases of occurrence of wind intensity at the level of the considered bin respectively.

The wind intensity is stronger over the northern Sahel due to the impact of mid-latitude synoptic features and probably to lower friction (no vegetation). The strongest winds over 5 min (>8 m $s^{-1}$ and 14 m $s^{-1}$ for southern and northern Sahel respectively) are mainly due to convective events. Such events contribute significantly to dust uplift.

At a finer scale (darker grey curve) at Bamba, maxima of 3 s mean wind over 5 min can reach 30 m $s^{-1}$ and correspond to microbursts.

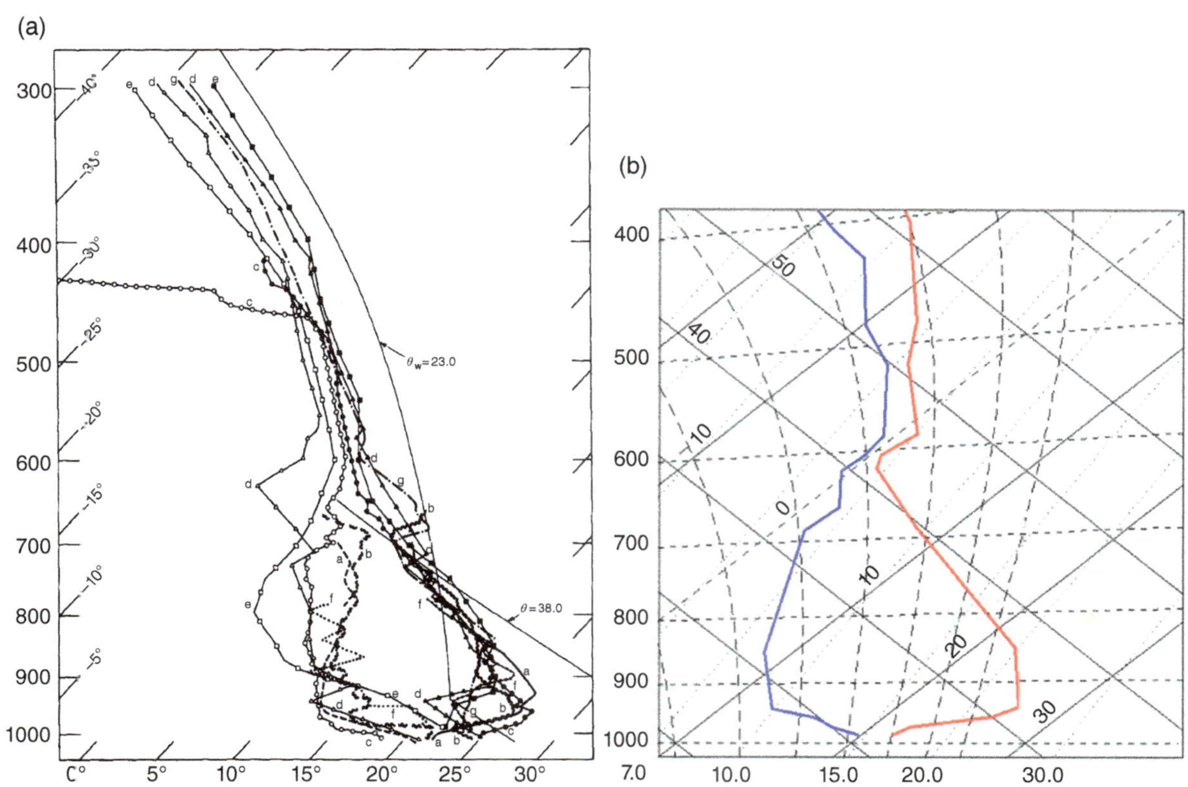


**Figure 3.9** Characteristic soundings in post-squall regions illustrating the impact of mesoscale subsidence. (a) For Caribbean and Global Atmospheric Research Program Atlantic Tropical Experiment (GATE) squall-lines. *Source*: Zipser (1977). © American Meteorological Society. Used with permission. (b) An African one observed at Niamey during the African Monsoon Multidisciplinary Analyses (AMMA) Special Observing Periods on 22 July 2006 at 1200 UTC.

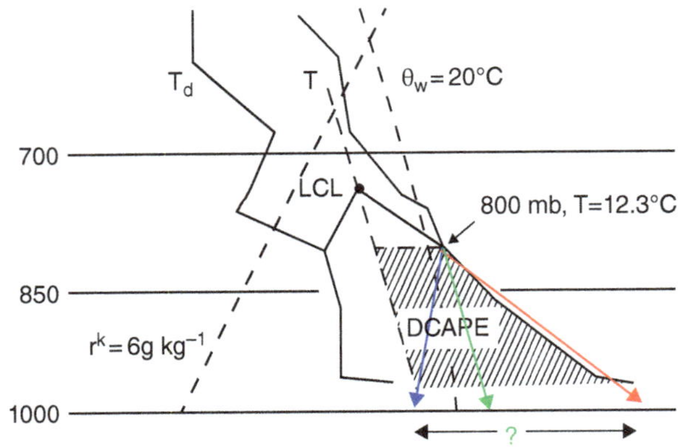


Figure 3.10 Illustration of the computation of the DCAPE index on a tephigram, for a subsiding parcel taken at 800 hPa. The blue arrow indicates the descent of air that reaches the ground completely saturated, the red arrow denotes dry adiabatic descent, and the green arrow represents the most common occurrence, lying between the moist and dry extremes. Therefore, there is considerable uncertainty in the temperature of the downdraught at ground level.

convection can be better organised, stronger, long-lasting and more efficient. It is thus necessary to analyse favourable conditions for the three types of subsidence occurring within convective areas (in (1) above). The downdraught convective available potential energy (DCAPE) proposed by Emanuel (1994) provides an index to evaluate those thermodynamic favourable conditions as illustrated by Figure 3.10. Taking an air parcel at a given level (here 800 hPa – the correct choice of origin level for DCAPE adds some uncertainty to the calculation, but generally it is taken at the first level where $\theta_w'$ reaches its minimum), it is first saturated at constant pressure by evaporation of rain, and then brought down pseudoadiabatically to the surface with just enough rain evaporation to keep it saturated. The energy released by this process is the parcel's downdraught available convective potential energy (DCAPE), proportional to the hatched area between the parcel trajectory and the environmental temperature profile as for CAPE or CIN. This trajectory is theoretical, as rain evaporation cooling will generate a deficit of buoyancy and thus a subsiding motion warming the parcel, so that DCAPE is a maximum, or upper bound, for the energy available to the downdraught. In reality, depending on the amount of rain evaporation, the subsiding parcel can reach the surface with a temperature that lies somewhere between its dry potential temperature $\theta$ (red arrow) and its moist potential temperature $\theta_w'$ (blue arrow). As the difference between these values can be greater than 10 K (in a 5–15 K range, from an analysis of atmospheric radiation measurement data in Niamey[2]), the thermodynamics of downdraughts is much more complicated than that for updraughts, leading to substantial uncertainty and variability in estimates of downdraught intensity.

Despite the significant uncertainties, some rules can be derived to guide the forecaster. Downdraughts are favoured by drier mid-level layers (e.g. 700–500 hPa), corresponding to lower $\theta_w'$, as dry mid-levels favour rain evaporation and increase the DCAPE area (Figure 3.10). Such dry conditions at mid-levels occur over the Sahel, where the AEJ lies to the south of a layer of very dry, elevated Saharan air with low $\theta_w'$ sometimes known as the SAL (see Section 2.1.2). 'Dry intrusions' (Section 2.1.5) of mid-level air may increase the likelihood of intense downdraughts on synoptic timescales. There is evidence that in the early and pre-monsoon periods, the first rains may have particularly strong downdraughts due to the dry mid-level conditions occurring before the full monsoon is established.

Elevated cloud bases and a deeper PBL also increase the DCAPE area, by lifting the origin level for the downdraught and the DCAPE calculation, which is increasingly the case as we move northward over West Africa. A well-mixed, deep PBL close to an adiabat, with a uniform vapour mixing ratio profile, favours intense downdraughts, because of the absence of ambient stability to oppose the downward flow. This is typically the case in the afternoon, where microsubsidence can form below cumulus clouds on the sub-kilometre scale, even with weak precipitation, due to dry air, fed by precipitation evaporation with lower buoyancy, penetrating within the deep Sahelian PBL. Such microsubsidences may generate microbursts near the surface, which are highly dangerous for aviation during take-off and landing periods (Box 3.1).

Rain evaporation within downdraughts generates cold air masses called 'cold pools' that spread over the surface like density currents. Figure 3.11 (left column) provides a schematic view of a density current. Its head is deeper – typically 1–2 km in depth in the Sahel – and characterised by a rotor circulation. Its propagation lifts ambient warmer air in the form of a roll cloud that can trigger new convective cells. The discontinuity between the two air masses is sharp (~500 m), as confirmed by the signature of its passage at the surface (Figure 3.11-right column). It is characterised by a sharp drop of temperature and of specific humidity (relative humidity may rise or fall, depending on the case), rotation of winds, intense gusts, a pressure jump (~1 hPa) and the arrival of rain a little later. Behind the head region, the tail of the cold pool is shallower, the winds slow down, and over a period of hours the low levels recover. The upper part of the density current is turbulent with intense wind shear, especially in the

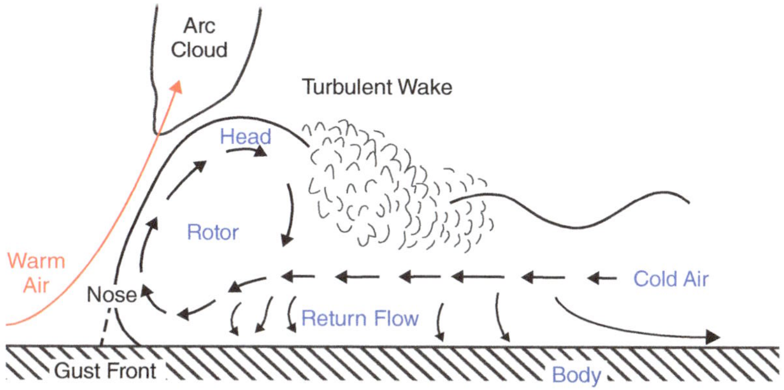

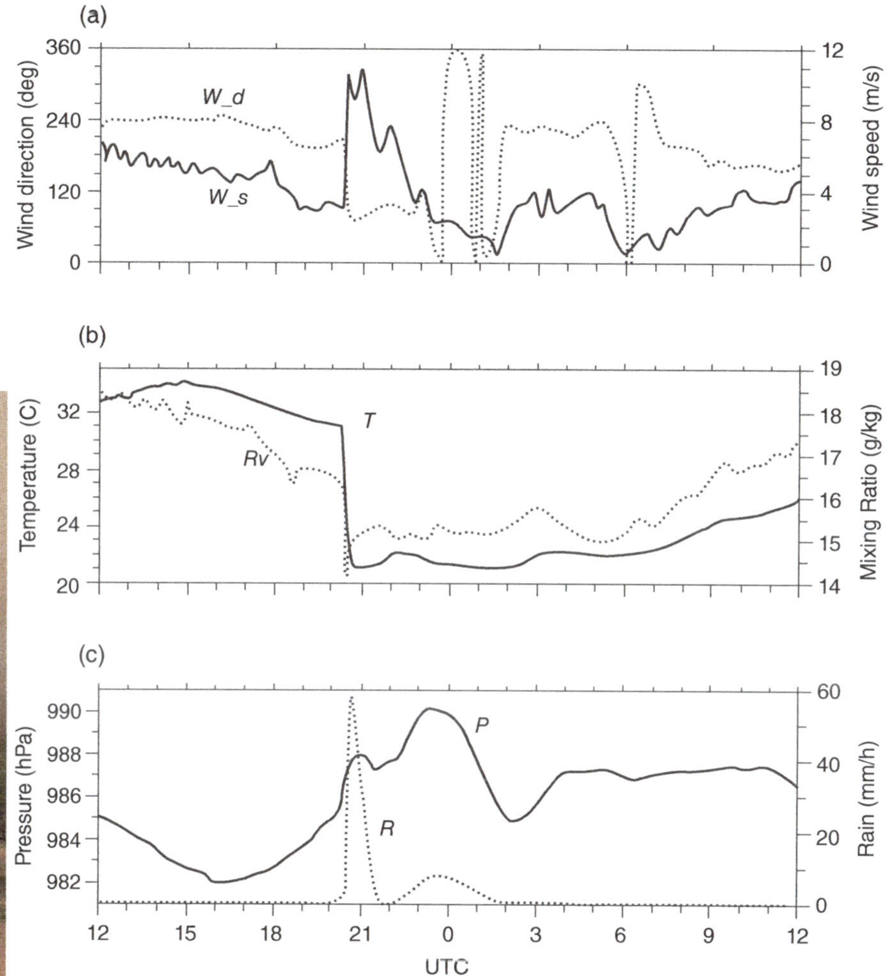


**Figure 3.11** Left column: schematic view of a cold pool, and a picture of the arrival of a convective cold pool lifting dust, a phenomenon often referred to as haboob. *Source*: © CNRS Photothèque, by F. Guichard and L. Kergoat, Mali, August 2004. Right column: an example of the signature of a cold pool passage at the surface for a Sahelian squall line for (a) wind intensity and direction, the temperature $T$ and vapour mixing ratio $R_v$, and (c) the surface pressure $P$ and the precipitation intensity $R$. *Source*: Redelsperger *et al.* (2002). Reproduced with permission of the Royal Meteorological Society.

head region. Over Africa cold pools can be intense (surface temperature drops larger than 10 °C) and deep (1–2 km), representing an important source of dust lifting over bare soil and allowing the visualisation of the cold pool (Figure 3.11 picture). Cold pools are a key feature causing the triggering of new, secondary convective cells in situations of strong inhibition, as occur over the Sahel. As shown in the following Section 3.1.2.3, this process of secondary triggering helps to organise convection, propagate it and enhance its duration.

#### 3.1.2.3 Wind Shear

The variation of the horizontal wind (i.e. shear) in the vertical, both in intensity and direction, is the third factor to consider in order to understand and forecast the convective activity. Early in the 1960s Ludlam pointed to strong shear as a key feature for severe storms (Ludlam, 1980). A tilting updraught enables precipitation to fall out of it – thus reducing the drag – and to feed a cold downdraft by rain evaporation. A tilted updraught and a strong downdraught may work side by side to produce intense and long-lasting convection. Now we have different theories explaining the role of the shear, but it is outside our present goal to detail them exhaustively. Here, we just present a few simple physical explanations that are relevant for Africa. Figures 3.12a and b illustrate the impact of the shear on a density current. In the case of no shear (Figure 3.12a), the circulation is symmetric[3] and the density current formed by convection spreads in all directions, taking a circular shape. Forced ascents are spread all along its leading edge and are thus weak. In contrast, in the case of linear shear (Figure 3.12b) (here easterly shear, as over West Africa), mid-level dry air feeding the density current transports momentum and the density current spreading is no longer symmetric. It is faster and deeper on its downshear (here western) edge, where the convergence is concentrated and stronger. The general rule is thus: *in the case of shear, the density current tends to trigger new convective cells on its downshear side.*

Arguments based on horizontal vorticity considerations by Rotunno *et al.* (1988) result in a similar conclusion. More robust upright updraughts occur at the head of a density current when the shear (vorticity) of the impeding ambient flow is opposed to the vorticity generation by the density current (Figure 3.12f); that is, the downshear side of the density current is more efficient.

Vertical motion in a sheared flow induces non-hydrostatic pressure deviations (~1 hPa). In simple terms, a convective updraught (downdraught) behaves as a 'soft obstacle'. A horizontal dipole of pressure opposes the ambient horizontal wind change and reduces the momentum difference between the draught and its environment. Such fluctuations of pressure at different levels generate vertical accelerations that favour new updraughts on the downshear side. Over West Africa the shear is moderate compared with mid-latitudes, but stronger than over the tropical oceans, contributing to differences in convective activity between these tropical regions.

For the sake of completeness, the *rotation* of the shear should also be considered, in order to understand specific organised convection such as supercells, as developed by Weisman and Klemp (1986). As shear rotation is weak over West Africa, supercells are uncommon, explaining why tornadoes have not been recorded in West Africa. Section 3.1.3.1.3 provides a brief description of a supercell storm.

#### 3.1.2.4 Discussion

The convection type and intensity results from the combination of the aforementioned three factors; namely, updraughts, downdraughts and wind shear (Figure 3.7). This allows us to partially respond to the question raised in Section 3.1.1: 'Why is convection different over Africa?' For instance, Jorgensen and LeMone (1989) and Lucas *et al.* (1994) compiled and compared the magnitude of convective draughts as observed during field experiments over oceans (GATE, Taiwan Area Mesoscale Experiment (TAMEX), Equatorial Mesoscale Experiment (EMEX)) and continent (Thunderstorm Project) (Figure 3.13), and for different types of system. It appears that convective draughts over the ocean are, on average, much less vigorous than those over the US Great Plains. Although updraughts are stronger than downdraughts everywhere, the symmetry of profiles between up- and downdraughts is to be noticed. Similar observations for the Tropical Ocean Global Atmosphere Coupled Ocean Atmosphere Response Experiment (TOGA-COARE) for the west Pacific warm pool confirm this difference between oceanic and continental regions. For West Africa we only have three-dimensional wind fields retrieved from the Convection Profonde Tropicale (COPT) 1981 and AMMA experiments that cannot be directly compared with the above direct airborne measurements. Nevertheless, vertical draught extrema are typically intense and similar to the ones over the US Great Plains (Lafore *et al.*, 1988), consistent with the distribution of the most extreme storms already shown by Figure 3.13.

The factors favourable to updraughts (CAPE and PW) cannot alone explain the difference in their intensities between oceanic and continental regions. PW is indeed much weaker over West Africa than over the tropical oceans. CAPE can reach large values in the

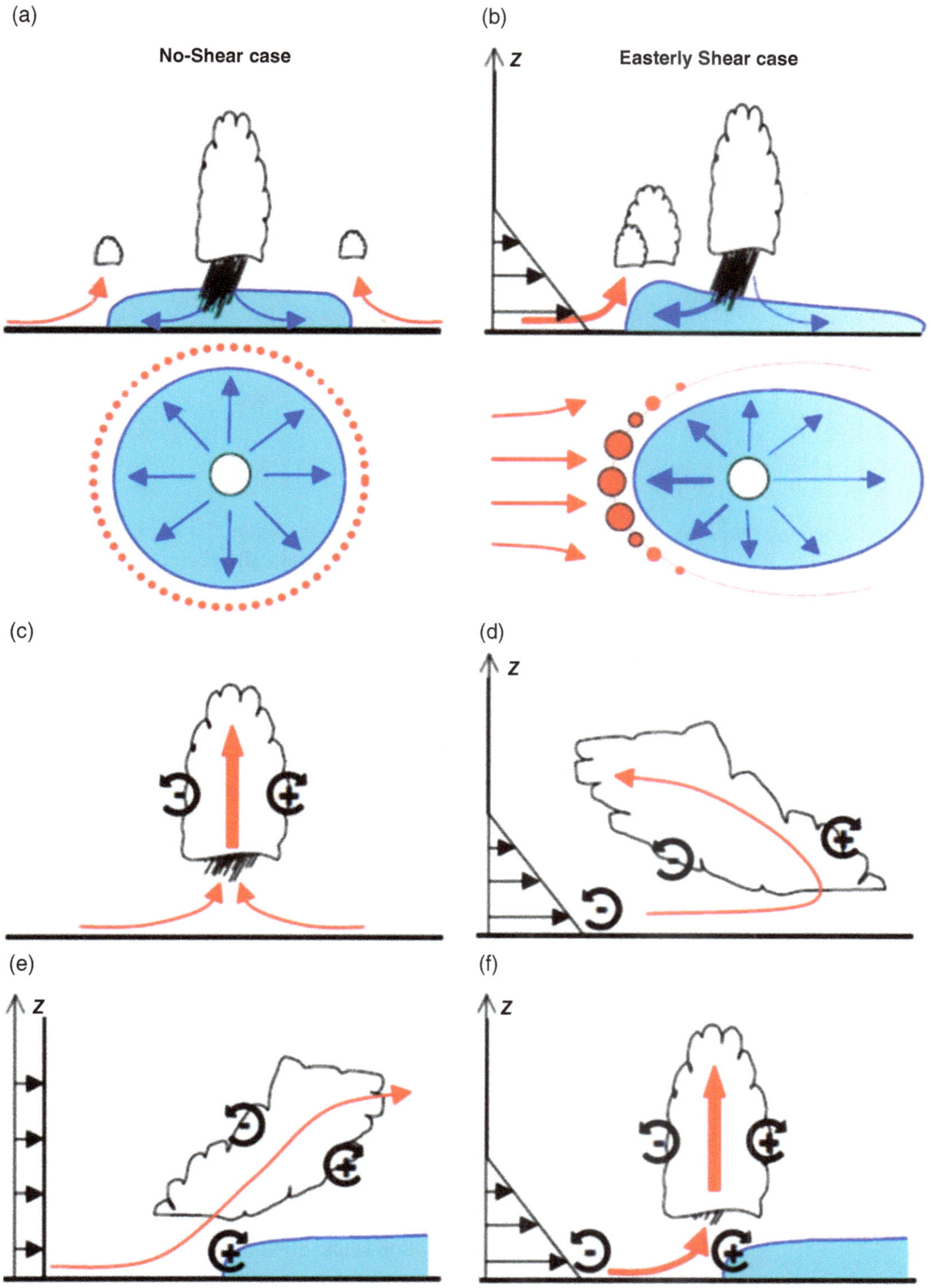


**Figure 3.12** Schematic view of the impact of the shear on convection organisation in cases of no shear (left column) and of unidirectional easterly shear (right column). Wind profiles in the environment are relative to the convection–cold pool system propagation. The upper row, (a) and (b), shows the impact of shear on a cold pool and its efficiency to lift new convective cells.

The lower two rows are a schematic diagram based on vorticity arguments showing how a buoyant updraught may be influenced by wind shear and/or a cold pool. (c) With no shear and no cold pool, the updraught axis produced by the thermally created, symmetric vorticity distribution is vertical. (d) With only shear, the distribution is biased toward negative vorticity and this causes the updraught to lean downshear. (e) With only a cold pool and no shear, the distribution is biased by positive vorticity of the underlying cold pool and causes the updraught to lean behind the cold pool. (f) With both a cold pool and shear, the two effects may negate each other and allow an erect updraught. *Source*: Rotunno *et al.* (1988). ©American Meteorological Society. Used with permission.

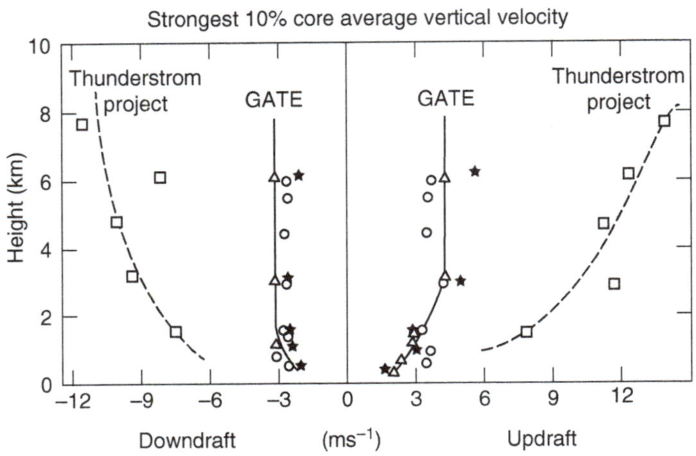


**Figure 3.13** Observed strongest 10%-level average of vertical velocity in updraught and downdraught cores as function of height for GATE, TAMEX, hurricanes, EMEX and the Thunderstorm Project, corresponding to triangles, stars, circles, plus signs and squares respectively. Adapted from Jorgensen and LeMone (1989) and Lucas *et al.* (1994). *Source:* Lucas *et al.* (1994). ©American Meteorological Society. Used with permission.

afternoon, higher than over the ocean and conducive to explosive cumulonimbus convection if the CIN can be overcome, but the clouds remain short-lived and localised without the action of other processes. The analysis of the other factors ((2) and (3) of Figure 3.7) provides some explanation. Compared with oceanic regions, over West Africa we have drier mid-level air, an elevated cloud base and a deep PBL, all favouring downdrafts and intense, deep cold pools. Also, owing to the AEJ occurrence, the shear is important in helping to organise pre-existing convection, whereas over the ocean shear is typically weak, except during specific periods such as the westerly bursts over the Pacific warm pool.

In short, the occurrence of convection cannot be understood (and forecasted) with the convective instability concept alone. The forecaster has to consider all of the four factors summarised by Figure 3.7. The formation of convection is only one half of the forecasting problem: the need to determine the type of convection that will result, and its subsequent life cycle, remains. This is essentially the problem of convective organisation, and is described in the next sections.

### 3.1.3 Organisation of Deep Convection

#### 3.1.3.1 Different Types of Organisation

Before speaking of convection organisation, we need to define the elementary convective structure. This task is not simple, owing to the turbulent nature of convection exhibiting a large range of scales down to a few metres, so that convection can appear as a 'fractal' object. In practice the definition of elementary convective structure depends on the available observations. For our purpose, we use the concept of a 'convective cell', regarded as an updraught region of a few to ten kilometres in size, extending vertically through a significant fraction of the troposphere. Each updraught cell is associated with a precipitation region identified by radar. Past studies of radar observations of convective cells, mainly over the US Great Plains, complemented by idealised modelling studies, have proposed three types of deep convective storms and conceptual models. Further recommended reading relevant to organised deep convection, and also other aspects addressed in this chapter, can be found in the book chapter by Moncrieff (2013).

##### *3.1.3.1.1 The Single-Cell Storm*

The single-cell storm is typically short-lived (30–50 min) with a life cycle consisting of three stages. The formation stage, with an ascending warm bubble, corresponds to a growing cumulus (passing then through the cumulus congestus phase). At the mature stage, glaciation usually occurs, and the storm extends over the whole tropospheric depth as a cumulonimbus cloud. Once glaciation has occurred, precipitation forms rapidly: the precipitation falls to the surface and initiates downdraughts, both through water loading and rain evaporation. Finally, during the dissipation stage, downdraughts dominate at mid and low levels and the resulting density current spreads across the surface, whereas the updraughts vanish at upper levels, feeding a residual anvil cloud.

The typical environment of the single-cell storm is characterised by a weak background wind shear, resulting in an upright updraught, so that precipitation falls within the updraught and cuts off the flow that is feeding the updraught. This type of short-lived storm is common over West Africa, particularly in regions far to the south of the AEJ core in summer, where the wind shear is light, but is usually not associated with severe weather, except occasionally microbursts. The single-cell storm propagates at about the mean wind speed of the convective layer. A key parameter to look at for the forecaster is the hodograph to visualise the vertical profile of the wind shear vector.

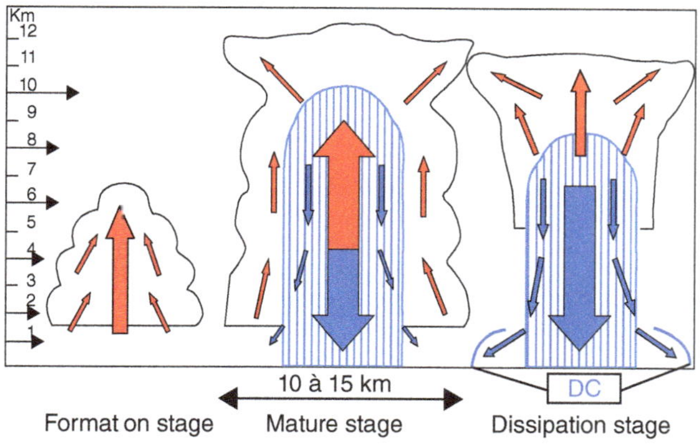


#### 3.1.3.1.2 *The Multicell Storm*

A multicell storm corresponds to an ensemble of shorter-lived single-cell storms at different stages of their life cycle that can persist over several hours up to a day or more. The key ingredient for this storm type is a moderate to strong wind shear in one direction (a linear hodograph). The schematic scenario is the following. Starting from a growing cell ('blue' cell) beginning to produce precipitation falling out (due to shear) of the updraught and initiating a downdraught, a new 'red' cell is initiated downshear (see Section 3.1.2.3 and Section 6.2.3). About 20 min later, the blue cell reaches its mature stage with an intense downdraught and a density current spreading downshear. This later reinforces the growing 'red' cell and favours the triggering of a new 'green' cell at the density current leading edge. The dissipation of the first 'blue' cell occurs 20 min later, thus contributing to feed the density current, now also fed by the 'red' cell reaching its mature stage, and so on. The interactions between cells contribute to form a long-lived convective system. The propagation of a multicell storm (a form of group velocity) is the combination of the propagation of each cell (mean wind speed in the convective layer) and the speed of the discrete triggering of new cells. So the propagation of the convective system as a whole may differ both in speed and direction from the mean wind. Multicell storms are frequent over West Africa, and are further classified in Sections 3.1.3.2 and 3.1.3.3. They may be associated with severe weather, in terms of heavy rainfall and squalls.

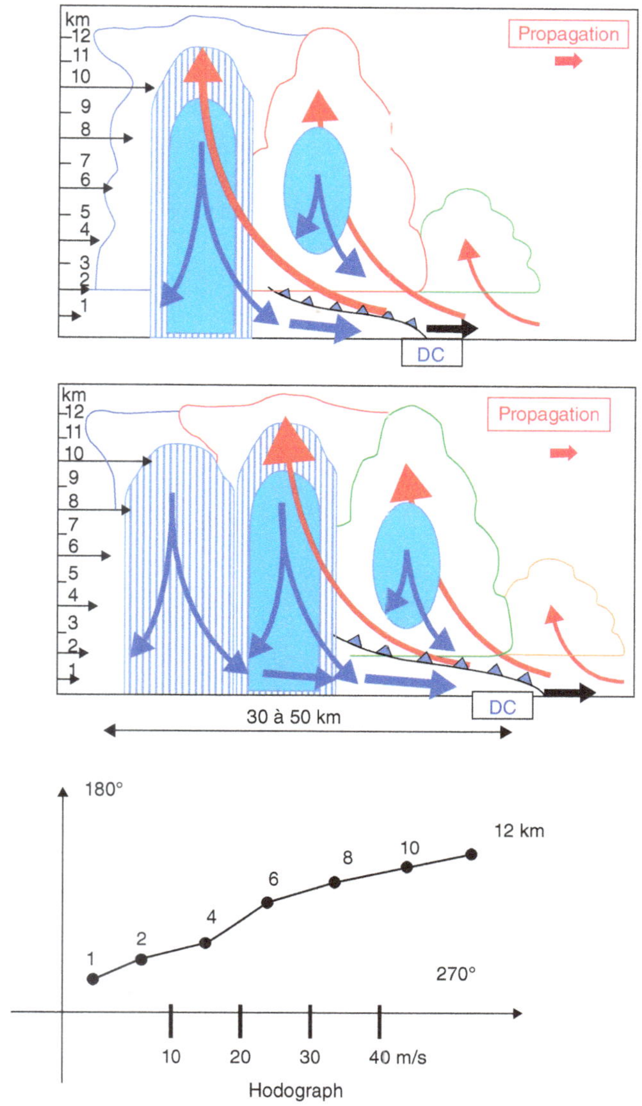


#### 3.1.3.1.3 *The Supercell Storm*

The supercell storm is potentially the most dangerous of the convective storm types. It may produce hail, tornadoes and strong winds; however, it has not been commonly recorded in West Africa, so it is only briefly mentioned here. The supercell consists of a single, powerful, quasi-steady rotating updraught. Apart from the required convective instability (CAPE), the key favourable ingredient for supercells is the vertical wind shear vector that needs both to be strong and to rotate at low levels. With this configuration, the supercell propagates differently from the mean wind toward the curvature of the hodograph at low levels. Owing to the lack of rotation of the shear, supercells are seldom observed over West Africa. Nevertheless, in some regions, such as in the lee of mountains or on the flank of squall lines, rotation of the shear can occur, so that such storms are not excluded, even if much less intense than over the US Great Plains.

Thus, forecasters are recommended to carefully examine hodographs to predict the storm type, propagation and so on and to evaluate risks.

#### 3.1.3.2 The Squall-line Conceptual Model

The fast-moving squall line is the predominant type of mesoscale convective system (MCS) over West Africa (see Section 3.1.3.3). It is a multicell, self-maintained system having three main interacting components (Figure 3.14).

- A curved line of powerful cumulonimbus occupies the convective region, a few tens of kilometres wide, that can extend for several hundred up to a thousand kilometres in length. The convective region is associated with a strong updraught extending over the whole troposphere, overshooting into the lower stratosphere as detected by cold IR satellite brightness temperature. Owing to the large latent heat release by convective cells, a low pressure (~1 hPa) ($L$ in Figure 3.14c) develops

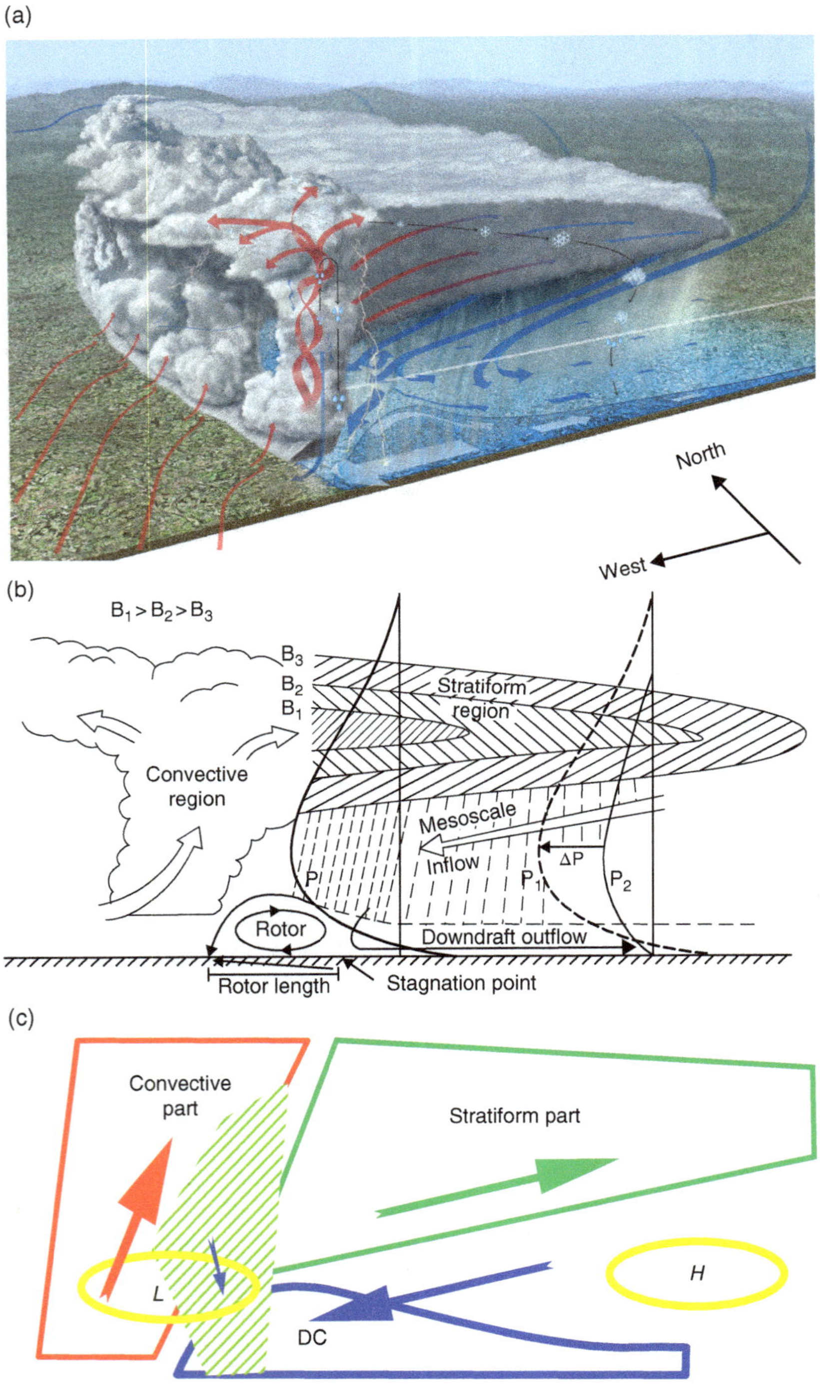


Figure 3.14 (a) Three-dimensional schematic view of a fast-moving squall line with the airflow circulation, the cold air mass forming the density current (DC) in blue shading, and some trajectories of hydrometeors. *Source*: Lafore (2004). Reproduced with permission of Météo-France. (b) Two-dimensional conceptual model of the squall line. *Source*: adapted from Lafore and Moncrieff (1989). ©American Meteorological Society. Used with permission. (c) Its box representation.

at mid-levels (~3 km), and intense precipitation is formed, feeding intense convective downdraughts. Owing to the elevated 0 °C isotherm (~4.5 km) over West Africa, a large part of the precipitation is liquid and falls rapidly (~7 m $s^{-1}$) underneath the convective region (see trajectories of hydrometeors in Figure 3.14a).

- A large amount of solid (ice) hydrometeors are also formed in the deep tropical troposphere (~16 km). Falling slower (~1–2 m $s^{-1}$), they feed a thick and widespread anvil called the stratiform region characterised by weak mesoscale ascent (a few tens of centimetres per second) in the anvil and a weak mesoscale subsidence (a few tens of centimetres per second) below the sloped elevated anvil cloud base. Owing to this sloped anvil cloud base and to the mid-level low pressure in the convective region, an inward horizontal pressure gradient develops. This favours the formation of a rear inflow that reinforces the AEJ behind the squall line and brings in drier air below the stratiform anvil, as illustrated by Figure 3.9b, with reinforcement of easterlies of up to 40 kt around the 700 hPa level.
- The mesoscale subsidence is forced by the rear inflow and maintained by both the rain evaporation and the melting of ice particles, contributing to feed the 'density current', corresponding to the third key component of the squall line. The deep head of the density current propagates downshear, where lifting of ambient air over this deep head helps to overcome the CIN barrier and trigger new convective cells in the convective region.

Owing to this efficient internal organisation, the squall line can maintain itself for a long time, even during the night, whereas diurnal convection (cumulus congestus and isolated, single-cell storms) mostly collapses around sunset. Further recommended reading directly relevant to MCSs is the review article by Houze (2004).

The Sahel is a favourable region for fast-moving squall lines during the monsoon, due to the coexistence of all the favourable and necessary factors; that is, strong CAPE, strong moisture in the monsoon layer resulting in elevated PW, easterly shear (~15 m $s^{-1}$ over ~4 km) between the AEJ and the monsoon, and dry air in the AEJ layer to feed the downdraughts and density current.

Figure 3.15 illustrates the reflectivity structure of a fast-moving squall line approaching Niamey on 11 August 2006 as seen by radar. In agreement with the theory, it moves westward in the shear direction and at a speed of about 13 m $s^{-1}$; about the AEJ speed. The narrow band of intense precipitation (red area on radar PPI plots) corresponds to the convective part, followed by a wide area (~200 km) of weaker reflectivity corresponding to the stratiform region part. The vertical cross-section (RHI) exhibits a forward anvil, new convective cells triggered ahead of the main convective part and the overshooting of the mature cells at its rear. Behind the bright band is a typical signature of the stratiform region, whose depth slowly decreases rearward. The IR Meteosat image (Figure 3.15d) allows us to identify the widespread cloud shield corresponding to the upper-level anvil. The coldest temperatures (−65 °C in red) indicate the wider overshooting areas, but the line structure of the convective part underneath the extended anvil cannot easily be detected from the IR image. The typical surface signature of the passage of a squall line has been already shown in Figure 3.11-right column. Apart from the sharp drop of temperature and of specific humidity, the rotation of winds, intense gusts and the pressure jump (~1 hPa) corresponding to the density current arrival, Figure 3.11 illustrates the occurrence of intense rainfall (which can reach 150 mm $h^{-1}$) over a short period of around 10–20 min during the passage of the convective part. After a relative break of precipitation, the stratiform region corresponds to weaker precipitation lasting a few hours, in which winds are weaker. A surface meso-high (~1 hPa) is often observed underneath the stratiform region, corresponding to the mesoscale subsidence and coupled with a weak divergent anticyclonic flow at the surface. Nevertheless, owing to the strong diurnal and semi-diurnal cycles of pressure in the tropics, such fluctuations are not easily detected on weather maps, except for the pressure change over 24 h.

#### 3.1.3.3 Climatology and Classification of Mesoscale Convective Systems

The high spatial and temporal resolution of geostationary satellites such as Meteosat (3 km and 15 min) allows the detection and tracking of MCSs for nowcasting (see Chapter 6), or for climatology and research purposes. Such algorithms detect convective areas from the IR imagery and search their temporal continuity to define convective systems for which global characteristics are provided, such as their duration $D$, propagation speed $V_p$, spatial extent and the trajectory of their 'centre of gravity'. A climatology of West African MCSs over a 25-year period (1986–2010) from Meteosat has been established using the algorithm described in Fiolleau and Roca (2013). The speed $V_p$ and duration $D$ parameters are used to describe MCSs in four classes, as proposed by Tomasini *et al.* (2006):

- The **C1** class ($D < 9$ h; $V_p < 10$ m $s^{-1}$) corresponds to small (between 5000 and 10 000 $km^2$) numerous diurnal and slow-moving MCSs.
- The **C2** class ($D > 9$ h; $V_p < 10$ m $s^{-1}$) includes *long-lived slow-moving* MCSs. They are bigger (up to 20 000 $km^2$) than those in the C1 class.
- The **C3** class ($D < 9$h; $V_p > 10$ m $s^{-1}$) corresponds to *diurnal and fast-moving* systems dissipating in the evening.
- The **C4** class ($D > 9$ h; $V_p > 10$ m $s^{-1}$) corresponds to *fast-moving long-lived* squall-line type systems. They are less numerous but are the largest (typically 30 000 $km^2$).

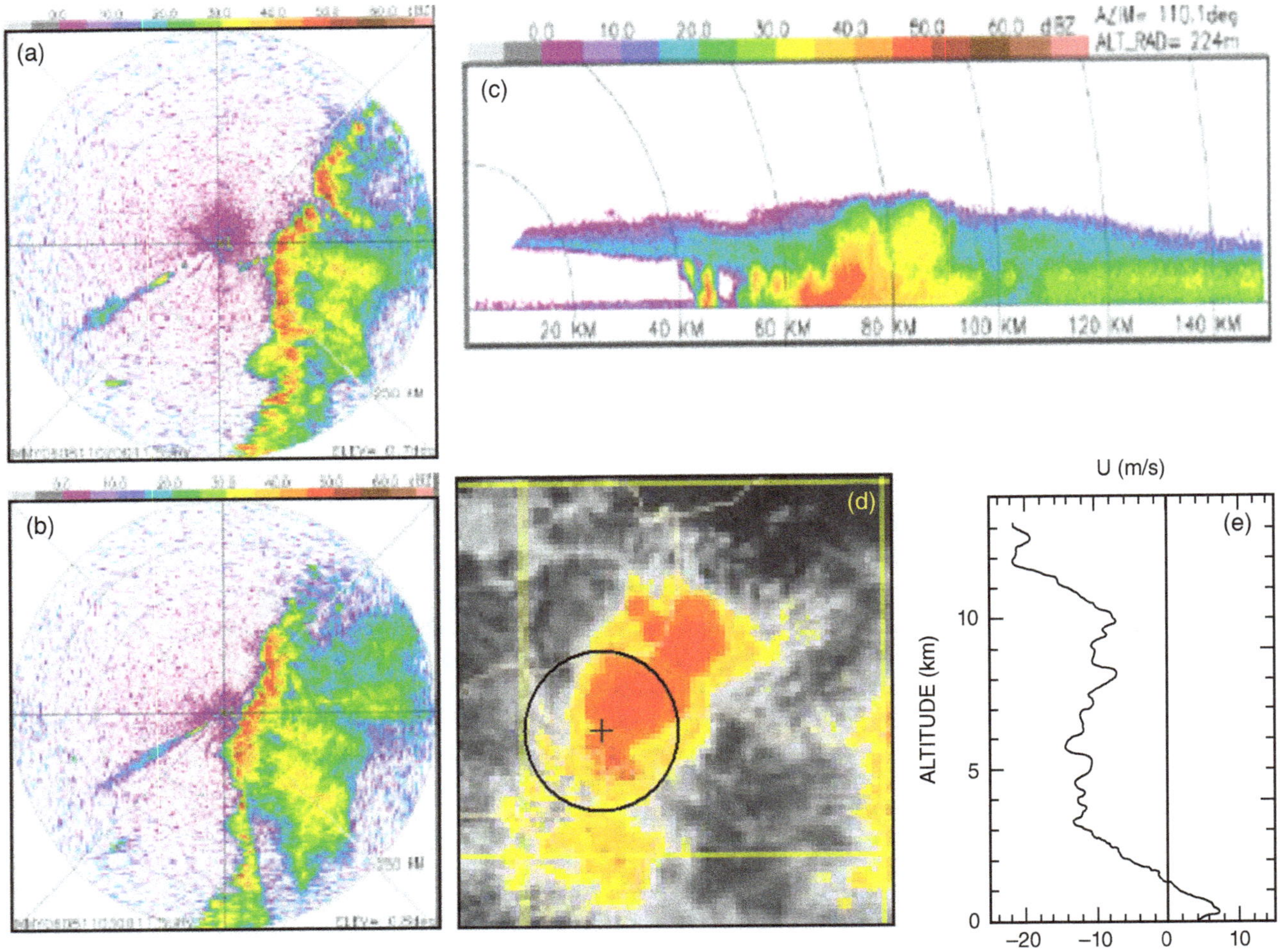


Figure 3.15 Squall line approaching Niamey (2.18°E, 13.48°N) on 11 August 2006 as seen by the MIT radar for reflectivity (dBZ) horizontal representation (plan position indicator (PPI)) at (a) 0200 and (b) 0300 UTC; (c) a vertical cross-section (range–height indicator (RHI) scan) at 0200 UTC. (d) The radar range (250 km) is superposed on the IR Meteosat image at 0300 UTC. (e) The vertical profile of zonal wind observed a few hours before the squall line passage. *Source*: Chong (2010). Reproduced with permission of the Royal Meteorological Society.

Figure 3.16 displays the spatial distribution of cloud cover JJAS climatology of MCSs for these four classes. This upper cloud cover (hours per month) is indicative of the convective rainfall.

- C1 systems are the more numerous but, with C3, have the weakest contribution to total rainfall. They are localised in the vicinity of elevated terrain (Fouta Djalon, Jos Plateau, Cameroon Mountains and Central Africa).
- In contrast, C2 systems have an important contribution to rainfall, especially in the Atlantic ITCZ, at the coast with maxima close to Conakry and the Niger Delta, and Central Africa.
- The contribution of C3 systems is also significant, especially in the Sahel band and over Central Africa.
- Fast-moving and long-lived C4 systems are concentrated in the Sahel band and strongly collocated with the AEJ, where they explain most of the rainfall. They correspond to fast-moving squall lines.

Favourable factors for C4 systems, as previously introduced (CAPE, PW, mid-level dryness, shear) are found in the Sahel band in agreement with the C4 climatology. Their maximum contribution to rainfall coincides with the mean location of the AEJ core. Dry air favours rain evaporation and intense density currents propagating westward at about the speed of the AEJ, and in the downshear direction. In a similar study, Mathon *et al.* (2002) defined an 'OCS' class corresponding to MCSs with temperature less than −60 °C, larger than 30 000 km$^2$ lasting more than 6 h and propagating with speed greater than 10 m s$^{-1}$, which is equivalent to the present C3+C4 fast-moving classes. They also make the largest contribution to the nebulosity (cloudiness) of the months

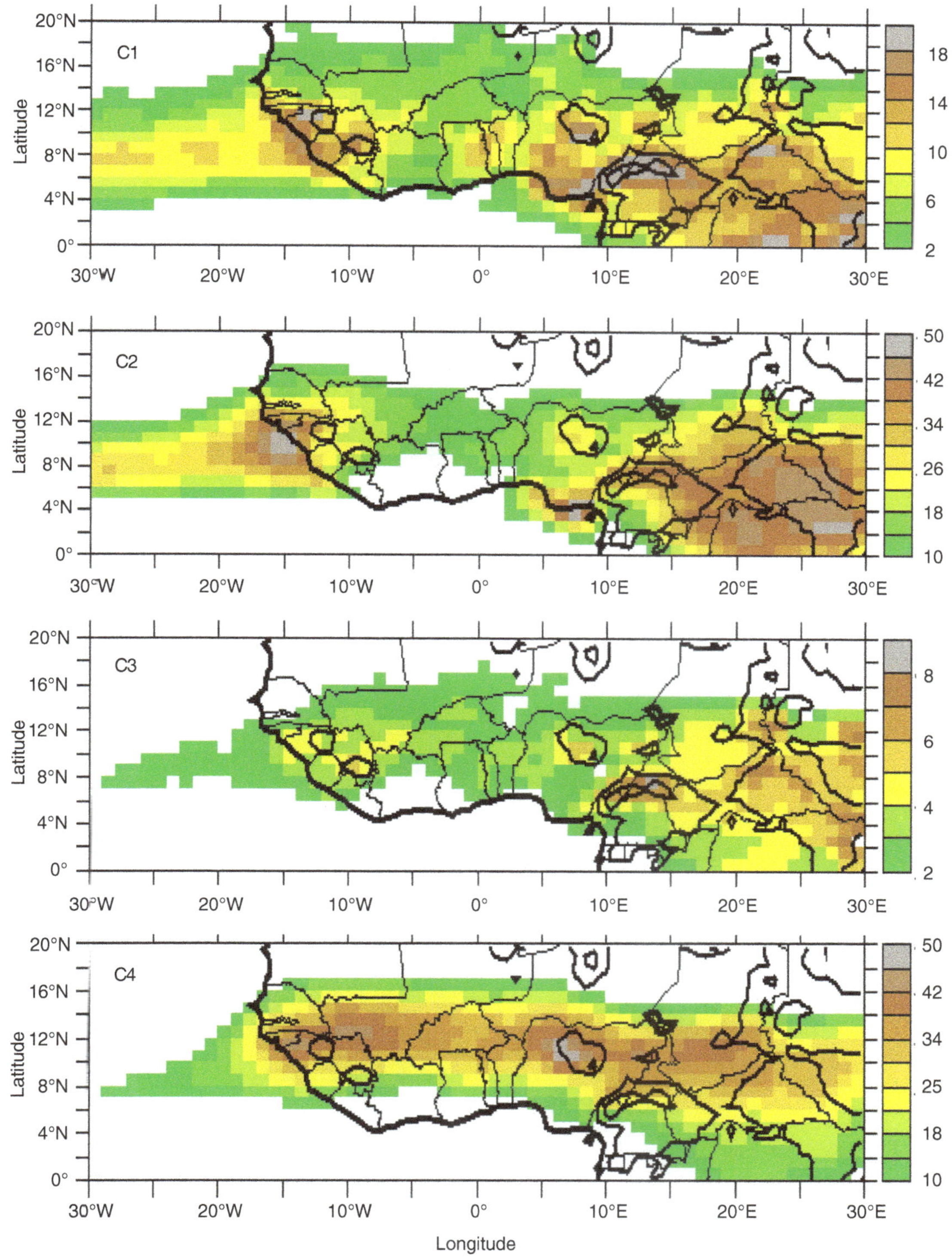


Figure 3.16 Spatial distribution of cloud cover JJAS climatology of MCSs over 25 years of Meteosat performed using the Fiolleau and Roca (2013) tracking algorithm for the four types of MCSs: classifications C1, C2, C3 and C4 (see text). Resolution is 1° square and the unit is hours per month. Heavy isolines represent surface altitude above 600 and 1200 m.

July and August 1993, especially in the Sahelian band, and are the most efficient in term of precipitation.[4] In contrast, south of the Sahel band and over the ocean, where the shear is weak and moister conditions prevail, MCSs propagate slower and stay close to their onset location, and other mechanisms are involved (such as forcing by the ocean/continent contrast, by the topography, and so on).

### 3.1.4 Life Cycle and Diurnal Cycle

#### 3.1.4.1 Triggering

Owing to the conditional character of the convective instability and to the strong convective inhibition (large CIN) over West Africa, especially over the Sahel, external forcings (factor (4) Figure 3.7) are necessary to trigger new convection. Figure 3.17 presents the climatology of triggering locations, to give insight into the mechanisms involved. Primarily, as short-lived convective systems are much more numerous than long-lasting (and well-organised) ones, the number of triggering events for C1 and C3 classes is greater than for C2 and C4 classes.

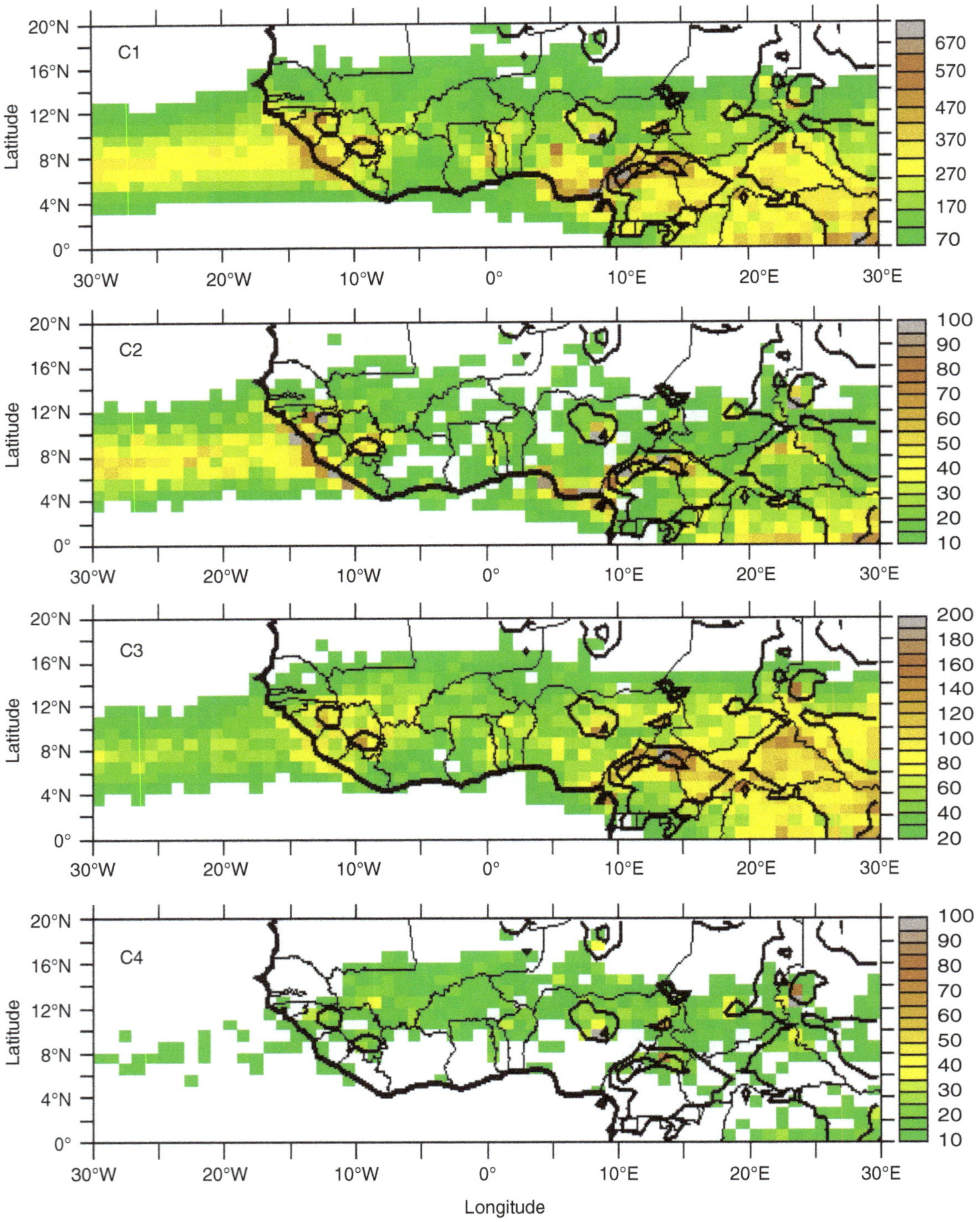


**Figure 3.17** As for Figure 3.16 for the location of the triggering. Unit is the total number of triggering events over the 25-year period of the climatology.

#### 3.1.4.1.1 Orographic Forcing

The triggering occurs most frequently in the vicinity of elevated terrain (above 600 m) for all classes; this significant result has been established for many years both in research and in forecasters' experiences. The areas south of the Jos Plateau (around 9°N, 8°E), and southwest of the Darfour (around 13°N, 23°E) and Cameroon mountains are very active triggering spots for all classes. Although a little weaker, especially for the C4 class, the Fouta Djalon (mainly in Guinea) is also active. The diurnal heating over higher terrain is the main cause of this triggering. In contrast, the dynamical effects over topography, such as orographic waves or blocking, are weak, because the flow is rather slow at low levels, the topography is moderate over West Africa, and the daytime convective boundary layer is neutrally stratified over a deeper layer than the hill tops (meaning that Froude numbers are high). Nevertheless, the southwestern flanks of topographic features are typically more favourable for triggering due to the upslope penetration of the moist southwesterly monsoon flow in these areas. A detailed examination of this triggering climatology brings useful information to local forecasters. For instance, the maxima over the Aïr Mountains (around 17°N, 8°E) or northern Benin confirm well-known local triggering spots.

#### 3.1.4.1.2 Coastal Circulations

Convective triggering occurs frequently in coastal regions, due to land–sea breezes induced by the surface temperature contrast between ocean and continent (see Section 4.1.4.2). In the summer season (JJAS), the Guinea–Sierra Leone and Nigerian coasts are very efficient to trigger MCSs in the C1, C2 and C3 classes. The combination of these two coastal forcings, with the orographic forcing due to the Fouta Djalon and Cameroon Mountains (as analysed by Vondou *et al.* (2010)) leads to the triggering of numerous MCSs of all types (except no C4 for South Nigeria and Cameroon), explaining the extrema of cloud cover (Figure 3.16) and rainfall in these two regions. Coastal circulations and weather conditions are discussed further in Sections 4.1.4.2 and 4.2.4.

#### 3.1.4.1.3 Surface–Atmosphere Coupling

Although less frequent, triggering is also detected over flat terrain below 600 m elevation over the western Sahel; for instance, over Mali for C3 and C4 MCSs (Figure 3.17). Based on observations during the AMMA campaign, and numerical simulations, recent studies suggest the surface–atmosphere coupling as a mechanism to trigger convection (these processes are discussed in more detail, in the context of surface microclimates, in Section 4.1.5.2). Taylor *et al.* (2011) analysed the initiation of over 3000 MCSs using satellite data and found a clear preference for the initiation of the first deep convective cells on strong soil moisture gradients induced by rain in previous days (see example in Figure 3.18). Figure 3.19 provides a conceptual model for this coupling mechanism. Previous rain events generate cold and moist soil patches at the surface, in turn affecting fluxes of sensible and latent heat, and the properties of the planetary boundary layer. Figure 3.19 sketches two moist patches (blue surface) surrounding a drier patch (red) over which the boundary layer is warmer and deeper. Breeze-type circulations, similar to but weaker than the sea breeze, force ascent over the edges of the dry patch. In the case of mean wind the forced ascent is stronger at the downwind edge of the dry patch where the preferred location for convective initiation is statistically observed. The effect is clearest for soil moisture patches on scales of 10–40 km, and was estimated to account for 1-in-8 MCS initiations in the western Sahel. As a result of this feedback, rain from single-cell storms and from MCSs in their early

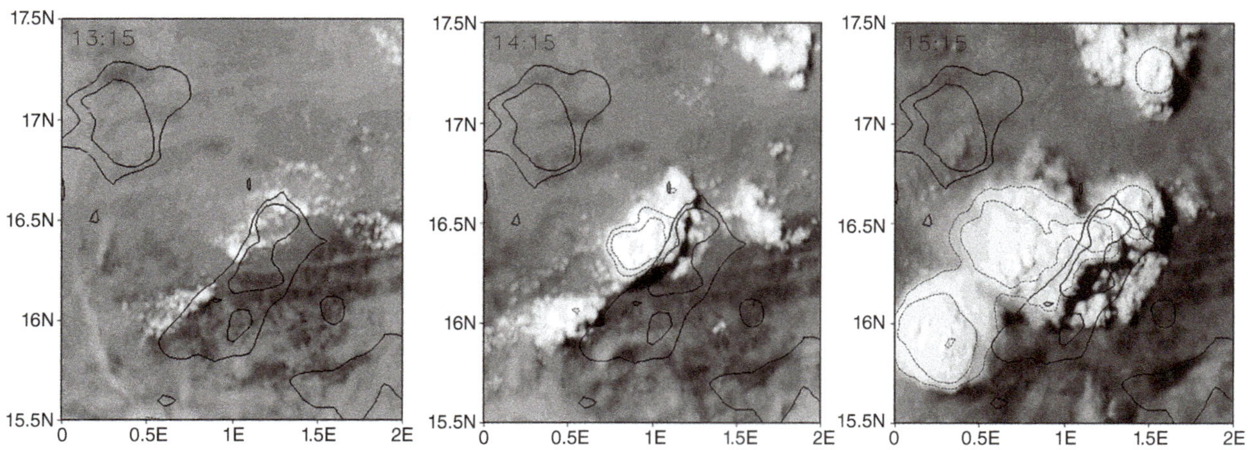


Figure 3.18 Example of the early afternoon development of an MCS in Mali from visible channel satellite data. Solid contours indicate wet patches created by rain the preceding day, and cloud top temperatures below −40 and −60 °C are enclosed by dashed contours. *Source*: Taylor *et al.* (2010). Reproduced with permission of the Royal Meteorological Society.

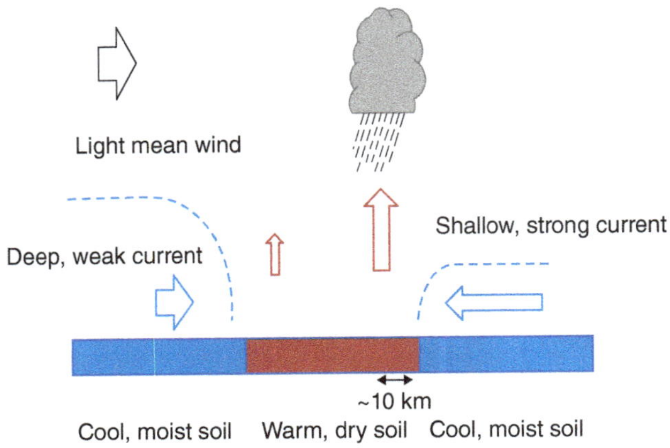


**Figure 3.19** Schematic depicting the impact of soil-moisture heterogeneity on convective initiation. Idealized soil-moisture-induced flows (blue arrows) under light synoptic winds (black arrow) create an ascent region (large red arrow) where the shallow, strong current opposes the mean wind. The preferred location for convective initiation coincides with the ascent region induced by the heating gradient at the downwind edge of the dry patch. Additional convergence over the dry patch is provided by a deep, weaker current at its upwind edge and cross-wind gradients in soil moisture. *Source*: Taylor *et al.* (2011). Reproduced with permission of *Nature Geoscience*.

stages is more likely over drier soil, in proximity to wet surfaces (Section 4.1.5.2).

Similar results from the Soudanian and Guinea Coast zones, where vegetation cover is much denser than in the Sahel, have shown that vegetation patterns, such as contrasts between forest cover and farmland, can also cause breeze-like circulations leading to convective triggering (Garcia-Carreras and Parker, 2011). A global analysis of satellite data (Taylor *et al.*, 2012) has confirmed this mechanism and highlighted Africa, and the Sahel in particular, as a region where the land-surface impact on precipitation is especially strong.

There is tentative evidence from case study events and modelling studies that mature MCS tracks will tend to follow the available moisture in the PBL. These results imply that, if an MCS approaches regions where there are differing amounts of PW, the MCS will move in the direction of the greater PW. These PW variations may be caused by differing moisture availability from the surface (due to synoptic-scale soil moisture patterns, or vegetation) as modelled by Gantner and Kalthoff (2010) and Wolters *et al.* (2010). Alternatively, the PW distribution can be modified by synoptic dynamics and advection: MCSs have been observed to propagate toward the moist nocturnal inflow of the monsoon (with similar behaviour observed in the USA).

In practice, forecasters may need soil moisture information such as Land SAF (http://landsaf.meteo.pt/) and VITO products and ASCAT (see Sections 9.1.5.9 and 9.2.1.3) to apply this theory. As such data are not generally available, forecasters can use the land surface temperature anomaly derived from Land SAF Meteosat product as a surrogate for soil moisture, with wet surfaces in the Sahel being cooler than dry surfaces. Indeed, negative land surface temperature is correlated with moister soil. Forecasters may also monitor the patterns of accumulated rainfall from the previous days to localise the wet/dry soil spots: regions where rainfall has fallen on a dry surface may be particularly favourable for convective initiation over the next 2–3 days.

#### *3.1.4.1.4 Synoptic Forcing and Convergence Lines*

In the absence of surface-based triggering, there are still dynamical patterns in the atmosphere that can trigger new convective cells. Synoptic patterns, particularly AEWs, are known to organise the distribution of convection, and the convective triggering (see the schematic of AEWs in Figure 2.10). However, it remains difficult to predict the exact location of triggering within a synoptic wave. It is known that the synoptic-scale vertical velocities within AEWs, on the order of a few centimetres per second, are not strong enough to cause convective triggering directly. Instead, it is thought that the AEWs create suitable conditions in which smaller, mesoscale 'coherent structures' may form, and these coherent structures provide the stronger vertical velocities, tens of centimetres per second or more, which are needed to lift air to its level of free convection at some location. In recent years, research field campaigns have led to an appreciation that such coherent structures are very common in the atmospheric PBL. Although they are still generally undetected by conventional observations, knowledge of their characteristics can help a forecaster to understand the likelihood of convection in some circumstances. An important family of features that tend to initiate cumulonimbus storms is that of 'convergence lines'. In the International $H_2O$ Project in the USA, almost all daytime storms were initiated at convergence lines (Wilson and Roberts, 2006; see also Sections 4.1.4.1 and 6.1.2.6).

Convergence lines may take several forms. Quite commonly a convergence line will form at a coastal headland and propagate hundreds of kilometres inland on the sea breeze, or monsoon flow. Similarly, convergence lines may be found as a line extending downwind from a range of hills. In the absence of topographic forcing, convergence lines may form spontaneously within a wider region of synoptic-scale convergence – for instance Bain *et al.* (2011) showed evidence of convergence lines embedded within the structure of an AEW. In this case, it seems that the synoptic-scale circulation at temperature gradients within the AEW have caused a small-scale convergence within the wave. Another likely process of generation of strong convergence in AEWs is through the dynamics of mid-level vortices. In some forecasting centres these vortices, which occur commonly on the monsoon trough at 850 hPa, are highlighted as likely locations for convective triggering. The exact nature of the dynamical triggering at such vortices is not well understood, but similar triggering by mesoscale vortices has been observed in the USA. Figure 3.20 provides illustrations of such triggering mechanisms as seen from the IR images. At 0900 UTC we have the residue of a previous MCS1 with characteristic high cloud debris rotating anticyclonically. Three hours later a new MCS4 is triggered in this same area well ahead the trough of the

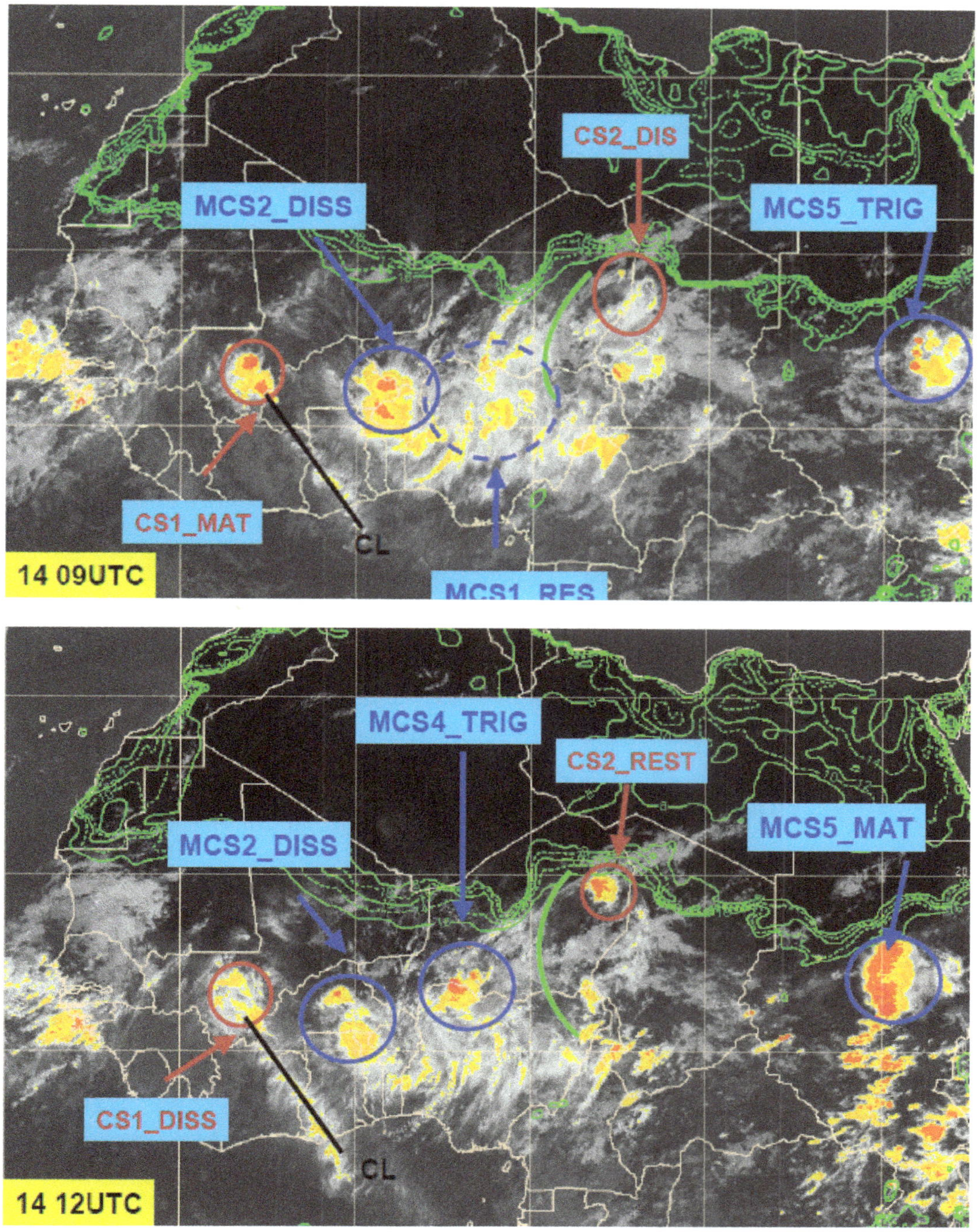


Figure 3.20 Illustrations of the dissipation and triggering of some MCSs on 14 August 2012, as extracted from the case study CS02 of an 'Archetype of AEW' observed between 13 and 16 August 2012 (for a detailed analysis, see e-learning material available on the site www.cnrm.meteo.fr/waf_handbook_casestudies/). IR Meteosat images are shown at 0900, 1200 and 1500 UTC with yellow and red colours for coldest clouds (less than −40 °C and −65 °C respectively), and with green isolines of $T_d$ at 2 m every 2 °C between 10 and 16 °C as provided by the ARPEGE operational analysis, to outline the intertropical discontinuity (ITD) location. The main AEW trough is drawn with a green arc-shaped line. The different MCS triggering, growing, mature, dissipation, residual and restart stages are labelled with the subscripts TRIG, GROW, MAT, DISS, RES and REST respectively.

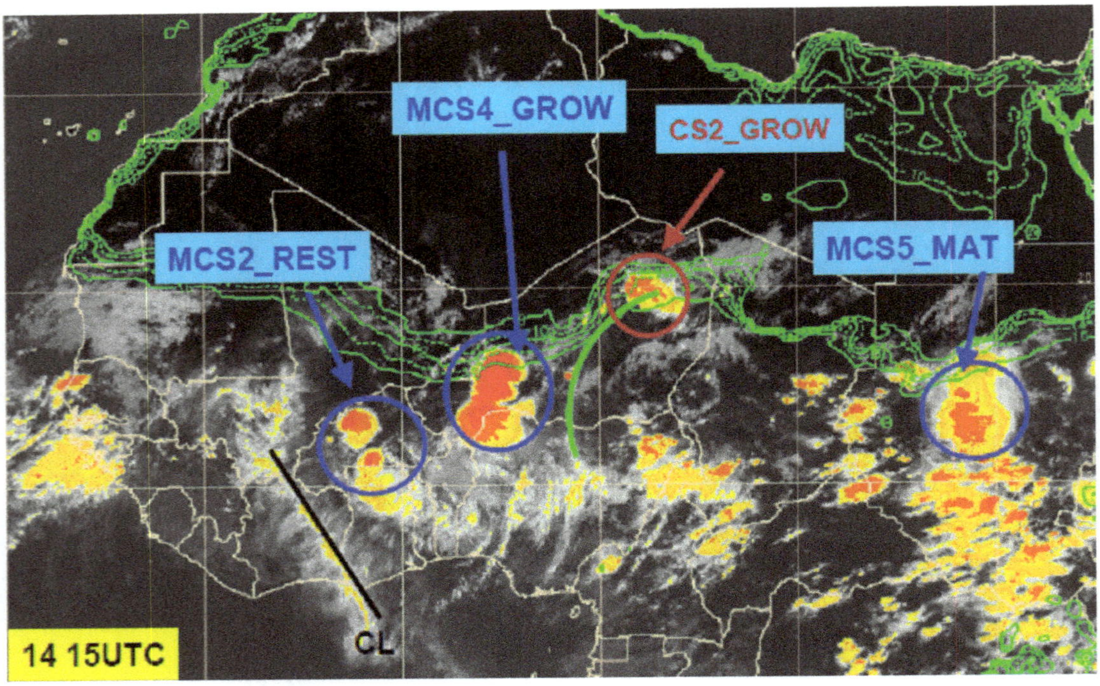


Figure 3.20 (Cont'd)

AEW (green arc-shaped line). Similarly, an intense fast-moving squall line, MCS2, is dissipating at 0900 UTC and restarts early in the afternoon. It is also the case of a convective system CS2 located behind the trough just south of the intertropical discontinuity (ITD) in the region of moist anomaly of the wave. The e-learning part of the Handbook[5] provides many such illustrations and detailed analysis for different case studies at different scales.

On smaller scales, the cold pools emerging from beneath MCSs have a strong gust-front at their head, and this may be regarded as a form of convergence line – indeed, as a cold-pool ages, at a long distance from its parent storm, it is mixed by boundary-layer turbulence and loses its gravity-current-like structure, in which case it may more generally be classified as a convergence line, and may be in evidence more than 24 h after its original genesis (e.g. Flamant *et al.*, 2009).

Convergence lines are generally difficult to detect in conventional observations, since they are narrow features (only a few kilometres across, if hundreds of kilometres long) with relatively small wind shift at the surface. However, in the afternoon the convergence lines can be revealed by shallow cumulus developing along their length, which may be apparent in visible satellite imagery, several hours before deep convection develops. It is thought that this shallow cumulus, by providing heating, can intensify a convergence line; for instance, as it extends downwind of a mountain. Figure 3.20 provides an illustration of such a convergence line (CL) detected from the IR imagery. The collapsing of MCS2 and the spreading of its density current far ahead in the shear direction (southwestward) generates a line of convergence visualised by a cloudy line, over Ghana and Ivory Coast, that lasted about a full day.

#### 3.1.4.1.5 Gravity Waves

The stably stratified troposphere supports gravity waves, which will radiate outward from a disturbance (Figure 3.6). In West Africa, intense and deep cumulonimbus storms generate strong sources of gravity waves, which propagate long distances. These gravity waves have already been mentioned in Section 3.1.1.2, in the way in which they spread the heating effects of convection into the wider environment. The gravity waves also have a role in triggering new convection: they can suppress or promote subsequent cumulonimbus storms in their environment.

A simple rule of thumb for gravity-wave propagation is obtained from the theoretical behaviour of gravity waves in a hydrostatic atmosphere of constant static stability $N$ and zero wind shear. In this model, two-dimensional (horizontal, vertical) gravity waves have a horizontal and vertical wavenumber ($k$, $m$), and their phase speed is given by $C_p = N/m$. Since $C_p$ is independent of $k$, these waves are non-dispersive. If a complex source of wave energy is initiated rapidly, or impulsively, as in the case of a cumulonimbus triggering event, waves of a given vertical wavenumber propagate together as a wave packet. The propagation of this wave packet will act to maintain mass conservation in the atmosphere: where the cumulonimbus storm generates a strong updraught, the wave packet will cause subsidence in the storm's environment to conserve mass, and this subsidence will also warm the environment, as described in Section 3.1.2.2.

A complex source of heating and cooling such as a cumulonimbus storm excites gravity waves with a range of vertical wavenumbers $m$. Generally, we expect these to be harmonics of the tropospheric depth; for example, if the tropopause is at 17 km, with $N = 10^{-2}$ s$^{-1}$, the deepest possible wave would be of vertical wavelength 34 km, and corresponding phase speed ~60 m s$^{-1}$, and the next mode would have vertical wavelength 17 km and phase speed ~30 m s$^{-1}$. However, shallow patterns of forcing exist within a cumulonimbus storm, for instance forced by evaporation of precipitation in the downdraught (Figure 3.6b), for which the vertical wavelength may be on the order of 6 km and the wave phase speed 10 m s$^{-1}$. Such a wave packet, forced by cooled, descending air in the storm, will force ascent in the environment and will tend to trigger new convection.

A good example of the role of linear gravity wave packets in triggering secondary convective storms is the case of 31 July 2006, described by Taylor *et al.* (2010) and modelled by Klüpfel *et al.* (2012) and Birch *et al.* (2013). In this case (Figure 3.21), a major secondary storm was initiated over a region of contrasting soil moisture, in a favourable synoptic environment, but the timing of the secondary initiation was determined by the arrival of a gravity wave that had been emitted from the parent storm further to the southeast.

In this example, and with the benefit of hindsight, the gravity wave could be seen in satellite imagery as a warm (approximately −10 °C) cloud band propagating away from the parent storm ahead of the dusty surface cold pool (Figure 3.21c). In other cases, propagating gravity waves may have no easily distinguished cloud signature, but forecasters should always be aware that there is a likelihood of wave packets propagating away from an existing storm, which can initiate daughter storms at remote locations.

#### 3.1.4.1.6 *Regeneration of Storms*

Multicell storms such as squall lines can have a remarkable persistence, in some cases being tracked for more than 24 h. More commonly, squall lines tend to decay in the morning hours. In such cases there is a possibility that convection, and a squall line, will be regenerated in the same place the following afternoon.

- The exact causes by which a new storm is regenerated are not absolutely understood, but a few mechanisms are likely. The MCS has a mesoscale circulation, a potential vorticity signature and some degree of rotational flow that has inertia and can persist for a few hours even if the convective region of the MCS ceases to be active. When the convective instability gets stronger again, later in the afternoon, this mesoscale flow can provide triggers, and organisation, to a regenerated storm. Figure 3.20 provides an illustration of this mechanism with MCS2 restarting at 1500 UTC after its dissipation stage in the morning and at noon. This kind of storm regeneration, through the maintenance of a mesoscale vortex, has been observed in the USA.
- The MCS wets the surface. The wet surface causes evaporation, to humidify the boundary layer, while the patchiness of soil moisture provides strong contrasts in the land–atmosphere forcing, which can trigger new storms (see Section 3.1.4.1.3).

#### 3.1.4.1.7 *Other Forcing*

The triggering of convection over Central Africa is weakly understood and needs further, specific studies. This region is wet and densely vegetated, with some topographic features, and may be partly forced by equatorial waves (Chapter 7).

Over the eastern Atlantic in the ITCZ, the triggering is similar for all MCS classes, except for C4 that disappear over the ocean (Figure 3.17). Convective triggering is easier due to weak inhibition (CIN) and lower cloud bases, and it appears that the large-scale convergence (in the ITCZ) forces the convection.

Crook and Moncrieff (1988) showed that the effects of large-scale convergence are important both for the triggering and maintenance of squall lines and MCSs. This is potentially relevant for forecasters because detection of large-scale convergence should be possible without special observations. Furthermore, the convergence aspect is associated with gravity wave and cold pool aspects already addressed earlier. However, large-scale convergence should be treated with great caution when forecasting storm initiation: Birch *et al.* (2014) have shown that model analysis convergence fields have little relation to actual locations of MCS initiation over West Africa.

### 3.1.4.2 The Diurnal Cycle

The distribution of local time of the minimum of brightness temperature, corresponding to the maximum of deep convection (Figure 3.22a), indicates opposite behaviours between ocean and continent in the tropics. Oceanic deep convection tends to reach its maximum in the early morning (green areas), whereas continental convection generally peaks in the evening (purple areas). A detailed analysis of the diurnal cycle of the fractional cloud cover (%) colder than 213 K over West Africa indicates substantial differences between four subregions (Figure 3.22b). The minimum of convective activity occurs at about noon over Africa, but the maximum is

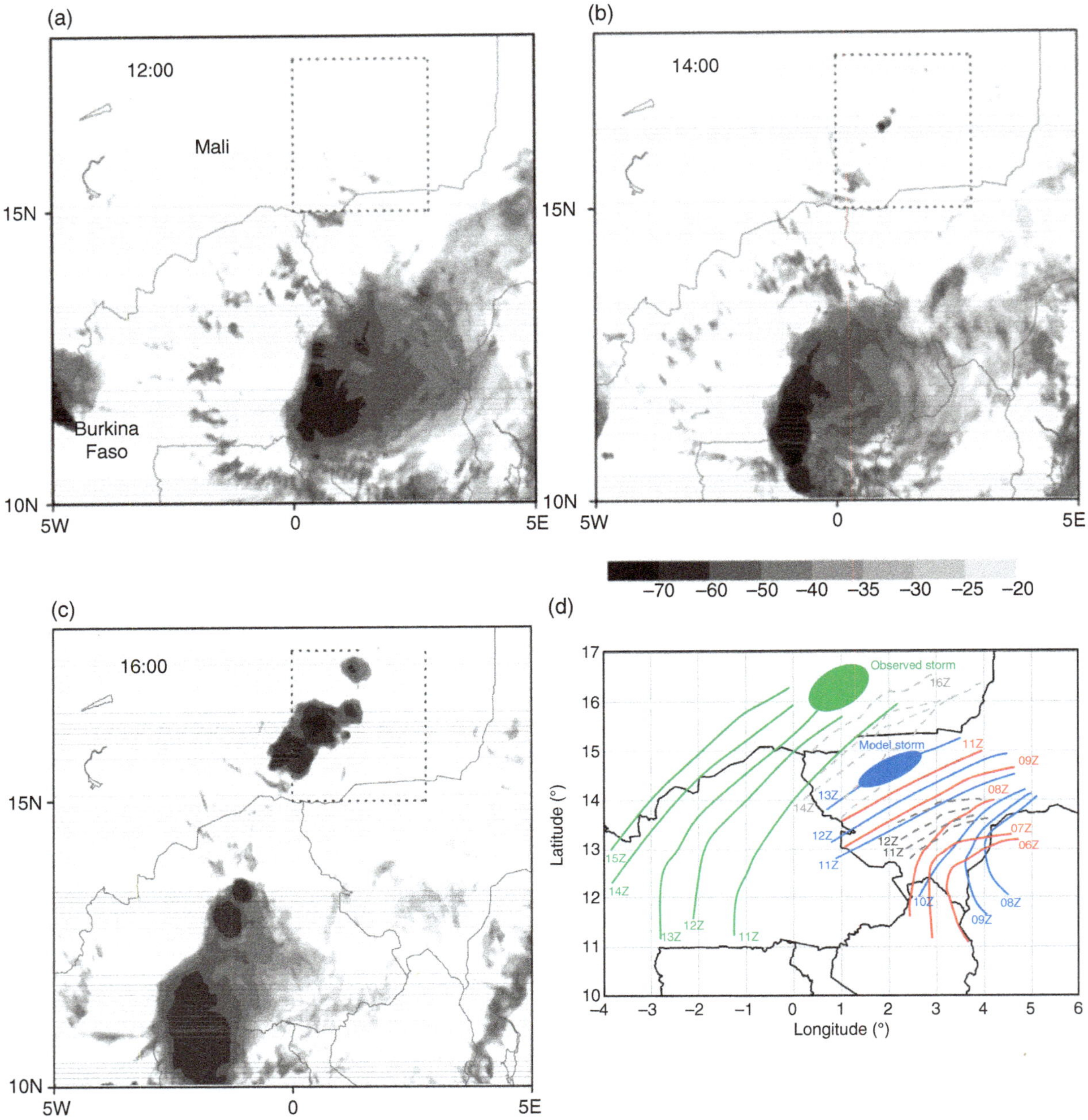


**Figure 3.21** Cloud structures from brightness temperature (°C) at (a) 1200, (b) 1400 and (c) 1600 UTC, on 31 July 2006, indicating gravity-wave cloud and secondary initiation. *Source*: Taylor *et al.* (2010). Reproduced with permission of the Royal Meteorological Society. (d) Observed gravity wave (solid green lines) and the observed cold pool (the light grey dashed lines). The first and second model gravity waves are shown by the solid red and blue lines respectively, and the dark grey dashed lines show the model cold pool. *Source*: Birch *et al.* (2013). Reproduced with permission of the Royal Meteorological Society.

dependent on the subregion; that is, at the end of the afternoon (~1700 LST, or local solar time) south of 9°N, and in the evening (~2000 LST) over the Sahel north of 9°N. Also, the nocturnal deep convection is more frequent over the central Sahel than to the south, consistent with more cloud coverage by long-lived C4 MCSs over the Sahel.

#### 3.1.4.3 Dissipation

For short-lived MCS classes such as C1 and C3, the distribution of the location of their dissipation (Figure 3.23) is similar to the distribution of their triggering (Figure 3.17) due to the short travel distance. Nevertheless, the patterns are smoother with a weak westward shift due to their mean propagation.

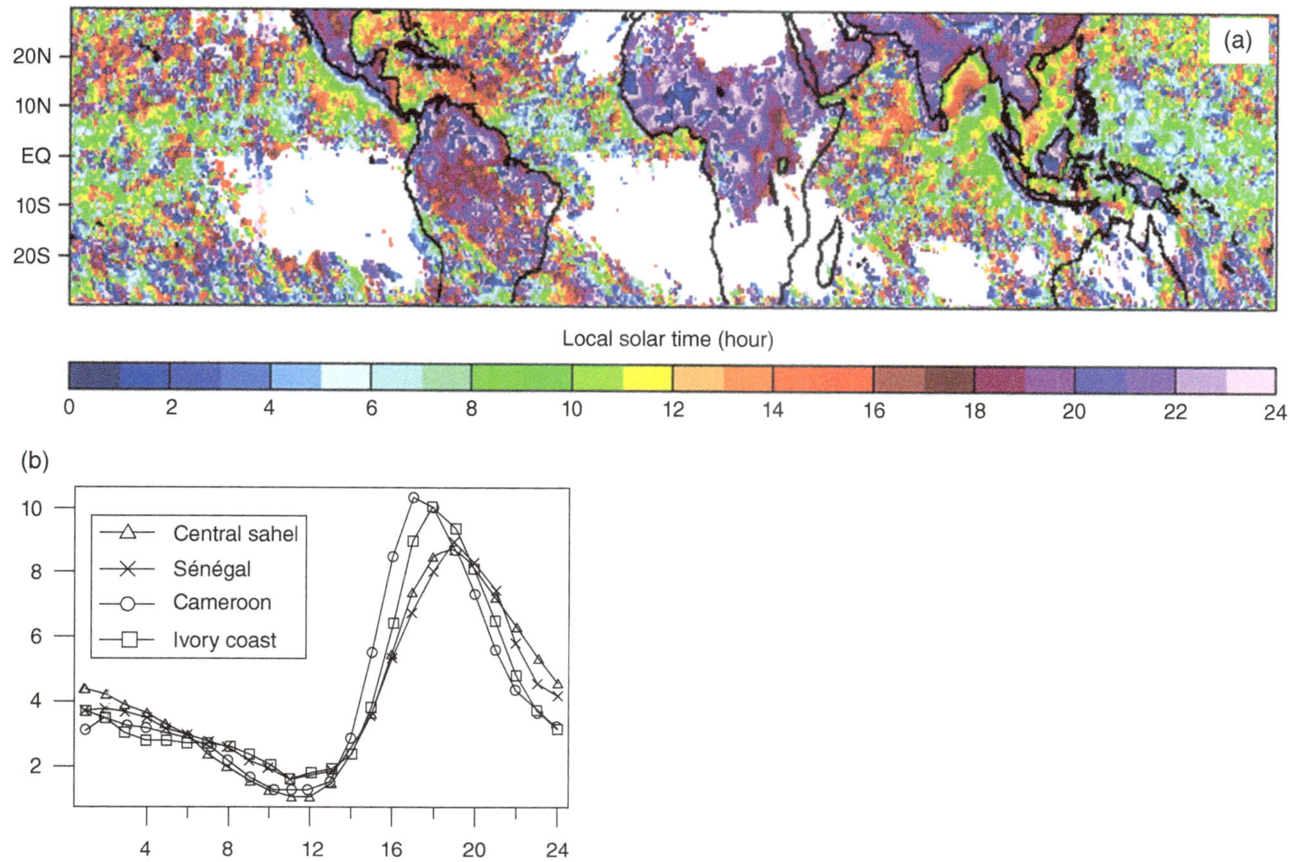


Figure 3.22 (a) Distribution of local time of the minimum of brightness temperature $T_b$ from CLAUS (adapted by J-M Piriou). *Source*: Yang and Slingo (2001). ©American Meteorological Society. Used with permission. (b) Composite diurnal cycle of the fractional cloud cover (%) colder than 213 K for four West African regions. *Source*: Mathon (2001).

Long-lived and slow-moving C2 MCSs dissipate farther to the west from their origin, with some interesting maxima over the ocean, offshore of Nigeria to the south of the Niger Delta, and also offshore of Guinea and Sierra Leone. The lack of dissipation and the weak occurrence of C1 and C2 systems in the [8°W–0°] band covering Ivory Coast, Ghana, Burkina Faso and Mali are remarkable. Indeed, this region is too far to the west of the main areas of triggering associated with the orography. The dissipation of long-lived and fast-moving C4 MCSs occurs within a large zonal band shifted to the south of the triggering area, and westward, so that C4 storms often dissipate over the eastern Atlantic, reaching 30°W in some cases. This is coherent with their fast westward propagation with a weak southward component toward moister regions, and in agreement with the shear direction. In contrast, especially west of 5°E, some systems move northward and can dissipate at latitudes of up to 20°N. Local maxima of dissipation can be noticed and can be confirmed by local forecasters based on their long experience. For instance, we can note well-known dissipation spots in central Nigeria, over Benin and offshore of Guinea Bissau. Owing to the secondary generation of MCSs (see Section 3.1.4.1.6), some triggering locations may be associated with previous dissipation locations such as the west of Benin.

## 3.2 Operational Methods

Predicting convection is one of the most important and, at the same time, most difficult tasks facing the forecaster. In this section we summarise the factors and diagnostics that should be taken into account when making a forecast of deep convective activity. These concepts are further developed in Chapter 11, in which systematic rules are provided for making and plotting convection on regional forecast charts.

Forecasting convection is difficult because there is inherent randomness in the behaviour of storms. The explosive nature of deep convection means that small changes in the weather conditions, or small-scale triggering mechanisms that may not be detectable by routine observations, can lead to violent storms which substantially change the larger scale evolution of the atmosphere. Once formed, each storm therefore changes the likelihood of future storms. This randomness, or 'stochastic' behaviour, means that forecasts of

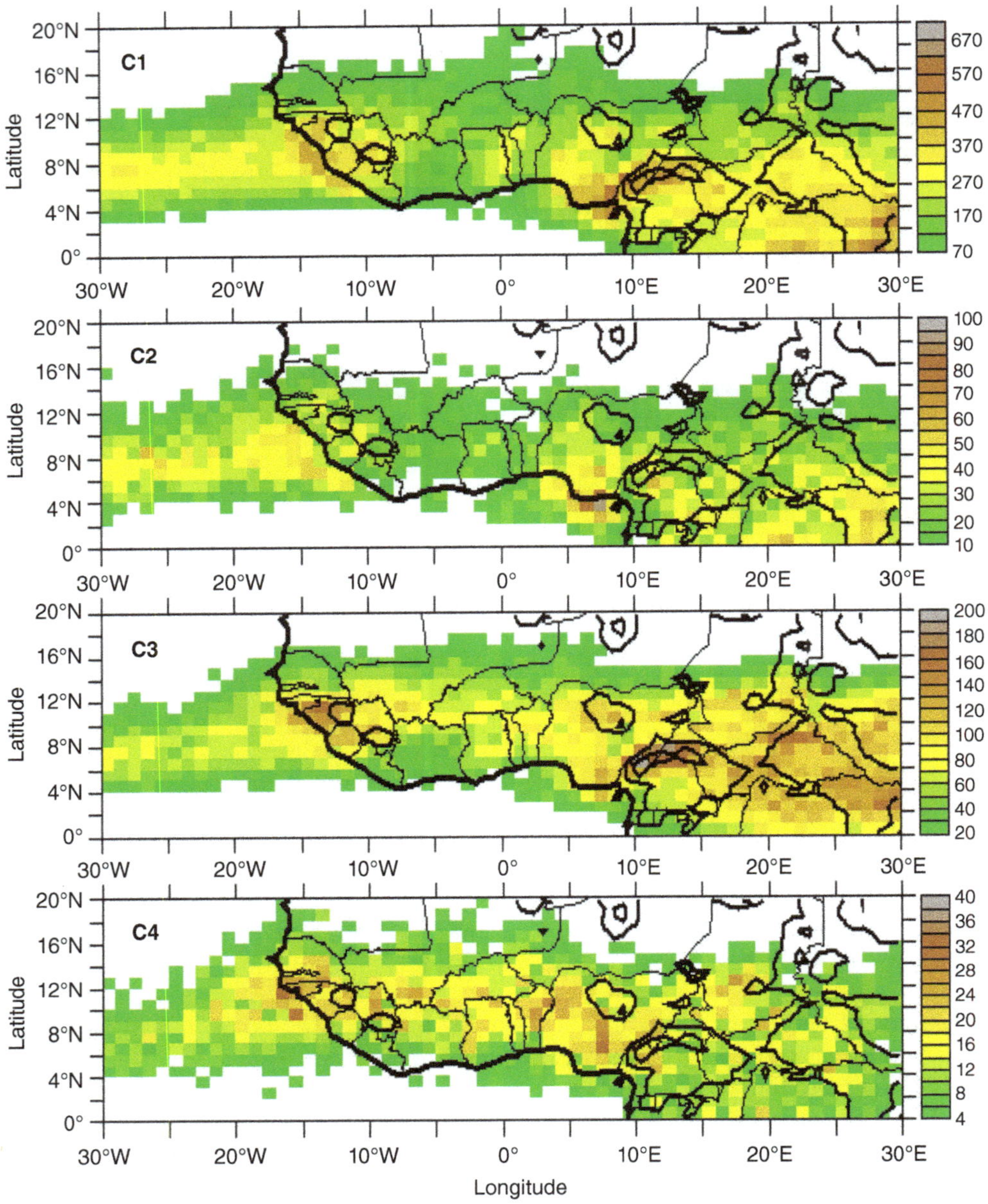


Figure 3.23 As for Figure 3.16 for the location of the dissipation. Unit is the total number of triggering events over the 25-year period of the climatology.

convective storms need to be considered as probabilistic forecasts, and the forecasts need to be evaluated statistically. Furthermore, forecasters cannot be trained by considering just one or two classical events; rather, they need to build up experience by considering many kinds of convective event, so that they gain an appreciation of the stochastic characteristics of deep convection. In Sections 3.2.1 and 3.2.2 we show examples of typical convective events; more cases are provided in the online supporting case-studies.[6] These case studies are numbered CS01, CS02, and so on for consistent referencing.

Forecasting of convection can be separated into three tasks:

- *Prediction of locations at which convection will initiate* (usually in the afternoon over land), and regions that will be free of convection (Section 3.2.3)
- *Prediction of the form of convection that will result,* once it has started; for instance, whether the convection will be isolated and short-lived, or organised, and the likely movement of any organised convection (Section 3.2.4).
- *Prediction of the decaying stage,* which occurs in the evening for the smallest systems, or for the

well-organised MCSs at the end of the night or in the morning. The morning is the less favourable period for deep convection, but specific attention must be paid at this period to MCS debris providing favourable conditions for the triggering of new systems in the afternoon (Section 3.2.5).

Once storms have formed, then their subsequent behaviour should be the subject of nowcasting (Chapter 6).

### 3.2.1 Life Cycle of Convection on 12–16 August 2012 (CS02)

To illustrate examples of the main phenomena described, we use a sequence of satellite images (Figure 3.24) and of West African Synthetic Analysis (WASA) maps (Figure 3.25) from case study CS02 for which the synoptic environment is described in Section 2.2.2.1.

#### 3.2.1.1 MCS1

This system appears on 13 August between 0900 and 1200 UTC from the debris of a previous storm, ahead of the trough of the AEW that is located over southern Chad. It is located south of the AEJ axis and seems to be linked to a low- and mid-level vortex (WASA, Figure 3.25). This system stays close to the trough of the AEW and keeps a moderate propagation speed. It grows until 2100 UTC and then quickly decreases in the first part of the night. It collapses on 14 August between 0600 and 1200 UTC, and generates a wide vortex that can be identified from the residual cloud structure and from the wind field (see vortex centres C600 and C600_850 on WASA at 0000 UTC and 1200 UTC respectively on 14 August (Figure 3.25). MCS4 will be triggered from this convective residue.

#### 3.2.1.2 MCS2

This system appears, like MCS1, ahead of the AEW trough, a little more to the north (~13°N, around the western side of Lake Chad). It quickly grows during the afternoon and evening, to become a big MCS overnight (2100 to 0000 UTC). It is located close to the AEJ axis and moves westward faster than the trough of the AEW. It quickly becomes independent from the AEW. It is developing further north, so in a drier area than MCS1, and reaches bigger dimensions and faster propagation. We have here a typical fast-moving squall line evolving in a favourable environment characterised by low-level shear, moisture availability (PW >50 mm) and dry air at mid-levels.

In the second part of the night MCS2 begins to decrease, with a southwest displacement, and then dissipates between 0900 and 1200 UTC on 14 August. When it collapses, the spreading of a density current, or possibly a gravity wave (Sections 3.1.4.1.5 and 4.1.4.1), far ahead to the southwest, generates a line of convergence visualised on the IR image by a cloudy line in the mid-troposphere, over Ghana and Ivory Coast. The resulting convergence line (CL) can explain the triggering of convective system 1 (CS1; see later).

In the afternoon of 14 August, MCS2 regenerates (not shown) over Burkina Faso (where a high PW anomaly is observed), then grows and moves westward, before decaying the following night. Its residue can be identified over Fouta Djallon on 15 August at 0900 UTC. So MCS2 lasted 2 days and travelled about 25° in longitude.

A simple conceptual model of the convection life cycle over the Sahel is exemplified in this case study. The same cycle is observed during a 24-h cycle for MCS2 and MCS4 (not shown):

- In the beginning of the afternoon, isolated convective cells appear ahead of the trough of the AEW.
- During the afternoon, these cells increase and merge to form an MCS that moves in the evening and the first part of the night faster than the AEW and becomes independent.
- In the second part of the night and in the morning, the MCS decreases.
- New isolated convective cells appear ahead the trough of the AEW in the beginning of the afternoon if the atmosphere is still convectively unstable, or the low levels moist enough.

#### 3.2.1.3 MCS3

This system appears on 13 August at 1200 UTC, at the south Mauritania–Mali frontier, a well-known geographical 'hot-spot' for convective triggering. The initiation occurs close to the mid-level vortex C600 on the southern flank of the AEJ core of the first perturbation (see corresponding WASA of Figure 3.25).

This system grows during the afternoon and becomes mature in the evening (2100 UTC). It covers the whole of Sénégal and the northwest of Guinea. The system decays during the night over Sénégal.

Precipitation over Sénégal as estimated from TRMM (Huffman *et al.*, 2007) and observed by the rain gauge network between 13 August at 0600 UTC and 14 August at 0600 UTC is shown in Figure 3.26. This example illustrates the footprint of MCS3, its high spatial variability of precipitation and the permanent challenge for the forecaster to forecast such events.

#### 3.2.1.4 CS1

Although of smaller size and shorter duration, the convective system CS1 is interesting to analyse as it is triggered at the evening of 13 August at 2100 UTC in a region of suppressed convection. A detailed analysis of the synoptic conditions as summarised by the

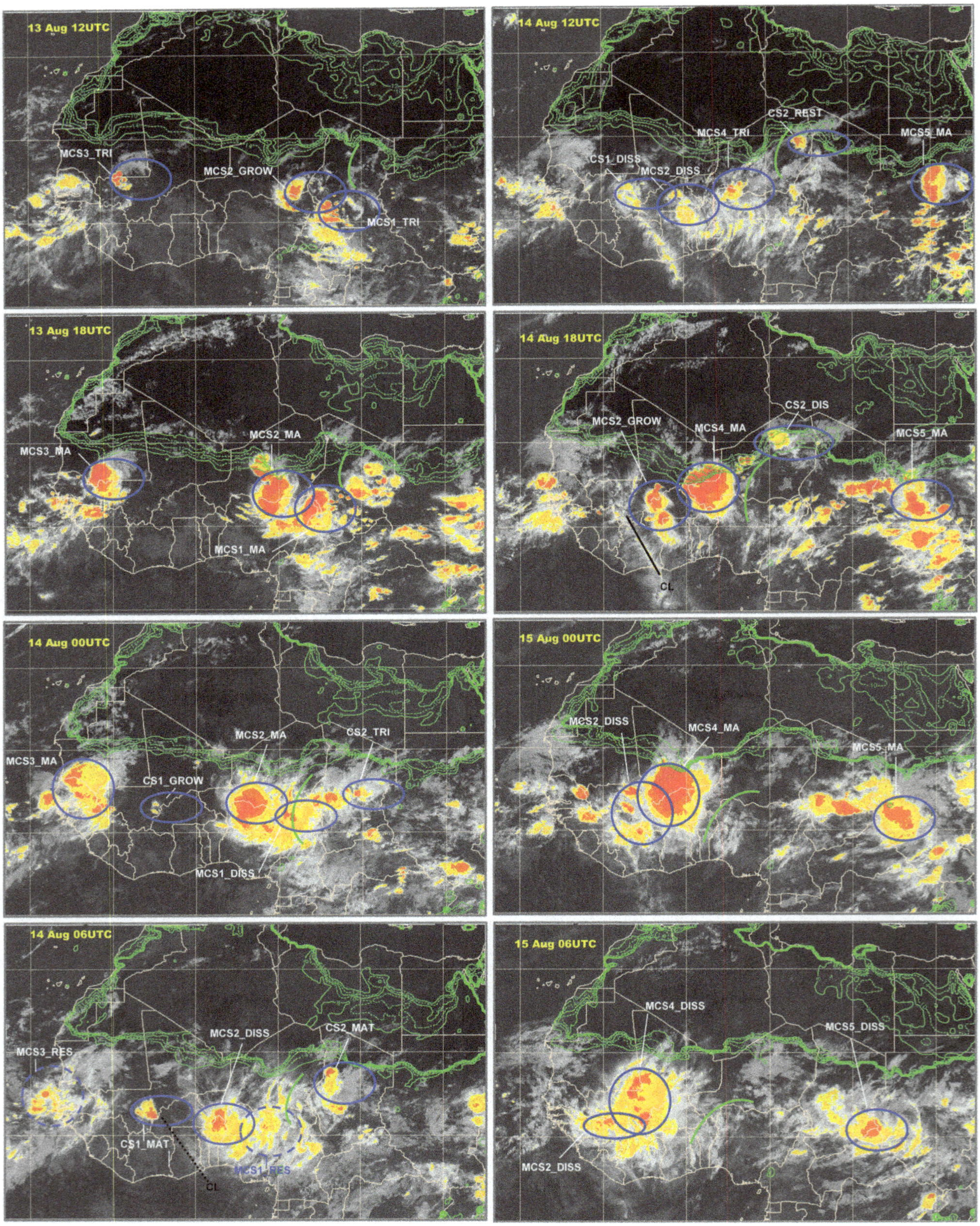


**Figure 3.24** Sequence of images showing evolution of MCSs in CS02 showing at a 6 h frequency, IR Meteosat image with yellow and red colours for coldest clouds (less than −40 °C and −65 °C respectively), and with green isolines of $T_d$ at 2 m every 2 °C between 10 and 16 °C, to outline the ITD location. The main AEW trough is drawn with a green arc-shaped line, and a convergence line is shown in black. The analysis is performed from 13 August at 12 UTC to 15 August at 1200 UTC. Convective systems (**CS**) and MCSs (**MCS**) are numbered in their order of appearance, starting from the east, with a suffix corresponding to the stage of their life cycle: TRIG for the triggering stage; GROW for the growing stage; MA for the mature stage; DISS for the dissipation stage; RES for a convective residue; REST for a restart from a convective residue.

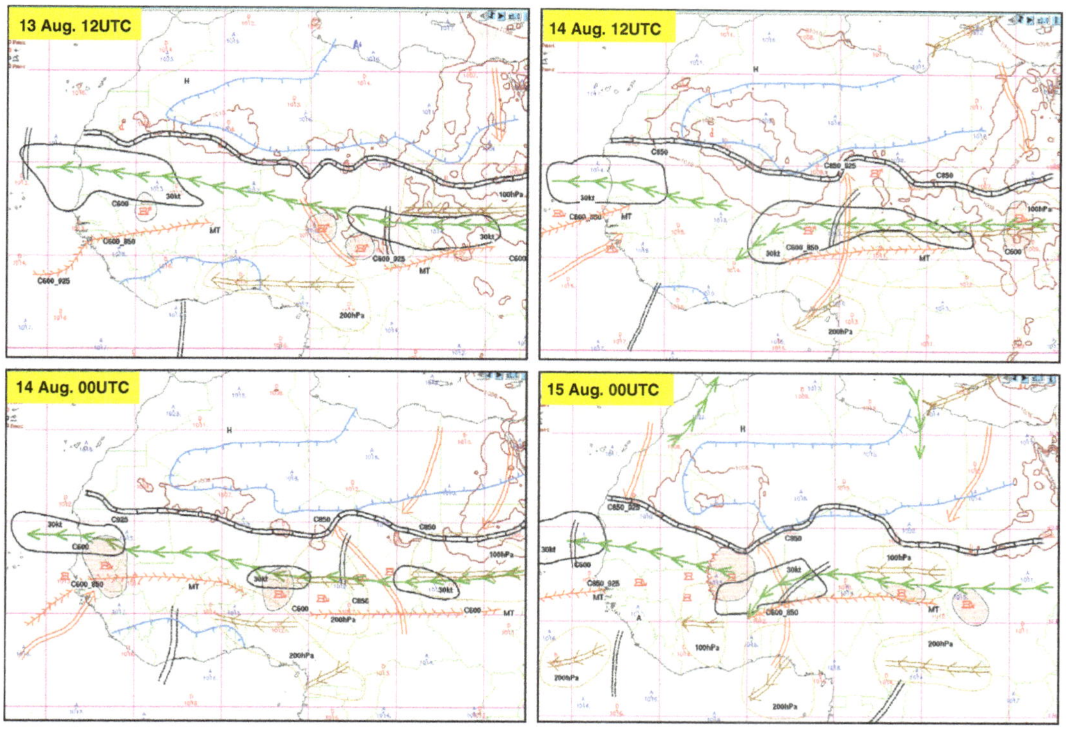


**Figure 3.25** Sequence of WASA corresponding to the sequence of IR images of Figure 3.24 at a 12 h frequency.

corresponding WASA (Figure 3.25) reveals that CS1 is triggered on the monsoon trough (MT) without any PW anomaly. Another possible contribution to its onset could be ascending motions due to the passage and the interference of gravity waves emitted by MCS2 and MCS3, as CS1 appears in between these two active MCSs. Also, local forcing, such as surface inhomogeneities or topographic features, cannot be excluded, as suggested by its 'V shape' and its weak propagation during the night.

In the morning of 14 August CS1 is passed by the convergence line CL. Their crossing corresponds to the reinforcement of CS1 at 09UTC. CS1 then continues at the speed of CL propagation, but dissipates before midday.

#### 3.2.1.5 CS2

This small convective system appears behind the AEW trough on 14 August at midnight over central Chad, within the cloudy area of the wet phase of the perturbation as indicated by the wet PW anomaly. It moves then northwestward during the night, toward the ITD penetrating northward. It dissipates in the morning but restarts at midday just south of the ITD over northeast Niger. It stays active during the afternoon along the ITD and dissipates in the evening.

Like the northward displacement of the ITD, these convective cells are due to the southerly flow behind the trough, which advects moisture into the northern Sahel, and where a high PW anomaly is usually observed.

#### 3.2.1.6 Suppressed Convection

We observe during this period an area of suppressed convection, from 1200 UTC on 13 August to 0000 UTC on 14 August, particularly over Mali and Burkina, where the sky is mainly clear, associated with a ridge at mid-levels and a dry PW anomaly. In this area, deep convection is not present, but some convective cells can develop, like CS2.

We observe another area of suppressed convection on 16 August at 1200 UTC (not shown) associated with a new ridge over the Gulf of Guinea, between two trough axes, one over western Mali and the other over Sudan.

### 3.2.2 A Cold Pool Case Study, 27 September 2014 (CS14)

Figure 3.27 illustrates a case of a squall line triggered by a cold pool as seen from a sequence of IR images on 27 September 2014. In the morning (left column) a previous MCS collapses in the region between Niger, Benin and Togo. From 1100 UTC a thin arc of clouds can be detected over north Burkina Faso, corresponding to the leading edge of a cold pool generated by this MCS and spreading out of it. At 1300 UTC the cloud signature is much clearer and its northern part reaches the Mali–Burkina Faso frontier. The visible channel allows a better detection of such cold pool leading edges, but the animation of Meteosat images also helps in the identification of

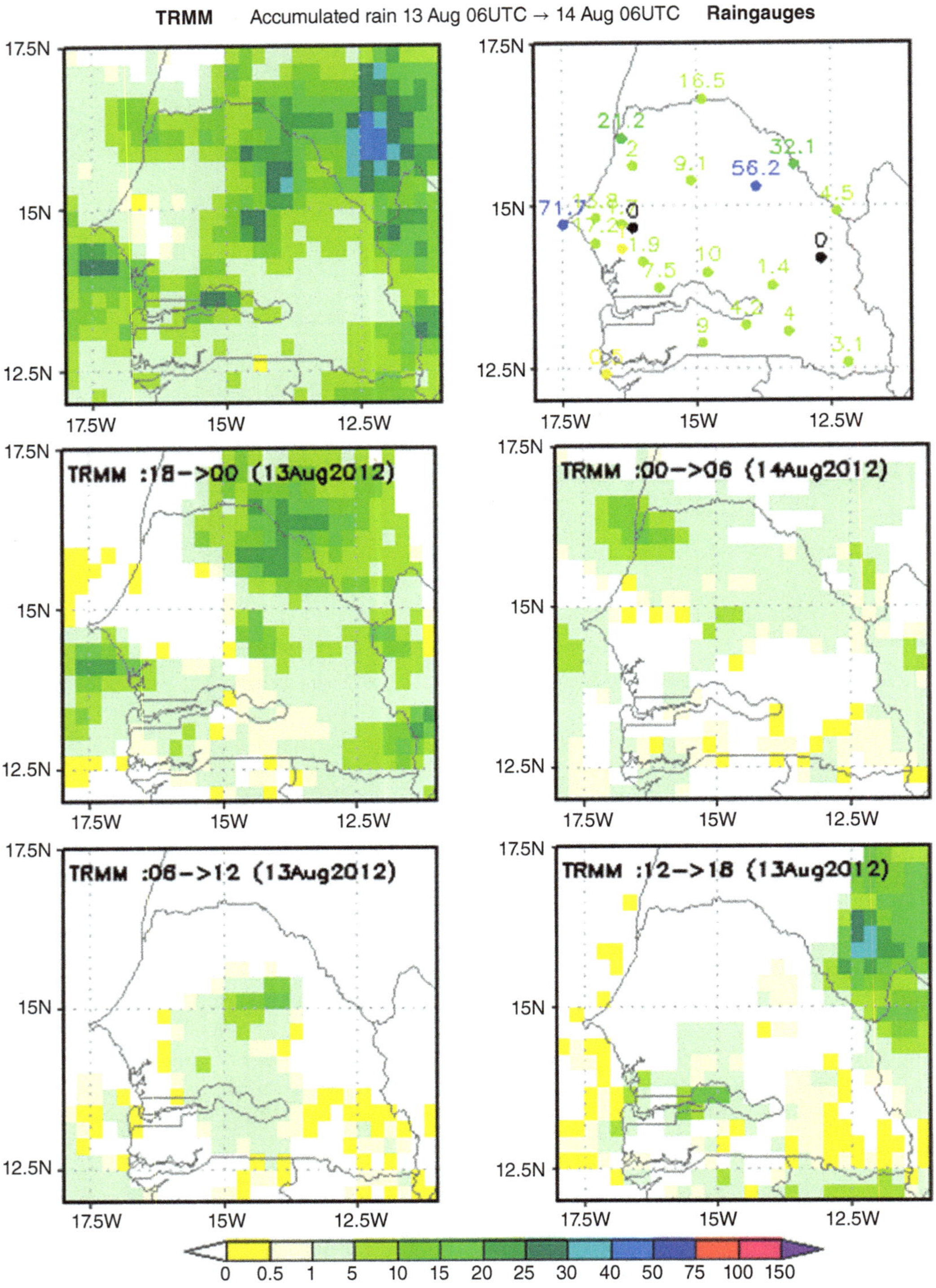


Figure 3.26 Precipitation over Sénégal as estimated from TRMM and observed by the rain gauge network between 0600 UTC on 13 August and 0600 UTC on 14 August (as available on the MISVA site http://isv.sedoo.fr/source/produit-sansup.php?current=20120814&nav=2012LS-Precip-Sénégal_OBSvsTRMM), illustrating the footprint of MCS3.

this fast-moving, arc-shaped line. One hour later, at 1400 UTC, a line of deep convective cells has been triggered corresponding to the convective part of a large squall line quickly moving westward. The squall line continues in the evening with two distinct parts north and south of 10°N. At midnight only the southern part stays very active, reaches Guinea and merges with the convection along the coast.

Figure 3.28 describes the synoptic situation on that day in late September. The monsoon has begun its

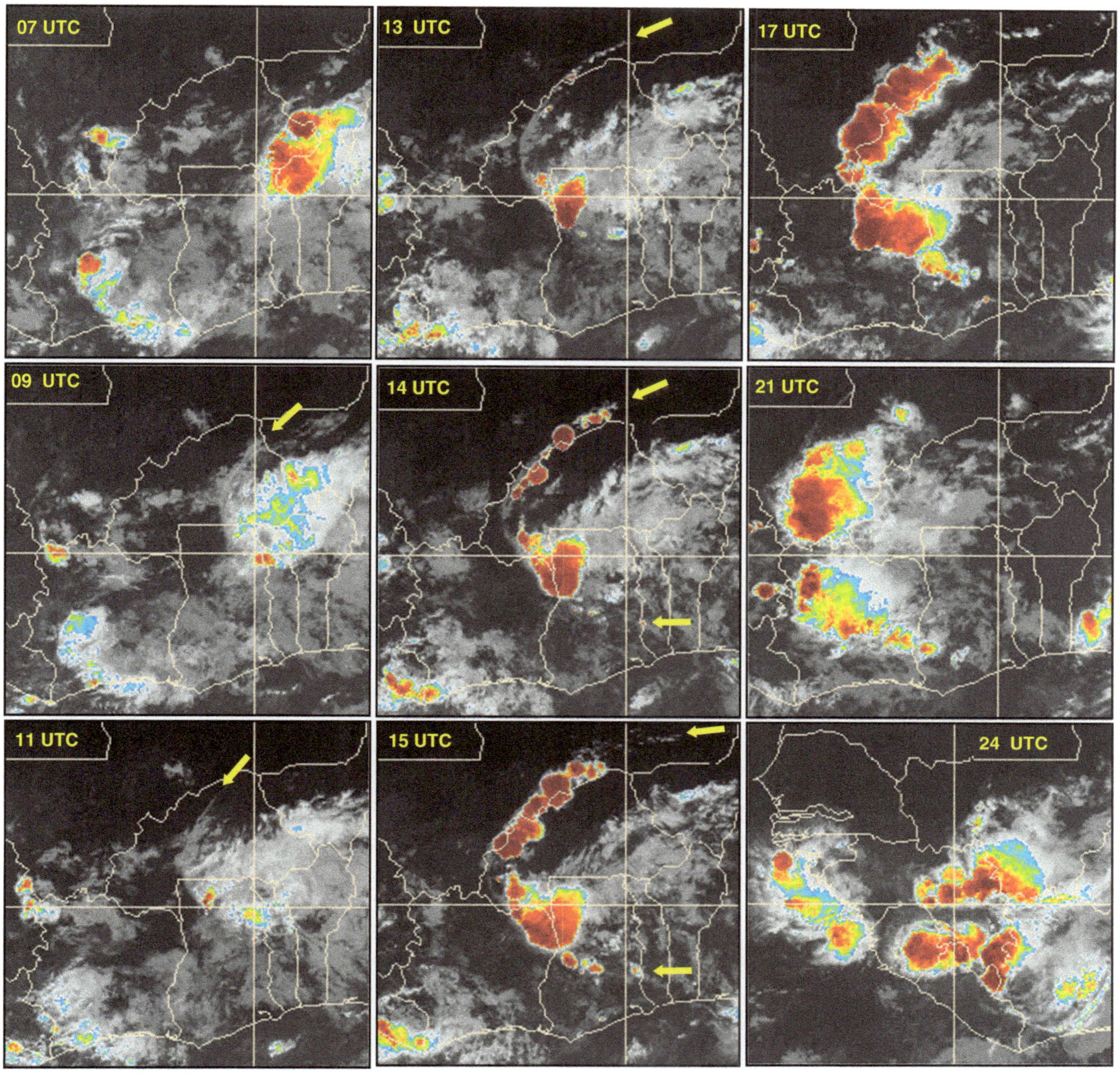


Figure 3.27 Sequence of IR images illustrating the triggering of an MCS by a cold pool generated by a previous MCS on 27 September 2014. Yellow arrows indicate the limits of the arc-shaped leading edge of the cold pool as detected by the low clouds.

retreat and the moist air is along the coast (Figure 3.28c and f) due to strong northerlies (Figure 3.28d). Consistently, the AEJ axis is at about 10°N with a core above 30 kt east of the Greenwich Meridian (Figure 3.28b). A trough is detected over Benin and a ridge over Sénégal–Guinea. The situation corresponds to a dry/moist dipole centred over northern Mali to Mauritania (dry) and Ivory Coast to Benin (moist) respectively (Figure 3.28e). In the morning, the MCS collapses in a dry region at mid-levels, characterised by strong easterly shear (above 20 m $s^{-1}$), as shown by Figure 3.28c. As explained in Section 3.1.2.3, all the conditions necessary to generate intense cold pools, sufficient to trigger convection, have been met, as illustrated by Figure 3.27. North of the AEJ, due to drier conditions, the squall line is triggered later and dissipates earlier in the night, contrary to the squall line part located south of the AEJ. It is to be noticed that the squall line moves at about the speed of the AEJ (15 m $s^{-1}$), faster than the trough, so that the squall line reaches the ridge region in the night. Also, the AEJ is reinforced behind the squall line.

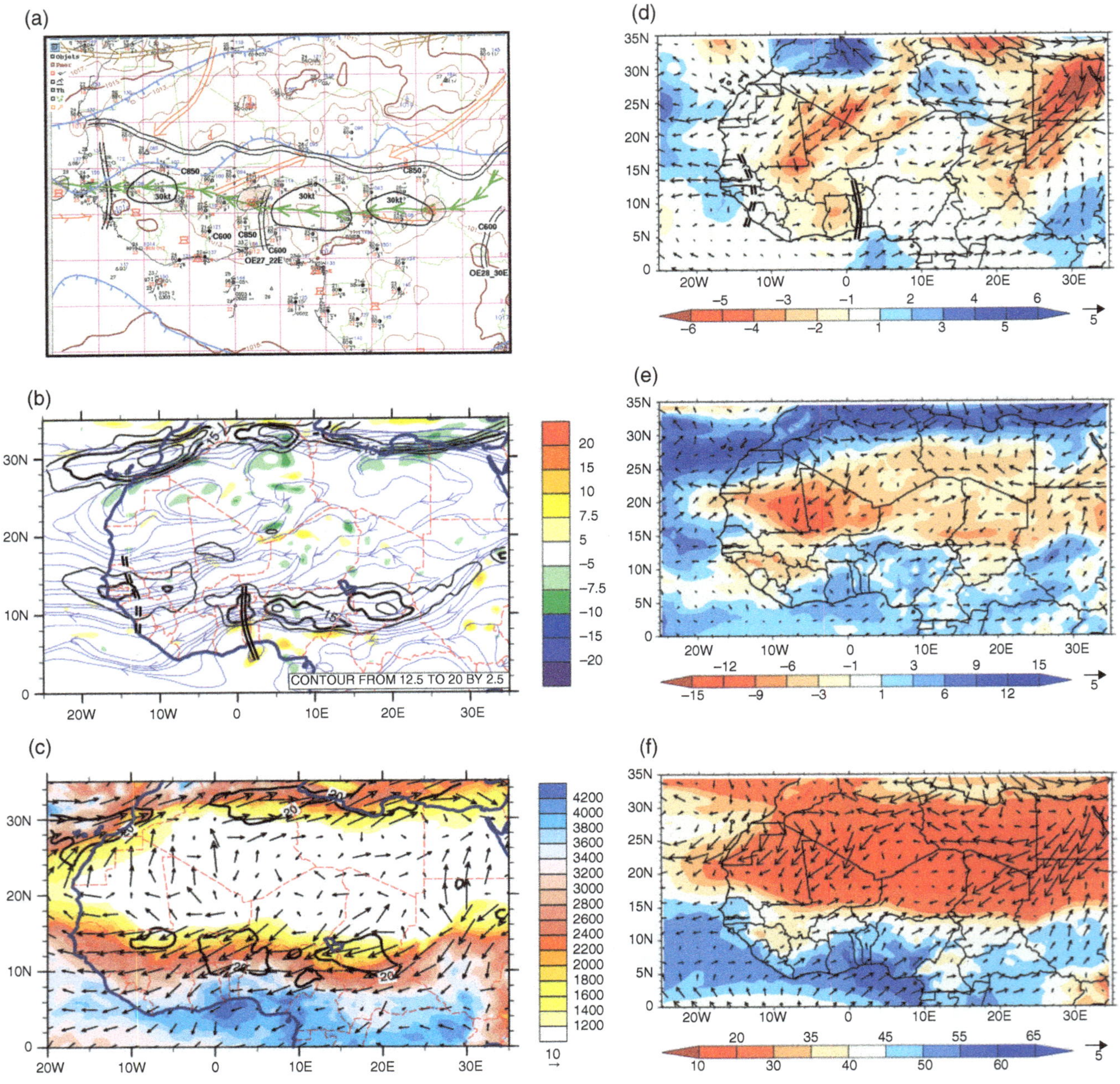


**Figure 3.28** (a) WASA map on 17 September 2014 at 0600 UTC. The following panels are for the analysis at 0000 UTC. (b) Streamlines, wind intensity (thick black line above 15 m $s^{-1}$) and vorticity (colour, $\times 10^{-5}$ $s^{-1}$) at 600 hPa. (c) Monsoon depth (colour, m) with wind shear vector (m $s^{-1}$) in the 950–600 hPa layer, areas with shear larger than 20 m $s^{-1}$ are outlined. (d) Mean meridional in the 950–600 hPa layer (colour, m $s^{-1}$) and mean wind vector in the 950–850 hPa layer. PW (colour) and wind vector at 950 hPa (m $s^{-1}$) for (e) the anomaly and (f) total fields respectively. See Chapter 11 for more explanations on these diagnostics and West African Synthetic Analysis and Forecast (WASA-F).

### 3.2.3 Forecasting Areas of Storm Initiation or Suppression

#### 3.2.3.1 Synoptic Conditions and Stability

##### *3.2.3.1.1 Climatological Controls*

1) *Zones A to D of the monsoon system*

As the ITD moves north and south, zones A to D (see Chapter 1, Figure 1.1) of the monsoon system follow it.

- *Zone A.* Precipitation is rare to the north of the ITD. Although deep convective systems do occur, especially if strongly forced by mid-latitude interactions or by orography, most of the precipitation will evaporate before reaching the ground. This evaporation can drive intense squalls, and a likelihood of dust uplift.

- *Zone B.* Infrequent isolated and short-lived but intense convection may occur in the zone 200 km south of ITD. One example is CS2 in case study CS02. Most cells die during the night; but some cells may persist, merge and move into zone C. These cells can be particularly nasty as there is dry air feeding the downdraughts, and there can be significant squalls and dust lift, but it is hard to determine which cells will develop intensely. In AEW conditions or other moist surges the situation is not so clear, and moisture surges behind the trough can allow strong convection close to the ITD.
- *Zone C.* Most deep convection occurs in zone C associated with the maximum rain zone of the monsoon. In the southern part of zone C, systems are typically slow moving; in the north of zone C they are faster moving. Precipitation is higher in this zone than in zone B, but squalls are generally less severe. The dry, mid-level air toward the north makes systems more dynamic, and precipitation is reduced. Also, due to this evaporation, the mid-level rear inflow generated behind squall lines may accelerate the AEJ locally as for MCS2 and MCS3 (Figures 3.24 and 3.25).
- *Zone D.* This is the stratocumulus zone corresponding to the little dry season.

2) *Convective life cycle, and the diurnal cycle*
   - Initiation of convective storms is really dominated by the production of new cells around 1500 LST. Over mountains (e.g. Jos Plateau or Aïr), the initiation may be a few hours earlier in the day.
   - Although there is an afternoon maximum, storms may initiate at any time of day.
   - Organised systems typically have maximum intensity overnight; for example, 13 August in CS02.
   - Systems may last up to 2 days (e.g. MCS2 in CS02), but, given the daily cycle of convection, with storms decaying in the morning and leaving only a residual system, or perhaps an 850 hPa vortex, it can be difficult to track a given MCS consistently over such a long period.
   - Quite commonly, an MCS may decay completely for a period of 12–24 h, but then regenerate again the following day. Some aspect of the dynamic and thermodynamic structure, which is not easily detected in routine observations, is preserved and continues to propagate consistently, to allow the system to regenerate. A Hovmöller plot of IR imagery may help to project systems (see Figures 2.14 and 2.15), and to extrapolate their location forward in time, indicating the locations of possible storm regeneration a few hours later.
   - Climatologically, systems propagate to the west and southwest, but systems to the north of the AEJ may move to the northwest, especially towards the west coast of the continent, where systems move to northern Mali and Mauritania, and Sénégal to a lesser extent.
   - The shear vector from the surface to the AEJ level (see Figure 3.28c and 11.16b) gives direction of propagation of squall-line-type systems. The speed of the squall lines is typically the AEJ speed. An example is CS02 MCS4_MAT on 15 August.
   - Decay of storms is often in the morning hours, 0600–1200 LST (see Figure 3.27).

3) *General geographical patterns*

A good forecaster needs local knowledge of climatology in relation to local features. This knowledge can be acquired through experience and through communication with colleagues. We recommend that weather services compile training material regarding their local climatologies of convective rainfall, relating to the prevailing season and synoptic state. Important local controls include the following.

- *Coastline:* sea breezes may initiate local storms. Sea breezes may be very variable day to day: sometimes the convection initiates along the coast; on other days it initiates inland (see Section 4.1.4.2 and Figure 4.18).
- *Hills:* orography is the primary surface control on convective initiation, which is very much enhanced over hills. Local features of the terrain, including steepness of slope relative to the prevailing winds, land surface type, albedo and vegetation, may all lead to local 'hot spots' of initiation. See the triggering climatology provided by Figure 3.17, with the Jos Plateau as a typical illustration.
- *Surface state:* hot spots for convective initiation may occur over flat terrain, due to soil type (albedo, infiltration); vegetation and land-use heterogeneity; and soil moisture anomalies due to previous storms, and their interactions (see Section 4.1.5).

#### 3.2.3.1.2 Synoptic State: African Easterly Waves, Mid-Latitude Troughs, Convergence Lines

Although convective activity is highly stochastic, the controls on convection forced by the synoptic state offer some predictability, which can increase confidence in a given forecast. It must be remembered that even when synoptic structures are clearly analysed, high variability in the resulting convection remains (see Section 2.1.3.6 and Figure 2.17), and exceptions to the general rules always seem to be possible.

1) *Convective indices*

A number of indices can be diagnosed from radiosonde profiles or from NWP data to indicate the stability of the profile to deep convection at that location. As discussed in Section 3.1.2.1, high values of CAPE and PW are favourable and necessary for deep convection, but not sufficient, and CIN tends to suppress the development of convection.

The lifted index (Galway, 1956) is calculated by subtracting the temperature of a parcel of air lifted from the surface to a mid-level height (defined to be a pressure of 500 mb) from the existing in-situ temperature at that height. If the observed 500 mb temperature is colder than the lifted air parcel then the lifted index is negative and the atmosphere is unstable to deep convection. The lifted index can be regarded as a crude measure of the CAPE of a parcel lifted from the surface, since it is a measure of the mid-level buoyancy (while the CAPE is the integral of this measure over the cloud layer). As the lifted index is conceptually related to CAPE, its variations match those of CAPE rather closely, and when data are being analysed computationally, CAPE would normally be preferred.

The *K*-index (George, 1960) is useful in determining the probability of occurrence of thunderstorms. The *K*-index is computed from various differences between temperatures and dew point temperatures at the 850, 700 and 500 mb levels:

$$K\text{-index} = (\mathrm{T850} - \mathrm{T500}) + \mathrm{TD850} + (\mathrm{TD700} - \mathrm{T700})$$

where T850, T700 and T500 are the 850 hPa, 700 hPa and 500 hPa temperatures respectively, and TD850 and TD700 are the dew point temperatures at 850 hPa and 700 hPa respectively. All temperatures are in Celsius.

The first term on the right hand side (T850 – T500) is a measure of the lapse rate of the 850 to 500 hPa layer and indicates an increased chance of convection when the difference is a large positive number. The second term (TD850) represents the amount of moisture in the lower layers. The last term represents the negative of dryness of the 700 hPa layer: a large difference between the dew point temperature and temperature at 700 hPa (which will tend to reduce *K*-index) indicates that the air at that level is dry and will tend to suppress the early development of cumulonimbus. It is this last term that makes *K*-index fundamentally different to CAPE. Dry air at mid-levels of the atmosphere, which will tend to suppress the early development of cumulonimbus, reduces *K*-index through the final (TD700 – T700) term, but CAPE is not affected by this dryness. However, unlike CAPE, *K*-index does not account for low-level moisture, aside from 850 hPa; therefore, it may not be sensitive to variations in the humidity (and $\theta_e$) of parcels, caused by soil moisture for instance. Note that *K*-index is intended for stations near sea level.

Forecasters in West Africa are taught to use the maximum of *K*-index at 0600 UTC to for the analysis and forecast of convection after 1200 UTC, with critical values of:

0–15: stable
15–25: small convective potential
25–40: moderate convective potential
>40 high convective potential

The "total totals index" (TT; Miller, 1972) is a simple index derived from the temperature lapse rate between 850 and 500 mb and moisture content at 850 mb:

$$\mathrm{TT} = \mathrm{T850} + \mathrm{TD850} - 2 \times \mathrm{T500}$$

Low values (around 30) are indicative of isolated moderate thunderstorms while high values, around 50 or greater, are associated with numerous severe thunderstorms.

2) *AEWs and convection*
Section 3.2.1 illustrates a case study of typical convective events associated with an AEW (Figures 3.24 and 3.25). Some general principles can be stated as follows.

- Long-lived MCS are often formed in or just ahead of the AEW trough close to the AEJ axis, and move to the southwest.
- To the south of the AEJ, on the AEW trough, slower-moving systems may be formed.
- Behind (to the east of) the trough, less organised convection may form in air of high PW; for example, in MCS1 of CS02 on 14 August 00UTC at 20°E.
- There is suppression of convection typically in and ahead of the ridge, especially where there is northerly flow above the monsoon layer. Exceptions can occur; for example, the 1200 UTC 14 August example of formation along the monsoon trough within the ridge of the wave. Often intense, fast-moving squall lines may move away from the trough and reach the ridge area.

3) *Mid-latitude troughs and convection*
Convection may be initiated in or ahead of (to the east of) upper-level, eastward-moving mid-latitude troughs, especially in the winter and springtime, as developed in Section 2.1.4.

In summer, mid-latitude troughs are weaker but may also force the AEJ and its oscillation (AEWs). For instance, the penetration of a mid-latitude trough over West Africa shifts the heat low southward, resulting in a monsoon reinforcement 1–2 days later, favourable to AEW and convection. At a larger scale of 2 weeks, the mid-latitude Rossby waves modulate the heat low location and intensity, resulting in periods of reinforcement and weakening of convection moving westward in the Sahel band.

4) *Mesoscale features controlling convection*
Various mesoscale features, which can sometimes be inferred from the synoptic state, and sometimes observed in satellite imagery, can initiate and interact with convection.

- Convergence lines, which are a very important mechanism of convective initiation (see Sections 3.1.4.1.4, 4.1.4.1 and 6.1.2.6), are very common but generally difficult to identify over land. Sometimes a convergence line can be apparent in shallow cumulus (visible imagery) or mid-level cloud (IR image; see Figure 3.24). Often such convergence lines correspond to the leading edge of a cold pool generated by a previous MCS, as illustrated in Section 3.2.2 (see Figure 3.27). Over northern regions with dry and bare soils, dust lifted by cold pools can be detected by Meteosat images (see Section 5.2.2.3), providing a way to detect cold pool convergence indirectly. A convergence line may be easier to diagnose in moisture flux convergence fields than the raw wind-convergence. Convergence lines may be more apparent over the sea. It is not uncommon for convergence lines, despite their short length scales, to be captured in the NWP model, especially when initiated by topographic features.

- Mid-level (850–600 hPa) vortices as indicated on the WASA-F maps (see Section 11.10 for the Monsoon Trough at 850 hPa) may commonly be associated with the initiation of convection and may themselves be generated by a convective storm. These vortices may be diagnosed from NWP model fields, although some caution is needed because a model with the wrong convective activity is likely to generate false vortices associated with that incorrect convection.

#### 3.2.3.2 History of Recent Storms

The various examples from CS02, shown in Section 3.2.1, illustrate how the decay and regeneration of storms is commonly associated with previous convective activity, from the past 24 h or so. Certain processes are well documented, as follows.

- A mature MCS will lead to a stabilisation of the atmospheric profile in its wake, decreasing CAPE and increasing CIN. This means that we would normally expect convective initiation to be suppressed in the hours following a significant MCS. However, the restoring time is about 12 h for CAPE, and CIN may recover quicker. Given that the air mass is typically moving too, synoptic-scale advection may contribute to a recovery of the instability of the atmospheric profile to deep convection within only a few hours.
- As seen in the case of MCS2 (Section 3.2.1, CS02), a decayed storm from day $D$ may regenerate on $D+1$. The exact mechanisms for the preservation of a 'memory' of the previous storm are not well understood, but in some cases it is thought that a vortex in the 850–600 hPa layer remains after the storm decays and forces the initiation of new activity the next day.
- Initiation of storms commonly occurs in connection with secondary triggering associated with a pre-existing 'parent' storm. A common sequence is for a parent storm to move to the southwest, but to emit gravity currents and gravity waves toward the north, where new storms will be initiated. These new storms may themselves then move to the southwest. In this way, a 'saw-tooth' sequence of storms is possible, with each successive storm moving southwest, and initiating a new system to the north.
- Soil moisture heterogeneity may help to initiate storms in the Sahel, where vegetation is sparse, typically north of 12°N. Soil moisture features may persist for 2 or even 3 days after rain. Therefore, forecasters should be alert to the fact that areas which have received rain on a dry surface may be favourable for convective initiation in the next 2–3 days.

#### 3.2.3.3 Use of Numerical Weather Prediction to Forecast Convective Initiation

At the current time, NWP models (Chapter 10) are not reliable for the prediction of convective events in West Africa, and great caution must be applied to their use.

The NWP data are helpful, to establish the synoptic environment for the prediction of initiation. Some measures of convection from the model, such as the northern limit of convection, may be useful.

There is good evidence that cases in which there is very strong synoptic control (e.g. very strong and robust AEW patterns) exhibit increased predictability of the NWP model convection. In such cases the forecaster might pay more attention to the model forecasts of convective initiation and suppression. However, on a day-to-day basis, the evidence for this increased predictability is not so clear, and care must be taken.

### 3.2.4 Forecasting Storm Types

#### 3.2.4.1 Organisation, Longevity, Speed and Direction

- Isolated storms are expected when the surface-to-AEJ-level wind shear is weak; this is typically in the south of zone C.
- Organised, propagating storms are expected when the surface-to-AEJ-level wind shear is strong[7] and the mid-levels dry – to the north of zone C and in zone B.
- Propagating storms typically move with a direction corresponding to the mean surface-to-AEJ-level wind vector. Their speed is typically similar to that of the AEJ.

#### 3.2.4.2 Weather Phenomena

- The conceptual model of a squall line is shown in Figure 3.14. From this model, the locations of squalls, wind shear, heavy convective and lighter stratiform rain, and low visibility can be inferred, relative to the system movement.

- Squalls depend on the degree of mid-level (say 700–500 hPa) dryness in the profile. This may be assessed, for example, by dew point depression in this layer. When the mid-levels are very dry, strong evaporation of hydrometeors occurs in the downdraught and high surface gusts are observed. However, the evaporation is also dependent on the microphysics of the downdraught, so squalls may on some occasions be weaker than expected.
- There is normally an intensification of the AEJ, and a lowering of its altitude, behind a propagating MCS.
- Precipitation is most intense at a squall line, and weaker in the more extensive stratiform region following the line.
- Flooding events are usually related to persistent moisture advection. Floods rarely come from squall lines, as these are fast moving; flood events are more normally connected with rainfall associated with closed vortices, or slow-moving organised systems (see the flood case study CS04 of Ouagadougou on 1 September 2009 in Section 2.2.2.2).
- Hail is very rare, although not completely unheard of, with about one case reported every year, and a favourable spot is over the Jos Plateau.
- Dust uplift associated with the gust front is commonly observed, but generally increases to the north, where there is more bare soil. During the season the limit of dust moves north due to the development of vegetation. Therefore, in predicting convective uplift of dust, the forecaster can take into account the northern limit of vegetation.

### 3.2.5 Forecasting Suppression of Convection

Models have difficulties in predicting areas of suppressed convection; a common problem of convective parameterisations for the tropics is for the precipitation to be too light and too widespread. Nevertheless, in some cases the forecaster can identify features that contribute to inhibit convection. In such cases one can localise regions where the probability of convection is weak. We can distinguish four types of factor to examine.

1) *Diurnal cycle*
Owing to the strong diurnal cycle of convection over West Africa, the probability of dissipation is large in the evening for isolated or weak organised convection (C1 and C3 type), whereas most organised MCSs (C2 and C4) dissipate during the night or in the morning.

But caution! Temporary dissipation during the morning can be followed by a reactivation during the afternoon.

2) *Climatologies*
Climatologies such as provided in this chapter (Figures 3.22 and 3.23) provide useful guidelines to forecasting suppression of convection, depending on the region and the type of MCS.

3) *Convection factors*
The examination of the first three types of environmental factors that govern the convective activity (Figure 3.7) allow the identification of regions favourable for the suppression of convection:

- weak CAPE and weak PW;
- large CIN, often associated with large areas of dry air advected over the monsoon layer;
- shear too weak to sustain fast-moving squall lines.

4) *Synoptic factors*
Anticyclonic circulation at low levels is the most efficient way to suppress convection. It can be identified in different ways:

- Streamlines at 925 and 850 hPa to detect anticyclonic curvature associated with a divergent flow. It is to be noticed that the little dry season over coastal regions is characterized by such an anticyclonic low-level circulation.
- Diagnostics explained in Chapter 11 (Figure 11.13b and c), such as the mean circulation in the 950–850 hPa or 950–600 hPa layers, help to identify these ridge axes, as illustrated in Section 2.2.2.3 (see case study 1–10 September 2014), where convection is suppressed.

Some other large-scale features can contribute to suppress convection, such as:

- A subsiding region typically between 500 and 200 hPa.
- Large-scale dry intrusions south of the ITD (except at their leading edge as drawn on WASA-F maps, see Section 11.6).
- A cyclonic circulation at the upper level (200 hPa). It can be a trough at 200 hPa associated with a northern tropical easterly jet (TEJ) that can prevent convection over a large part of West Africa as it moves slowly westward.

There are also occasions of sudden dissipation of big MCSs. It is often difficult to find an explanation, but some of the causes may be:

- strong shear due to a strong TEJ;
- drying of the low levels – for example, when an MCS crosses the ITD;
- drying of the mid-levels – for example, when an MCS propagates toward a region with drier mid-levels and with anticyclonic divergent low-level circulation;
- some west or northwest circulation near the Atlantic coast seems to dissipate MCSs.

## Acknowledgements

GPCP data were provided by the NOAA/OAR/ESRL PSD, Boulder, Colorado, USA, from their Web site at http://www.esrl.noaa.gov/psd/.

## Appendix: Acronyms

| | |
|---|---|
| AEJ | African easterly jet |
| AEW | African easterly waves |
| ARPEGE | Operational forecasting global model of Météo-France |
| CAPE | convective available potential energy |
| CIN | convective inhibition energy |
| DCAPE | downdraught convective available potential energy |
| GPCP | Global Prediction Climatology Project |
| ITCZ | intertropical convergence zone |
| ITD | intertropical discontinuity |
| LST | local solar time |
| MCS | mesoscale convective system |
| NWP | numerical weather prediction |
| PBL | planetary boundary layer |
| PPI | plan position indicator |
| PW | precipitable water |
| TEJ | tropical easterly jet |
| TRMM | Tropical Rainfall Measuring Mission |
| WASA/F | West African Synthetic Analysis/and Forecast |

## Notes

1 The flow vorticity $\xi$ and curvature R must be accounted for a better estimation:
2 M. Provod and J. Marsham, personal communication.
3 The Coriolis term is negligible at the convective time-scale and smaller spatial scale than the Rossby radius $\lambda$.
4 Study with the EPSAT-Niger rain gauge network.
5 www.cnrm.meteo.fr/waf_handbook_casestudies/.
6 www.cnrm.meteo.fr/waf_handbook_casestudies/.
7 Although if the shear is too strong the storm will decay.

## References

Adler RF, Huffman GJ, Chang A, *et al.* 2003. The Version 2 Global Precipitation Climatology Project (GPCP) monthly precipitation analysis (1979–present). *J. Hydrometeorol.* **4**: 1147–1167.

Bain CL, Parker DJ, Dixon N, *et al.* 2011. Anatomy of an observed African easterly wave in July 2006. *Q. J. R. Meteorol. Soc.* **137**: 923–933. doi: 10.1002/qj.812.

Birch CE, Parker DJ, O'Leary A, *et al.* 2013. Impact of soil moisture and convectively generated waves on the initiation of a West African mesoscale convective system. *Q. J. R. Meteorol. Soc.* **139**: 1712–1730. doi: 10.1002/qj.2062.

Birch CE, Marsham JH, Parker DJ, Taylor CM. 2014. The scale dependence and structure of convergence fields preceding the initiation of deep convection. *Geophys. Res. Lett.* **41**: 4769–4776. doi: 10.1002/2014GL060493.

Chong M. 2010. The 11 August 2006 squall-line system as observed from MIT Doppler radar during the AMMA SOP. *Q. J. R. Meteorol. Soc.* **136**: 209–226. doi: 10.1002/qj.466.

Crook NA, Moncrieff MW. 1988. The effect of large-scale convergence on the initiation and maintenance of squall lines. *J. Atmos. Sci.* **45**: 3606–3624.

Emanuel KA. 1994. *Atmospheric Convection*. Oxford University Press: New York.

Fiolleau T, Roca R. 2013. Composite life cycle of tropical mesoscale convective systems from geostationary and low Earth orbit satellite observations: method and sampling considerations. *Q. J. R. Meteorol. Soc.* **139**: 941–953. doi: 10.1002/qj.2174.

Flamant C, Knippertz P, Parker DJ, *et al.* 2009. The impact of a mesoscale convective system cold pool on the northward propagation of the intertropical discontinuity over West Africa. *Q. J. R. Meteorol. Soc.* **135**: 139–159.

Galway JG. 1956. The lifted index as a predictor of latent instability. *Bull. Am. Meteorol. Soc.* **37**: 528–529.

Gantner L, Kalthoff N. 2010. Sensitivity of a modelled life cycle of a mesoscale convective system to soil conditions over West Africa. *Q. J. R. Meteorol. Soc.* **136**: 471–482.

Garcia-Carreras L, Parker DJ. 2011. How does local tropical deforestation affect rainfall? *Geophys. Res. Lett.* **38**: L19802. doi: 10.1029/2011GL049099.

George JJ. 1960. *Weather Forecasting for Aeronautics*. Academic Press

Houze RA. 2004. Mesoscale convective systems. *Rev. Geophys.* **42**: RG4003. doi: 10.1029/2004RG000150.

Huffman GJ, Bolvi DT, Nelkin EJ, *et al.* 2007. The TRMM Multisatellite Precipitation Analysis (TMPA): quasi-global, multiyear, combined-sensor precipitation estimates at fine scales. *J. Hydrometeorol.* **8**: 38–55.

Jorgensen DP, LeMone MA. 1989. Vertical velocity characteristics of oceanic convection. *J. Atmos. Sci.* **46**: 621–640.

Klüpfel V, Kalthoff N, Gantner L, Taylor CM. 2012. Convergence zones and their impact on the initiation of a mesoscale convective system in West Africa. *Q. J. R. Meteorol. Soc.* **138**: 950–963. doi: 10.1002/qj.979.

Lafore J-P. 2004. Orages en fanfare. *Atmosphériques* no. 21, Janvier.

Lafore J-P, Moncrieff MW. 1989. A numerical investigation of the organization and interaction of the convective and stratiform regions of tropical squall lines. *J. Atmos. Sci.* **46**: 521–544.

Lafore JP, Redelsperger JL, Jaubert G. 1988. Comparison between a three-dimensional simulation and Doppler radar data of a tropical squall line: transport of mass, momentum, heat and moisture. *J. Atmos. Sci.* **45**: 3483–3500.

Largeron Y, Guichard F, Bouniol D, *et al.* 2015. Can we use surface wind fields from meteorological reanalyses for Sahelian dust emission simulations? *Geophys. Res. Lett.* **42**: 2490–2499. doi: 10.1002/2014GL062938.

Lucas C, Zipser EJ, LeMone MA. 1994. Vertical velocity on oceanic convection off tropical Australia. *J. Atmos. Sci.* **51**: 3183–3193. doi: 10.1175/1520-0469(1994)051<3183:VVIOCO>2.0.CO;2.

Ludlam FH. 1980. *Clouds and Storms: The Behavior and Effect of Water in the Atmosphere*. Pennsylvania State University Press: University Park, PA; 405 pp.

Liu C, Zipser EJ, Cecil DJ, *et al.* 2008. A cloud and precipitation feature database from nine years of TRMM observations. *J. Appl. Meteorol.* **47**: 2712–2727. doi: 10.1175/2008JAMC1890.1.

Marticorena B, Chatenet B, Rajot JL, *et al.* 2010. Temporal variability of mineral dust concentrations over West Africa: analyses of a pluriannual monitoring from the AMMA Sahelian Dust Transect. *Atmos. Chem. Phys.* **10**: 8899–8915. doi: 10.5194/acp-10-8899-2010.

Mathon V, Laurent H, Lebel T. 2002. Mesoscale convective system rainfall in the Sahel. *J. Appl. Meteorol.* **41**: 1081–1092. doi: 10.1175/1520-0450(2002)041<1081:MCSRIT>2.0.CO;2.

Mathon V. 2001. Etude climatologique des systèmes convectifs de méso-échelle en Afrique de l'Ouest. Thèse de l'Université Paris 7; 238 pp.

Miller RC. 1972. Notes on analysis and severe storm forecasting procedures of the Air Force Global Weather Central. Technical Report, 200(R). Headquarters, Air Weather Service, USAF; 190 pp.

Moncrieff MW. 2013. The multiscale organization of moist convection and the intersection of weather and climate. In: *Climate Dynamics: Why Does Climate Vary?*, D-Z Sun, F Bryan (eds). Geophysical Monograph Series, vol. 189. American Geophysical Union; pp. 3–26. doi: 10.1029/2008GM000838.

Mougin E, Hiernaux P, Kergoat L, *et al.* 2009. The AMMA-CATCH Gourma observatory site in Mali: relating climatic variations to changes in vegetation, surface hydrology, fluxes and natural resources. *J. Hydrol.* **375**(1): 14–33.

Neelin JD, Peters O, Hales K. 2009. The transition to strong convection. *J. Atmos. Sci.* **66**: 2367–2384. doi: 10.1175/2009JAS2962.1.

Redelsperger J-L, Diongue A, Diedhiou A, *et al.* 2002. Multi-scale description of a Sahelian synoptic weather system representative of the West African monsoon. *Q. J. R. Meteorol. Soc.* **128**: 1229–1257. doi: 10.1256/003590002320373274.

Riehl H, Malkus JS. 1958. On the heat balance in equatorial trough zone. *Geophysica* **6**: 503–538.

Roehrig R, Bouniol D, Guichard F, *et al.* 2013. The present and future of the West African monsoon: a process-oriented assessment of CMIP5 simulations along the AMMA transect. *J. Climate* **26**(17): 6471–6505. doi: 10.1175/JCLI-D-12-00505.1.

Rotunno R, Klemp JB,Weisman ML. 1988. A theory for strong, long-lived squall lines. *J. Atmos. Sci.* **45**: 463–485. doi: 10.1175/1520-0469(1988)045<0463:ATFSLL>2.0.CO;2.

Taylor CM, Harris PP, Parker DJ. 2010. Impact of soil moisture on the development of a Sahelian mesoscale convective system: a case study from the AMMA Special Observing Period. *Q. J. R. Meteorol. Soc.* **136**: 456–470. doi: 10.1002/qj.465.

Taylor CM, Gounou A, Guichard F, *et al.* 2011. Frequency of Sahelian storm initiation enhanced over mesoscale soil-moisture patterns. *Nat. Geosci.* **4**(7): 430–433. doi: 10.1038/ngeo1173.

Taylor CM, de Jeu RAM, Guichard F, *et al.* 2012. Afternoon rain more likely over drier soils. *Nature* **489**(7416): 423–426.

Tomasini M, Lafore J-P, Piriou C, *et al.* 2006. ATLAS on a climatology of West African mesoscale convective systems. Deliverable Du2.1.3.a of the Europen project AMMA-EU, no. 004089; 33 pp.

Vondou DA, Nzeukou A, Lenouo A, Mkankam Kamga F. 2010. Seasonal variations in the diurnal patterns of convection in Cameroon–Nigeria and their neighboring areas. *Atmos. Sci. Lett.* **11**: 290–300. doi: 10.1002/asl.29.

Weisman ML, Klemp JB. 1986. Characteristics of isolated convective storms. In: *Mesoscale Meteorology and Forecasting*, PS Ray (ed.). American Meteorological Society; pp. 331–358.

Wilson JW, Roberts RD. 2006. Summary of convective storm initiation and evolution during IHOP: observational and modeling perspective. *Mon. Weather Rev.* **134**: 23–47. doi: 10.1175/MWR3069.1.

Wolters D, van Heerwaarden CC, de Arellano JV-G, *et al.* 2010. Effects of soil moisture gradients on the path and the intensity of a West African squall line. *Q. J. R. Meteorol. Soc.* **136**(653): 2162–2175. doi: 10.1002/qj.712.

Yang C-Y, Slingo J. 2001. The diurnal cycle in the tropics. *Mon. Weather Rev.* **129**: 784–801. doi: 10.1175/1520 0493(2001)129<0784:TDCITT>2.0.CO;2.

Zipser EJ. 1977. Mesoscale and convective-scale downdrafts as distinct components of squall-line structure. *Mon. Weather Rev.* **105**: 1568–1589. doi: 10.1175/1520-0493(1977)105 <1568:MACDAD>2.0.CO;2.

Zipser EJ, Liu C, Cecil DJ, *et al.* 2006. Where are the most intense thunderstorms on Earth? *Bull. Am. Meteorol. Soc.* **87**: 1057–1071. doi: 10.1175/BAMS-87-8-1057.

# 4

# Local Weather

**Lead author:** *Douglas J. Parker*[a],

**Contributing authors:** *Abdou Kassimou*[b], *Bernard N. Orji*[c], *David Perry Osika*[d], *Ibrahim Hamza*[e], *Mariane Diop-Kane*[f], *Andreas Fink*[g], *Jim Galvin*[h], *Françoise Guichard*[i], *Benjamin L. Lamptey*[j], *Hama Hamidou*[e], *Roderick van der Linden*[k], *Robert Redl*[k] *and Thierry Lebel*[l]

**Reviewers:** *Chris Tubbs (UK Met Office/SWFDP) and Paul Hutcheon (UK Met Office)*

[a] *Institute for Climate and Atmospheric Science, School of Earth and Environment, University of Leeds, Leeds, UK*
[b] *Agence Nationale de l'Aviation Civile du Niger (ANAC-NIGER), Niamey, Niger*
[c] *Regional Training Centre, Nigerian Meteorological Agency, Oshodi–Lagos, Nigeria*
[d] *Ghana Meteorological Agency, Takoradi Airport, Takoradi, Ghana*
[e] *Ecole Africaine de la Météorologie et de l'Aviation Civile (EAMAC), Niamey, Niger*
[f] *Agence Nationale de l'Aviation Civile et de la Météorologie du Sénégal (ANACIM), Dakar, Sénégal*
[g] *Karlsruhe Institute of Technology (KIT), Karlsruhe, Germany*
[h] *Met Office, Exeter, UK*
[i] *CNRM-UMR3589, CNRS & Météo-France, Toulouse, France*
[j] *African Centre of Meteorological Applications for Development (ACMAD), Niamey, Niger*
[k] *University of Cologne, Cologne, Germany*
[l] *LTHE–IRD, Grenoble, France*

## 4.1 Scientific Background

### 4.1.1 Introduction

In this chapter we consider a set of challenges in weather forecasting for specific geographic locations, such as a particular village, town or airport. We consider phenomena on timescales from a few hours to 1–2 days, within which such localised forecasts are needed. The weather phenomena to be considered all tend to be linked, scientifically, through the dynamics of the planetary boundary layer (PBL), and are influenced by the local topography of hills, coastlines and land cover, on scales of a few tens of kilometres or even less – known as the mesoscale. This chapter is concerned, therefore, with predicting a family of weather patterns within the general areas of *mesoscale and boundary-layer meteorology*. We will not discuss rainfall prediction here in great detail, as this aspect is covered in Chapters 2, 3 and 6. However, initiation of local showers will be discussed.

The lower part of the atmosphere, which is influenced directly by the surface of the Earth on timescales of an hour or so, is known as the PBL. It normally has a depth varying from hundreds of metres up to 1 or 2 km: over the Sahara, the deepest boundary layers in the world are observed, extending up to depths of 5–6 km, or around 550 hPa in summer. Most of the thermal energy driving atmospheric circulations originates in the form of heating from the surface, which is warmed by the Sun's short-wave radiation: the lower part of the PBL mediates this flow of heat into the atmosphere. The PBL is also the most important part of the atmosphere in terms of human impacts: it is the region where we live, produce our food and release the bulk of our anthropogenic pollution.

The nature of the PBL is somewhat different from that of the 'free' atmosphere, in that the dynamics are dominated by the occurrence of *turbulence*. Turbulence is the random eddying motion of air – manifested as 'gustiness'. Gustiness of the wind increases its potential to damage trees and property, and turbulence is a hazard, occasionally fatal, for aircraft. Therefore, forecasting of turbulence is a high priority for a forecaster. The random turbulent motions of the PBL are accompanied by fluctuations in temperature and humidity on similar timescales. Turbulence is generated by two primary physical processes:

1) Under conditions of radiative heating of the land surface, the hot surface warms the air from below, leading to turbulent convection and mixing in the

*Meteorology of Tropical West Africa: The Forecasters' Handbook*, First Edition. Edited by Douglas J. Parker and Mariane Diop-Kane.

PBL. In contrast, strong surface cooling at night stabilises the boundary layer.

2) The wind flowing over the surface becomes turbulent as it flows over obstacles and the wind shear becomes unstable.

In terms of atmospheric circulations, the importance of turbulence is that it acts to cause rapid diffusion, or mixing, of heat, moisture, chemicals/aerosols or momentum; effectively, turbulence 'stirs' the air close to the ground. The fact that turbulence also mixes wind velocity means that it transfers the friction of the air flowing over the ground into the upper air. This effect is very important in determining the mean daily cycle of winds over West Africa.

Like all tropical environments, West Africa is characterised by a very strong diurnal cycle in the surface energy balance, driven by the diurnal cycle of solar heating. Indeed, one climatological definition of the 'tropics' is those regions for which the diurnal cycle is stronger than the seasonal cycle. The intense diurnal cycle tends to lead to high daytime surface temperatures, but the possibility also of cold surface temperatures by night. In the winter season, cold air moving over West Africa from the Sahara can cause significant health problems for local people. For instance, in the Magaria area of eastern Niger, winter weather conditions, known locally as 'Jamra', can bring night-time temperatures below 0°C, leading to human and livestock mortality, and the loss of crops.

The diurnal cycle of heating also leads to a strong cycle in the boundary-layer winds, and in cloud cover. As in other hot continental regions, such as Australia or parts of North America, nocturnal low-level jets are common and lead to significant wind shear, a hazard for aircraft. The nocturnal winds also transport and mix moisture, leading to substantial areas of stratus and stratocumulus over the Guinea Coast zone. Stabilisation of the lower boundary layer by night-time cooling can also be associated with fog formation, particularly over forested areas.

### 4.1.2 The Surface Energy Balance and the Atmospheric Surface Layer

The diurnal pattern of the surface energy balance is the primary control on local weather conditions in West Africa. The surface energy balance represents a budget of energy fluxes into a layer of the surface (whether ground or sea) underlying the boundary layer.

In many parts of the tropics, the large-scale winds are light, and in these conditions of low wind, radiative transfer at the surface of the Earth dominates the production or suppression of turbulence. The heating of the lower metres of the atmosphere depends on the heating of the surface, and the transfer of heat from the surface into the air. These processes can be quantified through the *surface energy balance.*

The 'atmospheric surface layer' is a region, around 30–50 m in depth, in which the fluxes are found to be approximately constant with height. The behaviour of this layer, and its exchange of heat or momentum with the underlying surface, controls the process of land–atmosphere, or air–sea, interaction.

#### 4.1.2.1 Surface Energy Balance Principles

The exchange of energy between the Earth's surface layer and its overlying atmosphere involves three important processes.

1) *Radiative fluxes*

The surface absorbs and reflects solar (or shortwave) electromagnetic radiation, and emits and absorbs terrestrial (or longwave) radiation. Most of the solar energy is associated with the wavelength range 0.3–2.0 μm, with a maximum located within the visible band, of 0.4–0.7 μm. The *irradiance* of a surface normal to the direct beam of solar radiation, at the top of the atmosphere, is about 1366 W $m^{-2}$. For a given area of the Earth's surface, the intensity of this solar radiation reduces as the Sun becomes lower in the sky, according to the *solar zenith angle* (see Figure 4.1).

Some of the direct solar beam is scattered and reflected by air molecules, aerosol particles and clouds in the atmosphere, so that the Earth's surface receives a reduced direct beam of solar radiation, as well as *diffuse solar radiation* associated with *scattering* from the atmosphere. The *net solar irradiance* of the surface can be represented by

$$S_n = S\downarrow - S\uparrow \quad (4.1)$$

where $S\downarrow$ is the total (direct plus diffuse) downwelling shortwave radiation (SWin, in Figure 4.1), and $S\uparrow$ is the total upwelling shortwave radiation. Because the Earth does not emit shortwave radiation, $S\uparrow$ is entirely associated with *reflection* of some of the downwelling radiation, and this is a constant fraction $\alpha$ of $S\downarrow$, depending on the surface type. It follows that

$$S_n = (1-\alpha)S\downarrow = (1-\alpha)(S_b + S_{diff}) \quad (4.2)$$

where $\alpha$ is known as the surface albedo, $S_b$ is the direct beam solar irradiance and $S_{diff}$ is the diffuse solar irradiance. $S_b$ and $S_{diff}$ are also strongly influenced by clouds and by atmospheric aerosol. The albedo varies from about 0.45 for bright desert sand, to 0.3 for short grasses, 0.2 for forest or dark soils and 0.05 for wet soils.

*Terrestrial radiation* refers to electromagnetic energy emitted by the Earth's surface and its surrounding atmosphere. The Earth's surface emits like a black body, and most of the energy is associated with the wavelength range 3–30 μm. This is often referred to as *longwave* radiation. For most surface materials, the longwave

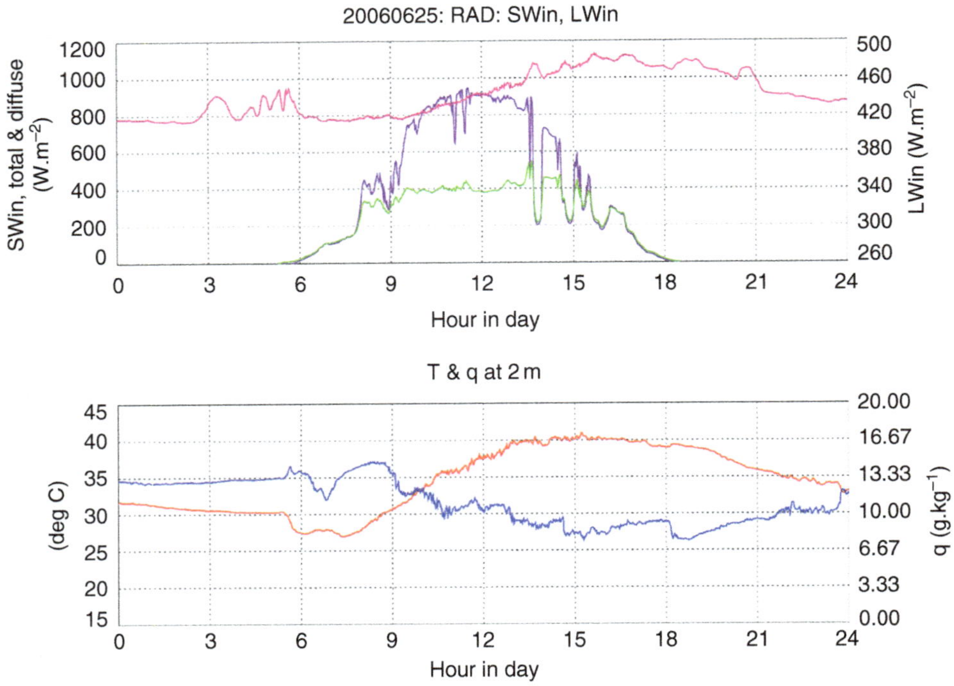


**Figure 4.1** A typical diurnal cycle of the surface energy balance, with 2 m temperature and specific humidity at Niamey, taken from the Atmospheric Radiation Measurement (ARM) Mobile Facility during the African Monsoon Multidisciplinary Analysis (AMMA) campaign, 25 June 2006. The incoming shortwave radiation (SWin, upper plot, purple line) follows the solar elevation, with a peak around 1200 UTC, but is interrupted by cloud events, especially in the afternoon, when its value falls substantially. The diffuse fraction (green line) on this day is around 50% of the total when the sky is clear; this fraction depends very much on the cloudiness and aerosol loading of the prevailing conditions, and can be higher or lower than shown here. Incoming longwave radiation is relatively constant over the day, and follows the surface temperature, with some nocturnal peaks due to cloud. The 2 m temperature (red line in lower panel) reaches its maximum in the afternoon, around 1500 UTC, while the specific humidity (blue line) is at its minimum at this time. It is possible to infer small reductions in 2 m temperature associated with the cloud events in the afternoon of around 1 °C.

energy emitted per unit area is predicted accurately by the Stefan–Boltzmann law:

$$L = \varepsilon\sigma T^4 \tag{4.3}$$

where $\varepsilon$ is an effective emissivity. From Kirchhoff's law, it follows that a surface material *absorbs* a fraction $\varepsilon$ of incident *longwave* radiation. For most land surface types, the effective emissivity is high; typically $\varepsilon$ is ~90%.

The atmosphere does not behave like a black body, because its emissivity varies strongly with wavelength. When the air is reasonably dry, the atmosphere is almost transparent to radiation in the 'atmospheric window' region of 8–12 μm, but nearly opaque ($\varepsilon$ is close to one) for wavelengths of 3 to 8 μm and > 12 μm. Its emission, and so too absorption, characteristics also vary with the concentrations of *greenhouse gases*, such as water vapour and carbon dioxide.

Clouds – even thin clouds such as thin cirrus – act like black-body radiators and absorbers, with an effective emissivity close to one. Therefore, the presence of clouds has an important influence on the longwave radiation. When clouds are absent and atmospheric humidity is low, longwave radiation in the atmospheric window of 8–12 μm can be transmitted to space; when clouds are present this band of radiation is absorbed by the clouds, and partly re-emitted back to the surface, with the atmospheric window effectively 'closed'. In conditions where the clear air is reasonably dry, the appearance of clouds can amount to 30 W m$^{-2}$ in increased downward longwave flux at the surface; therefore, clouds substantially increase the greenhouse effect locally. This effect is particularly important at night, when surface cooling tends to be much faster under clear skies than cloudy skies.[1]

The net surface irradiance $R_n$ is the sum of short- and longwave contributions:

$$R_n = R\downarrow - R\uparrow = (S\downarrow + L\downarrow) - (S\uparrow + L\uparrow) = S_n + L_n \tag{4.4}$$

Section 9.1.2 gives much more detailed explanation of radiative processes in general.

2) *Thermal conduction of heat energy within the ground*

There is also a significant heat flux through exchange with layers below the surface, known as the ground heat flux $G$. This term follows relatively simple physical principles, with heat flowing from the hot surface to the cooler, deep layers of the Earth during the afternoon, and flowing upward at night when the surface becomes colder than the deeper layers. The rate of heat flow depends on the thermal conductivity of the surface material, which is different for different soil types, and is strongly affected by factors such as the levels of moisture in the soil.

3) *Turbulent transfer of heat energy toward or away from the surface within the atmosphere*

The heat flowing from the surface into the atmosphere has two components:

The sensible heat flux $H$ is the heat flux as internal energy flows upward into the air (positive flux), by conduction and convection ('sensible' as opposed to 'latent' heat). It depends on the temperature difference between the surface and the air, as well as the wind speed and the strength of the turbulence, with stronger turbulence increasing the flow of heat. When warm air flows over a cold surface, $H$ can be negative, as sensible heat flows from the air into the ground.

The sensible heat flux is sometimes modelled as a function of the stability of the air in the surface layer:

$$H = -\rho c_{\mathrm{p}} K_{\mathrm{H}} \frac{\mathrm{d}\theta}{\mathrm{d}z} \tag{4.5}$$

where $K_{\mathrm{H}}$ is a 'turbulent diffusivity' or 'eddy conductivity' (Stull, 2009), which increases with the intensity of turbulent eddies. Therefore, the sensible heat flux increases as the turbulence becomes more intense and as the vertical gradient of temperature in the surface layer becomes more negative (hot surface below cooler boundary layer). At night, as the surface cools through upward longwave radiation, $H$ is negative (downward) because a stable inversion layer develops in which temperature increases with height, and $\theta$ increases strongly with height.

The latent heat flux $LE$, where $E$ is the rate of evaporation of water (or ice) and $L$ is the latent heat of condensation (around $2.5 \times 10^6$ J kg$^{-1}$ at 20°C) is the flow of latent heat, which occurs when water evaporates off the surface, or condenses onto it (recall that evaporation cools the ground, whereas condensation warms it). The latent heat flux is generally large over open water, but can also be large over wet soil, and is significantly affected by 'evapotranspiration' from the leaves of plants. In areas of substantial vegetation, the plants will dominate the flux, but in areas of sparse vegetation, such as the Sahel (e.g. in the vicinity of Niamey), bare-soil evaporation controls the latent heat flux. In well-vegetated regions, plants can access deep layers of water through their roots, but in the Sahel, where bare-soil evaporation is important, the latent heat flux is very sensitive to recent rainfall.

There is a strong control of African weather by the surface energy balance, and in particular by the sensible and latent heat fluxes of energy from the surface to the atmosphere. It is useful to define the ratio of the sensible and latent heat fluxes, termed the Bowen ratio

$$\beta = \frac{H}{LE} \tag{4.6}$$

or the 'evaporative fraction'

$$E_{\mathrm{F}} = \frac{LE}{LE + H} \tag{4.7}$$

to characterise the balance of turbulent fluxes. The Bowen ratio and evaporative fraction vary according to the surface state; for instance, a wet soil will tend to have high evaporation and low sensible heat flux, while a very dry soil, with sparse vegetation, will have high sensible heat fluxes. Such differences in the ratio of sensible to latent heat flux then have a profound influence on the PBL and the local weather conditions. For instance, if a surface becomes wet its Bowen ratio will be reduced, and the turbulent fluxes will be dominated by the latent heat flux. Then, the sensible heating of the atmosphere is lower, and the air temperatures are lower, but humidity is high.

Combining all of the energy flux terms, we can determine the energy balance for a ground layer of finite depth, as

$$C \frac{\mathrm{d}T}{\mathrm{d}t} = R_{\mathrm{n}} - G - H - LE \tag{4.8}$$

where $C$ is the heat capacity of the layer of the ground (Figure 4.2). In the special case of an infinitesimally thin ground layer, there can be no heat storage, and Equation 4.8 then reduces to a surface energy balance equation of the form

$$R_{\mathrm{n}} - G_0 - H - LE = 0 \tag{4.9}$$

or

$$R_{\mathrm{n}} - G_0 = H + LE \tag{4.10}$$

However, for many practical applications we need to consider surface temperature, both to understand the longwave emission and because surface temperature is, itself, important to certain users. Therefore, it is often desirable to work with the finite-layer form (Equation 4.8).

The surface energy balance offers a framework for understanding the relationships between the surface forcing of the boundary layer and the response of other aspects of the system, such as the surface temperature. Given that solar radiation is the main driver of the surface energy balance on the daily timescale, variations in

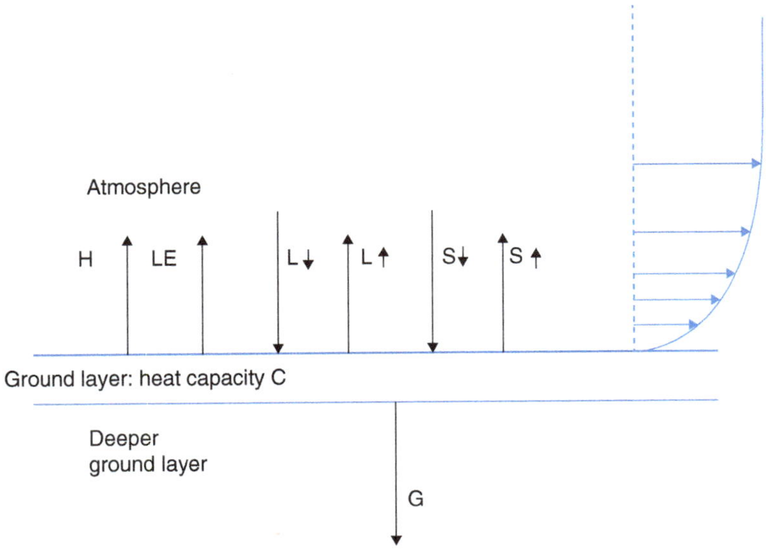


**Figure 4.2** Schematic of surface energy balance terms used in Equation 4.8, and the typical surface wind profile. Note that in the surface layer the mean wind direction is approximately constant with height.

the net solar flux $S_n$ lead to substantial changes in the local climate. Notably, cloud cover, by reducing the downward radiation, leads to cooler conditions near the surface; mineral dust in the atmosphere has a similar effect. However, the other energetic terms also have important roles to play. When the land surface is very dry, its temperature will tend to rise more rapidly (lower heat capacity $C$) and the air above it will become very hot, due to a high Bowen ratio (high sensible heat flux), relative to moist or vegetated regions.

Whilst the relationships that determine the surface energy balance are, conceptually at least, simple, using them accurately to predict quantities such as temperature can remain difficult. The difficulties arise because the system is a balance between various time-dependent terms, and there are some parts of the surface energy balance which are hard to constrain. The ground heat flux and the ground surface temperature are both sensitive to the soil type, which is often variable over a short distance, and to soil moisture, which varies considerably over short time and space scales. In turn, uncertainty in the surface temperature makes the upwelling longwave radiation uncertain. The turbulent fluxes are complex, and dependent on the atmospheric winds and stability, as well as the surface state.

Although the surface air temperature, for instance the 2 m temperature which forecasters need to predict, is linked to the ground surface temperature, their values can differ substantially – by perhaps 10°C, depending on the surface roughness and the sensible heat flux. At night the surface is usually cooler than the air, while by day it is warmer. Therefore, there can be a large difference between the air temperature and the radiation temperature, which appears in the surface energy balance (e.g. Equation 4.3). The gradient in temperature from the ground surface to 2 m is related to the heat flux, through Equation 4.5 for example, and to the intensity of the winds and turbulence in the air (which control the magnitude of $K_H$ in Equation 4.5). In Section 4.2.1 we will use empirical rules to relate changes in 2 m temperature to various processes that change the surface energy balance.

#### 4.1.2.2 Surface-layer Winds

Apart from the effect of surface heating on production and suppression of turbulence, there is also a significant process of wind shear generation of turbulence. The surface of the Earth is 'rough', on all scales, from the roughness associated with grains of sand, to trees, buildings and mountains protruding into the air. As the wind flows over any of these obstacles, the wind-shear is unstable and produces turbulence on a similar horizontal scale to the obstacle. Strong winds will generate turbulence to produce a well-mixed boundary layer, even at night under clear skies.

A standard result is that, in neutral conditions (for instance, zero sensible heat flux), the average wind profile in the atmospheric surface layer has a logarithmic form:

$$u(z) = \frac{u^*}{k} \ln\left(\frac{z}{z_0}\right) \tag{4.11}$$

where $u^*$ is the 'friction velocity', which is a measure of the strength of the turbulent variations in wind speed (the 'gusts') and $k$ is the von Kármán constant, and has a value of ~0.40. $z_0$ is the roughness length, and is regarded to be a constant for a given surface, related to the size, shape and distribution of the obstacles over which the surface air is flowing.

When the boundary layer is not neutral, there is some departure from the logarithmic wind profile, generally described with Monin–Obukhov theory (e.g. Stull, 2009).

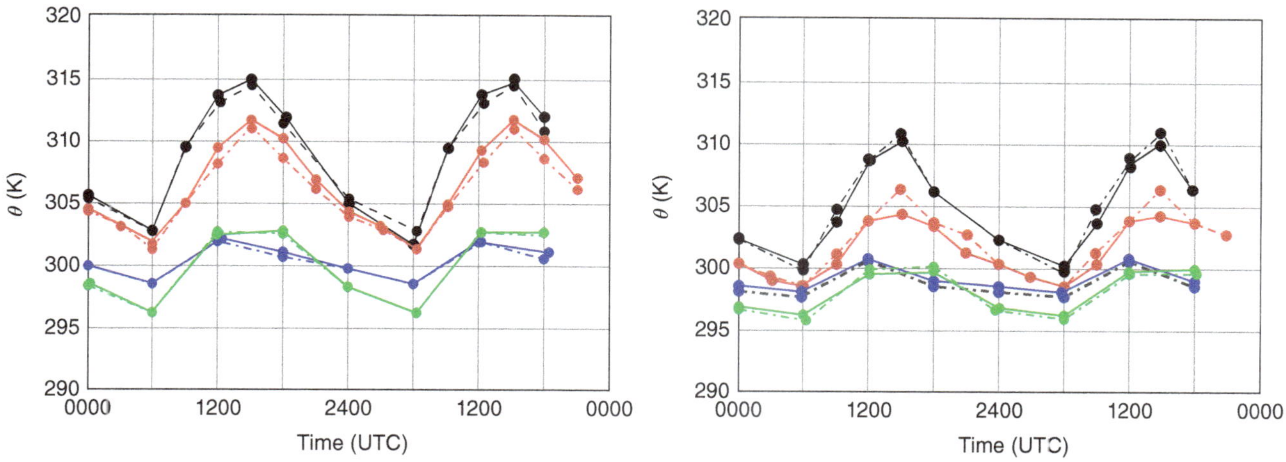


Figure 4.3 The diurnal cycle of 2 m potential temperatures for June (left panel) and August (right panel) for the stations at Agadez (black line, 17°N), Niamey (red line, 13.5°N); Parakou (green line, 9.5°N) and Cotonou (blue line, 6.5°N). For clarity, two diurnal cycles are shown. Solid lines show data from 2006; dashed lines are for August 2005–2008. *Source*: Gounou *et al.* (2012). Reproduced with permission of Springer.

During the day, when the land surface is hot and potential temperature decreases with height in the surface layer, the profile is unstable. In these unstable conditions, turbulent eddies are deeper and the low-level wind-shear is reduced.

In daytime, convective conditions, 'dust devils' may very commonly be observed (e.g. Figure 5.9). For instance, on research flights over Mauritania in the Fennec experiment of 2011 and 2012, in the afternoon hours dust devils would always be visible somewhere within sight of the aircraft (~10 km radius). More generally, forecasters report that this phenomenon occurs most frequently north of latitude 10°N during the transition period before the rainy season. In this period, the sensible heat flux is strong, the boundary layer is relatively deep and dry (Gounou *et al.*, 2012), and the wind shear is reasonably strong. Dust devils are observed to rotate both clockwise and anticlockwise; since the Coriolis force is relatively weak and the horizontal scale of the dust devils is small, they occur at very high Rossby number, and the effects of planetary rotation are negligible.

#### 4.1.2.3 The Surface Energy Balance over West African Climatic Zones

Chapter 1 describes the large-scale patterns of some of the terms of the surface energy balance across West Africa, according to time of year. These seasonal and geographic patterns cause different low-level weather conditions at the local scale. Figure 4.3 shows mean diurnal cycles in temperature at four West African stations, in the pre-monsoon period of June and the post-onset period of August.

In the pre-monsoon period, the air over the northern stations of Agadez and Niamey is dry, with relatively little cloud cover, leading to strong daytime heating and strong longwave cooling at night, and therefore a strong diurnal cycle in temperature.[2] In contrast, at the stations to the south, more extensive cloud cover and a humid atmospheric profile (relatively high PW), as well as a lower Bowen ratio due to greater vegetation cover, leads to a reduced amplitude of the diurnal cycle of 2 m temperature. At Agadez, and at Niamey in June, temperatures rise rapidly through the morning, but then in early afternoon rise more slowly, as the heating is dispersed into a very deep boundary layer. At Parakou and Cotonou, and Niamey in August, the temperatures warm more steadily in the morning, and fall less quickly in the afternoon, with higher mean temperature at 1800 UTC than at 1500 UTC at Parakou. At Cotonou the sea breeze dominates the afternoon conditions.

#### 4.1.2.4 Variation of Surface Energy Balance and Surface Temperatures in Response to Physical Processes

##### *4.1.2.4.1 Cloud Cover*

First we consider a cloudy day with a cloudy night. Incoming solar radiation is attenuated by the cloud canopy due to absorption and scattering. This results in lower maximum temperatures. At night, the outgoing longwave (terrestrial) radiation is absorbed and re-emitted back to the Earth's surface leading to surface warming. Hence, we have higher minimum temperatures. This scenario results in small diurnal temperature ranges, and would typically be associated with a high-pressure cell from the surface through 850 hPa.

Next we discuss a cloudless day and night. During the day, maximum solar radiation reaches the Earth's surface. At night, the terrestrial radiation escapes into the atmosphere without inhibition so we have a higher maximum temperature during the day and a lower minimum temperature by early morning. The result is a large diurnal temperature range.

It should, however, be recognised that when the air is very humid the influence of cloud on the night-time cooling is reduced; the atmospheric humidity means that there is relatively little cooling even when skies are clear. Cloud cover has a significant effect on night-time temperatures when the PW is less than around 20 kg $m^{-2}$ (Stephens *et al.*, 2012). From Figure 1.10, it can be inferred then that the effects of cloud cover on night-time cooling are most significant in zones A and B of the monsoon system (Figure 1.1), north of, and just to the south of the intertropical discontinuity (ITD).

#### *4.1.2.4.2 Weather*

Rain or thunderstorms preceding the time at which maximum temperatures are recorded has an important effect. For a few hours after the passage of a squall line, cold air is brought down to the ground in the mesoscale downdraught, while increased cloud cover reduces the solar radiation. In the subsequent hours, rainfall has a direct relation to the maximum temperature values, through a change in the ground heat flux and heat capacity (reducing the surface temperature change for a given solar heating), and a reduction in the Bowen ratio (or equivalently an increase of the evaporative fraction), decreasing the sensible heating for a given total heat flux. An increase in intensity of a precipitation event causes an appreciable decrease in the value of the maximum temperature with a decrease of 2–4°C in the daily maximum temperature following rain being common. As precipitation ceases by evening, cooling of the soil surface continues, leading to lower minimum temperatures by early morning.

Nocturnal temperature minima are strongly influenced by wind speed, especially in arid regions where the nocturnal low-level temperature inversion can be strong. High winds generate turbulence, which mixes out the inversion, lifting up the cool air away from the surface and bringing warmer air down to the ground. For example, inspection of the SYNOP data from Bilma shows that very cold temperatures of 0–5°C typically occur only when there is zero wind speed recorded, and that light winds from 2 to 10 kt at the surface can increase the nocturnal temperature minimum by 5–10°C.

#### *4.1.2.4.3 Dust*

Incoming solar radiation is attenuated by the dust canopy due to scattering by dust particles. Two crucial factors involved are dust concentration and its vertical extent. During Harmattan dust haze periods, there is a large concentration of dust particles with vertical extent to about 1000–3000 m. In such cases, the maximum temperature for the day may be reduced by about 2–5°C from the previous day's value. Also, the minimum temperature may increase by about 1–2°C, because the dust is an absorber of longwave radiation, acting in some way like cloud cover.

It should be recognised that dust events are generally also accompanied by mesoscale or synoptic temperature changes: the dust is generally lifted and transported by a significant weather system. For instance, dust may be raised by the cold-pool outflow from a convective storm, which brings cold air along with the dust; this air may also be moist, which would tend to increase nocturnal temperatures. Alternatively, when dust is generated by a mid-latitude trough (see Sections 2.1.4 and 5.1.2), cold advection follows the trough's passage, even though the dust is clearing. For these reasons, it is not easy to use observations to establish clear rules as to the effect of dust on surface temperatures: dust events are usually accompanied by synoptic and mesoscale changes that will mask the radiative changes in the surface energy balance.

The current generation of numerical weather prediction (NWP) models includes seasonal average dust concentrations, but does not include interactions with dust, and therefore they do not represent the radiative effects of unusually dusty events on surface temperatures. However, dust models are being used in research now and are likely to become operational in the coming years. Milton *et al.* (2008) evaluated NWP model simulations with and without dynamic dust, against surface measurements from the AMMA/Radiative Atmospheric Divergence using ARM Mobile Facility, GERB data and AMMA Stations deployment of the ARM Mobile Facility in 2006 (Miller and Slingo, 2007), for the very intense dust event of 8 March 2006 at Niamey. In their paper, Milton *et al.* (2008) show that the dust event led to a reduction in the surface shortwave fluxes of around 200 W $m^{-2}$, and a much smaller increase in the net longwave radiation. Daytime surface temperatures fell due to cool advection associated with the event, but were also more than 3°C colder in observations than the no-dust model simulation, implying that this is a typical magnitude of the dust-radiative influence on surface temperatures.

#### *4.1.2.4.4 Surface Type and Soil Moisture*

Areas with dark, rocky surfaces where the albedo is low typically have the highest daytime maximum temperatures, due to the high absorption of solar radiation in the surface.

In regions where vegetation is sparse, soil moisture can have a strong influence on surface temperatures, and daytime maxima in particular. Soil moisture tends to darken a bare-soil surface, reduce its albedo and lead to a higher input of solar energy to the surface. However, this

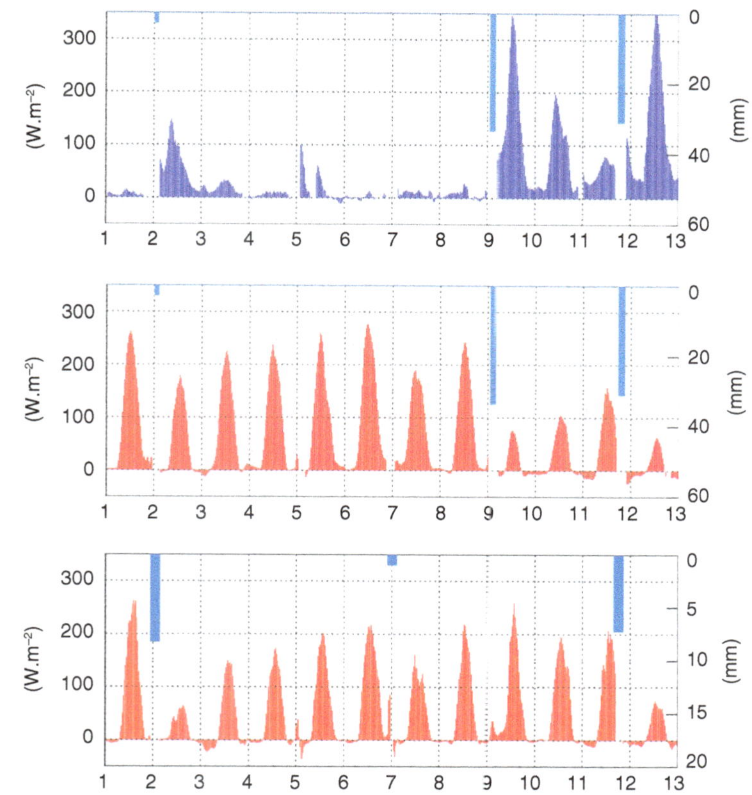


**Figure 4.4** Spatio-temporal variability of surface heat fluxes in measurements made in the vicinity of Niamey; the horizontal axis indicates dates in July 2006. The top two plots show time series of surface latent (blue) and sensible (red) heat fluxes in Wankama, with precipitation events indicated with thick blue bars. The bottom plot shows the sensible heat fluxes in Banizoumbou, a few tens of kilometres away. *Source*: Guichard *et al.* (2012). Reproduced with permission of *La Météorologie*.

is often a transient effect in the Sahel (around 1 or 2 days), as the top soil layer dries quickly after a rain event and the net effect of the surface energy balance over a wet soil is to reduce the surface temperatures, because the wet soil changes the Bowen ratio. The turbulent heat flux to the atmosphere is dominated by latent heat flux over a wet surface, with a weaker sensible heat flux, and the atmospheric temperature therefore rises more slowly. A wet surface also tends to have a higher heat capacity, so that the surface temperature rises more slowly for a given heat input (Equation 4.8).

The effects of soil moisture are strongest in semi-arid parts of the Sahel, where there are few deep-rooted trees, typically, north of around 12.5°N, and where there is plenty of bare soil. Soil moisture effects on surface temperature are thought to be particularly strong early in the rainy season, where there is little moisture in the deeper layers of the ground. Figure 4.4 shows the surface fluxes recorded on a sequence of days at two stations in the vicinity of Niamey (13.6°N, 2.2°E) in 2006. It can be seen that on the 1 or 2 days after rain, the latent heat fluxes rise from a few tens of watts per metre squared to 200 or 300 W $m^{-2}$, while the sensible heat fluxes are reduced by a factor of 3 or 4 after rainfall. The lower two panels of Figure 4.4, showing sensible heat fluxes at stations a few kilometres apart, illustrate the significant differences in surface energy balance over these spatial scales.

Airborne measurements around Niamey, conducted during the AMMA Special Observing Periods in 2006, have provided further evidence of the response of the surface and atmosphere to the soil moisture patterns. Aircraft measurements of afternoon surface temperatures, using infrared radiometry, have suggested surface temperature contrasts of around 8°C over neighbouring wet and dry surfaces, with boundary-layer temperatures, measured at the flight level of around 200 m, varying by around 4°C (Taylor *et al.*, 2007). From these 'synoptic surveys' it is reasonable to estimate that the 2 m maximum surface temperatures will be depressed by around 4–8°C over a wet surface, relative to a neighbouring dry surface. Such effects are significant over very short distances, down to around 10 km.

Even stronger changes in 2 m temperatures have been documented in case studies by Schwendike *et al.* (2010), where a drop of 10°C was observed from one day to the next, as a result of increased soil moisture following the passage of a mesoscale convective system (MCS) over the Dano site in Burkina Faso (11.2°N, 3.1°W). A weaker average response in 2 m temperatures to rainfall is usually found in surface station data, probably because the station data average is complicated by synoptic and cloud cover changes from day to day (Figure 4.5).

In regions with plenty of deep-rooted trees, the effects of soil moisture on surface temperatures are not thought to be so strong, because the trees maintain

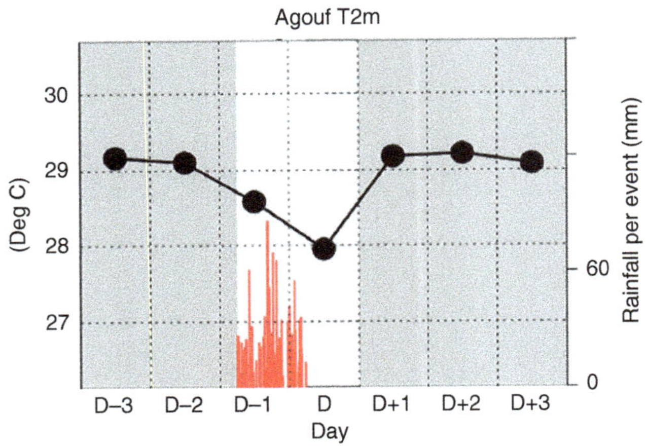


**Figure 4.5** Rainfall events (greater than 5 mm) and composite 2 m temperature observations from the Agoufou site (15.3°N, 1.6°W) in the Hombori mesoscale site of the AMMA–Couplage de l'Atmosphère Tropicale et du Cycle Hydrologique (CATCH) network, in the period 2002 to 2006.

evapotranspiration for many days after rain, even when the upper parts of the soil have become dry.

#### 4.1.2.4.5 Vegetation

Vegetation plays a significant role in modulating the temperature of the environment. The vegetation increases evapotranspiration, especially when deep-rooted trees are able to bring moisture from lower levels of the soil. Increased evapotranspiration in vegetated regions increases the latent heat flux; this causes the sensible heat flux to be reduced, through the surface energy balance, and results in a cooler atmosphere.

The vegetation cover provides a particularly strong control on the surface energy balance and surface temperatures in regions where there is water stress, such as the Sahel. Timouk *et al.* (2009) show comparisons of the sensible heat flux $H$ over different vegetation types in the Hombori AMMA–CATCH mesoscale site in Mali (15–17°N) from 2006. In the rainy season, the average daily maximum in $H$ is only around 30 W m$^{-2}$ over a forested site, in comparison with around 100 W m$^{-2}$ over grassland, 200 W m$^{-2}$ over semi-desert and 300 W m$^{-2}$ over bare soil. These differences are not due to the radiative balance, which has the opposite effect ($R_{net}$ is higher over forest than bare soil, due to reduced albedo over forest). Instead, the vegetated areas are reducing the sensible heat flux substantially, by increasing the evaporative fraction. Deep-rooted trees, with access to deeper soil moisture, have a stronger influence on the evaporation than grasses do.

Airborne measurements over heterogeneous vegetation in Benin (Garcia-Carreras *et al.*, 2010) indicate that in this climatic zone boundary-layer temperatures drop by about 1°C over forest relative to adjacent cropland; 2 m temperatures are expected to have a greater sensitivity than this.

In the southern parts of West Africa, nocturnal dew fall can be significant in vegetated areas, and influences the surface energy balance the next day, by making water available for evapotranspiration. In this way, increased humidity in the air of zone B, just to the south of the ITD (Figure 1.1), can increase evaporation, and latent heat fluxes, as the ITD advances northward just before the monsoon onset.

#### 4.1.2.4.6 Urban Temperatures

In most parts of the world, the so-called 'urban heat island' is primarily caused by the relative dryness of the urban surface, so that evapotranspiration is low and most of the available energy appears in sensible heating and in heating of the surface materials. The roughness of the urban surface (relative to typical rural areas) also means that there tends to be more turbulence at night, and therefore the nocturnal inversion can be weaker over cities than over rural areas.

Although quantitative evidence is not available for most West African cities, these effects are reckoned to amount to the urban temperatures in built-up and paved areas being perhaps 2–3°C warmer than rural areas. However, for Sahelian cities, there is also a possibility of a 'nocturnal cool island' in parts of the city where vegetation is maintained. Lindén (2010) described measurements from Ouagadougou (during the dry season) showing that night-time temperature differences between built-up and densely vegetated areas within the city amount to about 7°C, with the built-up areas some 2°C warmer than rural areas and the vegetated areas some 5°C cooler. This finding presumably occurs because the rural parts of the Sahel in the dry season have a surface energy balance and Bowen ratio not dissimilar to a built-up area, dominated by sensible heat flux, so that there are greater contrasts between vegetated and non-vegetated areas than there are between rural and urban. In the wet season and in countries to the south, where there is more rural vegetation even in the dry season, we would expect an urban heat island to prevail, as in most parts of the world, but we are not aware of systematic evidence of this.

## 4.1.3 Boundary Layer over West Africa: Diurnal Cycle

As the large-scale winds over most tropical regions are reasonably light, the production and suppression of turbulence in the boundary layer is dominated by surface heating by day and by cooling by night. These effects influence profoundly the boundary layer and surface layer winds, temperatures and humidity, with very different dynamics by day and by night.

### 4.1.3.1 Daytime Convective Boundary Layer

During the day the PBL over land is characterised by upward turbulent heat fluxes at the surface, an unstable surface layer with a depth of around 30–50 m, and a deep boundary layer that is well mixed by convective turbulence – this is termed a convective boundary layer (CBL) and typically has a depth of 1000–2000 m in the afternoon. The CBL may be capped by a layer of shallow cumulus or deeper cumulus congestus clouds. Over the Sahara, the CBL may reach depths of up to 5000 m – probably the deepest boundary layer in the world. Above the well-mixed layer is a stable layer, possibly an inversion (in which temperature increases with height) into which the layer of shallow cumulus or congestus clouds develops (Figure 4.6).

The well-mixed boundary layer has near-constant potential temperature throughout its depth, and, typically, a sharp 'lid' at its top. The convective turbulence tends to cause vertical mixing of the winds during the day, so that, in the absence of cumulonimbus clouds, horizontal warming and cooling by advection is relatively low (in contrast with the nocturnal regime described in Section 4.1.3.2).[3] For this reason, a one-dimensional model of the vertical boundary layer growth during the day can be used. This model was developed in papers by Betts (1973), Carson (1973) and Tennekes (1973) on the basis of a budget of energy from the surface, with a simple estimate of entrainment from the upper lid. The model yields good estimates of the mixed-layer depth, temperature and humidity, and can be used to say something about convective cloud cover (Haiden, 1997).

The model is constructed on the basis of a constant ambient static stability of the troposphere $\gamma$ into which the PBL grows in the morning. The profile is assumed to have a sharp inversion (a jump in potential temperature with height), as shown in Figure 4.7. Turbulence in the PBL entrains air from the free troposphere as the PBL grows upward, and this entrained air, which is warmer than the PBL air, leads to a downward heat flux. Here, the entrainment heat flux is assumed to be a constant fraction $A$ of the surface heat flux. By equating this entrainment with the amount of free-tropospheric air gained as the mixed layer grows in height, this height $h$ is found to follow a square-root relationship with the total surface heat flux over the day:

$$h^2 = h_0^2 + \frac{2(1+2A)}{\rho c_p \gamma} \int_{t_0}^{t} H(t')\,dt' \tag{4.12}$$

(a)

Figure 4.6 Photographs of cumulus convective clouds: (a) shallow cumulus (cumulus humilis), observed at Niamey; (b) cumulus congestus, observed from the UK BAe146 research aircraft during AMMA, with deeper cumulonimbus in the background; (c) isolated, precipitating cumulus cloud observed in much drier tropospheric conditions near Hombori. Photos: Douglas Parker (a), (b) and Françoise Guichard (c).

(b)

(c)

Figure 4.6 (Cont'd)

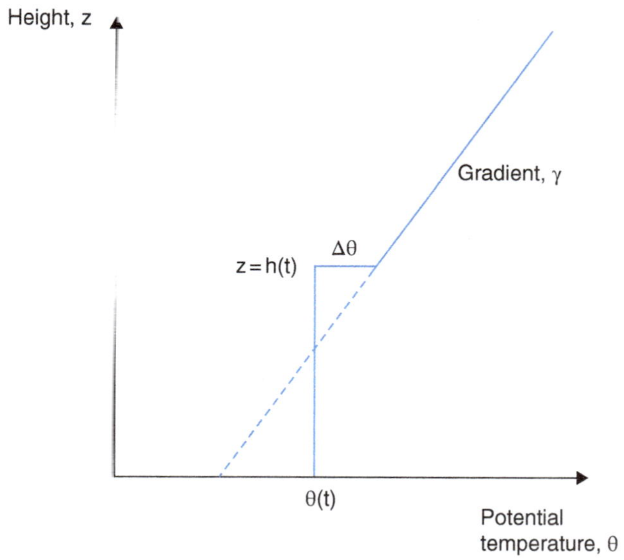


Figure 4.7 Schematic of the bulk boundary-layer model yielding (Equations 4.12 and 4.13). The well-mixed CBL has constant potential temperature $\theta$ up to height $z = h$ and is capped by a jump in $\theta$, at this height, above which there is constant stratification. As the CBL is heated by sensible heat fluxes from the surface, $h$ and $\theta$ both increase in time.

Similarly, the potential temperature follows

$$\theta = \theta_0 + \gamma\left(h - h_0\right)\frac{1+A}{1+2A} \tag{4.13}$$

where $h_0$ is the initial height of the well-mixed boundary layer at time $t_0$. In practice, the value of the entrainment constant $A$ is found to vary from 0.2 to 0.4, possibly with the higher values associated with cloud-topped boundary layers. Equations 4.12 and 4.13 are not particularly sensitive to this range of $A$.

The boundary-layer development described in Equations 4.12 and 4.13 turns out to be a rather good approximation in West Africa, because the meridional monsoon circulation tends to stabilise the boundary-layer profile at night, through cool southerly advection at low levels and a warmer southerly return flow higher in the boundary layer (Parker *et al.*, 2005b). In consequence, the CBL every morning tends to grow into an ambient profile that is quite stable. We can use the model to describe the typical growth of the boundary layer. Since in Equation 4.12 $h$ increases as the square root of the right-hand-side terms, growth is fastest in the late morning, when the boundary layer is still quite shallow and as the sensible heat flux is becoming large. In the afternoon, when the boundary layer is deep, its growth slows down – the heating is dispersed into a deeper layer, which therefore warms more slowly. In the north of the region, in 'zone B' of the monsoon system, it is also common for the boundary layer to break through into the elevated 'Saharan air layer' (SAL) aloft, which is a dry layer of low stability, into which the boundary layer can again grow very rapidly. Gounou *et al.* (2012) provide a helpful summary of the boundary-layer growth in different West African climatic zones, using radiosonde data from AMMA, which support these general conclusions (e.g. Figure 4.3).

Equations 4.12 and 4.13 also allow us to estimate some influences of meteorological controls, on the mixed-layer state. For instance, if we take $h_0$ to be negligible, a 50% reduction in the surface heat flux as seen in Figure 4.4, associated with soil moisture, would reduce the boundary-layer depth by 30%, and also reduce the warming of mixed-layer potential temperature by 30%.

The response of the boundary-layer humidity to surface heating is more subtle, because it depends on the entrainment of dry air at the top of the mixed layer, from the free troposphere (Betts and Ball, 1995). Over West Africa, this air can be extremely dry if it originates from over the Sahara. A high Bowen ratio (strong sensible heat flux) leads to a hot, deep boundary layer, which is dry due to the entrainment of dry air from aloft, while a low Bowen ratio (strong latent heat flux) provides a shallow, cool boundary layer that is more humid, with a lowered lifting condensation level (LCL). Given a net total surface heat flux, the surface flux of equivalent potential temperature $\theta_e$ is independent of Bowen ratio, so the $\theta_e$ is typically lower over the hot, dry surface than over the wet (due to dry entrainment from above). Given that convective available potential energy (CAPE) is generally quite closely related to the boundary layer $\theta_e$ (Williams and Renno, 1993; Parker, 2002), the likelihood of convective clouds is different over hot-dry (lower CAPE) and cool-wet surfaces with higher CAPE. However, as discussed in Section 4.1.5.2, deep convective clouds respond not only to levels of CAPE, but also to the convective inhibition (CIN), LCL and the mesoscale circulations induced by warm and cold air in proximity. In practice, we find that cumulus and cumulonimbus clouds preferentially form on the warm side of boundaries between hot-dry and cool-moist surfaces (Section 4.1.5.2).

The African easterly jet (AEJ) system is to some measure in thermal wind balance with the temperature gradient from the Guinea Coast to the Sahara (Holton, 1979; Cook, 1999; Parker *et al.*, 2005a). This means that, due to the south–north temperature gradient, there is vertical wind shear; from the southwesterly monsoon winds near the surface, to the AEJ with a maximum around 600–700 hPa. The sub-cloud CBL grows into this region of wind shear, typically reaching a depth at Niamey of around 1000–2000 m/850 hPa in the afternoon during the monsoon in summer, and 2000–3000 m/750 hPa in the pre-monsoon period (Gounou *et al.*, 2012). Therefore, the afternoon boundary layer mixes momentum between the AEJ and the monsoon flow, especially in the zone to the north of the AEJ core. In this northern region,

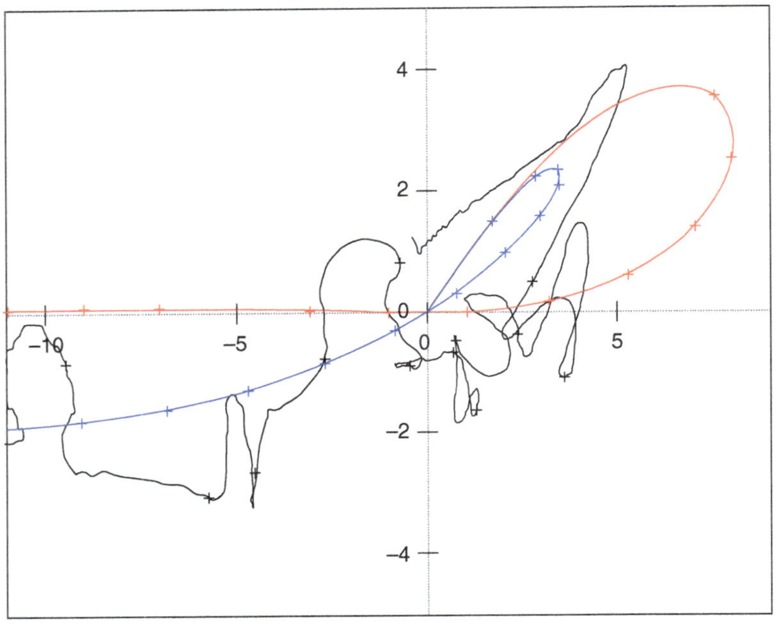


**Figure 4.8** Hodograph (locus of the $(u, v)$ wind vector) of the modified Ekman solution, as described by Equation 4.14, with surface geostrophic wind of 15 m s$^{-1}$ westerly, and an AEJ of 15 m s$^{-1}$ at 3 km height. The two curves show solutions with scale-height $\delta = 500$ m (red) and 1500 m (blue). Both solutions have $(u,v)=0$ at the surface, by definition, and the crosses mark heights at 200 m intervals above the surface. Note that both curves show a near-constant southwesterly wind direction over the lowest 200–300 m, and a maximum wind speed in the lower boundary layer at around 400–500 m altitude. The thin black line shows the hodograph for the Niamey sounding at 1200 UTC on 21 June 2006.

which represents zone B of Figure 1.1, the mixing can lead to northeasterly winds near the surface in the afternoon (Hamilton and Archbold, 1945).

The wind profiles in the daytime boundary layer are usually described, at least qualitatively, with an 'Ekman layer' model, in which the Coriolis force balances the pressure gradient and turbulent drag on the winds. The classic Ekman solution is

$$\begin{aligned} u &= u_g\left[1-e^{-z/\delta}\cos(z/\delta)\right]-v_g e^{-z/\delta}\sin(z/\delta) \\ v &= v_g\left[1-e^{-z/\delta}\cos(z/\delta)\right]+u_g e^{-z/\delta}\sin(z/\delta) \end{aligned} \quad (4.14)$$

where

$$(u_g, v_g) = \frac{1}{\rho f}\left(-\frac{\partial p}{dy}, \frac{\partial p}{dx}\right) \quad (4.15)$$

is the geostrophic wind, and the Ekman scale height is given by

$$\delta = \sqrt{\frac{2K}{f}} \quad (4.16)$$

in which $K$ is the turbulent diffusivity. It turns out that this solution can easily be applied to a linear profile of geostrophic wind

$$(u_g, v_g) = \left(u_{g0} + z\frac{du}{dz}, 0\right) \quad (4.17)$$

as a model for the wind shear beneath the AEJ. The geostrophic wind at the surface $u_{g0}$ can be surprisingly strong at these low latitudes; say 20 or 30 m s$^{-1}$, due to the low value of Coriolis parameter $f$. The solution to this model (Equations 4.14 and 4.17) is illustrated in Figure 4.8, with an example profile from the AMMA soundings of 2006 at Niamey. These curves show a couple of features that are not normally considered in relation to the standard 'Ekman spiral'. Notably, while the lower boundary-layer wind is to the left of the surface geostrophic wind, the hodograph tends to reverse back on itself in the lower boundary layer, toward the point where $u_g = 0$ in the geostrophic wind profile, at which $u = v = 0$. This leads to a maximum wind speed in the lower CBL, at around 500–1000 m, with a southwesterly direction at this level. Furthermore, if the boundary layer is deeper than the level where $u_g = 0$, and the Ekman scale $\delta$ is also comparable to or deeper than this height, there is northerly flow in the upper part of the Ekman boundary layer (where $u_g$ is easterly).

#### 4.1.3.2 Night-time Planetary Boundary Layer

The surface heat flux $H$ falls to zero in late afternoon – often this happens before sunset due to the strong longwave cooling and downward ground heat flux (Equations 4.3 and 4.9), at a time when the solar heating is falling fast. For example, over a forest site in the Sahel, Timouk *et al.* (2009) show $H$ falling to zero as early as 1500 UTC. After this time, the unstable surface layer rapidly stabilises, accompanied by falling surface temperature.

With the stabilisation of the surface layer around sunset, the boundary-layer turbulence becomes much weaker, over a short space of time; consequently, the turbulent stresses, which have during the day been retarding the winds, become rapidly reduced. When the balance between pressure gradient force, drag force and

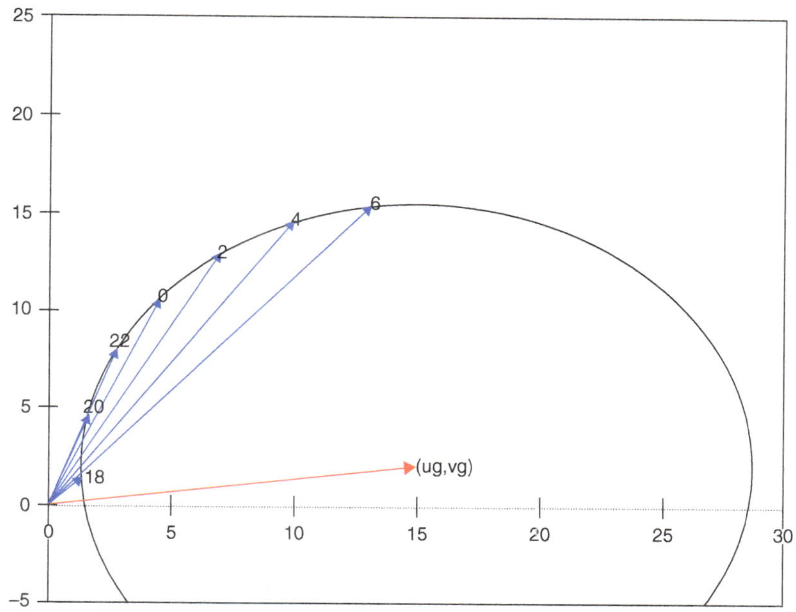


**Figure 4.9** (a) Hodograph of ($u, v$) vectors, showing an example of the solution of Equation 4.19 for an inertial osc llation in the nocturnal winds, indicating the wind vectors (blue) at 2-h intervals (labelled in local time), assuming the oscillation starts at 1800 UTC. The geostrophic wind vector is shown in red.

(rather small) Coriolis force is broken, the winds are free to accelerate down the pressure gradient. This acceleration often leads to the formation of a nocturnal jet in the lower part of the boundary layer, around 300 m in altitude. The nocturnal jet is a phenomenon that is observed in many regions of the world, notably those where the daytime CBL is deep and there is a pressure gradient on the larger scale.

The nocturnal jet is described by the model of Blackadar (1957), under the assumption of horizontally homogeneous, inviscid flow, in which the horizontal wind components, $u$ and $v$, vary only with height, and the large-scale horizontal pressure gradient is constant in time. The horizontal momentum equations then can be written as

$$\frac{du}{dt} - fv = -fv_g; \qquad \frac{dv}{dt} + fu = fu_g \qquad (4.18)$$

These equations can now be written in terms of the ageostrophic wind ($u_a, v_a$) as

$$\frac{du_a}{dt} - fv_a = 0; \qquad \frac{dv_a}{dt} + fu_a = 0 \qquad (4.19)$$

Equation 4.19 represent an 'inertial oscillation' of period $2\pi/f$, of the winds, around the geostrophic vector wind (see Figure 4.9).

The idealised model of Blackadar (1957) is complicated by the presence of the large-scale north–south temperature gradients in the monsoon system (as discussed by Rácz and Smith (1999) in the context of the Australian heat low). Therefore, on the continental scale the nocturnal jet is also part of a thermally direct circulation that advects cool air northward toward the Saharan heat low, where the warm desert air rises to compensate, and an elevated return flow carries warm, dry air toward the south. Locally, the nocturnal jet also has some characteristics of a gravity-current flow, and may in fact exhibit a gravity-current head at its leading edge, for instance at the ITD (Bou Karam *et al.*, 2008; Burton *et al.*, 2013).

The physics of the nocturnal jet are subtle, and detailed observations have been lacking. It seems that for hours at a time, turbulence in the night-time PBL (NBL) is very light, and nocturnal jet winds may be 'decoupled' from the surface, and almost frictionless. Very commonly at night, this decoupling occurs, and winds at the surface are very light; there may be no surface evidence of the rather strong NBL winds, which exist around 300–400 m above the ground (Lothon *et al.*, 2008). However, it must be recognised that the accelerating winds generate increasing wind shear, which also tends to become unstable. Therefore, the NBL represents a balance between surface cooling, which acts to stabilise the profile (and suppress turbulence), and accelerating winds acting to destabilise the profile, generating shear-driven turbulence. In this environment, turbulence tends to be intermittent (occurring in bursts of gusty turbulent activity with longer periods of calm), and characterised by coherent structures (see Section 4.1.4), such as propagating bores and gravity currents.

Although the formation of the nocturnal jet, and its rotation with time during the night, is well explained by the basic theory of Blackadar (1957), the evolution of the winds in the levels below the jet is less well understood, and there may be considerable local variations in the low-level winds from one night to another. Beneath the jet, there can be turbulence, driven by the wind

Table 4.1 Jet maxima and altitudes at different stations according to season. Special Observing Period (SOP)-1 is the pre-monsoon period, May–June 2006, and SOP-2 is the full monsoon period of July–August.

| | SOP-1 | | | SOP-2 | | |
|---|---|---|---|---|---|---|
| | Freq (%) | Speed (m $s^{-1}$) | Altitude (m) | Freq (%) | Speed (m $s^{-1}$) | Altitude (m) |
| Agadez | 35 | 9.9 (5.9–15.2) | 230 (110–370) | 80 | 9.4 (5.2–17.0) | 260 (130–670) |
| Niamey | 60 | 10.9 (5.1–19.0) | 320 (90–670) | 85 | 8.5 (5.2–15.6) | 360 (170–730) |
| Parakou | 55 | 8.6 (5.3–19.3) | 300 (130–590) | 50 | 8.1 (6.0–12.0) | 255 (110–430) |

*Source*: Gounou *et al.* (2012). Reproduced with permission of Springer.
The minimum and maximum values are indicated in parentheses.

shear, which tends to limit the wind speeds and can reduce the veering of the wind vector with time. For instance, Bain *et al.* (2010) showed observations of NBL in the Hombori region of Mali, in which the wind directions in the lowest 400 m, on several nights, were found not to be rotating, and their orientation was controlled by synoptic and local pressure gradients, not well captured by NWP models. Some theoretical extensions of Blackadar's (1957) theory have even suggested that, in the very lowest levels, there may be a reversed (anticlockwise) rotation of the wind vector with height (van de Wiel *et al.*, 2010). Suffice to say, that while inertial oscillations seem to be observed above the nocturnal surface layer (e.g. in pilot balloon data at 600 m; McGarry and Reed, 1978; Parker *et al.*, 2005b), the winds in the lowest 50 m or so may not show such a clear rotating cycle during the night.

The diurnal cycle of winds is also very sensitive to the time of year and climatic zone. Analyses by Parker *et al.* (2005b), Lothon *et al.* (2008) and Gounou *et al.* (2012) (see Table 4.1) indicate that the strength of the diurnal cycle of low-level winds varies according to season, and tends to be strongest when a station lies within 'zone B' of the monsoon system (Figure 1.1). In zone B there is a strong temperature (and therefore pressure) gradient, meaning a strong geostrophic wind vector at low levels, as well as strong daytime heating to suppress the winds during the day, which initiate the nocturnal jet around sunset (see Figure 4.9: light winds around sunset lead to a stronger acceleration of the nocturnal winds). It has also been suggested that the strong nocturnal jets in zone B around the time of monsoon onset have an important role in the transport of water vapour, to set up the initiation of the first rains (see Figure 1.29).

In addition to the seasonal evolution in the nocturnal jet, there seems to be synoptic and subseasonal variability. Forecasters in the Soudanian and Guinea Coast zones look to the temperatures and pressures in the Saharan and Sahelian regions to predict the strength of the nocturnal winds in the coming days, and this relationship was confirmed in observational data by Parker *et al.* (2005b): high temperatures in the northern region are related to strong nocturnal winds and diurnal cycle in the countries further south.

The nocturnal jet, and the moist advection which it brings in the monsoon season, are also related to the regular occurrence of nocturnal stratus and stratocumulus decks in the Guinea Coast zone (Schuster *et al.*, 2013; also see Section 2.1.6). These cloud decks further interact with the stratification, surface energy balance and turbulence. Schrage and Fink (2012) have argued that the onset of the nocturnal jet in this region leads to turbulence below the jet that mixes up moisture, resulting in the formation of a cloud layer. The stratocumulus can have a very shallow base, at night, and may be associated with fog on high ground; it usually dissipates between mid-morning and noon, through the mixing of the CBL.

The classic diurnal cycle of low-level winds can be inferred from Figure 4.10 in which there is very commonly a daily wind speed maximum in the morning hours before dawn. One significant detail in this figure is the occurrence of a maximum in *surface* winds in the morning hours; for example, on 14, 23 and 25 July 2006. This surface wind event appears to have first been noted in Australian observations by May (1995), and is caused by the mixing-down of the nocturnal jet momentum in the growing early-morning CBL. The phenomenon leads to a surface wind-speed maximum in the morning hours over wide spatial regions – this can lead to a morning peak in dust uplift for instance.

In terms of local weather forecasting, the nocturnal jet causes significant wind shear that may require notifications being issued for aviation. Abdou *et al.* (2010) provided a short review of different criteria used to categorise strong wind shear associated with nocturnal jets. From sources in the USA and existing practice at Agence pour la Sécurité de la Navigation Aérienne en Afrique et à Madagascar (Africa), Niamey, it seems that a value of

$$\frac{du}{dz} = 3-5\ \mathrm{m\ s^{-1}}/100\ \mathrm{m} = (3-5)\times 10^{-2}\ \mathrm{s^{-1}} \quad (4.20)$$

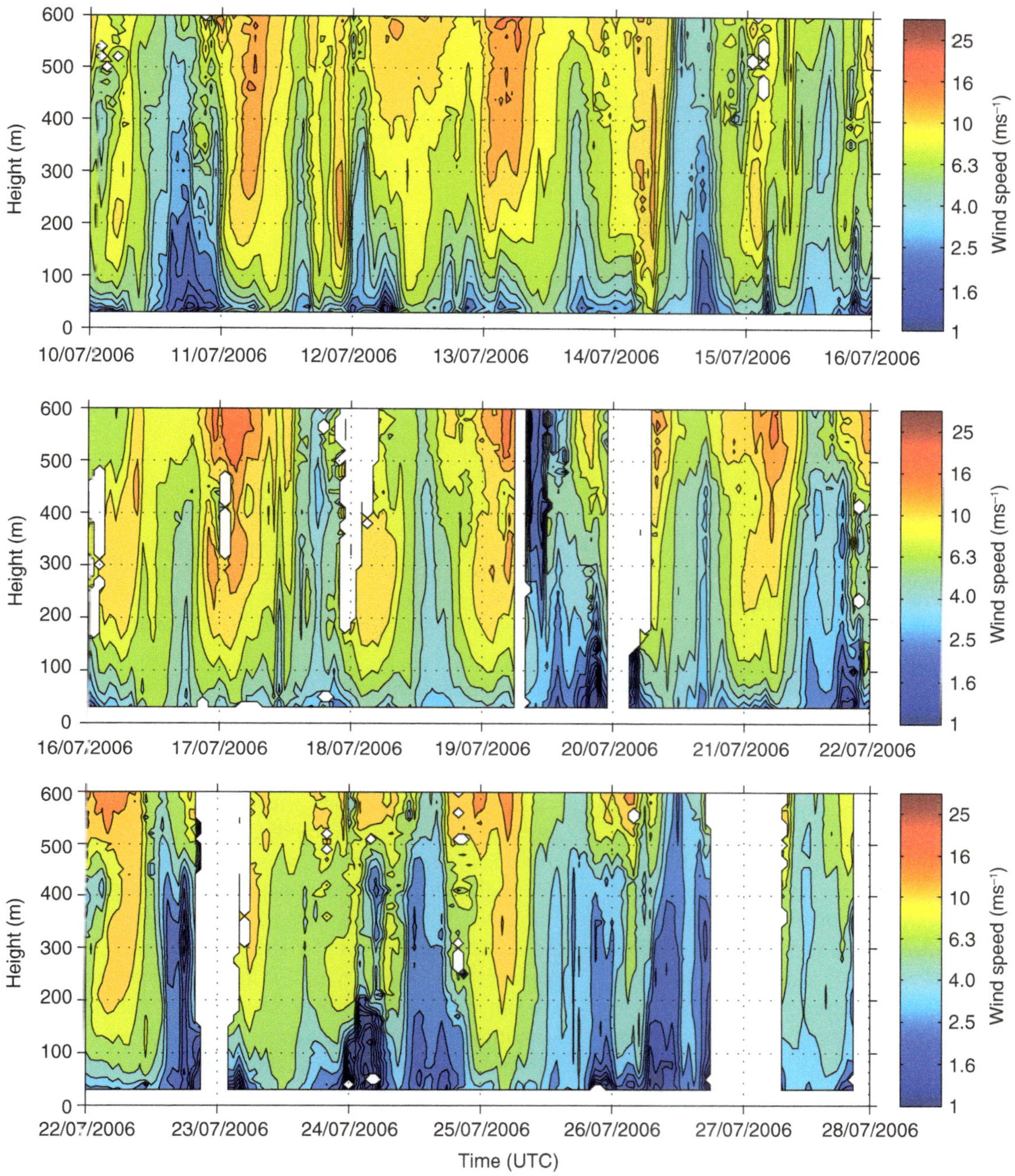


**Figure 4.10** Time–height contours of wind speed, averaged over a set of four sodars deployed around Niamey, in the period 10–27 July 2006. *Source*: Abdou *et al.* (2010). Reproduced with permission of the Royal Meteorological Society.

is a range of commonly used thresholds for strong wind shear, corresponding to a nocturnal jet of around 10–15 m s$^{-1}$ at 300 m altitude (around 20–30 kt at 1000 ft).

Figure 4.11 shows the diurnal statistics of low-level (30–300 m) wind shear, obtained from sodars deployed in the vicinity of Niamey for around 40 days in the summer of 2006. It can be seen that moderately strong wind shear is more common in the nocturnal hours, and weak wind shear is rarely observed in the period 1800–0900 UTC.

It is generally accepted that the NBL is less well understood than the daytime CBL, and that the NBL is rather poorly represented in NWP and climate models. During the day, the CBL characteristics are quite well described by bulk models such as that of Equations 4.12 and 4.13 in Section 4.1.3.1, given the surface heating, and therefore the boundary layer may be relatively insensitive to the details of the modelling schemes. At night, the NBL involves a delicate balance between surface cooling acting to stabilise the profile and accelerating winds acting

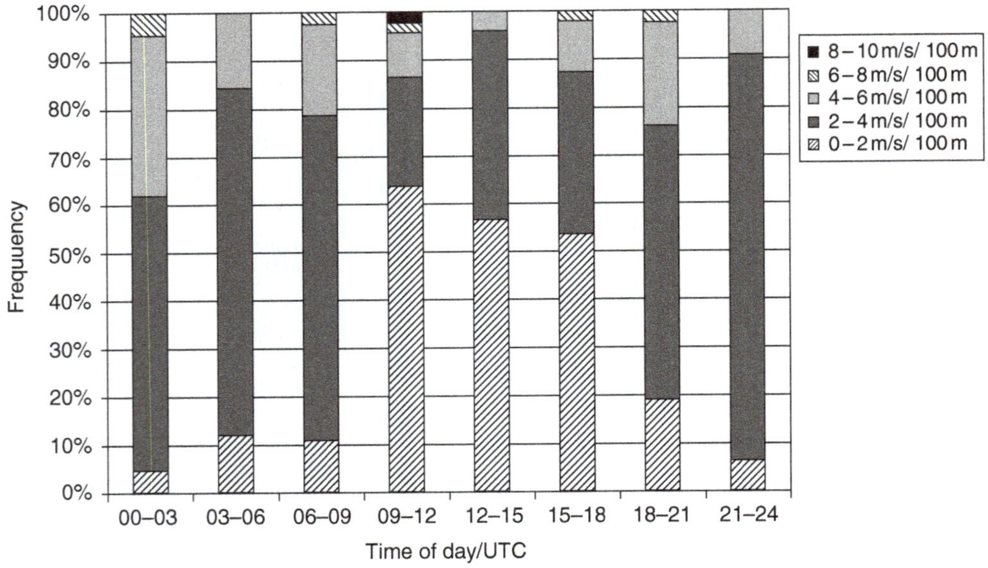


**Figure 4.11** Frequencies of most intense 15 min wind shears from a set of sodars deployed for the AMMA SOPs in the Niamey region in July–August 2006. The central category, of 4–6 m $s^{-1}$/100 m, is the one that would commonly require the notification of high wind shear to pilots. Note that these observations do not represent a reliable climatology of wind shear from convective squalls, because the sodars do not sample under conditions of falling precipitation. *Source*: Abdou *et al.* (2010). Reproduced with permission of the Royal Meteorological Society.

to destabilise it. Representation of these requires very high vertical resolution in models, and may be handled quite poorly by operational NWP systems (Bain *et al.*, 2010; Sandu *et al.*, 2013).

#### 4.1.3.3 Fog Formation

When the lower part of the NBL cools at night, through longwave radiative cooling of the ground, the relative humidity of the air rises and the formation of fog is possible, with a horizontal visibility value of less than 1000 m. Fog is notoriously difficult to predict, because its generation is a balance of various competing physical processes, including mixing with the layers above the surface. Condensation nuclei are also factors in fog formation; these are hygroscopic particles that include combustion products, organic salts and dust particles. When significant amounts of smoke contribute to intense fog events, the conditions are referred to as 'smog'.

There are several mechanisms for fog formation as follows.

##### *4.1.3.3.1 Radiation Fog*

This is by far the most common form in West Africa, and is associated with prolonged cooling of the ground surface at night. This cooling is much greater on clear, cloudless nights than cloudy nights. Radiation fog initially forms in the coldest air close to the surface, which is cooled to its dew point temperature. In order for a substantial depth of fog to form, there must be some mixing of air from the lowest metres of the atmosphere with air higher up. This mixing is suppressed by the stability of the NBL, but enhanced by light winds. As a fog layer deepens over the night, radiative cooling will occur from its upper surface, and this will tend to naturally destabilise the layer, providing the mixing needed to sustain a deep, cold fog layer.

Under calm conditions with no wind, cooling will be confined to the air immediately in contact with the ground, and in this case dew is most likely to be formed. Although light winds are thought to be necessary for the initial formation of a fog layer, the mixing also tends to dry the air, since the warmer air above the surface is usually of lower relative humidity, and if the wind is too strong fog will not form, or will disperse, possibly resulting in the formation of clouds (e.g. stratus or stratocumulus).

It should be noted when using observations of the very light winds needed for radiation fog to occur that old or poorly serviced anemometers may not respond to very light winds, so that a synoptic station may incorrectly report zero wind when the winds are light.

##### *4.1.3.3.2 Advection Fog*

Advection fog occurs due to the cooling of a warm air mass moving over a cold underlying surface. This is not thought to occur over land in West Africa since the surface is rarely significantly colder than the advecting air. However, the advection of moisture is generally accepted as a significant aid in the formation of radiation fog (e.g. Galvin, 2004). Sometimes, fog is formed by radiative cooling and can be transported wholly downstream within a moving air mass. In this case, radiation fog has been advected. However, strengthening winds will generate turbulence that will tend to mix the fog with drier air from aloft, and disperse it. On occasion, in the winter and spring months, stations on the Atlantic coast, such as Dakar or St Louis, can experience advection fog due to warm air moving over very cold ocean currents; the advection of this sea fog onshore can lead to visibility falling to a few hundred metres.

##### 4.1.3.3.3 Hill Fog

Hill fog is formed as a result of air streaming upward over a hill or mountain and consequently the air is subjected to adiabatic cooling. This type of fog is observed to be 'sitting' near the top of the hill or mountain and can explain increased occurrence of fog at higher altitude stations. In the summer months in the southern part of West Africa, when nocturnal stratocumulus decks prevail, the base of the stratocumulus can intersect the tops of hills, leading to the occurrence of fog at stations of high elevation.

##### 4.1.3.3.4 Steam Fog

Steam fog can be observed over the sea. This forms when cold air moves over warm water. The water vapour escaping from the water surface saturates on mixing with colder air and condensation takes place. However, owing to the lack of very cold air masses over West Africa, this kind of fog is not observed.

Once fog has been formed, it persists for as long as the ground surface remains cool (i.e. at low temperature). As incoming solar radiation hits the ground surface in the morning (at sunrise), the fog begins to lift or clear, giving way to improved surface visibility. Sometimes, the presence of a low-level inversion enables fog to persist for a while, but later, once the inversion is broken by solar heating, the fog clears.

Occurrence of fog is prevalent in the Guinea Coast zone, extending some distance inland, and to western coastal areas (see Figures 4.12a and 4.13). Note that in a given latitude band in Figure 4.12a there are stations that have frequent records of fog reports and stations that do not. Therefore, local conditions of surface and topography must be important. Fog occurrence tends to be favoured over and around hills and in forested areas.

The period of occurrence extends from the rainy season into the first dry-season months, in the period June–December, with a minimum in February and March (Figure 4.12b). The peak in fog occurs in September at the end of the rainy season, when sea-surface temperatures are comparatively low (such that air relatively close to saturation will be brought inland when winds have a southerly component), but soils are moist, following the summer monsoon, and the temperature range increases as nights lengthen. The seasonal cycle is also influenced by the increased nocturnal radiative cooling around the end of the rainy season, as the atmosphere dries out and cloud cover diminishes. Since the advection of moisture is generally accepted as a significant aid in the formation of radiation fog, it is very unlikely to form in the dry northeasterly Harmattan and is rarely observed in these SYNOP data between January and May, before the onset of the summer monsoon.

Note that stations such as Dakar and Ziguinchor on the west coast of the continent do experience fog events more typically in the early months of the year, February–May, advected into coastal locations from the sea.

Forecasters in West Africa reckon that fog does not occur north of 9°; however, there are some occasional fog reports recorded north of this latitude. Figure 4.12a also shows, quite remarkably, that there have been occasional synoptic reports of fog at several remote Saharan stations, such as Bordj Badji Mokhtar (21.33°N, 0.95°E). It appears that such events are not strictly fog in the conventional sense of the term, although the use of this weather code by the observers can be understood. In these remote desert regions, very occasional events of high humidity (near saturation) and low visibility can occur immediately after rainfall, compounded by elevated dust (Azzedine Saci, personal communication). Such desert conditions are perhaps a unique weather phenomenon.

#### 4.1.3.4 Marine Planetary Boundary Layer

We are only able to make a limited discussion of offshore meteorology, but some basic principles relating to the offshore environment are described here. Owing to the high heat capacity of the ocean mixed layer, and the strong mixing, relative to the heat capacity and ground heat flux over land, the ocean surface temperature has a much weaker diurnal cycle than the land surface (typically, the SSTs vary by less than 1°C over the daily cycle). Therefore, the boundary-layer structure over the ocean tends to be closer to equilibrium, evolving over longer space and timescales than the boundary layer over land, with the air temperatures quite similar to the SST, and high relative humidity.

Although the marine boundary layer is much less variable than that over land, it can still exhibit synoptic and mesoscale structures that contain strong gradients. For instance, deep convective cold pools, land breezes and coastal cyclones can all contain sharp frontal features, associated with strong gradients in winds and temperatures.

### 4.1.4 Local Winds: Mesoscale Variability and Circulations

The existence of local winds has been known for many years. Mariners have been aware of coastal wind effects for centuries, while in recent years pilots have become familiar with the influence of the land surface in generating local circulations of air, including thermals over hills and hot surfaces. Winds on these spatial scales may also

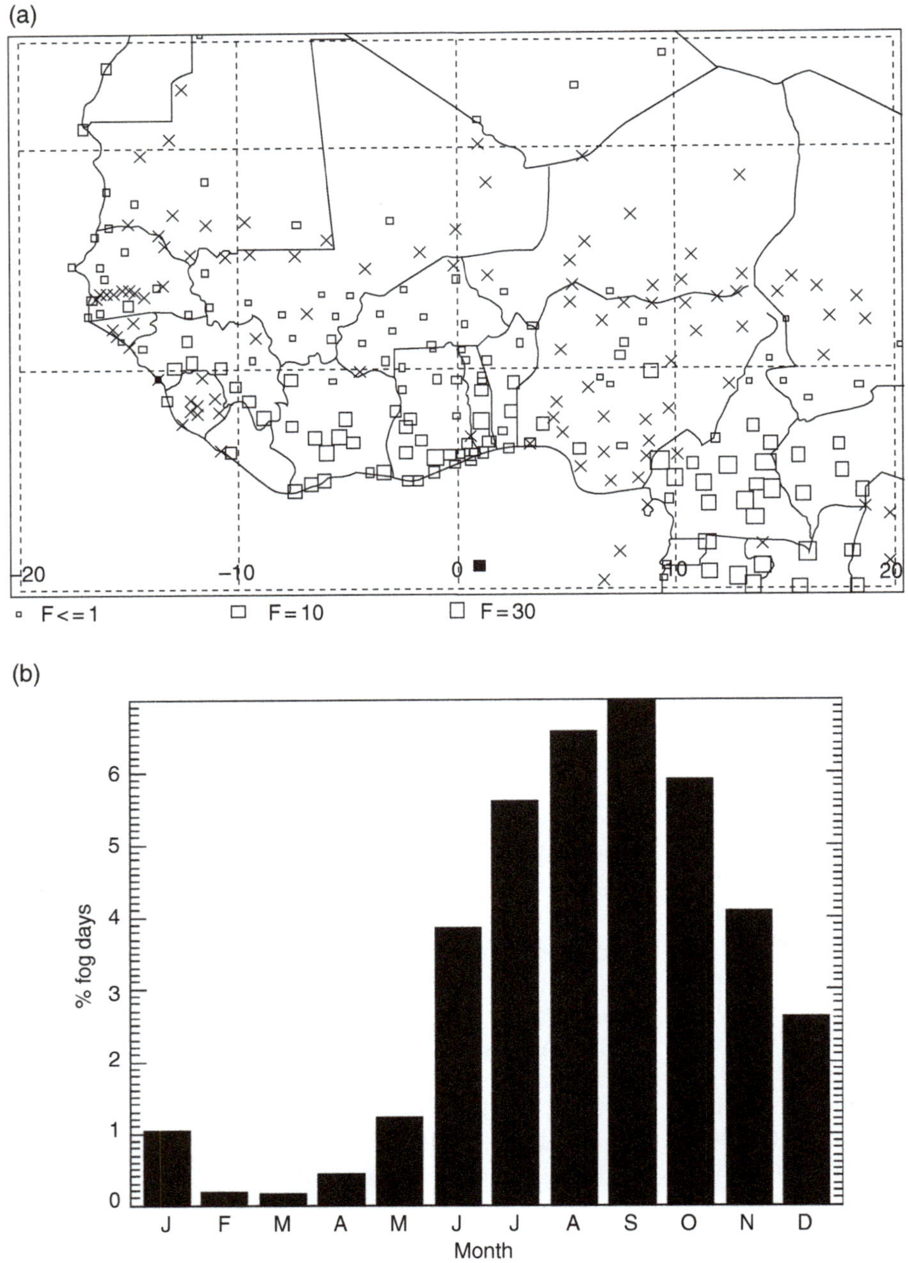


**Figure 4.12** Fog climatology. (a) Statistics of 0600 UTC fog reports: numbers denote *F*, the mean annual frequency of reports of fog at each station, while stations with no fog reports are marked 'x'. Some stations with frequencies above *F* = 20 have very low numbers of total reports and may have large errors on the exact value of *F*, but are retained here to show the geographic coverage of reports. (b) Monthly frequency of fog reports at 0600 UTC, as a percentage of all reports at 0600 UTC, for stations in the geographic zone 10°W–10°E, 7–10°N. Data are taken from West African synoptic station reports, for the years 1984 to 2012, obtained from the Met Office Integrated Data Archive System (MIDAS) database at British Atmospheric Data Centre (BADC).

be generated by the internal dynamics of the atmosphere, which tends to support gravity waves, through which disturbances can travel large horizontal distances, and other 'coherent structures', such as gravity currents, in which a mesoscale 'front' can be generated quite rapidly out of a much broader temperature gradient.

Local winds are important to many sectors, including aviation and marine applications. Because of their small scales, and subtle dynamics that are not always well captured by coarse-resolution weather prediction models, forecasters need to have an appreciation of the dynamics and statistics of local wind features.

#### 4.1.4.1 Coherent Structures in the Winds: Convergence Lines and Gravity Currents

The atmosphere, and especially the PBL, is a chaotic system, with an infinite variety of possible configurations. Given the complexity of much of the atmospheric

Figure 4.13 Photograph of fog approaching Takoradi airport, around one nautical mile from the sea, on 8 January 2012, 0630 UTC.

dynamics, it is quite remarkable that very repeatable 'coherent structures' exist, which can be categorised and described by reasonably simple conceptual models.

Some coherent structures take on the form of a horizontal line feature. 'Convergence lines' are a class of features that have been found to be prevalent in most situations over land, sometimes extending over many hundreds of kilometres, but with cross-line scales of a few kilometres, or even hundreds of metres. Although these convergence lines may not be easy to identify in conventional observations, they commonly act as the trigger for deep convective storms in the afternoon (see also Sections 3.1.4.1.4 and 6.1.2.6, and Figure 6.21).

An excellent example of a coherent structure that also causes convergence lines is that of a gravity current. A gravity current (also known as a density current) is the flow that develops as a result of horizontal gradients in density, at a horizontal surface such as the ground or sea surface (Figure 4.14). Under suitable conditions, the sea breeze is a classic example of an atmospheric gravity current (Simpson, 1999), developing in the morning hours as the land heats up faster than the sea. Cold pools (Section 3.1.2.2) emerging at the surface from cumulonimbus storms are another phenomenon commonly described as a gravity current.

From laboratory experiments (e.g. Figure 4.15), the propagation speed of the gravity current front is given by the formula

$$U_{\mathrm{f}} = k\sqrt{g\frac{\Delta\theta}{\theta}h} \tag{4.21}$$

where $U_{\mathrm{f}}$ is the front speed, $k$ is a constant of order 1, $\Delta\theta$ the drop in potential temperature at the front, and $h$ the depth of the gravity current flow, or the alternative form in terms of pressure behind the head:

$$U_{\mathrm{f}} = k\sqrt{\frac{\Delta p}{\rho}} \tag{4.22}$$

where $\Delta p$ is the pressure increase at the front and $\rho$ is the air density. Note that the speed of the flow behind the head (the 'feeder flow') is faster than the front-speed: the convergence of fluid at the head is balanced by upward motion and mixing with environmental fluid,

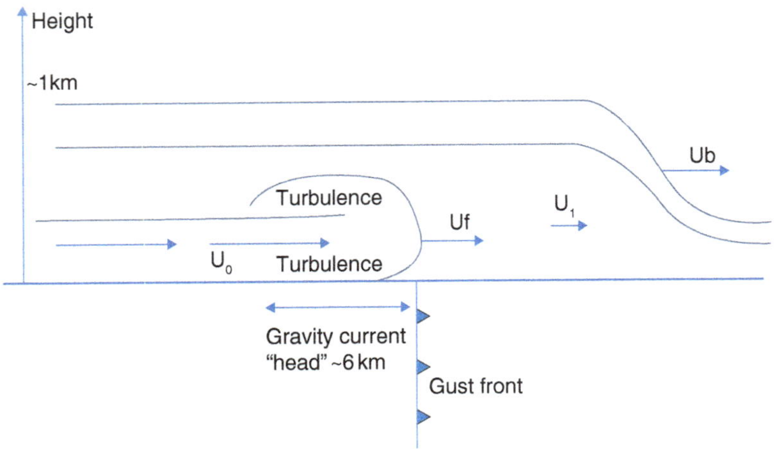


**Figure 4.14** Imagined situation of a gravity current penetrating into a nocturnal inversion, with a bore moving ahead of that. Contours indicate lines of potential temperature, increasing upwards. The feeder flow for the gravity current has a speed $U_0$ that is greater than the gravity current speed $U_f$. In contrast, the speed of the bore $U_b$ is greater than the speed of the flow behind it, $U_1$, so the bore moves as a wave through the flow. The excess air moving into the gravity current head is mixed out by strong turbulence in this region and forms a mixed layer that is left behind the head, above the feeder flow.

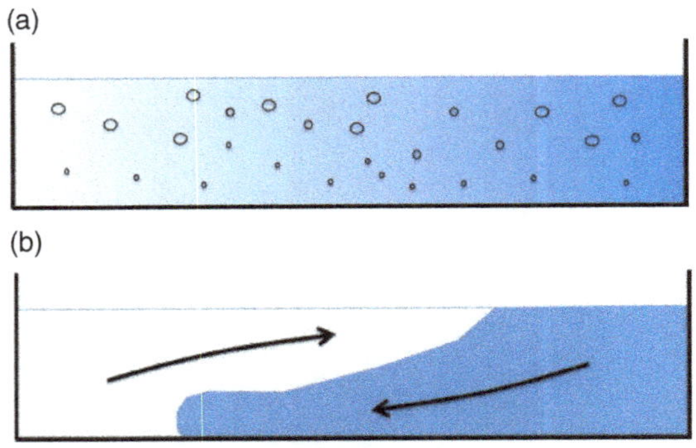


**Figure 4.15** A sketch indicating the general results of the laboratory fluid dynamics experiments of Linden and Simpson (1989), which suggest how mesoscale density gradients may be sustained in the atmosphere by day and by night. The dark fluid is denser than the ambient and represents cold air in the atmosphere. When the flow is turbulent (panel (a)), then a horizontal density gradient can be maintained, with relatively weak horizontal flow. When the turbulence diminishes (in the atmosphere, at night – panel (b)) then the dense fluid slumps below the ambient, in the form of a gravity current.

with the mixture moving behind the head to form a transition zone between the current and its environment. In the atmosphere, this relative flow of air from behind the head is one distinguishing feature of a gravity current (as opposed to a bore, for instance – see Section 4.1.4.5).

In the atmosphere, the characteristics of gravity-current-like flow are often substantially modified by the effects of wind shear, stratification and surface heating or cooling. While the gravity current model is often a useful one, to estimate the speed of propagation of mesoscale features, for example, a number of refinements are needed in different specific contexts.

When an atmospheric gravity current moves over a hot land surface, the current is heated from below and tends to be mixed with the environmental air by turbulence. Both of these effects tend to weaken the circulation. Linden and Simpson (1986) modelled this influence of turbulence on gravity current propagation in the laboratory by generating turbulence from the bubbling of air through the flow. They showed two situations that appear to describe the atmospheric context rather well. When the flow is well mixed by turbulence (Figure 4.15a), it can support relatively smooth horizontal temperature contrasts, where any tendency for the flow to overturn is balanced by vertical mixing of the temperature contrasts. Over West Africa, it seems that horizontal temperature contrasts can be sustained, in the afternoon hours, over quite short distances, say 10 km.

If the turbulence reduces (an analogue for the reduction of convective turbulence over land at nightfall) then the flow becomes free to respond to the temperature contrasts. Quite rapidly, the smooth temperature gradient evolves into a gravity current, with a strong temperature contrast at its head and strong convergence there (Figure 4.15b). This process of gravity current formation in a larger-scale thermal gradient at nightfall, as first described by Linden and Simpson (1986), is extremely important in tropical and subtropical continental dynamics. In particular, gravity currents will tend to form in the evening wherever there has been a pre-existing, larger-scale temperature gradient, and these may appear to be spontaneously generated out of the large-scale flow. The formation of gravity currents in such situations appears to be very common. The process has the result of generating very strong local convergence, sufficient to trigger cumulonimbus storms for instance, out of a much broader and weaker zone of large-scale convergence. One example of this process is the generation of gravity-current-like flows at the ITD, observed by Bou Karam *et al.* (2008).

The nocturnal stabilisation of the boundary layer means that gravity-current-like convergence interacts with a stable layer of air near the surface, typically surmounted by a rather neutrally stratified, residual layer above. This configuration of a stable layer, which supports gravity waves, lying beneath a neutral layer in

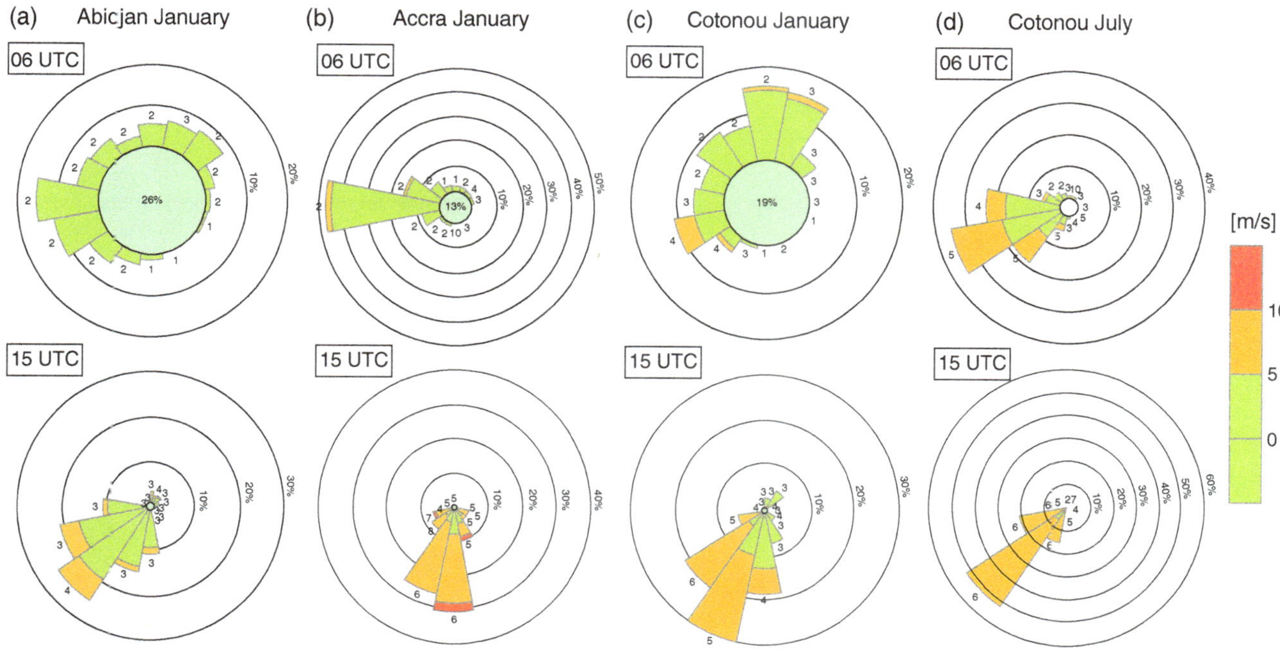


**Figure 4.16** Wind roses showing the patterns of land and sea breezes for three stations on the Guinea Coast. The top row shows plots from 0600 UTC, roughly corresponding to the time of maximum land breeze, and the lower row shows plots from 1500 UTC, when the sea breeze is typically a maximum. Panels (a) to (c) show wind roses for each station in January for Abidjan, Accra and Cotonou respectively, with (d) showing conditions at Cotonou in July (other stations show very similar behaviour in July).

which waves cannot propagate represents a *wave-guide*, on which a variety of wave structures may appear. There has been some work on such waves in other parts of the world, notably the USA and Australia, and there is some evidence of the effects in West Africa (Section 4.1.4.5).

In addition to gravity currents, there are other wind patterns that lead to the formation of convergence lines, such as the flow of air onshore over a headland, or the flow downstream of a mountain.

#### 4.1.4.2 The Sea Breeze and the Land Breeze

The sea breeze is a classic local wind system, which arises as a result of the rapid heating of the air over land during the daytime, in contrast with relatively little change in the temperature of the maritime air (Section 4.1.3.4). The warmer air over land leads to an enhanced 'heat low' (through increased atmospheric 'thickness'), resulting in a pressure gradient force from sea to land. In the morning hours, the winds accelerate toward the land, down this pressure gradient; if the Coriolis force is significant, the wind direction will later be turned to the right.

The strength of the breeze is greater on hot days, and may be weaker under cloudy conditions, when the land temperatures are not so high.

The *land breeze* is a reversal of the flow at the coast at night, leading to offshore winds as the land cools faster than the sea surface and the temperature gradient in the PBL is reversed. Because the NBL is stable, and much shallower than the daytime CBL, the land breeze is also shallower than the sea breeze, and less intense. Figure 4.16 shows observations from three stations on the Guinea Coast, illustrating the presence of the sea and land breezes. In January there is a significant contrast between land- and sea-breeze conditions at all the stations. However, while the sea breeze consistently blows from the southwestern sector at all of the stations, the strength and direction of the land breeze is much more variable from case to case, probably dependent on the details of local topography and coastline, to which we expect the land breeze to be quite sensitive. In July (Figure 4.16d) the flow is typically southwesterly throughout the day and night, dominated by the large-scale monsoon pressure gradient, and perhaps somewhat enhanced in the afternoon.

##### *4.1.4.2.1 Sea Breeze Characteristics*

The sea breeze brings cool air as a low-level intrusion (typically 500–1000 m deep) into the warmer air inland, with enhanced convergence at its leading edge. The sea breeze flow also carries humid air of high $\theta_e$ inland, and is progressively warmed by the surface fluxes. Although the sea breeze is often described as a gravity current (e.g. Figure 4.17a), during the day its dynamics are much more controlled by boundary-layer mixing than would be the case for a pure gravity current. In practice, the land over which the cool maritime air moves is relatively hot, and will weaken the sea breeze both by heating it from below and by generating convective turbulence that

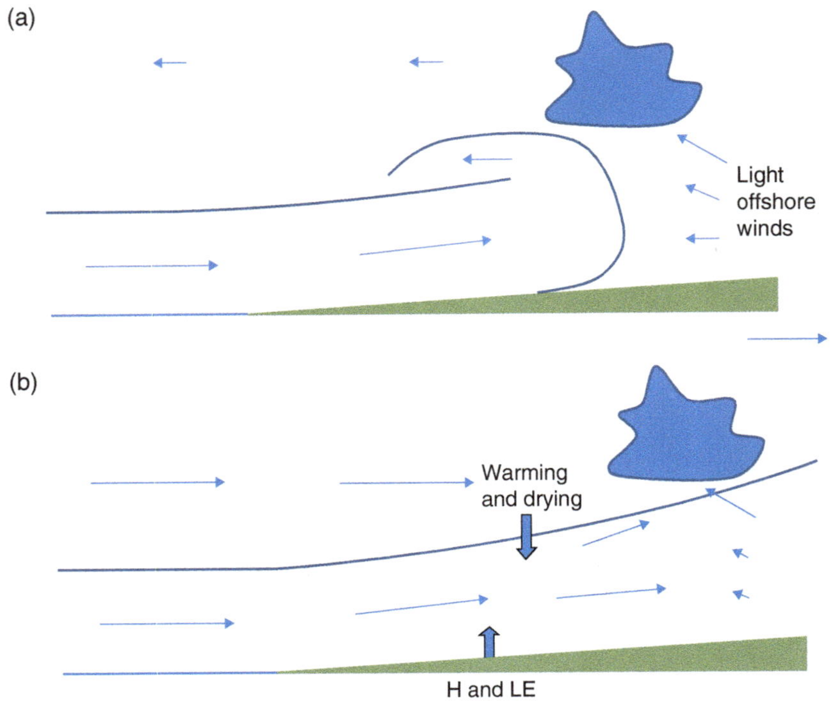


**Figure 4.17** Schematics of gravity current/PBL transition models for the sea breeze during the day. In (a) there are light offshore winds and the sea breeze progresses inland as a coherent gravity current, once the thermal contrast between land and sea is established in the morning. In (b) there is a robust onshore wind, as seen on the Guinea Coast when the monsoon is active. There is little diurnal signal in the winds at the coast, but in the morning, as the land warms up and the surface fluxes increase, the maritime boundary layer air is warmed as it moves inland, until cumulus can be formed at some distance inland.

mixes the maritime air into its surroundings, as in the model of Linden and Simpson (1986) (Figure 4.15). For these reasons, the penetration of the sea breeze inland may often be rather limited during the daytime hours, with the temperature gradient remaining within a few tens of kilometres of the coast.

On the Guinea Coast in the summer months, when the sea breeze is supported by a robust southwesterly monsoon flow, the local signature of the sea breeze winds may be hard to detect (with little change in the winds from day to night; Bajamgnigni Gbambie and Steyn, 2013). In these circumstances, the coastal airflow and thermodynamics may be regarded as a more passive modification of the marine boundary layer as it is carried inland by the southwesterly monsoon (Figure 4.17b). In this case, the evolution of the boundary layer moving inland may be modelled by the simple energy budget model of Section 4.1.3.1. In contrast, a stronger sea breeze convergence may occur in conditions where the large-scale winds have a very light offshore component (typically outside the monsoon months).

In the afternoon and evening, when the boundary-layer turbulence diminishes over land, a gravity current may form at the sea breeze front, which may propagate inland for distances of up to 100 km, and the propagation speed may be approximated by Equation 4.21. Taking a temperature difference across the front of 5°C and a depth of 500 m, Equation 4.21 gives a speed of around $3\ \mathrm{m\ s^{-1}}$, consistent with the observations of Figure 4.18.

Figure 4.18 shows a climatology of cloud cover over West Africa, and emphasises the significance of the sea breeze in modulating cloud cover in the coastal zone. At 1000 UTC (Figure 4.18a) there is a band of cloud along the coast of Ivory Coast, associated with the convergence of the cool oceanic boundary layer and the warming air over land; however, at this time the skies over the southwest–northeast-oriented coastline of Ghana are, on average, clear. Between 1200 and 1800 UTC the band of cloud intensifies all along the coast and moves inland. Note, however, that particularly strong convergence seems to be maintained over the Niger delta (Figure 4.18d). Note also the significant cloud patterns over Lake Volta.[4] The inland rate of advance of the peak cloud (in a statistical mean) is only around 60 km in 6 h, or around $3\ \mathrm{m\ s^{-1}}$. However, increased cloud frequency does extend further inland, suggesting that on some days the inland penetration of the sea breeze extends further and more rapidly.

Offshore, the panels of Figure 4.18 also show a general clearing of cloud upwind (to the south of) the Guinea Coast, perhaps due to the propagation of gravity waves (Love *et al.*, 2011).

Along the coastal regions in the early morning, slight rain or thundery activity could occur as the coastal convergence first develops. Later in the day, the cool sea breeze moves inland and regulates temperature around the coastal region and stabilises the atmospheric profile, generally leading to clear skies. As the flow moves inland, the sea breeze increases the frequency of afternoon

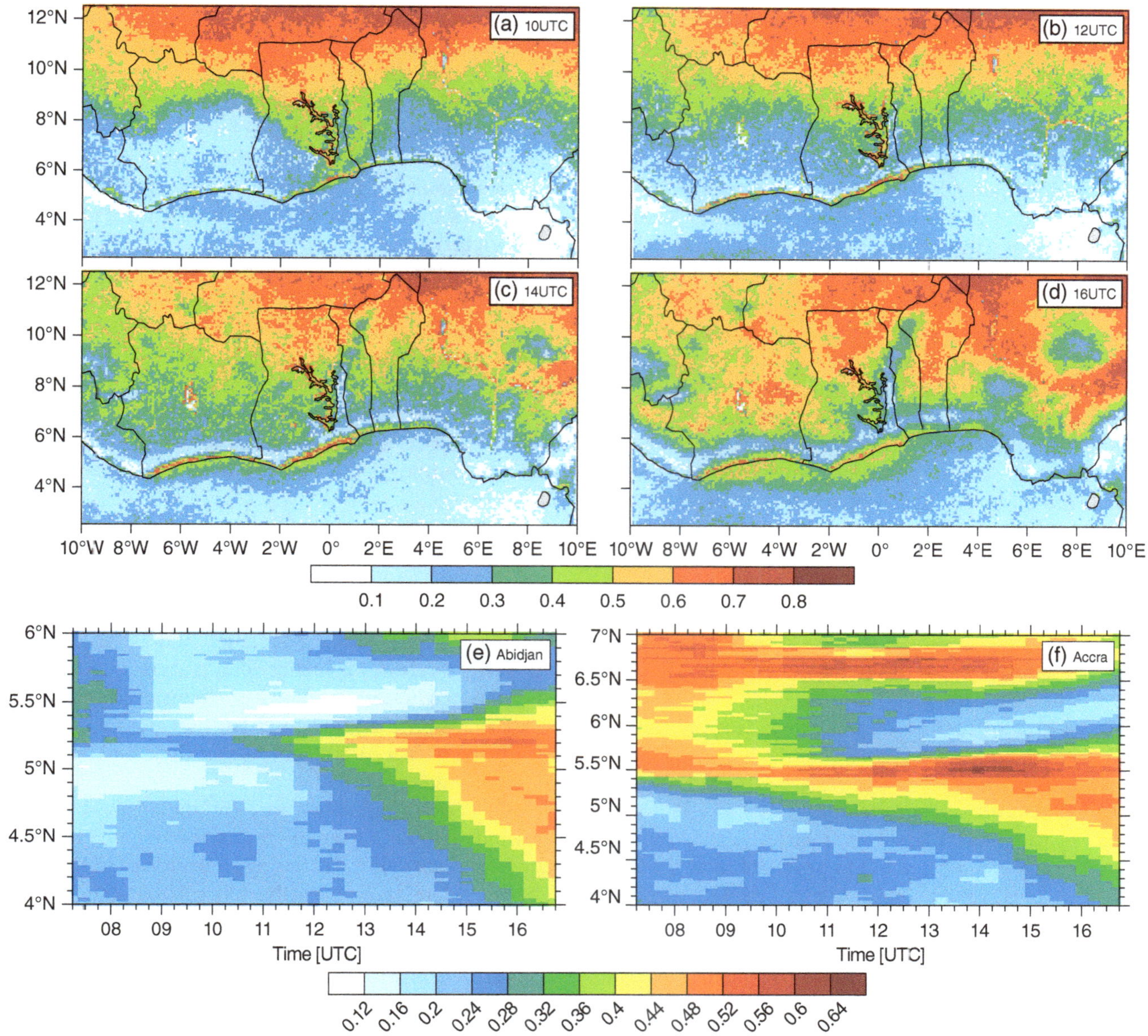


**Figure 4.18** Satellite-derived frequency of clear skies at (a) 1000 UTC, (b) 1200 UTC, (c) 1400 UTC and (d) 1600 UTC determined by the Satellite Application Facility on Support to Nowcasting and Very Short Range Forecasting cloud-type classification for October 2007–2011 (Derrien and Le Gléau, 2005). Red colours show clear skies and the land surface; blue and white colours indicate cloud fields. The final panels show the time–latitude sections of these data, for lines of longitude running through (e) Abidjan and (f) Accra.

showers at its leading edge, typically a few tens of kilometres (up to around 100 km) inland.

In the Mauritanian region in the summer months, Grams *et al.* (2010) have shown how the sea breeze forms part of a larger, synoptic-scale 'Atlantic inflow' circulation. By day, the sea breeze remains close to the Atlantic coast, with its inland penetration being suppressed by PBL convective mixing, but in the evening the cool marine air is free to flow as a gravity current inland, on occasion penetrating as far as the Tagant Mountains, some 400 km inland. The following day, the cool sea-breeze air is mixed into the deeper Atlantic inflow air, and thereby acts to ventilate the Saharan boundary layer. The inland penetration of the cool marine air in the Atlantic inflow is dependent on the position of the Saharan heat low core, and in turn probably related to the subseasonal modes of the heat low (Roehrig *et al.*, 2011; and see Section 7.1.5.1). When the heat low core is located to the east, the Atlantic inflow is strong, and the Saharan boundary layer remains relatively stratified and humid – conditions that can persist for many days.

In the Guinea Coast region, the sensible heating over land is lower, due to increased cloud cover and increased vegetation (low Bowen ratio), so the PBL mixing is weaker than at the Atlantic coasts of Sénégal and Mauritania, and a sea breeze gravity current penetrates inland more easily during the daytime, as implied by Figure 4.18.

There are other examples of surface forcing patterns that will lead to sea-breeze-like local winds. Lake breezes are well documented in Africa – there is a very strong diurnal cycle of winds and of clouds over Lake Victoria for example, and the signature of Lake Volta on the cloud fields of Figure 4.18 is dramatic.

Coastal areas have lower temperature ranges than inland stations because the sea breeze modulates temperatures, cooling the daytime maxima in particular (see Figure 4.3). Also, for the same reasons, areas around Lake Volta (Ghana) and other lakes have their temperature modulated by the microclimate of the environment, and small diurnal temperature ranges mostly are observed. For these reasons, an interruption in the sea breeze can have dramatic consequences locally. Along the Guinea Coast, a heat wave (up to 37°C) was observed in the period 4–6 March 2004, leading to temperatures in Cotonou higher than any maximum in the 1961–1990 period. A strong Harmattan surge (northeast wind) inhibited the onset of the cooling sea breeze until early afternoon, and allowed the temperatures to soar to record values (Knippertz and Fink, 2006).

#### 4.1.4.2.2 Land Breeze Characteristics

Because the land breeze is relatively shallow and relatively weak, it is regarded to be less reliable than the sea breeze. However, the observations (Figure 4.16) demonstrate that the flow can occur with regularity in some locations. Furthermore, at night there is evidence of propagating cloud and precipitation features moving offshore, which may be linked to propagating gravity waves and mesoscale winds (see Love *et al.* (2011) for an analysis of such flows in the Indonesian region). These offshore waves and flows can also cause nocturnal rainfall events over lakes; they are well-documented for Lake Victoria.

Fishermen make practical use of land and sea breezes for their activities. In unpowered boats, the land breeze carries fishermen out to sea at night, and the sea breeze brings them home during the day (see Box 4.1).

### 4.1.4.3 Topographically Forced Winds

Hills and mountains have an important control on the local meteorology, through the local circulations they generate.

#### 4.1.4.3.1 Anabatic Wind

This is a gentle upward flow on hill slopes on a fine, warm day. Under such conditions, the ground slope becomes warmed by solar radiation and air adjacent to the surface reaches a higher temperature than the air over the neighbouring valleys or plains. The warm air tends to become unstable and rises up the slope of the hill, being replaced by the surrounding air, which is cooler and denser.

If the background winds are light, the upslope, anabatic winds converge over the top of the hill, and cause a 'convective core' of warm, rising air over the hill summit. The boundary layer over the hill can be significantly

**Box 4.1 Q&A regarding land breeze, with Mr. Donkris Mevuta, Executive Director of Friends of the Nation, a non-governmental organisation involved in supporting community fisheries groups in Ghana**

- *What time does the land breeze start (e.g. before or after sunset)?*
  The land breeze, known in Fante as Daadzemframa, and referred to as Böfon, starts mostly before daybreak, normally around 12:00 midnight.
- *Are there any visual or other indications of the start of the land breeze?*
  The appearance of the sky, especially when it appears to be cool indicates possibility of land breeze occurrence. Additionally, the scent/smell that precedes land breeze can also give an indication. The smell is mostly a combination of vegetation (farmlands) and the dirt/garbage on land.
- *How far offshore does it extend?*
  It may range between 300 and 800 m.
- *How far onshore?*
  What is so obvious onshore is that the land breeze is stronger here than offshore
- *What time of day does the land breeze cease?*
  There isn't a specific time the breeze may cease, but ... if it starts at about midnight, it may probably cease around 7 am.
- *Is the land breeze stronger in different months or seasons?*
  The strongest land breeze season is after Christmas, somewhere between January and February.
- *Are there any indications that the land breeze will be strong or weak on a certain day?*
  The intensity of the coldness of the wind as well as intense pains/itches in the nostrils indicates strong land breeze, and vice versa.

warmer than that over neighbouring regions, partly because a shallower layer of air is being warmed over the hill. In this manner, the anabatic winds enhance orographic rainfall, both by warming the air and by providing upward motion for air parcels to overcome CIN.

#### 4.1.4.3.2 *Katabatic Wind*

This occurs during the night, when the slope surface loses heat by longwave radiation, especially under clear skies. The air that is cooled by contact with the cold slope surface becomes denser than the surrounding air. The force of gravity helps in directing the dense air toward the valley, with light downslope winds.

'Cold air pooling' in valleys can generate minima 10°C colder than neighbouring sites, as cold air formed by radiative cooling drains into the valley. Figure 4.19 shows satellite-derived surface temperatures for a case study night documented by Pospichal *et al.* (2010). This figure shows dramatically how the surface temperatures in the dry air to the north of the ITD are dominated by the patterns of topography, with river valleys clearly delineated over northern Nigeria and southwestern Niger. These valley cold pools amount to at least 10°C in amplitude, despite topographic differences of only 100 m in some of these areas. In contrast, to the south of the ITD the radiative signal of the surface is much weaker and the image is much more uniform; in part, the surface patterns are masked by radiation from water vapour and clouds, and in part the thermal contrasts are simply much weaker to the south of the ITD, due to stronger winds there, which erode the nocturnal inversion and break up any katabatic winds that may form.

Owing to this role in cold air pooling in valleys, katabatic winds are the major cause of fog formation during early mornings, which leads to poor visibility in the valleys.

#### 4.1.4.3.3 *Dynamic Influences on the Winds Around Hills*

Owing to the relatively low heights of the hills in West Africa, relative to the depth of the well-mixed boundary layer (1000–2000 m), downslope winds are not found to be significant in the region. However, there has been some evidence of downslope winds over the Hoggar mountains in Algeria, which may have contributed to the late onset of the monsoon in the region to the southwest of the Hoggar in 2006 (Drobinski *et al.*, 2009).

Probably more significant for the region is the possibility of funnelling of winds around mountains or along valleys, especially at night when the low-level stability is high and the nocturnal jet is strong. In such instances, the nocturnal jet may become intensified, running parallel to the side of the hills. This process has been seen to be significant in the generation of strong nocturnal jets over the Bodele Basin in Chad in the spring months, and therefore contributes to strong dust uplift in that region when the jet is mixed down to the surface in the morning (Section 5.1.2).

### 4.1.4.4 Local Winds and Convergence Associated with Soil Moisture and Vegetation Boundaries

In Section 4.1.2.4.4 it was shown how, in environments of sparse vegetation, patterns of soil moisture can be associated with highly contrasting surface temperatures,

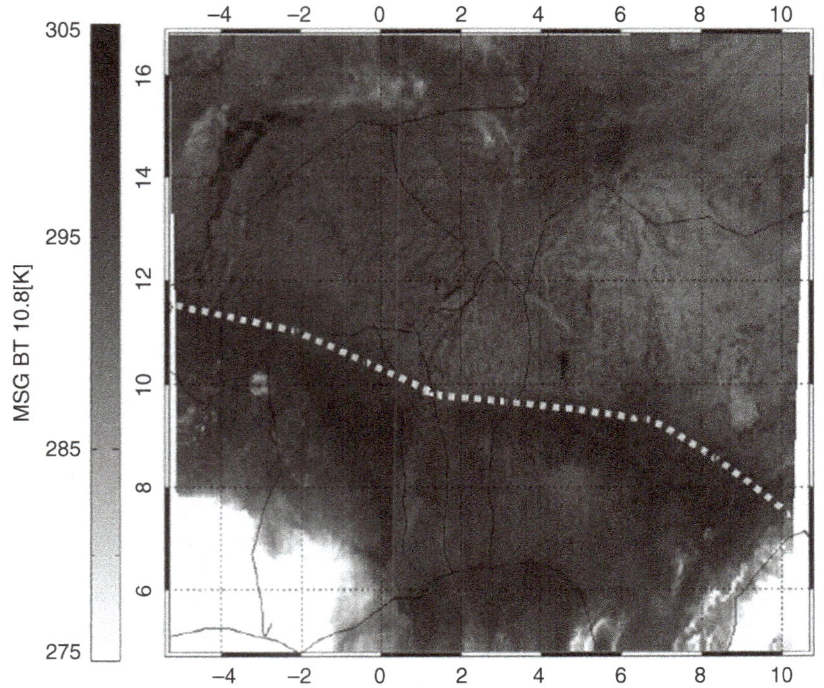


Figure 4.19 MSG SEVIRI 10.8 μm brightness temperature observations from 11 April 2006 at 0200 UTC. The white dotted line corresponds to the approximate ITD position from model analysis fields, using the 15°C dew point criterion. *Source*: Pospichal *et al.* (2010). Reproduced with permission of the Royal Meteorological Society.

Bowen ratio and heat fluxes into the boundary layer. Similar influences can result from vegetation patterns in the Sudan and Guinea Coast zones, where areas of high tree cover cause stronger evapotranspiration and lower sensible heat fluxes than areas of lower tree density (typically cropland). AMMA airborne measurements have demonstrated that these surface effects lead to comparable thermal patterns in the boundary layer, shown schematically in Figure 4.20. Over wet soil, the boundary layer is cooler, more humid and shallower than over dry soil (Taylor *et al.*, 2003). The cooler boundary layer will tend to subside and intrude, perhaps as a gravity current, into warmer neighbouring air. As in the case of the sea breeze, the circulations are weaker during the day, when CBL mixing is strong, and the thermal patterns remain quite closely coupled to the surface features that generated them; cool air is observed above cool/wet surfaces, rather than being advected substantially away.

Although an ambient wind will tend to weaken the PBL thermal patterns observed over patterns of surface state, these PBL features are observed in background winds of up to 5 m s$^{-1}$ (Dixon *et al.*, 2013), and the background wind tends to strengthen convergence where it opposes the slumping cool air (i.e. on the upwind side of a cool surface anomaly). Such effects seem to be significant on spatial scales down to about 10 km. On smaller scales than 10 km, boundary-layer turbulence mixes out the response to the surface rapidly. More precisely, Garcia-Carreras *et al.* (2011) showed that the thermal anomalies are maximised on the boundaries between hot, dry and cool, wet surfaces on the downwind side of a hot, dry surface feature.

#### 4.1.4.5 Trapped Waves, Bores and Solitary Waves

Section 3.1.4.1.5 described how pulses of gravity waves, forced by convective storms, can propagate large distances and influence the triggering, or suppression, of convection at remote locations. There is also evidence of *nonlinear* waves propagating across the region. The stratified atmosphere, in which the horizontal winds are sheared with height, can support a variety of coherent disturbances. A variety of types of wave may propagate, and these may lead to significant weather events, such as localised winds, dust uplift and the triggering of cumulonimbus storms. The majority of studies of these phenomena have been conducted in the USA and Australia. However, the West African environment appears to be suitable for the occurrence of these features, and some limited observational examples exist.

In favourable conditions, typically when a stable layer lies beneath an elevated layer of low stability (low Brunt–Väisälä frequency $N$), gravity wave energy can be trapped at lower levels. In such conditions, the profile is said to exhibit a 'waveguide', and gravity waves can propagate

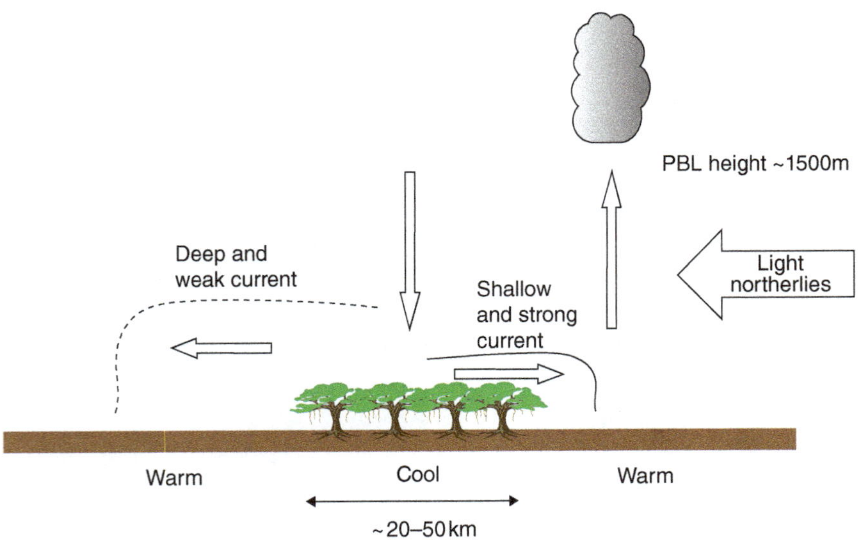


**Figure 4.20** Schematic of land-surface influences on temperatures, local winds and local initiation of convection. A cool surface, in this case resulting from a region of deep-rooted trees, leads to a relatively cool boundary layer at that location. Thermally direct local circulations, as indicated by the arrows, lead to convergence on the upwind side of this cool region, and a preference for convective clouds to form there in the afternoon. Equivalently, afternoon convective clouds form on the downwind side of warm surface features. The same response – probably a stronger response – has been observed over areas of soil moisture variability in the Sahel, especially where vegetation is sparse and in marshy areas. In such conditions, convective clouds are more likely to occur on the downwind edge of dry surfaces, adjacent to regions of higher soil moisture. *Source*: Garcia-Carreras *et al.* (2010). Reproduced with permission of John Wiley and Sons.

over long distances without loss of energy. Such waves have been observed in various studies in other parts of the world (Christie, 1992; Koch and Clark, 1999; and see references cited in Thomsen *et al.* (2009)).

A typical configuration for these waves is linked to a gravity current intruding horizontally into the stable layer (Figure 4.14): the gravity current may be a cold pool emerging below a cumulonimbus storm, or may be spontaneously generated by frontogenesis in a larger-scale thermal gradient. Ahead of the gravity current, we may find bores or solitary waves (Koch and Clark, 1999).

A bore is a propagating solitary wave in which the thickness of an atmospheric layer is increased permanently by the wave. At the passage of the wave, there is upward motion and the stable air layer is stretched to form a deeper layer behind the wave. Associated with the deeper layer is a higher surface pressure, through hydrostatic balance: for a bore of depth 500 m and temperature anomaly 3°C, this pressure change may amount to 0.5 hPa. The winds at the ground may be higher behind the bore, but the bore propagates faster than the winds both ahead and behind (relative to the bore, the wind is always in the same direction, but slower in the deeper layer – see Figure 4.14).

It can be difficult to distinguish bores from gravity currents, since both involve a deepening stable layer near the ground, a localised front at which there is convergence and a pressure rise on the passage of the event. Formally, these features are distinguished by their relative flow – in the frame of reference of the convergence line, a gravity current has inflow on both sides (and is therefore a material change in the air mass), while a bore has relative flow in the same direction on both sides (and therefore is not an air mass transition). However, in practice this distinction can be hard to make. Bores may be turbulent or undular –in which case they are followed by a series of waves.

Solitons represent another kind of nonlinear coherent structure that will propagate on a waveguide in the right conditions. A soliton is an isolated, large-amplitude wave; it involves lifting of the air as it passes, and then the descent of air after it has moved on. Therefore, solitons will pass through the air mass leaving it unchanged after the passage of the wave (rather like a linear wave packet as described in Section 3.1.4.1.5), unlike a bore, for which the air is lifted as the event passes and remains lifted. Solitons commonly occur as an ordered sequence of waves.

One example of a surface-based wave in West Africa, associated with large-scale convergence, is described by Burton *et al.* (2013), based on AMMA observations from Niamey made at the ARM Mobile Facility in January to February 2006. On several nights in this period, increased dustiness was observed in the lowest 500 m of the atmosphere in the early hours of the morning, associated with the passage of a reasonably sharp front and northerly flow above the surface. In observations and high-resolution numerical model results (Figure 4.21), it appears that this event is caused by the collapse of the large-scale temperature gradient (in this season with cold air to the north) into coherent structures, on at least one occasion corresponding to a propagating bore, in which there was a high dust loading. In the model simulations, solitary waves and gravity currents were also detected.

The monsoon layer, lying between an elevated SAL of low stability (Section 2.1.2.2) also provides a waveguide, and it is likely that waves may propagate large distances on the interface between the monsoon layer and the SAL. Observations of this occurrence have not conclusively been found, because they require high-resolution data (e.g. from radars), and because the waves, lying at some altitude, have a weaker signal in surface data than waves propagating on a nocturnal surface layer.

### 4.1.5 Convective Rainfall Distribution on the Local Scale

West Africa is characterised by deep convective rainfall, and in most regions a large fraction of this rainfall comes from organised convective systems that propagate large distances across the continent. The behaviour and forecasting of this organised convection is covered in Chapter 3. However, there is also a significant amount of locally generated convection, and local processes can significantly influence the location of initiation of organised storms, and the rainfall received from organised convection.

#### 4.1.5.1 Cumulus Congestus Clouds

The cumulus congestus represents an intermediate form of cloud, being taller than the shallow cumulus at the top of the boundary layer, but not so deep as fully developed cumulonimbus (see Figure 4.6). The congestus clouds are characterised by warm (liquid water) droplets, without much ice, and therefore they commonly have a smooth or 'tight' appearance in their outline, in contrast with cumulonimbus, whose ice particles give a 'wispy' appearance to the cloud at upper levels. Because the congestus clouds do not contain ice, they are typically non-precipitating, although some may produce a shower. This precipitation can be missed from some satellite estimates of rainfall, because the congestus cloud tops are quite warm and may not be detected by infrared or ice-scattering algorithms.

Congestus clouds are an important and universal feature of the tropics (Johnson *et al.*, 1999), and can represent a preliminary stage before the development of deep convection (cumulonimbus), or can occur without cumulonimbus occurring. Various studies (Johnson et al., 1996; Stein *et al.*, 2011; see Figure 1.13) have shown that the cumulus congestus clouds are usually limited by the freezing level, at around 500 hPa over West Africa in

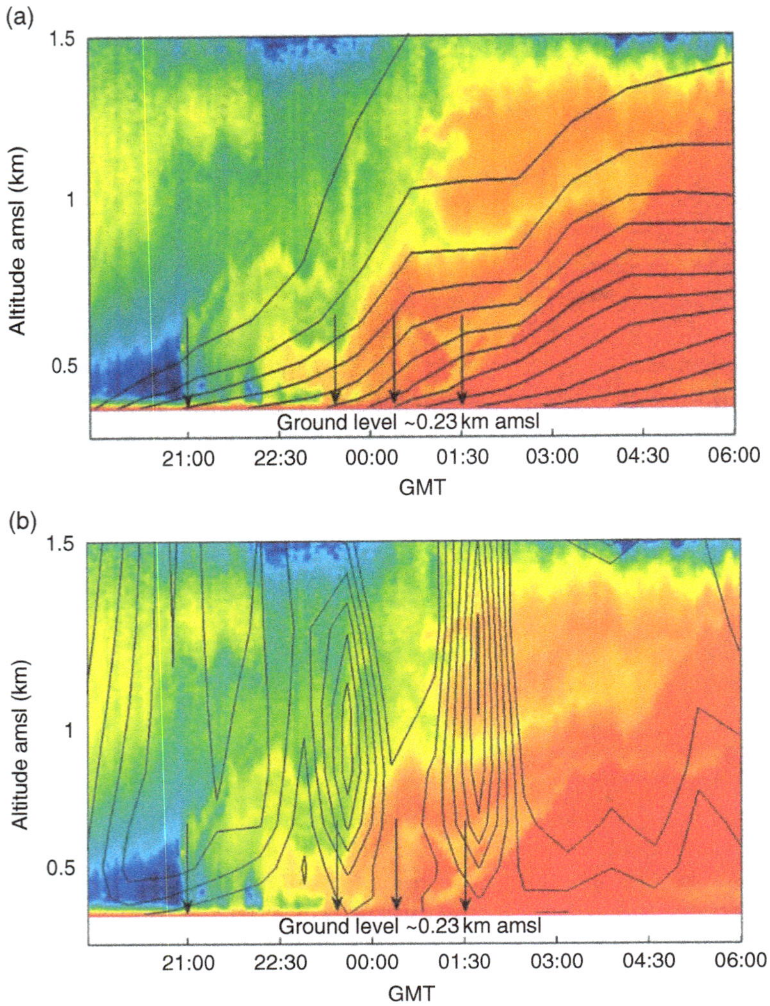


**Figure 4.21** Niamey dust data and overlain model potential temperature and vertical velocity fields. Time–height diagram at Niamey Airport for (a) potential temperature (contours with interval 1°C) and (b) vertical velocity (contours with interval 0.5 cm s$^{-1}$), both from a simulation with the Met Office Unified Model at 4 km horizontal resolution. In both, synchronous lidar observations of the dust loading are also shown. *Source*: Burton *et al.* (2013). Reproduced with permission of the Royal Meteorological Society.

the summer. The congestus clouds carry heat and moisture from the boundary layer into the lower troposphere; in this way, they can moisten the lower troposphere, which commonly consists of overrunning Saharan air, and precondition the profile for deep cumulonimbus development later in the day. Therefore, congestus clouds are important, both for their occasional production of warm rain and for their role as a precursor for the initiation of deep convection.

#### 4.1.5.2 Land Surface Controls on Rainfall: Soil Moisture and Vegetation

In Section 3.1.4.1 it has been shown how convective storms tend to initiate over hilly areas and in areas of low albedo (dark surfaces). In addition to these topographic controls on rainfall, West Africa is also characterised by important dynamic controls in the initiation of rainfall, by soil moisture and vegetation. Given that the surface patterns of soil moisture and vegetation can change with time, on timescales of hours (soil moisture), weeks, seasons and decades (vegetation), these surface–rainfall feedbacks are ways in which the fast rainfall dynamics are linked to slower processes in the climate system.

Although soil moisture and vegetation both influence the surface energy balance in a number of different ways, including surface albedo, roughness and heat capacity, we can understand their influence on subsequent rainfall through the patterns of the Bowen ratio (or evaporative fraction). The Bowen ratio tends to be lower (i.e. have a higher evaporative fraction) over wet soil than over dry. Comparable differences are also observed between surfaces with deep-rooted trees and those with shallower vegetation (e.g. cropland), with a lower Bowen ratio (high evaporative fraction) associated with the forested areas.

When the Bowen ratio is high, over a dry surface, a boundary layer will grow quickly, whereas for a low Bowen ratio, over a wet surface, the boundary layer will remain quite shallow. Given that the air above the boundary layer is typically quite dry (for instance in the SAL), this rapid growth of the boundary layer over a dry surface will tend to reduce $\theta_e$ relative to the boundary layer

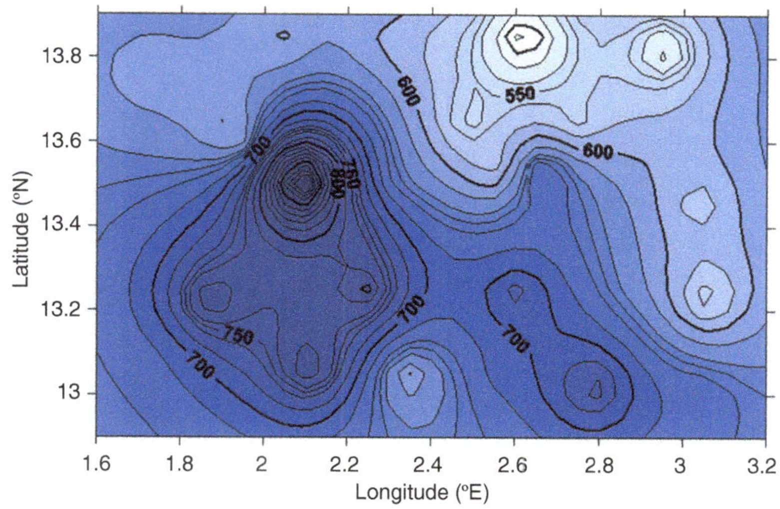


Figure 4.22 Seasonal rainfall totals for 1998 in the vicinity of Niamey. In parts of the domain there are gradients of 400 mm in seasonal rainfall over a distance of 30 km. This represents a factor of 2 in seasonal rainfall on the village-to-village scale. Data collected by the AMMA-CATCH observatory (Lebel *et al.*, 2010).

over the wet surface. There is, therefore, higher CAPE over a wetter surface.[5]

Triggering of convective clouds is also influenced by the vertical circulations, which can lift air parcels through a layer of CIN. The effect of boundary-layer convergence (forced by the thermal patterns) means that the strongest $\theta_e$ and CAPE, and lowest CIN, are moved to a location on the boundary between hot-dry and cool-wet surfaces, on the downwind side of a hot-dry surface feature (Garcia-Carreras *et al.*, 2011). This is also a region of maximum convergence between the cool and warm air (overlying the cool and warm surfaces), and therefore the downwind edge of a warm-dry surface feature appears to be a preferred location for the formation of moist convection in the afternoon. This effect has been confirmed in satellite statistics. More significantly, Taylor *et al.* (2011) showed in satellite data that this location, on the downwind side of surface 'hotspots', favours the initiation of MCSs – it was shown that 1-in-8 MCSs over the Sahel are initiated over surface hotspots. Conversely, an idealised modelling study of Garcia-Carreras and Parker (2011) indicated that convective initiation is very much suppressed over cool/moist surfaces adjacent to hotter land. For these reasons, it makes sense for forecasters to consider surface state contrasts – where there are strong mesoscale gradients in soil moisture or vegetation – as possible areas for afternoon convective triggering.

As a result of the generally higher $\theta_e$ and CAPE over cool/moist surfaces, it is thought that large, organised convective systems, such as squall lines and, more generally, MCSs, deliver more precipitation to a wet surface than to a dry one, within their swath (Clark *et al.*, 2003). This process, however, is not entirely understood, and it likely to be sensitive to climatic zone and to time of day. MCSs may also propagate preferentially over wetter surfaces, on the larger scale (e.g. 300 km scale, comparable to the size of the system – Gantner and Kaltoff, 2010; Wolters *et al.*, 2010). This proposed behaviour of MCSs, delivering more rain to surfaces that are already wet, may explain the observed tendency of rainfall patterns on the mesoscale (e.g. 10–100 km scale) to persist throughout the rainy season in the Sahel (where MCSs provide over 80% of the annual rain); Figure 4.22, taken from the 1998 rainy season, exemplifies this behaviour. Taylor and Lebel (1998) first showed this effect using Hydrology–Atmosphere Pilot Experiment in the Sahel 1990–1992 rain gauge data, and the behaviour has been observed in the Niamey region rain gauge network (AMMA-CATCH) in subsequent years.

In summary, these recent studies have indicated some coherent relationships between patterns of land-surface state, local winds and rainfall.

1) Over high soil moisture, or dense vegetation, during the daytime the atmosphere is cooler and more humid than over neighbouring areas, with higher CAPE.
2) Circulations develop, with maximum convergence on the downwind side of a warm anomaly.
3) Owing to convergence, and the reduction of dry-air entrainment, maximum $\theta_e$, maximum CAPE and minimum CIN are found in the convergence zones close to land-surface boundaries.
4) Convergence, close to land-surface boundaries, tends to increase cloud cover and convective triggering over the hotter surface. In the Sahel, this effect may account for 1-in-8 convective initiation events.
5) Mature MCSs appear to respond to the thermodynamics of the surface. There is higher moisture and CAPE over a cool, moist surface than a hot, dry one; therefore, MCSs may preferentially propagate over moist or forested surfaces, and may deliver more rain to such moist surfaces within their swath, on the smaller scale (e.g. 10 km scale).

Overall, areas of significant spatial variability in the surface fluxes, through soil type, soil moisture or vegetation patterns, may be favourable for the initiation of new convective cells.

### 4.1.6 Marine Weather

Marine weather is a large topic that can only superficially be discussed here.

SSTs are important for fisheries and have important seasonal variations, with cooler waters developing in the Gulf of Guinea in summer around the time of monsoon onset, from late June. There is very strong cooling of the SST along the Ghana–Togo coast, starting in June, with a 7–8°C drop in SST from early June to August. This phenomenon is important for fisheries, as well as for coastal circulations and rainfall.

Sea fog may be observed during the dry season off the west coast of the continent; for example, off Sénégal (see also Sections 4.1.3.3.4 and 4.2.2.2).

Cumulonimbus storms can propagate from land to sea in the Gulf of Guinea, in May to early July, and again in September to late October. Storms are rarely seen to initiate over the sea. In the morning, storms moving over the sea may persist; at night they may decay.

Propagating thunderstorms, moving from the land with a trajectory toward the marine environment, have the potential to cause serious damage to structures and ships. Also, thunderstorms that develop over the sea constitute a potential hazard to coastal infrastructure, including fishing boats. In both cases, the hazard signature is high-velocity downdrafts and heavy downpours. Owing to strong downdrafts (≥40 kt), wave heights can increase greatly and affect marine operations adversely. Coastal erosion and flooding cannot be ruled out. Wetlands and low-lying lands can be inundated by rising waves, causing environmental degradation as well.

Large variations of wave height over the coastal waters of West Africa (Gulf of Guinea and Sénégalo-Mauritanian coast) constitute a serious hazard for mariners and fishermen. Common sources of this hazard can be attributed to two drivers: thunderstorms and the long fetch of modulated waves from the St Helena High and Azores High. During intense building of the St Helena High, strong winds are generated over the sea surface off the coast of the Gulf of Guinea. This situation accentuates the wave height of pre-existing waves. Thus, these modulated waves arrive at the coastal areas with much increased heights, which can impact negatively on fishing activities as well as structural damage to ships (due to collision amongst them). The same principle applies to coastal regions off Sénégal and Mauritania, in association with the Azores High, or with the presence of a very deep low between Greenland and the British Isles.

## 4.2 Operational Methods

### 4.2.1 Forecasting of Maximum and Minimum Temperatures (Tx, Tn)

Forecasting of daily maximum (Tx) and minimum (Tn) temperatures is a priority in the West African region, for various economic sectors and for the general public. Quantitative temperature prediction involves the integration of different sources of information. Experience is important to the forecaster, and this is a key area of human intervention in the weather forecast.

Maximum and minimum temperatures in West Africa typically occur around 1400–1500 UTC and 0500–0600 UTC respectively. In the absence of significant mesoscale or synoptic weather events, over the West African domain the gradient of surface temperature is rather slack: temperature values between adjacent inland stations rarely show a difference of 1–2°C. However, weather events, sea breezes and significant changes in the surface state (e.g. soil moisture) can lead to bigger differences.

Warnings are generally issued for maximum temperatures exceeding certain thresholds, including high mean temperatures or nocturnal temperatures. However, the thresholds for extreme temperature alerts depend on the customer and geographical region. Inland regions have different expectations, sensitivities and, therefore, thresholds to coastal regions. Therefore, certain considerations have to be taken into account in judging whether extreme high or low temperatures are likely to cause problems for public health.

In Niamey, people are quite used to dealing with maximum temperatures of 40–43°C at the end of the dry season, although above 45°C there is thought to be a statistical increase in mortality. In regions of the Guinea Coast zone or within 100 km of the Atlantic coast, temperatures above 35°C, associated with high humidity and pollution in the cities, cause discomfort and health problems.

In regard to temperature minima, for public health, temperatures lower than 15°C are widely regarded to be uncomfortable; cattle can die at temperatures below 9–10°C.

The effect of high temperatures on public well-being is also sensitive to humidity. Recently, some success has been obtained in applying the National Oceanographic and Atmospheric Administration Weather Prediction Center 'heat index' (HI) measure over West Africa, to determine the likely impact of high temperatures. HI is calculated as an empirical fit to temperature and relative humidity data (Steadman, 1979), which is summarised in Figure 4.23.

Once HI has been computed, its significance can be assessed as

- HI >39°C: danger
- HI >51°C: extreme danger.

Temperature/°C

| RH/% | 27 | 28 | 29 | 30 | 31 | 32 | 33 | 34 | 36 | 37 | 38 | 39 | 40 | 41 | 42 | 43 |
|---|---|---|---|---|---|---|---|---|---|---|---|---|---|---|---|---|
| 40 | 27 | 27 | 28 | 29 | 31 | 33 | 34 | 36 | 38 | 41 | 43 | 46 | 48 | 51 | 54 | 58 |
| 45 | 27 | 28 | 29 | 31 | 32 | 34 | 36 | 38 | 40 | 43 | 46 | 48 | 51 | 54 | 58 | |
| 50 | 27 | 28 | 29 | 31 | 33 | 35 | 37 | 39 | 42 | 45 | 48 | 51 | 55 | 58 | | |
| 55 | 27 | 29 | 30 | 32 | 34 | 36 | 38 | 41 | 44 | 47 | 51 | 54 | 58 | | | |
| 60 | 28 | 29 | 31 | 33 | 35 | 38 | 41 | 43 | 47 | 51 | 54 | 58 | | | | |
| 65 | 28 | 29 | 32 | 34 | 37 | 39 | 42 | 46 | 49 | 53 | 58 | | | | | |
| 70 | 28 | 30 | 32 | 35 | 38 | 41 | 44 | 48 | 52 | 57 | | | | | | |
| 75 | 29 | 31 | 33 | 36 | 39 | 43 | 47 | 51 | 56 | | | | | | | |
| 80 | 29 | 32 | 34 | 38 | 41 | 45 | 49 | 54 | | | | | | | | |
| 85 | 29 | 32 | 36 | 39 | 43 | 47 | 52 | 57 | | | | | | | | |
| 90 | 30 | 33 | 37 | 41 | 45 | 50 | 55 | | | | | | | | | |
| 95 | 30 | 34 | 38 | 42 | 47 | 53 | | | | | | | | | | |
| 100 | 31 | 35 | 39 | 44 | 49 | 56 | | | | | | | | | | |

Figure 4.23 Values of heat index (°C) as a function of 2 m temperature (°C) and relative humidity (%) (http://www.hpc.ncep.noaa.gov/heat_index/about_hi.html and Steadman, 1979).

Several predictive methods for daily temperature maxima and minima are available to forecasters, including statistical tools, NWP model products, information regarding the weather situation (such as cloud cover, dust and winds) and, finally, the use of personal experience at the location for which the forecast is being prepared. The combination of these tools can be termed 'heuristic' forecasting methods.

In the USA and EU, the large forecasting centres use the same blend of measurements, NWP and heuristic methods, but typically rely most heavily on NWP model data. Predictions from the NWP model are improved computationally using an automated post-analysis system that relates model data to historical statistics, called model output statistics (MOS; see Met Office (1996)). In major NWP centres, a forecaster will typically start with the Tx/Tn forecast provided in the model or MOS data, and then correct this with knowledge of the processes that the model may not capture correctly, such as cloud cover, aerosol and local surface conditions. At West African sites, the NWP models are less well tested and *a priori* confidence in the NWP prediction may be quite low; in such cases, a forecaster may begin by inspecting the current and recent observations and use the tendencies in the NWP prediction to guide the forecast, among other sources of information. The typical sequence of steps involved in the Tx/Tn forecast is as described in the following subsections.

#### 4.2.1.1 Statistically Based Forecasts

For a given station, the mean diurnal cycle of temperature is constructed for a given month (or shorter period), based, for example, on a 30-year climatology (Figure 4.24). This mean diurnal cycle corresponds to a climatological mean set of hourly temperature differences. In conditions where synoptic changes are not prevailing, these hourly changes can be used to project daily temperatures on the basis of the latest observations. For example, based on Figure 4.24, the maximum temperature, at 1500 UTC, is 9.5°C warmer than the temperature at 0900 UTC. If the temperature at 0900 is 22°C, then an afternoon maximum of 31.5°C can be expected from the climatology. Similarly, a night-time minimum of 15°C can be predicted.

This kind of statistical method is reckoned to be good in the absence of significant mesoscale or synoptic changes over the forecast region. From experience, temperatures in the dry season are easier to predict with this climatological method, while the rainy season is harder to predict due to the increased atmospheric and land-surface variability. Note that this type of forecast has zero bias relative to observations, since it automatically fits climatology by definition.

In other parts of the world, more complex statistical functions have been employed. Such methods typically use a linear combination of temperature and dew point measurements from one or more observation times, to

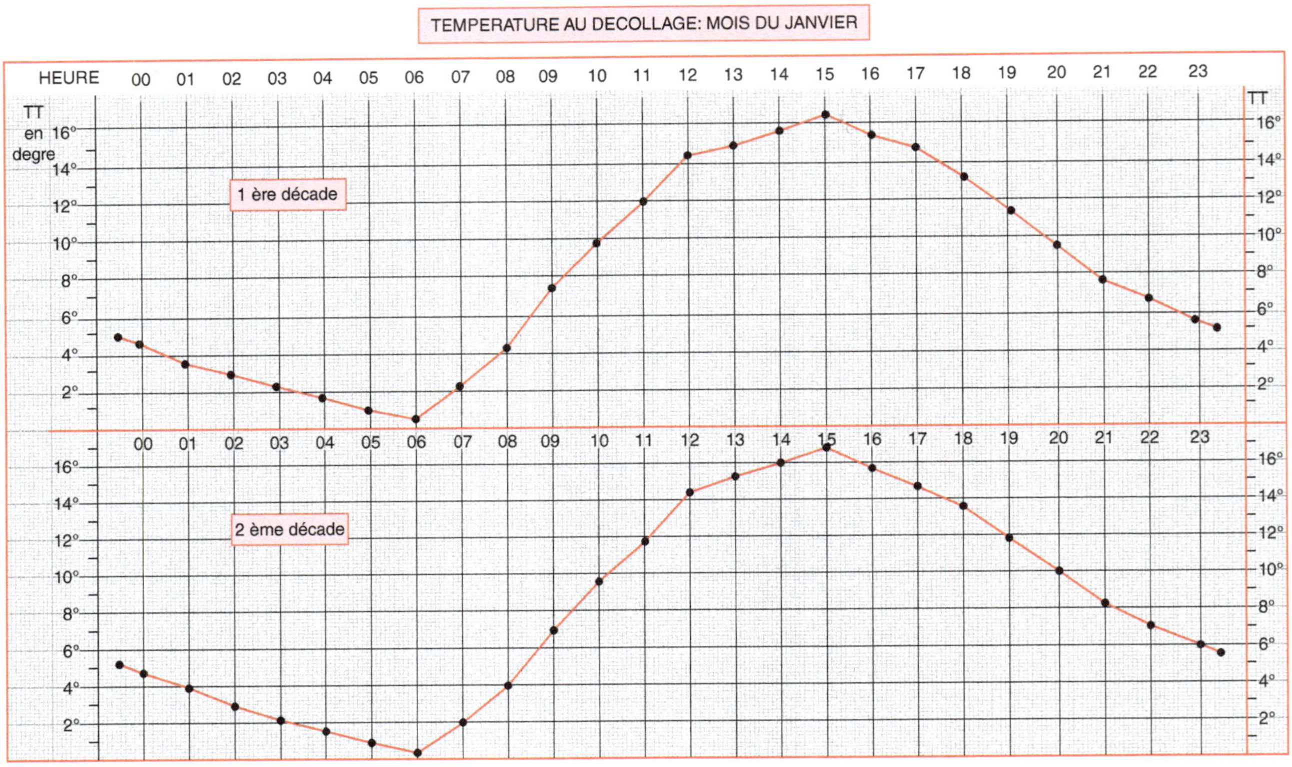


**Figure 4.24** Climatological daily variation of surface temperature as used to predict Tn and Tx. This example comes from Niamey in the first decad of January, and is used for the prediction of ambient temperatures for take-off of aircraft. Note that the values are relative, not absolute.

predict a future value of Tx or Tn; an example is MacKenzie's method for prediction of Tn, given by

$$\mathrm{Tn} = 0.5(\mathrm{Tx} + \mathrm{Td}) - A \tag{4.23}$$

where Td is the dew point at the time of maximum temperature and $A$ is a constant depending on forecast surface wind and cloud amount, which must be computed from climatology and tabulated for every forecast location (Met Office, 1996). Although such methods have been tried in research mode in West Africa, they seem not to be used at all in operational forecasting. The constant parameters in the predictions, such as $A$ in Equation 4.26, need to be calibrated for each location, and reliable versions of such formulae are not, to our knowledge, available for West Africa.

#### 4.2.1.2 Physical Methods Based on Sounding Data

The physical processes determining surface temperatures amount to a thermodynamic budget, involving the upward and downward energy fluxes (Section 4.1.2.1), thermal advection by the winds and the profile of temperature. Therefore, some physically based methods can be used that apply these terms quantitatively. Note that these methods, which consider a vertical boundary layer evolution and/or large-scale advection, are not suitable for areas influenced by strong mesoscale effects, such as sea breezes.

The 'thickness'[6] of the lower part of the atmosphere, typically the 1000–850 hPa layer, can be used as the basis for a forecast of daily temperature maximum, since it represents a measure of the average temperature of the boundary layer at the time of observation (or model time). It is then possible to generate a climatology of Tx according to thickness, cloud cover and time of year. Effectively, the time of year and cloud cover determine the net radiation at the surface (Section 4.1.2.1) and, therefore, the input of heat to the boundary layer. This climatology should be computed for specific locations, and the results will depend on other factors influencing the surface energy balance, such as soil moisture (Section 4.2.1.4).

A related method, currently used in West Africa to predict Tx, is to consider the observed nocturnal surface inversion in a radiosonde profile (or in a model profile). A strong and deep inversion implies that there will be slower and weaker mixing in the lower levels in the morning; this effect amounts to a cooler layer-mean temperature, so it is physically consistent with the use of thickness described earlier.

A practical, quantitative use of the method follows the UK Met Office Forecasters' Reference Book (Met Office,

1996) and is illustrated in Figure 4.25. The pressure difference $\Delta p$ is found from a lookup table according to time of year, and effectively defines the expected energy input to the boundary layer (through surface heat flux $H$), at that time of year. Given $\Delta p$, the point I is adjusted so that there are equal areas (shaded) to the left and right of the isotherm, H–I. Once I is determined in this way, the surface temperature can be taken as point F, where the adiabat meets the surface isobar.

Given that the thermodynamic method illustrated in Figure 4.25 relies on the specification of $\Delta p$, climatological values of $\Delta p$ need to be derived for each station or climatic region. These will typically be comparable to the normal CBL depth, so they can be estimated from soundings. Figure 4.26 shows boundary layer depths in metres, for the pre-monsoon and monsoonal periods for representative West African stations, based on AMMA data. Since $\Delta p$ depends on the heat input to the boundary layer at the surface, the method is sensitive to the surface conditions and cloud cover. As shown in Section 4.1.2.1, the value of sensible heat flux $H$ on any given day can depend strongly on the soil moisture, especially in regions of sparse vegetation, as in the Sahel. Corrections should be applied to reduce $\Delta p$ according to these factors. For instance, in applying this method in the UK, $\Delta p$ would be reduced by 10% for 8/8 cirrus, or 50% for 8/8 stratocumulus.

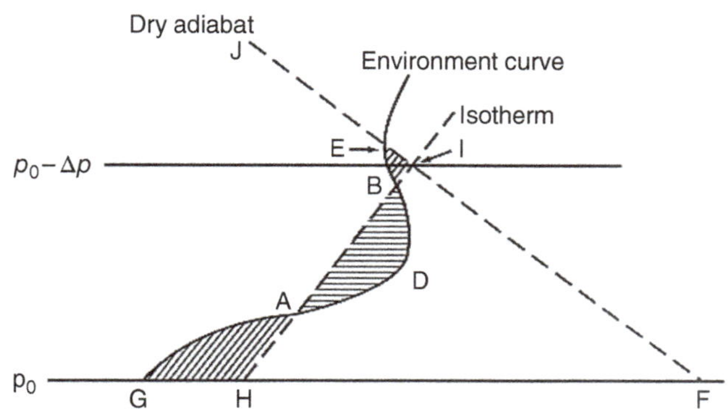


Figure 4.25 Use of a tephigram construction to predict daytime temperatures. The thickness of the layer, $\Delta p$, is found from a lookup table, according to time of year and time of day. The point I is then constructed so that the shaded areas add up to zero (positive and negative areas). F is the forecast surface temperature. *Source*: Met Office (1996). Reproduced with permission of Met Office.

Some forecasters use the rotation of the wind vector with height, derived from a local radiosonde observation, to predict temperature changes based on the geostrophic thermal advection (Holton, 1979). For instance, forecasters at Dakar use the rotation of the wind to show the likely temperature change. Veering of the wind (clockwise rotation with height) indicates warm advection, while backing of the wind (anticlockwise rotation with height) indicates cold advection. The 850–700 hPa layer is used in the rainy season and 700–500 hPa in the dry season. This method is then used to infer the sign of the larger-scale temperature trend (warming/cooling), relative to a simple climatological prediction. For example, if the radiosonde profile indicates warm advection, then the predicted temperature changes may be higher than indicated simply by the mean climatology.

While these profile methods for temperature forecasting are designed for use with local radiosonde data, they can also be applied to model analysis and forecast profiles. However, in applying model data caution must be taken, because the methods are typically sensitive to the fine structure in the thermodynamic sounding, and particularly in the nocturnal inversion. Often the model

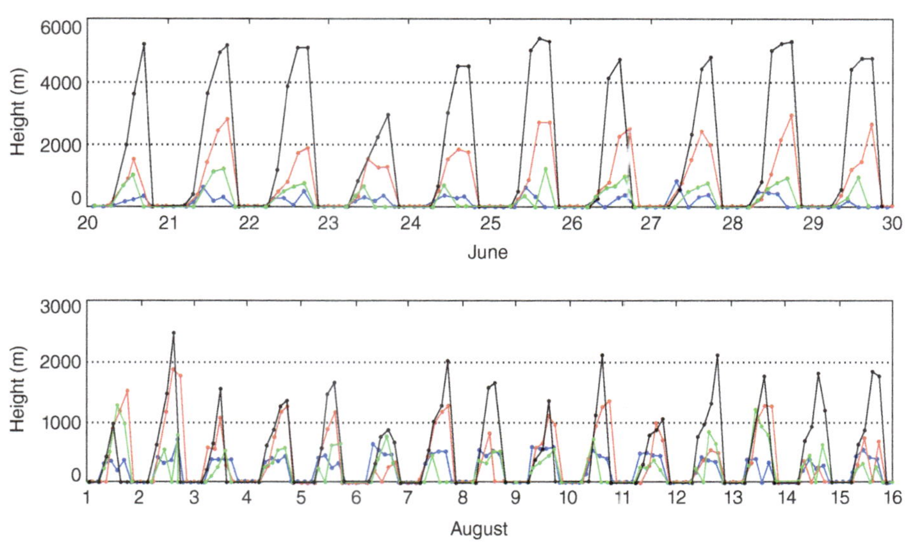


Figure 4.26 Boundary layer depths (m) for days in the pre-monsoonal period of June, and the monsoonal period of August 2006, from AMMA soundings, for the stations at Agadez (black line, 17°N), Niamey (red line, 13.5°N), Parakou (green line, 9.5°N) and Cotonou (blue line, 6.5°N). *Source*: Gounou *et al.* (2012). Reproduced with permission of Springer.

fields supplied to forecasters do not have sufficient levels (vertical resolution) in the boundary layer to apply the profile methods accurately.

#### 4.2.1.3 Integration with Numerical Weather Prediction Data

NWP data can be used directly to provide a specific forecast of 2 m temperatures at a given location. However, the NWP forecast has some uncertainty due to low resolution of the NWP model in representing local-scale variations in surface conditions, on scales of hundreds of metres. Although increasing resolution of NWP models may improve the local-scale representation of surface state in NWP models in the coming years, there is always likely to be uncertainty, due to the short horizontal scales of variability in the surface, and the sensitivity of the surface energy balance to rather unpredictable phenomena such as isolated showers.

It is possible to use NWP products in conjunction with local observations and weather conditions to generate a combined, heuristic forecast. For instance, the hourly trend in 2 m temperatures can be taken from the NWP model instead of the local climatology in the statistical method of Section 4.2.1.1. In this case, the climatological curve of Figure 4.24 is replaced by the temperature curve from the model forecast. Use of the model in this way takes into account large-scale synoptic changes, which we expect to be reasonably well modelled by the NWP system. However, the biases in the NWP model are typically not constant over time, and may be different by day and night, unlike the observational climatology, so the NWP system needs to be carefully evaluated relative to local observations before being used in this way.

The NWP model data can also be used in conjunction with local observations to derive estimates for surface temperatures; typically, the model is used to give information about changes in the vertical profile of the atmosphere, which will influence the PBL growth and therefore its temperature.

Finally, the NWP system is increasingly likely to offer statistical measures of possible surface temperatures, obtained from a forecast ensemble. These can give a measure of confidence in the forecast, especially if they can be linked to the prevailing meteorological conditions (e.g. the likelihood of a convective event, which will suppress daytime temperatures if it occurs, and may only exist in some of the ensemble members).

#### 4.2.1.4 Physical Processes Influencing Daily Temperatures

While a statistical tool as described in Section 4.2.1.1 provides a first estimate of daily temperatures, this can be improved by considering the particular synoptic and weather conditions prevailing on a given day.

The altitude is crucial when making forecasts of maximum and minimum temperatures in the short range. Places like Jos (Nigeria) and Amedzofe (Ghana) have lower temperatures than some other cities owing to their altitude. This effect will be reflected in the climatologies and mean diurnal cycle at each station.

In addition to large-scale synoptic changes in temperature, a number of other factors that should be considered when forecasting maximum and minimum temperatures include cloud cover, dust, weather element, recent rainfall, vegetation and topography. Each of these factors influences the surface energy balance, and therefore the way in which the air in the lowest metres of the atmosphere responds to the large-scale conditions. Table 4.2 summarises our knowledge of the likely effects of various phenomena in modifying surface temperatures.

### 4.2.2 Visibility

#### 4.2.2.1 Basic Principles of Visibility Forecasting

In meteorology, visibility refers to the transparency of the atmosphere in relation to human vision. Meteorological personnel are concerned with horizontal visibility, which is expressed in metres or kilometres. This may vary in direction, in which case the lowest value is usually reported. It should also be realised that visibility from an aircraft is different from the horizontal visibility, which meteorologists measure near the Earth's surface.

In daylight, meteorological visibility is defined as the greatest distance at which a black object of suitable dimensions can be seen and recognised against the horizon sky. This definition cannot be used at night; hence, the general definition is the greatest distance at which a black object of suitable dimensions could be seen and recognised, if the general illumination were raised to the normal daylight level.

In practice, the most suitable objects for determining visibility at night are unfocused lights of moderate intensity at known distances. The silhouettes of hills or mountains against the sky may also be used.

It is desirable to determine as close as possible the conditions that would be experienced by a pilot on landing or on take-off: the 'runway visual range'. Hence, observations are made of the visual range of objects or light provided for the guidance of the pilot. Runway visual range is defined as: 'The maximum distance in the direction of take-off or landing, at which the runway or the specified lights or markers delineating it can be seen from a portion above a specified point on its centre line, at a height corresponding to the average eye-level of pilots at touch-down' (Retallack, 1970: 397). Runway visual range observations are only made when the ordinary visibility is a kilometre or less. A height of approximately 5 m is regarded as corresponding to the average eye-level of pilot at touch-down. In daylight, the objects are markers placed alongside the runway, and at night runway lights are used.

The main causes of reduced visibility are elaborated upon in the following subsections.

Table 4.2 Summary of the effects of various phenomena on daily minimum and maximum temperatures, Tn, Tx, relative to a similar location not influenced by that phenomenon.

| Phenomenon (see Section 4.1.2.4) | Effect on minimum temperature, Tn | Effect on maximum temperature, Tx | Comments |
|---|---|---|---|
| Arrival of moist monsoonal air in Sahel | +5 to +10°C | | Humid night warmer than dry night |
| Cu and stratiform cloud: 1–5 okta | +1 to +2°C | −1 to −2°C | |
| Stratiform cloud: 6–8 okta | +1 to +2°C | −1 to −6°C | Weaker influence in humid air/ Guinea Coast zone. Stronger effect in dry air/over desert |
| Dust event | +1 to +2°C | up to −3°C | Milton *et al.* (2008). Cold advection also likely to be active |
| After MCS event | −1 to −2°C | −2 to −4°C | |
| Strong low-level winds | +1 to +5°C | −1 to −3°C | Stronger nocturnal effect in conditions of clear skies |
| Soil moisture | Not known | −4 to −8°C | In semi-arid zone |
| Dense vegetation/ deep-rooted trees | +1 to +2°C | −1 to −2°C | Nocturnal warming in late winter and spring |
| Topography | −10°C in valleys if cold air pools form under light winds | −1°C per 100 m of altitude | |
| Urban | −5°C to +2°C | −2°C to +2°C | Highly dependent on vegetation cover of urban and adjacent rural surfaces |

#### 4.2.2.1.1 *Precipitation*

The obscurity may be caused by either water droplets or ice particles or both. Visibility in rain depends on both the size of the droplets and their number in a given volume of air. Light rain has little effect, but heavy rain can reduce the visibility to as low as 50–500 m. In drizzle, the visibility will depend on the intensity and may range from about 500 m to 3 km.

#### 4.2.2.1.2 *Fog and Mist*

Fog is a cloud of microscopic small drops of water condensed from and suspended in the air at the surface of the Earth in sufficient numbers to reduce the horizontal visibility to less than 1000 m. It can be classified as stratus cloud with its base at the Earth's surface. Mist is almost the same feature, except that mist is reported when the visibility is above 1 km.

#### 4.2.2.1.3 *Smoke and Pollution*

A thick haze is often produced by smoke from industrial and domestic fires, as well as bush fires. Most materials which are burnt produce minute particles of carbon. The carbon from smoke produces the black haze that is characteristic of large industrial cities. If an inversion is present, it checks rising currents and the smoke is confined to the lower layers of the atmosphere. In light winds and humid conditions, fog may also be present. The combined effect of smoke and fog is often called 'smog', as mentioned in Section 4.1.3.3. During this situation, visibility can be reduced to less than 1000 m. The visibility in some cities may be somehow reduced by the presence of oil particles in the air. The escape of oil fumes from motor vehicles and industries pollutes the air, hence causing drastic reduction in visibility to about 1000 m at times.

#### 4.2.2.1.4 *Dust and Sand*

Dust and sand may be raised from the ground by the wind and carried upward. When the visibility is reduced below a kilometre, the phenomenon may be referred to either as a sandstorm or as a dust storm. Sandstorms consist of relatively coarse sand that is too heavy to be lifted far. The sand particles rarely extend to a height of more than 20 or 30 m. Dust storms consist of minute particles of fine dust, and may on occasions be distributed to heights of several kilometres above the Earth's surface. During the dry season the northeasterly winds transport large quantities of dust particles across West Africa, reducing the visibility most drastically (typically <100 m to above 5000 m); such weather conditions are called the Harmattan dust haze. Dust and sand are described in much more detail in Chapter 5.

#### 4.2.2.1.5 *Wind-Blown Spray*

As the speed of wind increases over the sea, the crests of the waves become higher, and eventually spray begins to form from the breaking waves, and the spray reduces visibility. With stronger winds (Beaufort 6, or 25 kt or greater), streaks of foam are blown through the air.

Sea spray often evaporates after being blown into the atmosphere. Each droplet then leaves behind a particle of salt to become a condensation nucleus at a later time. The haze produced by sea salt has a whitish appearance.

Here, visibility values can be reduced to as low as 2000–5000 m, or even 1000 m along the coastal strip, within, say, 100 m of the sea.

#### 4.2.2.2 Fog Forecasting

##### 4.2.2.2.1 Synoptic Considerations

Fog forecasting starts with the synoptic forecast; radiation fog results from prolonged radiative cooling of the Earth's surface at night, and requires conditions of clear skies and light winds. High relative humidity is necessary since it determines saturation. However, during fog conditions, relative humidity is not always 100%. In most cases, the value is above 95%, but sometimes fog can form when RH is 90%.

Factors involved in fog formation are as follows.

1) Positive pressure tendency, associated with fair weather and clear skies at night, leading to radiation loss.
2) High RH >80%. Wet bulb depression: ≤2°C.
3) Surface dry bulb temperature threshold ≤23°C.
4) Light or calm winds overnight.
5) In the coastal region, light onshore winds.
6) Stable PBL profile; presence of high-pressure cells both at surface and 850 hPa level.
7) Local conditions of surface (e.g. topography or forest cover) modify the likelihood of fog. Forests are a relatively humid environment where fog may be favoured. Fog may also be more likely over hills (hill fog) and in valleys, due to katabatic drainage of cold, humid air downslope.
8) Soil moisture may also influence fog formation by raising the humidity of the air close to the surface. For example, over Nigeria it is found that fog may follow rain if a ridge forms overnight and the skies clear.

Cloudy conditions (e.g. medium clouds) are counterproductive in relation to fog formation. This is due to the fact that the cloud layers reradiate the outgoing longwave (terrestrial) radiation back to the Earth's surface, thereby preventing the surface from cooling to the fog-point temperature.

The most probable time of occurrence of fog is 0330 to 0530 UTC during winter. However, fog has been observed sometimes during summer after a long, heavy precipitation event during the day.

On the western coast, fog may develop in conditions of clear sky, weak winds (2–3 kt), and air close to saturation; typically this occurs in the months of February to April, when SSTs are cold. For instance, at Dakar, on occasion (less than once per annum; Figure 4.12a) a sea breeze can lead to <2000 m visibility in the morning. The first indicator for the afternoon forecaster is a change of wind direction from weak northeasterlies to weak westerlies. One will then forecast fog if saturation is likely. This is well illustrated in Figure 4.27a and b showing the time series of synoptic parameters during a dense fog episode in Dakar: very light winds off the Atlantic are associated with a falling dew point depression until the arrival of fog abruptly at 0200 UTC. Patches of fog can be seen in the RGB night microphysics image of 0600 UTC in Figure 4.27c. Visibility then improves steadily after sunrise.

##### 4.2.2.2.2 Fog Forecasting Tools

1) The humidity in the lower part of the troposphere is significant in determining the likelihood of fog. Specifically, considering the depression of the surface temperature $T_0$ below the wet-bulb potential temperature at 850 hPa,

$$D = \theta_w^{850} - T_0 \tag{4.24}$$

the probability of fog increases with $D$.

2) The 'Saunders method' is an empirical approach, devised to forecast radiation fog using temperature and humidity profiles in the boundary layer, from forecast tephigrams or actual ascents (Saunders, 1950). The method involves the following procedure:
   - From the upper-air sounding, the LCL of the air with maximum temperature is found, using a standard parcel ascent (Normand's construction). If there is a super-adiabatic layer near the surface, construct the parcel ascent using an adiabat representative of the well-mixed boundary layer, instead of the surface temperature. (See Met Office (1996) for more details and exceptions.)
   - The humidity mixing ratio at the LCL is followed back to the surface isobar. The point where these meet is the expected fog-point temperature.

An assessment of the performance of Saunders' method, for 16 fog events at Parakou, Benin, between September 2006 and December 2008, has been conducted using radiosonde profiles for 1200 UTC on the day preceding fog events, from Abidjan. There is very good agreement in the predictions of the fog-point temperature $T_f$ by Saunders' method for these cases, with the overnight minimum temperature reaching, or falling below, the predicted fog-point temperature in 14 out of the 16 events (and within 1°C for the other two cases). Parakou is an inland site in the uplands of the centre of the country with ground rising to the north and west, as well as a river valley to the west, incising the plateau. This brings a number of factors,

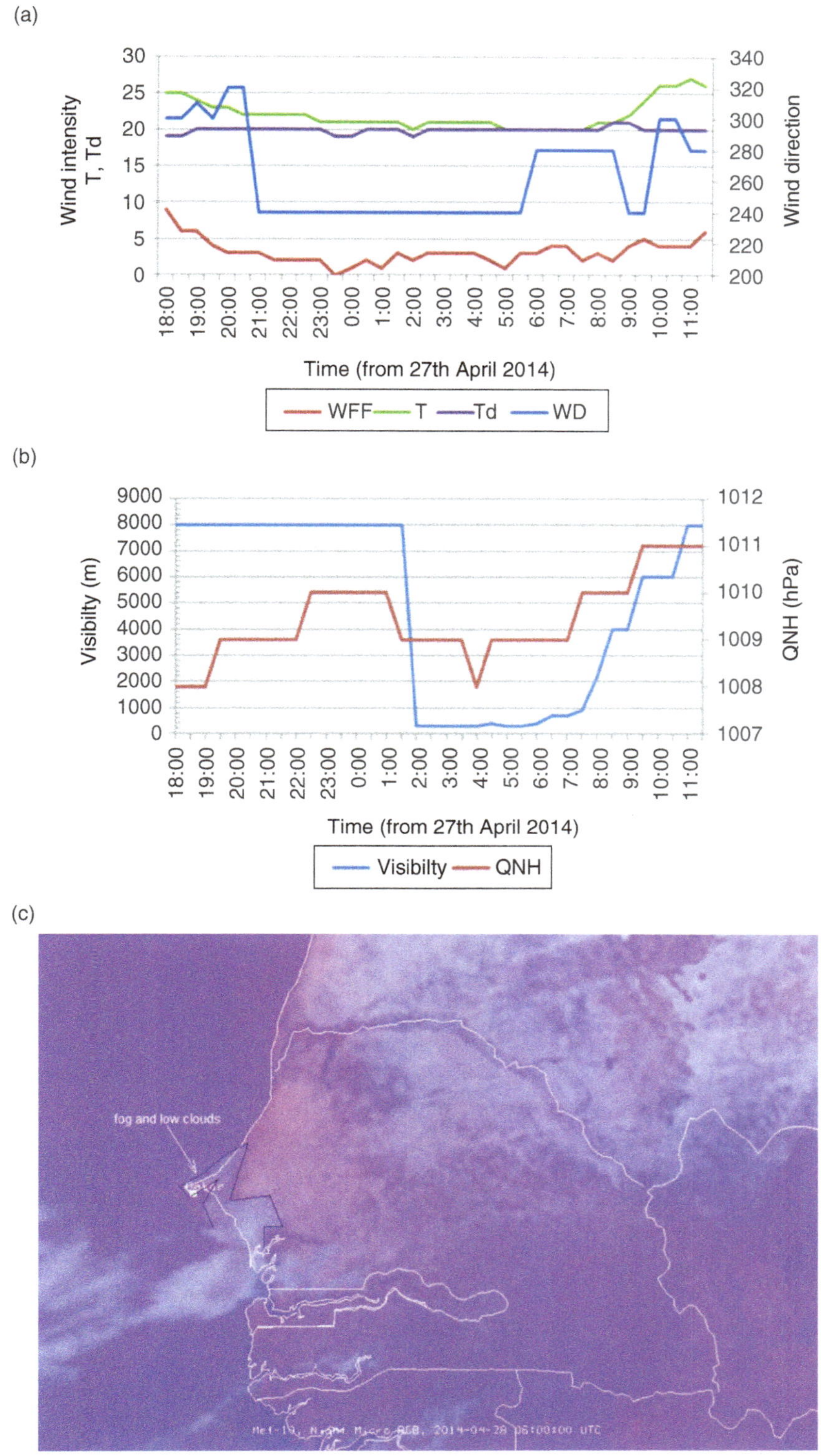


Figure 4.27 Example of coastal fog on the night of 27–28 April 2014 on the coast of Sénégal, showing (a, b) synoptic measurements from Dakar and (c) RGB night microphysics on 28 April 2014 at 0600 UTC. Details for the RGB image can be found in Section 9.1.4.4.6. *Source*: figure courtesy Jochen Karl Kerkmann. Reproduced with permission of EUMETSAT.

increasing the likelihood of radiation-fog formation: (i) a relatively high diurnal temperature range, aided by distance from the sea; (ii) the potential for condensation from moist low-level airflows; (iii) possibility of katabatic drainage from higher ground that would aid the fall of temperature overnight.

The role of moist advection in favouring fog formation appears to be significant in much of West Africa, and radiosonde profiles from Abidjan suggest a higher fog-point temperature at 0000 UTC than at 1200 UTC. Winds with a southerly component at Parakou favour fog formation. On at least three occasions in the study, fog formed with winds from the west, northwest or northeast, which may indicate a drainage component in fog formation at this site. In six of the events, winds were calm at 0600 UTC and it is not possible to confirm the advection of moisture in these cases, except from the increase in $T_f$ overnight, evident in Abidjan radiosonde data on 17 September 2008.

3) Other statistical methods may be possible, based on analysis of the conditions of wind, humidity and temperature preceding fog events in the observational record.

##### *4.2.2.2.3 Fog Clearance*

Forecasting the time of fog clearance is very important. Fog clearance occurs shortly after insolation sets in at sunrise, almost always by 0900 UTC. However, at hill stations or forested areas, fog can on occasion persist after 0900 UTC.

### 4.2.3 Forecasting Wind Shear and Turbulence

Wind shear is defined in the Glossary of the American Meteorological Society as 'The local variation of the wind vector or any of its components in a given direction'. In the context of local forecasting, wind shear is an important phenomenon for aviation, potentially representing a hazard to aircraft. Wind shear can also cause damage to structures such as buildings and offshore platforms, and is typically associated with high turbulence.

It is possible to forecast meteorological events and conditions that render significant wind shear more likely, but not to forecast the wind shear itself with any confidence. For example, it should be possible to forecast a likelihood that there will be wind shear based on a forecast of cumulonimbus convection. More precise warnings of wind shear are supplied through a process of nowcasting, which is presented in Sections 6.2.3 and 6.2.4.

Wind shear conditions are normally associated with some particular meteorological phenomena as follows.

1) Cumulonimbus storms and thunderstorms are a well-known source of wind shear hazard, due to the rapid release of kinetic energy within the systems and the violent airflows that may develop. At low levels the wind shear hazard is mainly due to the effects of the convective downdraught, when it impacts on the surface and generates a horizontal outflow.
   - *Squalls* typically form at the gust front of an organised cumulonimbus system such as a squall line. Almost all large-scale propagating convective storms over West Africa can be expected to exhibit squalls at their leading edge.
   - *Microbursts and downbursts* are more short-lived events, occurring at the time when the convective downdraught first impinges on the surface. At this time, the outflow winds, and wind shear, can be particularly intense. Surprisingly, intense microbursts can be formed from relatively small cumulonimbus storms.

   Forecasting of thunderstorms is covered in more detail in Section 3.2.
2) The nocturnal jet is a source of wind-shear that on occasion can be hazardous. As described in Section 4.1.3.2, the nocturnal jet forms when the nocturnal boundary layer becomes stable at night and the winds are decoupled from the surface. Therefore, the nocturnal jet, at around 400 m altitude, may be impossible to detect in surface measurements (Lothon *et al.*, 2008). It is reckoned by forecasters that the hottest days are often followed by the strongest nocturnal low-level jets. Although we do not have clear evidence for this, there are some possible mechanisms which would explain hot days being followed by high wind shear at night. The synoptic conditions of clear skies and a dry profile would favour both a hot afternoon and a stable nocturnal boundary layer. At the same time, hotter daytime temperatures typically lead to strong turbulence and weaker daytime winds, following which a stronger inertial oscillation and stronger nocturnal jet can be expected.
3) *Local topography* can cause different airflows to develop, as outlined in Sections 4.1.4.2 and 4.1.4.3; in some cases these airflows can cause strong wind shear. This is another area of local weather forecasting where the experience of the forecaster in appreciating the topography and climatology at the local scale is extremely valuable.
   - Sea breeze fronts typically correspond to strong gradients in the winds. They may penetrate 100 km inland during the afternoon (Section 4.1.4.2; Figure 4.18). The front is likely to be sharper under conditions of a light headwind in the synoptic-scale flow.
   - Hilly perturbations to the airflow can cause strong low-level jets or wind shear regions. Mountain waves can exist downwind of a hilly area, when the atmospheric profile is suitable; in extreme cases, mountain waves can lead to low-level rotors in the terminal area. Alternatively, there is a possibility of channelling of low-level winds around hills and in gaps between hills, especially under stable atmospheric conditions at night, leading to locally enhanced wind shear.

There is a requirement for aviation forecasters to consider the likelihood of clear air turbulence (CAT). This phenomenon is notoriously difficult to predict in any part of the world, and the conditions for its occurrence are not unique in West Africa.

Compared with the mid-latitude regions, CAT is relatively rare in the tropics (Galvin, 2007), but it can be encountered in association with the subtropical jet or tropical easterly jet. Synoptic indicators of CAT include the following.

1) Significant wind shear, in the vertical or the horizontal, which can provide a source of instability, leading to the generation of turbulence.
2) The flanks of jets, which are locations for horizontal wind shear, but also tend to show thermodynamic anomalies.
3) In flow showing significant curvature of the streamlines.

### 4.2.4 Forecasting Land, Sea and Lake Breezes

The sea breeze has some typical features.

1) Moderate/fresh wind speeds, typically ~10 kt at Cotonou, with onset at around 1100–1400 local time, attaining a maximum speed by early afternoon, and terminating around 2000 local time (Bajamgnigni Gbambie and Steyn, 2013).
2) The flow extends vertically to approximately 1 km.
3) The flow can penetrate to approximately 100 km inland.
4) It develops best during the dry season.
5) Weak pressure and temperature gradients are in evidence; there is an absence of major synoptic features.
6) Light, offshore, ambient winds are present.

The land breeze has some characteristic features.

1) Weak wind speeds, usually ≤5kt.
2) Extends just a few hundred metres vertically.
3) It is best developed near dawn when the water surface is relatively warm.

Lake breezes may also be significant, for instance in the vicinity of Lake Volta. Vegetation breezes have been shown to exist, in relation to strong contrasts in surface vegetation cover (e.g. from forest to cropland), but these are likely to be relatively weak, of a few knots, and intermittent.

### 4.2.5 Marine Forecasting

Marine forecasting is a big topic, which cannot be treated in great detail here. However, a basic summary of forecasting priorities and principles is given.

Marine forecasts are essential for the safety of life at sea and part of the Global Maritime Distress and Safety System. Forecasts may also be required for inland seas such as Lake Chad. Key forecast parameters for the marine environment are winds and squalls, sea state (waves and swell) and visibility.

#### 4.2.5.1 Winds over the Sea

Winds are generally described by the Beaufort scale (Table 4.3).

As in the case of wind shear warnings for aircraft, different warnings of high winds are needed for different kinds of shipping. In West Africa, warnings are usually issued for force 6 winds (22–27 kt, including gusts); or, for example, if the mean wind speed is 15 kt and it gusts to 25 kt or more, the situation calls for a warning. In the high seas, at great distances from land, marine agencies are mainly concerned with forecasting storms (force 10), but thresholds are lower for inshore waters.

Knowledge of background climatology of the forecast domain is imperative for good forecast skill. In synoptic terms, variations in the St Helena and Azores high-pressure systems are monitored. Cyclonic circulations (sometimes blocked and/or slow moving) are often associated with extreme weather just off the coast of Sénégal. Ocean winds should also be monitored using remote-sensing products, as described in Section 9.1.5.10.

NWP products are the primary tool for forecasting winds over the sea. In fact, without NWP products accurate ocean forecasts are extremely difficult. Experience has shown that the models are generally very good at representing wind velocities in the maritime environment, and are reasonably effective at capturing local circulations such as sea breezes. The main challenge in the use of NWP models is to forecast the presence of deep convection and associated squalls, which are poorly represented in low-resolution models and may be overemphasised in convection-resolving models (see Sections 10.1.2.4 and 10.2.2.3).

Thunderstorm development should also be monitored in remote-sensing products (see Sections 3.2, 6.2, 9.2.3.2 and 9.2.3.4) because of the risk of convective squalls. Additional tools of thermodynamic variables may be integrated in the case of thunderstorms. For instance, studying squalls in the waters offshore of western equatorial Africa, Berkes *et al.* (2012) found that on a daily basis:

- the probability of squalls increases with CAPE, downdraught CAPE and 925–700 hPa wind shear (with CAPE being the strongest influence), and
- the probability of squalls is diminished by high CIN

where these profiles are taken from the nearest oceanic gridpoint in reanalysis datasets.

#### 4.2.5.2 Ocean Waves

'Wind waves' or 'sea waves' are the waves generated by the winds currently being experienced, while the 'swell waves' are those waves generated by winds remote from the location. The sea state is therefore the combination of both wind waves and swell, each of which can be very significant. NWP model outputs are used in most cases in forecasting wave heights as well as speeds. They serve

Table 4.3 The Beaufort scale for winds over the sea, and associated wind conditions.

| Beaufort wind scale | Mean wind speed | | Limits of wind speed | | Wind descriptive terms | Wave height[a] (m) | | Sea state | Sea descriptive terms |
|---|---|---|---|---|---|---|---|---|---|
| | kt | m s$^{-1}$ | kt | m s$^{-1}$ | | Probable | Probable max. | | |
| 0 | 0 | 0 | <1 | <1 | Calm | — | — | 0 | Calm (glassy) |
| 1 | 2 | 1 | 1–3 | 1–2 | Light air | 0.1 | 0.1 | 1 | Calm (rippled) |
| 2 | 5 | 3 | 4–6 | 2–3 | Light breeze | 0.2 | 0.3 | 2 | Smooth (wavelets) |
| 3 | 9 | 5 | 7–10 | 4–5 | Gentle breeze | 0.6 | 1 | 3 | Slight |
| 4 | 13 | 7 | 11–16 | 6–8 | Moderate breeze | 1 | 1.5 | 3–4 | Slight to Moderate |
| 5 | 19 | 10 | 17–21 | 9–11 | Fresh breeze | 2 | 2.5 | 4 | Moderate |
| 6 | 24 | 12 | 22–27 | 11–14 | Strong breeze | 3 | 4 | 5 | Rough |
| 7 | 30 | 15 | 28–33 | 14–17 | Near gale | 4 | 5.5 | 5–6 | Rough to Very rough |
| 8 | 37 | 19 | 34–40 | 17–21 | Gale | 5.5 | 7.5 | 6–7 | Very rough to High |
| 9 | 44 | 23 | 41–47 | 21–24 | Strong gale | 7 | 10 | 7 | High |
| 10 | 52 | 27 | 48–55 | 25–28 | Storm | 9 | 12.5 | 8 | Very High |
| 11 | 60 | 31 | 56–63 | 29–32 | Violent storm | 11.5 | 16 | 8 | Very High |
| 12 | — | — | >64 | >33 | Hurricane | >14 | — | 9 | Phenomenal |

*Source*: http://www.metoffice.gov.uk/guide/weather/marine/beaufort-scale.

[a] These values refer to well-developed wind waves of the open sea. The lag effect between the wind getting up and the sea increasing should be borne in mind; wave heights will be lower in inshore winds, when the winds are offshore or if there is a passing weather system, which might not have winds that last long enough in one place to form a fully developed sea.

as a guide in making forecasts both over the short and medium range. In the West African region, the forecaster has to monitor

1) trends in the variation of the intensity of the St Helena and Azores Highs, using NWP and observational mean sea-level pressure and wind charts, and
2) South Atlantic deep depressions producing swells that reach the Gulf of Guinea

in order to make an estimate of the likelihood of significant swell and wind waves.

Wave period and height are generally computed using the average values of the better formed waves. Unlike the consistency and uniformity of swell waves, sea waves are characterised by a size irregularity. Care should be taken not to split the smaller and larger waves of a given sea into different wave systems of sea and swell. The higher components of a sea can resemble swell by their longer crests and periods. Once in a while, sea and swell do come from the same direction. When this happens, the period of the swell will be several seconds longer than that of the sea.

Certain measures of the wave state are used:

1) Wave height is the vertical distance between a wave crest and the following trough.
2) Significant wave height $H_{sig}$ is the mean of the highest third of the waves, over a period of 10–20 min.
3) Maximum wave height $H_{max}$ is the highest wave to be expected in every 100 consecutive waves. For wind waves it is found (Met Office, 1996) that

$$H_{max} = 1.67 H_{sig} \tag{4.25}$$

The maximum wave height in conditions with both wind waves and swell is given by

$$H_{max} = \sqrt{\left(H_{max} \text{ for wind waves}\right)^2 + \left(\text{swell}\right)^2} \tag{4.26}$$

4) Extreme waves may occur when the fetch is long, and the high winds persist for many hours, leading to significant impacts. The height of extreme waves may be 50% greater than the maximum wave height.

#### 4.2.5.3 Visibility over the Sea

Visibility ranges are:

| | |
|---|---|
| <1000 m | very poor (i.e. fog); |
| 1000–3900 m | poor; |
| 4–9 km | moderate; |
| 10 km or more | good. |

These are reflected in SHIP code visibility ranges.

At sea, visibility cannot be observed as accurately as at land stations, since marker objects are not available.

In a large ship, it is possible to use objects aboard ship for estimation when visibility is low. Errors can arise, however, since the air is affected by the ship. For higher ranges of visibility, the appearance of the land when coasting is a useful guide. In addition, if fixes can be obtained, the distances of landmarks, just as they are appearing or disappearing, may be measured from the chart. Similarly, in open sea, when other ships are sighted and their distances are known, visibility can be obtained. Sometimes the distances can be obtained by radar.

In the absence of other objects, the appearance of the horizon, as observed from different levels, may be used as a basis of the estimation. Care is needed, however, as abnormal refraction may introduce errors. At night, the appearance of navigation lights can give a useful indication of the visibility.

Sometimes the visibility varies in different directions. It should then be estimated or measured in the direction of least visibility excluding reduction of visibility due to ship's smoke. A suitable entry should be made in the log.

### 4.2.6 Marine Services

#### 4.2.6.1 Forecasts/Warnings

Inshore Waters forecasts (with warnings of force 6 winds or stronger) differ from those for the Open Sea areas (with warnings of force 8 or more) and High Seas forecasts (with warnings of force 10 or more). The first is aimed at small vessels, the second is aimed mainly at fishermen and commercial shipping, and the last is aimed mainly at commercial shipping. There are four sets of forecasts/warnings that could be issued:

1) *Shipping forecast/gale warning.* The forecast contains details of gale warnings in force, a general synopsis and sea area forecast containing wind direction and force, weather and visibility. This is updated four times a day and covers the next 24 h.
2) *Inshore Waters forecast/strong winds.* These include strong coastal winds, the coastal forecast and an outlook. The forecast is made for coastal areas up to 12 miles offshore. It is updated four times daily.
3) *High Seas forecast/storm warnings.* These include storm warnings in force, coastal forecasts and an outlook for the next few days. This is updated twice daily. The High Seas forecast will include an outlook, to cover the period 24–48 h ahead of the issue time.
4) *Extended outlook.* These are for sea areas, starting 48 h ahead. They are updated once a day.

#### 4.2.6.2 Routeing of Ships

The International Convention for the Safety of Life at Sea 1974, as amended in 2000, includes 'Ships' routeing' in its Regulation 10. Routeing is the art of achieving a safe, economic passage across an ocean, taking into consideration all available meteorological and oceanographic factors. Ship routeing leads to greater safety and more economy in ship operations. The advantages include less chance of heavy weather damage, less chance of cargo shifting, more comfort for the people on board and a faster passage, resulting in time and fuel savings.

The factors that influence routeing are distance, ocean currents, wind and waves, ice at sea, fog and very low temperatures.

1) *Climatological routeing.* Until recently routeing was performed solely by the use of Pilot books and monthly or seasonal charts of the winds and currents of the oceans.
2) *Weather routeing.*

   Weather routeing is done in geographical regions where changes in the weather are so drastic that seasonal routeing is inadequate. The routeing must be done based on the prevailing weather. In tropical areas, climatological routeing is quite successful, as seasonal variation is fairly regular and systematic.

   Weather routeing of a ship is more difficult than that of an aircraft. This is because routing of a ship requires long-term forecasts, days in advance; this poses several problems. Aviation routing requires short-term forecasts based on recent observations, made possible because of the very fast, modern methods of communications that have been established by international cooperation through the World Meteorological Organisation. This same international cooperation is what has made weather routeing of ships possible.
3) *Ships' performance curves.* Apart from ocean currents, waves are the biggest factors that cause a reduction of a vessel's speed. The waves can be sea, swell or a combination of both. This makes it very important to know a vessel's performance under various wave conditions before attempting to weather-route a particular vessel. Forecasts of sea state and weather are produced and Ships' Masters must apply the forecasts to their own vessels, of which they have the best knowledge. Performance curves can be constructed for this purpose by inspecting the records of previous voyages. If the records are not readily available, the curves of a similar vessel may be used in the meantime. Separate performance curves can be drawn for light and load conditions.

## Acknowledgements

We are grateful to the US Department of Energy for data obtained from the Atmospheric Radiation Measurement (ARM) Mobile Facility (AMF), which operated at Niamey during the AMMA SOPs in 2006, and to the BADC for surface station data which have been obtained from the MIDAS. We benefitted from helpful remarks and ideas from many colleagues.

## Appendix: Acronyms

| | |
|---|---|
| AEJ | African easterly jet |
| AMMA | African Monsoon Multidisciplinary Analysis |
| ARM | Atmospheric Radiation Measurement |
| CAPE | Convective available potential energy |
| CAT | Clear air turbulence |
| CBL | Convective boundary layer |
| CIN | Convective inhibition |
| CATCH | Couplage de l'Atmosphère Tropicale et du Cycle Hydrologique |
| HI | Heat index |
| ITD | Intertropical discontinuity |
| LCL | Lifting condensation level |
| MCS | Mesoscale convective system |
| MOS | Model output statistics |
| NBL | Night-time, or nocturnal, PBL |
| NWP | Numerical weather prediction |
| PBL | Planetary boundary layer |
| PW | Precipitable water |
| RADAGAST | Radiative Atmospheric Divergence using ARM Mobile Facility, GERB data and AMMA Stations |
| SAL | Saharan air layer |
| SST | Sea-surface temperature |
| SOP-1, SOP-2 | Special Observing Periods (of AMMA) |

## Notes

1 It is important to note that when the precipitable water (PW) vapour is high, more than 20 kg $m^{-2}$, the longwave cooling is reduced, regardless of the state of cloud cover (Stephens *et al.*, 2012), and surface temperatures are then much less sensitive to cloudiness.

2 Earlier in the year, when the profile is even drier, the diurnal temperature range at these stations is even stronger.

3 Exceptions to these conditions will occur close to strong horizontal temperature gradients, such as the ITD or within African easterly waves, and possibly in the early morning when the nocturnal jet can be mixed down to the surface by the growing CBL – see Section 4.1.3.2.

4 There is a clear land–sea breeze cloudiness pattern over Lake Volta. However, immediately east and northeast of Lake Volta is the 500–900 m high Togo-Atakora mountain chain that is responsible for the 'downwind' cloud pattern.

5 The convective initiation is also controlled by the CIN, the LCL and the level of free convection (Section 3.1.2), which exhibit more complex behaviour (Haiden,1997; Findell and Eltahir, 2003).

6 Defined as the difference in height between two specified pressure surfaces, and exactly equal to the layer-mean temperature for a hydrostatic atmosphere.

## References


Abdou K, Parker DJ, Brooks B, *et al.* 2010. The diurnal cycle of lower boundary-layer wind in the West African monsoon. *Q. J. R. Meteor. Soc.* **136**: 66–76. doi: 10.1002/qj.536.

Bajamgnigni Gbambie AS, Steyn DG. 2013. Sea breezes at Cotonou and their interaction with the West African monsoon. *Int. J. Climatol.* **33**: 2889–2899. doi: 10.1002/joc.3637.

Bain CL, Parker DJ, Taylor CM, *et al.* 2010. Observations of the nocturnal boundary layer associated with the West African monsoon. *Mon. Weather Rev.* **138**: 3142–3156. doi: 10.1175/2010MWR3287.1.

Berkes F, Knippertz P, Parker DJ, et al. 2012. Convective squalls over the eastern equatorial Atlantic. *Weather Forecast.* **27**: 770–783. doi: 10.1175/WAF-D-11-00086.1.

Betts AK. 1973. Non-precipitating cumulus convection and its parametrization. *Q. J. R. Meteorol. Soc.* **99**: 178–196.

Betts AK, Ball JH. 1995. The FIFE surface diurnal cycle climate. *J. Geophys. Res.* **100**: 25679–25693.

Blackadar AK. 1957. Boundary layer wind maxima and their significance for the growth of nocturnal inversions. *Bull. Am. Meteorol. Soc.* **38**: 283–290.

Bou Karam D, Flamant C, Knippertz P, *et al.* 2008. Dust emissions over the Sahel associated with the West African monsoon inter-tropical discontinuity region: a representative case study. *Q. J. R. Meteorol. Soc.* **134**: 621–634.

Burton RR, Parker DJ, Dixon N, *et al.* 2013. The Harmattan over West Africa: nocturnal structure and frontogenesis. *Q. J. R. Meteorol. Soc.* **139**: 1364–1373. doi: 10.1002/qj.2036.

Carson DJ. 1973. The development of a dry inversion-capped convectively unstable boundary layer. *Q. J. R. Meteorol. Soc.* **99**: 450–467.

Christie DR. 1992. The morning glory of the Gulf of Carpentaria: a paradigm for non-linear waves in the lower atmosphere. *Aust. Meteorol. Mag.* **41**: 21–60.

Clark DB, Taylor CM, Thorpe AJ, *et al.* 2003. The influence of spatial variability of boundary-layer moisture on tropical continental squall lines. *Q. J. R. Meteorol. Soc.* **129**: 1101–1121.

Cook KH. 1999. Generation of the African easterly jet and its role in determining West African precipitation. *J. Climate* **12**: 1165–1184.

Derrien M, Le Gléau H. 2005. MSG/SEVIRI cloud mask and type from SAFNWC. *Int. J. Remote Sensing* **26**: 4707–4732. doi: 10.1080/01431160500166128.

Dixon NS, Parker DJ, Garcia-Carreras L, *et al.* 2013. The effect of background wind on mesoscale circulations above variable soil moisture in the Sahel. *Q. J. R. Meteorol. Soc.* **139**: 1009–1024. doi: 10.1002/qj.2012.

Drobinski P, Bastin S, Janicot S, *et al.* 2009.On the late northward propagation of the West African monsoon in summer 2006 in the region of Niger/Mali. *J. Geophys. Res. Atmos.* **114**: D09108. doi: 10.1029/2008JD011159.

Findell KL, Eltahir EAB. 2003. Atmospheric controls on soil moisture–boundary layer interactions. Part I: framework development. *J. Hydrometeorol.* **4**: 552–569.

Galvin JFP. 2004. Radiation fog on 9 and 10 December 2003. *Weather* **59**: 177–182

Galvin JFP. 2007. The weather and climate of the tropics part 2 – the subtropical jet streams. *Weather* **62**: 295–299.

Gantner L, Kalthoff N. 2010. Sensitivity of a modelled life cycle of a mesoscale convective system to soil conditions over West Africa. *Q. J. R. Meteorol. Soc.* **136**: 471–482. doi: 10.1002/qj.425.

Garcia-Carreras L, Parker DJ, Taylor CM, *et al.* 2010. Impact of mesoscale vegetation heterogeneities on the dynamical and thermodynamic properties of the planetary boundary layer. *J. Geophys. Res. Atmos.* **115**: D03102. doi: 10.1029/2009JD012811.

Garcia-Carreras L, Parker DJ, Marsham JH. 2011. What is the mechanism for the modification of convective cloud distributions by land surface-induced flows? *J. Atmos. Sci.* **68**: 619–634. doi: 10.1175/2010JAS3604.1.

Garcia-Carreras L, Parker DJ. 2011. How does local tropical deforestation affect rainfall? *Geophys. Res. Lett.* **38**, L19802. doi: 10.1029/2011GL049099.

Gounou A, Guichard F, Couvreux F. 2012. Observations of diurnal cycles over a West African meridional transect: pre-monsoon and full-monsoon seasons. *Bound.-Lay. Meteorol.* **144**: 329–357.

Grams C, Jones S, Marsham J, *et al.* 2010. The Atlantic inflow to the Saharan heat low: observations and modelling. *Q. J. R. Meteorol. Soc.* **136**: 125–140. doi: 10.1002/qj.429.

Guichard F, Kergoat L, Taylor CM, *et al.* 2012. Interactions entre surface et convection au Sahel. *La Météorologie* (Spécial AMMA): 25–32. doi: 10.4267/2042/48129.

Haiden T. 1997. An analytical study of cumulus onset. *Q. J. R. Meteorol. Soc.* **123**: 1945–1960. doi: 10.1002/qj.49712354309.

Hamilton RA, Archbold JW. 1945. Meteorology of Nigeria and adjacent territory. *Q. J. R. Meteorol. Soc.* **71**: 231–264.

Holton JR. 1979. *An Introduction to Dynamic Meteorology*. Academic Press.

Johnson RH, Ciesielski PE, Hart KA. 1996. Tropical inversions near the 0 degrees C level. *J. Atmos. Sci.* **53**: 1838–1855.

Johnson RH, Rickenbach TM, Rutledge SA, *et al.* 1999. Trimodal characteristics of tropical convection. *J. Climate* **12**: 2397–2418.

Knippertz P, Fink AH. 2006. Synoptic and dynamic aspects of an extreme springtime Saharan dust outbreak. *Q. J. R. Meteorol. Soc.* **132**: 1153–1177. doi: 10.1256/qj.05.109.

Koch SE, Clark WL. 1999. A nonclassical cold front observed during COPS-91: frontal structure and the process of severe storm initiation. *J. Atmos. Sci.* **56**: 2862–2890. doi: 10.1175/1520-0469(1999)056<2862:ANCFOD>2.0.CO;2.

Lebel T, Parker D, Flamant C, *et al.* 2010. The AMMA field campaigns: multiscale and multidisciplinary observations in the West African region. *Q. J. R. Meteorol. Soc.* **136**: 8–33. doi: 10.1002/qj.486.

Linden PF, Simpson JE. 1986. Gravity driven flows in a turbulent fluid. *J. Fluid Mech.* **172**: 481–497.

Lindén J. 2010. Nocturnal cool island in the Sahelian city of Ouagadougou, Burkina Faso. *Int. J. Climatol.* **31**: 605–620.

Lothon M, Saïd F, Lohou F, Campistron B. 2008. Observation of the diurnal cycle in the low troposphere of West Africa. *Mon. Weather Rev.* **136**: 3477–3500.

Love BS, Matthews AJ, Lister GMS. 2011. The diurnal cycle of precipitation over the Maritime Continent in a high-resolution atmospheric model. *Q. J. R. Meteorol. Soc.* **137**: 934–947.

May PT (1995) The Australian nocturnal jet and diurnal variations of boundary-layer winds over Mt. Isa in north-eastern Australia. *Q. J. R. Meteorol. Soc.* **121**: 987–1003.

McGarry MM, Reed RJ. 1978. Diurnal variations in convective activity and precipitation during phases II and III of GATE. *Mon. Weather Rev.* **106**: 101–113.

Met Office. 1996. *Source Book to the Forecasters' Reference Book*. Met Office: Bracknell; 277pp.

Miller MA, Slingo A. 2007. The ARM Mobile Facility and its first international deployment – Measuring radiative flux divergence in West Africa. *Bull. Am. Meteorol. Soc.* **88**: 1229–1244. doi: 10.1175/BAMS-88-8-1229.

Milton SF, Greed G, Brooks ME, *et al.* 2008. Modeled and observed atmospheric radiation balance during the West African dry season: role of mineral dust, biomass burning aerosol, and surface albedo. *J. Geophys. Res. Atmos.* **113**: D00C02.

Parker DJ. 2002. The response of CAPE and CIN to tropospheric thermal variations. *Q. J. R. Meteor. Soc.* **128**: 119–130.

Parker DJ,; Thorncroft CD,; Burton RR, Diongue-Niang A. 2005a. Analysis of the African easterly jet, using aircraft observations from the JET2000 experiment. *Q. J. R. Meteor. Soc.* **131**: 1461–1482. doi: 10.1256/qj.03.189.

Parker DJ, Burton RR, Diongue-Niang A, *et al.* 2005b. The diurnal cycle of the West African monsoon circulation. *Q. J. R. Meteor. Soc.* **131**: 2839–2860. doi: 10.1256/qj.04.52.

Pospichal B, Karam DB, Crewell S, *et al.* 2010. Diurnal cycle of the intertropical discontinuity over West Africa analysed by remote sensing and mesoscale modelling. *Q. J. R. Meteorol. Soc.* **136**: 92–106.

Rácz Z, Smith RK. 1999. The dynamics of heat lows. *Q. J. R. Meteorol. Soc.* **25**: 225–252.

Retallack BJ. 1970. 2.16 Runway Visual Range. In: *Compendium of Lecture Notes for Training Class IV Meteorological Personnel.* WMO no. 266, TP 150. World Meteorological Organization: Geneva.

Roehrig R, Chauvin F, Lafore J-P. 2011. 10–25 day intraseasonal variability of convection over the Sahel: a role of the Saharan heat low and midlatitudes. *J. Climate* **24**: 5863–5878.

Sandu I, Beljaars A, Bechtold P, *et al.* 2013. Why is it so difficult to represent stably stratified conditions in numerical weather prediction (NWP) models? *J. Adv. Model. Earth Syst.* **5**: 117–133. doi: 10.1002/jame.20013.

Saunders WE. 1950. A method of forecasting the temperature of fog formation. *Meteorol. Mag.* **79**: 213–219.

Schrage JM, Fink AH. 2012.Nocturnal continental low-level stratus over tropical West Africa: observations and possible mechanisms controlling its onset. *Mon. Weather Rev.* **140**: 1794–1809. doi: 10.1175/MWR-D-11-00172.1.

Schuster R, Fink A, Knippertz P. 2013. Formation and maintenance of nocturnal low-level stratus over the southern West African monsoon region during AMMA 2006. *J. Atmos. Sci.* **70**: 2337–2355. doi: 10.1175/JAS-D-12-0241.1.

Schwendike J, Kalthoff N, Kohler M. 2010. The impact of mesoscale convective systems on the surface and boundary-layer structure in West Africa: case-studies from the AMMA campaign 2006. *Q. J. R. Meteorol. Soc.* **136**: 566–582.

Simpson JE. 1999. *Gravity Currents: In the Environment and the Laboratory*. Cambridge University Press; 244pp.

Steadman RG. 1979. The assessment of sultriness. Part I: a temperature-humidity index based on human physiology and clothing science. *J. Appl. Meteorol.* **18**: 861–873.

Stein THM, Parker DJ, Delanoe J, *et al.* 2011. The vertical cloud structure of the West African monsoon: a 4 year climatology using CloudSat and CALIPSO. *J. Geophys. Res. Atmos.* **116**: D22205. doi: 10.1029/2011JD016029.

Stephens GL, Wild M, Stackhouse PW, *et al.* 2012. The global character of the flux of downward longwave radiation. *J. Climate* **25**: 2329–2340. doi: 10.1175/JCLI-D-11-00262.1.

Stull RB. 2009. *An Introduction to Boundary Layer Meteorology*. Springer.

Taylor CM, Lebel T. 1998. Observational evidence of persistent convective-scale rainfall patterns. *Mon. Weather Rev.* **126**: 1597–1607.

Taylor CM, Ellis RJ, Parker DJ, *et al.* 2003. Linking boundary-layer variability with convection: a case-study from JET2000. *Q. J. R. Meteor. Soc.* **129**: 2233–2253. doi: 10.1256/qj.02.134.

Taylor CM, Parker DJ, Harris PP. 2007. An observational case study of mesoscale atmospheric circulations induced by soil moisture. *Geophys. Res. Lett.* **34**: L15801. doi: 10.1029/2007GL030572.

Taylor CM, Gounou A, Guichard F, *et al.* 2011. Frequency of Sahelian storm initiation enhanced over mesoscale soil-moisture patterns. *Nat. Geosci.* **4**(7): 430–433. doi: 10.1038/ngeo1173.

Tennekes H. 1973. A model for the dynamics of the inversion above a convective boundary layer. *J. Atmos. Sci.* **30**: 558–567.

Thomsen GL, Reeder MJ, Smith RK. 2009. The diurnal evolution of cold fronts in the Australian subtropics. *Q. J. R. Meteorol. Soc.* **135**: 395–411.

Timouk F, Kergoat L, Mougin E, *et al.* 2009. Response of surface energy balance to water regime and vegetation development in a Sahelian landscape. *J. Hydrol.* **375**: 178–189.

Van de Wiel BJH, Moene AF, Steeneveld GJ, *et al.* 2010 A conceptual view on inertial oscillations and nocturnal low-level jets. *J. Atmos. Sci.* **67**: 2679–2689. doi: 10.1175/2010JAS3289.1.

Williams E, Renno N. 1993. An analysis of the conditional instability of the tropical atmosphere. *Mon. Weather Rev.* **121**: 21–36.

Wolters D, van Heerwaarden CC, Vilà-Guerau de Arellano J, *et al.* 2010. Effects of soil moisture gradients on the path and the intensity of a West African squall line. *Q. J. R. Meteorol. Soc.* **136**: 2162–2175. doi: 10.1002/qj.712.

# 5

# Dust

**Lead author:** *Peter Knippertz*[a]

**Contributing authors:** *Ibrahim Hamza*[b], *Abdou Kassimou*[c], *Benoit Laurent*[d], *Bernard N. Orji*[e], *David Perry Osika*[f], *Kerstin Schepanski*[g] *and Ina Tegen*[g]

**Reviewer:** *Andreas Fink (Institute of Meteorology and Climate Research, Karlsruhe Institute of Technology, Karlsruhe, Germany)*

[a] *Institute of Meteorology and Climate Research, Karlsruhe Institute of Technology, Karlsruhe, Germany*
[b] *L'Ecole Africaine de la Météorologie et de l'Aviation Civile (EAMAC), Agence pour la Sécurité de la Navigation Aérienne en Afrique et à Madagascar (ASECNA), Niamey, Niger*
[c] *Agence Nationale de l'Aviation Civile du Niger (ANAC-NIGER), Niamey, Niger*
[d] *Laboratoire Interuniversitaire des Systèmes Atmosphériques, UMR7583 CNRS, Université Paris Diderot, Université Paris-Est Créteil, Paris, France*
[e] *Regional Meteorological Training Centre, Nigerian Meteorological Agency, Lagos, Nigeria*
[f] *Ghana Meteorological Agency, Takoradi, Ghana*
[g] *Leibniz Institute for Tropospheric Research, Leipzig, Germany*

## 5.1 Scientific Background

### 5.1.1 The Dust Cycle

Atmospheric mineral dust is mainly produced by aeolian (wind-driven) erosion acting in arid and semi-arid areas. North African desert areas are the world's most important dust sources. Mineral dust emitted on local and regional scales can represent risks and nuisances for the exposed populations living close to the source areas. Dust storms strongly reduce the visibility in the atmosphere, causing traffic problems in regions with frequent dust events (Pauley *et al.*, 1996). Frequent exposure to high dust concentrations occurring during dust storms has also been related to health problems like silicosis and is suspected of playing a role in the spread of infections and outbreaks of meningitis during the dry season in northern Africa (Morman and Plumlee, 2014).

During their transport in the atmosphere, mineral dust particles have an impact on the Earth's radiative budget by absorbing and scattering incoming solar and outgoing terrestrial radiation. Mineral dust can be involved in heterogeneous and multiphase atmospheric chemistry, affecting photo-oxidant concentrations and the composition of precipitation. It can also contribute to biogeochemical cycles of nutrients like iron (Fe) and phosphorus (P) in deposition areas. The correct description of the spatial and temporal variability of emission, transport and deposition of dust is a prerequisite to estimate and forecast atmospheric dust concentrations and their impacts.

#### 5.1.1.1 Emission

Dust emission involves nonlinear processes that are governed by the meteorology and by the state and properties of the surface (see Marticorena (2014) for an overview). The number of fine particles present in a free state at the surface of wind-erodible soils is usually insignificant. These particles are often incorporated into coarser sand-sized aggregates. The force exerted on a soil aggregate by wind is the shear stress $\tau$ that can be expressed as $\tau = \rho_a u^{*2}$, $\rho_a$ being the air density and $u^*$ the wind friction velocity. For neutral conditions $u^*$ is defined by

$$u(z) = \frac{u^*}{k} \ln\left(\frac{z}{z_0}\right) \quad (5.1)$$

where $k$ is the von Karman constant, $u(z)$ the wind velocity at a height $z$ above the surface and $z_0$ the aerodynamic roughness length (see also Section 4.1.2.2, Equation 4.11).

The setting into motion of soil aggregates is caused by a wind stress, and this is a threshold process, in that a significant wind stress is needed before any dust will be lifted. For values of friction velocity greater than the threshold $u_t^*$, wind stress counterbalances weight, interparticle forces and capillary forces between grains that maintain aggregates on the ground. $u_t^*$ controls both the occurrence of dust emission and its intensity and is a

*Meteorology of Tropical West Africa: The Forecasters' Handbook*, First Edition. Edited by Douglas J. Parker and Mariane Diop-Kane.

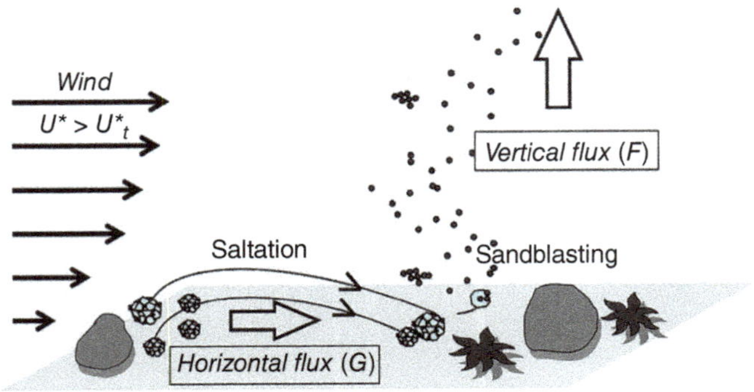


**Figure 5.1** Schematic diagram of dust production. The vertical flux is produced during the 'sandblasting' process, when the saltating grains (horizontal flux) hit the surface and break the interparticle bonds linking dust particles together or to the surface. Figure adapted from Laurent (2005).

function of the soil grain diameter (Bagnold, 1941). For large grains, $u_t^*$ increases with size due to gravity. For small particles, $u_t^*$ increases with decreasing size mainly due to the interparticle cohesive forces reinforcing grain bonds. These two effects lead to a minimum in $u_t^*$ for a grain size around 80 μm. These particles are too heavy to stay in suspension for longer periods and therefore saltate or creep, leading to a horizontal dust flux *G* (Figure 5.1). Non-erodible elements (pebbles, stones or vegetation) dissipate part of the wind momentum that will not be available to initiate particle motion. The presence of such non-erodible elements therefore leads to a global decrease of the wind shear stress acting on the erodible surface and to an apparent increase of $u_t^*$. Soil moisture can also increase $u_t^*$ by promoting the development of capillary forces between soil grains (Fécan *et al.*, 1999). In the same way, physical (Gomes *et al.*, 2003) and biological (Belnap *et al.*, 2004) crusts can enhance the binding of particles.

Theoretical and experimental work (Bagnold, 1941; Gillette, 1979) showed that *G* is proportional to the third power of $u_t^*$. The actual dust production occurs when the aggregates in saltation hit the ground. A part of their kinetic energy can be used to break the interparticle bonds linking dust particles together or to the surface and eject finer dust particles (a sandblasting process; Figure 5.1). Dust emission generally does not occur in the absence of saltation (Gillette, 1977; Sow *et al.*, 2009). Warren *et al.* (2007) noticed that in the Bodélé Depression (Chad) the breakdown of saltating diatomite flakes as they collide in saltation and with the surface is the most prolific mechanism of dust production. The vertical dust flux *F* is composed of the finest particles moving into suspension from the saltation layer. These particles can be transported over long distances.

Efforts have been made to develop dust emission schemes based on the physical processes at the interface between the atmosphere and the surface. Marticorena and Bergametti (1995) use the so-called drag partition $f_{eff}$ between the roughness elements (characterised by the aerodynamic roughness length $z_0$) and the erodible surface (characterised by the smooth roughness length $z_{0s}$). $u_t^*$ can then be parameterised in 'rough' or 'smooth' situations that are encountered in arid and semi-arid areas. *G* simulated by Marticorena *et al.* (1997) reproduces quite well the total mass and the size distribution of wind tunnel measurements for different soil substrates and wind friction velocities (Williams, 1964). Only sandblasting-process models allow simulation of both the mass and the size distribution of *F* explicitly. Alfaro and Gomes (2001) proposed a model based on typical dust particle populations. Dust of a given size is produced when the kinetic energy of the saltating soil particles exceeds the corresponding threshold. The dust size distribution varies therefore as a function of the saltating particle size and their velocity. The dust production can also be estimated as a function of the volume of soil removed by the saltating grains impacting on the surface. Shao (2004) developed an elaborated version of this model taking the saltation bombardment, the aerodynamic entrainment and the aggregate disintegration into account.

Regional and global input databases of desert surface roughness, soil dry size distribution and soil texture have been specifically developed alongside the advancements of dust emission models (Marticorena *et al.*, 1997; Tegen *et al.*, 2002; Prigent *et al.*, 2005; Laurent *et al.*, 2008a). However, large-scale application of sandblasting models remains conditioned to the determination of required input data from measurements at this scale. The issues of surface dynamics, as for instance seasonal vegetation, temporary surface crusts and supply limitation, and specific soils, such as the diatomite soil in the Bodélé Depression, are still largely unresolved. Kardous *et al.* (2005) point out the necessity to investigate the agricultural practices used in semi-arid areas. Future advances in dust modelling will also result from better-resolved and more accurate wind fields (Laurent *et al.*, 2008b).

There is no reliable quantitative observation of dust emission over the Sahara. However, the measurements of recent North African field campaigns (including Bodélé Dust Experiment (BoDEx), African Monsoon Multidisciplinary Analysis (AMMA), Saharan Mineral Dust Experiment (SAMUM) and Fennec) and indirect observations of emitted and transported dust give useful information to evaluate simulations, at least in terms of spatio-temporal variations. Using different approaches (surface stations, models and satellite products) it is nowadays possible to locate the most important North African source areas and to investigate the seasonal cycle of dust emission. Estimates of the Saharan dust load emitted range from 240 Tg $yr^{-1}$ using satellite observations (Kaufman *et al.*, 2005) to 760 Tg $yr^{-1}$ using modelling approaches (Laurent *et al.*, 2008a).

#### 5.1.1.2 Transport

Atmospheric processes on synoptic, regional and local as well as turbulent scales can generate dust emission and force atmospheric mixing of dust particles. The transport of dust from and to the surface, as well as within the atmosphere, is controlled by the characteristics of the planetary boundary layer (PBL; see Chapter 4).

The main long-range transport paths are westward over the North Atlantic to North or South America, northward across the Mediterranean to southern Europe, and along easterly trajectories across the eastern Mediterranean to the Middle East (Middleton and Goudie, 2001, and references cited therein). Stuut *et al.* (2005) point out the effect of the seasonal shift of the intertropical convergence zone (ITCZ), from 19°N during boreal summer to 5°S during winter, on the origin of dust and the pathways of the transported dust (Figure 5.2). In summer, strong vertical mixing of dusty air occurs over the Sahel and southern Sahara. Over the ocean, the Saharan dust layer is transported above the trade wind inversion (up to 5–7 km above sea level) (Kalu, 1979; Figure 5.3). In winter, vertical mixing is weaker and occurs further south, such that most of the dust is transported within the trade wind layer below 1.5–3 km (Figure 5.3).

#### 5.1.1.3 Wet and Dry Deposition

The mineral dust cycle is sensitive to the rate of removal of dust from the atmosphere back to the surface, through dry and wet deposition (see Bergametti and Forêt (2014) for an overview). Dry deposition is the dust flux from the atmosphere to the surface through molecular and turbulent diffusion and gravitational settling, while wet deposition is the dust transfer to the surface through precipitation. The dry deposition velocity of dust depends on particle size (large particles fall faster), flow properties and surface characteristics (gravitational settling, turbulent transfer, Brownian diffusion, impaction, interception, particle rebound; Zhang *et al.*, 2001). Wet deposition includes in-cloud and sub-cloud scavenging.

Experimental studies usually show a wide range in deposition values depending on the measurement techniques used (total, dry or wet, soluble or insoluble deposition). Mahowald *et al.* (1999) present a compilation of dust indicators and records from a terrestrial and marine palaeoenvironment database. Annual fluxes to marine sediments range over several orders of magnitude from 0.2 to 141.6 g $m^{-2}$ $yr^{-1}$, and from 0.8 to 27.4 g $m^{-2}$ $yr^{-1}$ for

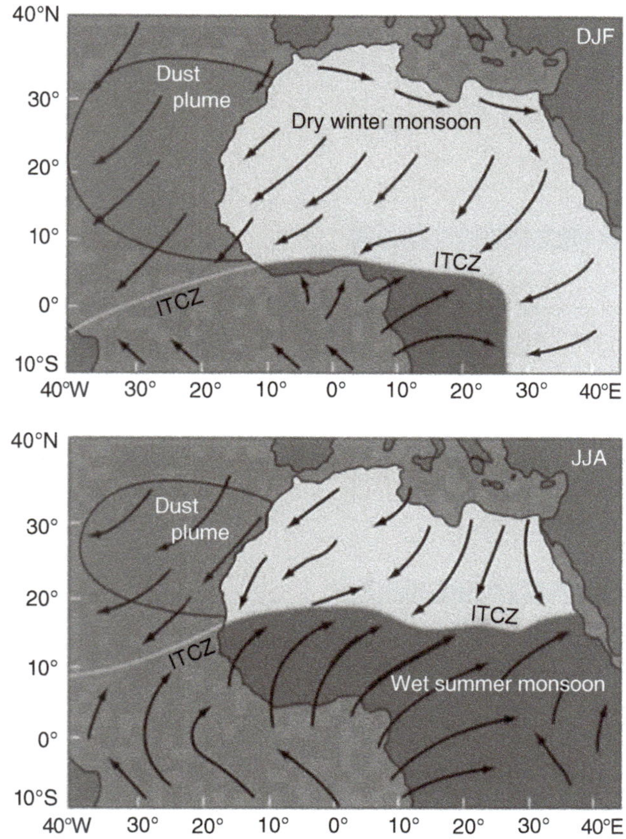


**Figure 5.2** Seasonal variation in latitudinal position of the ITCZ and its consequences for atmospheric conditions over northern Africa. Arrows indicate direction of trade winds; the extension of the dry and often dusty Harmattan air is marked by light grey shading over land and by a darker grey line over the ocean. December, January and February (winter) and June, July and August (summer) are shown. *Source*: Stuut *et al.* (2005). Reproduced with permission of John Wiley and Sons.

aeolian fluxes to sediment traps. Middleton and Goudie (2001) gathered estimates of rates of dust deposition for sites at varying distances from the Sahara. Annual deposition values range from less than 1 g $m^{-2}$ for western Europe up to 100–200 g $m^{-2}$ close to Saharan source areas.

Aerosol removal processes remain an important source of uncertainty in models, and their evaluation remains difficult due to the locations and periods for which in-situ observations are available. Global aerosol distributions are strongly controlled by the rate of removal of aerosols from the atmosphere by wet scavenging. The efficiency of the wet deposition depends on the type and intensity of the precipitation, and the size distribution of raindrops. It also depends on the size distribution and the chemical characteristics of the dust aerosol. The simulated annual deposition flux can range from ten to several hundreds of grams per square metre per year over the Sahara, and from five to several tens of grams per square metre per year over the Atlantic downwind from the North African dust sources (Mahowald, 2007; Figure 5.4a). Figure 5.4b indicates the regions

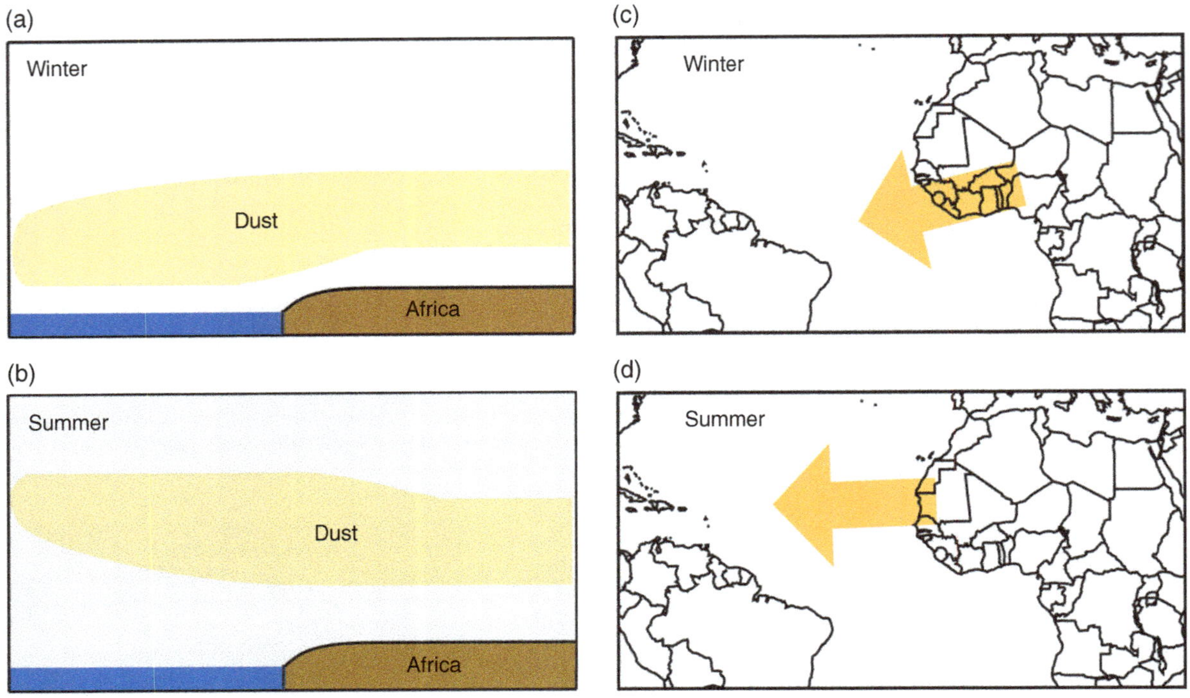


**Figure 5.3** Schematic diagram of vertical and horizontal dust export from North Africa toward the tropical East Atlantic for Northern Hemisphere winter (a, c) and summer (b, d). (See Section 5.1.1.2 for typical plume heights) © Schepanski *et al.* (2009a) CC Attribution 3.0 License. After Kalu (1979).

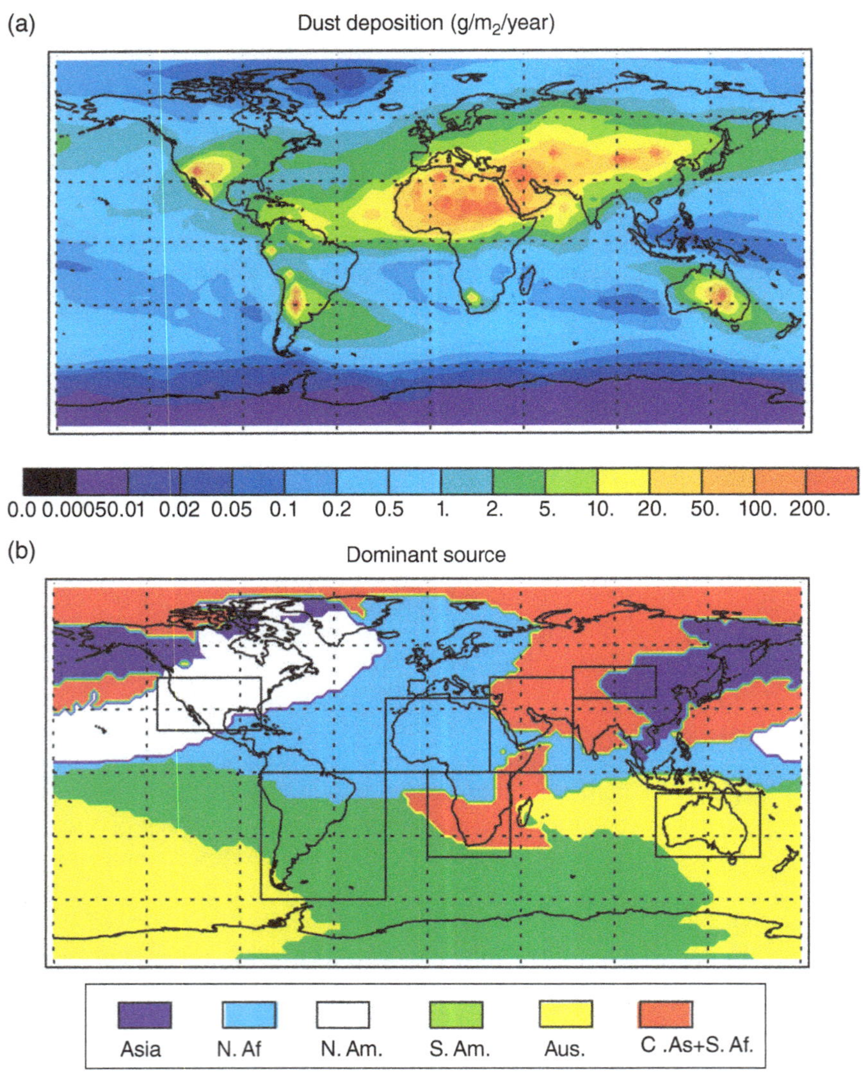


**Figure 5.4** (a) Annual deposition fluxes estimated from a dust model. (b) Dominance of source regions for dust deposition indicated by different colours. *Source*: Mahowald (2007). Reproduced with permission of John Wiley and Sons.

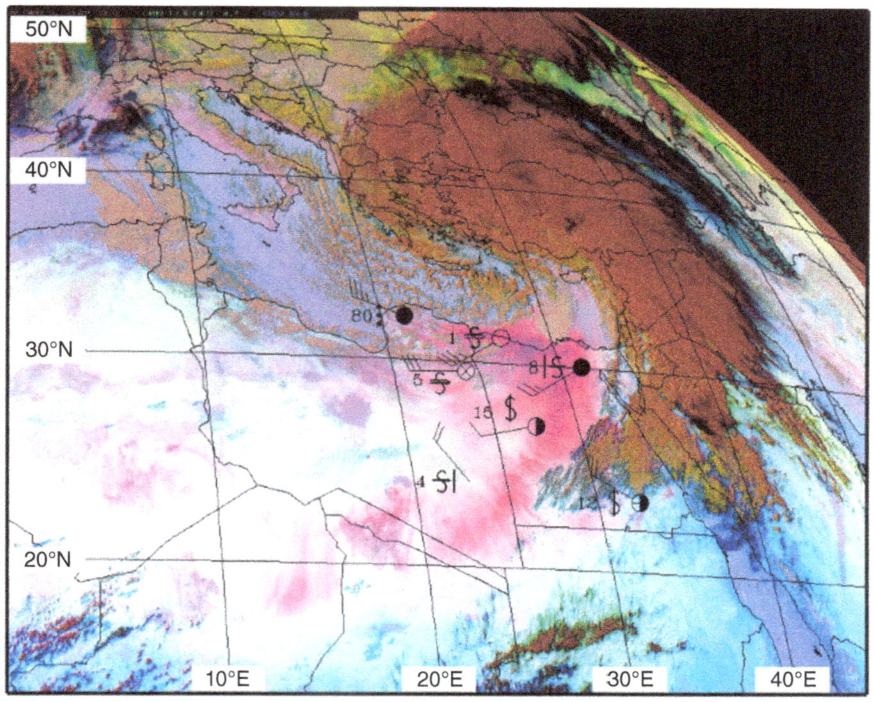


**Figure 5.5** Example of an intense Khamsin cyclone over the eastern Mediterranean Sea at 1200 UTC 22 January 2004. The underlying satellite image is a composite of three infrared (IR) channels from the Meteosat Second Generation (MSG) Spinning Enhanced Visible and InfraRed Imager (SEVIRI) instrument. High, optically thick clouds appear red in this image and dust in magenta. The synoptic station reports shown in this figure indicate severe dust storms with wind speeds of up to 35 kt and a visibility down to 100 m behind the cold front over Egypt and Libya. *Source:* Knippertz and Todd (2012). Reproduced with permission of John Wiley and Sons.

dominated by the deposition of mineral dust emitted from North Africa (mainly North Africa itself, the North Atlantic Ocean, the Mediterranean basin and Europe).

### 5.1.2 Meteorological Systems

Meteorological systems that cause near-surface winds strong enough to mobilise and transport dust occur on spatial scales from a few metres to spanning the whole of northern Africa (see Knippertz and Todd (2012) and Knippertz (2014) for an overview). The synoptic scale (hundreds to thousands of kilometres) is dominated by extratropical weather systems with marked regional and seasonal differences. In the northern and central Sahara, surface depressions regularly form in connection with southward-penetrating upper-level troughs or cut-off lows from Europe or the adjacent Atlantic Ocean, in particular during spring and winter (Thorncroft and Flocas, 1997). Typical examples are the so-called Khamsin, Sharav or desert depressions that form in the lee of the Atlas Mountains and then track east or northeastward causing dust storms along the Mediterranean coast (El Fandy, 1940; Pedgley, 1972; Alpert and Ziv, 1989; Figure 5.5). Dust mobilised along the trailing cold front of such systems can occasionally reach as far south as the Sahel.

In the Sahel and southern Sahara, intense dust storms frequently occur during the West African dry season (December to April) in connection with low-level pressure surges associated with an intensification of anticyclonic conditions over northern Africa (Kalu, 1979; Klose *et al.*, 2010). These are related to cold air intrusions from the mid-latitudes and are sometimes enhanced by evaporational cooling (Knippertz and Fink, 2006). The increased south–north pressure gradient across northern Africa associated with a strengthened subtropical anticyclone leads to an intensification of the Harmattan winds. In weak cases, this may only cause localised emissions from preferential sources (Knippertz *et al.*, 2011) such as the Bodélé Depression in Chad, where orographic channelling accelerates the low-level northeasterlies (Washington and Todd, 2005). In strong cases, long-lived, fast-moving dust fronts can form that become almost continental in scale (Knippertz and Fink, 2006; Shao *et al.*, 2010; Figure 5.6).

During the summer monsoon season, the large-scale pressure distribution differs markedly and is characterised by a pronounced heat low centred over northern Mali and fairly strong pressure gradients toward the Atlantic to the west, the Gulf of Guinea to the south and the Mediterranean Sea to the northeast (see Section 1.5). During the day, the PBL is usually deep and well mixed over large parts of the area, impeding the response of low-level winds to large-scale pressure gradients. During the night, when the low levels stabilise, a more coherent flow toward the heat low can be observed (Parker *et al.*, 2005). The southerly monsoon flow can develop a sharp leading edge that behaves almost like a gravity current and leads to dust emission (Bou Karam *et al.*, 2008). Dust storms have also been observed in cold-air surges from the Mediterranean Sea toward the heat low (Knippertz *et al.*, 2009).

An important phenomenon that occurs over cloud-free, dry regions with strong background pressure gradients during all seasons is the formation of nocturnal low-level

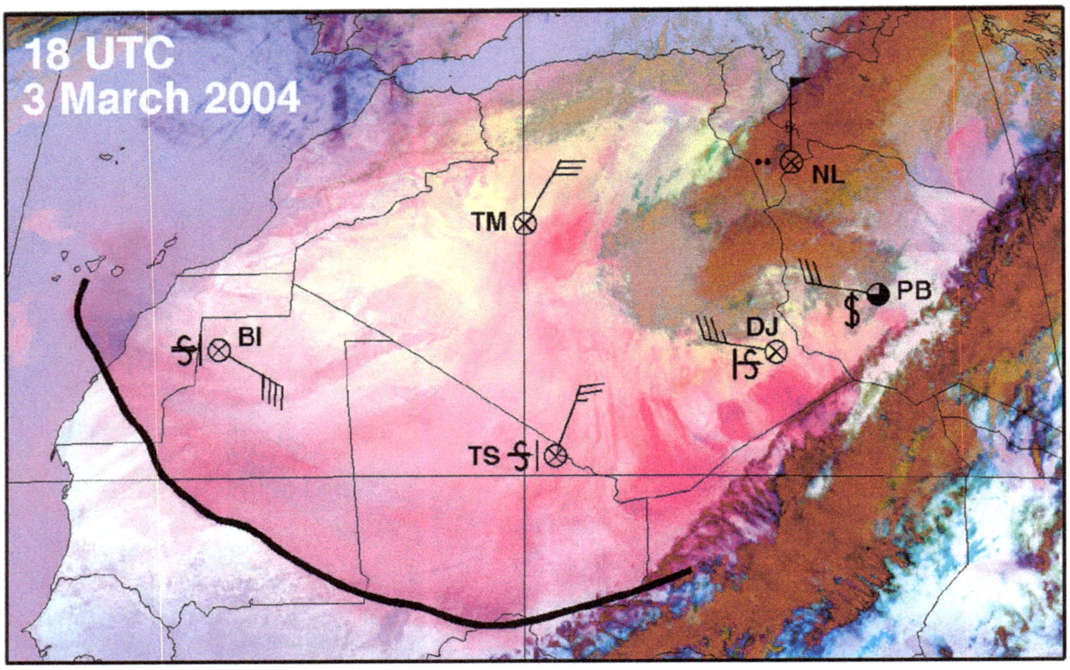


**Figure 5.6** Example of an intense large-scale dust outbreak caused by rapid anticyclogenesis over northwestern Africa at 1800 UTC 3 March 2004. The underlying satellite image and station reports are as in Figure 5.5. During the following days the dust front continued to spread until it formed a long arc from the Gulf of Guinea to the Canary Islands. See Knippertz and Fink (2006) for more details. *Source*: Knippertz and Fink (2006). Reproduced with permission of the Royal Meteorological Society.

jets (LLJs; Stensrud, 1996; Fiedler *et al.*, 2013; see Section 4.1.3.2). These are dynamically related to the decoupling of the layer above the radiation inversion from surface friction during the night. During the increasing stabilisation of the PBL after sunset, the wind in this layer performs an inertial oscillation around the geostrophic wind (or some other equilibrium wind) with a period of $2\pi$ divided by the Coriolis parameter $f$. This corresponds to about 28 h at 25°N and therefore can lead to flow that is highly supergeostrophic (faster than the geostrophic wind) around sunrise. Downward mixing of momentum from the LLJ during the build-up of the PBL in the morning leads to a peak in near-surface gusts and dust emission during these hours (Washington *et al.*, 2006; Schepanski *et al.*, 2009b). This effect has been demonstrated for the Bodélé Depression during BoDEx (Figure 5.7). Knippertz (2008) found that the gust parameterisation used by the European Centre for Medium-Range Weather Forecasts (ECMWF) is capable of representing this mechanism in the region of large pressure gradients to the north of the mean summertime heat trough. Using ECMWF data, Engelstaedter and Washington (2007) found a strong correlation between the annual cycle in gusts and dust over the heat low region.

On the mesoscale (a few to several hundred kilometres), density currents driven by the cooling associated with the evaporation of convective precipitation in the hot and dry desert air (so-called convective 'cold pools'; see Section 3.2.2) play an important role. Dust storms along the leading edge of a rapidly spreading cold pool are often called haboobs and were first documented for squall-line-type convection in the Sudan (Sutton, 1925; Freeman, 1952). Over northern Africa haboobs are most frequent at the margins of the Sahara where enough moisture is available to sustain the parent deep moist convection. In recent years a number of detailed observational studies on Sahelian haboobs using aircraft and ground-based radar have been published (Flamant *et al.*, 2007; Marsham *et al.*, 2008a; Williams *et al.*, 2008). Figure 5.8 shows a photograph of a typical haboob and radar data taken during the AMMA field campaign in 2006. The frontal nature of the leading edge of the cold pool, or density current, is evident with a pronounced updraft of more than 8 m $s^{-1}$. Particles or other small objects appear to be transported upward and rearward in the head of this gravity current, reaching heights of about 2.5 km. Winds are more turbulent and dust more equally distributed in the rearward parts of the system. Haboobs have also been discovered over Morocco and Algeria, where they are related to moist convection over the Atlas Mountains, particularly during the summer half year (Emmel *et al.*, 2010). Occasionally, haboob activity extends from the Sahel far into the central and even northern Sahara (Knippertz *et al.*, 2009; Cuesta *et al.*, 2010). There is evidence that such situations are connected to moisture transport in the southerly sectors of intense African easterly waves, particularly if these interact with disturbances in the subtropics (Knippertz and Todd, 2010).

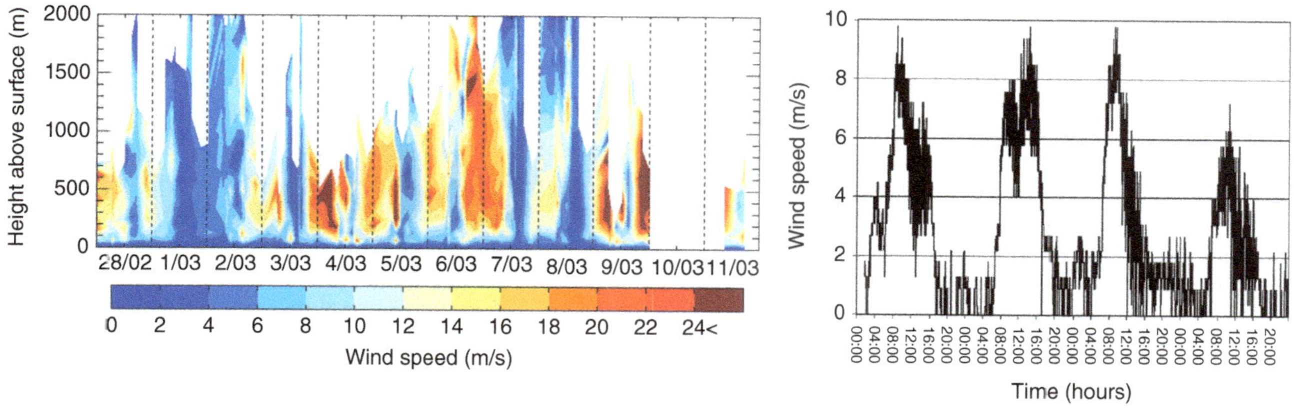


**Figure 5.7** Measurements of the effect of nocturnal LLJ formation on surface winds at Chicha (16°53′N, 18°33′E, Chad) during the BoDEx campaign. Left panel: time–height profile of wind speed from pilot balloon data for 28 February to 11 March 2005. Right panel: 1 min averages of 2 m wind speed for 5–8 March 2005. The data show the formation of nocturnal jets and corresponding peaks in surface winds during the morning for some days during the campaign. See also Figure 4.10. *Source*: Washington *et al.* (2006). Reproduced with permission of John Wiley and Sons.

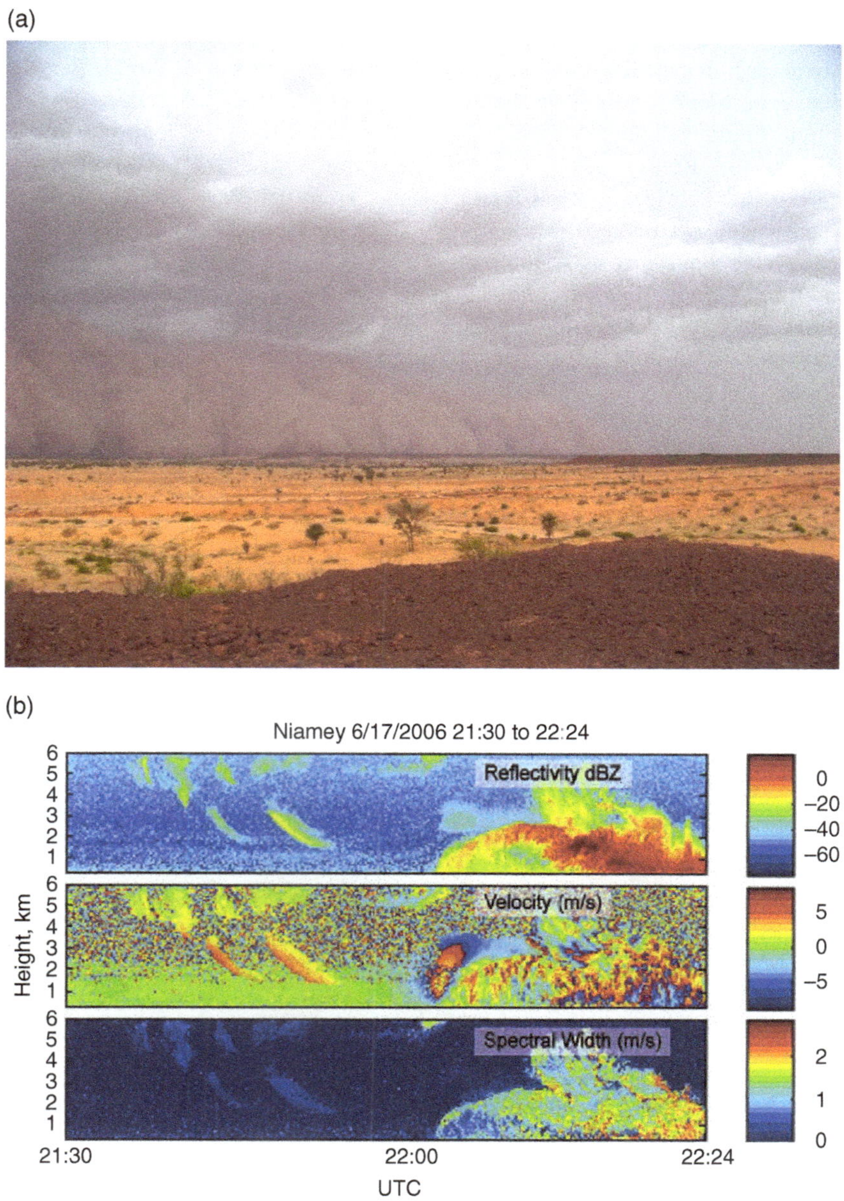


**Figure 5.8** Examples of haboob dust storms generated by gust fronts ahead of vigorous squall lines over the Sahel. (a) Photograph of a haboob arriving over Niamey on 26 June 2012. *Source*: Reproduced with permission of Laurent Labbé. (b) Vertical profiles of radar reflectivity, mean Doppler velocity and Doppler spectral width for a gust front on 17 June 2006 measured with the MIT Doppler radar at Niamey, Niger, at an operating frequency of 95 GHz. Note that insects, sticks and blades of grass are probably the main radar targets, rather than the dust itself. *Source*: Williams *et al.* (2008).

The dynamics, climatology and contribution to the total dust budget of haboobs are still not very well understood. This is caused on the one hand by a scarce observational network in the Sahara, combined with difficulties of detecting haboobs from space due to the cloudiness associated with the parent convection (Williams, 2008), and on the other hand by the challenges of numerically simulating the organised moist convection causing a haboob, which requires high resolution and a suitable trigger for convection. Examples have been documented where even large haboobs are not reflected at all in operational analysis (Knippertz *et al.*, 2009). Recent high-resolution modelling studies suggest an important contribution of haboobs to dust uplift during the summer monsoon, which is not reproduced by simulations using parameterised convection (Heinold *et al.*, 2013).

On the microscale (<2 km) dust is mobilised in connection with dry convective mixing in the PBL, part of which is realised through rotating dust devils and part through non-rotating convective plumes (see Figure 5.9 for examples). Estimating the contribution of these phenomena to global dust emission is challenging (Koch and Renno, 2005), but recent work suggests rather negligible values for most parts of northern Africa (Jemmett-Smith et al., 2015). Dust devils typically have diameters of 3–100 m, lifetimes of a few minutes and tangential winds of 3–15 m $s^{-1}$ (Sinclair, 1969). They are caused by convergence of ambient vorticity in regions of intense dry convective uplift. Dust plumes, in contrast, are characterised by larger diameters and lifetimes up to an hour, and also by weaker winds and smaller dust fluxes. Typical conditions for the formation of dust devils/plumes include moderate, near-surface wind shear and a large temperature contrast between the surface and the lowest few metres of the atmosphere (difference between temperature at 1 and 2 m above ground on the order of 1 K, with superadiabatic stratification reaching as high as several hundred metres above the ground (Ryan, 1972). The latter corresponds to high sensible heat flux and therefore, through the surface energy balance (Equations 4.5 and 4.9), requires intense downward solar radiation, a dry soil and low aerosol optical thickness (AOT; Ansmann *et al.*, 2009).

Very few studies have addressed dust devil/plume activity in and around the Sahara specifically. Ansmann *et al.* (2009) present lidar measurements from southern Morocco, which indicate coherent dust plumes of more than 2 km vertical extension. In recent years, first attempts have been made to simulate dust devils numerically, using large eddy models (Raasch and Franke, 2011). Another important aspect of the vigorous PBL mixing is the homogenisation of dust from different sources and the vertical transport to heights of more than 5 km over arid or semi-arid subtropical land surfaces during the daytime (Marsham *et al.*, 2008b; Cuesta *et al.*, 2009).

### 5.1.3 Climatologies

Dust source activity and resulting dust storms change with space and time. To describe their variability, time series from weather station observations, such as visibility records, weather reports and satellite images are

(a) (b)

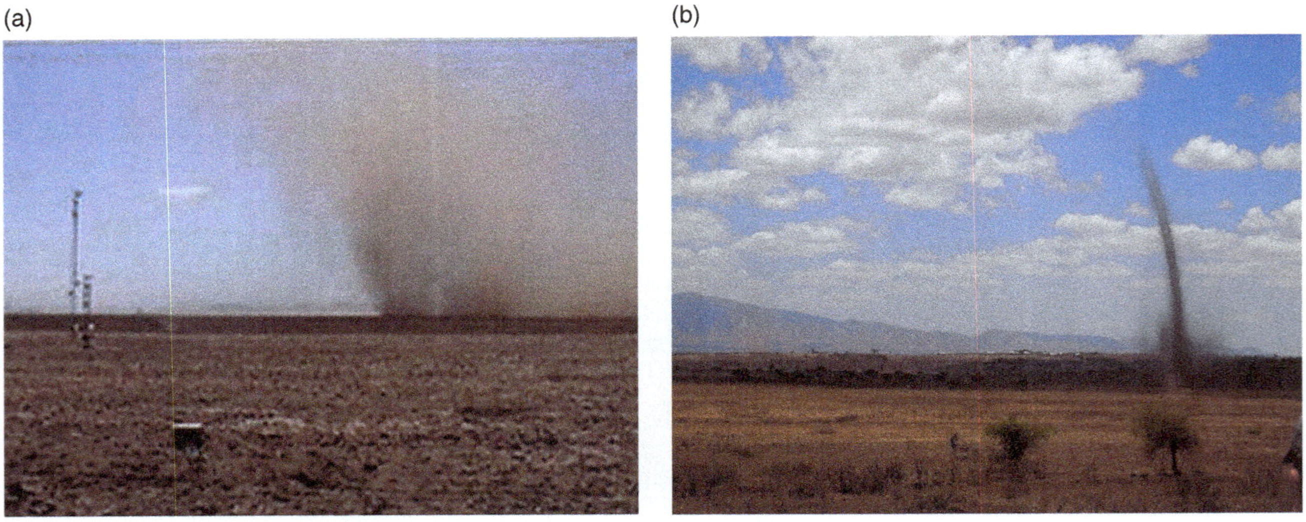

Figure 5.9 (a) Example of a non-rotating dust plume (taken by N. Renno during a 2002 field campaign near Eloy, Arizona, USA). *Source*: Koch and Renno (2005). Reproduced with permission of John Wiley and Sons. (b) Example of a rotating dust devil in Tanzania. *Source*: reproduced with permission of Philippe Peyrillé.

widely used. Records from some stations cover more than 60 years and are suitable to compile a local climatology of dust storm frequencies, their intensities during the year as well as interannual changes (N'Tchayi Mbourou *et al.*, 1997; Mahowald *et al.*, 2007; Klose *et al.*, 2010; Cowie et al., 2014). The frequency of days with reduced visibilities, as well as the time of day with minimum visibility, depends on the position of the station in relation to emitting dust sources and the wind regime transporting the dust toward the station. As wind regimes over North Africa change with the seasons, dust-emitting sources and types of dust transport change as well.

The study by N'Tchayi Mbourou *et al.* (1997), analysing weather reports from North African weather stations, illustrates a pronounced diurnal cycle in visibility reduction in the Sahel. Typically, minimum horizontal visibility occurs during late morning to mid-afternoon due to the development of the convective boundary layer. At Saharan stations the diurnal cycle is less pronounced, but minima in visibility are frequently observed during the late morning and mid-afternoon. Seasonal variation in the occurrence of reduced visibilities changes with the rainfall regime, and thus with distance from the equator (Figure 5.10). Generally, the dust season lasts longer as the latitude increases: south of the Sahel (below 15°N) dust is typically observed from October to April/May, peaking in December and January. Going further north the dust season extends, finally covering the whole year. At Saharan stations, the frequency of days with reduced visibility due to dust peaks in summer. At stations south of the Sahara, dust outbreaks are observed during the dry season. At Saharan and sub-Saharan stations, dust plumes generated by haboobs are also present during the rainy season.

Satellite observations provide better spatial coverage but often for much shorter time periods. Figure 5.11 shows monthly means of total atmospheric aerosol loading, retrieved from the MISR sensor flying on-board

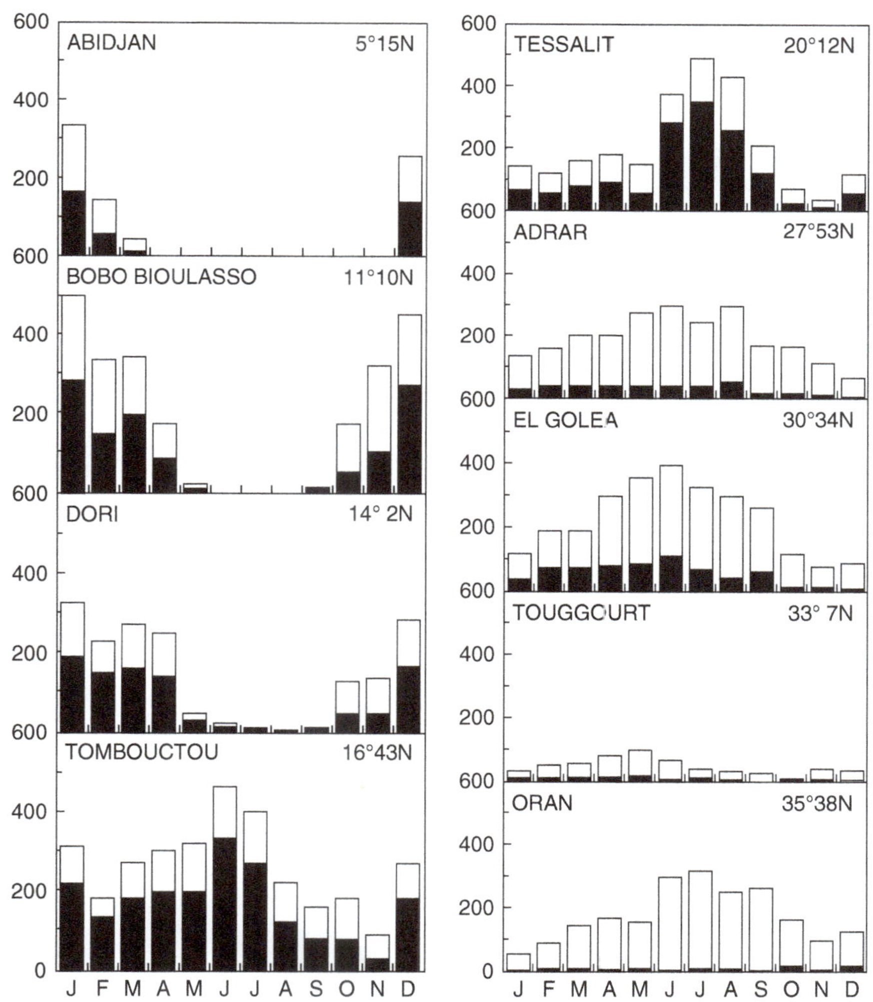


Figure 5.10 Mean monthly number of hours with visibility reduced to less than 5 km (shaded bars) and to less than 10 km (unshaded bars) during the period 1983–1987. Left: stations from the Guinea coast northward to the central Sahara. Right: stations from the central Sahara northward to the Mediterranean coast. *Source*: N'Tchayi Mbourou *et al.* (1997). © American Meteorological Society. Used with permission.

the polar-orbiting Terra satellite expressed through AOT. As AOTs represent the total atmospheric column, high aerosol loadings do not necessarily correspond to high surface concentrations and thus low horizontal visibilities. In addition to dust, aerosols such as soot from burning biomass contribute to the AOT, particularly over the Gulf of Guinea region during November to March. In general, two different seasons of dust loading can be identified (Figure 5.11). From October to February the atmospheric dust loading over North Africa is at its minimum, although two main regions of increased dust loading are evident: the Bodélé Depression–Ténéré region (Chad and Niger), and northwest Mali. Enhanced aerosol loadings over the Gulf of Guinea region are due to combustion. During March to May the atmospheric dust loading generally increases, forming a band of enhanced dustiness extending from the Red Sea to the Atlantic, centred at around 15–20°N, that persists until August and retreats in September. Areas of high AOT values may be linked not only to frequent dust emission but also to windblown transport, or topographic structures fostering the accumulation of dusty air (Schepanski *et al.*, 2012).

In addition to MISR, the MSG SEVIRI IR dust index or Moderate Resolution Imaging Spectroradiometer (MODIS) AOT images have been used to infer the location of individual dust sources (Schepanski *et al.*, 2007; Ginoux *et al.*, 2012), and, in the case of the former, also the diurnal cycle of dust source activations. Besides the Bodélé Depression, active dust sources are identified in the foothills of Saharan mountains, where wadis (*oueds*) and pluvial drainage systems act as dust sources. The local dust source activity changes during the course of the year, as reflected by the change in MISR AOTs (Figure 5.11).

### 5.1.4 Interactions with Weather and Climate

Dust particles influence the regional and global radiation balance, atmospheric dynamics, cloud microphysics, biogeochemical processes and atmospheric chemistry in several ways, which are only partly understood and remain largely unquantified (Figure 5.12). Emission, transport and deposition of atmospheric dust particles depend on meteorological and climate parameters like surface wind speed, precipitation and vegetation cover in dust source areas. Therefore, the impacts of dust on those parameters have the potential to feed back upon the atmospheric dust processes themselves. Because quantitative data of the atmospheric dust content cannot be obtained from observations alone, studies of such interactions between dust, weather and climate must rely on results of large-scale models of dust transport.

#### 5.1.4.1 Direct Radiative Effect

Dust impacts considerably on the energy balance of solar and thermal IR radiation (Solomon *et al.*, 2007). Incoming solar radiation is partly reflected and absorbed by atmospheric particles. With diameters of 2 μm or larger, dust particles can also absorb and emit outgoing terrestrial radiation. Dust particle size and complex refractive indices determine their optical properties, and thus the magnitude and sign of their radiative effects. They can be determined by measurements in the laboratory or field, or by remote sensing. There is a lack of information on the wavelength-dependent complex refractive indices of dust from different source regions. Optical properties, in particular single scattering albedo values that are derived with inversion techniques and those that are inferred from laboratory measurements of dust samples or specific minerals, often disagree. This is mainly due to different *a priori* assumptions about particle properties, in particular with respect to their shape.

The optical properties of dust are used to compute the fractions of absorbed and transmitted radiation passing through an atmospheric layer that contains dust. Net radiative forcing by dust is defined as the difference of net irradiance of the dust-containing atmosphere and a clear atmosphere. In addition to its optical properties, the magnitude of dust net radiative forcing depends on phenomena such as the presence of clouds, the vertical distribution of the dust particle concentration and on the albedo of the underlying surface. Values of global average net radiative forcing by dust at the top of atmosphere (TOA) range from −0.6 W $m^{-2}$ (Yoshioka *et al.*, 2007) to +0.35 W $m^{-2}$ (Balkanski *et al.*, 2007). Case studies of instantaneous dust net radiative forcing during individual dust events estimate negative TOA values above −6 W $m^{-2}$ (Christopher and Jones, 2007) or negative values up to −400 W $m^{-2}$ at the surface (Costa *et al.*, 2006). In the global annual average, dust net radiative forcing at TOA is most likely negative (Solomon *et al.*, 2007), but positive values may occur regionally over bright surfaces like the Sahara Desert. A negative dust forcing suggests that an enhancement of atmospheric dust aerosol would lead to a cooling of the climate.

#### 5.1.4.2 Impact on Atmospheric Stability

Radiative forcing by dust influences the energy distribution within the atmosphere. At the surface below a dust cloud incoming solar radiation is reduced, which results in a decrease in latent heating and therefore a weakening of the hydrological cycle (Miller *et al.*, 2004; Yoshioka *et al.*, 2007). The reduction of incoming radiation at the Earth's surface by dust can be substantial, while the estimates of the average TOA forcing are small. Thus, dust causes an increase of energy absorption within the

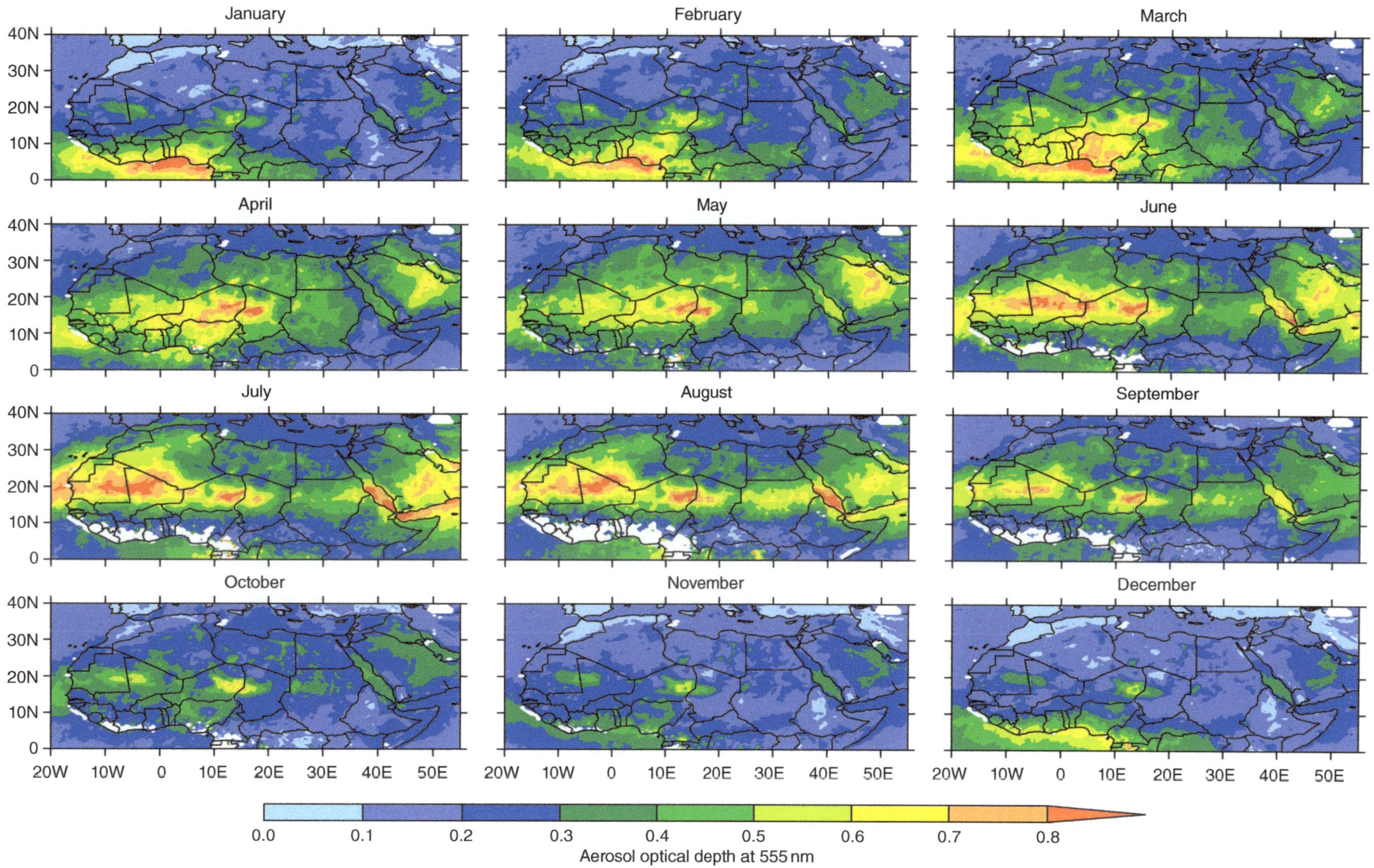


**Figure 5.11** Multi-angle Imaging SpectroRadiometer (MISR) AOT monthly means for 2001–2011. Note that during November to March soot contributes significantly to the AOT over the Gulf of Guinea region.

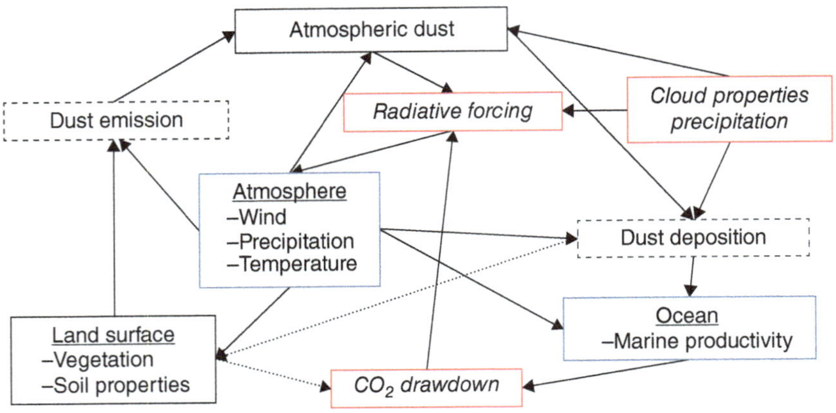


**Figure 5.12** Linkages between mineral dust aerosol and climate, indicating direct and indirect aerosol effects, and the role of dust deposition for oceanic ecosystems.

atmospheric column. This implies that, in the presence of dust aerosol, atmospheric heating rates are altered. Early estimates range between 1.6 and 15.6 K day$^{-1}$ for a dust AOT of 1.5 at 550 nm wavelength (e.g. Ackerman and Chung, 1992). Such processes may influence weather and climate patterns.

The presence of dust stabilises the PBL downwind of the dust source. Surface wind speed is slightly reduced below a dust cloud. Above the dust cloud the wind speed is enhanced, which can be explained by the changes in the vertical temperature structure. Figure 5.13 shows an example of a regional model result of the horizontal distribution of changes in surface temperature and 10 m wind speeds for a daytime and a night-time example in the Chad region (Tegen *et al.*, 2006). The maximum surface cooling by dust exceeds 20 K, corresponding to a decrease in net radiation flux by about 500 W m$^{-2}$. The maximum reduction in surface wind speeds at this time due to the dust radiative forcing is 3 m s$^{-1}$. For surface wind speeds that are near the emission threshold such a difference may lead to considerably reduced dust emission. At night-time, the thermal re-emission by the dust cloud causes surface warming, causing stronger mixing than the clear-sky case without dust forcing. Section 4.1.2.4.3 includes additional evidence of these effects of dust on surface temperatures.

Generally, results from regional models show that dust affects the atmospheric stability and thus the processes of dust emission in source areas. Atmospheric heating by absorbing aerosol particles, like mineral dust, may also lead to a reduction in cloud cover (the so-called semi-direct aerosol effect) (Hansen *et al.*, 1997). Including effects of dust in global and regional weather models generally improves predictions (Pérez *et al.*, 2006).

#### 5.1.4.3 Impact on Cloud Microphysics

The influence of aerosol particles on cloud evolution is a major uncertainty in climate research. Dust may influence the formation and lifetimes of clouds via the so-called indirect aerosol effect. A coating of dust particles by soluble material enhances the ability of those particles to form cloud condensation nuclei (Wurzler *et al.*, 2000), meaning that complex atmospheric chemistry may be needed to understand the effects of the dust on clouds. Simulations of aerosol–cloud processes with large-scale models contain great ambiguities.

There are considerable uncertainties in the estimates of the impact of mineral dust on the ice phase of clouds and precipitation. While mineral dust particles can be efficient ice nuclei, the extent of the influence of dust particles on the formation and development of ice clouds remains largely unquantified. An influence of dust particles on properties and the evolution of ice clouds can change the radiation budget as well as the hydrological cycle. Primary ice formation occurs either homogeneously (in other words, involving only the freezing of liquid droplets; at temperatures below −35 to −40 °C) or heterogeneously (in which the freezing is assisted by an ice or aerosol particle) in warmer regimes. Heterogeneous ice nucleation is subdivided into immersion, condensation, deposition and contact freezing. It is not yet clear under which conditions which of the identified ice nucleation mechanisms are of importance and thus have to be considered in studies of cloud development. From observations it is difficult to distinguish between the different possible freezing processes. Under certain conditions field measurements provide important information for understanding of freezing processes in clouds. Lidar measurements show that dust particles start to act as ice nuclei at a wide range of temperatures (Ansmann *et al.*, 2008). Laboratory studies are used to clarify the roles of freezing mechanisms (Möhler *et al.*, 2006). The description of ice nucleation processes, even in the most detailed microphysical models, is not complete.

#### 5.1.4.4 Fertilisations of Ecosystems by Dust

Soil dust particles add essential nutrients into terrestrial and marine ecosystems when they are deposited to the ground. Fertilisation experiments adding dissolved iron (Fe(II)) to the ocean show an enhanced productivity

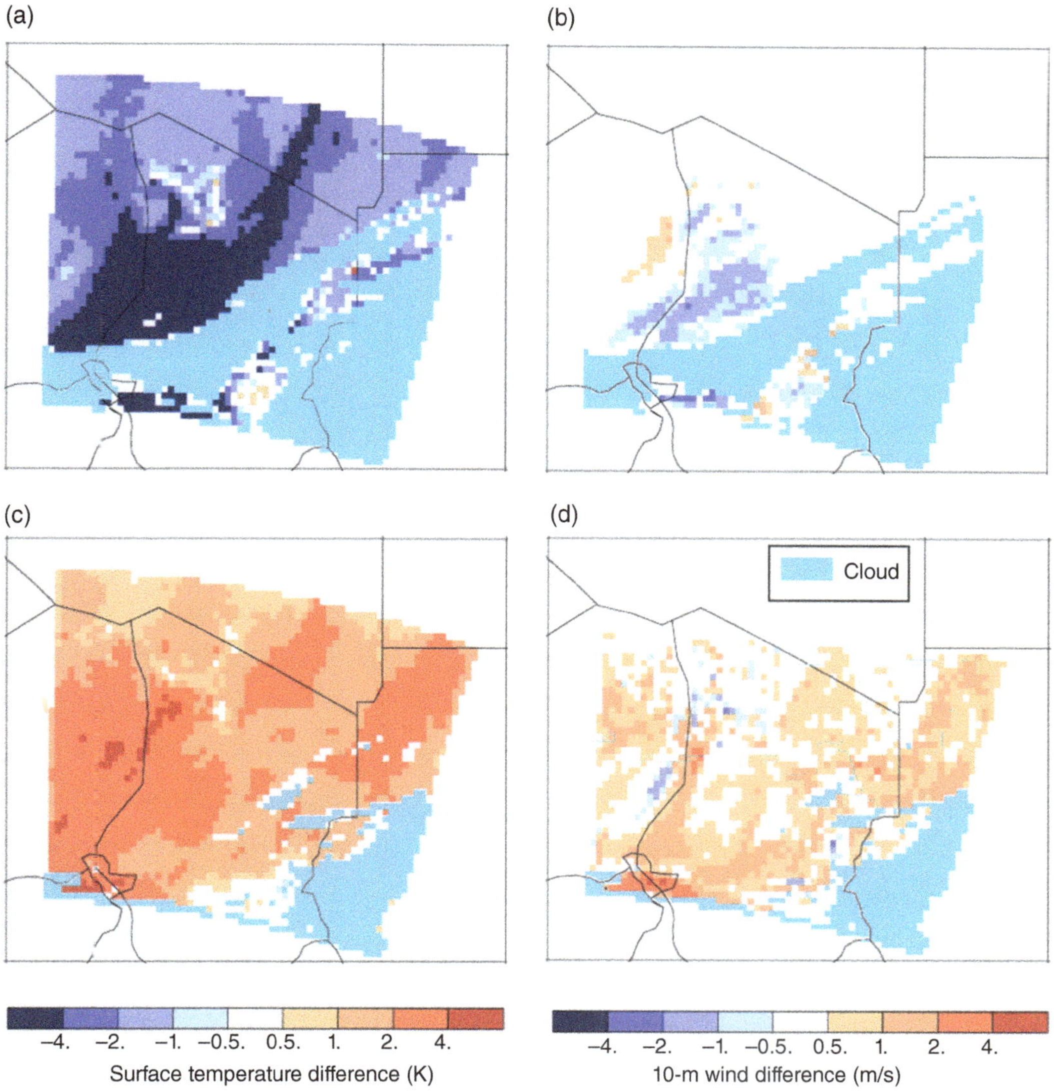


Figure 5.13 Regional model results of changes in 2 m temperatures (a, c) and 10 m wind speeds (b, d) for the Bodélé region in Chad. The results are shown as difference of the regional model results including dust forcing and results from the model without including dust, on 11 March 2005 for daytime (12 UTC, a, b) and night-time (00 UTC) (c, d) conditions.

of oceanic microorganisms (e.g. plankton) and increase in $CO_2$ drawdown in regions rich in macronutrients (Coale *et al.*, 1996). A summary of findings is published by Jickells *et al.* (2005). Microorganisms utilise iron in enzyme systems, including those for photosynthesis and nitrogen fixation, and therefore additions of dissolved iron enhance these processes. In nutrient-rich regions of the oceans the addition of the micronutrient iron can enhance primary production, species composition and $CO_2$ uptake, while in nutrient-poor regions, like the subtropical Atlantic, the iron supply from deposited dust aerosol particles can enhance nitrogen fixation (Moore *et al.*, 2006). The solubility of soil dust particles near their source areas is only on the order of 1% or less (Bonnet and Guieu, 2004). Higher local solubilities of aerosol iron, of up to 40%, have been found in atmospheric particles, either due to photochemical processes at the particle surface or aerosol cloud processing. At low pH values, of less than 2, Fe(III) is easily dissolved, such that coatings of dust particles with acid compounds (e.g. sulphuric acid) may greatly enhance its availability for oceanic microorganisms when they are deposited into the ocean. As yet it is unknown whether the iron fertilisation by dust is an important contributor to the oceanic $CO_2$ sink at a global scale. Transported desert dust deposited onto nutrient-poor soils is suspected to contribute micronutrients to terrestrial ecosystems. In particular, the transport of Saharan dust into the Amazon rainforest is suspected to support vegetation growth (Koren *et al.*, 2006), which would increase $CO_2$ uptake.

## 5.2 Operational Methods

### 5.2.1 Introduction and Structure

Dust and sand storms or dust haze are a part of day-to-day weather forecasting for many parts of West Africa. Dust deteriorates visibility, with adverse effects on aviation and road traffic. Dust is also detrimental to human health, fostering respiratory diseases, and being associated with meningitis episodes in the Sahel. The following section will provide guidelines to forecasters to predict dust storms and episodes of dust haze, typically caused by southward advection of large amounts of dust from the Sahara into the southern parts of West Africa. Section 5.2.2 provides background on the most useful sources of information to generate a successful dust forecast, including data from station observations, remote sensing and numerical models. Section 5.2.3 will explain how best to use these data in an operational forecast environment for the two main seasons (November–April Harmattan dust haze (HDH) and April–August convective dust storms in the Sahel).

### 5.2.2 Data

#### 5.2.2.1 Surface Station Data

An important source of information for forecasting dust storms or haze is the three-hourly surface synoptic observation (SYNOP), hourly meteorological aviation routine weather report (METAR) and SPECI airport reports network in West Africa, which provides instrumented and visual observations in near-real time through the Global Telecommunication System. The key parameter to look out for in dust forecasting is the current weather code (ww). A list of all codes relevant for dust forecasting is given in Table 5.1.

Of these reports, ww 06 is by far the most frequently reported, as the number of stations in active dust source areas is low. ww 07 is also commonly reported during HDH episodes in the Sahel during the period November to April, while ww 08 and ww 09 are used less. Reports of ww 30–35 are quite rare and only reserved for the most intense storms with very high winds and low visibility.

The direct eye (visual) observations of dust storms or haze as reported in the ww part of the SYNOPs can be complimented by indirect indicators such as reports of strong winds (typically 30 kt or more) or low horizontal visibility (typically below 5 km).

#### 5.2.2.2 Surface and Pressure-level Charts

A key element in dust forecasting is sufficiently high winds to raise dust and transport it quickly downstream. These winds can be assessed directly on model-generated analysis or forecast charts (typically 10 m to assess emission and 925 hPa to assess transport) or indirectly from charts of mean sea-level pressure (MSLP) or geopotential height (again at 925 hPa), as winds are very closely related to the horizontal gradient in pressure or geopotential height. Recall that the actual wind at the surface and within the lowest layers will blow at an angle to lines of constant pressure (isobars) or lines of constant geopotential height (isohypses) pointing toward the lower pressure (see Chapter 2 for more details).

**Table 5.1** SYNOP current weather codes (ww) related to dust storms or dust haze.

| ww | Description | |
|---|---|---|
| 05 | Haze (can be an indication of dust haze, but can also be caused by other types of aerosols) | |
| 06 | Widespread dust in suspension in the air, not raised by wind at or near the station at the time of observation | |
| 07 | Dust or sand raised by the wind at or near the station at the time of the observation, but no well-developed dust whirl(s), and no sandstorm seen | |
| 08 | Well-developed dust whirl(s) or sand whirl(s) seen at or near the station during the preceding hour or at the time of observation, but no dust storm or sandstorm | |
| 09 | Dust storm or sandstorm within sight at the time of observation, or at the station during the preceding hour | |
| 30 | | has decreased during the preceding hour |
| 31 | Slight or moderate dust storm or sandstorm | no appreciable change during the preceding hour |
| 32 | | has begun or increased during the preceding hour |
| 33 | | has decreased during the preceding hour |
| 34 | Severe dust storm or sandstorm | no appreciable change during the preceding hour |
| 35 | | has begun or increased during the preceding hour |
| 98 | Thunderstorm combined with dust storm or sandstorm at time of observation | |

#### 5.2.2.3 Remote Sensing

As the synoptic station network is sparse in the Sahara, satellite remote sensing provides a crucial source of additional information to monitor dust storms. Data from several space-borne instruments are available to detect dust plumes in different temporal and spatial resolutions,

**Table 5.2** Freely available satellite dust products for West Africa.

| | VIS AOT | DeepBlue AOT | Near-UV AOT | AAI | IR dust |
|---|---|---|---|---|---|
| Instrument (satellite) | MODIS (Aqua & Terra)[a] | MODIS (Aqua & Terra)[a] | OMI (Aura)[b] | OMI (Aura)[c] | SEVIRI (MSG)[d] |
| Wavelength | VIS | VIS/near-UV | near-UV | near-UV | IR |
| Spatial resolution | Re-gridded, 1° × 1° | Re-gridded, 1° × 1° | Re-gridded, 1° × 1° | Re-gridded, 1° × 1° | 3 km × 3 km to ~5 km × 5 km |
| Temporal resolution | Daily | Daily | Daily | Daily | 1 hourly |
| Surface | Dark | Bright land | Land & ocean | Land & ocean | Land |
| Quantitative | Yes | Yes | Yes | Yes[e] | No |
| Night | No | No | No | No | Yes |
| Near-real time | 2–4 days later | 2–4 days later | 3 days later | 3 days later | Yes |

VIS: visible; UV: ultraviolet; IR: infrared.
[a] http://giovanni.sci.gsfc.nasa.gov/giovanni/. Select 'Daily' and 'Aerosol Optical Depth'.
[b] http://giovanni.sci.gsfc.nasa.gov/giovanni/. Select 'Total Aerosol Optical Depth' and 'OMI'.
[c] http://giovanni.sci.gsfc.nasa.gov/giovanni/. Select 'Aerosol Index' and 'OMI'.
[d] http://oiswww.eumetsat.org/IPPS/html/MSG/RGB/DUST/WESTERNAFRICA/index.htm.
[e] The aerosol index is sensitive to the height of the aerosol plume. Index values are increasing with increasing level height assuming a constant total aerosol loading.

all with their own limitations and sensitivities. Three types of satellite-based dust products are common:

1) AOT providing quantitative information on the atmosphere aerosol amount, usually based on extinction in different parts of the electromagnetic spectrum (columns 1–3 in Table 5.2).
2) Absorbing aerosol index (AAI), or simply aerosol index, providing semi-quantitative information on atmospheric loading of UV-light-absorbing aerosol particles (e.g. mineral dust and soot) (column 4 in Table 5.2).
3) Thermal IR index providing qualitative information on the presence of atmospheric dust (column 5 in Table 5.2).

All products listed in Table 5.2 are freely available, but only the SEVIRI IR dust product is available in near-real time, and therefore the following explanations will concentrate on this type of data. The other products are currently available online after 2 or 3 days and can therefore only be used for post-analysis of specific events or forecast verification. Another important limitation for the use of satellite information is the prerequisite of cloud-free conditions. Thus, forecasters need to be aware that there may well be significant amounts of dust and very low visibility present underneath clouds, particularly during April–August in the Sahel.

As mentioned, the SEVIRI IR dust product is the most widely used for operational dust forecasting owing to its near-real-time availability, high time resolution and applicability during day and night. Thermal IR wavelengths are suitable for dust monitoring, as the thermal emissivity of dust is less than that of the underlying surface. Multispectral measurements at IR wavelengths allow separating between surface emissivity and dust emissivity by the split-window technique (e.g. Ackerman, 1997; Wald *et al.*, 1998; Sokolik, 2002) using differences in brightness temperatures of two adjacent channels within the same atmospheric window.

EUMETSAT (European Organisation for the Exploitation of Meteorological Satellites) offers a tri-spectral IR dust false-colour image using MSG SEVIRI measurements (see Figure 5.16 for an example). Dust is evident in these images by a magenta or pink colour and is not affected by soot plumes. The colour does not give a quantitative estimate of dust loading or AOT, but the high temporal resolution of 1 h allows tracking of individual dust plumes in cloud-free areas. The MSG dust product has several important limitations that forecasters should be aware of (see detailed discussion in Brindley *et al.* (2012)). Dust can become invisible in areas of high column water vapour (typically of ≥25 mm). These stand out in the images through a more intense blue component. Furthermore, the appearance of the dust in the images depends on the height of the dust layer, its thickness and the time of day (Figure 5.14). Thin dust clouds during the day will appear pink irrespective of height. Low-level thin dust clouds can become hard to see

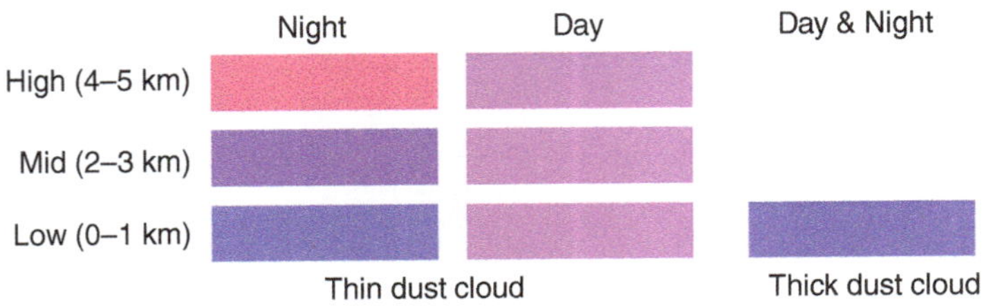


**Figure 5.14** Typical appearance of dust clouds in the MSG dust product depending on thickness of the dust cloud, time of day and height. *Source*: reproduced with permission of Jochen Kerkmann, EUMETSAT).

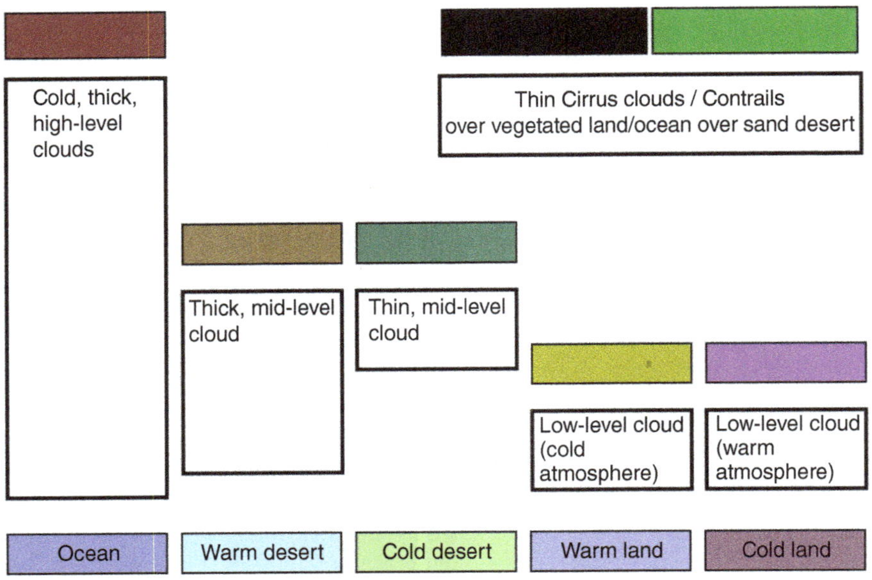


**Figure 5.15** Typical appearance of clouds and cloud-free surfaces in the MSG dust product. *Source*: reproduced with permission of Jochen Kerkmann, EUMETSAT.

during the night due to effects of near-surface temperature inversions. Thick dust clouds near the surface will appear more purple than pink.

Despite these limitations, the high time resolution of 15 min makes the MSG false-colour images a very important tool for dust storm monitoring and forecasting for the short term. If possible, animations should be used to track and extrapolate dust plume movement. Typically, the human eye is capable of linking outstanding features, even if colours change over time.

The MSG false-colour images also allow a monitoring of clouds (Figure 5.15), which is particularly important for dust storms related to deep moist convection. These will appear in dark red on the images, often with black areas around the edges where thin cirrus prevails. An example of a dust storm over northern Africa with strong dust emission behind the cold front is provided in Figure 5.16.

In addition to the qualitative near-real-time SEVIRI IR product, aerosol plumes can be identified more quantitatively by spectro-radiometer measurements at different wavelengths (see Table 5.2 for an overview). For example, in the visible part of the spectrum (VIS) dust particles scatter back solar radiation and appear bright compared with dark surfaces like the ocean, forest or dark rocks. This works less well for dust over bright surfaces like deserts. Radiative transfer models applied to measurements from satellite sensors are used to obtain quantitative information on dust amount. Along its way through the atmosphere, radiation gets absorbed and scattered by aerosol particles. AOT is the most widely used measure for this extinction (sometimes also called aerosol optical depth). It can be understood as a measure for the total aerosol loading of an atmospheric column. Over West Africa the most important aerosol types contributing to the total AOT are (i) mineral dust, (ii) soot from biomass burning, especially during November–March, and (iii) emissions from combustion of fossil fuels, particularly along the densely populated Guinea Coast.

In the visible part of the spectrum AOT retrievals using measurements from MODIS, Sea-viewing Wide Field-of-view Sensor (SeaWiFS), Advanced Very High Resolution Radiometer (AVHRR) and SEVIRI are widely used. These algorithms are limited to dark surfaces and therefore not applicable to large parts of the Sahara. Measurements at the deep blue part of the visible spectrum by Terra MODIS and SeaWiFS on-board GeoEye are used to compute AOT over bright surfaces ('Deep Blue'; Hsu *et al.*, 2004). IR-AOTs are computed from SEVIRI measurements, albeit only during daytime. At UV wavelengths, dust, soot particles and volcanic ash can be identified by their spectral contrast. As surface reflectivity is low at these wavelengths, particles can be identified over land and ocean. The widely used AAI

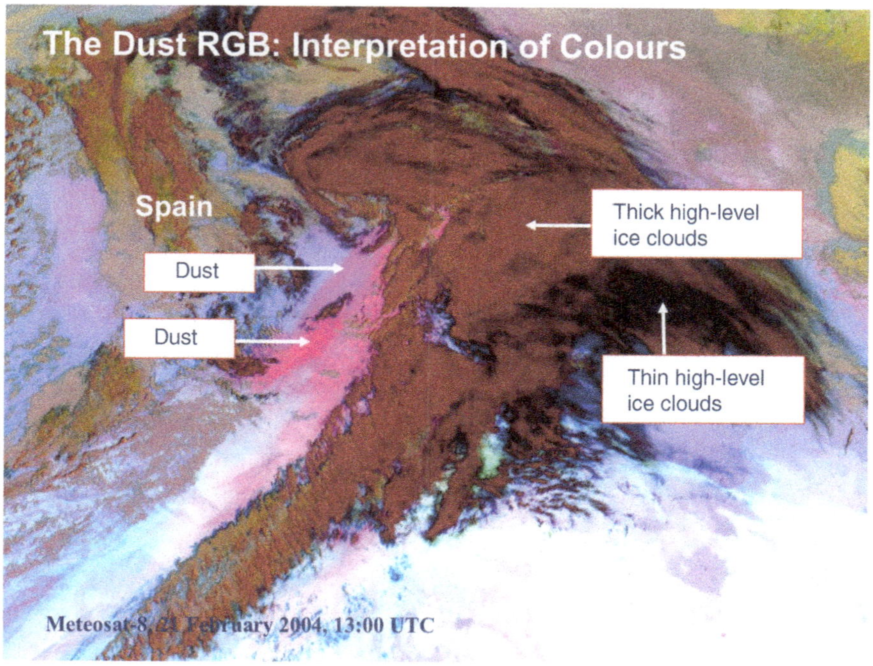


**Figure 5.16** Example of the appearance of a dust storm in the MSG dust product. *Source*: reproduced with permission of Jochen Kerkmann, EUMETSAT).

(Herman *et al.*, 1997) based on this technique and using data from the Ozone Monitoring Instrument (OMI) and its precursor, the Total Ozone Mapping Spectrometer (TOMS), however, cannot distinguish between mineral dust aerosol and soot. The aerosol index has also been reported to be biased toward elevated dust plumes (Hsu *et al.*, 1999). Near-UV AOTs are obtained directly by using an aerosol model (Torres *et al.*, 2007) or indirectly by rescaling the AAI (e.g. Chiapello *et al.*, 1999).

AOTs and information on aerosol type and characteristics can also be obtained from the global network of ground-based sun-photometer measurements, AERONET (AErosol RObotic NETwork; Holben *et al.*, 1998). Retrievals are limited to daytime cloud-free conditions. AERONET provides an online visualisation tool, which allows easy access to the measurements.[1] Three quality levels are available, and only Level 1 data, which might be contaminated by clouds in some cases, are available in near-real time. Unfortunately, the AERONET station network across West Africa is sparse and subject to changes. Amongst the stations with longer term data coverage are Banizoumbou (Niger), Cinzana (Mali), Dakar (Sénégal), Ilorin (Nigeria), Tamanrasset (Algeria) and Capo Verde (Sal Island).

#### 5.2.2.4 Numerical Model Predictions of Dust

There are now a rapidly increasing number of operational centres that run operational forecasts of dust concentrations and loading, globally or specifically for northern Africa (see Tables 5.3 and 5.4). A very useful overview over these models with real-time analysis and forecast products is available at the webpage hosted by the Northern Africa–Middle East–Europe Regional Centre of the World Meteorological Organization Sand and Dust Storm Warning Advisory and Assessment System (SDS-WAS).[2] All systems use numerical schemes for dust emission, transport and deposition driven by winds from meteorological models (global forecasts or regional models nested into global forecasts). Most centres also use radiative transfer codes to provide estimates of dust AOT.

The global products (Table 5.3) typically have coarser resolution (between 1 and 4°), but MetUM and GEOS-5 are now running at grid spacings well below 1°. The METEO FRANCE model focuses on air-quality predictions and therefore includes an extended chemistry module at rather coarse resolution. The MACC-ECMWF, NGAC, MetUM and GEOS-5 products have high numbers of vertical levels ($\geq$60). GEOS-5, MACC-ECMWF, MetUM and NAAPS run data assimilation systems that include satellite information for atmospheric composition with ongoing further development.

The regional models (Table 5.4) all cover a similar domain including northern hemispheric Africa and adjacent Europe, the Arabian Peninsula, the Middle East, and the tropical and subtropical Atlantic Ocean. BSC-DREAM, TAU/DREAM, DREAM-NMME-MACC, DREAMABOL and SKIRON are from the same 'model family' based on the Eta/NCEP model. Except for DREAM-NMME-MACC, all these models are driven by NCEP analysis and forecasting data at the lateral boundaries. NMMB/BSC-Dust is the regional version of the global model with the same name. The main differences for practical forecasting purposes are different forecast ranges, different horizontal resolution (0.24, 1/3 and 1°) and different graphical products for the output.

In general, forecasters should carefully compare the output from different centres, as fairly large deviations

Table 5.3 Selection of currently operational global modelling systems with dust modules (as of July 2016).

| Abbrev. | Full name | Centre | Range | Freq. | Output | Driven by | Resol. | Webpage |
|---|---|---|---|---|---|---|---|---|
| CAMS-ECMWF | Copernicus Atmospheric Monitoring Service | European Centre for Medium-Range Weather Forecasts (ECMWF) | 120 h | daily at 00 UTC | dust AOT | ECMWF | 1° × 1°<br>60 levels | http://www.gmes-atmosphere.eu/d/services/gac/nrt/nrt_opticaldepth!03!Dust |
| METEO FRANCE | PREV'AIR System | Official Air Quality Monitoring Associations, France | 72 h | daily at 00 UTC | daily peak and mean sfc concentrations | ARPEGE | 4° × 4° | http://www.prevair.org/en/prevision_pous_desert.php |
| MetUM | Met Office Unified Model | UK Met Office, UK | 144 h | daily at 00 UTC | dust AOT and sfc conc. | MetUM | 0.3516° × 0.2344°<br>70 levels | http://sds-was.aemet.es/forecast-products/dust-forecasts/u.k.-met-office |
| NAAPS | Navy Aerosol Analysis and Prediction System | Naval Research Laboratory, USA | 120 h | 4 times daily | dust AOT and sfc conc. | NOGAPS | 1° × 1°<br>25 levels | http://www.nrlmry.navy.mil/aerosol/ |
| NASA-GEOS-5 | Goddard Earth Observing System Model, Version 5 | National Aeronautics and Space Administration, USA | 240 h | Twice daily at 00 & 12 UTC | dust AOT | GEOS-5 | 0.5° × 0.666°<br>72 levels | http://portal.nccs.nasa.gov/cgi-fp/fp_2d_chem.cgi?region=global |
| NCEP-NGAC | NEMS GFS Aerosol Component | National Centers for Environmental Prediction (NCEP), USA | 120 h | daily at 00 UTC | dust load, dust AOT, deposition, emission, sfc conc. | NEMS GFS | 1° × 1°<br>64 levels | http://www.nco.ncep.noaa.gov/pmb/products/ngac/ |
| NMMB/BSC-Dust | Nonhydrostatic Multiscale Meteorological Model on the B grid | Barcelona Super-computing Center, Spain | 72 h | daily at 00 UTC | dust load, sfc conc., AOT, deposition | NCEP | 1.40625° × 1.0°<br>40 levels | http://www.bsc.es/earth-sciences/mineral-dust/nmmbbsc-dust-forecast |

Table 5.4 Selection of currently operational regional modelling systems with dust modules that cover West Africa (as of July 2016).

| Abbrev. | Full name | Centre | Range | Freq. | Output | Driven by | Resol. | Webpage |
|---|---|---|---|---|---|---|---|---|
| BSC-DREAM8b | Dust REgional Atmospheric Model | Barcelona Super-computing Center, Spain | 72 h | daily at 12 UTC | dust load, sfc conc., dust AOT, deposition, profiles, cross-sections | NCEP | 1/3° × 1/3° 24 levels | http://www.bsc.es/earth-sciences/mineral-dust-forecast-system/bsc-dream8b-forecast |
| DREAM-NMME-MACC | DREAM - Nonhydrostatic Mesoscale Model on E-grid – Monitoring Atmospheric Composition & Climate | South East European Climate Change Center (SEEVCCC), Serbia | 72 h | daily at 00 UTC | dust load, sfc conc., deposition | ECMWF | 0.2° × 0.2° | http://www.seevccc.rs/?p=8 |
| DREAMABOL | DREAM Bologna | Institute of Atmospheric Sciences and Climate (ISAC), Italy | 72 h | daily at 12 UTC | dust AOT, sfc conc. | BOLAM | | http://bolchem.isac.cnr.it/ |
| NMMB/BSC-Dust | Nonhydrostatic Multiscale Meteorological Model on the B grid | Barcelona Super-computing Center, Spain | 72 h | daily at 12 UTC | dust load, sfc conc., dust AOT, deposition, profiles, cross-sections | NCEP | 1/3° × 1/3° 40 levels | http://www.bsc.es/earth-sciences/mineral-dust/nmmbbsc-dust-forecast |
| NOA-WRF-CHEM | Weather Research and Forecasting model coupled with Chemistry | National Observatory Athens (NOA), Greece | 84 h | daily at 12 UTC | sfc conc. | NCEP | 0.2° × 0.2° | http://meteo.gr/meteomaps/ |
| SKIRON | | University of Athens, Greece | 120 h | daily at 12 UTC | dust load, AOT, wet/dry deposition, sfc conc. | NCEP | 0.24° × 0.24° 38 levels | http://forecast.uoa.gr/dustindx.php |
| TAU/DREAM-8b | Dust REgional Atmospheric Model | Tel-Aviv University, Israel | 72 h | daily at 12 UTC | load, concentrations | NCEP | 1/3° × 1/3° 24 levels | http://wind.tau.ac.il/dust8/dust.html |

can be expected in some situations. The SDS-WAS webpage provides direct model comparisons and some multi-model products (e.g. the median of all models) that facilitate the assessment of forecast uncertainty. Given the sometimes large model disagreement, a multi-model median is very likely the best estimate. The SDS-WAS page also provides near-real-time evaluation of models that help to better understand biases and systematic errors. It is recommended that forecasters should look at these evaluations for their specific region of interest. Model uncertainties are caused, among other things, by uncertainties in dust-source location, strength and characteristics as well as by the highly nonlinear relation between dust emission and peak winds near the surface, which are often badly represented in models.

Generally, advection of existing dust plumes will be better represented than fresh emission. As stated earlier, some centres have started assimilating satellite information into their models, which should improve the representation of existing dust plumes. However, most of this is currently restricted to oceanic areas and therefore of little advantage to West African forecasters. The UK Met Office also assimilates SEVIRI information over land, which can be expected to improve HDH forecasts for southern West Africa. Haboob storms during the rainy season are probably the most problematic, as they are closely tied to the prediction of convection and convective organisation.

### 5.2.3 Practical Guidelines for Forecasters

#### 5.2.3.1 Introduction

For dust forecasting in West Africa two main types/seasons with very distinct characteristics can be distinguished:

1) HDH affects large parts of West Africa during November–April. It is typically related to dust being lifted from sources over the Sahara or sometimes the Sahel, which is then mixed vertically within the PBL and transported horizontally with the prevailing northeasterly Harmattan winds. The weak precipitation activity over most parts of West Africa prevents wet removal of dust particles, and therefore HDH can occur far to the south, even reaching the Guinea Coast in some cases.
2) Convective dust storms occur in the Sahel during the pre-monsoon to peak monsoon period from April to August. They are related to the sinking and spreading at the surface of cooler air generated by evaporation, often in connection with mesoscale convective systems such as squall lines (see Section 3.1.3.2).

The following two sections will detail practical tools to best forecast these two situations. Dust activity in September and October is often weak. This is typically related to a weakening of the precipitation activity in the Sahel during the late monsoon season, weaker downdrafts due to moister mid-levels and weaker dust sources due to enhanced vegetation cover and soil moisture in the Sahel.

#### 5.2.3.2 Forecasting Harmattan Dust Haze

Generally, three phases are involved in HDH forecasting: dust raising, transport and clearance. Forecasters should concentrate on four aspects in their assessment of a given situation:

- Watch out for significant amounts of dust being raised over the Sahara/northern Sahel.
- Estimate the likelihood of the dust being transported to your region and the arrival time.
- Monitor the situation to estimate duration and severity of the HDH event.
- Watch out for indications of a clearing due to a change of air masses or ceasing dust emission.

##### *5.2.3.2.1 Raising of Dust*

The main source regions for HDH are over the southern Sahara and northern Sahel. As evident from Figure 5.17, there is some disagreement on the exact location of the main sources between different products and analysis techniques. The areas most common to many products are shaded grey in Figure 5.17 and numbered PSANAF-1 to -6. PSANAF-2 and -3 are most relevant for Senegambia, while PSANAF-5 affects Nigeria and neighbouring countries. The other main dust sources PSANFA-1, -4 and -6 are too far away to strongly affect southern West Africa, except for extraordinary large-scale and long-lived dust episodes.

The raising of dust is usually associated with the intensification of the subtropical anticyclone, which increases the south–north pressure gradient over the respective region (see Figure 1.19 for the background climatology of surface pressure). Frequent examples are an intensification of the Libyan High (eastern–central parts of West Africa) or the intensification/southeastward extension of the Azores High into northwestern Africa (central–western parts) as schematically depicted in Figure 5.18 (see also Section 1.7). Both are often preceded by the eastward passage of a mid-latitude depression across northern Africa. The high pressure often builds rapidly behind the cold front, which may penetrate far into the African continent. The two cases shown in Figure 5.18 are consistent with cases 2 and 1 of Klose *et al.* (2010), which together account for about 70% of dust episodes in the Sahel during the dry season. A third case that accounts for about 14% is characterised by a large low-pressure system centred over the Iberian Peninsula with a high over northeast Africa (not shown).

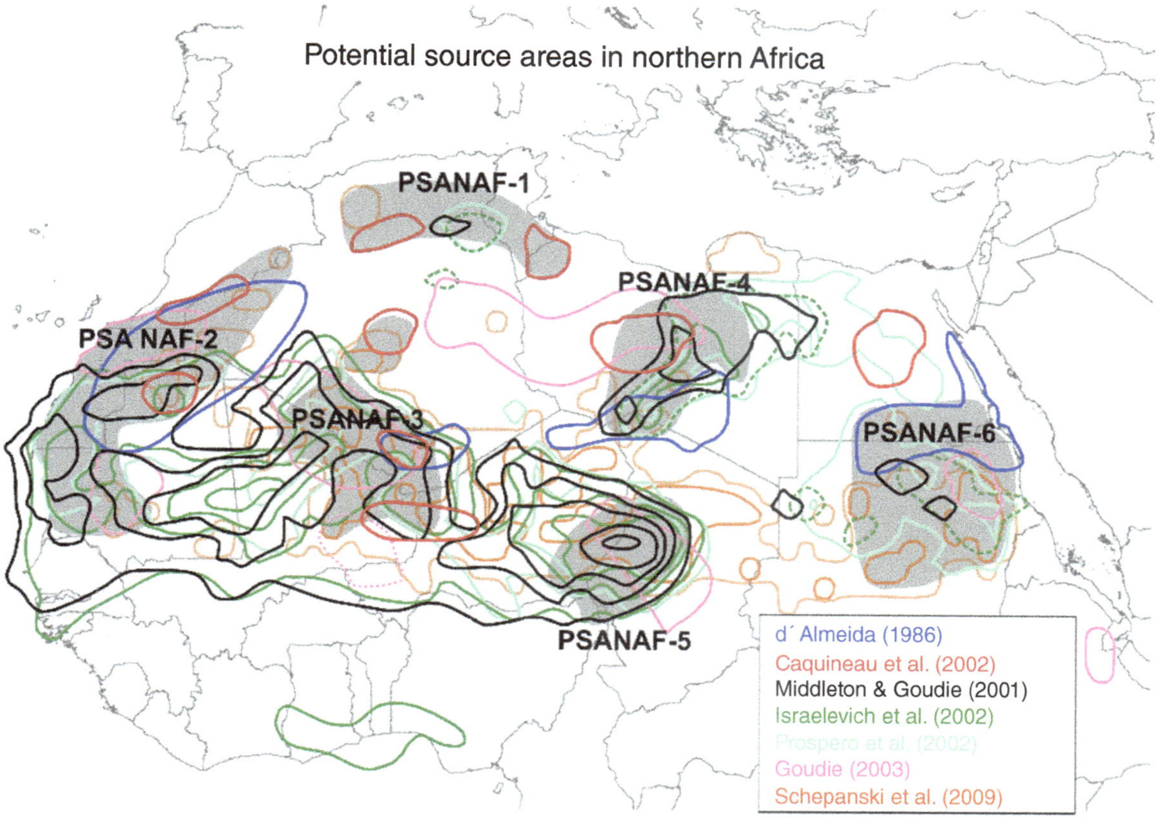


**Figure 5.17** Overview of potential dust source areas in northern hemispheric Africa (PSANAF) based on a synergy analysis by Formenti *et al.* (2011). The different lines and colours represent analyses by different authors based on different data and techniques (see Formenti *et al.* (2011) for references in the legend), illustrating the challenge of identifying dust sources with great accuracy. The most important regions are grey shaded: PSANAF-1: zone of chotts in Tunisia and northern Algeria; PSANAF-2: foothills of Atlas Mountains (PSANAF-2a) and western coastal region (PSANAF-2b; western Sahara, western Mauritania); PSANAF-3: Mali–Algerian border region; PSANAF-4: central Libya; PSANAF-5: Bodélé Depression (western Chad); PSANAF-6: southern Egypt, northern Sudan. Other source regions worth noting are the southern parts of the Azawagh, a dry basin in northwestern Niger/northeastern Mali (pink stippled line), and the Kaouar region in northern Niger (not specifically marked on the figure).

The unusual pressure distributions over northern Africa associated with intense HDH episodes are often accompanied by high-amplitude waves in the subtropical jet stream around the 250 hPa level.

Dust emission can occur at 10 m wind speeds as low as about 10 kt, but typically wind speeds need to exceed 30 kt to lift amounts of dust significant enough to create an HDH episode in the more populated southern parts of West Africa.

A number of techniques have been suggested to analyse the synoptic features, which will create these wind-speeds, from observations and output from forecast models:

- As a first orientation, the 1015 hPa isobar of MSLP often demarks the southern edge of the region where dust can be raised. This line should be highlighted on analysis charts (see thick dashed line in Figure 5.18). If no model data are available or as a quick check on the reliability of a given model analysis, the 1015 hPa isobar can be determined from SYNOP observations of MSLP, when sufficient data are available.
- As stated earlier, the key indicator for dust emission is a large pressure gradient, which can be estimated from the spacing of isobars (e.g. 1015 hPa crossing 15°N and 1020 hPa crossing 20°N). Alternatively, the pressure gradient may be estimated from suitable pairs of stations; for example, differences between Abeche (64756) and Sebha (60785) of more than ≥10 hPa indicate an HDH episode in downstream northern Nigeria. Forecasters should be aware that pressure gradients can change locally or areas of tight gradients can migrate without an apparent change of the centre of the parent high-pressure system.
- Another useful indicator is observations of dust, low visibility and strong winds at upstream SYNOP stations (see Section 5.2.2.1). Things to look out for are reports of severe dust storms (ww 30–35), dust being raised near a station (ww 07) and dust haze (ww 06), which allows an assessment of the severity

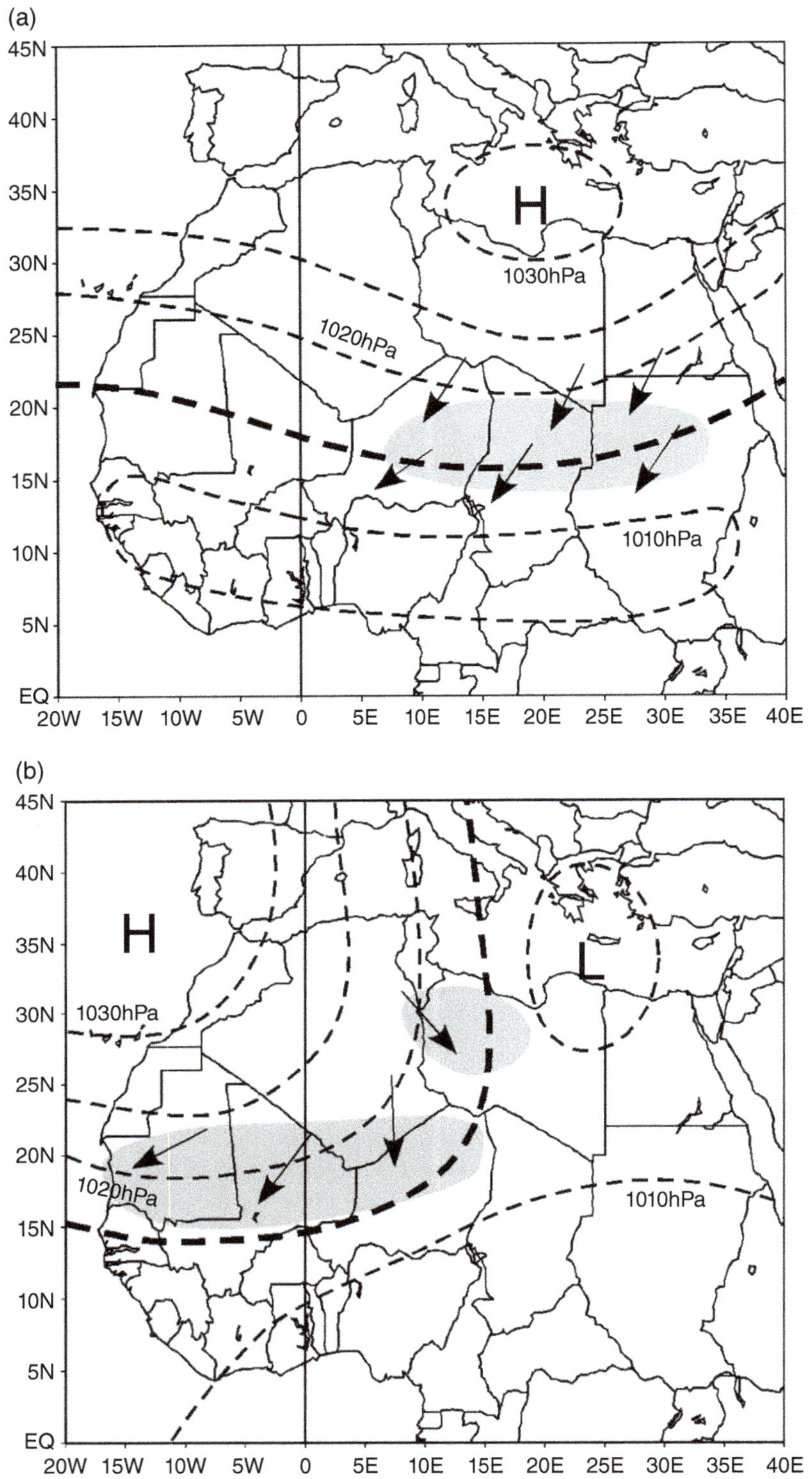


**Figure 5.18** Typical situations for the raising of dust during HDH periods. (a) Intensified Libyan High; (b) intensified/southeastward shifted Azores High accompanied by the passage of an extratropical trough across northern Africa. The dashed lines show typical isobars spaced every 5 hPa, with the 1015 hPa highlighted in bold. The arrows mark the regions of strongest 10 m winds indicating typical wind directions. The grey areas contain active dust sources. Situation (a) will typically precede an HDH episode in the eastern and central parts of West Africa, while situation (b) is more typical of HDH in the western parts.

of the dust event. The latter can be informed further by reports of low visibility (<5 km for haze down to less than 100 m for severe dust storms). As an additional source of information, the 10 m wind speed measured at the station should be considered with velocities above 30 kt indicative of significant dust events as mentioned earlier. Note that, owing to the sparse coverage in the Sahara, even significant dust storms might not show up in SYNOP observations.

- In addition, MSG IR composite imagery (see Section 5.2.2.2) should be used to identify significant dust plumes. The high temporal resolution allows an extrapolation of the movement of an identified plume.

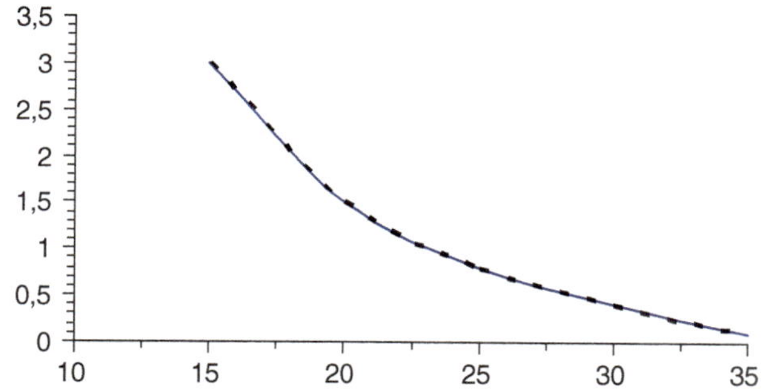


Figure 5.19 Climatological relationship between wind speed (horizontal axis in knots) and horizontal visibility (vertical axis in kilometres) as derived from many Sahelian station observations during different seasons.

Forecasters should be aware of the limitations of this product, including concealing of dust by clouds, qualitative nature of the images (no direct relation to forecast-relevant parameters like visibility or surface dust concentration) and day–night differences (see Section 5.2.2.3 for details).

- Be aware of a general tendency for dust emission to peak during the morning hours, as a large fraction of events during the Harmattan season are related to the breakdown of nocturnal LLJs.
- Forecasts of raising of dust for the following days should be based on monitoring the evolution of the 1015 hPa isobar in model-generated predictions (as in Figure 5.18). In addition, forecast fields of 10 m or 925 hPa winds should be considered. Note that both can have a strong diurnal cycle, such that looking at more than one time of day is advisable.

An important aspect for forecasters is to estimate the visibility in situations of active dust lifting. Figure 5.19 shows the climatological relationship for the Sahel as a first orientation. For a given situation, of course, the exact visibility will depend on the local characteristics of the soil and on more or less dust being advected from upstream.

#### 5.2.3.2.2 *Dust Transport and Plume Arrival*

The subsequent transport of the dust from its source (identified under Section 5.2.3.2.1) to the region of interest mainly relies on low-level wind information. A number of forecast techniques have been suggested:

- If available, use wind information from analysis and forecast charts for 950 and 925 hPa, where the bulk of the dust is typically transported. The predominant wind direction (typically northeasterly) should point from the identified active dust source to the region for which the forecast is made. The travel time can be estimated using typical wind speeds and the distance from the source (e.g. 10 m $s^{-1}$ corresponds to 864 km $day^{-1}$).
- If no model charts are available, 10 m winds from SYNOPs can be used, but be aware that these will tend to underestimate the progression of the plume. Where available, pilot balloon winds (PILOTS) may give a better representation of the speed of progression, for instance using the 600 m winds.
- In the main transport region, winds at 950 and 925 hPa should be on the order of 30 kt or more for significant HDH events. Slower winds will create longer travel times, and therefore a larger fraction of the original dust plume will be removed by gravitational settling and dry deposition. Because of this, the intensity of the HDH will be a combination of travel time and amount of the dust originally lifted.
- Plumes can also be tracked using the MSG IR composite described earlier (see Section 5.2.2.3). Dust also becomes increasingly difficult to detect in areas of high column water vapour (>25 mm), but this is a less relevant limitation during the dry Harmattan period.
- In addition, the forecaster is expected to watch out for low visibility and dust haze (ww 06) reports from upstream SYNOP stations (see Section 5.2.2.1) to make an informed decision for the forecast domain. Visibility reports from upstream are a useful basis for forecasting this parameter.
- The actual arrival of the dust will in most cases differ from day to night. After around midday the PBL is typically well mixed, and therefore dust concentrations will vary little from the surface up to 1–2 km. During the night and early morning hours, the formation of LLJs can transport dust in a vertically confined layer overhead, while horizontal visibility at the surface is still high. The breakdown of the jet in the morning will lead to an abrupt deterioration of visibility at the surface (see Figure 5.20). As the biggest particles tend to gravitate down from the dusty jet layer during the presence of a jet, a forecaster may be able to smell the arrival of the dust, and should warn for poor visibility in the morning.
- When the PBL grows deeper in the course of the first dust day, cleaner air from above gets entrained and visibility improves to some extent. If a dust episode persists for several days, dust particles will be mixed through a deeper layer, and therefore the afternoon improvement of visibility may be less pronounced.

#### 5.2.3.2.3 *Clearance*

Generally, there are two ways through which an HDH episode can come to an end.

1) *Forecast region stays in the northeasterlies, but dust emission ceases.*
   The weakening of the subtropical anticyclone is the first sign of the end of an HDH episode. This will be accompanied by a retreat of the 1015 hPa line to the northern Sahara or even farther. The reduction of the pressure gradient will weaken the winds at the surface

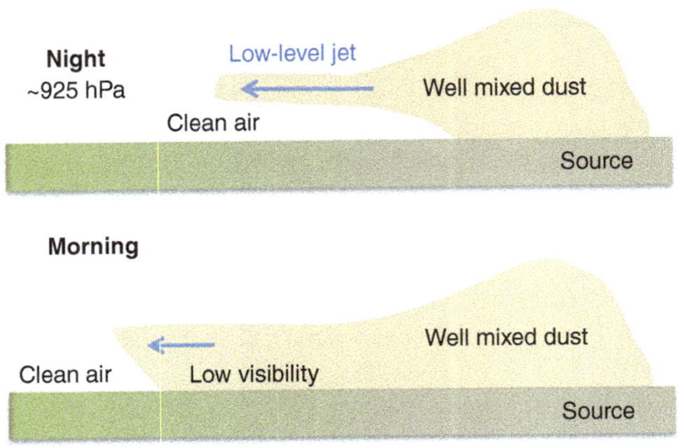


Figure 5.20 Schematic depiction of transport and mixing of dust associated with a nocturnal LLJ.

(typically below 15 kt) and, therefore, cut off the supply of dust into the atmosphere. At the same time, the southward transport will become weaker. If the forecast region will remain in northeasterly flow then the clearance will tend to be gradual as the HDH is only slowly replaced by clearer air from further north. Again, observations from stations upstream can be used to monitor the end of a dust-lifting period and increases in visibility.

2) *Forecast region comes under the influence of moister tropical air from the south.*
   A more abrupt clearance and improvement of visibility will occur if the HDH air is replaced by cleaner air from the south. To forecast such an event, the forecaster should look out for the presence of an eastward-propagating mid-latitude trough, which can extend equatorward to about 20°N in some cases. The lower pressure over the Sahara causes the intertropical discontinuity (ITD; see Section 1.1) to shift poleward. The area influenced by the southerly or southwesterly flow of cleaner, moister air from the Gulf of Guinea will expand across southern West Africa, leading to the clearance of dust in the atmosphere. This analysis should be based on pressure, wind and possibly moisture charts to monitor the progress of the ITD.

#### 5.2.3.3 Forecasting Convective Dust Storms

Convective dust storms are a feature of the Sahel region of West Africa (between 10 and 20°N). The dust storms mainly occur during the pre-monsoon season (April–June) within zone B of the ITD (see Figure 1.1). The number and intensity of cases decrease once the monsoon has set on. The storms are related to strong winds associated with cool outflow (cold pools) from deep convective storms. These can be more short-lived and localised, but also organised on the mesoscale into elongated squall lines (up to several hundred kilometres long; see Section 3.1.3.2). The leading edges of these cold pools are often very sharp and appear as a 'wall' of reddish-brown dust of up to several kilometres in height approaching at the horizon. The dry conditions over the Sahel in the pre-monsoon season associated with weak southwesterlies favour dust storms because of dry soils and strong evaporation at mid-levels. The latter can be so strong that no precipitation reaches the ground, leaving the dust suspended for many hours. If a rainstorm follows immediately after the initial dust storm, as is increasingly the case around the monsoon onset, the dust is washed out very quickly, leaving clear conditions behind.

A synoptic-scale factor for the occurrence of pre-monsoon dust storms can be interactions with upper-level troughs from the extratropics, which tend to shift the ITD northward (see Section 2.1.4). In these circumstances, the northward-propagating section of a large haboob, penetrating far into the Sahara, can create a long-lived, large dust plume that can subsequently create significant haze, when advected back into the Sahel for example.

Forecasting convective dust storms is a considerable challenge to the forecaster closely linked to the challenge of forecasting convection, as discussed in detail in Chapter 3. The following steps are recommended:

- Assess the state of the soil. If it has been dry for at least the last 2 days, the soil is typically dry enough to produce dust, provided sufficiently high wind speeds occur.
- Assess the potential for the development of deep convection (see Chapter 3 for details and definition of parameters). In particular, downdraft convective available potential energy (see Section 3.1.2.2) can be a useful indicator for the strength of associated downdrafts, and thus the potential intensity of the surface haboob.
- Assess the potential for convective organisation and downdraft formation. This typically includes an analysis of vertical wind shear and mid-level dryness (see Sections 3.1.2.2 and 11.6 for details). The intensity of the dust storm will be closely related to the intensity of the downdraft, and therefore the amount of evaporative cooling at mid and low levels. These factors will also allow an assessment of the likelihood of significant precipitation at the surface that could wash out dust after the initial storm.
- Owing to the difficulties in forecasting a storm before it has formed, nowcasting techniques are of great importance (see Chapter 6). Once a storm has formed and has been reported at SYNOP stations (see Section 5.2.2.1) upstream (i.e. typically to the east), a forecast can be made based on simple extrapolation of the storm track. Here, communications between stations are very crucial.

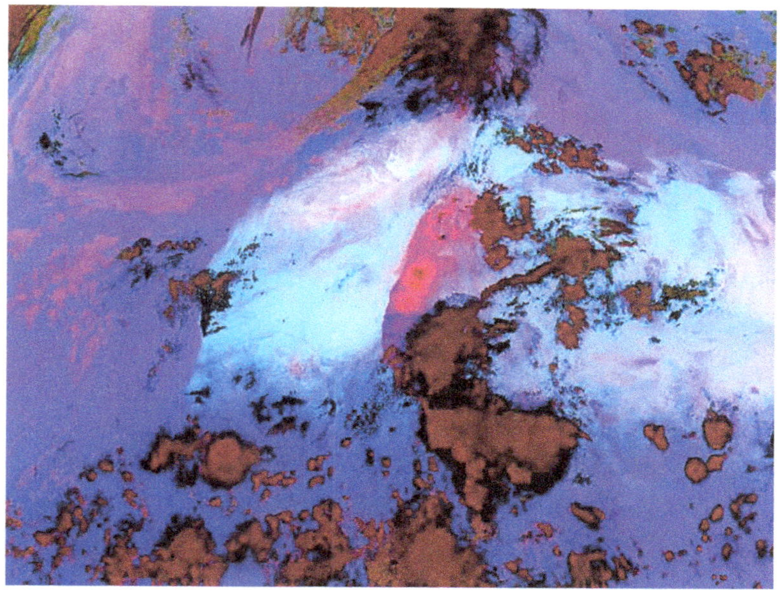

**Figure 5.21** Large convective dust storm covering parts of the Sahel and Sahara on 9 June 2010, 1700 UTC. The bright pink dust over the Sahara is from a previous convective development, while the darker purple dust is from the convective system to the east and south. Note that clouds conceal the southern extension of the cold pool into the Sahel, as is often the case.

- Meteosat satellite imagery (either standard IR or the IR dust product discussed in Section 5.2.2.3) is the best way to monitor the convection potentially leading to dust storms. The forecaster should be aware that a large fraction of dust storms over the Sahel will be concealed by the cirrus shield surrounding the convective core (see example in Figure 5.21), while the part of the cold pool stretching into the Sahara might be easier to see. After decay of the associated convective system, dust particles already suspended in the atmosphere can be tracked for some considerable distance downstream (see aged dust cold pool over the Sahara in Figure 5.21). Haze episodes associated with the advection of dust from haboobs can be forecast this way.

### 5.2.4 Case Study Examples Online

Two examples of typical HDH situations are presented in the online supporting material at http://www.cnrm-game-meteo.fr/waf_handbook_casestudies/, one with an intensified Libyan High, the other with an intensified Azores High. The 15–18 March 2012 and 4–7 February 2012 cases include MSLP and wind charts, satellite (MSG and MODIS) images, SYNOPs maps and dust model output.

## Appendix: Acronyms

| | |
|---|---|
| AAI | absorbing aerosol index |
| AERONET | AErosol RObotic NETwork |
| AMMA | African Monsoon Multidisciplinary Analysis |
| AOT | aerosol optical thickness |
| BoDEx | Bodélé Dust Experiment |
| ECMWF | European Centre for Medium-Range Weather Forecasts |
| EUMETSAT | European Organisation for the Exploitation of Meteorological Satellites |
| HDH | Harmattan dust haze |
| IR | infrared |
| ITCZ | intertropical convergence zone |
| ITD | intertropical discontinuity |
| LLJ | low-level jet |
| MISR | Multi-angle Imaging SpectroRadiometer |
| MODIS | Moderate Resolution Imaging Spectroradiometer |
| MSG | Meteosat Second Generation |
| MSLP | mean sea-level pressure |
| NCEP | National Centers for Environmental Prediction |
| OMI | Ozone Monitoring Instrument |
| PBL | planetary boundary layer |
| PSANAF | potential dust source areas in northern hemispheric Africa |
| SAMUM | Saharan Mineral Dust Experiment |
| SDS-WAS | Sand and Dust Storm Warning Advisory and Assessment System |
| SeaWiFS | Sea-viewing Wide Field-of-view Sensor |
| SEVIRI | Spinning Enhanced Visible and InfraRed Imager |
| SYNOP | surface synoptic observation |
| TOA | top of atmosphere |
| UV | ultraviolet |
| VIS | visible |

## Notes

1 http://aeronet.gsfc.nasa.gov/cgi-bin/type_piece_of_map_opera_v2_new?level=3.
2 http://sds-was.aemet.es/.

## References


Ackerman SA. 1997. Remote sensing aerosol using satellite infrared observations. *J. Geophys. Res.* **102**(D14): 17069–17079.

Ackerman SA, Chung H. 1992. Radiative effects of airborne dust on regional energy budgets at the top of the atmosphere. *J. Appl. Meteorol.* **31**: 223–233.

Alfaro S, Gomes L. 2001. Modeling mineral aerosol production by wind erosion: emission intensities and aerosol size distributions in source areas. *J. Geophys. Res.* **106**: 18075–18084.

Alpert P, Ziv B. 1989. The Sharav Cyclone: observations and some theoretical considerations. *J. Geophys. Res.* **94**: 18495–18514.

Ansmann A, Tesche M, Althausen D, *et al.* 2008. Influence of Saharan dust on cloud glaciation in southern Morocco during the Saharan Mineral Dust Experiment. *J. Geophys. Res.* **113**: D0421. doi: 10.1029/2007JD008785.

Ansmann A, Tesche M, Knippertz P, *et al.* 2009. Vertical profiling of convective dust plumes in southern Morocco during SAMUM. *Tellus B* **61**: 340–353. doi: 10.1111/j.1600-0889.2008.00384.x.

Bagnold RA. 1941. *The Physics of Blown Sand and Desert Dunes*. Methuen: New York; 265 pp.

Balkanski Y, Schulz M, Claquin T, Guibert S. 2007. Reevaluation of mineral aerosol radiative forcings suggests a better agreement with satellite and AERONET data. *Atmos. Chem. Phys.* **7**: 81–95.

Belnap J, Phillips SL, Miller ME. 2004. Response of desert biological soil crusts to alterations in precipitation frequency. *Oecologia* **141**: 306–316. doi: 10.1007/s00442-003-1438-6.

Bergametti G, Forêt G. 2014. Dust deposition. In: *Mineral Dust – A Key Player in the Earth System*, P Knippertz, J-B Stuut (eds). Springer Netherlands; pp. 179–200.

Bonnet S, Guieu C. 2004. Dissolution of atmospheric iron in seawater. *Geophys. Res. Lett.* **31**: L03303.

Bou Karam D, Flamant C, Knippertz P, *et al.* 2008. Dust emissions over the Sahel associated with the West African monsoon inter-tropical discontinuity region: a representative case study. *Q. J. R. Meteorol. Soc.* **134**: 621–634. doi: 10.1002/qj.439.

Brindley H, Knippertz P, Ryder C, Ashpole I. 2012. A critical evaluation of the ability of SEVIRI thermal IR RGB rendering to identify dust events. Part A: theoretical analysis. *J. Geophys. Res.* **117**: D07201. doi: 10.1029/2011JD017326.

Chiapello I, Prospero JM, Herman JR, Hsu NC. 1999. Detection of mineral dust over the North Atlantic Ocean and Africa with the Nimbus 7 TOMS. *J. Geophys. Res.* **104**(D8): 9277–9291.

Christopher SA, Jones T. 2007. Satellite-based assessment of cloud-free net radiative effect of dust aerosols over the Atlantic Ocean. *Geophys. Res. Lett.* **34**: L02810. doi: 10.1029/2006GL027783.

Coale KH, Johnson KS, Fitzwater SE, *et al.* 1996. A massive phytoplankton bloom induced by an ecosystem-scale iron fertilization experiment in the equatorial Pacific Ocean. *Nature* **383**: 495–501.

Costa MJ, Sohn BJ, Levizzani V, Silva M. 2006. Radiative forcing of Asian dust determined from the synergized GOME and GMS satellite data – a case study. *J. Meteorol. Soc. Jpn.* **84**: 85–95.

Cowie S, Knippertz P, Marsham JH. 2014. A climatology of dust emission events from northern Africa using long-term surface observations. *Atm. Chem. Phys.* **14**: 8579–8597. doi:10.5194/acp-14-8579-2014.

Cuesta J, Marsham JH, Parker DJ, Flamant C. 2009. Dynamical mechanisms controlling the vertical redistribution of dust and the thermodynamic structure of the West Saharan atmospheric boundary layer during summer. *Atmos. Sci. Lett.* **10**: 34–42. doi: 10.1002/asl.207.

Cuesta J, Lavaysse C, Flamant C, *et al.* 2010. Northward bursts of the West African monsoon leading to rainfall over the Hoggar Massif, Algeria. *Q. J. R. Meteorol. Soc.* **136**: 174–189. doi: 10.1002/qj.244.

El Fandy MG. 1940. The formation of depressions of the Khamsin type. *Q. J. R. Meteorol. Soc.* **66**: 323–336.

Emmel C, Knippertz P, Schulz O. 2010. Climatology of convective density currents in the southern foothills of the Atlas Mountains. *J. Geophys. Res.* **115**: D11115. doi: 10.1029/2009JD011819.

Engelstaedter S, Washington R. 2007. Atmospheric controls on the annual cycle of North African dust. *J. Geophys. Res.* **112**: D03103. doi: 10.1029/2006JD007195.

Fécan F, Marticorena B, Bergametti G. 1999. Parameterization of the increase of the aeolian erosion threshold wind friction due to soil moisture for semi arid areas. *Ann. Geophys.* **17**: 149–157.

Fiedler S, Schepanski K, Heinold B, *et al.* 2013. Climatology of nocturnal low-level jets over North Africa and implications for modeling mineral dust emission. *J. Geophys. Res. Atmos.* **118**: 6100–6121. doi: 10.1002/jgrd.50394.

Flamant C, Chaboureau J-P, Parker DJ, *et al.* 2007. Airborne observations of the impact of a convective system on the planetary boundary layer thermodynamics and aerosol distribution in the inter-tropical discontinuity region of the West African monsoon. *Q. J. R. Meteorol. Soc.* **133**: 1175–1189. doi: 10.1002/qj.97.

Formenti P, Schütz L, Balkanski Y, *et al.* 2011. Recent progress in understanding physical and chemical properties of African and Asian mineral dust. *Atmos. Chem. Phys.* **11**: 8231–8256. doi: 10.5194/acp-11-8231-2011.

Freeman M. 1952. *Duststorms of the Anglo-Egyptian Sudan*. Meteorological Reports Series, no. 11. Her Majesty's Stationery Office: London; 22 pp.

Gillette DA. 1977. Fine particulate emissions due to wind erosion. *Trans. Am. Soc. Agr. Eng.* **20**: 890–897.

Gillette DA. 1979. Environmental factors affecting dust emission by wind erosion in Saharan dust. In: *Saharan Dust*, C Morales (ed.). John Wiley & Sons, Ltd: Chichester; pp. 71–94.

Ginoux P, Prospero JM, Gill TE, *et al.* 2012. Global-scale attribution of anthropogenic and natural dust sources and their emission rates based on MODIS Deep Blue aerosol products. *Rev. Geophys.* **50**: RG3005. doi: 10.1029/2012RG000388.

Gomes L, Rajot JL, Alfaro SC, Gaudichet A. 2003. Validation of a dust production model from measurements performed in semi-arid agricultural areas of Spain and Niger. *Catena* **52**: 257–271.

Hansen JE, Sato M, Ruedy R. 1997. Radiative forcing and climate response. *J. Geophys. Res.* **102**: 6831–6864.

Heinold B, Knippertz P, Marsham JH, *et al.* 2013. The role of deep convection and low-level jets for dust emissions in West Africa. *J. Geophys. Res. Atmos.* **118**: 4385–4400. doi: 10.1002/jgrd.50402.

Herman JR, Bhartia PK, Torres O, *et al.* 1997. Global distribution of UV-absorbing aerosols form Nimbus7/ TOMS data. *J. Geophys. Res.* **102**: 16911–16922.

Holben BN, Eck TF, Slutsker I, *et al.* 1998. AERONET: a federal instrument network and data archive for aerosol characterization. *Remote Sens. Environ.* **66**: 1–16.

Hsu NC, Herman JR, Torres O, *et al.* 1999. Comparison of the TOMS aerosol index with Sun-photometer aerosol optical thickness: results and application. *J. Geophys. Res.* **104**: 6269–6280.

Hsu NC, Tsay SC, King MD, Herman JR. 2004. Aerosol properties over bright-reflecting source regions. *IEEE Trans. Geosci. Remote Sens.* **42**(3): 557–569.

Jemmet-Smith BC, Marsham JH, Knippertz P, Gilkeson CA. 2015. Quantifying global dust devil occurrence from meteorological analyses. *Geophys. Res. Lett.* **42**: 1275–1282. doi:10.1002/2015GL063078.

Jickells TD, An ZS, Andersen KK, *et al.* 2005. Global iron connections between desert dust, ocean biogeochemistry, and climate. *Science* **308**: 67–71.

Kalu AE. 1979. The African dust plume: its characteristics and propagation across West Africa in winter. In: *Saharan Dust*, C Morales (ed.). John Wiley & Sons, Ltd: Chichester; pp. 95–118.

Kardous M, Bergametti G, Marticorena B. 2005. Aerodynamic roughness length related to tillage ridges of the semi-arid regions. *Ann. Geophys.* **23**: 3187–3193.

Kaufman YJ, Koren I, Remer LA, *et al.* 2005. Dust transport and deposition observed from the Terra-Moderate Resolution Imaging Spectroradiometer (MODIS) spacecraft over the Atlantic Ocean. *J. Geophys. Res.* **110**: D10S12. doi: 10.1029/2003JD004436.

Klose M, Shao Y, Karremann MK, Fink AH. 2010. Sahel dust zone and synoptic background. *Geophys. Res. Lett.* **37**: L09802. doi: 10.1029/2010GL042816.

Knippertz P. 2008. Dust mobilization in the West African heat trough – the role of the diurnal cycle and of extratropical synoptic disturbances. *Meteorol. Z.* **17**(5): 553–563. doi: 10.1127/0941-2948/2008/0315.

Knippertz P. 2014. Meteorological aspects of dust storms. In: *Mineral Dust – A Key Player in the Earth System*, P Knippertz, J-B Stuut (eds). Springer Netherlands;, pp. 121–148.

Knippertz P, Fink AH. 2006. Synoptic and dynamic aspects of an extreme springtime Saharan dust outbreak. *Q. J. R. Meteorol. Soc.* **132**: 1153–1177. doi: 10.1256/qj.05.109.

Knippertz P, Todd MC. 2010. The central west Saharan dust hotspot and its relation to African easterly waves and extratropical disturbances. *J. Geophys. Res.* **115**: D12117. doi: 10.1029/2009JD012819.

Knippertz P, Todd MC. 2012. Mineral dust aerosols over the Sahara: processes of emission and transport, and implications for modeling. *Rev. Geophys.* **50**: RG1007. doi: 10.1029/2011RG000362.

Knippertz P, Ansmann A, Althausen D, *et al.* 2009. Dust mobilization and transport in the northern Sahara during SAMUM 2006 – a meteorological overview. *Tellus B* **61**: 12–31. doi: 10.1111/j.1600-0889.2008.00380.x.

Knippertz P, Tesche M, Heinold B, *et al.* 2011. Dust mobilization and aerosol transport from West Africa to Cape Verde – a meteorological overview of SAMUM-2. *Tellus B*, **63**: 430–447. doi: 10.1111/j.1600-0889.2011.00544.x.

Koch J, Renno NO. 2005. The role of convective plumes and vortices on the global aerosol budget. *Geophys. Res. Lett.* **32**: L18806. doi: 10.1029/2005GL023420.

Koren I, Kaufman Y, Washington R, *et al.* 2006. The Bodélé Depression: a single spot in the Sahara that provides most of the mineral dust to the Amazon forest. *Environ. Res. Lett.* **1**. doi: 10.1088/1748-9326/1/1/014005.

Laurent B. 2005. Simulation des émissions d'aérosols désertiques à l'échelle continentale: analyse climatologique des émissions du nord-est de l'Asie et du nord de l'Afrique. Thesis, University Paris 12, Créteil; 225 pp.

Laurent B, Marticorena B, Bergametti G, *et al.* 2008a. Modeling mineral dust emissions from the Sahara Desert using new surface and soil developments. *J. Geophys. Res.* **113**: D14218. doi: 10.1029/2007JD009484.

Laurent B, Heinold B, Tegen I, *et al.* 2008b. Surface wind accuracy for modeling mineral dust emissions: comparing two regional models in a Bodélé case study. *Geophys. Res. Lett.* **35**: L09804. doi: 10.1029/2008GL033654.

Mahowald NM. 2007. Anthropocene changes in desert area: sensitivity to climate model predictions. *Geophys. Res. Lett.* **34**: L18817. doi: 10.1029/2007GL030472.

Mahowald N, Kohfeld K, Hansson M, *et al.* 1999. Dust sources and deposition during the last glacial maximum and current climate: a comparison of model results with paleodata from ice cores and marine sediments. *J. Geophys. Res.* **104**: 15895–15916.

Mahowald NM, Ballantine JA, Feddema J, Ramankutty N. 2007. Global trends in visibility: implications for dust sources. *Atmos. Chem. Phys.* **7**: 3309–3339.

Marsham JH, Parker DJ, Grams CM, *et al.* 2008a. Uplift of Saharan dust south of the inter-tropical discontinuity. *J. Geophys. Res. Atmos.* **113**: D21102. doi: 10.1029/2008JD009844.

Marsham JH, Parker DJ, Grams CM, *et al.* 2008b. Observations of mesoscale and boundary-layer scale circulations affecting dust transport and uplift over the Sahara. *Atmos. Chem. Phys.* **8**: 6979–6993.

Marticorena B. 2014. Dust production mechanisms. In: *Mineral Dust – A Key Player in the Earth System*, P Knippertz, J-B Stuut (eds). Springer Netherlands; pp. 93–120.

Marticorena B, Bergametti G. 1995. Modeling the atmospheric dust cycle: 1 – design of a soil derived dust production scheme. *J. Geophys. Res.* **100**: 16415–16430.

Marticorena B, Bergametti G, Aumont B, *et al.* 1997. Modeling the atmospheric dust cycle: 2 – simulation of Saharan sources. *J. Geophys. Res.* **102**: 4387–4404.

Middleton NJ, Goudie AS. 2001. Saharan dust: sources and trajectories. *Trans. Inst. Brit. Geogr.* **26**: 165–181.

Miller RL, Tegen I, Perlwitz J. 2004. Surface radiative forcing by soil dust aerosols and the hydrologic cycle. *J. Geophys. Res.* **109**: D04203.

Möhler O, Field PR, Connolly P, *et al.* 2006. Efficiency of the deposition mode ice nucleation on mineral dust particles. *Atmos. Chem. Phys.* **6**: 3007–3021.

Moore JK, Doney SC, Lindsay K, *et al.* 2006. Nitrogen fixation amplifies the ocean biogeochemical response to decadal timescale variations in mineral dust deposition. *Tellus B* **58**: 560–572.

Morman SA, Plumlee GS. 2014. Dust and human health. In: *Mineral Dust – A Key Player in the Earth System*, P Knippertz, J-B Stuut (eds). Springer Netherlands; pp. 385–409.

N'Tchayi Mbourou G, Bertrand JJ, Nicholson SE. 1997. The diurnal and seasonal cycles of wind-borne dust over Africa north of the equator. *J. Appl. Meteorol.* **36**: 868–882.

Parker DJ, Burton RR, Diongue-Niang A, *et al.* 2005. The diurnal cycle of the West African monsoon circulation. *Q. J. R. Meteorol. Soc.* **131**: 2839–2860. doi: 10.1256/qj.04.52.

Pauley PM, Baker NL, Barker EH. 1996. An observational study of the 'Interstate 5' dust storm case. *Bull. Am. Meteorol. Soc.* **77**: 693–720.

Pedgley DE. 1972. Desert depression over northeast Africa. *Meteorol. Mag.* **101**: 228–244.

Pérez C, Nickovic S, Pejanovic G, *et al.* 2006. Interactive dust-radiation modeling: a step to improve weather forecasts. *J. Geophys. Res.* **111**: D16206.

Prigent C, Tegen I, Aires F, *et al.* 2005. Estimation of the aerodynamic roughness length in arid and semi-arid regions over the globe with the ERS scatterometer. *J. Geophys. Res.* **110**: D09205. doi: 10.1029/2004JD005370.

Raasch S, Franke T. 2011. Structure and formation of dust devil-like vortices in the atmospheric boundary layer: a high-resolution numerical study. *J. Geophys. Res.* **116**: D16120. doi: 10.1029/2011JD016010.

Ryan JA. 1972. Relation of dust devil frequency and diameter to atmospheric temperature. *J. Geophys. Res.* **77**: 7133–7137.

Schepanski K, Tegen I, Laurent B, *et al.* 2007. A new Saharan dust source activation frequency map derived from MSG-SEVIRI IR-channels. *Geophys. Res. Lett.* **34**: L18803. doi: 10.1029/2007GL030168.

Schepanski K, Tegen I, Todd MC, *et al.* 2009a. Meteorological processes forcing Saharan dust emission inferred from MSG-SEVIRI observations of subdaily dust source activation and numerical models. *J. Geophys. Res.* **114**: D10201. doi: 10.1029/2008JD010325.

Schepanski K, Tegen I, Macke A. 2009b. Saharan dust transport and deposition towards the tropical northern Atlantic. *Atmos. Chem. Phys.* **9**: 1173–1189.

Schepanski K, Tegen I, Macke A. 2012. Comparison of satellite based observations of Saharan dust source areas. *Remote Sens. Environ.* **123**: 90–97. doi: 10.1016/j.rse.2012.03.019.

Shao Y. 2004. Simplification of a dust emission scheme and comparison with data. *J. Geophys. Res.* **109**: D10202. doi: 10.1029/2003JD004372.

Shao Y, Fink AH, Klose M. 2010. Numerical simulation of a continental-scale Saharan dust event. *J. Geophys. Res.* **115**: D13205. doi: 10.1029/2009JD012678.

Sinclair PC. 1969. General characteristics of dust devils. *J. Appl. Meteorol.* **8**: 32–45.

Sokolik IN. 2002. The spectral radiative signature of wind-blown mineral dust: implications from remote sensing in the thermal IR region. *Geophys. Res. Lett.* **29**(24): 2154. doi: 10.1029/2002GL015910.

Solomon S, Qin D, Manning M, *et al.* (eds). 2007. *Climate Change 2007: The Physical Science Basis*. Cambridge University Press; 989 pp.

Sow M, Alfaro SC, Rajot JL, Marticorena B. 2009. Size resolved dust emission fluxes measured in Niger during 3 dust storms of the AMMA experiment. *Atmos. Chem. Phys.* **9**: 3881–3891.

Stensrud DJ. 1996. Importance of low-level jets to climate: a review. *J. Climate* **9**: 1698–1711. doi: 10.1175/1520-0442(1996)009<1698:IOLLJT>2.0.CO;2.

Stuut J-B, Zabel M, Ratmeyer V, *et al.* 2005. Provenance of present-day aeolian dust collected off NW Africa. *J. Geophys. Res.* **110**: D04202. doi: 10.1029/2004JD005161.

Sutton LJ. 1925. Haboobs. *Q. J. R. Meteorol. Soc.* **51**: 25–30.

Tegen I, Harrison SP, Kohfeld K, *et al.* 2002. Impact of vegetation and preferential source areas on global dust aerosol: results from a model study. *J. Geophys. Res.* **107**(D21): 4576. doi: 10.1029/2001JD000963.

Tegen I, Heinold B, Todd MC, *et al.* 2006. Modelling soil dust aerosol in the Bodélé Depression during the BoDEx campaign. *Atmos. Chem. Phys.* **6**: 4345–4359.

Thorncroft CD, Flocas HA. 1997. A case study of Saharan cyclogenesis. *Mon. Weather Rev.* **125**: 1147–1165.

Torres O, Tanskanen A, Veihelmann B, *et al.* 2007. Aerosol and surface UV products from Ozone Monitoring Instrument observations: an overview. *J. Geophys. Res.* **112**(D24): D24S47. doi: 10.1029/2007JD008809.

Wald AE, Kaufman YJ, Tanré D, Gao BC. 1998. Daytime and nighttime detection of mineral dust over desert using infrared spectral contrast. *J. Geophys. Res.* **103**(D24): 32307–32313.

Warren A, Chappell A, Todd MC, *et al.* 2007. Dust-raising in the dustiest place on earth. *Geomorphology* **92**(1–2): 25–37. doi: 10.1016/j.geomorph.2007.02.007.

Washington R, Todd MC. 2005. Atmospheric controls on mineral dust emission from the Bodélé Depression, Chad: the role of the low level jet. *Geophys. Res. Lett.* **32**: L17701. doi: 10.1029/2005GL023597.

Washington R, Todd MC, Engelstaedter S, *et al.* 2006. Dust and the low-level circulation over the Bodélé Depression, Chad: observations from BoDEx 2005. *J. Geophys. Res.* **111**: D03201. doi: 10.1029/2005JD006502.

Williams E, Nathou N, Hicks E, *et al.* 2008. The electrification of dust-lofting gust fronts ('haboobs') in the Sahel. *Atmos. Res.* **91**(2–4): 292–298. doi: 10.1016/j.atmosres.2008.05.017.

Williams ER. 2008. Comment on 'Atmospheric controls on the annual cycle of North African dust' by S. Engelstaedter and R. Washington. *J. Geophys. Res.* **113**: D23109. doi: 10.1029/2008JD009930.

Williams G. 1964. Some aspects of the aeolian saltation load. *Sedimentology* **3**: 253–256.

Wurzler S, Reisin TG, Levin Z. 2000. Modification of mineral dust particles by cloud processing and subsequent effects on drop size distributions. *J. Geophys. Res.* **105**: 4501–4512.

Yoshioka M, Mahowald NM, Conley AJ, *et al.* 2007. Impact of desert dust radiative forcing on Sahel precipitation: relative importance of dust compared to sea surface temperature variations, vegetation changes, and greenhouse gas warming. *J. Climate* **20**: 1445–1467.

Zhang L, Gong S, Padro J, Barrie L. 2001. A size-segregated particle dry deposition scheme for an atmospheric aerosol module. *Atmos. Environ.* **35**: 549–560.

# 6

# Nowcasting

**Authors:** Rita D. Roberts[a] and James W. Wilson[a]

**Contributors:** Henk Verschuur[b], Paul Kucera[a], David Johnson[a], Mariane Diop-Kane[c], Douglas J. Parker[d], James Galvin[e], Nancy Rehak[a], Earle Williams[f], Mary O. Iso[g] and Michael McGraw-Herdeg[f]

**Reviewers:** Charles Yorke, (GMet), Jim Galvin (UK Met Office), Nicolas Chapelon (Météo-France), Jean Philippe Lafore (CNRM-UMR3589, Météo-France and CNRS, Toulouse, France)

[a] National Center for Atmospheric Research, Boulder, CO, USA
[b] EUMETSAT, Eumetsat Allee 1, D-64295, Darmstadt, Germany
[c] Agence Nationale de l'Aviation Civile et de la Météorologie du Sénégal (ANACIM), Dakar, Sénégal
[d] School of Earth and Environment, University of Leeds, Leeds, UK
[e] Met Office, Exeter, UK
[f] Lincoln Laboratory, Massachusetts Institute of Technology (MIT/LL), Lexington, MA, USA
[g] Nigerian Meteorological Agency (NIMET), Ikeja, Lagos, Nigeria

## 6.1 Scientific Background

### 6.1.1 Introduction

Nowcasting, as defined by the World Meteorological Organization (WMO) Working Group on Nowcasting Research, is 'forecasting with local detail, by any method, over a period from the present to a few hours ahead'; this includes a detailed description of the present weather. The term nowcasting is used to emphasise that the forecasts are time and place specific for a period from 0–6 h. In the first published book on nowcasting, Browning (1982) described the great need for timely, location-specific predictions of high-impact weather that are of particular importance to commercial and general aviation for planning and routing air traffic, to the public for outdoor sporting events, and to the construction industry, power utilities and ground transportation organisations that conduct much of their work outdoors. Current numerical weather prediction (NWP) models lack the accuracy necessary for predicting high-impact weather on the time and spatial scales required, as discussed in Chapters 3 and 10. Thus, the role of the forecaster is crucial in providing the public with detailed information regarding local weather features and the expected evolution of the weather over the next few hours (Mass, 2003). In fact, it is expected that the role of the forecaster will undergo a significant transformation during the coming decades from spending most of their time on longer-term forecasts, to nowcasting on the 0–6 h time period (Mass, 2012). It is also expected that rapid communications of the nowcast to users, such as terminal aerodrome forecasts (TAFs) to pilots and controllers, and high-impact weather to emergency managers and the public, will be greatly improved by the use of the internet and social media (Mass, 2012).

Nowcasting high-impact weather can be conducted in a variety of ways, depending on the concentration of observations and computing resources available to the forecasters. The actual process of producing nowcasts is similar and intimately related to the processes involved in producing and amending TAFs, area forecast discussions, and mesoscale area weather updates throughout the day. Continual monitoring of the current weather and understanding how it evolves is critical for assessing the potential for nowcasting high-impact weather. Anticipating high-impact weather relies on forecasters' experience and knowledge of the climatology of their area, the dominant weather regimes, and their conceptual models of the weather processes and triggers that lead to high-impact weather.

Two particularly noteworthy nowcasting successes from the USA are the warning systems for tornadoes and for microbursts, each having saved many lives and reduced property damage. In the case of tornadoes. when severe weather is threatening, the forecaster continually monitors the radar looking for severe storm signatures, like rotation couplets, bow echoes and flare echoes. The forecaster also has automated tools looking for severe storm features, but they depend mostly on

*Meteorology of Tropical West Africa: The Forecasters' Handbook*, First Edition. Edited by Douglas J. Parker and Mariane Diop-Kane.

their own experiences. Microburst warnings at airports are completely automated, based primarily on radar features proposed in Roberts and Wilson (1989).

Historically, very short-term forecasting (or nowcasting) techniques have been based on hand-drawn extrapolations of current observations into the future, using the past motion of the weather feature. Automated, computer-based extrapolation techniques are now commonly in use in forecast offices around the world, and form the backbone to more complex nowcast systems that have been developed during the past two decades. These nowcast systems combine extrapolation of observations with NWP forecasts (blended systems) for 0–6 h nowcasts, or use heuristic techniques to combine observations with forecaster rules and conceptual models for 0–2 h nowcasts of severe weather. This is done to overcome the limitations in using extrapolation techniques or NWP forecasts alone. The steps of the nowcast process are described in Section 6.1.2.

The quality and usefulness of the nowcast products are dependent firstly on the available observations and secondly on the maintenance and data quality from the observation facilities, the communication networks and the training of the nowcasters. Appendix B provides guidance, in the form of a check list, from the WMO Working Group on Nowcasting Research to countries interested in developing a nowcasting capability. The guidance was developed during a panel discussion at the WMO Nowcasting Symposium in Vancouver, Canada, in 2010.

This nowcasting chapter focuses primarily on the nowcasting of convective storms and associated hazards. Nowcasting can also apply to other phenomena such as fog and dust haze; these are dealt with in Chapters 4 and 5 respectively. The chapter presents general methods and tools used for convective storm nowcasting in the international forecasting community. Some of the techniques and examples are taken from regions outside of West Africa, but the principles and procedures for nowcasting remain the same regardless of geographical location. In particular, it should be noted that a number of illustrations in this chapter utilise radar data. At the present time, ground-based meteorological radars are not very common in West Africa; however, most West African countries have plans to acquire such radars and operational forecasters aspire to use them. One of the most useful tools worldwide for nowcasting thunderstorms is radar detection and monitoring of boundary-layer convergence lines. This chapter will present data from non-operational radars in West Africa that clearly show that boundary-layer convergence lines are common and they play a very important role in determining where and when new convective storms will occur and how they may evolve. Satellite images and products are becoming very good, and increasingly easily accessible to the African community. While they generally cannot provide the detailed high-resolution data of radars, satellite imagery can provide excellent real-time data that can be used to nowcast convective storms, much of which is itemised in Chapter 9. It is noteworthy that the examples shown in this chapter are illustrative examples; in practice, forecasts were issued and will be issued based on all available observational data and NWP products, alongside the experience of the forecaster, which is a key element in the forecasting chain. This chapter will also give insights to basic research to develop statistical forecasts much needed for nowcasters.

The chapter is organised into two main sections (Sections 6.1 and 6.2). Section 6.1 discusses the nowcast process and describes the required activities for the on-duty nowcaster. This discussion also provides a background to the science behind the steps in the nowcasting process. Section 6.2 provides guidelines and examples for operational nowcasting and amending TAFs.

### 6.1.2 Nowcast Process

The nowcast process as described here begins when the forecaster first assesses the synoptic and mesoscale weather pattern, and continues to evolve, ideally, with continuous monitoring of the local high-resolution observations. The nowcasting steps discussed below include (a) becoming familiar with the synoptic-scale features that will affect the local area during the day, (b) considering the climatology and conceptual models for storm evolution that are most applicable for the expected local synoptic and mesoscale weather pattern, (c) conducting a stability analysis for the day, (d) determining the likely storm type for the day, (e) extrapolating existing precipitation fields and (f) nowcasting new storm initiation, growth, and decay of existing extrapolated storms.

Fundamental to the nowcast process is frequent examination of surface station data (temperature, moisture, pressure, visibility, rain gauge measurements) to identify and monitor approaching weather such as gust fronts, sea breezes and precipitation systems, or changing local conditions such as a localised high associated with an approaching convective system. Satellite visible and infrared (IR) imagery, lightning and radar data, when available, are particularly useful to track the movement of large-scale convergence lines or mesoscale convective systems (MCSs), including gust fronts and squall lines approaching the area and the timing for potentially severe weather.

#### 6.1.2.1 Convective Weather Outlook

The first step in the nowcast process is for the forecaster to examine synoptic weather maps, NWP model analysis and forecast fields, and all available real-time observations to assess the potential for high-impact weather and to provide a convective outlook for the day.

Current NWP model forecasts of convective storms and precipitation on the 0–6 h time frame lack the accuracy needed by end users in the prediction of site-specific locations of convective-scale precipitation, particularly for events without strong synoptic forcing during the warm season months (Olson *et al.*, 1995; Wilson and Roberts, 2006). The ability of NWP models to accurately predict severe weather on the small temporal (0–2 h) and spatial (~1 km horizontal) scales needed is extremely limited. Median timing errors of 2.5 h have been observed with high-resolution mesoscale model forecasts (Fowle and Roebber, 2003), and spatial offsets of as much as 250 km were observed 44% of the time with rapid update cycle 3-h operational forecasts (Wilson and Roberts, 2006), which is quite problematic for providing accurate, high-resolution nowcasts of weather that might impact approach and departure gates at major airports. These systematic NWP forecast errors are due in part to the lack of robust operational observations necessary to provide a complete representation of the atmospheric environment when models are initialised (Benjamin *et al.*, 2004; Stensrud *et al.*, 2009), insufficient model physics on the storm scale, and to the use of convective and land surface parameterisation schemes within NWP models. Furthermore, owing to limited computing resources, many NWP models update only twice per day, and are unable to predict rapidly changing weather or weather resulting from secondary convection.

Increasingly, NWP models run in the research community at higher horizontal grid resolutions are showing improvement in representing mesoscale storm precipitation structure and intensities (Done *et al.*, 2004), but as the resolution increases, objective verification scores are degraded due to spatial and timing errors in the forecasts (Mass *et al.*, 2002). It is anticipated that the assimilation of radar, satellite, profiler and other operational datasets into NWP models (Xue *et al.*, 2003; Sun, 2005; Benjamin *et al.*, 2009; Weygandt *et al.*, 2009), explicit specification of convection (Done *et al.*, 2004; Weisman *et al.*, 2008) and use of ensemble runs of NWP models to produce probabilistic forecasts will help improve the accuracy of quantitative precipitation forecasting (Stensrud *et al.*, 1999; Roebber *et al.*, 2004) and can be used to increase air traffic capacity in the future (Steiner *et al.*, 2010). Sun *et al.* (2014) have shown that the assimilation of radar data into advanced numerical models does significantly improve the forecasts particularly in the first 2–3 h. They list the primary challenges in using NWP for nowcasting applications as (a) the theoretical limit of predicting small-scale convective storm features, (b) the need for improved mesoscale observation networks and (c) the need for improvement of rapid-update NWP and data assimilation systems. They also discuss improved physical understanding of storm-scale processes and correspondingly improved NWP model physics.

Because of the lack of NWP accuracy in predicting convective weather on the nowcasting time and space scale, non-numerical methods are required, which use, primarily, high-resolution observations.

#### 6.1.2.2 Conceptual Models and Climatology

Essential to the entire nowcasting process is the fact that the forecasters must know their area well. This is achieved through daily forecast experience, and facilitated further by knowledge of the climatological frequency and location of severe weather, as compiled from statistical processing of data from their regional observational network. Examples of the types of climatological statistics useful for nowcasting thunderstorms and strong boundary-layer winds are shown in Figures 6.1 and 6.2 respectively. (See also various climatological examples in Chapters 1, 3 and 4). Figure 6.3 shows the monthly climatologies for frequency of lightning occurrence in Africa. Developing a detailed, diurnal climatology *for every grid point* within the forecast domain (as shown in Figure 6.4 using lightning data for Florida) is particularly beneficial for forecasters tasked with providing frequently updated, location-specific nowcasts.

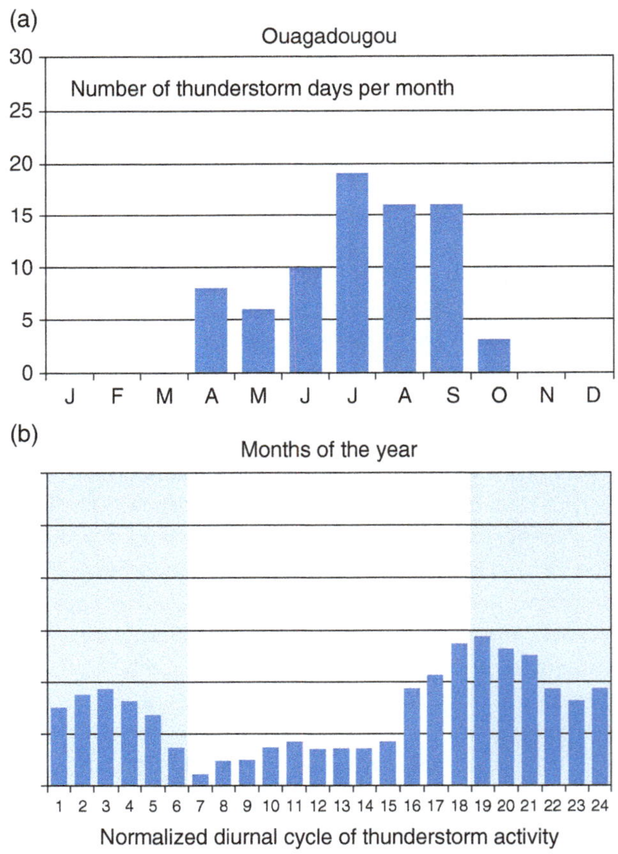


**Figure 6.1** A climatology of all thunderstorms occurring in the vicinity of the Ouagadougou, Burkina Faso airport during 2012, showing the frequency of thunderstorms (a) monthly and (b) diurnally (Johnson *et al.*, 2014). *Source*: Reproduced with permission of Dave Johnson.

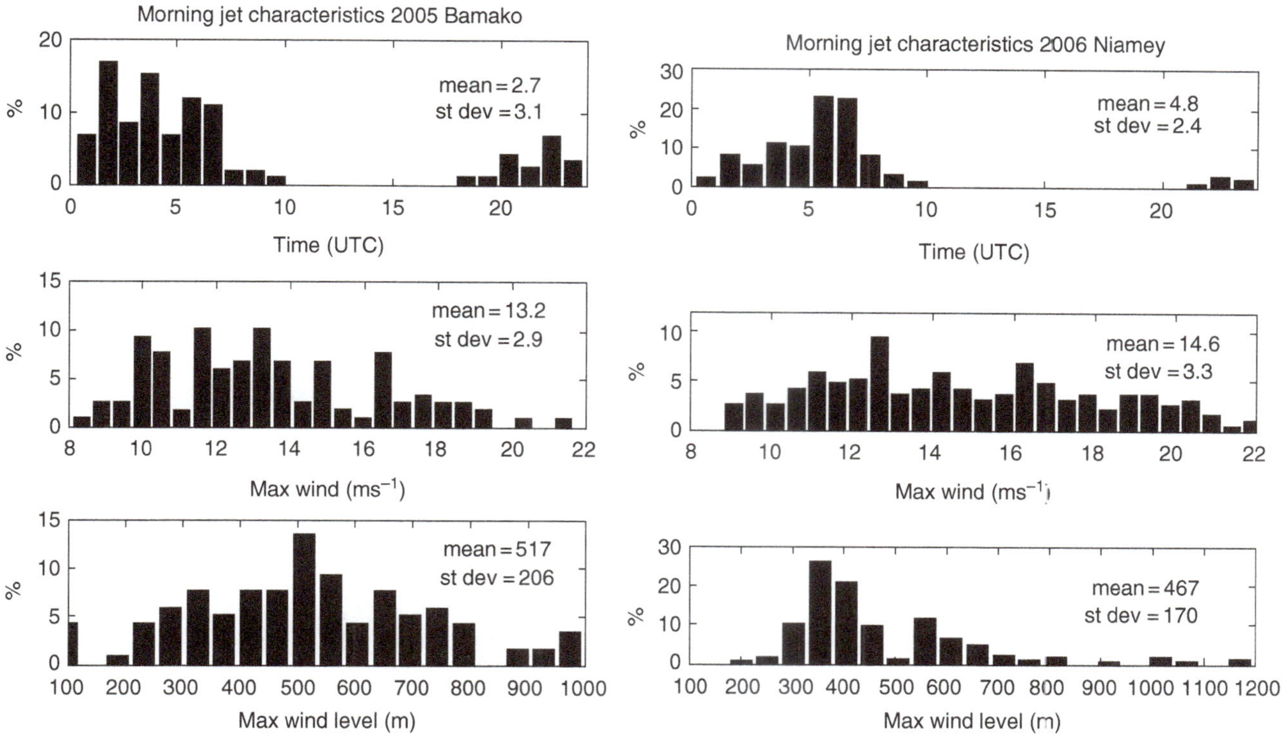


**Figure 6.2** Characteristics of the nocturnal low-level jets in Bamako and Niamey. *Source*: Madougou (2010). See also Figure 4.11.

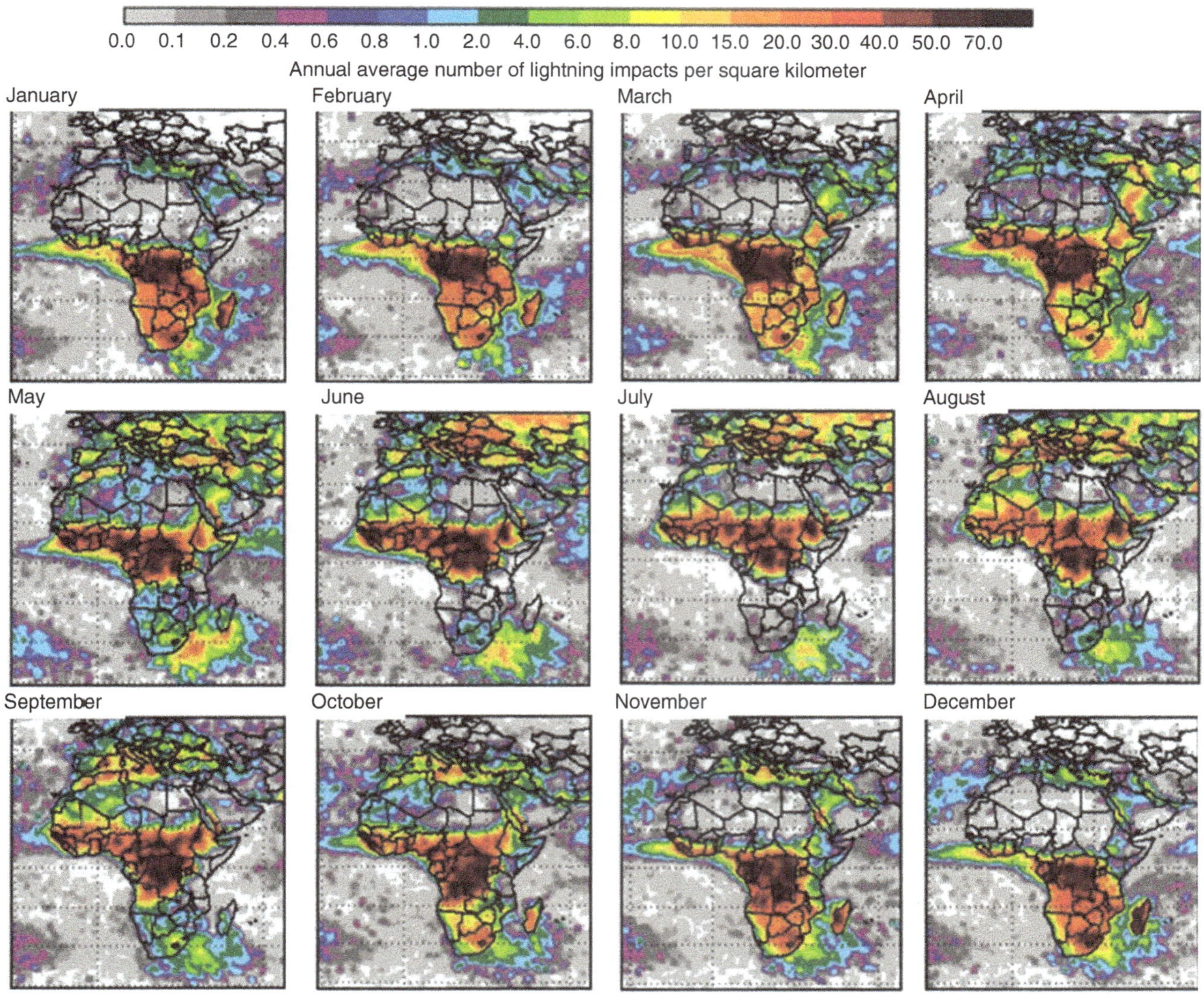


**Figure 6.3** Monthly lightning climatologies for Africa produced from a decade of total lightning flash rates collected on NASA's polar-orbiting satellites. *Source*: Johnson *et al.* (2014). Reproduced with permission of Dave Johnson.

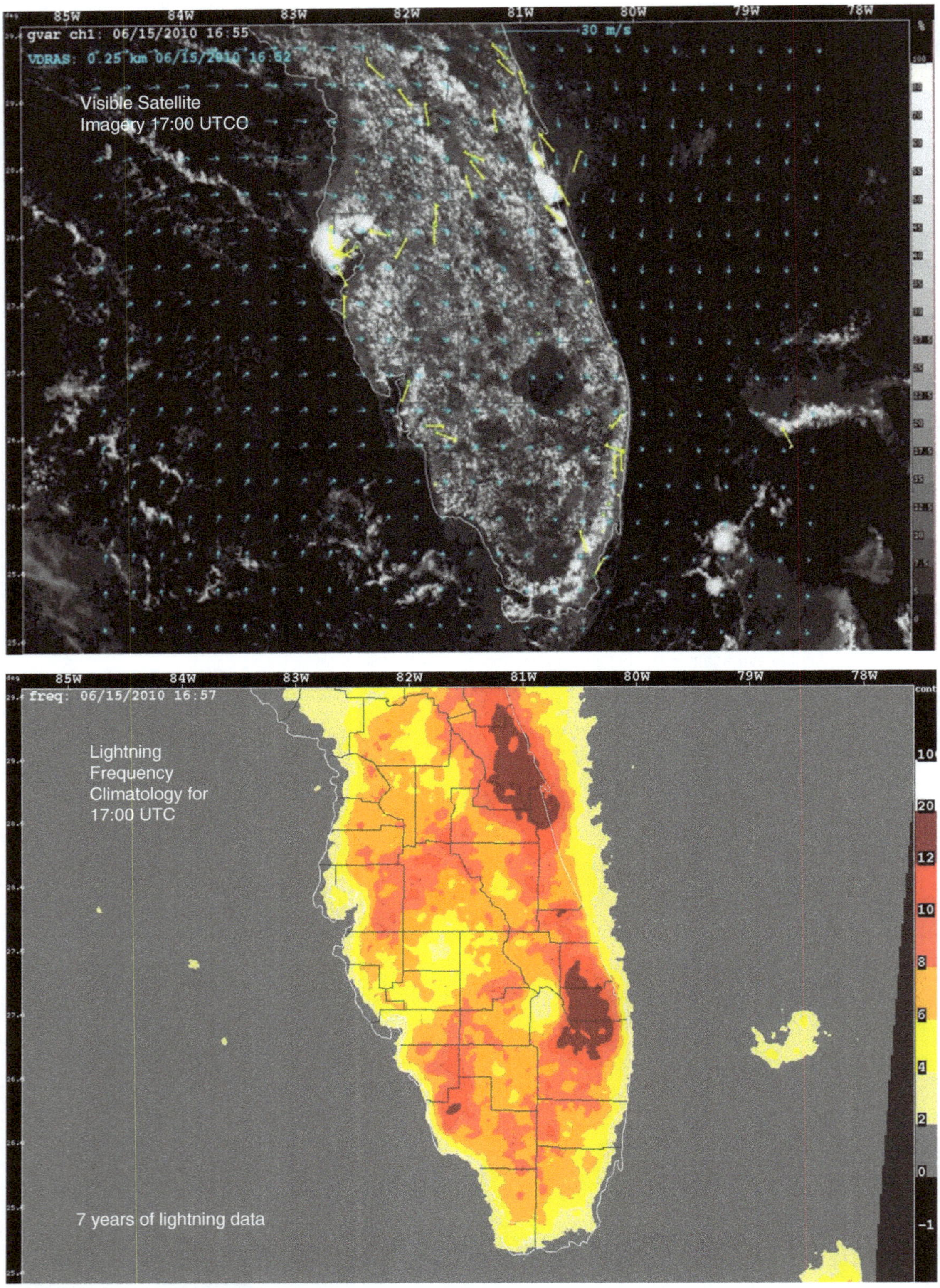


**Figure 6.4** High-resolution, gridded climatology used in nowcasting convective storms for the Florida peninsula in the USA. Top panel: satellite visible imagery of convective storm development at 1700 UTC (13:00 LT). Bottom panel: 7 years' climatology of lightning frequency on a 1 km resolution grid, valid for the same time period indicating locations of highest frequency of thunderstorms.

The forecaster uses the climatology information in conjunction with patterns observed in the different observational datasets to develop a conceptualised view of what is currently happening in the atmosphere and what is anticipated for the next 0–6 h. The forecaster skills needed specifically for nowcasting include local knowledge of the nuances of their own weather patterns of importance, good subjective mesoanalysis skills for rapid identification of the prevailing weather regime of the day, ability to anticipate the changes in atmospheric stability, ability to allow for limitations in NWP model performance when using model fields as guidance, and assessment of the quality of real-time observations. For example, the local forecaster, on seeing that the present satellite image was closely matching the lightning climatology in Figure 6.4 for the same time of day, and knowing the synoptic patterns were similar, would then be inclined to use the climatology for the remainder of the day as a major guide for nowcasting storm evolution for the next few hours.

Many of the conceptual models used or developed by forecasters are designed to represent specific attributes and forcing on the synoptic or mesoscale. Figure 6.5 is an example of a conceptual model developed specifically for anticipating and nowcasting daily storm initiation and evolution over the Amazon River basin in Brazil (Lima and Wilson, 2008). Studies showed that the first storms initiated, typically, near noon from solar heating along the slopes of small hills (~300 m high) or along boundary-layer convective rolls. Some of these storms would precipitate (Figure 6.5a) and generate cold pools with circular gust fronts along the leading edge of the cold pool. New storms would then initiate along the gust front that was downwind of the storm steering level flow (Figure 6.5b). Later in the afternoon, gust fronts would collide, causing the largest storms to form (Figure 6.5c). Nowcasters need to develop similar conceptual models for their region; examples are shown in Figures 4.17 and 4.20, relating to local initiation of storms by sea breezes and by vegetation patterns respectively.

#### 6.1.2.3 Stability Analyses

NWP forecasts of environmental stability or large-scale dynamic forcing are very important in the nowcast process, as they help the forecaster to anticipate the weather that may occur later in the day. Recent techniques for assessing the instability of the atmosphere using both NWP forecasts of environmental stability combined with satellite-derived stability indices provided by the Meteosat Second Generation (MSG) global instability index (GII) product (see Section 9.1.5.2.2) are showing promise for providing 6–9 h lead times for the convective potential of the atmosphere and regions with potential for severe convection (Koenig and de Coning, 2009; de Coning *et al.*, 2011). The satellite-retrieval method for deriving the GII products is currently calculated during pre-convective conditions; that is, only over cloud-free regions of the atmosphere. With the high-frequency, 15 min MSG update cycle, changes in atmospheric stability can be monitored continuously; a distinct advantage over using NWP model forecasts alone (Koenig and de Coning, 2009). Examples of the GII products available to forecasters from the European Organisation for the Exploitation of Meteorological Satellites (EUMETSAT) are shown in Figure 6.6. Figure 6.7 shows a $K$-index plot (see Section 3.2.3.1.2) for South Africa for the pre-convective environment on 26 October 2006, where regions of strong convective potential are highlighted in yellow. Approximate 2.5 h later, Figure 6.8, thunderstorms occur in one of these regions where the $K$ index values approached 40°C.

The South African Weather Service combines a regional version of the GII called the regional instability indices (RII) with a local mesoscale model, and factors for orographic lift, to produce a probability for convection in the form of a new combined instability index (CII) (de Coning *et al.*, 2011). Again, with the high update rate of the MSG imagery, this product can be displayed operationally every 15 min for forecaster viewing, providing a great benefit over radiosonde soundings that may only be available once or twice a day and covers only a limited area. An example of this product is shown in Figure 6.9.

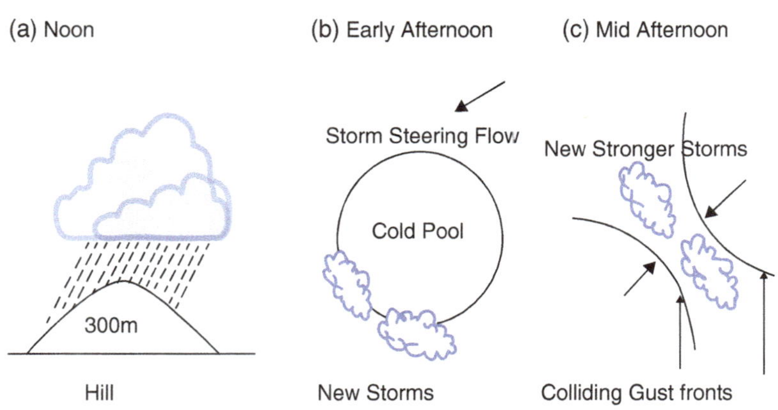


**Figure 6.5** Conceptual model of storm initiation and evolution over the Amazon River basin. See text for explanation. *Source*: Lima and Wilson (2008).

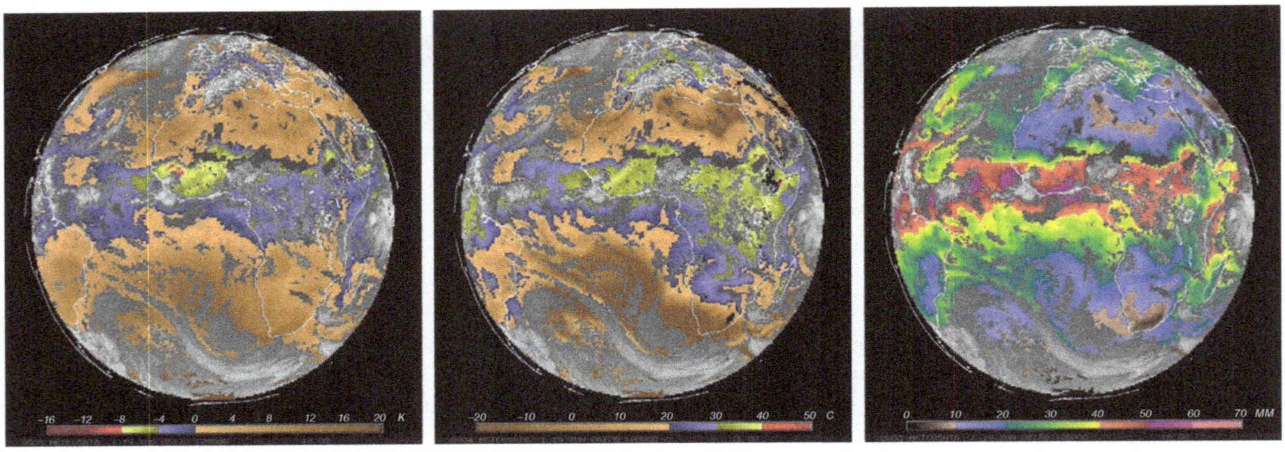


**Figure 6.6** Examples of the EUMETSAT GII products derived from the MSG satellite for Africa on 19 June 2006. Indices shown in the panels are lifted index (left), *K* index (centre) and total precipitable water (right). *Source*: Koenig and de Coning (2009). © American Meteorological Society. Used with permission.

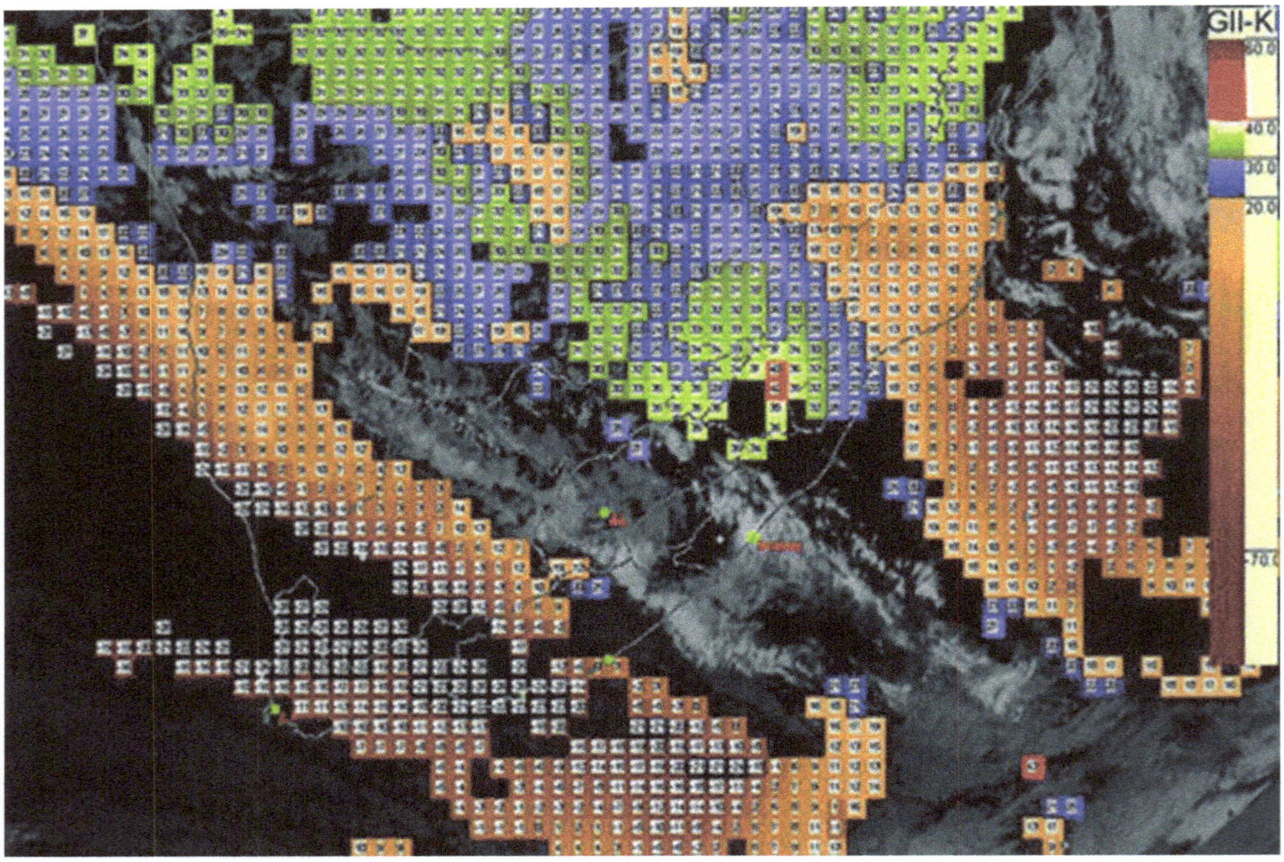


**Figure 6.7** The *K* index (values of instability are shown from blue to yellow to red) over South Africa, at 0800 UTC 26 October 2006. *Source*: Koenig and de Coning (2009). © American Meteorological Society. Used with permission.

A next step in assessing the potential for *locally triggered* convective weather over the next 0–6 h is to examine the most recent radiosonde tephigram plot for the computed convective available potential energy (CAPE), convective inhibition (CIN), lifted index (LI) and other stability indices (see Chapter 3 for definitions). Figure 6.10 shows the CAPE and CIN areas derived from lifting a surface parcel of air. The methodology for determination of CAPE and CIN is described in most basic radiosonde sounding manuals and online tutorials (e.g. see the COMET module 'Tephigram mastery' link in Appendix D). In brief, the procedure is to lift the surface parcel of

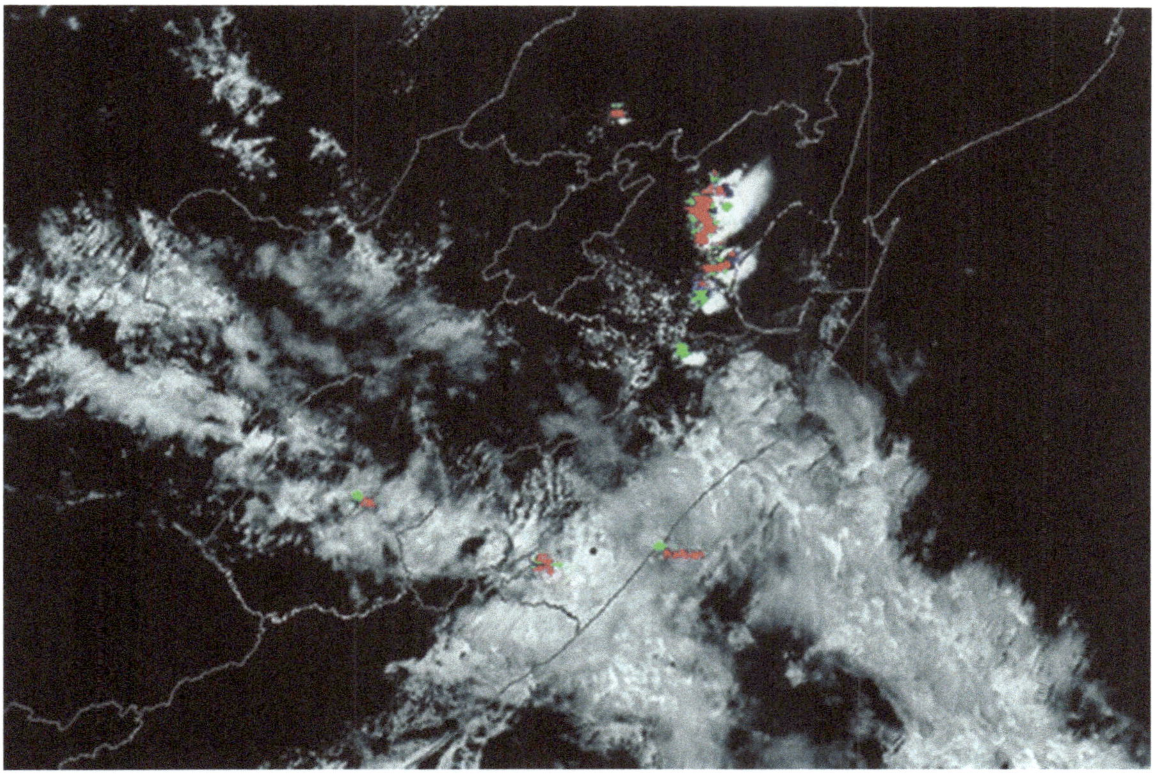

Figure 6.8 MSG high-resolution visible channel at 1030 UTC 26 October 2006. A strong storm has developed over the area of the previously high (near 40°C) *K* values. Lightning reports are overlaid. The red symbols are all strokes reported in the last 5 min, the blue strokes are those reports 5–10 min ago, and the green strokes are those reported between 10 and 15 min ago. *Source*: Koenig and de Coning (2009) © American Meteorological Society. Used with permission.

air, dry adiabatically to its condensation level (lifted condensation level) and then moist adiabatically to the tropopause. The height at which the lifted parcel becomes warmer than the environmental temperature is called the level of free convection (LFC), as indicated in Figure 6.10. The values typically used for the lifted parcel are the lowest 50 mb average of the dry bulb temperature and mixing ratio. The area on the tephigram below the LFC, where the environmental temperature is colder than the lifted temperature, represents the amount of energy needed to lift the air parcel vertically to the LFC; this is the called the CIN. Correspondingly, the area on the tephigram above the LFC, where the environmental temperature is warmer than the lifted parcel temperature, is the positive buoyant energy and is called the CAPE.

Often a low-level temperature inversion is present in morning soundings, indicating that some negatively buoyant air (CIN; Figure 6.10) will need to be overcome before convection will ensue. In the absence of any synoptic or mesoscale forcing in the area, locally driven convection will arise from surface heating as the surface temperature approaches the convective temperature $T_c$. Figure 6.11 illustrates how $T_c$ is derived from the sounding. In this case it has been assumed that the surface moisture will remain constant; thus, the near-surface mixing ratio value has been lifted to the level where it intersects the temperature profile: the convective condensation level (CCL). The convective temperature is then retrieved by following the dry adiabat to the surface. This is an important step in the nowcast process: to anticipate the timing and start of convection by monitoring surface station temperatures as they approach $T_c$.

Convective temperature will often be reached first along the southern- or eastern-facing slopes of elevated terrain. In the Brazilian Amazon region, solar heating of the shallow (<300 m) terrain is sufficient for convective temperature to be reached prior to 1200 local time, with convective clouds and precipitation showers ensuing by noon, as shown in the conceptual model in Figure 6.5. The same pattern seems to be seen over the Jos Plateau in Nigeria, with the first convective cells appearing over the southern slopes, probably also enhanced by the moist monsoonal inflow from that direction.

As the surface temperature and dew point temperature changes during the day, so will the atmospheric stability. The forecaster can anticipate changes in stability by lifting surface parcels using the current surface station mesonet measurements and noting changes in areas of CAPE and CIN compared with those from the morning. In addition, the forecaster should have an expectation of how and when temperatures in the upper atmosphere might change, based on examination of upper air plots

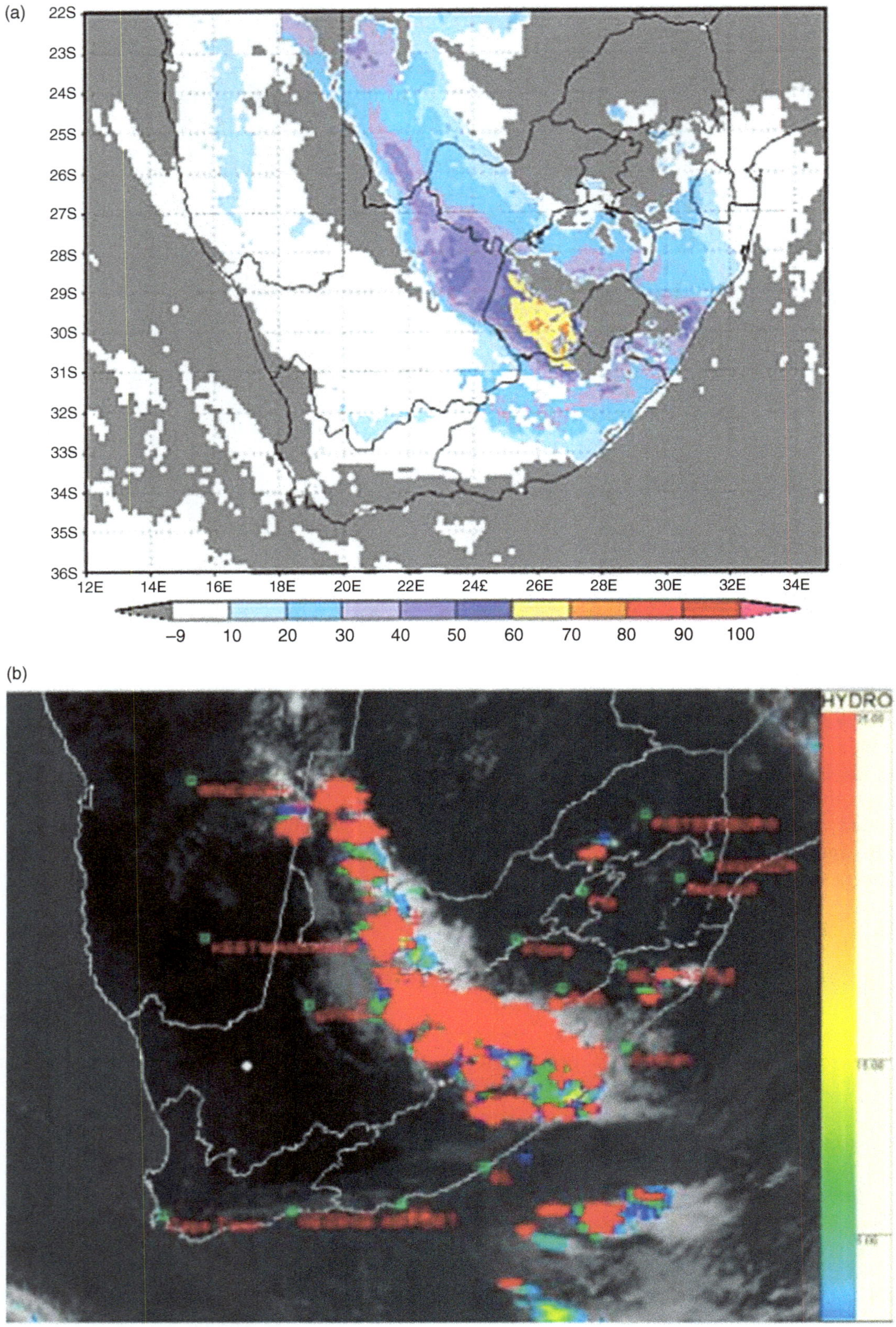


**Figure 6.9** Three-hourly time average of CII between 0600 and 0900 UTC (%) over South Africa (a) and lightning occurrence (indicated in red, green and blue crosses) as well as Hydroestimator (mm) at 1500 UTC (b) for 31 January 2010. Grey shading indicated cloud cover. *Source*: de Coning *et al.* (2011). © American Meteorological Society. Used with permission.

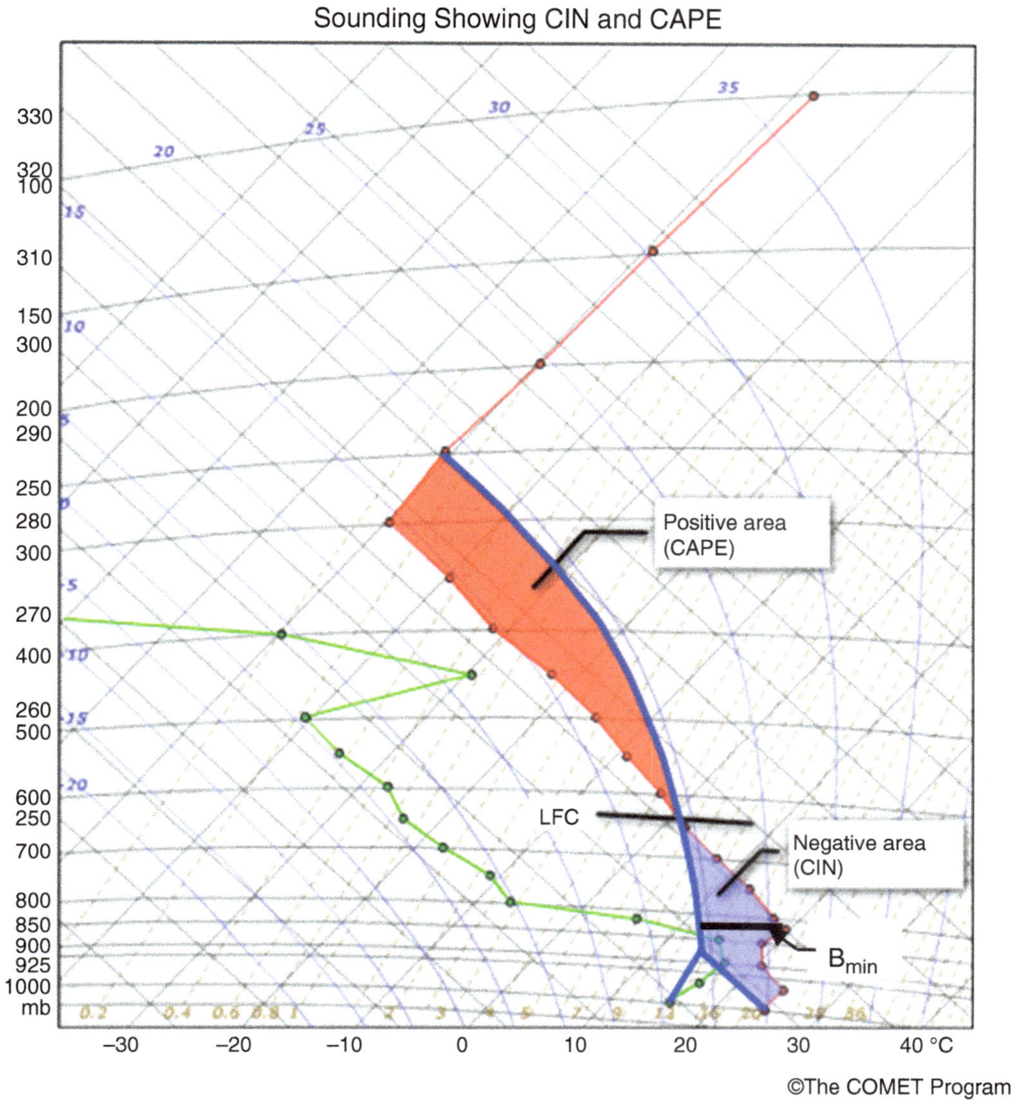


Figure 6.10 Example tephigram sounding for a given day and time, showing the negative and positively buoyant areas of the atmosphere for a parcel being lifted from the surface. *Source*: reproduced with permission of COMET.

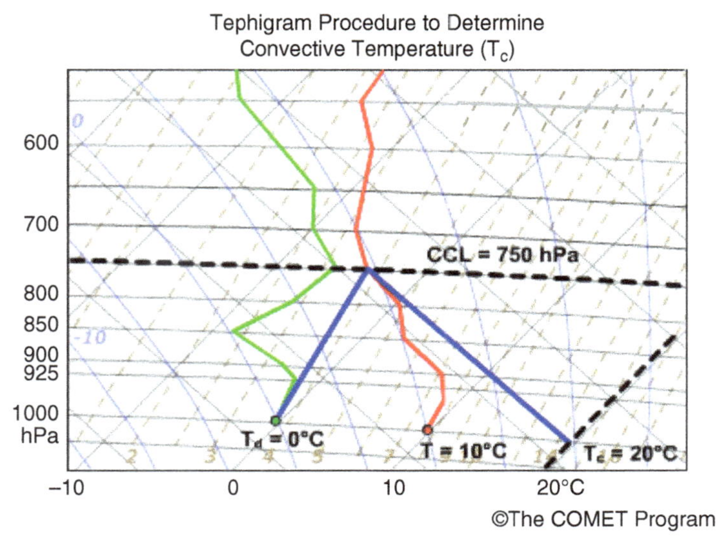


Figure 6.11 Process for computing convective temperature. *Source*: reproduced with permission of COMET.

and NWP forecasts for indications of warm or cold air advection, $\theta_w$ advection, or positive vorticity advection. Satellite images of visible, infrared and moisture fields are particularly useful in tracking the approach of different air masses. Some sounding analysis packages will automatically provide forecast thermodynamic profiles, but the best method to obtain the timing for when local convection will begin, which requires no computing resources, is to update the sounding profile manually by plotting the current surface temperature and dew point temperature values while keeping the rest of the sounding unchanged, and then recompute CAPE and CIN values.

Numerical simulations have shown that convection initiation is particularly sensitive to small changes in temperature and moisture in the boundary layer, and that these small variations are typically observed. These variations have a number of causes: they occur as the boundary layer changes during the diurnal cycle; they happen seemingly randomly, as a result of small-scale and mesoscale circulations; and they may be a coherent result of differing land-surface state. Section 4.1.4.4 has outlined various ways in which patterns of vegetation, land-use and soil moisture can have quite substantial effects on the boundary-layer temperature and humidity, on spatial scales of 10 km or larger. While a moisture drop-off of 1 g kg$^{-1}$ can make a difference in the strength of existing convection, a small temperature drop-off of 1°C in the boundary layer can make the difference between convection occurring and no convection at all (Crook, 1996). This is because the negatively buoyant (CIN) portion of the atmosphere is much more sensitive to temperature variations than to moisture variations. Thus, it is important to monitor the surface temperature change as the day progresses to determine if and when the CIN will be overcome and convection will occur. Once convection ensues, CAPE provides useful information on the potential strength and intensity of the convection, and many studies have stressed the importance of identifying significant CAPE thresholds for severe convection. However, widespread regions of CAPE can exist in the atmosphere and provide little indication on when and where convection will occur (Dai *et al.*, 1999; Tuttle and Davis, 2006).

A new methodology, based on parcel theory, for detecting changes in maximum negative buoyancy has been developed with direct application to nowcasting convection initiation (Trier *et al.*, 2014a). This method involves the calculation of a term $B_{min} = (T_{vp} - T_{ve})$, which is the minimum buoyancy (in temperature units) of a vertically displaced air parcel, where $T_{vp}$ and $T_{ve}$ are the parcel and the environmental virtual temperatures respectively. $B_{min}$ can be determined graphically from atmospheric soundings (see Figure 6.10) or calculated from NWP model-derived soundings to produce an NWP model-based, gridded $B_{min}$ field. Laing *et al.* (2012), in their simulations study of deep convection and MCS systems that form over the Ethiopian Highlands and propagate westward over the eastern and central Sahel region of northern Africa, calculated $B_{min}$ values for long-lived versus short-lived convective cases (Figure 6.12). They found that the 75th percentile value of the mid-afternoon $B_{min}$ from five long-lived cases was approximately −0.5 K, which supports local convection initiation, while the 75$^{th}$ percentile value from the seven short-lived cases was approximately −1.8 K, which presents a much more significant barrier to convection initiation. Additional simulation studies have recently been conducted using $B_{min}$ calculations to help quantify the differences in thermodynamic destabilisation in surface-based versus elevated-based MCSs (Trier *et al.*, 2014b). Similar studies and statistics should be compiled for West Africa.

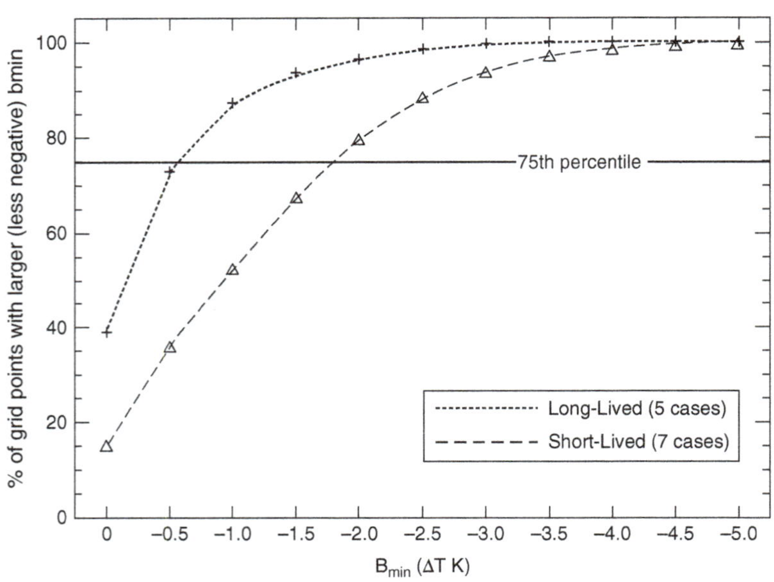


**Figure 6.12** Cumulative distribution functions for minimum negative buoyancy for the most unstable parcel over a 400 x 400 km$^2$ grid region for the long- and short-lived simulated cases at 1200 UTC (Source: Laing *et al.*, 2012; © American Meteorological Society. Used with permission.)

#### 6.1.2.4 Storm Type

The forecaster must determine whether there is sufficient CAPE to support the potential for convective weather. If the values are significantly large for their area, the forecaster must decide if there is potential for severe weather and what type of severe weather might occur. The vertical shear in the wind profile provides guidance on the type of storms that may occur (see Weisman and Klemp (1986) and also the discussion in Section 3.1.3 in this book). Vertical wind shear takes the form of unidirectional speed shear, directional shear, or may have attributes of speed and directional shear, which tend to support single-cell, multicell or a mix of multicell and supercell-type storms respectively. Based on studies from North American storms (Chisholm and Renick, 1972), Figure 6.13 is an example of what typical shear profiles may look like for West Africa and the type of storm structure that would likely result.

In addition to vertical wind shear, the buoyant energy or CAPE of the atmospheric environment also contributes to storm type and severity. The bulk Richardson number (BRN), which is an indication of the potential severity of storms, is the ratio of stability to the shear in the atmosphere and is given by

$$\mathrm{BRN} = \frac{\mathrm{CAPE}}{0.5\left(U_{6\mathrm{km}} - U_{500\mathrm{m}}\right)^2} \tag{6.1}$$

where $U_{6\mathrm{km}}$ is the mean wind speed in the lowest 6000 m and $U_{500\mathrm{m}}$ is the mean wind speed in the lowest 500 m. Figure 6.14 shows the BRN values computed from soundings and the observed storms, for a specific region of the USA. BRN values >45 in the USA typically support single-cell and multicell storms, whereas values <45 support supercell storms. Threshold values will be different for different areas in the world, and it is important to develop a plot like this using the CAPE and shear threshold values for severe weather for specific regions.

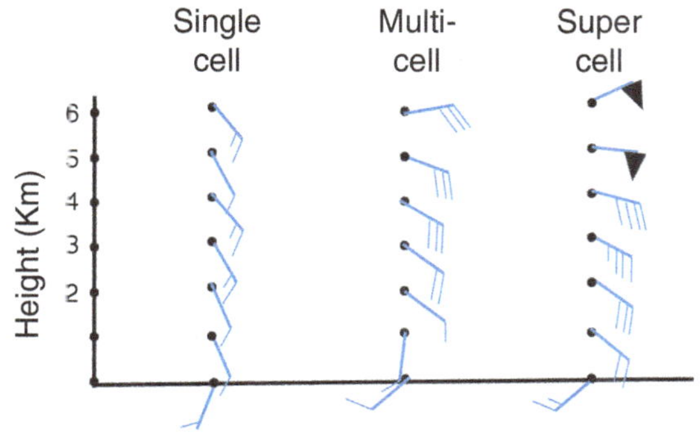


Figure 6.13 Proposed sample vertical wind shear profiles for single cell, multicell and supercell storms for West Africa. The multicell storm is characterised by much stronger rotational shear of the winds. These are educated guesses based on North American studies, and the reader should note that there is little or no evidence of supercell storms occurring in West Africa. A half barb is 5 kt, a full barb is 10 kt and a flag is 50 kt.

#### 6.1.2.5 Extrapolation of Existing Storms

Nowcasting the future location of convective storms and severe weather has, historically, been based on extrapolation of current radar, satellite or lightning observations into the future. Radar and satellites have been particularly well suited for use in short-term nowcasts as they provide large spatial coverage of the weather at high spatial and temporal resolutions. If one wishes to have a general idea of where storms might be located 1–2 h in the future, extrapolation of observations will provide a reasonable estimation of anticipated location. Figures 6.15 and 6.16 are examples of automated storm detection and extrapolation techniques applied to satellite and radar data respectively. Both of the tools shown in these figures have the ability to track single-cell and longer-lived, more organised, multicellular storms (e.g. MCSs and squall lines).

However, it is important to note that an implicit assumption of most extrapolation techniques is that

Figure 6.14 Types of observed storms for different BRN values derived from NWP model-retrieved soundings. *Source:* reproduced with permission of COMET and American Meteorological Society.

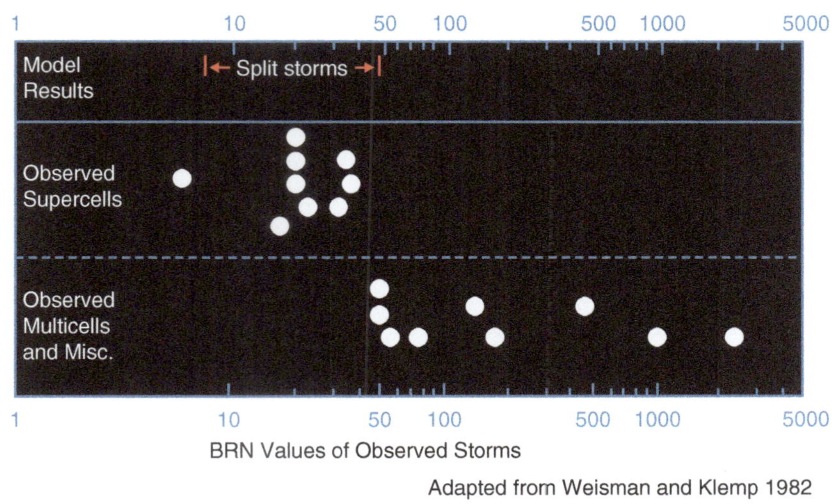

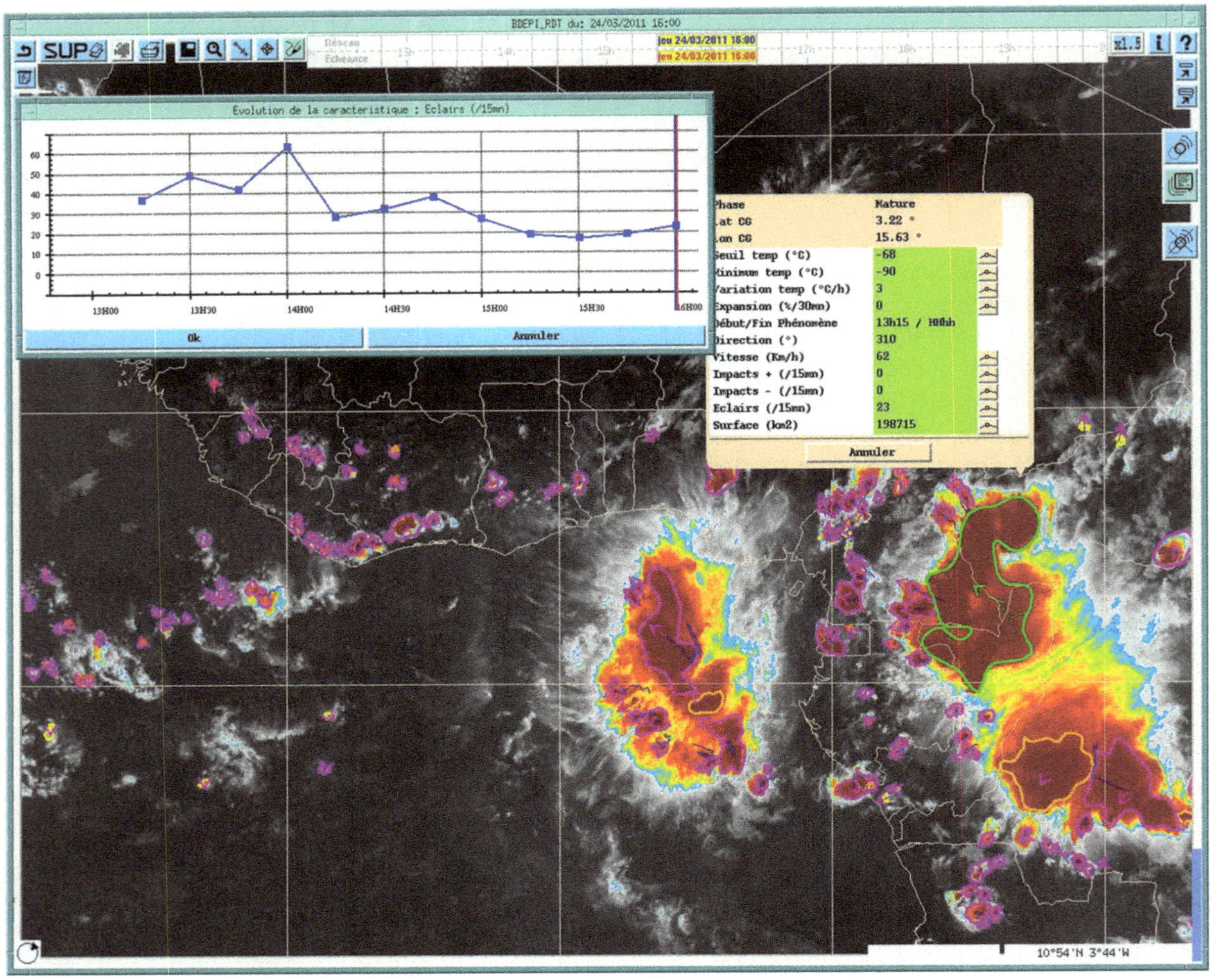


**Figure 6.15** Example of the RDT product for West Africa for tracking storms observed in satellite data. Coloured polygons and vectors overlaid on satellite imagery are automated storm detection and motion products. Specific information on storm and lightning characteristics and storm motion for the storm bounded by a green polygon are shown in the pop-up window. *Source*: Météo-France web site, 2014 (data from 2011).

storms do not change in size or intensity but maintain a steady state. Studies have long shown that the predictability of storms and convective storm motion is closely associated with storm size, evolution and duration (Kessler and Russo, 1963; Wilson, 1966). Multicellular storm complexes such as squall lines, MCSs and supercells are generally more organised, and are composed of storms that merge or split during their lifetimes and have much longer durations of >1 h (Figure 6.17). Because of the likelihood that these larger scale complexes will persist for longer than 1 h, short-term predictability is possible by extrapolation of these storm positions at least 1–2 h into the future. The largest systems, like MCSs, without concern for precipitation intensity, can have useful extrapolations, out to 6 h or more. In contrast, small-scale, single-cell storms are much more perishable in nature with mean durations of <25 min (Foote and Mohr, 1979; Henry, 1993), and thus generally are only predictable for a few tens of minutes using simple extrapolation techniques, as these techniques are unable to predict the dissipation of storms and tend to keep storms 'alive' well beyond their lifetimes.

Tsonis and Austin (1981) investigated the use of trends in echo size and intensity to improve nowcasts of convective cells that had already lived at least 30 min, but they found negligible improvement of skill even in elaborate nonlinear time-trending schemes. Wilson *et al.* (1998) reconfirmed this finding with an additional data set. This led to the conclusion by both sets of investigators that essential physical processes that dictate the change in rainfall with time are not necessarily observable in the past history of a particular radar echo development. In the case of convective storms, these physical processes are often events occurring in the boundary layer, such as convergence (Garstang and Cooper, 1981) or spatial variability of atmospheric instability. Figure 6.18 is an example of storm extrapolation where

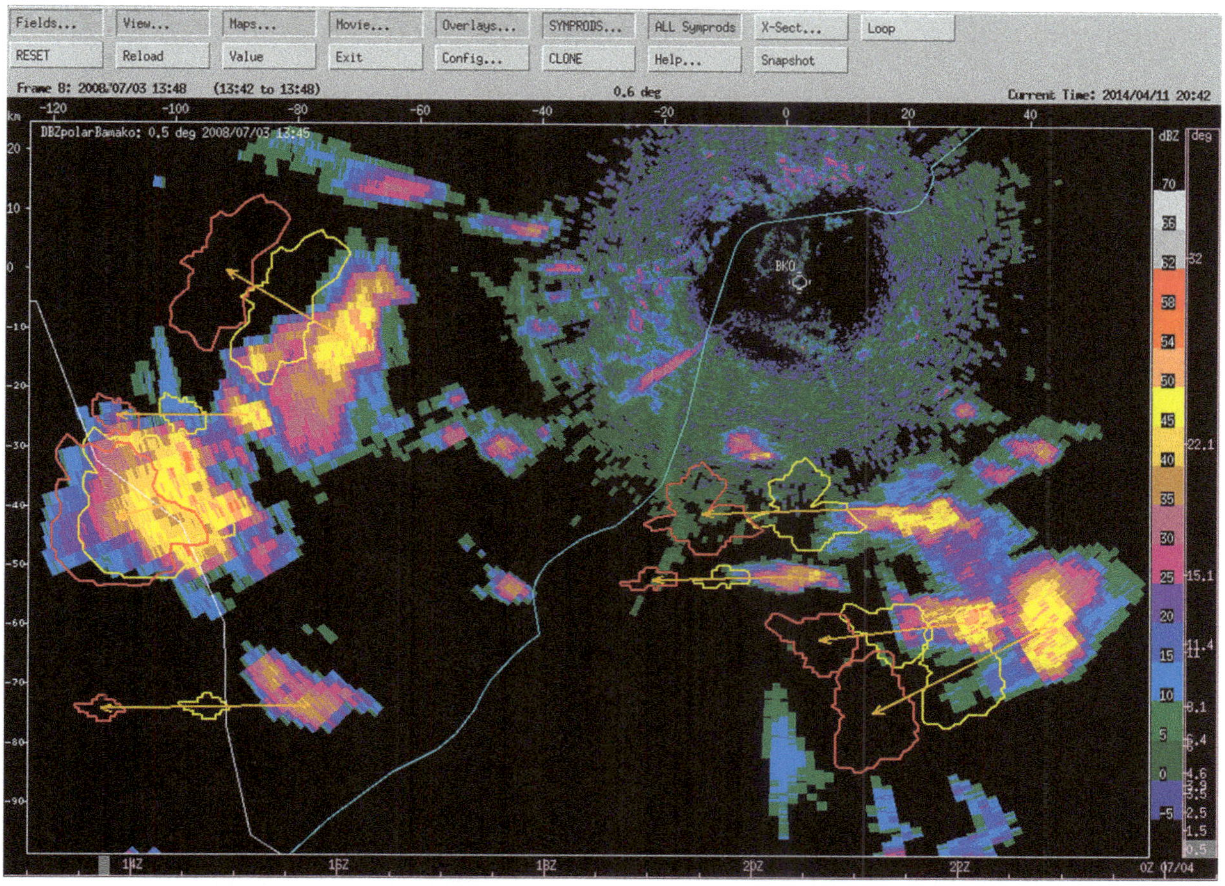


Figure 6.16 Example of automated storm extrapolations run on radar data from Bamako, Mali, using the TITAN algorithm (Dixon and Wiener, 1993). Shown are 30 min (orange polygons) and 60 min (red polygons) storm extrapolations and the storm motion vectors on 3 July 2008. Blue contour is the Niger River. Bamako-Senou International Airport is located at $x = 3$ km, $y = -2$ km.

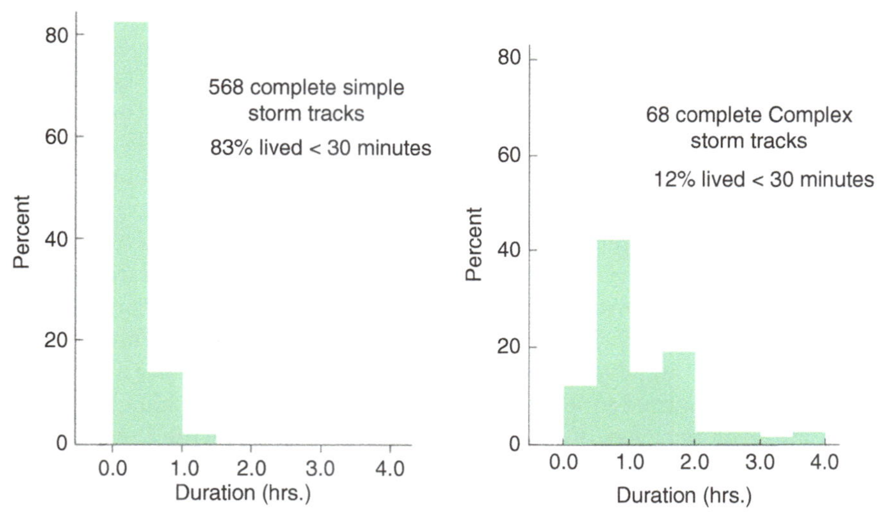


Figure 6.17 Histogram showing the lifetime of simple and complex storms observed during the summer of 1991 near Denver, Colorado (USA), based on data from an automated cell tracking system called TITAN. A simple storm is one that does not merge or split during its lifetime, and a complex storm is one that does. *Source*: Henry (1993). © American Meteorological Society. Used with permission.

major initiation, growth and decay took place within the 60 min extrapolation period. While this is an extreme example of evolution during a 60 min period, it nicely demonstrates the need to predict the evolving boundary-layer conditions.

#### 6.1.2.6 Nowcasting Storm Initiation and Evolution

This section is concerned primarily with the initiation and evolution of convective storms prior to the MCS stage. Chapter 3, particularly Section 3.1.3, discusses the evolution of MCSs.

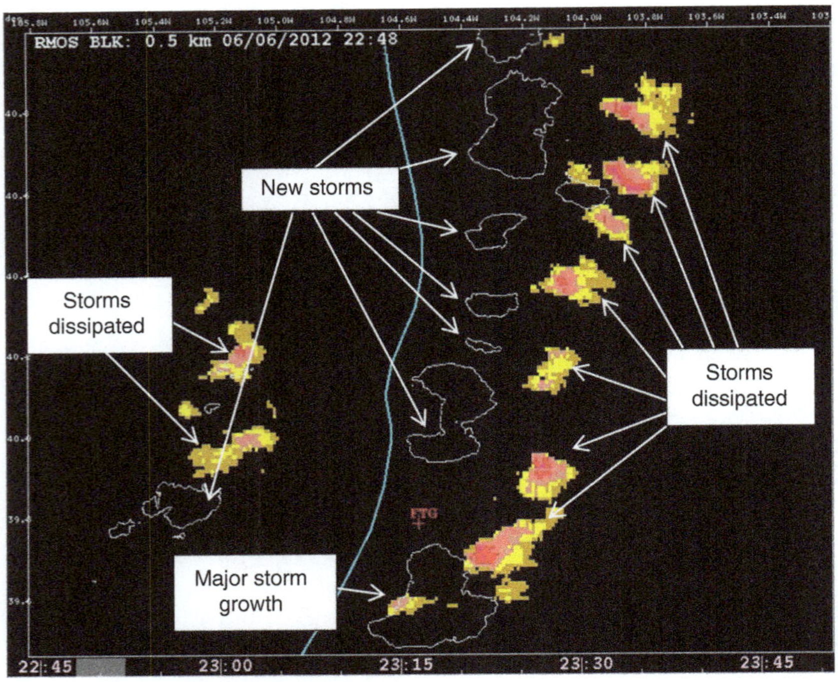


**Figure 6.18** Sixty-minute extrapolation nowcast of storm position and intensity (colour-filled objects) and actual location of storms at verification time (white contours). The outer colour is a radar reflectivity of 35 dBZ with each new colour increasing in 5 dBZ steps. The echoes were moving primarily from the southwest. The white contour is 35 dBZ. This is an extreme example where nearly all the extrapolated storms dissipated within 60 min even though many of the storms were >50 dBZ. The N–S line of new storms are mostly >50 dBZ that initiated just east of a boundary-layer convergence line (blue line).

As already discussed, even for nowcasting periods as short as 30 min there can be significant errors in location, timing and intensity of convective weather by extrapolating weather into the future using past storm motion and intensity trends. This is because new storm initiation and changes in intensity of existing storms are not taken into account. It has also been discussed earlier, and in Chapters 3 and 10, that NWP has not been able to provide the level of accuracy required for nowcasting convective storms and precipitation on the time and spatial scale needed. However, it has been demonstrated that forecasters have the skill for nowcasting convective storms. They accomplish this by using NWP model fields as guidance and subjectively monitoring observations from radar, satellite, surface station and profilers within the context of conceptual models of convective storm evolution in their forecast area (Wilson *et al.*, 2010; Roberts *et al.*, 2012).

Nowcasting of convective storm evolution began to develop in the 1970s and 1980s, beginning with examination of weather satellite imagery, and then Doppler radars started to show that thunderstorm initiation and evolution was often controlled by boundary-layer convergence lines – something that had been proposed years earlier by Byers and Braham (1949). Convergence lines, often referred to as 'boundaries', represent the interface between moist and dry low-level air, thermal gradients or changes in wind direction and speed (see also Section 4.1.4). Purdom (1973, 1976, 1982) showed that convergence lines were often visible on satellite imagery as lines of cumulus clouds and were observed to interact with one another and initiate thunderstorms. Examples of satellite-detected outflow boundaries are shown in the red, green, blue (RGB) dust composite images in Figures 6.19 and 6.20 (see Section 9.1.4.4 for explanation of RGB images). It is evident from this sequence of RGB satellite images that storm initiation occurs along the boundaries between the moist and dry air (Figure 6.19b) and evolves into an MCS (Figure 6.19c). Using the RGB products, some gust fronts can be tracked across northern Africa for many hundreds of kilometres.

Wilson and Carbone (1984) proposed that sensitive Doppler radars could be used to monitor convergence lines. Of particular significance, they showed that radar could be used to track convergence lines in the *absence* of clouds. Wilson and Schreiber (1986) showed that 80% of the thunderstorms in the Denver, Colorado, area formed along radar-detected convergence lines. Examples of radar-detected convergence lines, from different locations around the world, including West Africa, are shown in Figure 6.21. Other examples from West Africa are shown in Dione *et al.* (2014). Boundary-layer convergence lines on radar appear at low radar-elevation angles as thin lines of enhanced reflectivity and are observed in the Doppler radial velocity field as a region of converging radial winds. An example from West Africa of converging winds associated with gust front and squall-line passage is shown in Figure 6.22. With modern weather radars, convergence lines can be observed over land areas almost anywhere in the world. Their detection is not dependent on the presence of clouds because the backscattering mechanism for their detection is insects (Battan, 1973; Vaughn, 1985) that become concentrated in the updrafts along convergence lines (Wilson *et al.*, 1994).

Figure 6.19 Dust RGB plots highlighting low-level convergence boundaries between moist and dry air and storm initiation and evolution above these boundaries on 4 October 2005 at three time periods. *Source*: EUMETSAT web site, 2014. Reproduced with the permission of Jochen Kerkmann (EUMETSAT) and Daniel Rosenfeld (HUJI).

As long as temperatures within these convergence and updraft regions are >10°C, insects are generally present in the updrafts and act as effective backscatters of radar energy.

Based on a variety of research field experiments, mostly in the last 30 years, much has been learned about the types of boundary-layer convergence lines and their role in the initiation and evolution of convective storms. Today it is common practice for operational weather forecasters in many countries to closely monitor their location and motion to update and amend thunderstorm forecasts.

The most common type of boundaries are associated with sea breezes, land breezes, dry lines, synoptic fronts, gust fronts, bores and topographically induced boundaries. With the exception of synoptic-scale boundaries, most boundaries are about 1 km deep with widths of 1–3 km and generate vertical wind speeds of a few metres per second to >10 m $s^{-1}$ when boundaries collide (Mahoney, 1988). The ability of a boundary to initiate convective storms is obviously related to the magnitude of the vertical motion and the stability of the air being lifted. Thorpe *et al.* (1982) and Rotunno *et al.* (1988) showed that the low-level vertical wind shear profile (surface to 2.5 km) directed normal to a gust front is related to the extent and longevity of thunderstorms. The Rotunno *et al.* (1988) simulations show that the optimal conditions for long-lived intense convection occur when the import of positive vorticity associated with the low-level shear just balances the net-buoyant generation of negative vorticity by the cold pool. This allows deep, vertical updrafts at the gust front. In a very similar manner, Weisman and Klemp (1986) and Wilson and Megenhardt (1997) show that initiation, organisation and lifetime of convective storms is dependent on the relative motion between the cumulus clouds and the convergent line.

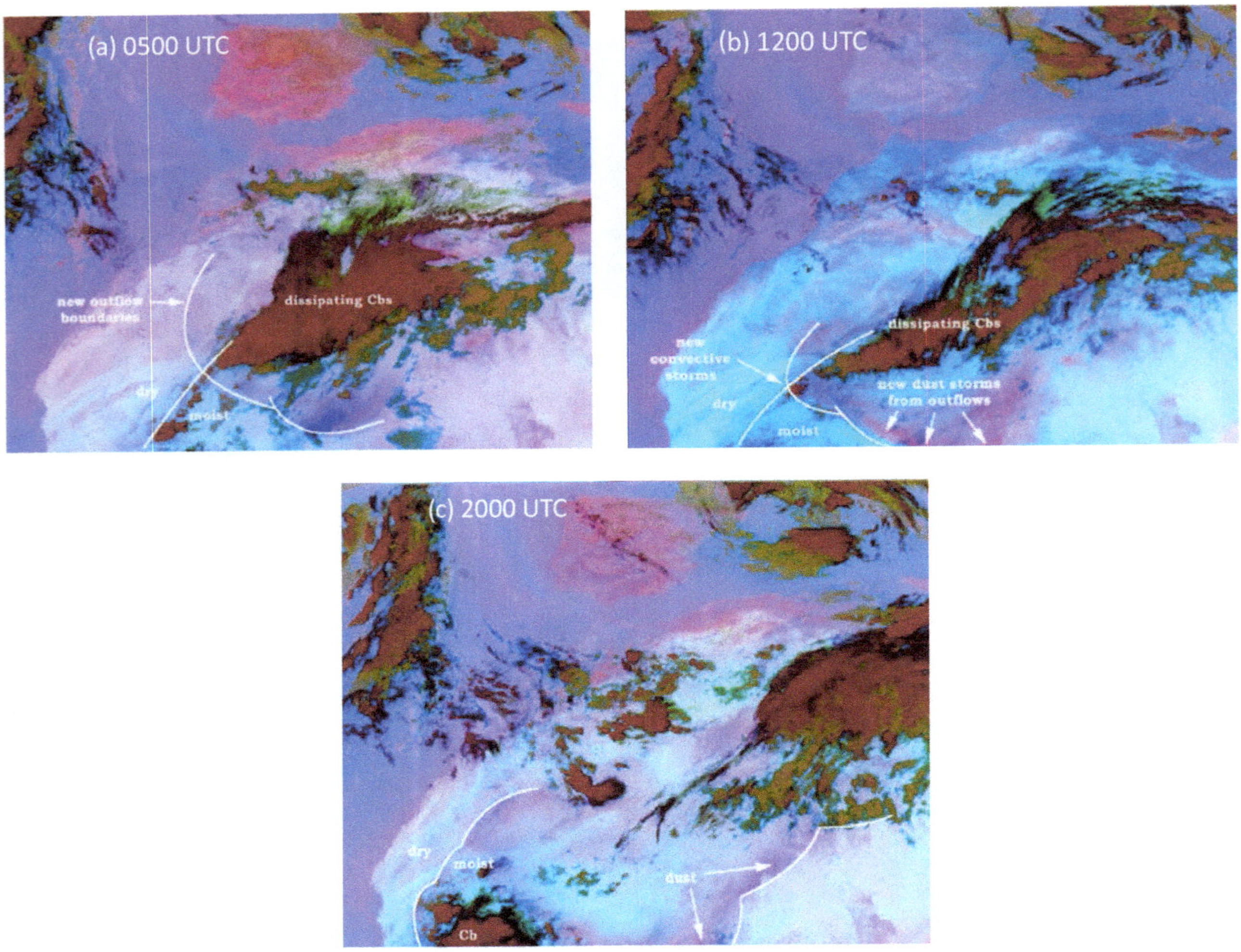


**Figure 6.20** Same as Figure 6.19, except from 5 October 2005 at (a) 0500 UTC, (b) 1200 UTC, and (c) 2000 UTC. *Source*: EUMETSAT web site, 2014. Reproduced with the permission of Jochen Kerkmann (EUMETSAT) and Daniel Rosenfeld (HUJI).

Figure 6.23 contrasts favourable and unfavourable conditions for initiation and sustenance of convection. In the favourable situation, the convergence at the leading edge of the gust front is maximised and the cloud steering-level winds will be similar to the gust-front motion, resulting in deep, erect updrafts. The exact opposite occurs with the unfavourable situation, where the convergence at the leading edge of the gust front is smaller, updrafts are tilted and the cells would tend to move away from the leading edge of the gust front. See also Figure 3.12.

Owing to the strong updrafts they can generate, colliding boundaries can be very proficient at initiating more intense storms (Mahoney, 1988). In addition, boundaries moving into a field of cumulus clouds may initiate storms (Wilson and Mueller, 1993) or intensify existing storms. Storm initiation along a boundary has been shown to occur where boundary-layer horizontal convective rolls (Weckwerth *et al.*, 1999) intersect a boundary (Atkins *et al.*, 1995; Wilson *et al.*, 1992). Roberts and Rutledge (2003) examined the rate of growth of cumulus clouds above different types of radar-detected, surface convergence boundaries by tracking satellite-derived cloud-top temperatures over time. The rates of cloud-top cooling above stationary convergence boundaries tend to be a bit slower following the first jump in cloud-top cooling (−10°C/15 min in Figure 6.24) than the rates observed with gust front interaction with horizontal convective rolls (Figure 6.25a) and the strong cooling rates associated with gust fronts in general (−10 to −16°C/15 min in Figure 6.25b), where stronger surface convergence and vertical lifting are more prevalent. Once storms form, storm intensification can occur fairly rapidly, as can be seen by the jump in reflectivity in Figure 6.25a. In the absence of available radar data, the forecaster should monitor the 15-min MSG imagery for pronounced cloud-top cooling of cumulus clouds in order to gain extra lead time in nowcasting for thunderstorms.

Recent automated nowcasting efforts have been directed at blending radar extrapolation with numerical models to produce a seamless 0–6 h nowcast. Nowcasting and Initialization for Modelling Using Regional Observation Data Systems (Golding, 1998) was likely the first such operational system. The typical procedure is to give full weight, for the first 2–3 h, to extrapolation of the

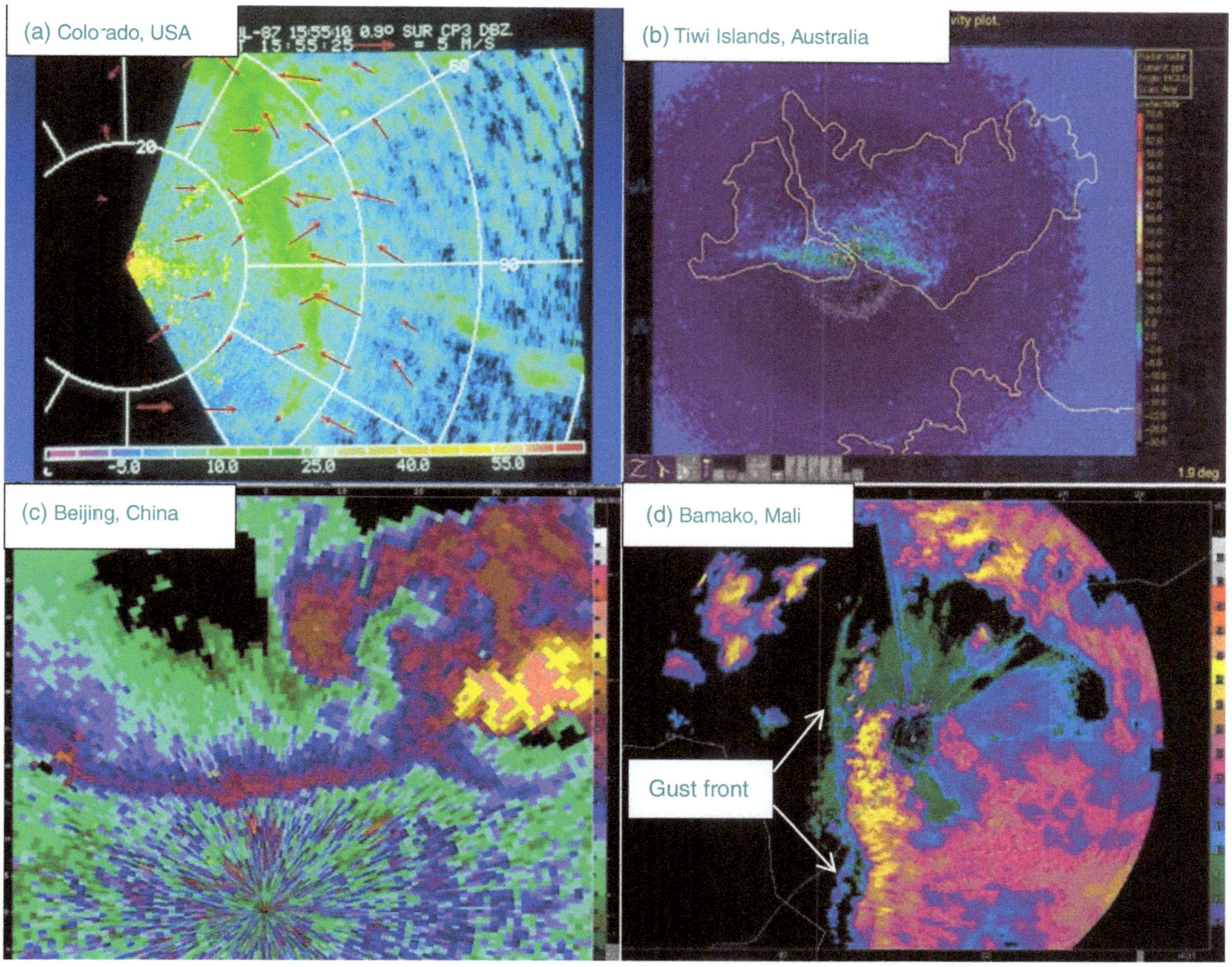


Figure 6.21 Examples of boundary-layer convergence lines ('boundaries') seen as thin lines of enhanced radar reflectivity. (a) Result of topographical-induced convergence, Colorado, USA; the red arrows are surface-wind vectors from surface mesonet stations. (b) Sea-breeze front, Tiwi Islands off the north coast of Australia., (c) Gust front, Beijing, China, during the Olympic Games. (d) Gust front at the leading edge of a squall line in Bamako, Mali, West Africa.

observed precipitation field and then to give increasing weight to the model until the model gets full weight by about the fifth to sixth hour. Figure 6.26 is an example of the forecast skill versus forecast length for extrapolation, NWP, intensity and positioning (phase-corrected) NWP and extrapolation, blended with phase-corrected NWP. The skill scores are for July 2012 for the eastern two-thirds of the USA. The primary points from Figure 6.26 are the rapid decrease in extrapolation skill with increasing forecast length, and the low skill by numerical techniques at all time periods, and the improvement in NWP when corrections are made to the intensity and positioning errors.

Table 6.1 is a compilation of nowcasting rules for convective storm evolution associated with boundary-layer convergence lines. These rules are used in a nowcasting system developed at the National Center for Atmospheric Research called the Autonowcaster (Mueller *et al.*, 2003; Roberts and Rutledge, 2003; Saxen *et al.*, 2008). The Autonowcaster is a 0–1 h precipitation nowcasting system that extrapolates precipitation echoes and nowcasts the initiation, growth and dissipation of precipitation, with nowcast updates approximately every 5 min. Sections 3.2.3 and 11.12 provide additional information for nowcasting other types of convection, like those associated with the African easterly wave (AEW).

## 6.2 Operational Methods for Nowcasting Severe Weather

Rainy days and thunderstorms are frequent occurrences at many of the West African aerodromes during the monsoon season, as can be seen in Table C6.1 in Appendix C. Clusters of thunderstorms and multicellular storms comprise many of the MCSs that track from east to west over West Africa. They bring heavy rain, lightning, strong winds and the potential for flash floods. As

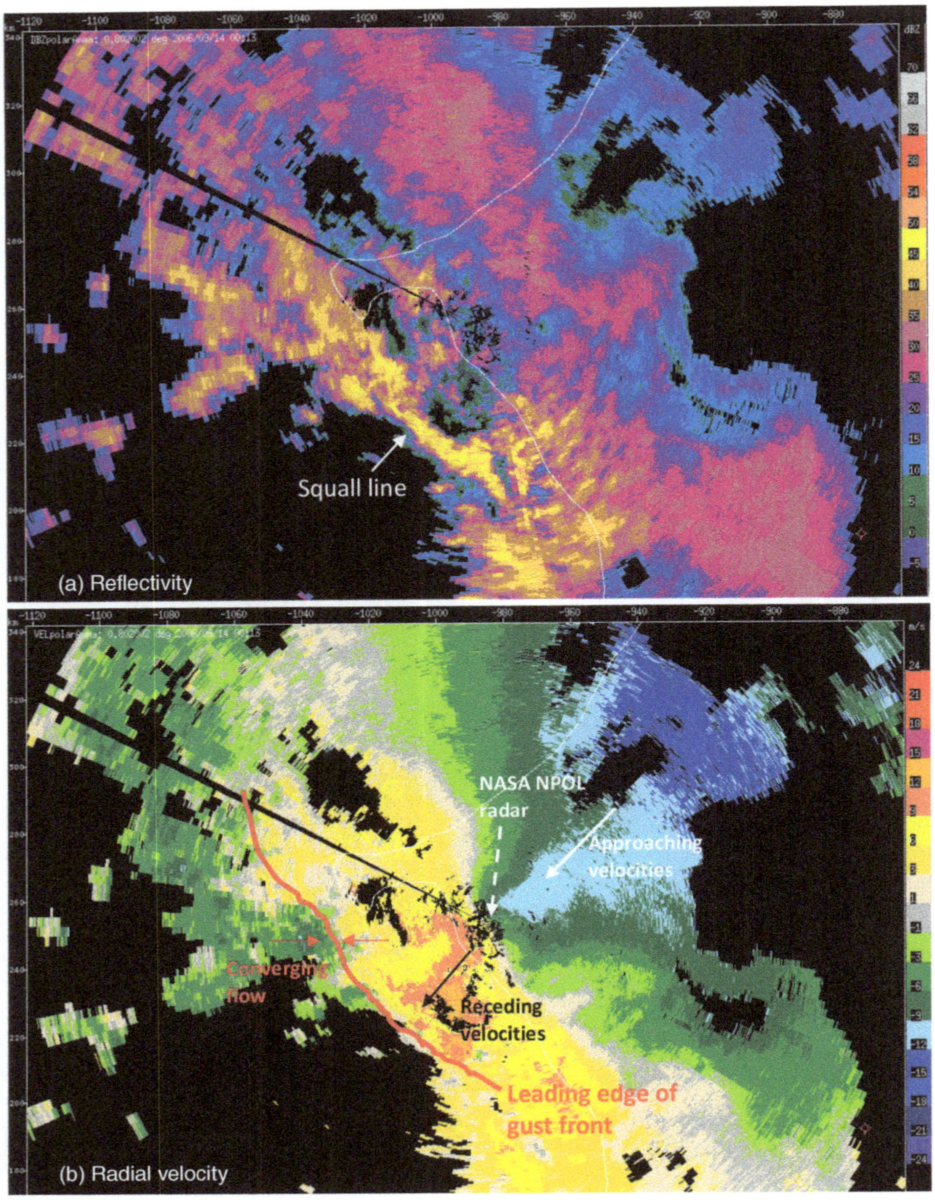


**Figure 6.22** NASA NPOL detection of (a) squall line (reflectivity in decibels relative to *Z*) and (b) associated gust front (radial velocity in metres per second) near the Léopold Sédar Senghor International Airport in Dakar on, 14 September 2006 at 0013 UTC.

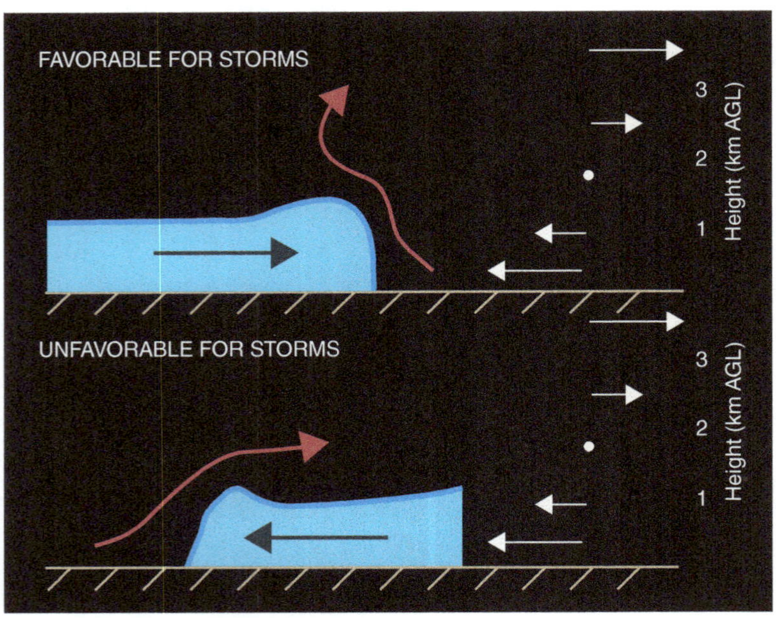


**Figure 6.23** Illustration of favourable and unfavourable situations for convective storm development triggered by a convective storm outflow boundary. The red arrows show the airflow over the relative cool convective storm outflow (blue colour). The black arrow shows the direction of motion of the outflow. The environmental winds are shown by the white vectors. This is an example for mid-level westerly winds. Similar plots can be made for mid-level easterlies.

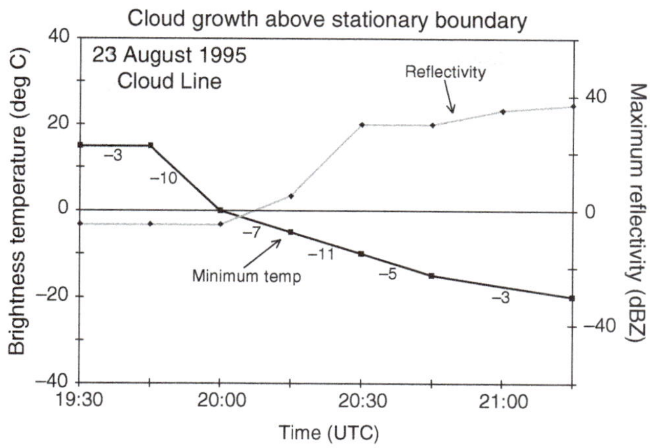


Figure 6.24 Time-series plot of minimum IR brightness temperature (black curve) in Celsius and maximum radar reflectivity (grey curve) in decibels relative to *Z* associated with growth of a cloud line above a semi-stationary convergence line on 23 August 1995 in Colorado, USA. Rate-of-change values determined over 15 min intervals are listed below the temperature curve. *Source*: Roberts and Rutledge (2003). © American Meteorological Society. Used with permission.

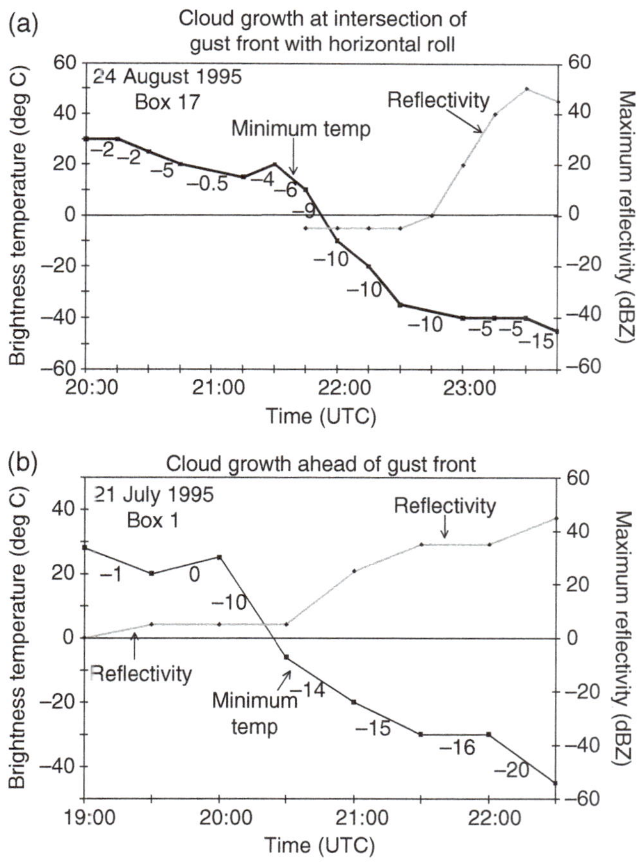


Figure 6.25 Same as Figure 6.24, but for (a) cloud growth above gust front interactions with horizontal rolls on 24 August 1995 and (b) cloud growth ahead of an approaching gust front on 21 July 1995. *Source*: Roberts and Rutledge (2003). © American Meteorological Society. Used with permission.

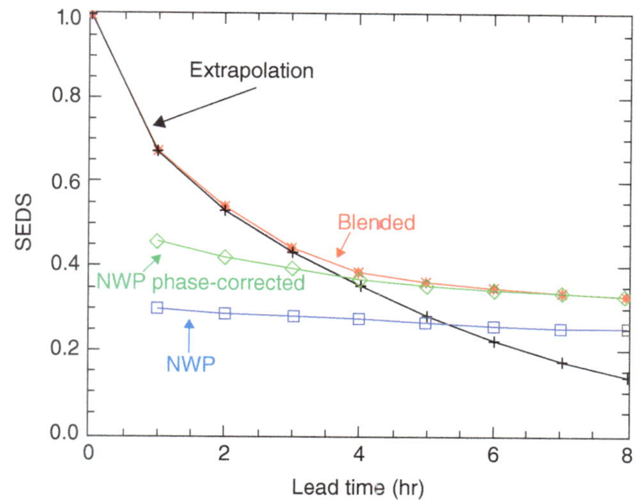


Figure 6.26 Example of skill score (SEDS, symmetric extreme dependency score; Hogan *et al.*, 2009) versus forecast lead time for a blended nowcasting system. This blended system is being tested in the USA for aviation purposes.

Table 6.1 Nowcast rules[a] for convective storm initiation, growth and dissipation associated with boundary-layer convergence lines ('boundaries').

***Semi-stationary boundaries***

- Nowcast thunderstorm initiation when IR cloud-top temperature falls below 0°C.
- Nowcast thunderstorm initiation when the precipitation reflectivity aloft increases to >20 dBZ.
- Nowcast thunderstorm initiation where convective rolls intersect a convergence boundary and radar indicates increasing precipitation reflectivity aloft.

*Moving boundary*

- For a circular (ring) gust front, nowcast thunderstorms only along the portion of the gust front downwind of the steering-level winds.
- Nowcast thunderstorms if a boundary is moving into a field of cumulus clouds, particularly if boundary and storms have a similar motion. Do not nowcast thunderstorms if boundary and cumulus clouds are moving rapidly in opposite directions.
- If boundary is moving into an existing thunderstorm, nowcast thunderstorm intensification, particularly if boundary and storm are moving in the same direction.
- Nowcast thunderstorm intensification if the storm is moving about the same speed and direction as the boundary.
- Nowcast thunderstorm dissipation if boundary is moving away from the thunderstorm.

*Colliding boundaries*

- Nowcast thunderstorm initiation if cumulus clouds or storms are present in the vicinity of collision. There is potential for these thunderstorms to be particularly intense.

*Storm moving from mountains to plains*

- Nowcast thunderstorm dissipation if there is no indication of instability over the plains based on a recent sounding and/or lack of cumulus clouds on the plains.

[a] These rules are based on Roberts and Rutledge (2003), Wilson and Mueller (1993), Atkins *et al.* (1995), Rotunno *et al.* (1988), Wilson and Megenhardt (1997), Mahoney (1988) and Wilson *et al.* (2010).

individual thunderstorms, they have the potential to produce significant wind shear events, such as microbursts, downbursts and gust fronts, and bring lightning and occasionally hail. The detail that can be provided in the nowcast for these events, at any given location such as an airport, depends on the quality of the datasets available. If the RGB products from satellites are available (see Sections 9.1.4.4 and 9.2 for examples), a general short-period forecast, using extrapolation, can be made for a few hours in advance for the likelihood of rainfall, lightning and/or wind shifts associated with an MCS. Chapters 3 and 9 discuss forecasting MCSs using satellite data. More precise timing of heavy rainfall, lightning, strong winds, wind shifts, downbursts and hail require Doppler weather radar and preferably dual-polarisation radar. Even with radar, these high-impact weather events can usually be nowcast only for periods less than 60 min. The following sections provide procedures for nowcasting these events and updating TAFs based on weather radar and satellite information.

### 6.2.1 Thunderstorms, Hail and Flash Floods

At the beginning of the forecaster's shift, the forecaster needs to determine the type of weather that will be impacting their area over the next several hours. If the weather will be driven by large-scale synoptic forcing, the forecaster will need to determine the timing and onset for the convective weather associated with these larger scale systems, such as MCSs, caught up in the easterly wave. By extrapolating weather features evident in satellite-visible, IR or RGB imagery, the forecaster should estimate the time of arrival of precipitation at specific locations. If the weather looks to be more locally driven or forced by mesoscale processes, the forecaster will need to be diligent in monitoring the available observational data for changes occurring in the atmosphere, to provide location and time-specific nowcasts of high-impact weather. As mentioned earlier, much of the latter detail requires storm-scale information available from weather radars. Table 6.2 provides a recommended daily routine for nowcasting convective storm activity and potential for severe weather.

**Table 6.2** Daily operational guidelines for short-term forecaster or nowcaster.

1. Short-term forecaster (nowcaster) arrives on shift and receives briefing from outgoing short-term forecaster.
2. Within first hour, decide on dominant synoptic and mesoscale weather scenario for the day using upper air plots, surface weather analysis plots and NWP model forecasts.
3. Determine the potential for convective instability and severe weather over the next 0–6 h for the forecast area using:
   - NWP forecast and satellite-derived stability indices
   - recent nearby radiosondes to determine CIN, CAPE and anticipated convective temperature, and time it will be reached
   - amount of total precipitable water vapour
   - height of the freezing level and the LFC of the atmosphere
   - potential for cold or warm air advection
4. Use satellite data, sounding, surface observations, radar data, lightning data, as available, to determine the following:
   - locations of surface convergence boundaries (gradients in temperature, moisture or wind)
   - locations of cumulus clouds and growing clouds
   - local areas approaching convective temperature
   - advection of storms into the forecast area

### 6.2.2 Procedure for Nowcasting Thunderstorms, Hail and Flash Floods

On 9 June 2008, numerous thunderstorms formed in the vicinity of the Bamako-Sénou International Airport in Bamako, Mali. Attributes of the environment early in the day are shown in the following soundings.

From Figure 6.27, the forecaster will have determined that the atmosphere on this day is conditionally unstable, with sufficient low- and mid-level moisture, vertical shear and plenty of CAPE to support vigorous convection. Steering-level winds are from the east. The forecaster will now need to monitor all available observations for the onset of convection and thunderstorm development and for nowcasting the potential for severe weather within their forecast domain. Particular attention should be given to examining the satellite imagery and RGB products for any evidence of approaching MCSs, gust fronts, stationary convergence boundaries and cloud growth.

In the IR satellite imagery in Figure 6.28 at 13:00 UTC, it is evident that the only appreciable cloudiness is located approximately 600 km to the E–SE of Bamako, where some middle- and high-level clouds are present. No MCSs are evident at this time, so the forecaster will need to focus on the possibility for more locally driven convection initiation. Detecting the initiation of convective storms will help the forecaster to extend the lead time for issuing watches and warnings.

In the early afternoon on 9 June the forecaster will have been monitoring surface station information anticipating when $T_c$ will be reached locally. Examination of the satellite infrared imagery at 13:30 UTC shows that there are no convective storms present in the Bamako area at this time (Figure 6.29a) but a few cumulus clouds are present ~60 km southeast of Bamako. Examination of the Bamako radar data at 13:27 UTC (i.e. 13:27 LT) similarly shows no precipitation echoes present in the Bamako area at this time (Figure 6.30a).

Sensitive radars, such as the one located in Bamako, are able to see both storm precipitation echoes (with reflectivities from 10–60 dBZ) and non-precipitation echoes, which are referred to as 'clear air' features (with reflectivities generally from −10 to +20 dBZ). Although

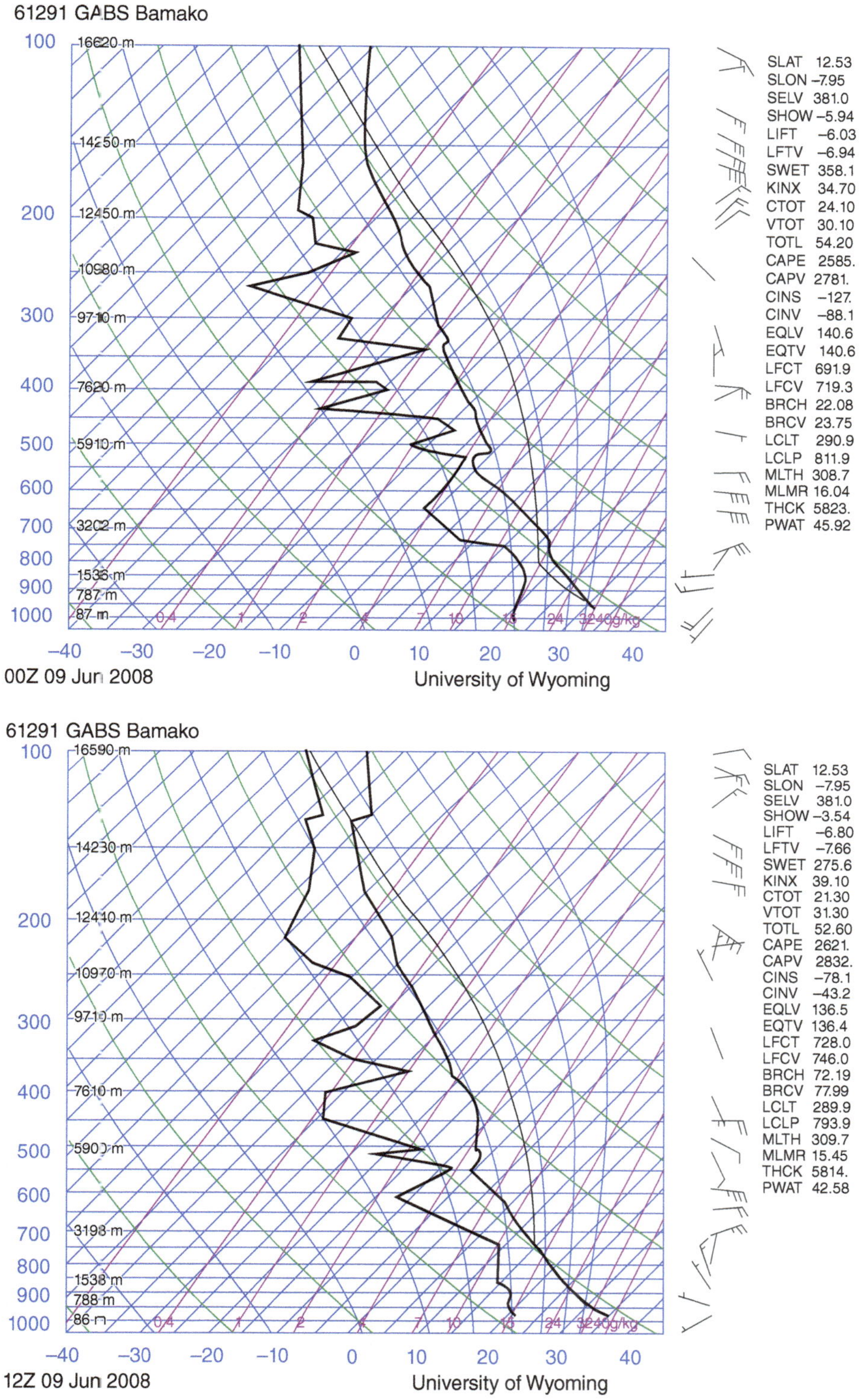


**Figure 6.27** Bamako, Mali, soundings from 9 June 2008 at 0000 UTC (top panel) and 1200 UTC (bottom panel) plotted on Skew-T plots. *Source*: University of Wyoming web site, 2014.

no storms are present at this time, two stationary convergence boundaries are evident in the 'clear air' return of the radar in the form of ENE–WSW thin lines of reflectivity. As discussed in the previous section, detection of these thin lines is important as they represent regions of surface convergence that concentrate moisture and provide vertical lifting of air, and are thus potential areas for storm initiation and intensification.

Table 6.1 indicates that the nowcaster should be monitoring the satellite data for cloud-top cooling of the

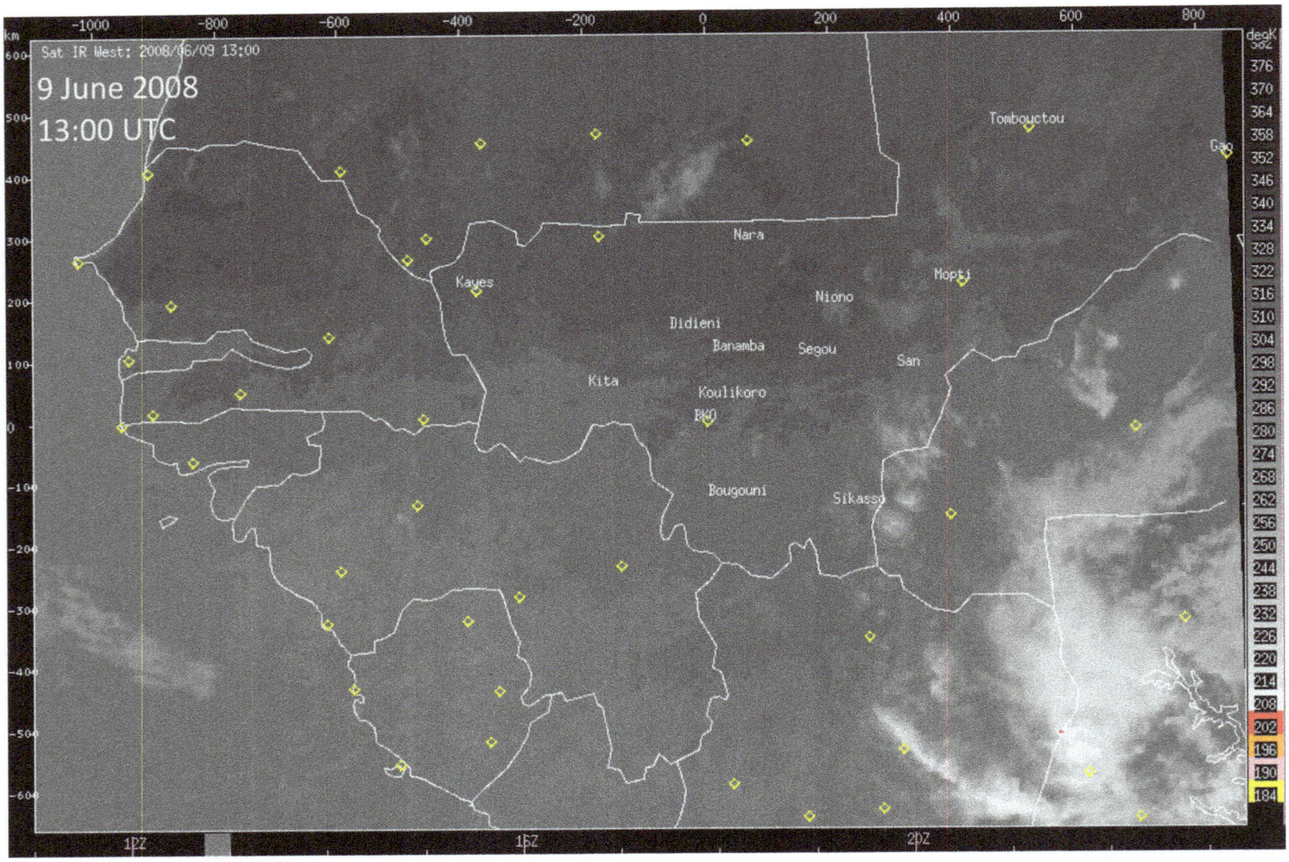


Figure 6.28 Satellite imagery of IR brightness temperatures at 13:00 UTC on 9 June 2008 for a portion of western Africa. Yellow diamond-shaped icons of the locations of major airports are overlaid and include the location of Bamako airport (BKO). White contours are country borders.

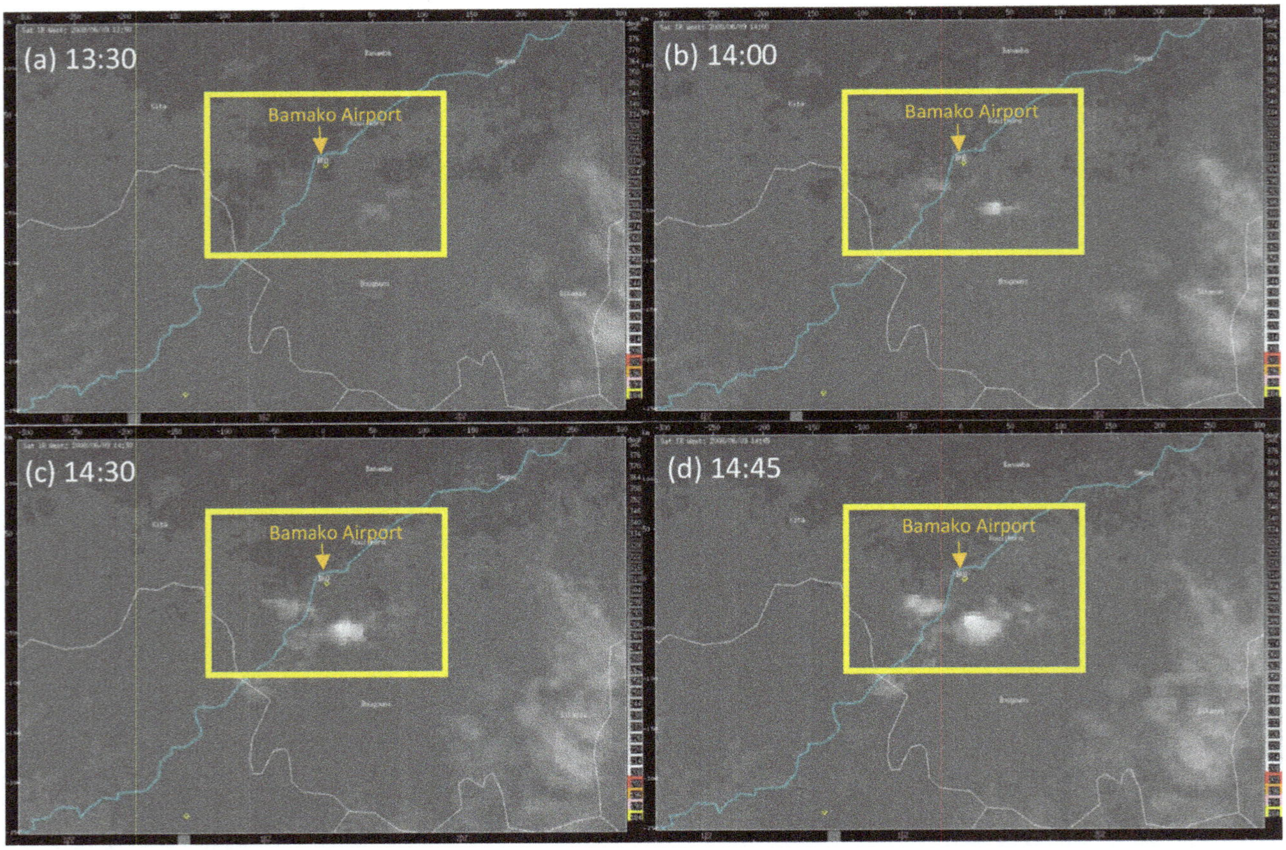


Figure 6.29 Sequence of IR satellite images centred near the Bamako, Mali, airport on 9 June 2008 at the four time periods (in UTC) shown. The colour scale, in brightness temperature, is the same as in Figure 6.28. The yellow box shows the domain covered by the Bamako radar images in Figure 6.30. White contours are country borders and the blue contour is the Niger River.

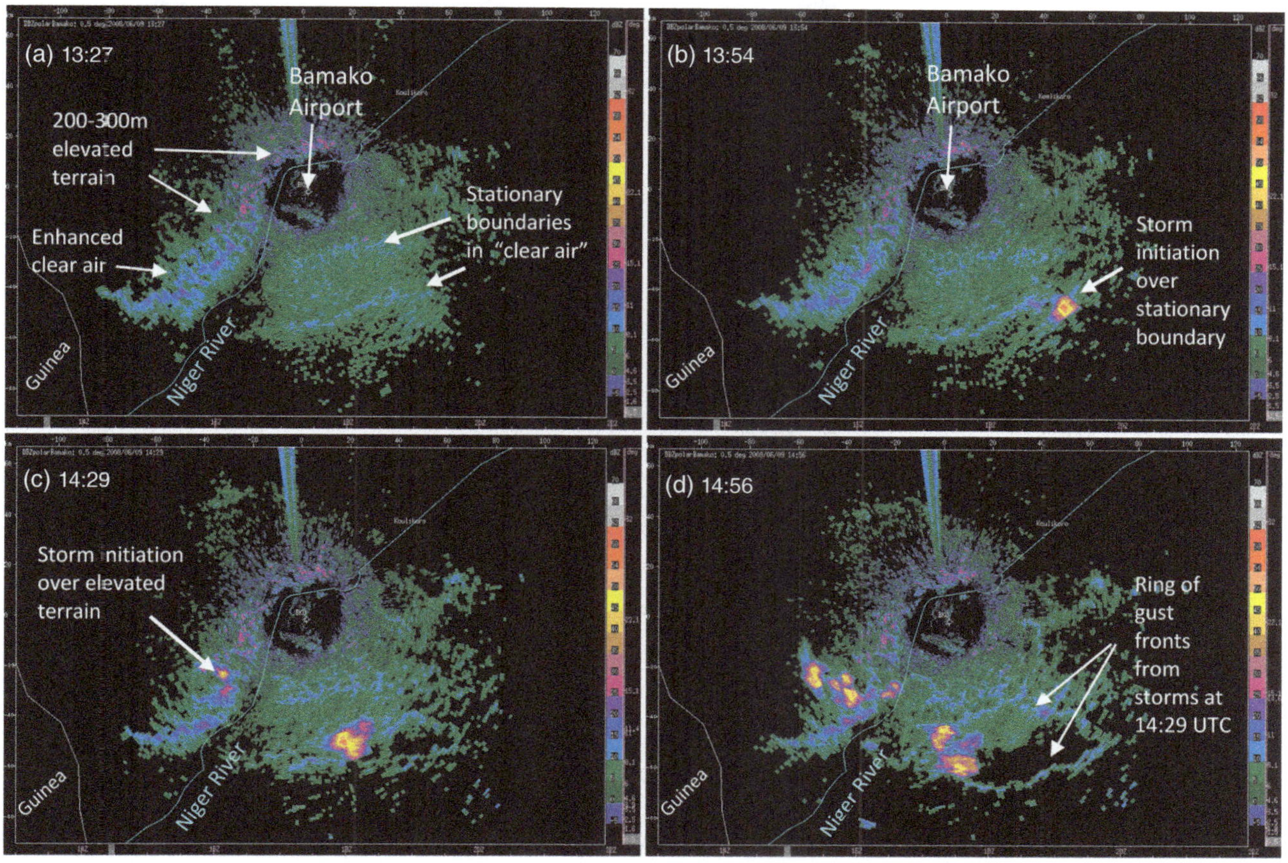


**Figure 6.30** Sequence of radar reflectivity images from the Doppler radar located near the Bamako, Mali, airport on 9 June 2008 at the four time periods (in UTC) shown. The colour scale is in 5 dBZ increments, with most intense thunderstorms having reflectivities of 50–60 dBZ (orange/red colours). The green and light blue colours in the images represent regions of clear air (no storms). The radar is able to detect clear air because of the backscattering of radio electromagnetic energy from insects in the clear air.

cumulus clouds and evidence of precipitation reflectivity aloft on the radar, above these stationary convergence boundaries. As can be seen ~30 min later, the first convective storm has initiated S–SE of Bamako Airport above one of the stationary boundaries (Figure 6.30a), with the satellite imagery showing a cumulus congestus cloud with a taller and colder cloud top (Figure 6.29b) and a corresponding precipitation echo in the radar imagery (Figure 6.30b); $T_c$ has clearly been reached. The appearance of this storm should put the forecaster on alert for additional convection to ensue. During the next hour (Figure 6.29c and d), additional cumulus cloud development is evident on satellite imagery. Comparison of the satellite and radar imagery (Figure 6.30c and d) indicates that these clouds and precipitating storms have developed above the stationary boundary observed in the radar imagery in Figure 6.30a, and above the slightly elevated terrain located to the southwest of Bamako along the Niger River, a likely climatologically favoured area for storm initiation. (*Note to forecaster:* whenever possible, it is best to use animation for viewing radar images, as animation makes it much easier to monitor gust fronts and storm evolution, rather than viewing still figures as are shown in this and the following figures.) No weather is currently impacting the airport, but the nowcaster should be wary of the possibility for convection developing nearby or immediately above the airport. The radar does not scan directly overhead, so it is particularly important for the forecaster to monitor the satellite imagery in the vicinity of the airport for growing cumulus clouds, as evidenced by colder cloud tops.

During the next 2 h period, numerous storms have developed in the area south of the Bamako Airport (Figure 6.31) and appear to have a somewhat linear orientation to them in the satellite imagery and quite cold cloud tops. The satellite IR imagery does not provide any guidance on where the new storms will initiate. Examination of the Bamako radar imagery shows that most of these storms were initiated by the ring gust front (i.e. a thunderstorm outflow that moves away from the initial storms in all directions) seen in Figure 6.32a and b, and by the gust front produced by storms located over the slightly elevated terrain. Tracking the motion of the storms by hand, by animation of the images over time or by running an automated tracking algorithm like Rapid Development Thunderstorm (RDT) or TITAN (see Figures 6.15 and 6.16), the forecaster will notice that most of the storms are moving to the west, posing no threat to

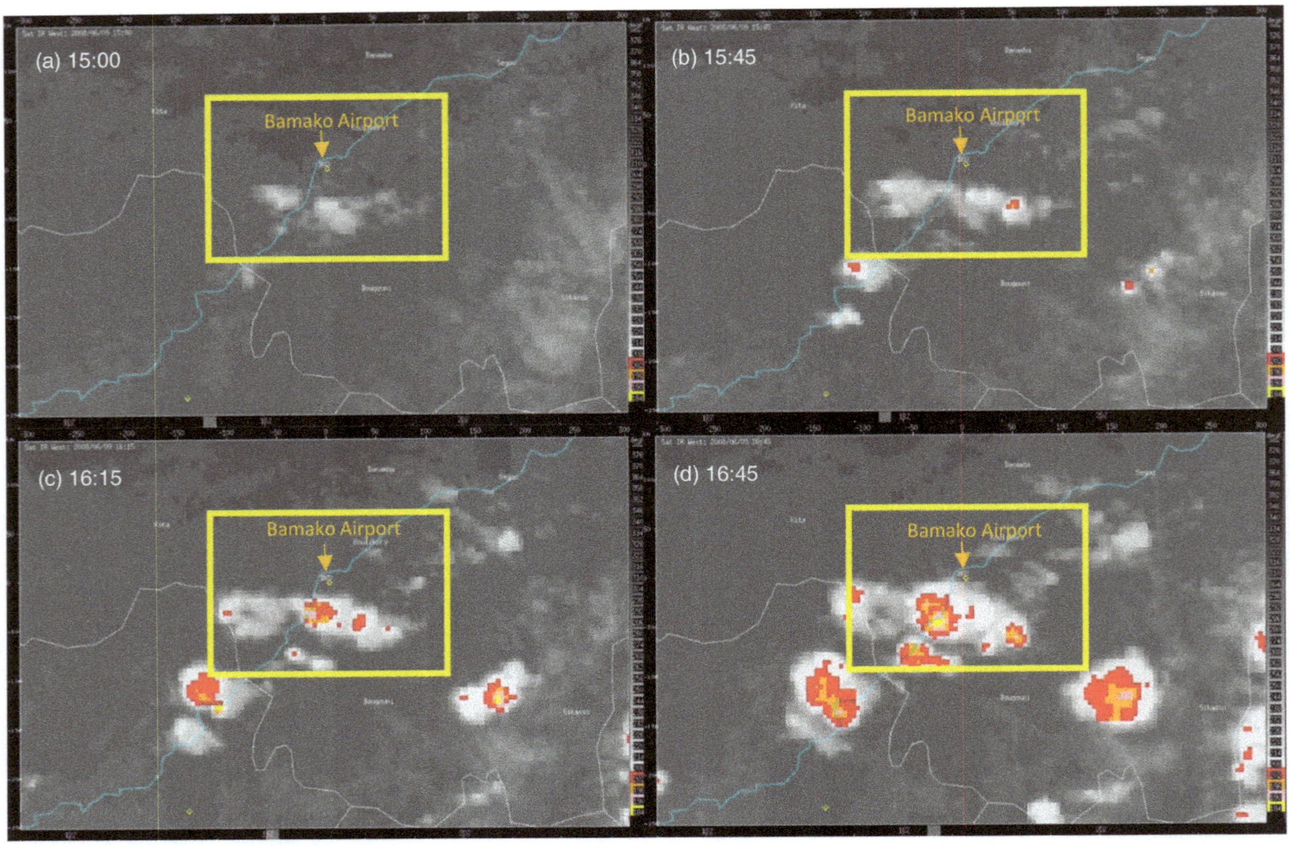


Figure 6.31 Same as Figure 6.29, except at four later times from 15:00 to 16:45 UTC.

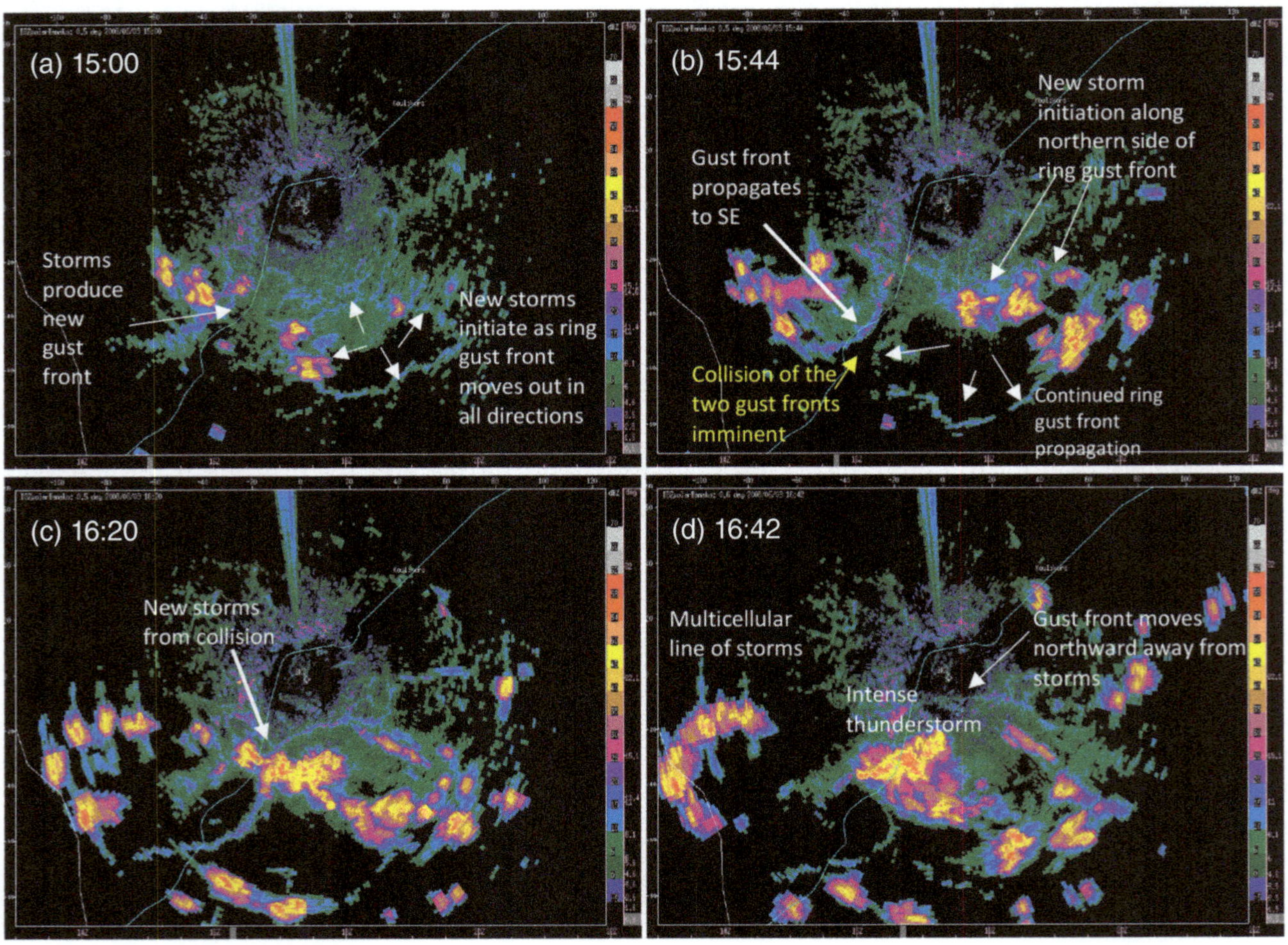


Figure 6.32 Same as Figure 6.30, but for four later time periods from 15:00 to 16:42 UTC.

the airport currently. However, at 15:44 UTC (Figure 6.32b) it is clear that the northern portion of the ring gust front continues propagating to the north and is initiating new storms as it moves. Also at 15:44 UTC, the nowcaster should note that two gust fronts will soon collide southwest of Bamako and new intense storms are likely (see Table 6.1 for nowcasting rule on colliding boundaries). Of course, these detailed observations of the gust fronts are only possible if the radar data are available.

**At this point, the nowcaster should be considering whether to issue an advanced warning for the possibility of thunderstorms over the airport in the next 30–60 min.** Without any other additional information or independent observations, the nowcaster should assume that the possibility exists for new thunderstorms to move over the airport within an hour, based purely on extrapolation of the existing situation. A possible advanced warning could be:

GABS AD WRNG 1 valid 091600/091800
CB/TS OBS 30KM SSW, MOV N 20KT INTSF

A potential amendment to the TAF could be made as follows:

TAF AMD GABS 091544Z 0916/1018 16010KT 7000 BKN030 FEW040CB TEMPO 0916/0918 VRB30G40KT 1000 TSRA BECMG 0919/0921 NSW SCT040

Figure 6.32c and d shows that the atmospheric activity is actually more complicated than what pure extrapolation would predict to happen. Although there continued to be some storm initiation along the northern portion of ring gust front during the next hour, most of the storms moved west, away from the gust front, and subsequently dissipated as they were no longer tied to their source of low-level moist air being lifted up by the gust front. This is an important rule for nowcasting thunderstorms – that storms will dissipate if they do not move in the same direction as the surface convergence feature that triggered it (see Table 6.1 for rule on moving boundaries). Instead, what actually happened was much more important, as two gusts fronts collided to the southwest of the airport (Figure 6.32c) and produced a particularly intense thunderstorm of >55 dBZ, with lightning and potentially hail (see Figures 6.32d and 6.33a). Examination of the satellite imagery in Figure 6.34 shows the very cold cloud tops associated with this storm.

Without spotter reports, there is no way to be absolutely certain that hail existed in this storm just because of the high values of reflectivity observed in Figure 6.33. Statistically, hail is a relatively infrequent event in West Africa, given the number of hail days that occur annually at African aerodrome sites (see Table C6.1, Appendix C). However, Table C6.1 shows that the potential for hail does exist. Table 6.3 lists some of the signatures that a forecaster can watch for that are indicative of hail in

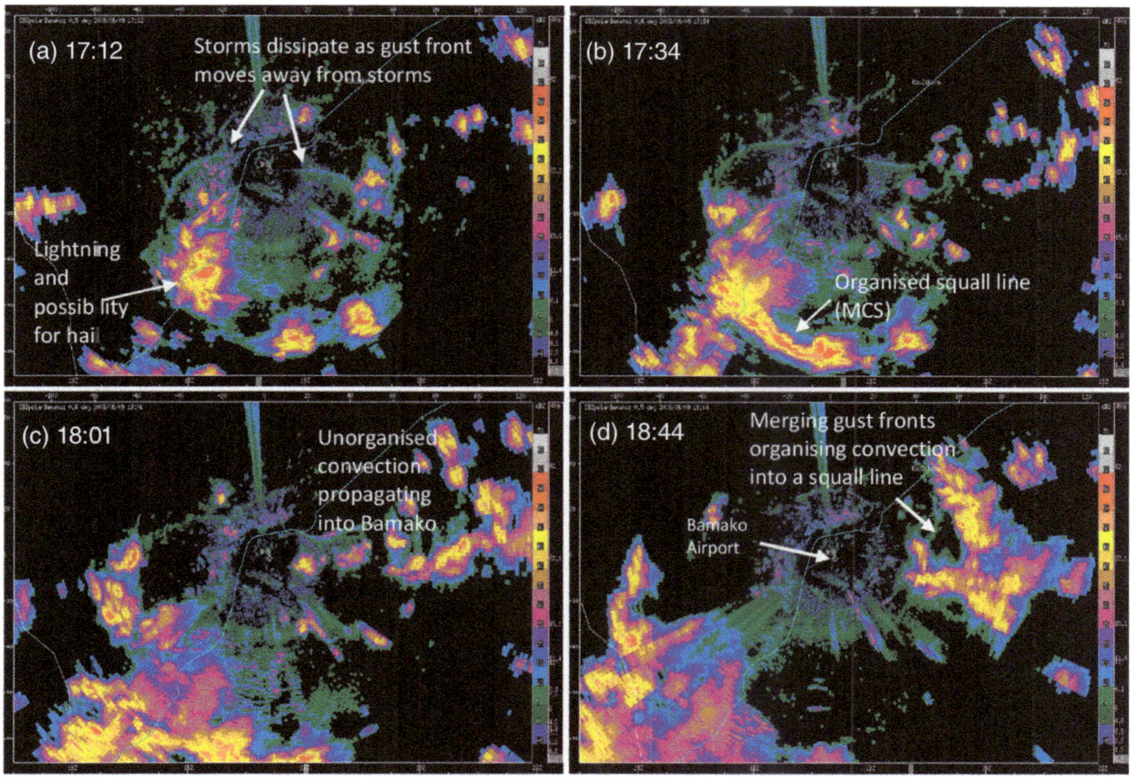


Figure 6.33 Same as Figure 6.32, but for four later time periods from 17:12 to 18:44 UTC.

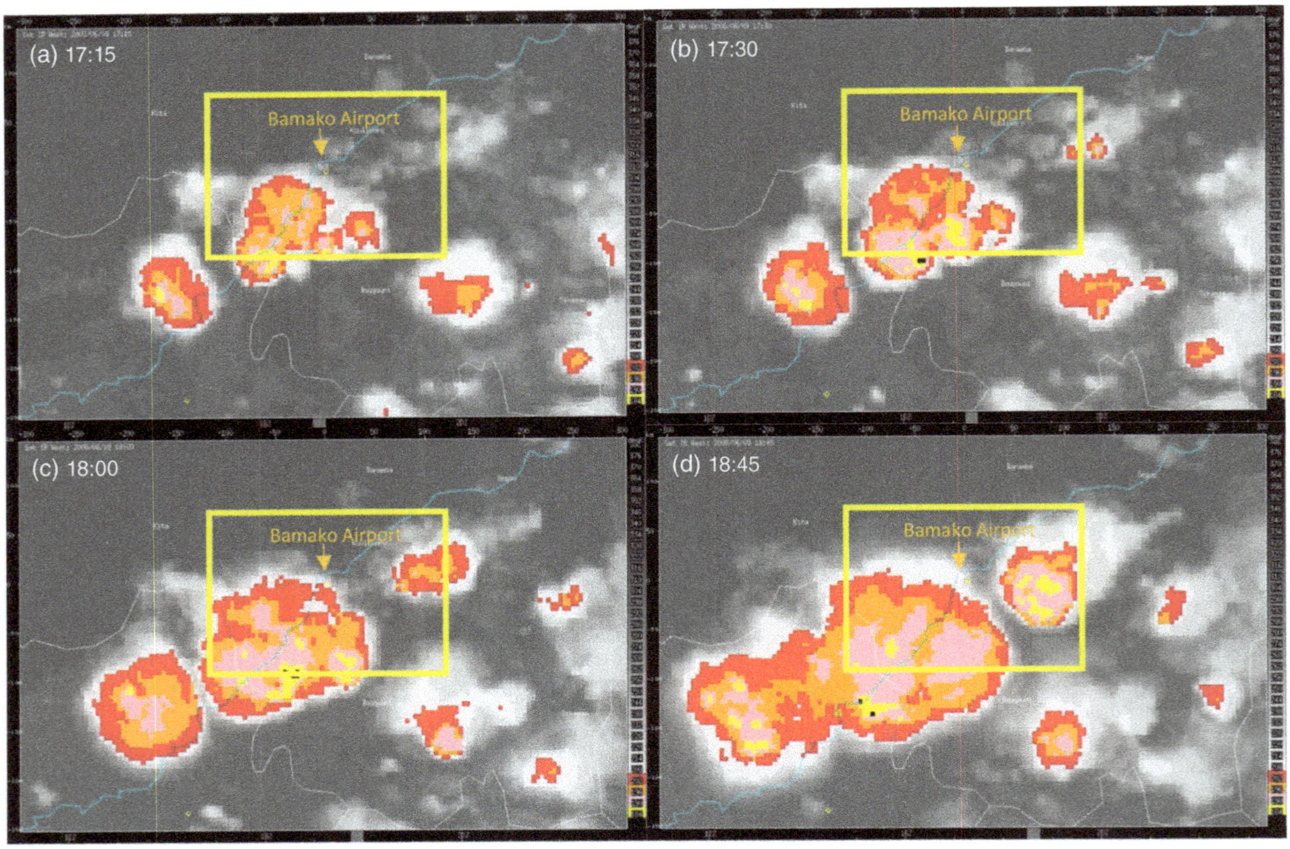


**Figure 6.34** Same as Figure 6.31, except at four later time periods from 15:15 to- 18:45 UTC.

**Table 6.3** Hail signatures in convective storms.

**Single polarisation radar signatures**

*Radar reflectivity (Z)* is dependent on the size of the particle; that is, $Z \approx D^6$, where $D$ is the diameter of the hydrometeor particle. Thus, reflectivity values ≥60 dBZ usually indicate the presence of hail within the storm. Radar reflectivity of 70 dBZ indicates the presence of large hail. Radar reflectivity >45 dBZ above the freezing level is a useful indicator of the presence of hail (Waldvogel *et al.*, 1979) in the storm; the larger the 45 dBZ area above the freezing level the greater the likelihood hail will reach the ground and the greater the likelihood the hail will be large (Joe *et al.*, 2004). In addition, the closer the height of the 0°C isotherm (in the sounding) is to the ground, the greater the likelihood of hail reaching the ground without melting.

*V-notch shaped reflectivity echo* indicates a region where large hydrometeor particles are being swept in a V-pattern around a strong updraft that is blocking the flow. Strong updrafts that can block the flow are strong enough to support the production of hail. Reflectivity values within the V-notch are generally ≥55 dBZ.

*Radar flare echo* (Wilson and Reum, 1988), a backscattering signature from hail within storms, observable with S-band and C-band radars. The hail size associated with this signature is ≥3/8″ (0.94 cm).

**Dual polarisation radar signatures**

*Differential reflectivity (ZDR)* with values between 0.0 and 1.0 dB for reflectivities >50 dBZ (Bringi *et al.*, 1986).

*Automatic hail classification* using a combination of dual polarisation fields (Vivekanandan *et al.*, 1999) and atmospheric sounding information.

**Satellite signatures**

*Overshooting tops*, evidence of strong updrafts that can support hail, are evident in the visible and IR imagery, and in the IR brightness temperature difference products. The lowest cloud-top brightness temperatures are found within an overshooting top.

*Cold-U/V and cold-ring shape signatures* are severity-related cloud-top features found in enhanced brightness temperature imagery, representing the cold parts of a regular storm anvil, surrounding longer-lived and larger-sized, embedded warm areas (from M. Setvák, Czech Hydrometeorological Institute, 2010).

**Favourable environmental conditions**

*Large environmental CAPE* to support strong updrafts.

*Large total precipitable water.*

*Large speed and/or directional vertical wind shear.*

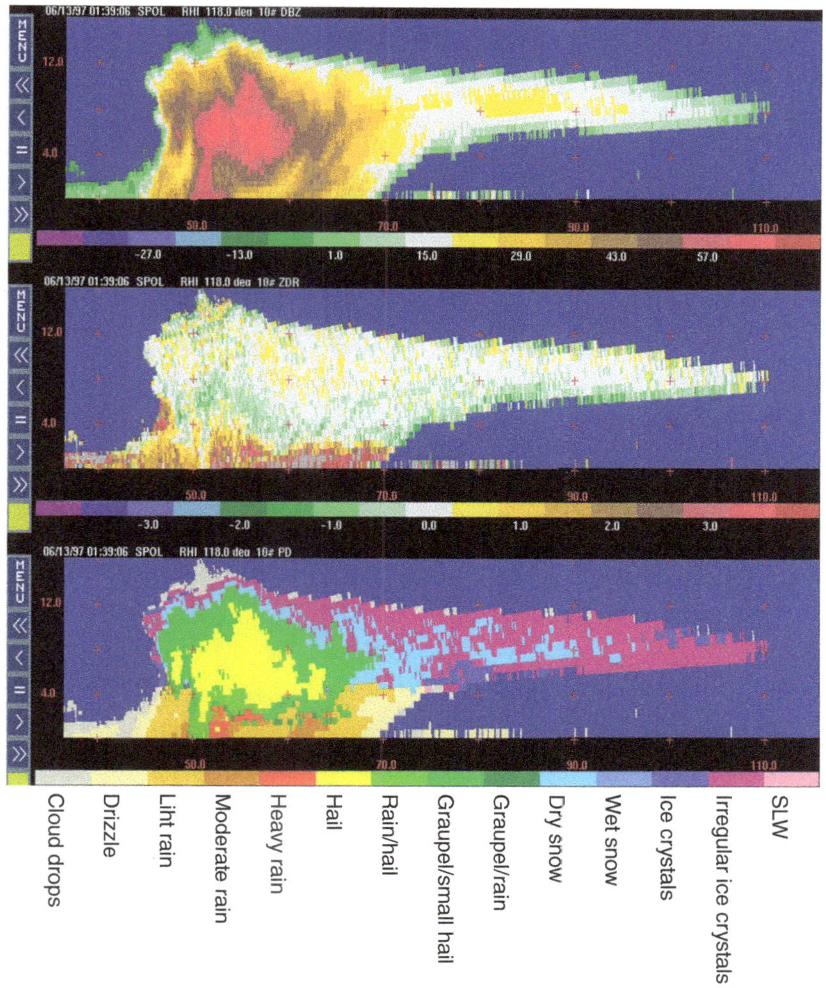


**Figure 6.35** Range–height indicator (RHI) scans of reflectivity (top panel), differential reflectivity (ZDR; middle panel), and corresponding particle classification (PID) results. Measurements were collected by the NCAR S-Pol research radar in 1997. *Source*: Vivekanandan *et al.* (1999). © American Meteorological Society. Used with permission.

severe storms. In addition, there is great value in having a dual-polarisation radar, which has the ability to detect the different hydrometeor particle types in storms and the presence of hail within storms (Vivekanandan *et al.*, 1999), as is illustrated in Figure 6.35.

With this new intense storm less than 30 km from the airport (Figure 6.33a), the nowcaster will need to decide if it will impact the airport. Since the storm steering-level flow has been moving the storms westward, the nowcaster needs to closely monitor the observations to confirm this is still the case and to also make sure that the storm is not 'back building' toward the airport. **If animation of the radar indicates the storm is moving west, then the nowcaster should not issue a TAF amendment.** It is interesting to note that because the storms are moving to the west in the same approximate direction as a portion of the ring gust front, the storms not only continue to live (Table 6.1, rule for moving boundaries), but intensify into a well-organised squall line (Figure 6.33b) propagating southwest away from Bamako. This is also evident in growth of the anvil cloud associated with the subsequent squall line (MCS) development at 17:30 UTC (Figure 6.34b) and the expansion of the anvil to the southwest (Figure 6.34c and d).

At 18:01 UTC, the nowcaster notes a gradual increase in unorganised convection east of the airport (Figure 6.33c). By 18:44 UTC, merging gust fronts appear to be organising the storms (Figure 6.33d). It is most likely that this weather is not locally driven but, rather, that it is associated with a larger scale event, possibly an easterly wave. Figure 6.36 illustrates how the approaching, somewhat unorganised, cluster of thunderstorms (MCS) becomes more organised into a multicellular line of storms through the merging of individual thunderstorm outflows into one large gust front. This front then initiates new storms along the leading edge of the MCS. Of particular concern to the nowcaster is to determine (1) whether the storm and associated lightning will reach the airport, and at what time, and (2) whether any other severe weather associated with this MCS will impact airport operations and when the impact will occur. The satellite data starting at 19:00 UTC are shown in Figure 6.37; lightning data that are typically available to the nowcaster would normally be overlaid onto the

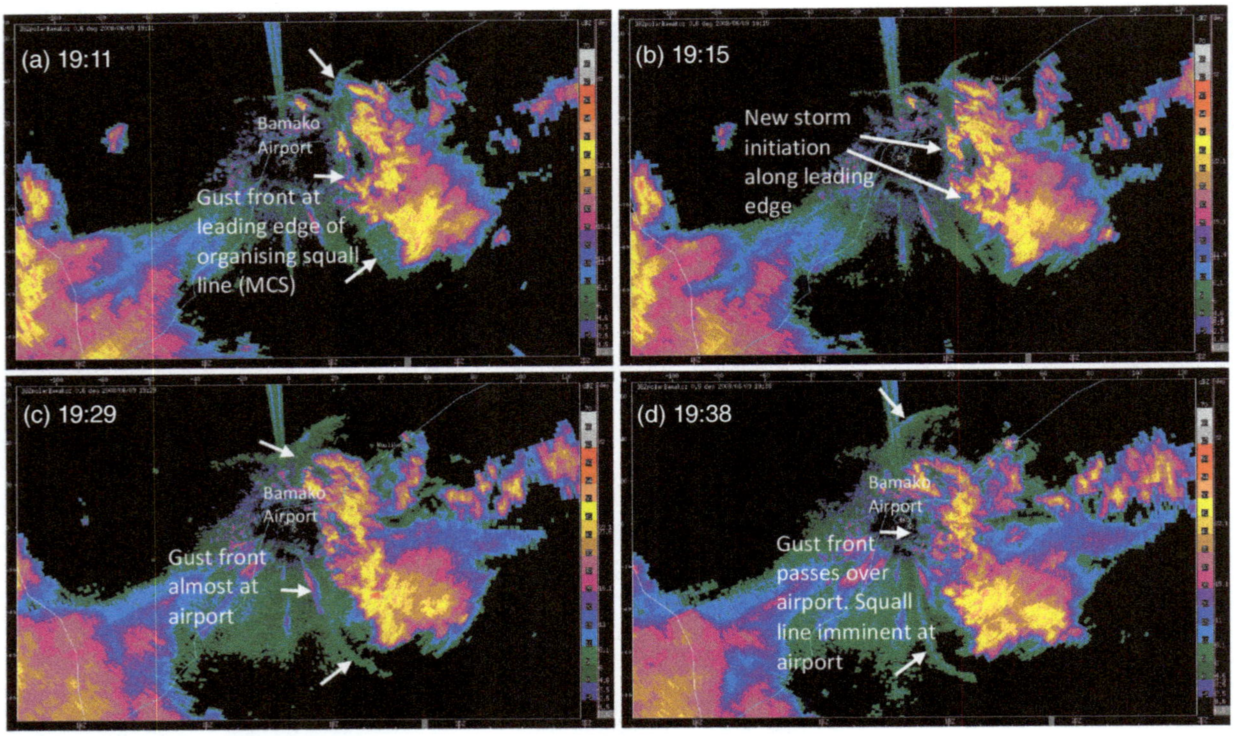


Figure 6.36 Same as Figure 6.33, but for four later time periods from 19:11 to 19:38 UTC. Small white arrows show leading edges of gust fronts.

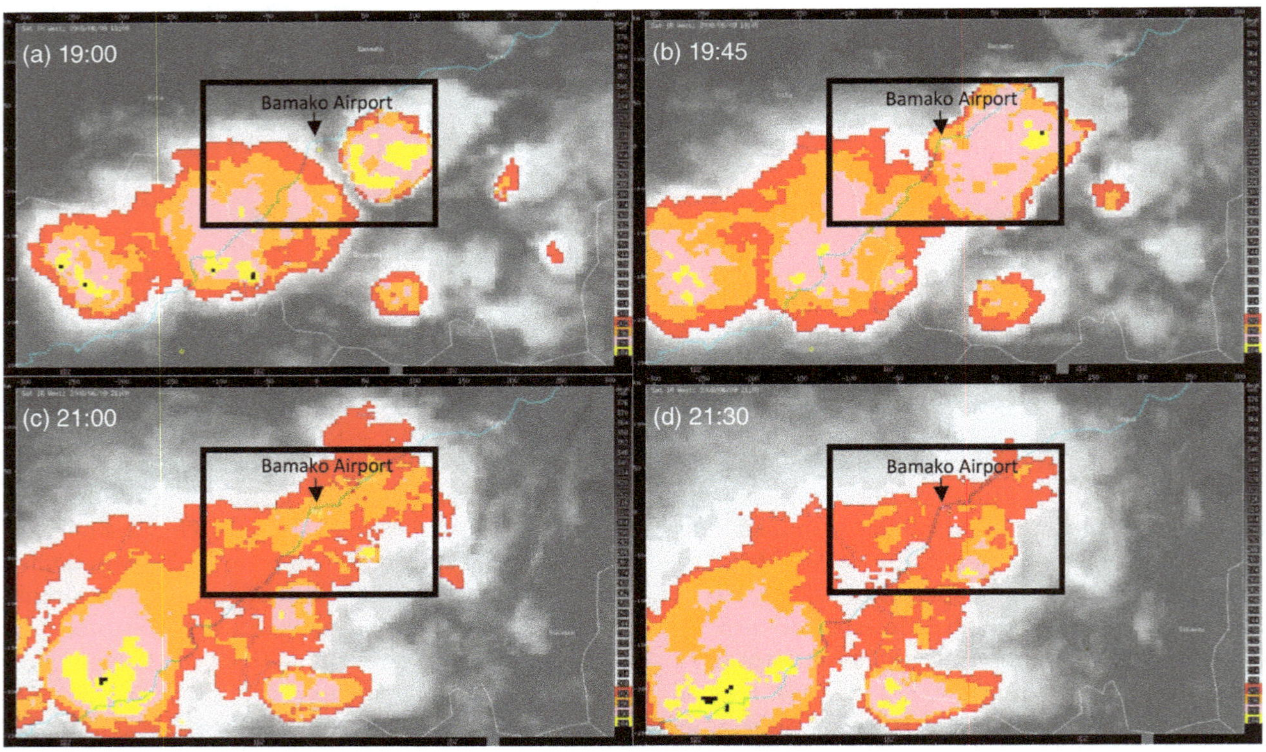


Figure 6.37 Same as Figure 6.34, except at four later time periods from 19:00 to 21:30 UTC.

satellite imagery. If surface station data exist, it will be important to know how much cooling has already occurred at the surface with the passage of the earlier gust front, as it is likely this will have affected the stability of the atmosphere in that area (i.e. it is probable that the atmosphere has been made more stable, and therefore less likely to support continued intense convection that is surface based). If the nowcaster has access to all of this information, this should affect the composition and content of the TAF, with emphasis less on heavy rain and more on lightning and strong winds. **At 19:11 UTC, based on examination of all observations and extrapolation of key features, the nowcaster should consider putting out a TAF amendment for this new weather.** A possible advanced warning for the next hour period could be:

GABS AD WRNG 2 valid 091900/2030
CB/TS OBS 25KM ENE, MOV W 22KT INTSF=

The TAF can also be amended as follows:

TAF AMD GABS 091911Z 0920/1024 22010KT 7000 TS BKN030 FEW040 CB TEMP 0919/0921 10030G40KT 1000 TSRA BECMG 0921/0923 9999 SCT040 =

As can be seen in Figure 6.36 and Figure 6.38a and b, thunderstorms and associated lightning do pass over the airport within the hour, but storm intensities have significantly weakened during the period and cloud tops are not as cold (Figure 6.37). It is likely that the MCS has diminished in intensity due to it moving into an area that has already seen substantial cooling from numerous thunderstorm outflows during the afternoon, creating a much more stable atmospheric environment at this time. It is also getting late in the day and, based on the climatology for the area, dissipation should be expected. In fact, that is exactly what is observed over the next couple of hours as the active convection associated with the MCS evolved into widespread stratiform precipitation (Figures 6.37d and 6.38d). Aside from the lightning associated with the MCS when it passed directly over the airport, the other significant weather impact for the airport is gust front passage over the airport at ~19:38 UTC (Figure 6.36d) causing possible crosswinds for the aircraft (as confirmed by Doppler radial velocities not shown here). Careful inspection of the radar data in Figures 6.36 and 6.38 shows the movement of this gust front (the thin line in reflectivity) as it moves westward over the airport.

This example demonstrates that, with access to high-resolution observations from a sensitive weather radar, it is possible for a nowcaster to understand better the physical mechanisms influencing the evolution of the convection, and so provide more precise nowcasts.

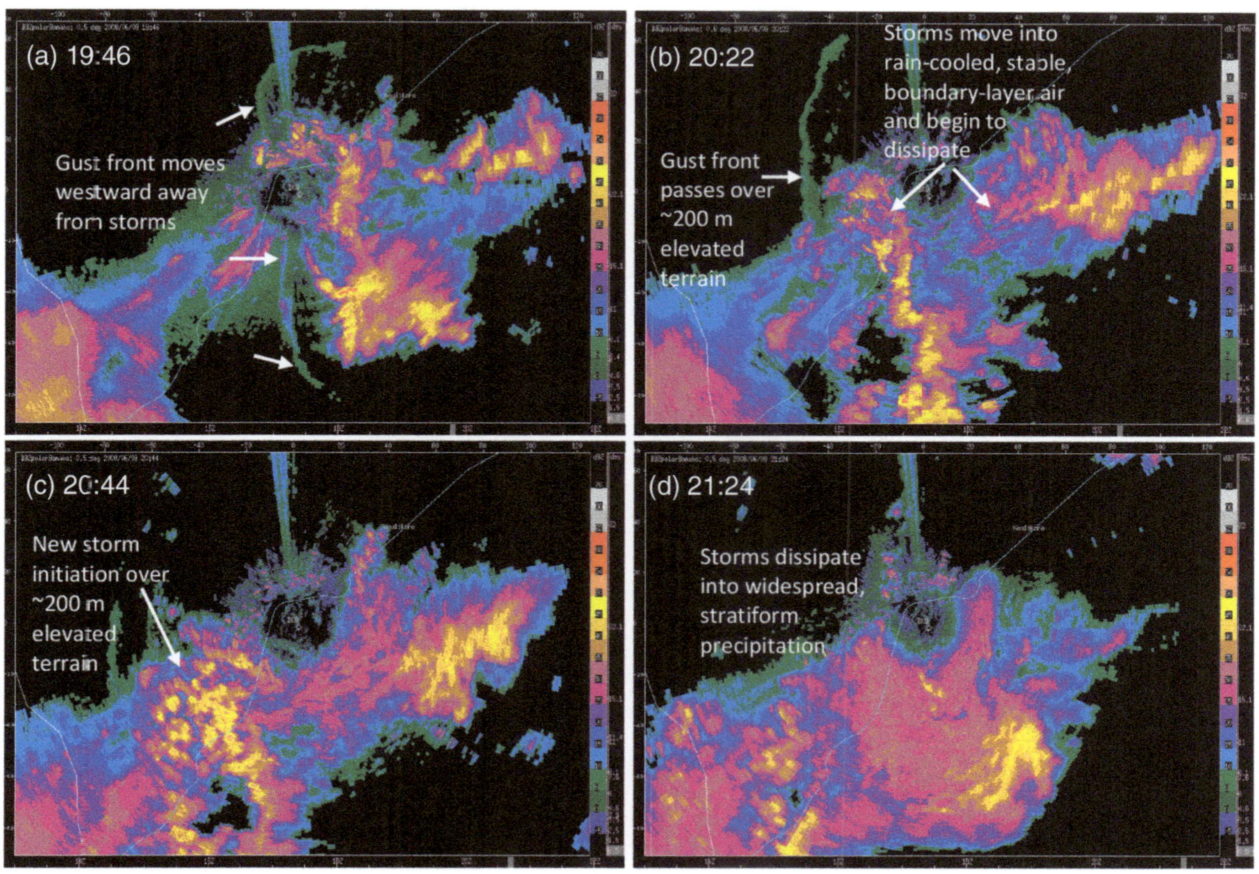


Figure 6.38 Same as Figure 6.36, but for four later time periods from 19:46 to 21:24 UTC.

### 6.2.3 Wind Shear

Wind shear poses one of the great hazards to the aviation community. In the 1970s, wind shear associated with thunderstorm density currents (gust fronts) was well documented in the scientific literature, as was the significant hazard it can present, in the form of turbulence and crosswinds, to aircraft taking off and landing. In 1975, while investigating an aircraft accident, Fujita (1976) and Fujita and Byers (1977) defined the term 'downburst' as a strong downdraft that induces an outburst of damaging winds on or near the ground, as seen in the schematic in Figure 6.39. They found that these high-intensity downdrafts delivered speeds comparable to or greater than the approximate rate of descent or climb of a jet aircraft on final approach or take-off, at 91 m (300 ft) above the surface (see Figure 6.40). The outflow winds associated with downbursts have horizontal extensions >4 km and can persist for 5–30 min. During the Joint Airport Weather Studies project (McCarthy *et al.*, 1982), Doppler radar examination of numerous thunderstorm downdrafts and outflows indicated that wind shear particularly hazardous to aircraft occurred on much smaller temporal and spatial scales. Fujita classified these smaller scale, but more hazardous, downdrafts/outflows as 'microbursts' (Figure 6.41). The damaging outflow winds associated with microbursts have horizontal extensions of ≤4 km and can persist for 2–10 min, typically (Wilson *et al.*, 1984). Microbursts and downbursts have been cited as the cause for numerous aircraft accidents. Microbursts are difficult to detect by ground-based anemometers, unless surface anemometers are spaced very close together, as is commonly done at airports using the Low-Level Wind-Shear Alert System (LLWAS; see web link in Appendix D). Doppler radars are very accurate for detecting the diverging outflows associated with microbursts and downbursts due to the high-resolution spatial coverage and rapid-update frequencies of radars. Terminal Doppler weather radars, designed specifically to detect microbursts, scan over airports every 1 min and have a horizontal spatial resolution of 150 m.

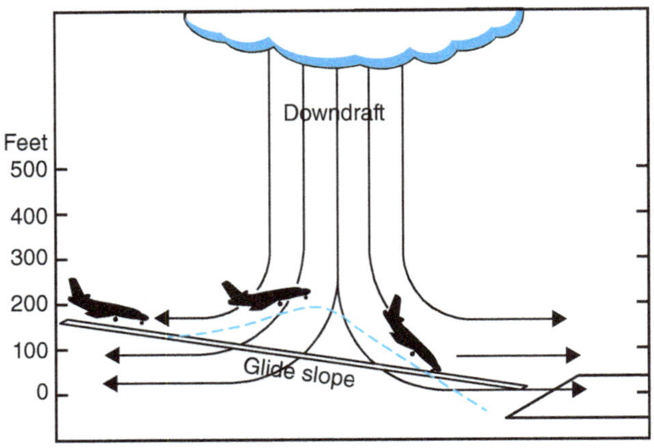


Figure 6.39 Schematic representation of the impact of a downburst on an aircraft on final approach to the runway. *Source*: McCarthy *et al.* (1982). © American Meteorological Society. Used with permission.

The earliest microburst-related aircraft accident documented by Fujita (1985) occurred at Kano, Nigeria, on

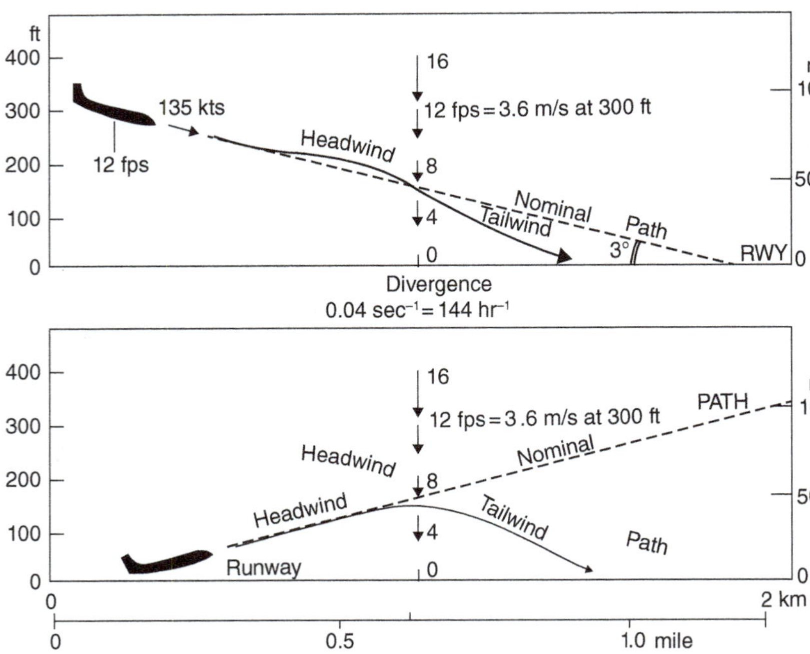


Figure 6.40 Schematic diagrams of flight paths under the influence of a downburst cell. Downward arrows represent downdraft with speeds of 3.6 m $s^{-1}$ at an altitude of <100 m (300 ft). *Source*: Fujita and Caracena (1977). © American Meteorological Society. Used with permission.

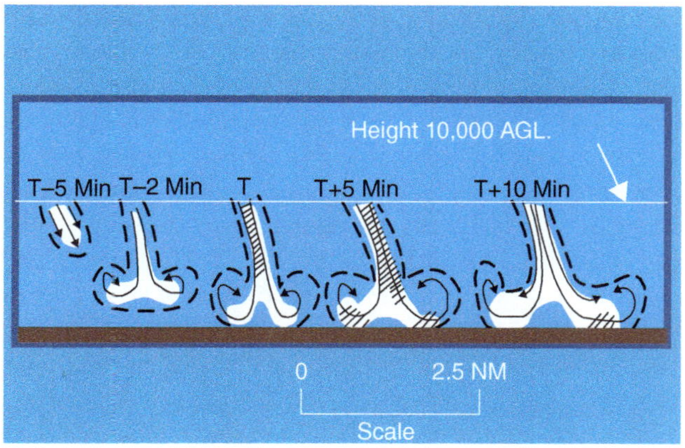


**Figure 6.41** Schematic representation of the short-lived, temporal evolution and very small spatial scale of a microburst occurring below 3000 m (10 000 ft). Scale is in nautical miles. *Source*: Wilson *et al.* (1984). © American Meteorological Society. Used with permission.

24 June 1956. The microburst was 3 km in horizontal extent, located in front of the departure end of the runway (Figure 6.42) and was associated with a thunderstorm and heavy rain. During take-off, the aircraft encountered the intense downdraft and tailwind and began to lose airspeed and altitude very quickly (Figure 6.42, bottom panel). When it corrected to level flight, the aircraft's altitude was only 4–6 m above ground! The aircraft hit a tree and then crashed; 45 people were killed.

Downbursts and microbursts can be caused by thunderstorms that have low cloud bases and heavy rain or from high-based convective clouds with mid-level dry air entrainment. Various areas in West Africa are susceptible to both types of thunderstorms; thus, wind shear events pose a very real hazard for aviation. In addition to thunderstorm-generated wind shear, wind shear conditions in West African airports can be associated with squalls (at the gust fronts), strong surface winds coupled with local topography, sea breeze fronts, hilly perturbations or mountain waves (including low-level rotors in the terminal area) and low-level temperature inversion. Detection of wind shear at airports can be obtained from LLWAS, Doppler radars, an array of surface wind and pressure sensors along approach and departure paths, or by aircraft pilot reports during climb-out or approach phases of flight.

In West Africa, when wind shear is observed or reported with high probability, it is considered to be severe. However, when reporting wind shear, pilots may use the qualifying terms 'moderate', 'strong' or 'severe', based, to a large extent, on their subjective assessment of the intensity of the wind shear encountered. Table 6.4 provides a summary of the probability of wind shear. It should be noted that these guidelines apply to operations in the airport vicinity. The clues are considered cumulative. The

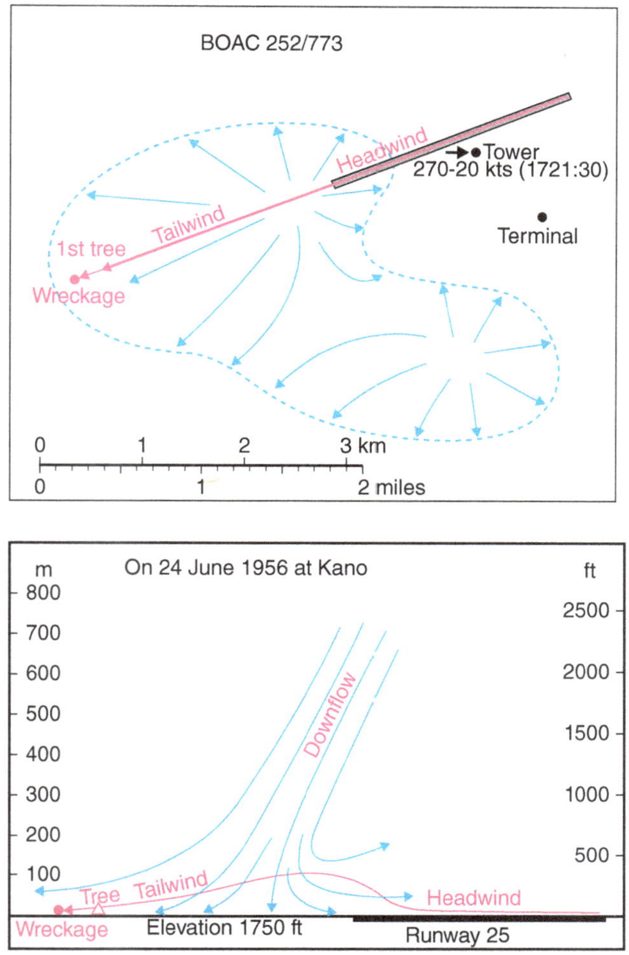


**Figure 6.42** Vertical and horizontal views of the microburst that caused the crash of BOAC Flight 252/773 accident at Kano, Nigeria. Wind patterns were reconstructed based on observations and information received from the Ministry of Transport and Civil Aviation, Her Majesty's Stationary Office, London, UK. *Source*: Fujita (1985).

hazard increases with proximity to the convective weather; hence, when those features in Table 6.4 are reported/observed, a wind shear alert/warning should be issued.

### 6.2.4 Procedure for Nowcasting Wind Shear

An event that took place on 27 August 2006 at the Léopold Sédar Senghor International Airport in Dakar, Sénégal, is discussed here to illustrate the procedure for nowcasting wind shear. Figure 6.43 gives the current weather at 06:00 UTC and the anticipated weather that was expected for 27 August 2006 at 18:00 UTC over West Africa, as issued by the African Monsoon Multidisciplinary Analysis (AMMA) team based in Niamey. Isolated convection was forecast for Sénégal for late in the afternoon, with more intense convection possible in northern Sénégal.

**Table 6.4** An example of wind shear probability guidelines for West Africa.

| S/n | Observations | | Wind shear probability |
|---|---|---|---|
| 1. | Presence of convective weather near aerodrome with: | | |
| | (a) | localised strong winds (reports of observed squall, blowing dust, rings of dust, etc.) | High |
| | (b) | heavy precipitation (observed or radar indications of contour, red or attenuation shadow) | High |
| | (c) | lightning (based on the rate of count; i.e. number of cloud to cloud or cloud to ground strikes per second) | Medium to high |
| | (d) | moderate turbulence (reported or radar indications). | Medium to high |
| 2. | On-board wind shear detection system alert (reported or observed by pilot) received from ATS unit. | | High |
| 3. | Pilot report of airspeed loss or gain: | | |
| | (a) | 15 kt or greater | High |
| | (b) | less than 15 kt | Medium |
| 4. | LLWAS alert (continues red signals) | | High |
| 5. | Abrupt change in wind speed/wind direction: | | |
| | (a) | 20 kt or greater (strong surface wind) | High |
| | (b) | less than 20 kt to 15 kt or more | Medium |
| | (c) | surface wind directional variation of 60° or more | High |
| | (d) | surface wind directional variation of 30° or more to less than 60° | Medium |
| 6. | Low-level temperature inversions. | | Medium |
| 7. | Hilly perturbations or mountain waves (including low-level rotors in the terminal area). | | High |
| 8. | Wind shear forecast from model charts. | | Low |

The sounding launched at 11:00 UTC from Dakar, Sénégal (Figure 6.44), indicated there was plenty of CAPE available for afternoon convection and no CIN. The sounding showed that the 700–600 hPa steering winds were southeasterly, and limited vertical shear was present to support multicell storms. Deep cyclonic circulation was present. Based on satellite-derived wind imagery, it was possible to see that, by late afternoon, the winds between 400 and 800 hPa had increased to 15–20 kt. It is likely that $T_c$ had already been reached by 11:00 UTC and that the atmosphere was very moist. The potential for thunderstorms in Dakar was favourable. However, on satellite imagery there is no convective activity occurring in Sénégal (Figure 6.45a) at 11:00 UTC.

From the afternoon into the evening, the satellite imagery shows convective storms developing S–SE of Dakar, Sénégal (Figure 6.45), that evolve into an MCS that moves northwestward toward Dakar Airport by 21:30 UTC (Figure 6.45d). Figure 6.46 shows higher time and space resolution satellite imagery centred on Sénégal. While individual storms can be seen south and southeast of Dakar before 21:00 UTC, by 21:30 UTC the satellite imagery shows one large anvil region of cold cloud-top temperatures associated with the MCS to the southeast of Dakar. Corresponding data from the Dakar radar at 21:00 (Figure 6.47a) show that, underneath the extensive cloud shield, there are numerous individual cells of moderate (45 dBZ) intensity over land, and even some new small storm cells just off the coast of Sénégal. These new cells grow (Figure 6.47b), and by 23:29 UTC have mostly dissipated, producing a ring gust front (Figure 6.48a) while the MCS further south continues to grow.

Animation of the radar images would have indicated to the nowcaster that the storms that produced the ring gust front were initiated by a land breeze that had moved off the coast of Sénégal. Although it is unlikely that the airport would be impacted by either the storms along the coast or the ring gust front produced by these storms, the fact that the storms produced an outflow should alert the nowcaster that the atmosphere would support gusty winds and downbursts.

At 23:29 UTC the MCS is close to Dakar: both satellite and radar show an area of widespread precipitation south of Dakar (Figure 6.48a and c). The satellite images show an extensive MCS with an overshooting top (orange and pink shades from −67 to −88°C in Figure 6.48c). However, examination of the NASA NPOL radial velocity field indicates that, in addition to potentially heavy rainfall, there are further reasons for concern. As discussed in Section 6.2.3, downbursts are composed of strong downdrafts that spread out in all directions upon reaching the surface. Radars are able to detect these diverging outflows quite accurately. Diverging outflow can be seen (Figure 6.48b) in association with a 45 dBZ embedded cell. Vertical cross-sections through the storm (Figure 6.48d and e) show the classic structure of flow converging into storm at mid-levels, supporting the downdraft, and the flow spreading out radially once it reaches the ground (Figure 6.48e).

**The nowcaster should decide, after initial observation of this downburst, whether to issue a TAF amendment for the possibility of a downburst.**[1] An advanced warning will be issued as follows:

GOOY AD WRNG 1 valid 272344/280300
CB/TS OBS 45 KM SSW MOV NW 25KT INTSF

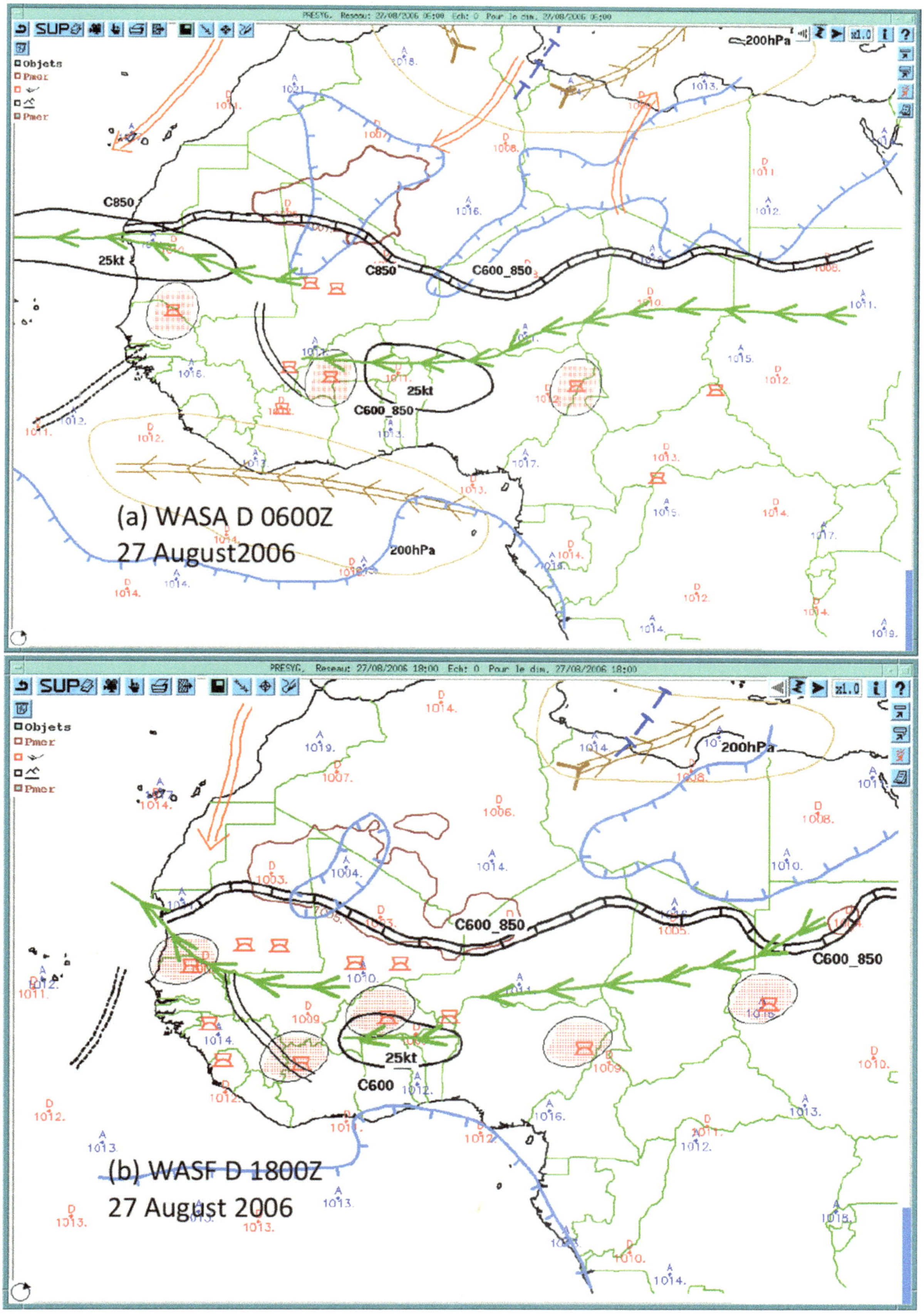


**Figure 6.43** Weather analysis (WASA) and forecast (WASF) maps for 27 August 2006 over Western Africa. Shown are the intertropical discontinuity (double-black barred contours); the African easterly jet locations (green line contours with arrowheads); the AEW trough (double-black contours); the subtropical jet (orange double-line contours with arrowheads); shaded polygons representing areas of most intense convection (MCS) estimated from the IR imagery (orange corresponding to −65°C); and isolated convective cells (identified by a Cb symbol only). See Chapter 9 for further explanations. (a) WASA at 06:00 UTC and (b) WASF for 18:00 UTC. *Source*: Météo-France, 2014.

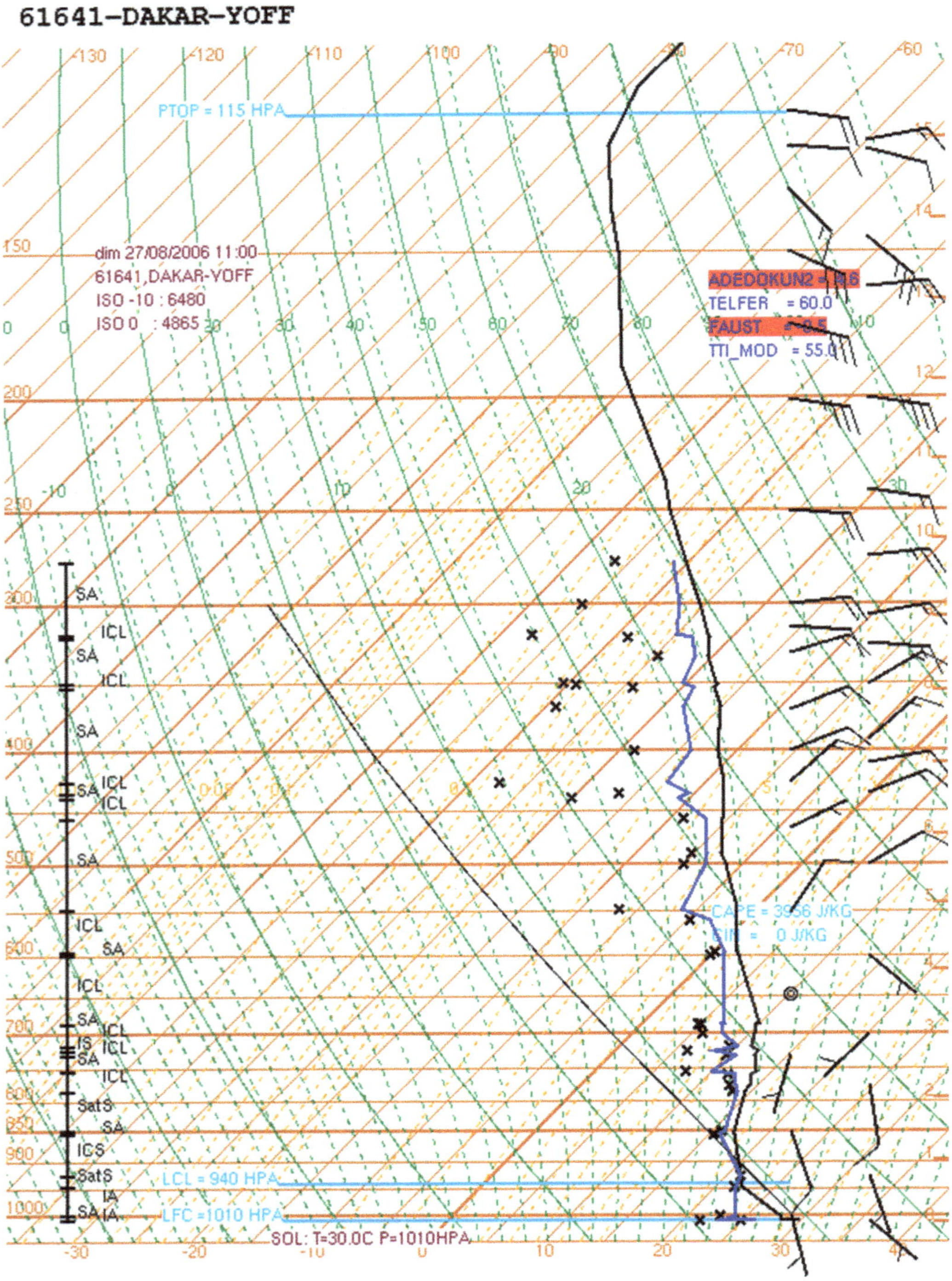


Figure 6.44 Tephigram sounding at 11:00 UTC on 27 August 2006 for Dakar, Sénégal. Sounding calculated values for CAPE and CIN were 3956 J kg$^{-1}$ and 0 J kg$^{-1}$ respectively. *Source*: AMMA Experiment web site, 2014.

If a TAF had already been issued, a possible amendment could be

TAF AMD GOOY 272344Z 2800/2906 35010KT 3000 TS BKN030 FEW040 CB TEMP 2801/2803 17020KT 800 TSRA BECMG 2803/2806 9999 SCT040 =

The diverging flow became stronger over time, and also became larger in scale (Figures 6.49 and 6.50), eventually evolving into a gust front by 23:44 UTC. Figure 6.51 shows the evolution of both the radar Doppler velocity and the satellite IR cloud-top temperatures. The radar has the advantage of being able to show the progression of the gust front toward the Dakar Airport. Had this downburst occurred closer to the Dakar Airport, the nowcaster should definitely issue a TAF amendment for a downburst. The nowcaster should also consider the

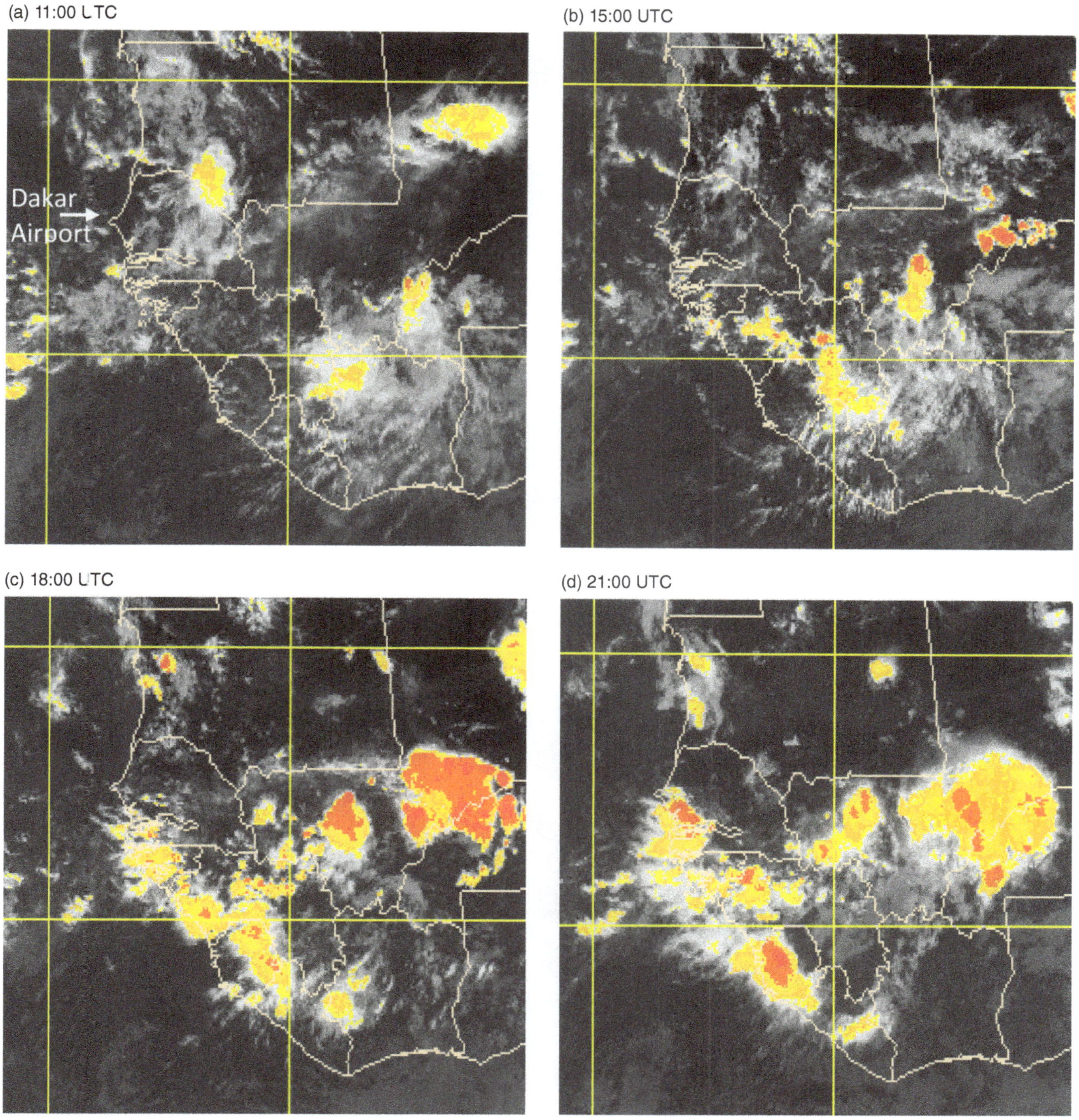


**Figure 6.45** Meteosat IR (10.8 μm) imagery on 27 August 2006 at (a) 11:00 UTC, (b) 15:00 UTC, (c) 18:00 UTC and (d) 21:00 UTC. Orange areas represent cloud-top temperatures of −65°C used for tracking intense MCSs. The approximate location of the Dakar, Sénégal, airport is indicated in (a). *Source*: AMMA Experiment web site, 2014.

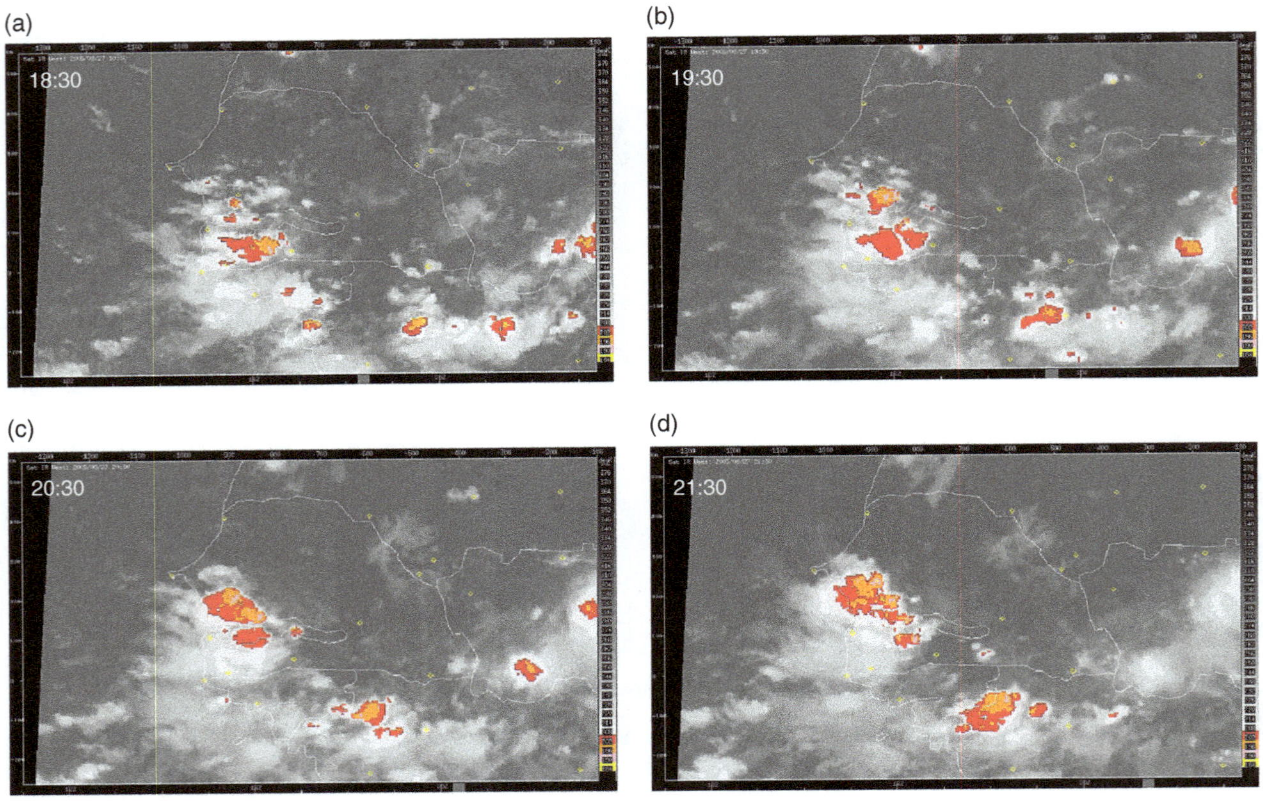


**Figure 6.46** Sequence of satellite IR imagery from 18:30 to 21:30 UTC centred on Sénégal. Orange shades represent cloud-top temperatures colder than −67°C. Country borders are outlined in white.

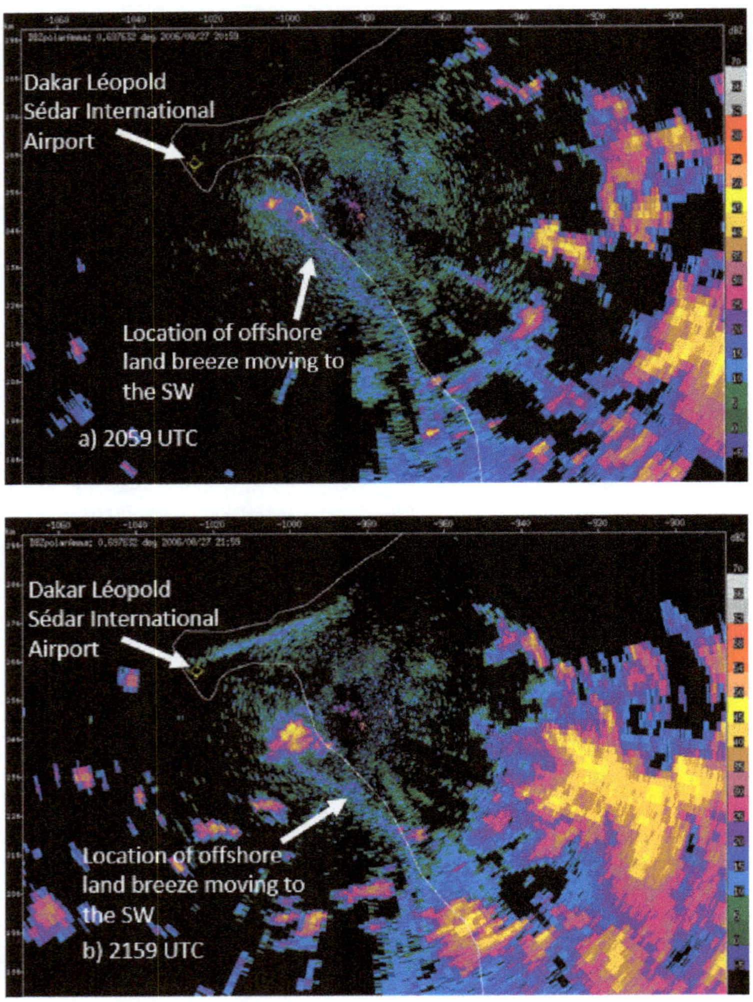


**Figure 6.47** NASA NPOL radar reflectivity image at 20:59 UTC on 27 August 2006. The white contour is the coastline of Sénégal and the yellow icon represents the location of the Dakar Airport. The radar is located 40 km to the southeast of the airport. (a) 20:59 UTC; (b) 21:59 UTC.

(a)

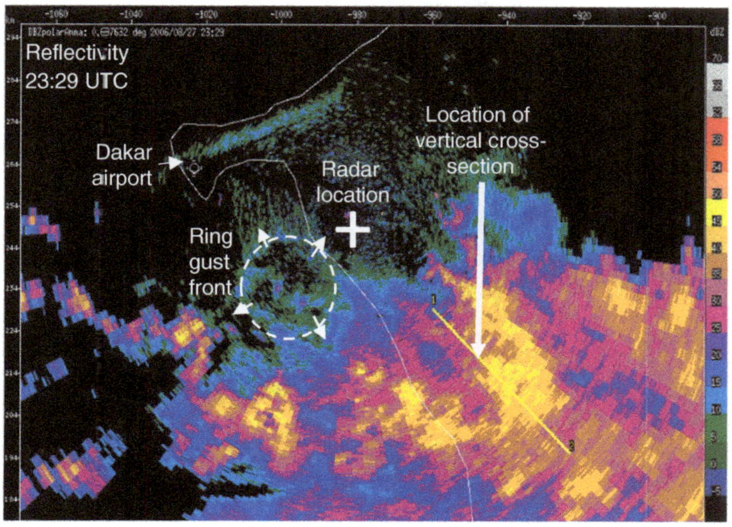


(b)

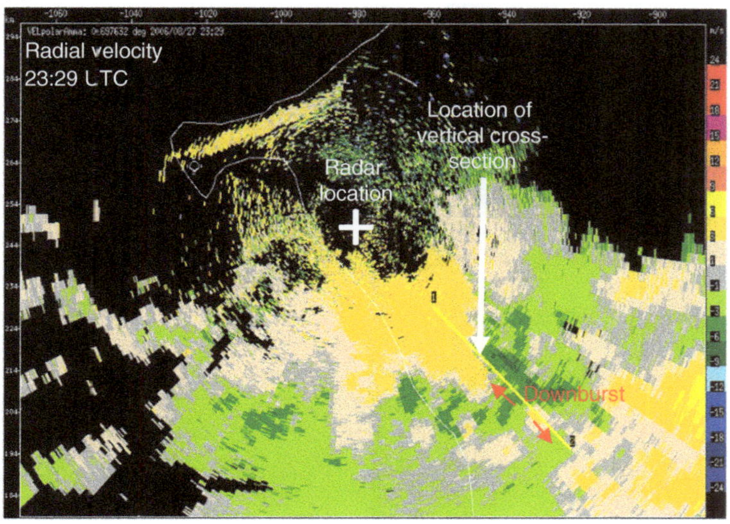


(c)

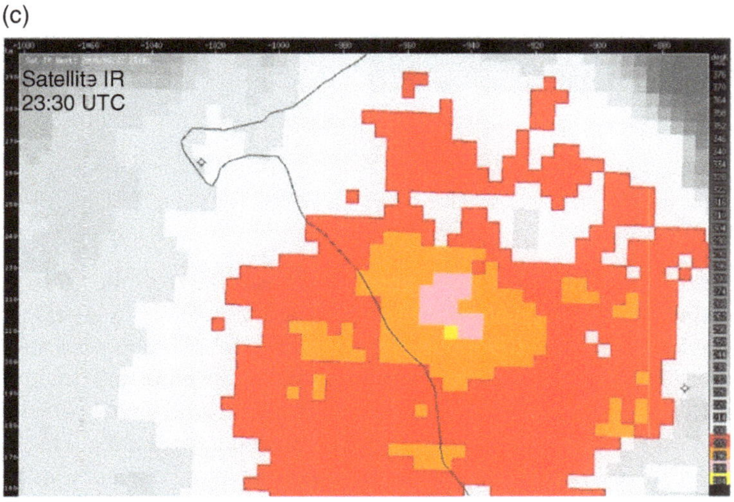


**Figure 6.48** Radar horizontal and vertical cross-sections through a downburst-producing storm in Sénégal on 27 August 2006 at 23:29 UTC. (a) Horizontal radar reflectivity (dBZ). Higher reflectivity values represent more intense storms and precipitation. (b) Horizontal radar radial velocity (m $s^{-1}$). Blue and green shades are regions of flow toward the radar, and yellow and red shades are flow away from the radar. Yellow line shows the locations of vertical cross-sections shown in (d) and (e). The red arrows in (b) show the Doppler wind directions associated with the downburst. (c) Satellite IR image at 23:30 UTC. Orange shade is −67°C and pink shade is −88°C. (d) Radar reflectivity (vertical) cross-section (dBZ). (e) Radar radial velocity (vertical) cross-section (m $s^{-1}$). Blue and green shades are regions of flow toward the radar, and yellow and red shades are flow away from the radar. Radar position is located to the left of the image.

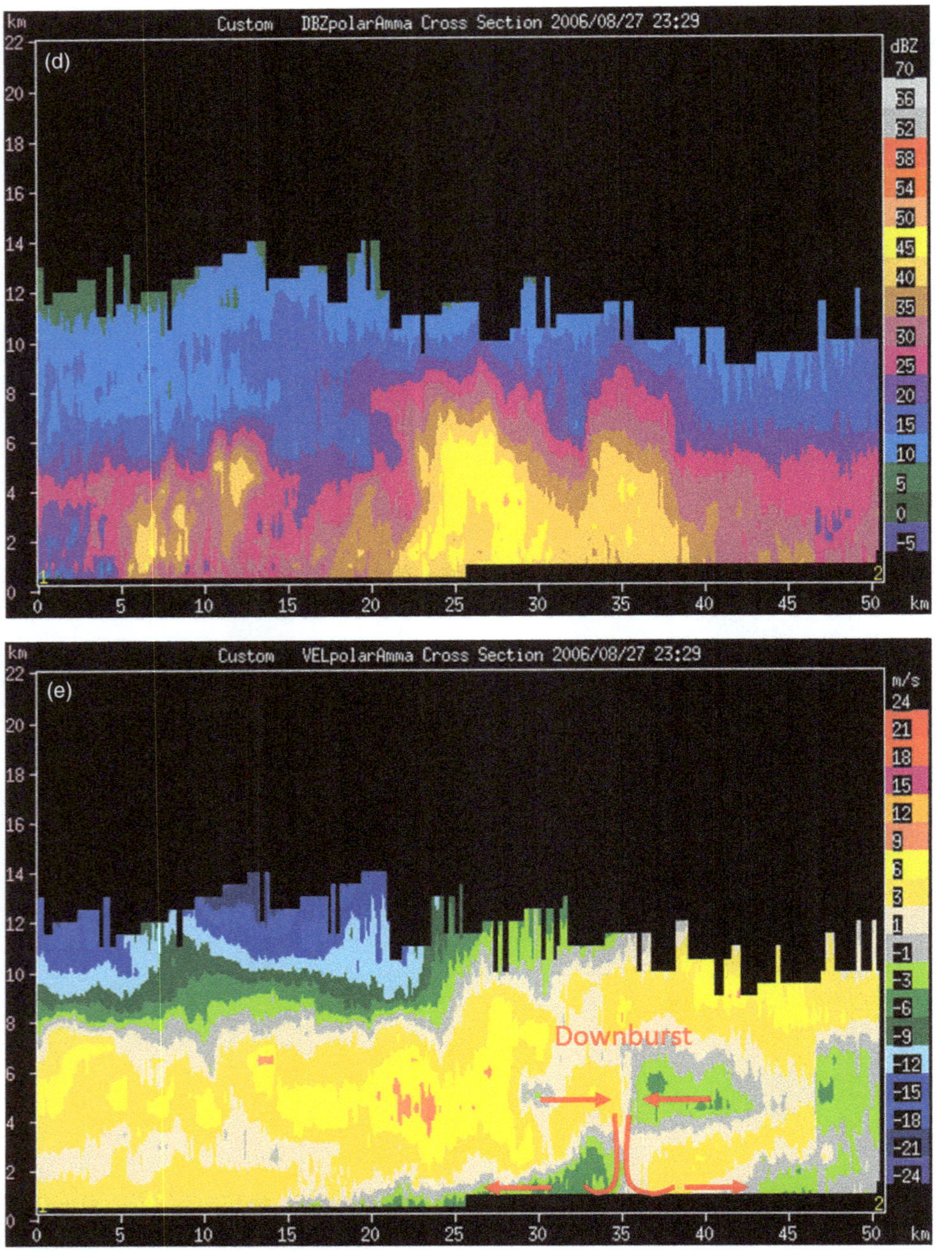


Figure 6.48 (Cont'd)

impact the gust front approaching from the southeast may have on airport operations. **At this point, the nowcaster needs to decide if the gust front will reach the airport, and what the wind direction and potential strength of the winds might be.** This information will be dealt with using METAR and SPECI.

The examples shown in Figures 6.48–6.51 demonstrate how Doppler radar can be used to observe remarkable small-scale details associated with gust fronts and downbursts, making it possible to provide life-saving nowcasts in the aerodromes.

This case is a good demonstration of the greater nowcasting capabilities that are possible if a working, calibrated Doppler radar is available. It is the particular ability of radars, to observe small-scale wind and precipitation features underneath an MCS cloud shield or anvil cloud, that enables more advanced nowcasting. This is especially true with respect to nowcasting the initiation of individual storms by boundary-layer convergence lines, detection of gust fronts and detection of downbursts.

The Dakar surface station meteogram in Figure 6.52 shows the timing of gust-front passage.

(a)

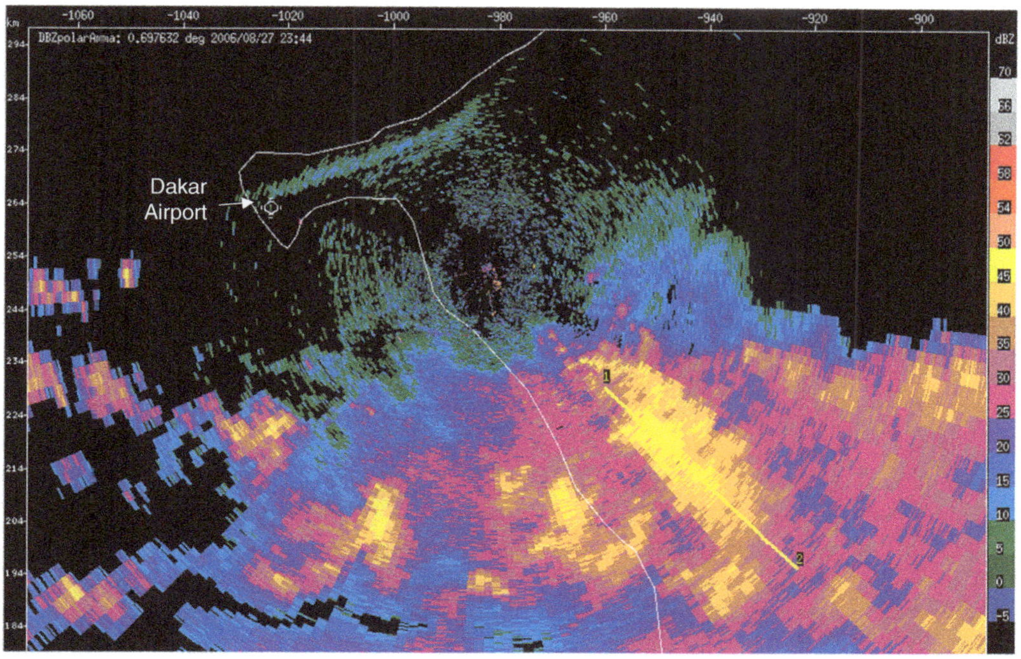


(b)

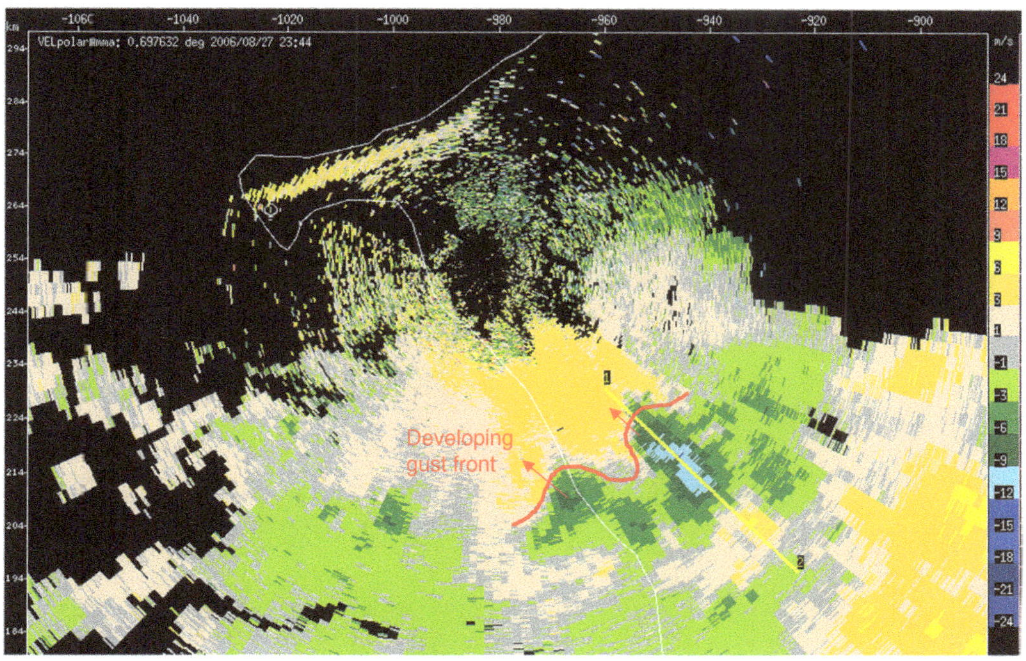


**Figure 6.49** Radar horizontal and vertical cross-sections through a downburst-producing storm in Sénégal on 27 August 2006 at 23:44 UTC. (a) Horizontal radar reflectivity (dBZ) and b) horizontal radar radial velocity (m $s^{-1}$). Blue and green shades are regions of flow toward the radar, and yellow and red shades are flow away from the radar. Yellow line shows the location of the vertical cross-sections in (c) and (d). (c) Radar reflectivity (vertical) cross-section (dBZ). (d) Radar radial velocity (vertical) cross-section (m $s^{-1}$). Blue and green shades are regions of flow toward the radar, and yellow and red shades are flow away from the radar. Radar position is located to the left of the image.

(c)

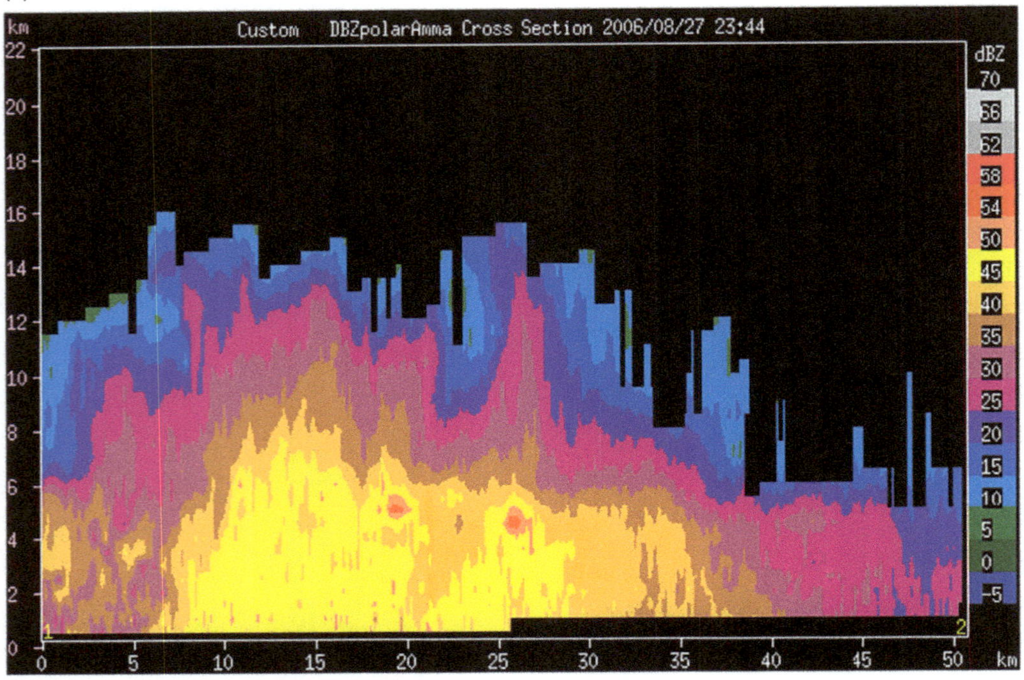


(d)

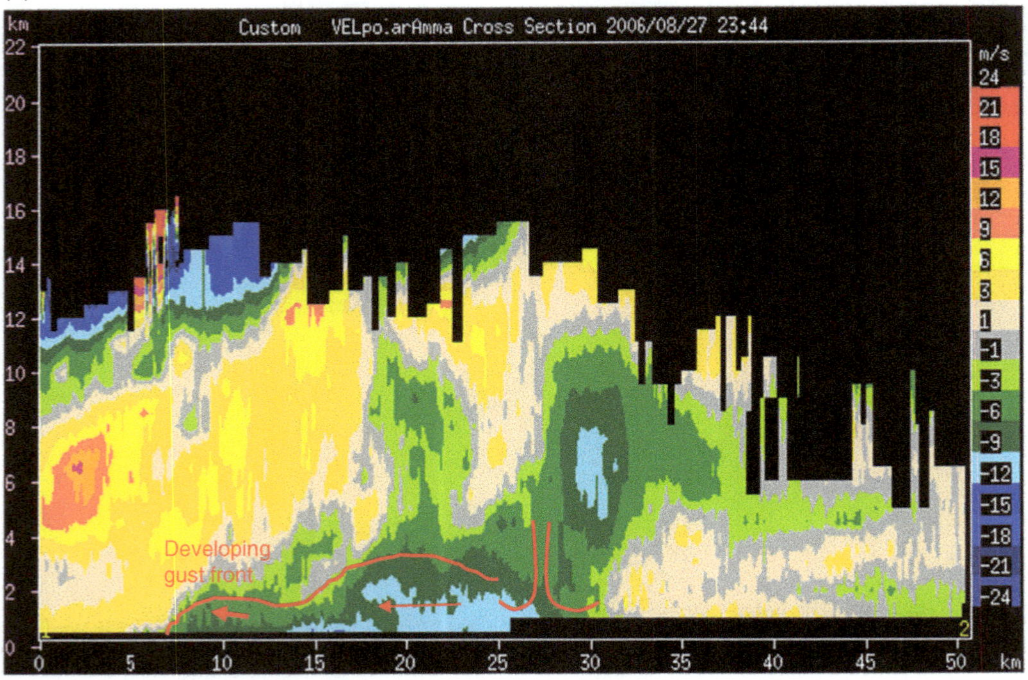


**Figure 6.49** (Cont'd)

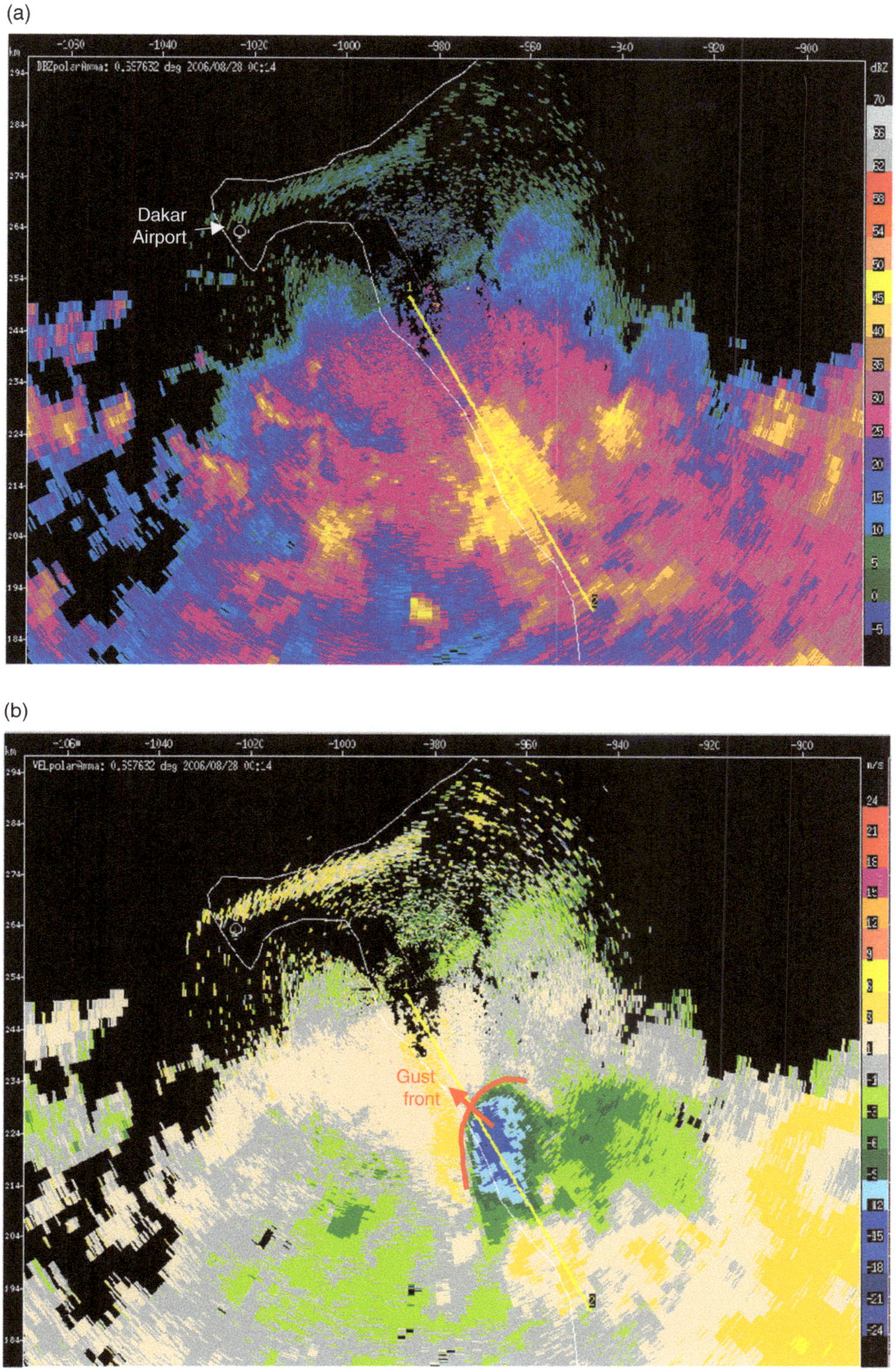


**Figure 6.50** Radar horizontal and vertical cross-sections through a downburst-producing storm in Sénégal on 27 August 2006 at 00:14 UTC. (a) Horizontal radar reflectivity (dBZ) and (b) horizontal radar radial velocity (m $s^{-1}$). Blue and green shades are regions of flow toward the radar, and yellow and red shades are flow away from the radar. Yellow line shows the location of the vertical cross-sections in (c) and(d). (c) Radar reflectivity (vertical) cross-section (dBZ). (d) Radar radial velocity (vertical) cross-section (m $s^{-1}$). Blue and green shades are regions of flow toward the radar, and yellow and red shades are flow away from the radar. Radar position is located to the left of the image.

(c)

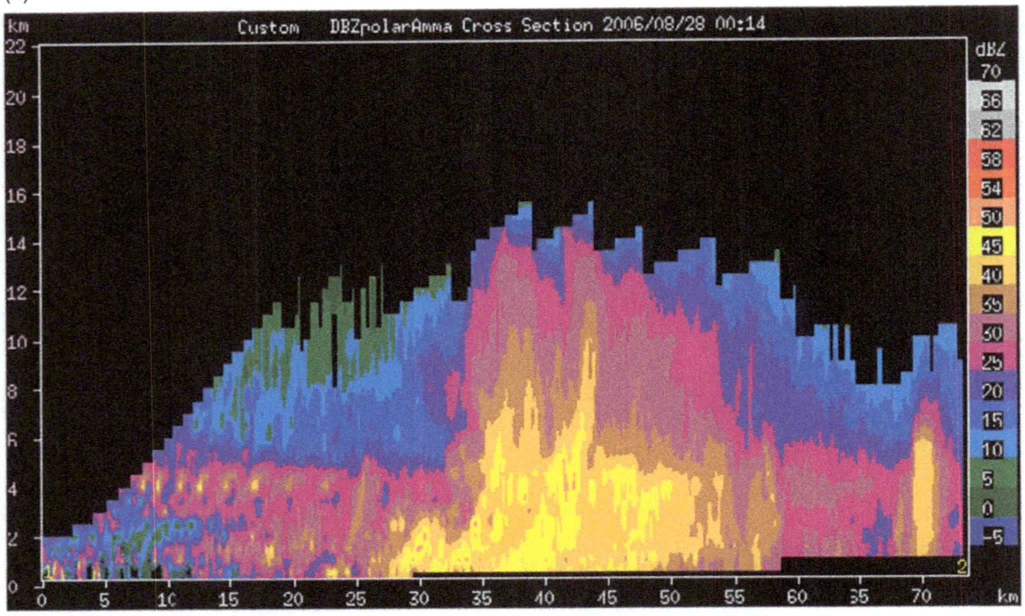


(d)

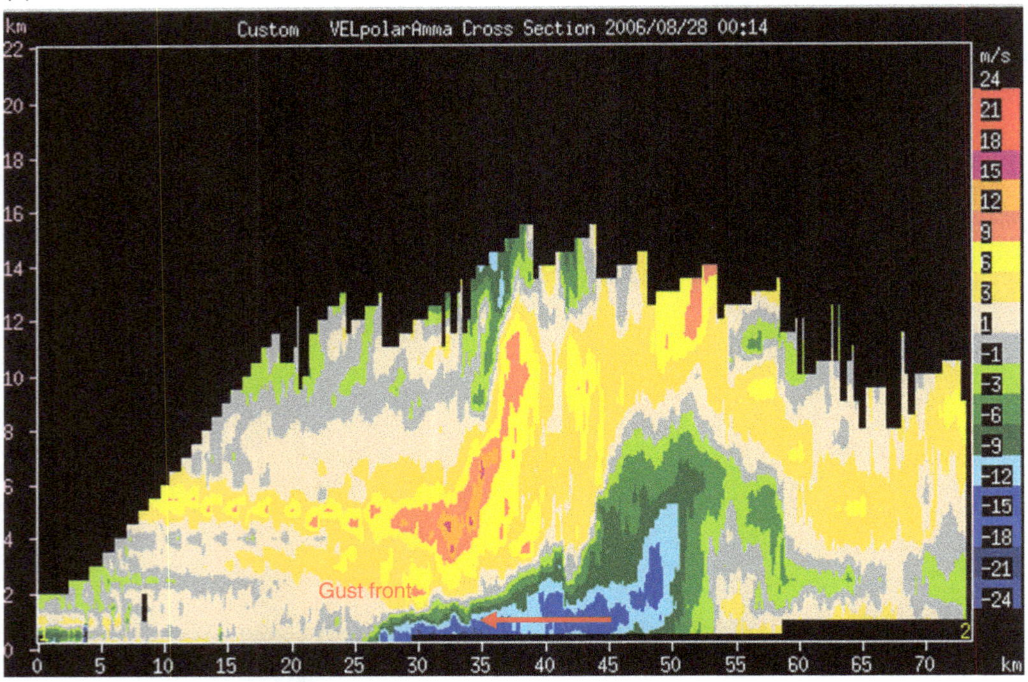


**Figure 6.50** (Cont'd)

### 6.2.5 Lightning

In the absence of radar data, lightning detection systems are particularly useful for identifying the location of thunderstorms and for tracking thunderstorm motion. In Africa, lightning-strike data are available at 5 min intervals through EUMETCAST, at 1 min resolution or 'bundled' into data packages delivered at 15 min intervals, and can be displayed on local workstations (Figure 6.53) and overlaid onto other datasets such as satellite IR imagery (Figure 6.54). The lightning

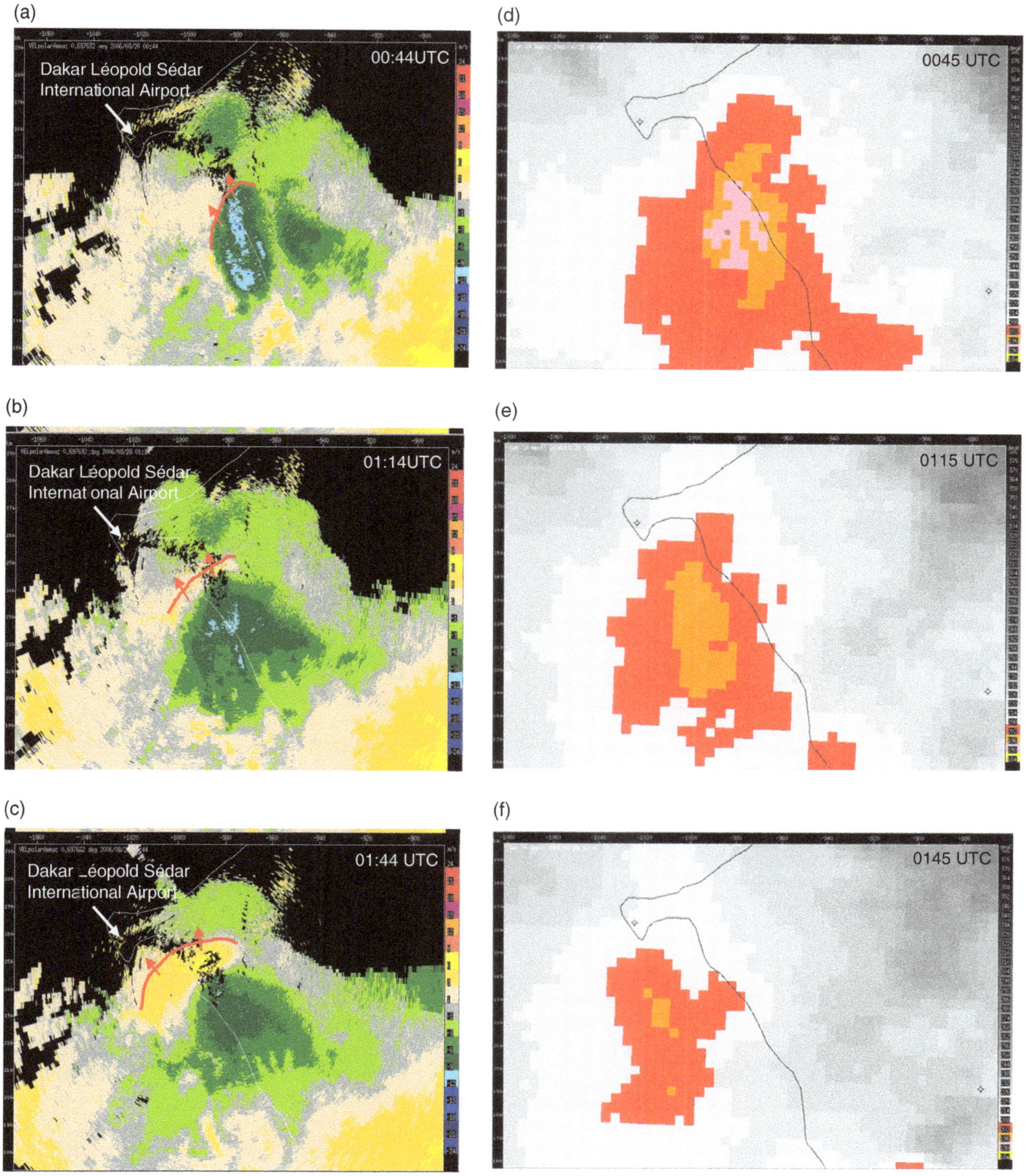


**Figure 6.51** (a)–(c) Gust-front propagation to the northwest over a 90 min period as observed in the NASA NPOL radar radial velocity field. Green and blue shades represent radial winds approaching the radar; yellow and red shades are radial winds moving away from the radar. The radar is located 40 km to the southeast of the airport. The red contour and arrows indicate the location of the leading edge of the gust front and its direction of propagation. (d)–(f) Satellite IR corresponding to the radar radial velocity data for the same time periods.

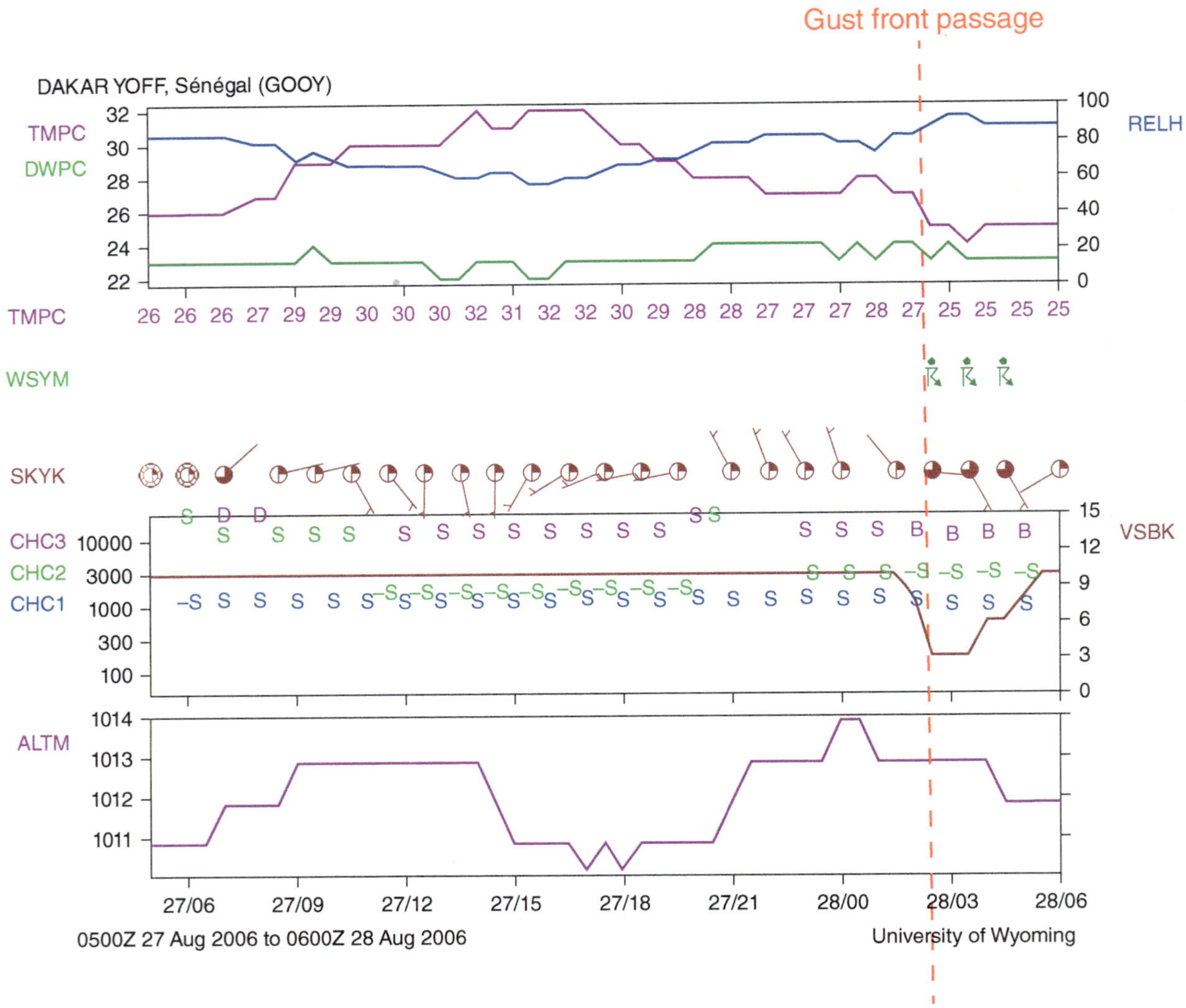


**Figure 6.52** Meteogram from surface station at Dakar, Sénégal on 27–29 August 2006. Time is shown on the horizontal axis in the form of day/hour. Gust-front passage occurred at approximated 02:15 UTC based on meteogram and NPOL radar radial velocity data. In the top panel, temperature (°C) is plotted in purple, dew point temperature (°C) is plotted in green and relative humidity is plotted in blue. In the middle panel, sky conditions are shown in red; the visibility is plotted in brown. The altimeter reading is plotted in purple in the lowest panel. Note that 7.2 mm of rains were recorded at Dakar station. *Source*: AMMA Experiment web site, 2014.

data are obtained from the UK Met Office ATD sferics system (Hooper, 1981; Lee, 1986; Nash *et al.*, 2006). The ATD system can detect about one-third of the cloud-to-cloud and cloud-to-ground lightning strikes.

### 6.2.6 Procedure for Nowcasting Lightning

ATD is particularly useful as it provides a measure of the area covered by cumulonimbus clouds. Using these data, it is possible to locate areas where one might expect to see occasional cumulonimbus clouds; this is especially so when combined with satellite imagery (e.g. Bader *et al.*, 1995: chapter 6). This aspect of the system works particularly well in the tropics. Using the lightning data to identify areas of deep convection, and combining the output with satellite imagery, it is also possible to detect MCSs.

If a forecaster knows the attributes of an MCS (its movement, speed, organisation and intensity) and the stability of the environment into which it is moving, they can predict when an MCS will move into their forecast domain using extrapolation techniques, and can nowcast when lightning is likely to occur. Using ATD sferics data, plotted onto a regional or national background, the forecaster can quickly estimate the advection of lines or areas of thunderstorms, reinforcing and modifying the weather developments they expected in longer-period forecasts. In this way, sferics data can be used by forecasters to predict the arrival and cessation of thunderstorms at a particular location using a trend that will modify and improve upon the forecast produced in a TAF.

In the absence of radar data, the sferics system also has great value in identifying developing storms in their early stages when cumulonimbus clouds are too

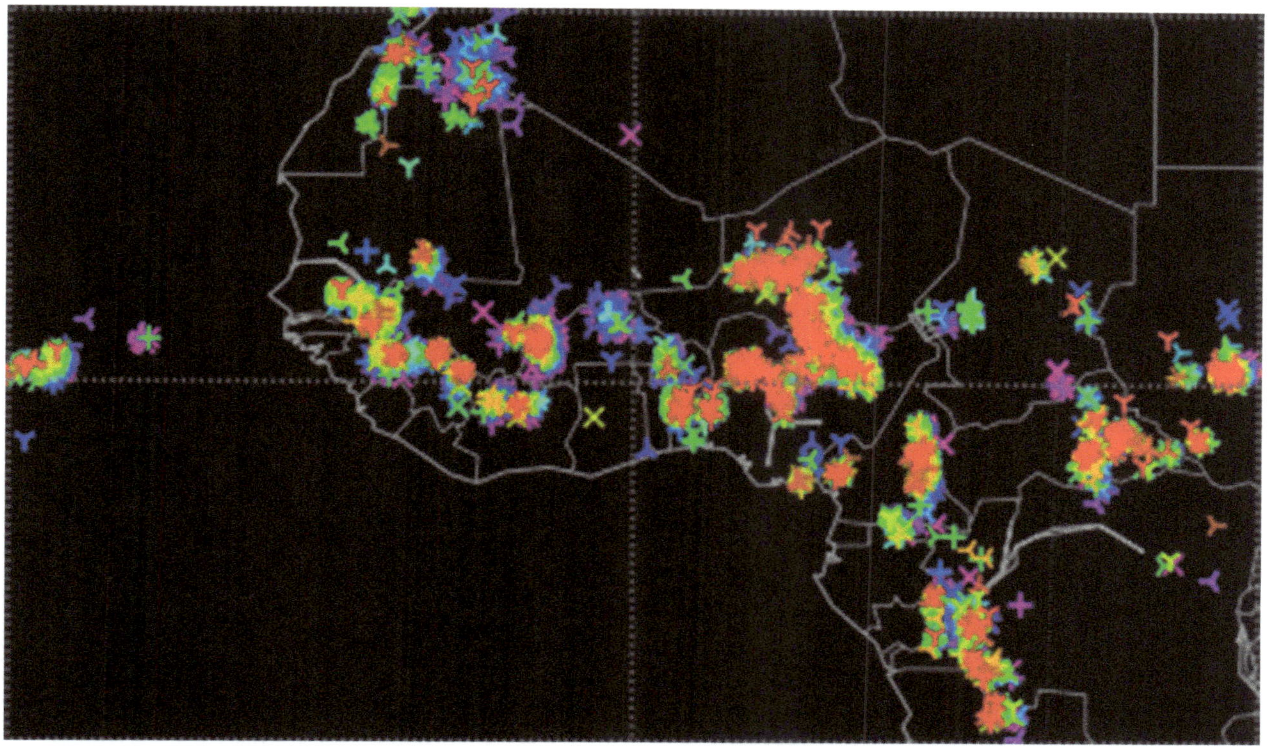

**Figure 6.53** Arrival time delay (ATD) sferics returns plotted for a 2 h period, showing the westward progress of MCS bands over West Africa on 28 April 2005. The most recent sferics (at 18:20 UTC) are shown in orange, the oldest (at 16:25 UTC) in purple. *Source*: J. Galvin, 2014.

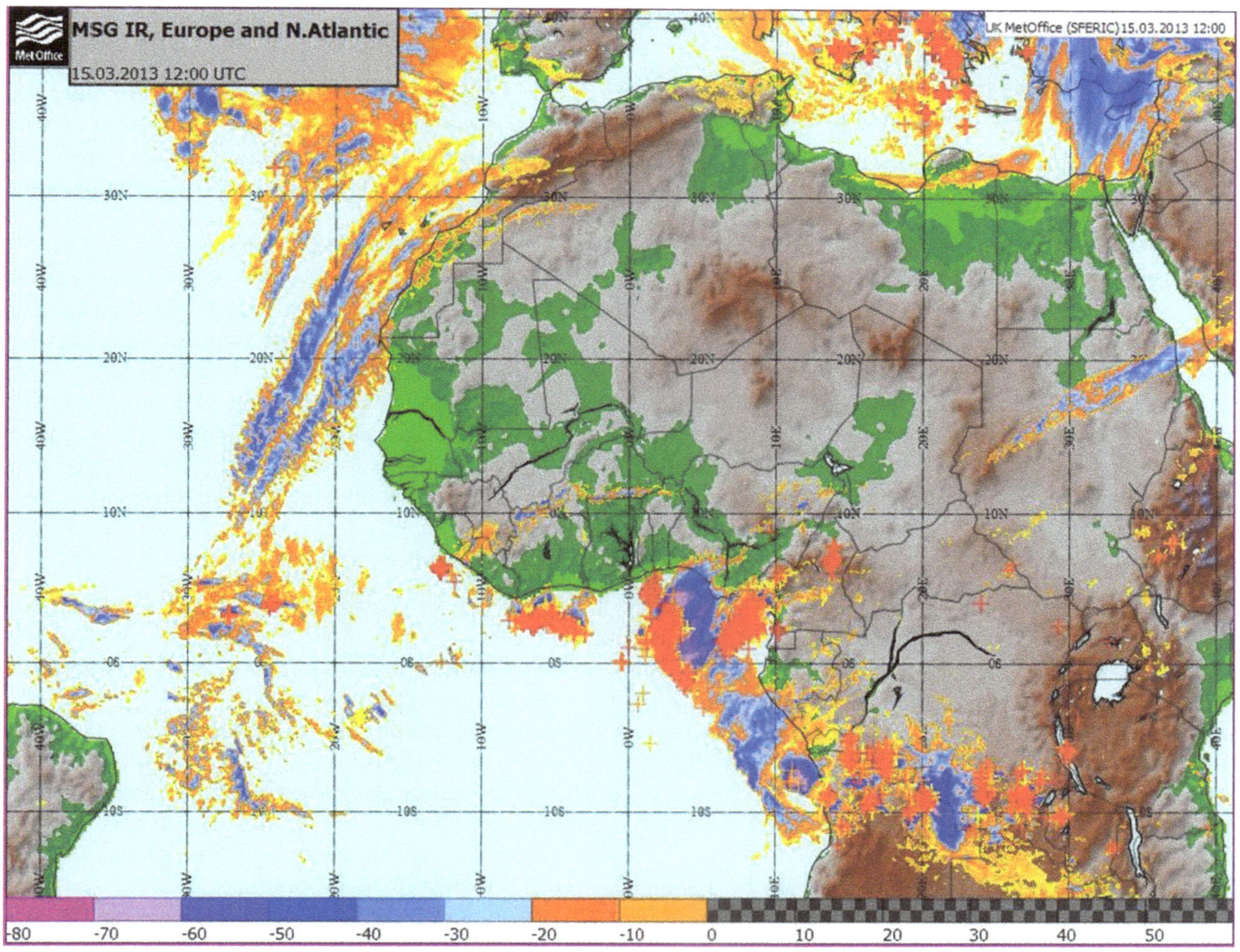


**Figure 6.54** (a) Returns from the ATDNET sferics system (crosses) over North Africa overlaid on IR cloud imagery (cloud-top temperature colour scale is shown at the bottom of the image) by the Swift/Visual Weather display system. Areas of active deep convection can be seen clearly in the extensive regions of cold cloud-top temperatures (blue shades) and high density of lightning strikes. The sferics plots are coded from red to yellow at 5 min intervals over a total of 60 min. *Source*: Met Office. © Crown Copyright.

small in horizontal extent to be clearly shown by satellite imagery, particularly at night, when the cold top of a cumulonimbus cloud must be several kilometres in diameter before it can be detected by IR sensors on weather satellites. The lightning data are also useful in the daytime when higher clouds obscure growing cumulus clouds.

## Appendix A: Acronyms

| | |
|---|---|
| BRN | bulk Richardson number |
| CAPE | convective available potential energy |
| CIN | convection inhibition |
| LFC | level of free convection |
| LLWAS | Low-Level Wind-Shear Alert System |
| MCS | mesoscale convective system |
| METAR | aerodrome routine meteorological report |
| NWP | numerical weather prediction |
| RDT | Rapid Development Thunderstorm |
| RGB | red, green, blue |
| RHI | range–height indicator |
| RMDCN | Regional Meteorological Data Communications Network in Europe |
| TAF | terminal aerodrome forecast |
| SPECI | aerodrome special meteorological report |
| WMO | World Meteorological Organization |
| ZDR | differential reflectivity |

## Appendix B: Checklist for Improving Operational Nowcasting Capability

The WMO provides several options and avenues for countries to receive assistance in education, training and research in very short-term forecasting. Responding to the question: "What else can be done?", the following recommendations were put forward by the WMO Weather Group on Nowcasting Research Group during a panel discussion at the WMO Nowcasting Symposium in Vancouver Canada in 2009:

1) Plan in advance what your equipment needs are before purchasing an expensive instrument measuring system.
2) Define and plan the reception of necessary datasets for the forecasting process via Internet, GEONETCAST (data sharing via space-based digital video broadcast (DVB) technology), or RMDCN (Regional Meteorological Data Communications Network in Europe).
3) If you cannot afford it all, at least begin by gaining access to satellite data '...it is an efficient poor man's nowcasting system'.
4) Plan how you will support the maintenance, provision of spare parts for, and repair of these systems.
5) Plan how you will develop the expertise and technical support staff in your service necessary to keep these systems operational.
6) Develop a plan for incorporating very short-term forecasting in the operational setting.
7) Document the process to make it sustainable.
8) Decide on the best sustained approach for training your forecast staff.
9) Develop a guidance system for forecasters.
10) Forecasters must assume ownership of the process of disseminating products.
11) Provide end-to-end training; this is crucial for both forecasters and emergency managers and responders.
12) Document emergency manager and end-user requirements and needs.
13) Cooperate with neighbouring countries in a regional scale, like in the WMO Virtual Laboratory regional focus group discussions.

A significant nowcasting issue is the need for rapid and easy communication of the nowcast to the users. Fortunately this problem is being solved by the advent of the internet and social media communications (Mass, 2012). In addition to the need for rapid and easy dissemination to the public, other examples are dissemination of TAFs to pilots and controllers, and high-impact weather conditions to city and emergency managers.

Overall, it is stressed that developing countries need to create an end-to-end plan; that is, a 'recipe for success' for short-term forecasting of high-impact weather for their countries. For developing countries, deficiencies in observational networks, slow internet connectivity for data transmission, and computing resources are important to overcome for providing timely nowcasts, product dissemination and warnings to end users. Know what you need and carefully add the ingredients to achieve that goal.

Common needs for achieving these goals are:

1) Need for forecasters to be trained in nowcasting concepts.
2) Need for single visualisation display for viewing all nowcasting-related datasets.
3) Common data formats for sharing of datasets and use of common applications.
4) Need to communicate with and nurture emergency managers.
5) Forecasters need to understand emergency managers' needs to provide useful products.
6) Design of nowcast guidance systems to aid the forecast process and dissemination.

# Appendix C

Table C6.1 Climatology statistics compiled at African aerodrome sites. Note: the web link listed with this table has since been updated to http://en.tutiempo.net/climate/africa.html. *Source*: Johnson *et al.* (2014).

| Country | Aérodromes | Rain Days | Thunderstorm Days | Hail Days | Fog Days |
|---|---|---|---|---|---|
| Sénégal | Cap-Skiring GOGS | 79.2 | 90.2 | 0.4 | 1.6 |
| | 2 Dakar/Yoff GOOY | 53.8 | 51.4 | 0.0 | 1.6 |
| | Saint-Louis GOSS | 35.4 | 42.2 | 1.0 | 2.4 |
| Mauritanie | 20 Nouakchott GQNN | 23.2 | 23.0 | 0.0 | 2.6 |
| | Nouadhibou GQPP | 17.4 | 7.2 | 0.2 | 9.6 |
| Mali | 8 Bamako/Senou GABS | 82.2 | 131.8 | 0.2 | 1.8 |
| | Sikasso GASK | 67.0 | 123.0 | 2.0 | 1.0 |
| | Mopti GAMB | 37.0 | 68.0 | 0.0 | 0.0 |
| | Tombouctou GATB | 26.0 | 29.0 | 0.0 | 0.0 |
| Burkina Faso | 13 Ouagadougou DFFD | 83.2 | 122.0 | 0.0 | 0.4 |
| | Babo Dioulasso DFOO | 67.0 | 93.3 | 0.5 | 1.0 |
| Niger | 17 Niamey-Aero DRRN | 58.4 | 74.4 | 0.2 | 0.0 |
| | Zinder DRZR | 29.6 | 58.0 | 0.0 | 0.0 |
| | Agadés DRZA | 18.2 | 42.0 | 0.0 | 0.0 |
| Tchad | 16 N`Djamena FTTJ | 61.8 | 96.4 | 0.2 | 0.0 |
| | Sarh FTTA | 58.5 | 115.0 | 1.5 | 1.0 |
| Guinée-Bissau | Bissau GGOV | 84.0 | 86.0 | 0.0 | 9.0 |
| Côte d`Ivoire | 7 Abidjan DIAP | 166.8 | 140.8 | 0.0 | 11.8 |
| | Yamoussoukro DIYO | 73.0 | 106.5 | 0.0 | 25.5 |
| Togo | 12 Lomé DXXX | 90.9 | 115.2 | 0.2 | 6.6 |
| | Niamtougou DXNG | 55.3 | 107.3 | 0.0 | 3.5 |
| Bénin | 10 Cotonou DBBB | 110.4 | 127.4 | 0.0 | 5.8 |
| | Parakou DBBP | 64.3 | 77.7 | 0.7 | 13.0 |
| Cameroun | 6 Douala FKKD | 210.8 | 272.2 | 0.0 | 7.4 |
| | 18 Yaoundé FKYS | 134.0 | 210.7 | 0.0 | 24.0 |
| | Garoua FKKR | 53.0 | 106.0 | 0.5 | 0.5 |
| Guinée Equatoriale | 5 Malabo FGSL | 151.0 | 234.5 | 0.0 | 0.0 |
| | 9 Bata FGBT | 117.0 | 166.0 | 0.0 | 0.0 |
| Gabon | 1 Libreville FOOL | 207.2 | 199.6 | 0.4 | 1.8 |
| | 14 Port Gentil FOOG | 135.3 | 112.3 | 0.0 | 0.0 |
| Centrafrique | 19 Banqui M`Poko FEFF | 122.7 | 189.0 | 0.3 | 27.0 |
| Congo | 4 Brazzaville FCBB | 135.2 | 169.4 | 0.4 | 3.0 |
| | 3 Pointe Noire FCPP | 153.4 | 116.2 | 0.2 | 4.2 |
| Comores | 15 Moroni FMCH | 137.0 | 51.0 | 2.3 | 0.8 |
| Madagascar | 11 Antananarivo FMMI | 137.6 | 119.4 | 0.6 | 39.6 |
| | Mahajanga FMNM | 80.4 | 70.4 | 0.4 | 1.2 |
| | Toamasina FMMT | 187.6 | 41.8 | 0.6 | 0.2 |
| | Tolagnaro FMSD | 131.2 | 29.2 | 0.2 | 2.0 |
| | Nosy-Bé/Fascene FMNN | 51.0 | 37.3 | 0.0 | 0.3 |
| Maroc | Agadir GMAD | 27.8 | 7.8 | 0.0 | 29.2 |
| Gambie | Banjul/Yundum GBYD | 76.0 | 86.8 | 0.4 | 3.0 |
| Libéria | Monrovia GLRB | 193.0 | 148.0 | 0.0 | 22.3 |
| São Tomé & Principe | São Tomé FPST | 80.0 | 119.0 | 0.0 | 0.0 |

REFERENCE DATA SOURCE: http:/www.tutiempo.net/en/climate/

Most of the data entries above (shown in a bold font) are -year annual averages. Regular (non-bold) font entries indicate that data for all years are not available for that airport (i.e. 2–4 years of data only). The entries shown in a light blue font indicate that only data for a single year is available.

11 The dark, numbered squares on many of the airports' data line identify the 20 busiest ASECNA airports, in terms of their annual number of aircraft operations ( as obtained from a recent ASECNA Annual Report).

## Appendix D: Online Web Sites Containing Weather-related Training Materials

Tephigram mastery: https://www.meted.ucar.edu/training_module.php?id=1018#.Ux-G84XYNc4

Review of aeronautical meteorology – Africa: https://www.meted.ucar.edu/training_course.php?id=36

Forecasting fog for aviation – Kenya case study: https://www.meted.ucar.edu/asmet/e_africa_a7

Nowcasting for aviation in Africa: https://www.meted.ucar.edu/avn_int/nowcasting

WRF-EMS aviation products: https://www.meted.ucar.edu/avn_int/wrf_avnprods

Writing TAFs for convective weather, 2nd edition: https://www.meted.ucar.edu/dlac2/mod2_v2/navmenu.php?tab=1&page=1.0.0&type=flash

Writing TAFs for ceilings and visibility, Africa edition: https://www.meted.ucar.edu/avn_int/taf_africa_cv

Virtual Laboratory for Training and Education in Satellite Meteorology (VLab): http://www.wmo-sat.info/vlab

For information on wind shear, see the following publications and web link:

ICAO Manual on Low-Level Wind Shear, Doc 9817 AN/449 (English version)

ICAO Manual on Low-Level Wind Shear, Doc 9817 (French version)

*ICAO Journal*, **62**(2), 2007, 16–20.

http://llwas-technology.net/references/references-to-llwas-literature

## Note

1 The TAF amendment should include wording about the possibility that downbursts could occur in the vicinity of the airport. Experience has shown that if one downburst occurs, then there could easily be more downbursts in the area. Thus, the forecaster should put out a TAF amendment or advisory for possibility of downbursts. This is important, as it will get the pilot's immediate attention.

## References

Atkins NT, Wakimoto RM, Weckwerth TM. 1995. Observations of the sea breeze front during CaPE. Part II: dual-Doppler and aircraft analysis. *Mon. Weather Rev.* **123**: 944–969.

Bader MJ, Forbes GS, Grant JR, *et al.* 1995. *Images in Weather Forecasting. A Practical Guide for Interpreting Satellite and Radar Imagery*. Cambridge University Press; 499 pp.

Battan LJ. 1973. *Radar Observations of the Atmosphere*. University of Chicago Pres; 324 pp.

Benjamin SG, Dévényi D, Weygandt SS, *et al.* 2004. An hourly assimilation–forecast cycle: the RUC. *Mon. Weather Rev.* **132**: 495–518.

Benjamin S, Hu M, Weygandt S, Devenyi D. 2009. Integrated assimilation of radar, satellite, and METAR cloud data for initial hydrometeor/divergence fields to improve hourly updated short-range forecasts from the RUC, Rapid Refresh, and HRRR. In: *WMO WWRP Symposium on Nowcasting and Very Short Term Forecasting*, Whistler, British Columbia, 31 Aug–4 Sept.

Bringi VN, Vivekanandan J, Tuttle JD. 1986. Multiparameter radar measurements in Colorado convective storms. Part II. Hail detection studies. *J. Atmos. Sci.* **43**: 2564–2577.

Browning KA. 1982. *Nowcasting*. Academic Press: London; 256 pp.

Byers HR, Braham RR, Jr. 1949. *The Thunderstorm*. US Government Printing Office: Washington, DC; 187 pp.

Chisholm AJ, Renick JH. 1972. The kinematics of multicell and supercell Alberta hailstorms. Alberta Hail Studies, Research Council of Alberta Hail Studies, Rep. 72-2, Edmonton, Canada; pp. 24–31.

Crook NA. 1996. Sensitivity of moist convection forced by boundary layer processes to low-level thermodynamic fields. *Mon. Weather Rev.* **124**: 1767–1785.

Dai A, Giorgi F, Trenbreth KE. 1999. Observed and model-simulated diurnal cycles of precipitation over the contiguous United States. *J. Geophys. Res.* **104**: 6377–6402.

De Coning E, Koenig M, Olivier J. 2011. The combined instability index: a new very short-range convection forecasting technique for southern Africa. *Meteorol. Appl.* **18**: 421–439.

Dione C, Lothan M, Badiane D, *et al.* 2014. Phenomenology of Sahelian convection observed in Niamey during the early monsoon. *Q. J. R. Meteorol. Soc.* **140**: 500–516.

Dixon M, Wiener G. 1993. TITAN: thunderstorm identification, tracking, analysis and nowcasting – a radar-based methodology. *J. Atmos. Oceanic Technol.* **10**: 785–797.

Done J, Davis CA, Weisman ML. 2004. The next generation of NWP: explicit forecasts of convection using the Weather Research and Forecasting (WRF) model. *Atmos. Sci. Lett.* **5**: 110–117.

Foote GB, Mohr CG. 1979. Results of a randomized hail suppression experiment in northeast Colorado. Part IV: post hoc stratification by storm type and intensity. *J. Appl. Meteorol.* **18**: 1589–1600.

Fowle MA, Roebber PJ. 2003. Short-range (0–48 h) numerical prediction of convective occurrence, mode, and location. *Weather Forecast.* **18**: 782–794.

Fujita TT. 1976. Spearhead echo and downburst near the approach end of a John F. Kennedy Airport runway, New York City. SMRP Research Paper 137, University of Chicago, Chicago, IL; 51 pp.

Fujita TT. 1985. *The Downburst.* Satellite and Mesometeorology Research Project, University of Chicago; 122 pp.

Fujita TT, Byers HR. 1977. Spearhead echo and downburst in the crash of an airliner. *Mon. Weather Rev.* **105**: 129–146.

Fujita TT, Caracena F. 1977. An analysis of three weather-related aircraft accidents. *Bull. Am. Meteorol. Soc.* **58**: 1164–1181.

Garstang M, Cooper HJ. 1981. The role of near surface outflow in maintaining convective activity. In: *Proceedings, Nowcasting-I Symposium*, Copenhagen, ESA SP-165; pp. 161–168.

Golding BW. 1998. Nimrod: a system for generating automated very short range forecasts. *Meterorol. Appl.* **5**: 1–16.

Henry SG. 1993. Analysis of thunderstorm lifetime as a function of size and intensity. In: *26th Conference on Radar Meteorology*, Norman, OK. American Meteorological Society; pp. 138–140 (preprint).

Hogan RJ, O'Connor EJ, Illingworth AJ. 2009. Verification of cloud-fraction forecasts. *Q. J. R. Meteorol. Soc.* **135**: 1494–1511.

Hooper AH. 1981. An arrival-time difference sferics system. In: *Second WMO Technical Conference on Instruments and Methods of Observation (TECIMO-II)*, Mexico City, 13–17 October; pp. 59–66.

Joe P, Burgess D, Potts R, *et al.* 2004. The S2K severe weather detection algorithms and their performance. *Weather Forecast.* **19**: 43–63.

Johnson DB, Chapman M, Roberts R, Kucera P. 2014. A technical and scientific evaluation of wind shear and rainstorm hazard detection systems for L'Agence pour la Sécurité de la Navigation Aérienne en Afrique et à Madagascar (ASECNA). Report #1.

Kessler E, Russo JA. 1963. Statistical properties of weather radar echoes. In: *Proceedings, 10th Weather Radar Conference*, Washington, DC. American Meteorological Society; pp. 25–33 (preprint).

Koenig M, de Coning E. 2009. The MSG Global Instability Indices product and its use as a nowcasting tool. *Weather Forecast.* **24**: 272–285.

Lee ACL. 1986. An operational system for the remote location of lightning flashes using a VLF arrival time difference technique. *J. Atmos. Oceanic. Technol.* **3**: 630–642.

Laing AG, Trier SB, Davis CA. 2012. Numerical simulation of episodes of organized convection in tropical northern Africa. *Mon. Weather Rev.* **140**: 2874–2886.

Lima MA, Wilson JW. 2008. Convective storm initiation in a moist tropical environment. *Mon. Weather Rev.* **136**: 1847–1864.

Madougou S. 2010. Etude du potential éolien du jet nocturne dans la zone sahélienne à partir des observations de radars, profileurs de vent. PhD thesis, l'Université de Toulouse; 214 pp.

Mahoney WP, III. 1988. Gust front characteristics and the kinematics associated with interacting thunderstorm outflows. *Mon. Weather Rev.* **116**: 1474–1491.

Mass C. 2012. Nowcasting the promise of new technologies of communication, modeling, and observations. *Bull Am. Meteorol. Soc.* **93**: 797–809.

Mass CF. 2003. IFPS and the future of the National Weather Service. *Weather Forecast.* **18**: 75–79.

Mass CF, Ovens D, Westrick K, Colle B. 2002. Does increasing horizontal resolution produce more skillful forecasts? *Bull. Am. Meteorol. Soc.* **83**: 407–430.

McCarthy J, Wilson JW, Fujita TT. 1982. The Joint Airport Weather Studies Project. *Bull. Am. Meteorol. Soc.* **63**: 15–22.

Mueller C, T Saxen, Roberts R, *et al.* 2003. NCAR Auto-Nowcast system. *Weather Forecast.* **18**: 545–561.

Nash J, Atkinson NC, Hibbett E, *et al.* 2006: The new Met Office ATDNET lightning detection system. http://wmo.int/pages/prog/www/IMOP/publications/IOM-94-TECO2006/2(7)_Nash_UK.pdf (accessed 27 July 2016).

Olson DA, Junker NW, Korty B. 1995. Evaluation of 33 years of quantitative precipitation forecasting at the NMC. *Weather Forecast.* **10**: 498–511.

Purdom JFW. 1973. Satellite imagery and the mesoscale convective forecast problem. In: *8th Conference on Severe Local Storms*, Denver. American Meteorological Society; pp. 244–251 (preprint).

Purdom JFW. 1976. Some uses of high resolution GOES imagery in the mesoscale forecasting of convection and its behavior. *Mon. Weather Rev.* **104**: 1474–1483.

Purdom JFW. 1982. Subjective interpretations of geostationary satellite data for nowcasting. In: *Nowcasting*, K Browning (ed.). Academic Press: London; pp. 149–166.

Roberts RD, Rutledge S. 2003. Nowcasting storm initiation and growth using GOES-8 and WSR-88D data. *Weather Forecast.* **18**: 562–584.

Roberts RD, Wilson JW. 1989. A proposed microburst nowcasting procedure using single-Doppler radar. *J. Appl. Meteorol.* **28**: 285–303.

Roberts RD, Anderson ARS, Nelson E, *et al.* 2012. Impacts of forecaster involvement on convective storm initiation and evolution nowcasting. *Weather Forecast.* **27**: 1061–1089.

Roebber PJ, Schultz DM, Colle BA, Stensrud DJ. 2004. Toward improved prediction: high-resolution and ensemble modeling systems in operations. *Weather Forecast.* **19**: 936–949.

Rotunno R, Klemp JB, Weisman ML. 1988. A theory for strong, long-lived squall lines. *J. Atmos. Sci.* **45**: 463–485.

Saxen T, Mueller CK, Warner TT, *et al.* 2008. The Operational Mesogamma-Scale Analysis and Forecast System of the U.S. Army Test and Evaluation Command. Part IV: the White Sands Missile Range Auto-Nowcast system. *J. Appl. Meteorol.* **47**: 1123–1139.

Steiner M, Bateman R, Megenhardt D, *et al.* 2010. Translation of ensemble weather forecasts into probabilistic air traffic capacity impact. *Air Traff. Control Q.* **18**: 229–254.

Stensrud DJ, Brooks HE, Du J, *et al.* 1999. Using ensembles for short-range forecasting. *Mon. Weather Rev.* **127**: 433–446.

Stensrud DJ, Wicker LJ, Kelleher KE, *et al.* 2009. Convective-scale warn-on-forecast system: a vision for 2020. *Bull. Am. Meteorol. Soc.* **90**: 1487–1499.

Sun J. 2005. Convective-scale assimilation of radar data: progress and challenges. *Q. J. R. Meteorol. Soc.* **131**: 3439–3463.

Sun J, Xue M, Wilson JW, *et al.* 2014. Use of NWP for nowcasting convective precipitation: recent progress and challenges. *Bull. Am. Meteorol. Soc.* **95**(3): 409–426.

Thorpe AJ, Miller MJ, Moncrieff MW. 1982. Two-dimensional convection in nonconstant shear: a model of midlatitude squall lines. *Q. J. R. Meteorol. Soc.* **108**: 739–762.

Trier SB, Davis CA, Ahijevych DA, Manning KW. 2014a. Use of the parcel buoyancy minimum (Bmin) to diagnose simulated thermodynamic destabilization. Part I: methodology and case studies of MCS initiation environments. *Mon. Weather Rev.* **142**: 945–966.

Trier SB, Davis CA, Ahijevych DA, Manning KW. 2014b. Use of the parcel buoyancy minimum (Bmin) to diagnose simulated thermodynamic destabilization. Part II: composite analysis of mature MCS environments. *Mon. Weather Rev.* **142**: 967–990.

Tsonis AA, Austin GL. 1981. An evaluation of extrapolation techniques for the short- term prediction of rain amounts. *Atmos. Ocean* **19**: 54–65.

Tuttle JD, Davis CA. 2006. Corridors of warm season precipitation in the central United States. *Mon. Weather Rev.* **134**: 2297–2317.

Vaughn CR. 1985. Birds and insects as radar targets: a review. *Proc. IEEE* **73**: 205–227.

Vivekanandan J, Zrnic DS, Ellis SM, *et al.* 1999. Cloud microphysics retrieval using S-band dual-polarization radar measurements. *Bull. Am. Meteorol. Soc.* **80**: 381–388.

Waldvogel A, Federer B, Grim P. 1979. Criteria for the detection of hail cells. *J. Appl. Meteorol.* **18**: 1521–1525.

Weckwerth TM, Horst TW, Wilson JW. 1999. An observational study of the evolution of horizontal convective rolls. *Mon. Weather Rev.* **127**: 2160–2179.

Weisman ML, Klemp JB. 1986. Characteristics of isolated convective storms. *Mesoscale Meteorology and Forecasting*, PS Ray (ed.). American Meteorological Society; chapter 15.

Weisman ML, Davis C, Wang W, *et al.* 2008. Experiences with 0–36-h explicit convective forecasts with the WRF-ARW model. *Weather Forecast.* **23**: 407–437.

Weygandt S, Benjamin S, Smirnova T, *et al.* 2009. Evaluation of the High Resolution Rapid Refresh (HRRR): an hourly updated convection resolving model utilizing radar reflectivity assimilation from the RUC/RR. In: *WMO WWRP Symposium on Nowcasting and Very Short Range Forecasting*, Whistler, British Columbia, 31 Aug–4 Sept.

Wilson J, Carbone RE. 1984. Nowcasting with Doppler radar: the forecaster–computer relationship. In: *Nowcasting II*, K Browning (ed.). ESA SP-208. European Space Agency: Noordwijk, Netherlands; pp. 177–186.

Wilson JW. 1966. Movement and predictability of radar echoes. Tech. Memo ERTM-NSSL-28, National Severe Storms Laboratory; 30 pp (available from National Information Service, Operations Division, Springfield, VA 22161.).

Wilson JW, Megenhardt DL. 1997. Thunderstorm initiation, organization and lifetime associated with Florida boundary layer convergence lines. *Mon. Weather Rev.* **125**: 1507–1525.

Wilson JW, Mueller CK. 1993. Nowcasts of thunderstorm initiation and evolution. *Weather Forecast.* **8**: 113–131.

Wilson JW, Reum D. 1988. The flare echo: reflectivity and velocity signature. *J. Atmos. Oceanic Technol.* **5**: 197–205.

Wilson JW, Roberts RD. 2006. Summary of convective storm initiation and evolution during IHOP: observational and modeling perspective. *Mon. Weather Rev.* **134**: 23–47.

Wilson JW, Schreiber WE. 1986. Initiation of convective storms by radar-observed boundary layer convergent lines. *Mon. Weather Rev.* **114**: 2516–2536.

Wilson JW, Roberts RD, Kessinger CK, McCarthy J. 1984. Microburst wind structure and evaluation of Doppler radar for airport wind shear detection. *J. Appl. Meteorol.* **23**: 898–915.

Wilson JW, Foote GB, Fankhauser JC, *et al.* 1992. The role of boundary layer convergence zones and horizontal rolls in the initiation of thunderstorms: a case study. *Mon. Weather Rev.* **120**: 1758–1815.

Wilson JW, Weckwerth TM, Vivekanandan J, *et al.* 1994. Boundary layer clear-air radar echoes: origin of echoes and accuracy of derived winds. *J. Atmos. Oceanic Technol.* **11**: 1184–1206.

Wilson JW, Crook NA, Mueller CK, *et al.* 1998. Nowcasting thunderstorms: a status report. *Bull. Am. Meteorol. Soc.* **79**: 2079–2099.

Wilson JW, Feng Y, Chen M, Roberts R. 2010. Nowcasting challenges during the Beijing Olympics: successes, failures and implications for future nowcasting systems. *Weather Forecast.* **25**: 1691–1714.

Xue M, Wang D-H, Gao J-D, *et al.* 2003. The Advanced Regional Prediction System (ARPS), storm-scale numerical weather prediction and data assimilation. *Meteorol. Atmos. Phys.*, **82**: 139–170.

# 7

# Subseasonal Forecasting

**Lead author:** *Wassila M. Thiaw*[a]

**Contributing authors:** *Serge Janicot*[b], *Endalkachew Bekele*[a,c], *Kerry H. Cook*[d], *Bernard Fontaine*[e], *Ademe Mekonnen*[f], *Ousmane Ndiaye*[g], *Pierre-Honoré K. Tamo*[b] *and Edward K. Vizy*[d]

**Reviewer:** *George Kiladis (NOAA ESRL, Boulder, CO, USA)*

[a] *Climate Prediction Center, National Centers for Environmental Predictions, National Oceanic and Atmospheric Administration, College Park, MD, USA*
[b] *Institut de Recherche et de Developpement, LOCEAN/IPSL, Paris, France*
[c] *Joint Office for Science Support, University Corporation for Atmospheric Research, Boulder, CO, USA*
[d] *Department of Geological Sciences, Jackson School of Geosciences, The University of Texas at Austin, Austin, TX, USA*
[e] *Departement Biogeosciences/CRC, Centre National de Recherches Scientifiques and Universite de Bourgogne, Dijon, France*
[f] *Energy & Environmental Systems, North Carolina A&T State University, Greensboro, NC, USA*
[g] *Division Recherches et Developpement, Agence National de l'Aviation Civile et de la Meteorologie du Sénégal, Dakar, Sénégal*

## 7.1 Scientific Background

### 7.1.1 Introduction

Rainfall forecasting at the subseasonal timescale is extremely important in advance risk planning in various socio-economic sectors, including agriculture, water, health and food security. We define subseasonal as variability roughly from 10 days to less than 90 days. This timescale helps close the gap between short- to medium-range weather forecasting and seasonal to interannual climate predictions, thereby enabling 'seamless' predictions. Advances in numerical weather predictions (NWPs) over the past several decades have resulted in tremendous improvements in short- and medium-range weather forecasting up to 5 days. In more recent decades, enormous strides in the prediction of seasonal climate anomalies in the tropics have been made through the understanding of the global climate as a coupled ocean–land–atmosphere system, especially (1) the El Niño Southern Oscillation (ENSO) and its effects on the global climate, and (2) other modes of variability, such as the Atlantic dipole mode (Lamb, 1978), which has a more direct effect on the West African monsoon (WAM) variability. Up until the last few years, however, little progress has been made in predictions at the subseasonal timescale, due in part to the fact that internal variability in the atmosphere hinders any advantages gained from predictability associated with boundary conditions, including sea surface temperature (SST) or soil temperature and moisture.

However, it is at the subseasonal timescale where a number of users of climate information are looking to the science community to provide a greater understanding of the onset of the seasonal rains, the dry and wet spell frequencies, and the probability of extreme events occurring. Fortunately, recent research has indicated important potential sources of predictability for this time range that can be realised through better representation of coupled ocean–atmosphere phenomena, such as the Madden–Julian oscillation (MJO) in numerical weather and climate models, and the development of tools to monitor and predict the MJO, and its impacts, on the global climate. Better understanding of these potential sources of predictability, together with improvements in model development, data assimilation and computing resources, should result in more accurate forecasts. In particular, the representation of the MJO in models has improved substantially in recent years, and some models now have skill beyond 20 days. This has important implications globally due to links between the MJO and other major modes of variability, such as ENSO and the North Atlantic oscillation. Models are beginning to represent these links and processes better, but there is still a need for further improvements. This chapter will focus primarily on modes of variability most relevant to the subseasonal timescales and, therefore, will not discuss synoptic phenomena.

Note: some high0frequency signals such as Kelvin waves can have profound implications on the subseasonal timescale. These will be described in this chapter, with additional information provided in Chapter 2.

The chapter is organised as follows: Section 7.1.2 describes the most relevant datasets for subseasonal forecasting and Section 7.1.3 discusses the main modes of intraseasonal variability of convection. Section 7.1.4 describes the convectively coupled equatorial waves (CCEWs). Other convectively coupled signals, including Kelvin waves, signals between 10 and 25 days, and the MJO are presented in Section 7.1.5. Section 7.1.6 discusses the mechanisms associated with wet and dry spells, while the onset of the rains is presented in Section 7.1.7. Finally, Section 7.2 focuses on operational subseasonal forecasting: Section 7.2.1 discusses the operational prediction tools, while Section 7.2.2 describes a step-by-step guideline for operational subseasonal forecasting.

### 7.1.2 Data

In this section we describe the data that are used primarily in subseasonal forecasting of precipitation over Africa. They include precipitation, outgoing longwave radiation (OLR), brightness temperature, SST, and reanalysis, including winds and other derived fields such as velocity potential (VP).

#### 7.1.2.1 Precipitation

Real-time operational subseasonal forecasting consists of predicting weekly climate anomalies with lead times ranging from 1 day to approximately 2–3 weeks. This requires knowledge of the weekly climate base state. For this purpose, climatological weekly running-means of rainfall are constructed to help (1) locate areas of seasonal precipitation and (2) determine current and past climate anomalies, critical for the forecast verifications. Preference is given to rain gauge measurements. However, owing to the sparse distribution of the observational network, forecasters often rely on satellite rainfall estimates for a more accurate representation of important precipitation features.

There are several real-time daily satellite rainfall estimates available, including the National Centers for Environmental Prediction (NCEP), Climate Prediction Center's African Rainfall Estimation Algorithm Version 2 (CPC RFE 2.0) and the African Rainfall Climatology 2.0 (ARC2) (Novella and Thiaw, 2013) and Tropical Rainfall Measuring Mission (TRMM) (Huffman *et al.*, 2007). In the following, we provide a brief description of the CPC RFE2 and ARC2, which serve as a basis for operational subseasonal forecasting of precipitation for Africa at NCEP.

The operational daily RFE2 is on a 10 km grid and incorporates gauge data, geostationary infrared (IR) and polar orbiting microwave Special Sensor Microwave/Imager (SSM/I) and Advanced microwave sounding unit (AMSU-B) satellite data. However, its climatological record is short, extending back only to 2001. ARC2 is a subset of RFE2 and consists of a blend of daily gauge data and Meteosat IR data. It has a longer climatological record from 1983 to the present and therefore is used to assess current precipitation anomalies.

In the following we provide a brief analysis of the precipitation data for a week-1 outlook case valid during the period 19–25 June 2012. Figure 7.1 shows the weekly rainfall totals for 12–18 June 2012 (Figure 7.1a) and anomalies (Figure 7.1b), representing the most recent period before the forecast period of 19–25 June. These help assess recent rainfall trends and how they relate to the atmospheric circulation. They are also useful for verification purposes. The week of 12–18 June was marked by enhanced rainfall activity over the central sector of the Gulf of Guinea region with a northward extension into Burkina Faso and southwestern Mali. However, the western sector of the Sahel and the eastern sector of the Gulf of Guinea region experienced below-average rainfall.

The long-term mean of weekly rainfall totals for 19–25 June is shown in Figure 7.1c to document the rainfall climatology of the target period, and to locate areas that climatologically receive rainfall during this period. Knowledge of the climatology of the target period will guide the forecaster to focus on areas that are climatologically wet. The north to south precipitation gradient across West Africa is clearly depicted, with the highest estimated rainfall totals in the Gulf of Guinea region and the lowest in the Sahel. This period roughly coincides with the onset of the monsoon season in the Sahel.

#### 7.1.2.2 Sea Surface Temperature

The National Ocean and Atmospheric Administration (NOAA) optimum interpolation (OI) SST analysis, produced weekly on a 1° grid, is used in the preparation of the subseasonal forecasts. The SST analysis uses in situ and satellite-derived SST adjusted for biases using the method of Reynolds (1988) and Reynolds and Marsico (1993). SST simulated by sea-ice cover is also used. The OI analysis is done over all ocean areas and the Great Lakes. There is no analysis over land. The ocean and land areas are defined by a land–sea mask.

Understanding the annual cycle of Atlantic SST is essential for subseasonal forecasting for West Africa. Climatologically, SST is warmer in the Gulf of Guinea than over the tropical northern Atlantic during the Northern Hemisphere winter and spring. However, by mid spring, and ahead of the rainfall season in the Sahel, SST begins to decrease in the Gulf of Guinea and the equatorial cold tongue begins to emerge. The cold tongue is a band of relatively low SST that forms along coastal central Africa and expands westward as it deepens throughout the summer season. It is a response to the amplification of the St Helena anticyclone and its associated southeasterly trade winds. The seasonal cooling of the equatorial SST enhances northward gradients of temperature and moist static

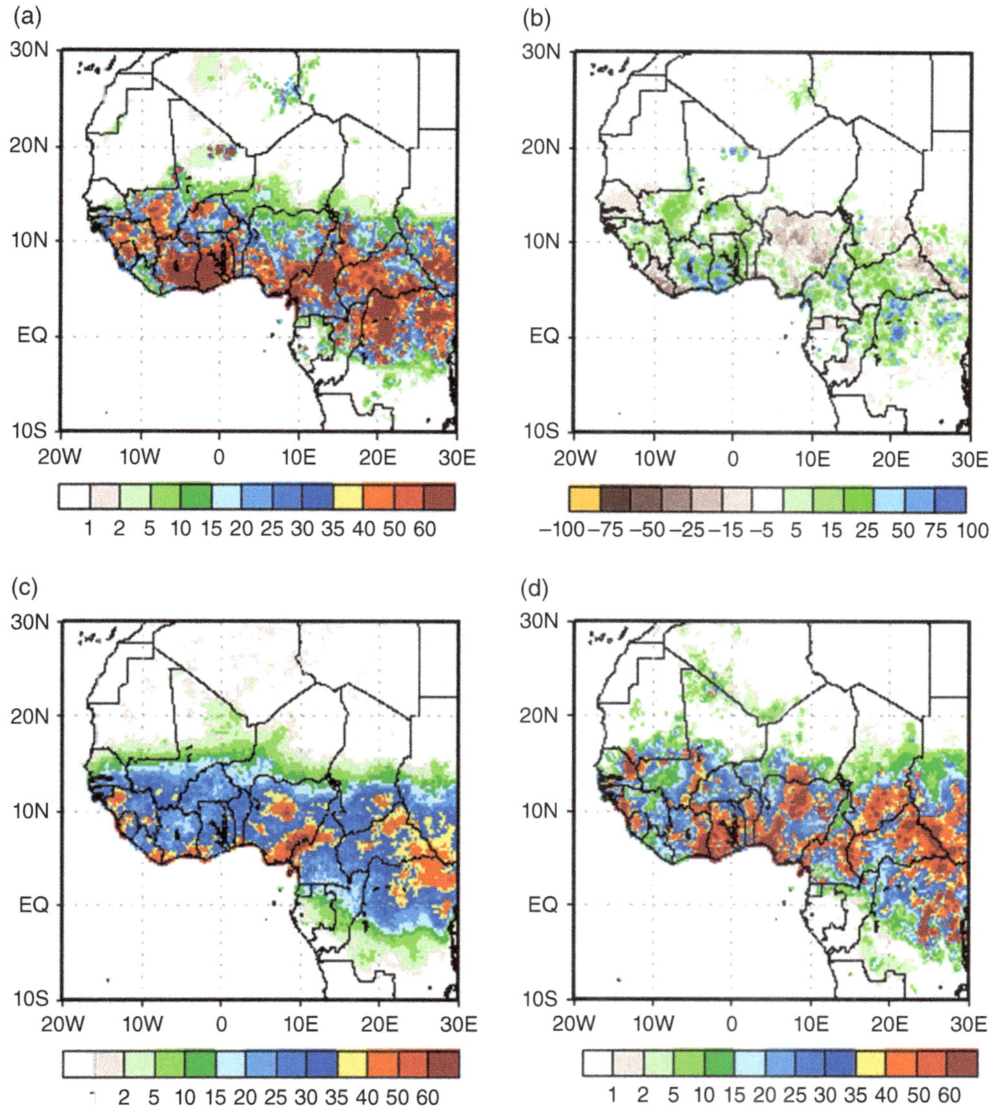


**Figure 7.1** Satellite-derived daily rainfall estimates (mm) from the blended gauge-satellite IR African Rainfall Climatology (ARC2; 1983–2012) dataset for (a) rainfall total for 12–18 June 2012, (b) rainfall anomaly for 12–18 June (1983–2012), (c) rainfall climatology for the week of 19–25 June and (d) rainfall total for 19–25 June 2012.

energy in the atmospheric boundary layer (Okumura and Xie, 2004). The resulting surface heat flux gradients increase wind speeds northward of the Equator, to shift convection over land. During this time, the West African heat low (WAHL; Lavaysse *et al.*, 2009; see also Sections 1.1.5, 1.1.7 and 2.1.2.1) amplifies and expands north and west.

During the Northern Hemisphere summer in July and August, as the WAM intensifies, low geopotential heights associated with the northward excursion of the marine Atlantic intertropical convergence zone (ITCZ) superimpose with low geopotential heights associated with the WAHL to result in the low-level westerly jet (LLWJ), also known as the West Africa westerly jet (WAWJ), which forms in the eastern Atlantic off the west coast of Africa, north of the Equator, and expands westward as it strengthens during summer (Pu and Cook, 2010, 2012). The WAWJ converges over West Africa with the south-southwesterly flow from the Gulf of Guinea. Tropical Atlantic SST, which has a direct influence on the WAM system, can change rapidly over a relatively short period of time. Conversely, however, SST anomalies when prominent, combined with the low-level wind anomalies, can be used as a tool for forecasting weekly precipitation anomalies on subseasonal timescales. For instance, an increased meridional SST gradient in the eastern Atlantic can act to enhance the LLWJ that converges over West Africa with the south-southwesterly flow from the Gulf of Guinea, creating favourable conditions for enhanced rainfall activity.

#### 7.1.2.3 Outgoing Longwave Radiation

OLR is often used as a proxy for deep convection and precipitation, especially in the tropics. It is a powerful tool for the monitoring of global tropical precipitation anomalies.

OLR data at the top of the atmosphere are observed from the Advanced Very High Resolution Radiometer (AVHRR) instrument aboard the NOAA polar orbiting spacecraft. The Earth radiation budget is made up of the incoming solar flux and the outgoing top-of-the-atmosphere (TOA) radiative fluxes. The outgoing radiative fluxes consist of the reflected part of the incoming solar flux, as well as the thermal flux emitted by the Earth–atmosphere system. It is the thermal flux that is often referred to as OLR. Thus, OLR is defined as the emission to space of terrestrial radiation from the TOA (Salby, 2012). In physical terms, it is strongly controlled by three main meteorological variables; namely, the temperature of the Earth and the atmosphere above it, the presence of water vapour in the atmosphere (which strongly absorbs IR radiation attenuating the terrestrial signal) and the presence of clouds (which may completely block all outgoing IR radiation from the surface). Thus, global maps of OLR averages and anomalies reveal information on the temperature, humidity and cloudiness of the atmosphere, making it very suitable for the monitoring of global climate anomalies. In the tropics, OLR values are strongly influenced by cloudiness and vary directly with the cloud-top temperature (Chelliah and Arkin, 1992). Hence, regions of intense convection and precipitation exhibit low OLR values, while the cloudless regions are associated with relatively high OLR values. These properties of the OLR data led to their extensive use in the qualitative estimation of the intensity of tropical convection and convective precipitation (Arkin, 1983) despite the fact that caution must be used in the interpretation of OLR anomalies, especially over arid and semi-arid regions, and on daily timescales because of non-precipitating high cloudiness. Thus, for climate monitoring purposes, OLR raw data are converted into a standardised anomaly index such that negative OLR is indicative of enhanced convection and hence more cloud coverage, while conversely, positive OLR indicates suppressed convection and less cloud cover.

#### 7.1.2.4 Brightness Temperature

The characteristics of convection can also be investigated using brightness temperature (TB) data from the Cloud Archive User Service (CLAUS) project of the European Union (see also Section 9.1.2.2.2). CLAUS TB is compiled from multiple satellite observations and it is a resample at 0.5° global horizontal grid on a three-hourly time step. The CLAUS TB dataset is compiled by Hodges *et al.* (2000) based on the International Satellite Cloud Climatology Project (ISCCP) level B3 products in a 10–12 μm window. We note that the original ISCCP B3 data has a spatial resolution of 25–30 km. CLAUS TB has been successfully applied in several studies (e.g. Yang and Slingo, 2001; Mekonnen *et al.*, 2006; Kiladis *et al.*, 2009). The data are available for the period July 1983–June 2009.

#### 7.1.2.5 Reanalysis

The NCEP Reanalysis (Kalnay *et al.*, 1996), referred to as the Climate Data Assimilation System (CDAS), and the European Centre for Medium Range Weather Forecasting (ECMWF) ERA-Interim reanalysis (Dee *et al.*, 2011) are datasets that are widely utilised to monitor and diagnose circulation features associated with the current and/or past climate. They use a state-of-the-art analysis/forecast system to perform data assimilation to integrate in situ, satellite and aircraft observations with NWP model output from the present back to 1949 for CDAS and back to 1979 for ECMWF. These data are available in a four-times daily format and also as daily averages, and provide in real time an adequate representation of the state of the Earth's atmosphere. Variables relevant to subseasonal forecasting include land surface and 2 m air temperature, mean sea-level pressure, wind and geopotential height at various altitude levels of the atmosphere, and derived fields such as the VP. The VP field, especially at 200 hPa, provides an excellent approximate measure of the global mass circulation at the upper troposphere, such that areas with negative VP values are generally associated with divergence and enhanced convection and precipitation, while positive values of VP generally suggest convergence and suppressed convection and precipitation. Since this is true only on the largest scales (waves 1 and 2; Ventrice *et al.*, 2013), VP and OLR are often used together in the monitoring of the MJO.

### 7.1.3 Detection of the Main Modes of Subseasonal Variability of Convection

When forecasters are considering operational weather maps, what appears clearly are the synoptic features like jets, easterly waves, intertropical front (ITF) or heat-low centres, as described in Chapters 1 and 2. However, it is well-known that the African monsoon is also characterised by dry sequences lasting several consecutive days or longer fluctuations of weakened or enhanced convective activity at local to regional scales. These regimes are not easily detectable from daily weather maps as an integration of information over longer time scales is needed, and spatial structures may be obscured by the synoptic weather patterns. So, specific computations of the meteorological data (OLR, rainfall, wind, geopotential fields) must be performed in order to extract the relevant patterns.

Several approaches exist for defining periodicities and increasing the signal-to-noise ratio in space and time.

When the data are noisy and/or multivariate, principal component (PC) analyses and/or singular value decomposition are often useful for isolating the leading empirical orthogonal functions (EOFs). The main scales of variability must be then identified through a wavelet transform on time series of spatial indices defined over regions of interest or through spectral analyses; for example, a multivariate singular spectrum analysis performed on their associated time coefficients. The resulting time series can be then used for prediction; for example, through the maximum entropy method (Keppenne and Ghil, 1992). As an example, Figure 7.2 (Janicot and Sultan, 2001) shows the time series from 1 May to 31 October 1968 of daily rainfall averaged over 10°W–10°E/12.5–15°N, and the associated wavelet diagram. The wavelet decomposition enables us to evaluate the spectral distribution of variance at every time step of the time series, providing a 'local' spectrum. The smooth seasonal cycle of rainfall is clear (it is represented by the blue line superimposed on the rainfall time series) with weak rainfall amounts in May and October, and the highest amounts during the summer in July and August. Day-to-day fluctuations are also evident, characterising the influence of synoptic-scale weather systems like easterly waves and mesoscale convective systems with a timescale of less than 10 days. Moreover, rainfall sequences longer than 10 days of persistent high and low rainfall amounts can also be noticed. The wavelet diagram highlights this intraseasonal timescale variability as intermittent signals with more variance within two main periodicity intervals, 10–25 days and 25–60 days.

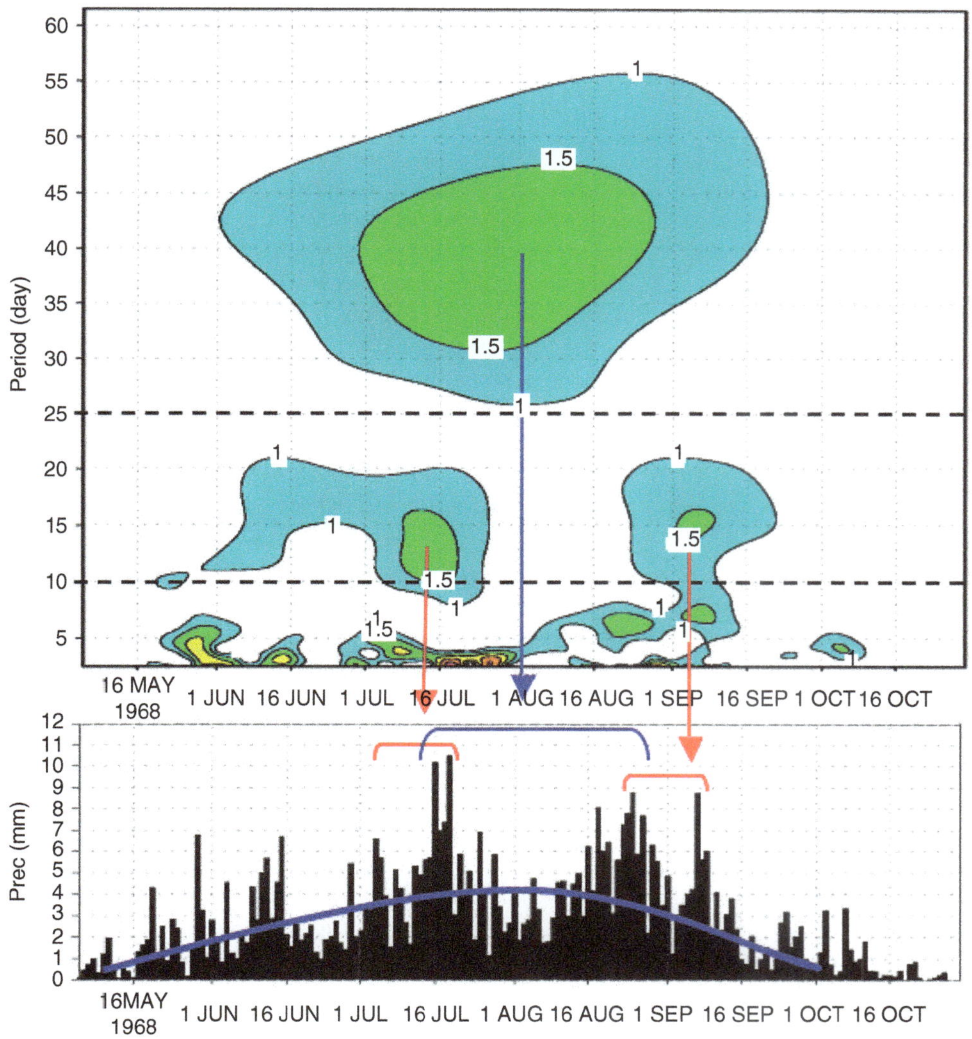


Figure 7.2 Bottom: daily rainfall time series (mm) from 1 May to 31 October 1968 averaged over 10°W–10°E/12.5–15°N. The smooth blue line represents the low-passed rainfall data. Top: wavelet analysis of the daily rainfall time series. Only periods lower than 60 days are presented. Values lower than one are not displayed. Sequences of 15-day and 40-day high variance are highlighted in red and blue respectively. *Source*: Janicot and Sultan (2001). Reproduced with permission of John Wiley and Sons.

When the bandwidth of the signal is known *a priori*, or when it is deduced from spectral analyses, different band-pass filters can be utilised for extracting the signal within the desired cut-off frequencies. One of the most efficient ways (see Christoph *et al.* (1995) for more details) is the recursive Butterworth filter proposed by Murakami (1979). In this procedure the band-pass filter is applied twice after removing the mean and linear trend in the time series, once forward and then backward, so that there is no phase change in the filtered signal. The practical way of handling missing entries at the boundaries of the signal is to repeat the first and last value, perhaps 20 or 100 times before filtering the series, to fill the window of the filter

More theoretical bases can be used to define the mode of filtering. As shown in Section 7.1.4, the shallow water theory applied to the analysis of the atmospheric equatorial waves enables us to define space–time domains specific to eastward- or westward-propagating waves like Kelvin and equatorial Rossby (ER) waves.

Interactions with other variables (precipitation, temperature, three-dimensional wind, moisture flux) can then be diagnosed through lag correlation or lag covariance analyses.

Applications of these procedures on climate data are shown below, applied in northern spring and summer. From filtering based on the shallow water theory, three types of convectively-coupled waves will be highlighted: Kelvin, ER and mixed Rossby–gravity (MRG) waves. Also, three main timescale ranges have been detected: 2–10 days, 10–25 days and 25–90 days. Janicot *et al.* (2011) propose some synthesis of these intraseasonal dominant modes.

### 7.1.4 Convectively Coupled Equatorial Waves

The structure of CCEWs over West and Central Africa is discussed in this section. It is well known that a significant part of the variability in convection in the tropics is associated with synoptic-scale wave disturbances, broadly referred to as equatorial waves. These waves are coupled with convection (dynamically and thermodynamically) to form coherent structures, and hence their name (see Takayabu (1994a,b) and Wheeler and Kiladis (1999)). Note that there are also equatorial waves that are not coupled with water vapour or convection in the atmosphere. These are referred to as 'free' or 'dry' internal waves and they propagate through the atmosphere without interacting with latent heat through cloud formation or precipitation (see the review by Kiladis *et al.* (2009) for details).

The structure and dispersion characteristics of CCEWs have been predicted based on the shallow water equations on an equatorial beta-plane. In the late 1950s it was realised that the low-latitude region of a rotating planet could give rise to a particular class of 'trapped' motions along the Equator. In his classic paper on equatorial motions, Matsuno (1966) solved for the zonally propagating wave solutions of a system of equations known as the 'linearised shallow water equations on an equatorial beta plane', which govern motions in a layer of constant-density fluid when the restoring forces are gravity and a linearly varying Coriolis parameter. He was the first to derive the complete set of zonally propagating wave solutions to these equations, isolating the normal modes of this system and their dependence on zonal wave number, and thereby introducing zonal wave number $k$ and frequency $\omega$. Solutions that decay from the Equator provide relations between $k$ and $\omega$ for each positive integer) (called meridional mode number), thus defining the horizontal dispersion relation for the waves.

The leading modes are presented in Figure 7.3. The CCEWs that conform to Matsuno's solutions include Kelvin waves, ER waves, and MRG waves. Note that other convectively coupled waves that do not conform to Matsuno's solution include the African easterly waves (AEWs), described in Chapter 2.

A first step toward investigating various wave structures and their impact on convection is to carry out filtering or scale separation. As mentioned earlier, different methods are available to filter the different wave regimes. One way is to filter data in space and time. Briefly, the space–time filtering method, as performed by, for example, Wheeler and Kiladis (1999), involves decomposition of a data field into wave number–frequency components for eastward- and westward-propagating wave types, and the resulting spectral density is plotted as a function of wave number and frequency. The decomposition is performed by using 96-day segments such that each segment overlaps the previous segment for 31 days. Prior to the decomposition, the data series is detrended and the ends of the time series are tapered to zero to control spectral leakage. As a result, this process identifies regions of a significant wave number–frequency domain for each wave type (see Figure 7.4). Note that the observed phase speeds of the CCEWs are considerably slower than their dry counterparts. Also, note that the wave number–frequency decomposition method separates the filtered field into symmetric and antisymmetric components as shown in Figure 7.4 based on brightness temperature data. The spectrum is the result of dividing the raw power by an estimate of its red noise background. The background spectrum is determined by smoothing the raw spectrum many times. Detailed descriptions of symmetric and antisymmetric parts of the spectrum can be found in Wheeler and Kiladis (1999). Briefly, the total spectrum is a combination of symmetric and antisymmetric components. At a given latitude, symmetric/antisymmetric components for brightness temperature $T_b$ are defined as

$$\mathrm{Tb}(\text{symmetric}) = \frac{\mathrm{Tb}(\phi) + \mathrm{Tb}(-\phi)}{2}$$

$$\mathrm{Tb}(\text{antisymmetric}) = \frac{\mathrm{Tb}(\phi) - \mathrm{Tb}(-\phi)}{2}$$

where $\phi$ is the latitude.

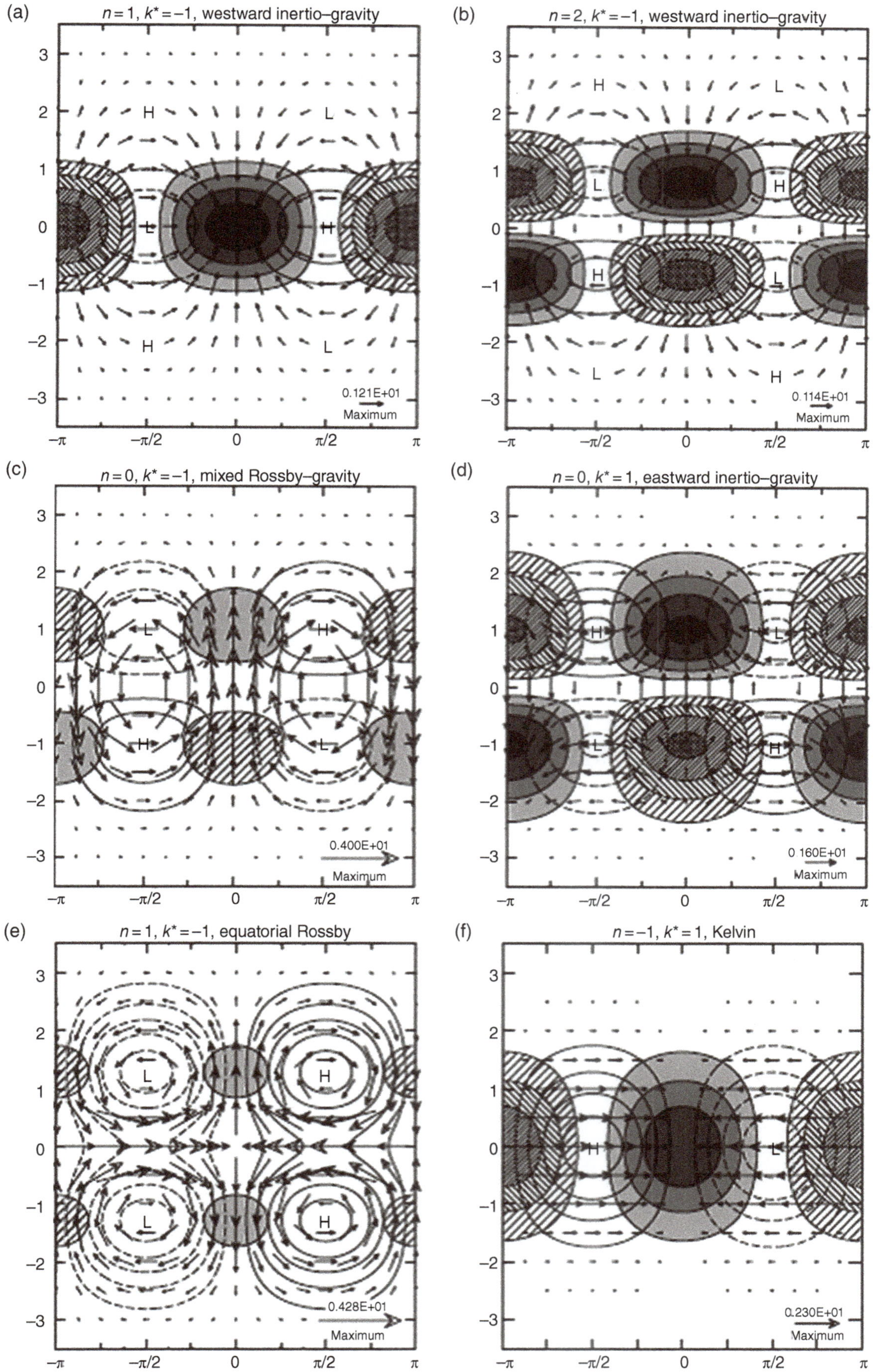


**Figure 7.3** Horizontal structures of a subset of the zonally propagating wave solutions to the shallow water equations. Each is shown for a non-dimensional zonal wave number, $k^* = \pm 1$. All scales and fields have been non-dimensionalised. The Equator runs through the centre of each diagram. Hatching is for divergence and shading is for convergence, with a 0.6 unit interval between successive levels. Unshaded contours are geopotential, with a contour interval of 0.5 units. Negative contours are dashed and the zero contour is omitted. The maximum wind vectors in each panel are specified in the bottom right corner (see Kiladis *et al.* (2009: figure 3) for more details). *Source*: Kiladis *et al.* (2009). Reproduced with permission of John Wiley and Sons.

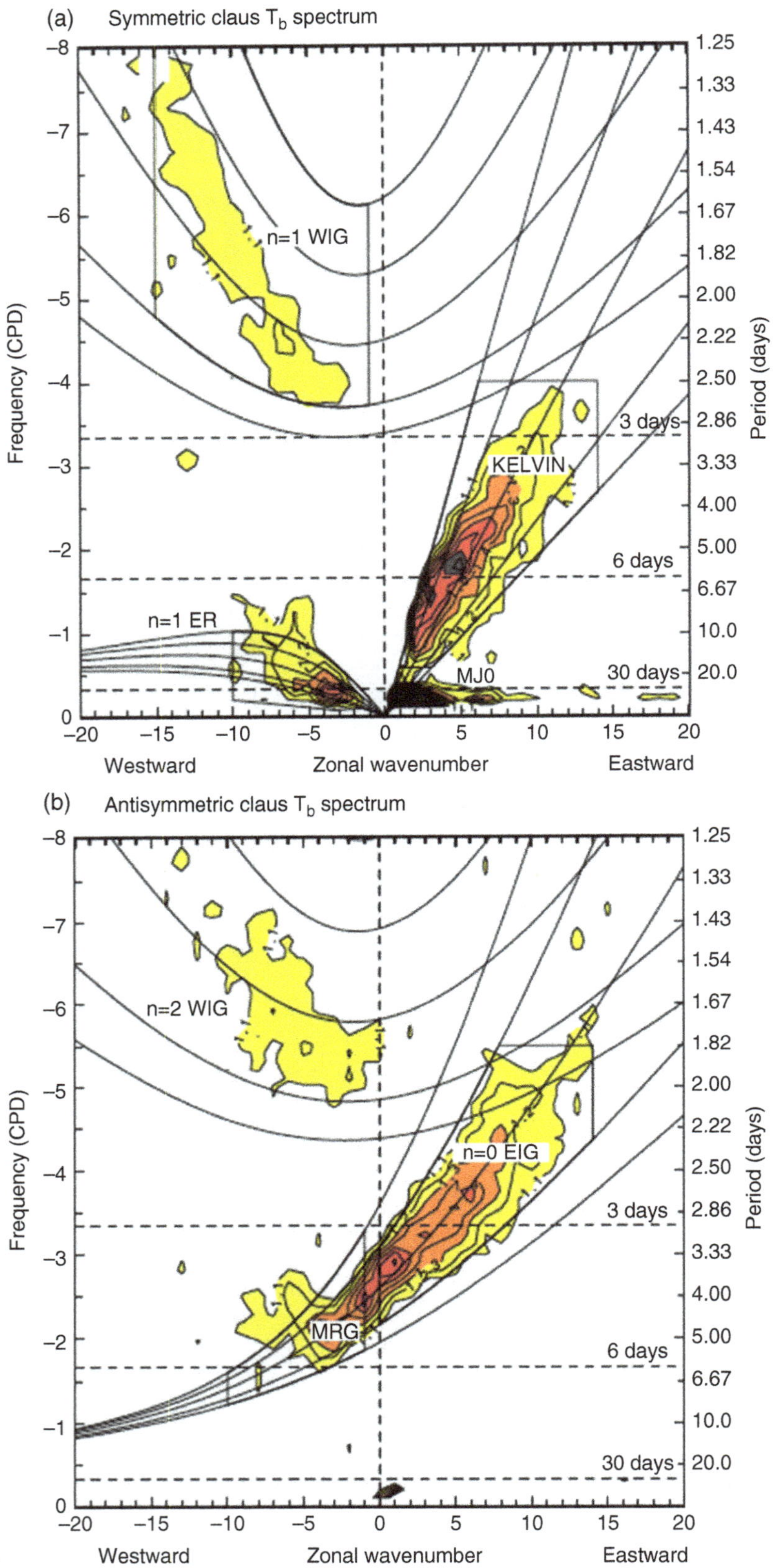


**Figure 7.4** Wave number–frequency power spectrum for symmetric (top) and antisymmetric (bottom) spectrum of brightness temperature $T_b$ from the European CLAUS for July 1983–2005, averaged over 15°N, 15°S. *Source*: Kiladis *et al.* (2009). Reproduced with permission of John Wiley and Sons.

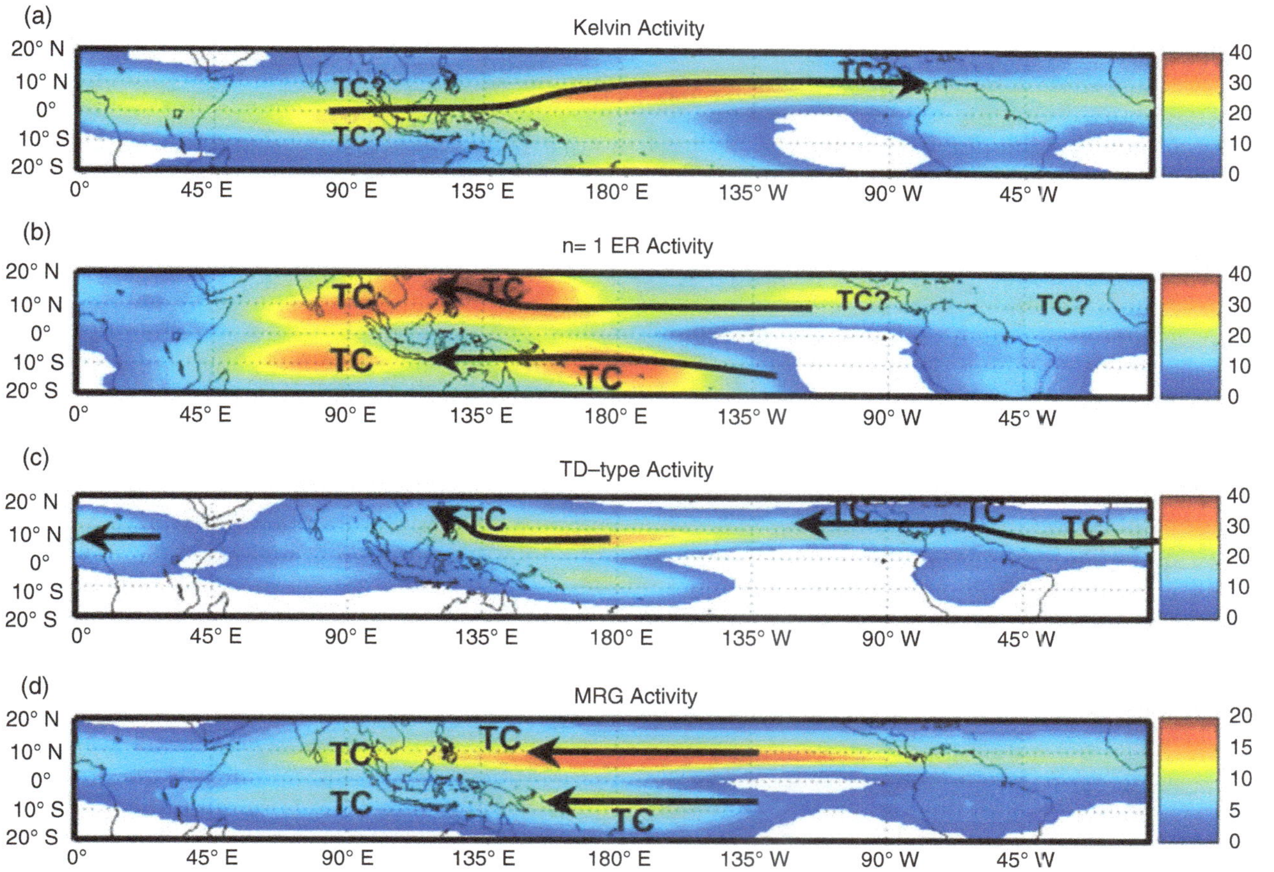


**Figure 7.5** Distribution of annual mean variance of CLAUS brightness temperature $T_b$ filtered for the (a) Kelvin, (b) $n = 1$ ER, (c) tropical depression (TD) and (d) MRG wave bands. Also shown are preferred direction of propagation and preferred location of tropical cyclone (TC) genesis associated with each mode. *Source:* Kiladis *et al.* (2009). Reproduced with permission of John Wiley and Sons.

We now briefly discuss three CCEW types; namely, Kelvin waves, ER and MRG waves.

#### 7.1.4.1 Kelvin Waves

Theoretical Kelvin waves are eastward-moving non-dispersive wave types (Figure 7.7), and are mainly characterised by zonal wind fluctuations that peak at the Equator and decrease poleward. The mean variance (implying a measure of wave activity) associated with the Kelvin waves peaks near the Equator over the western Indian Ocean and Pacific warm pool region, but slightly to the north of this over the central and eastern Pacific (Figure 7.5a). The spatial structure of a Kelvin wave over West Africa (Figure 7.6) has been obtained by the computation on May-October of composites of TB (shading), geopotential height (contours), and 850 hPa winds (vectors) fields associated with a perturbation of −10K of the Kelvin-filtered TB index at the reference location 2°N–10°E. The circulation field is nearly symmetric about the Equator, and is characterised broadly by a wavelength of about 5000 km and an eastward phase speed of about 15 ms$^{-1}$ (e.g., Mounier *et al.*, 2007; Mekonnen et al., 2008). Several studies (e.g., Mekonnen *et al.*, 2008; Mounier *et al.*, 2007; Ventrice *et al.*, 2012; Laing *et al.*, 2011) showed that these waves have an important impact on local convection and on AEWs in Africa, and are implicated in MJO initiation over the Indian Ocean. Matthews (2004) suggested that a dry intraseasonal Kelvin wave tied to the MJO that starts in the western Pacific and travels around the globe reaches Africa 20 days later, and enhances moisture flux from the equatorial Atlantic toward West Africa. Also, Nguyen *et al.* (2011) and Laing *et al.* (2011) suggested that convective events over equatorial Africa are triggered or modulated in association with Kelvin wave activity during the March–May season. Results from a case study are displayed in Figure 7.7 and reveal that four AEWs were initiated in East Africa in association with the Kelvin wave passage in August 1987 (Mekonnen *et al.*, 2008). These AEWs are seen propagating across the Atlantic and, some of them, into the East Pacific.

#### 7.1.4.2 Equatorial Rossby Waves

Atmospheric ER waves are westward-moving waves whose variance peaks off the Equator along 10°S near the dateline and along 10°N near 135°E over the West Pacific (Figure 7.5b). As shown in Kiladis *et al.* (2009), ER

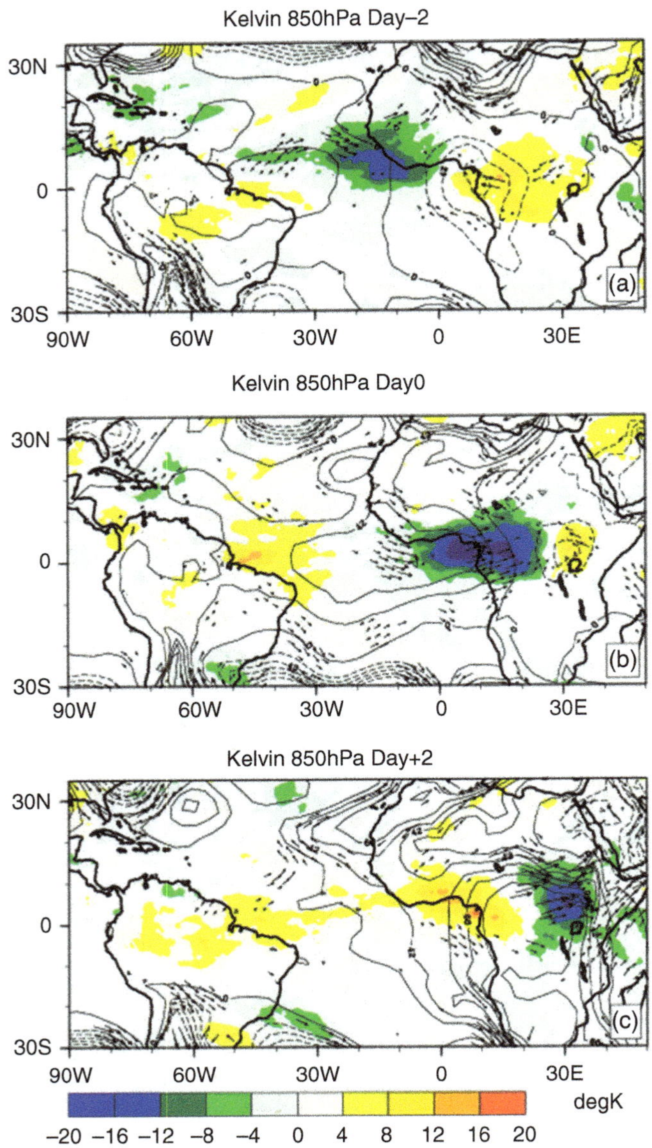


**Figure 7.6** Composite maps of May to October anomalous brightness temperature $T_b$ (shading), geopotential height (contours) and wind (vectors) associated with a −10 K perturbation in Kelvin wave $T_b$ at the base point 2°N, 10°E, for (a) day −2 at 850 hPa, (b) day 0 at 850 hPa and (c) day +2 at 850 hPa. The contour interval is 12 mgp, with negative contours dashed. Cold shading is for negative and warm shading is for positive $T_b$ perturbations. $T_b$ and wind vectors are locally significant at the 95% level, with the largest vectors around 2 m s$^{-1}$.

variance amplitude also appears to peak over the northern and southern Indian Ocean along 10°N and 10°S. Convectively coupled ER waves are characterised by periods in the range of 10–15 days and phase speeds in the range 5–10 m s$^{-1}$ in the Pacific Ocean (Wheeler and Kiladis, 1999). Studies have suggested that, in addition to local convection, ER waves sometimes impact the genesis of the MJO (e.g. Roundy and Frank, 2004). Figure 7.8 shows a typical structure of convectively coupled ER waves over West Africa computed over the months May to October. A suppressed–enhanced convective pair associated with circulation features propagates westward between day −5 and day +5. Note that the easterly flow to the east and westerly flow to the west of the enhanced convective centre is seen. ERs contribute partly to the Sahel mode (see Sections 7.1.5.1.2 and 7.1.5.1.1) at a biweekly timescale and to the westward propagation of the MJO signal from the Indian basin (Janicot *et al.*, 2010).

#### 7.1.4.3 Mixed Rossby–Gravity Waves

Distinct from TD-type disturbances, the MRG waves are identified in wave numbers 1–5 in the wave number–frequency spectrum (Figure 7.4) and propagate westward with phase speed of 15–25 m s$^{-1}$ (eastward group velocity of about 5 m s$^{-1}$) and wavelength of about 9000 km (Takayabu and Nitta, 1993; Kiladis *et al.*, 2009). The strongest signals of MRG waves are observed over the western and central Pacific during the boreal summer and autumn seasons (Figure 7.5d; Kiladis *et al.*, 2009), although they are present nearly all year round in the tropical oceanic convergence zones. Convection is enhanced off the Equator and suppressed on the opposite side of the Equator; that is, it has an antisymmetric signal. Figure 7.9 shows MRG wave structure over West Africa computed over the months May to October. The structure is similar to the theoretical pattern but with a weaker structure on its southern part. MRG waves have an important impact on convection and tropical cyclones (Kiladis *et al.*, 2009).

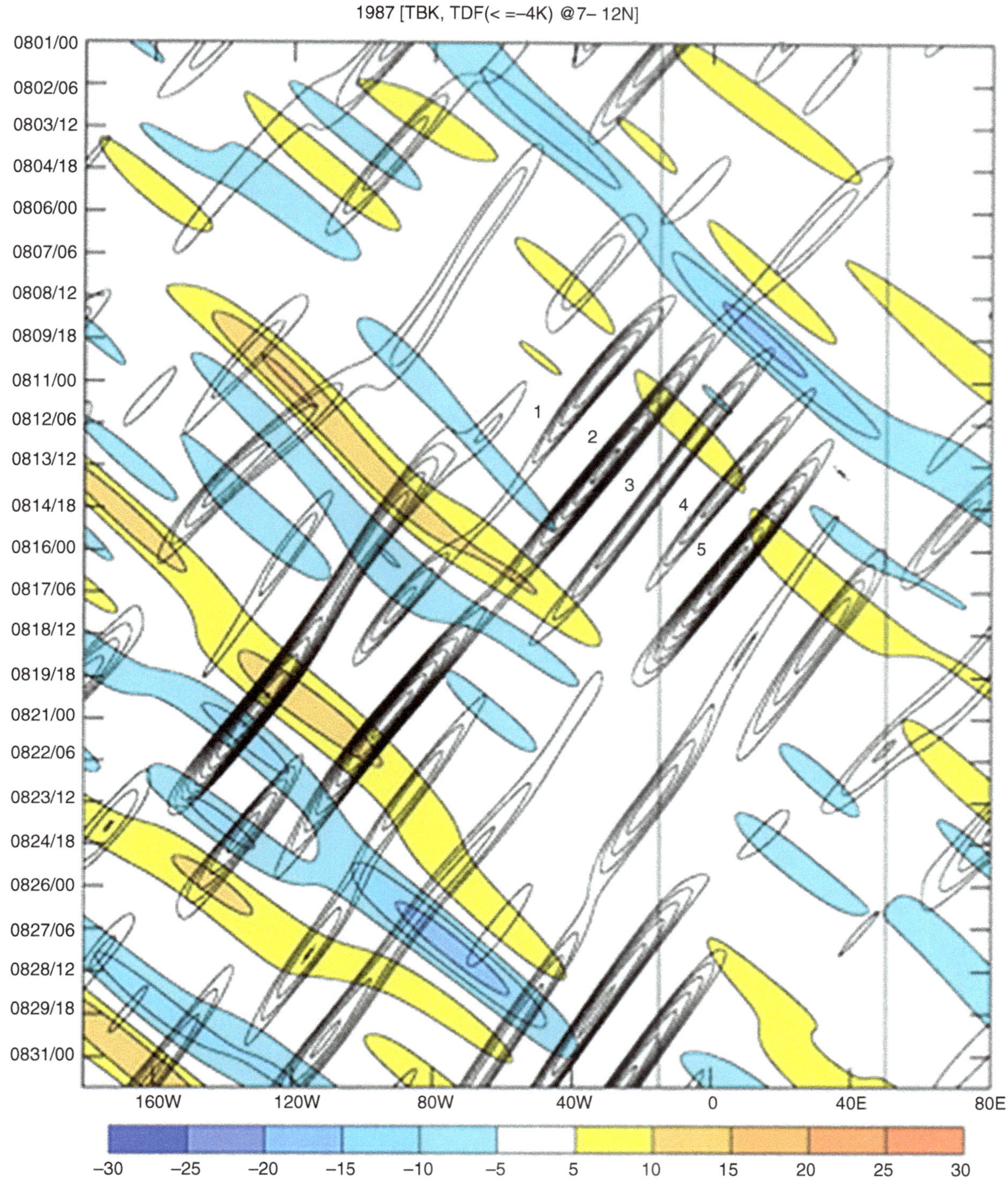


Figure 7.7 Kelvin-filtered cloud brightness temperature $T_b$ (shaded) and AEWs (contoured every 2 K from −4 K; only negative values are shown for clarity) for August 1987, averaged between the 7°N and 12°N range. AEWs are determined by filtering $T_b$ in the period of 2–7.5 days and westward wave number 6–20. Numbers 1–5 denote the AEWs initiated or enhanced in association with the enhanced phase of the Kelvin wave. *Source*: Mekonnen *et al.* (2008). ©American Meteorological Society. Used with permission.

### 7.1.5 Other Convectively Coupled Signals and Links with Equatorial Waves

#### 7.1.5.1 Signals Between 10 and 25 Days

The 10–25-day oscillation occurs mainly during the summer season and has been identified as one of the two primary intraseasonal modes of the WAM. The second is the 25–90-day periodicity, which is rather intermittent and discussed in Section 7.1.5.2. Figure 7.10 shows deseasonalised time series of OLR indices in summer 2006, averaged over the Sahel (top right), the Guinea coast (top left) and West Africa (bottom), as well as superimposed 10–25-day filtered signal (top) and 25-90-day

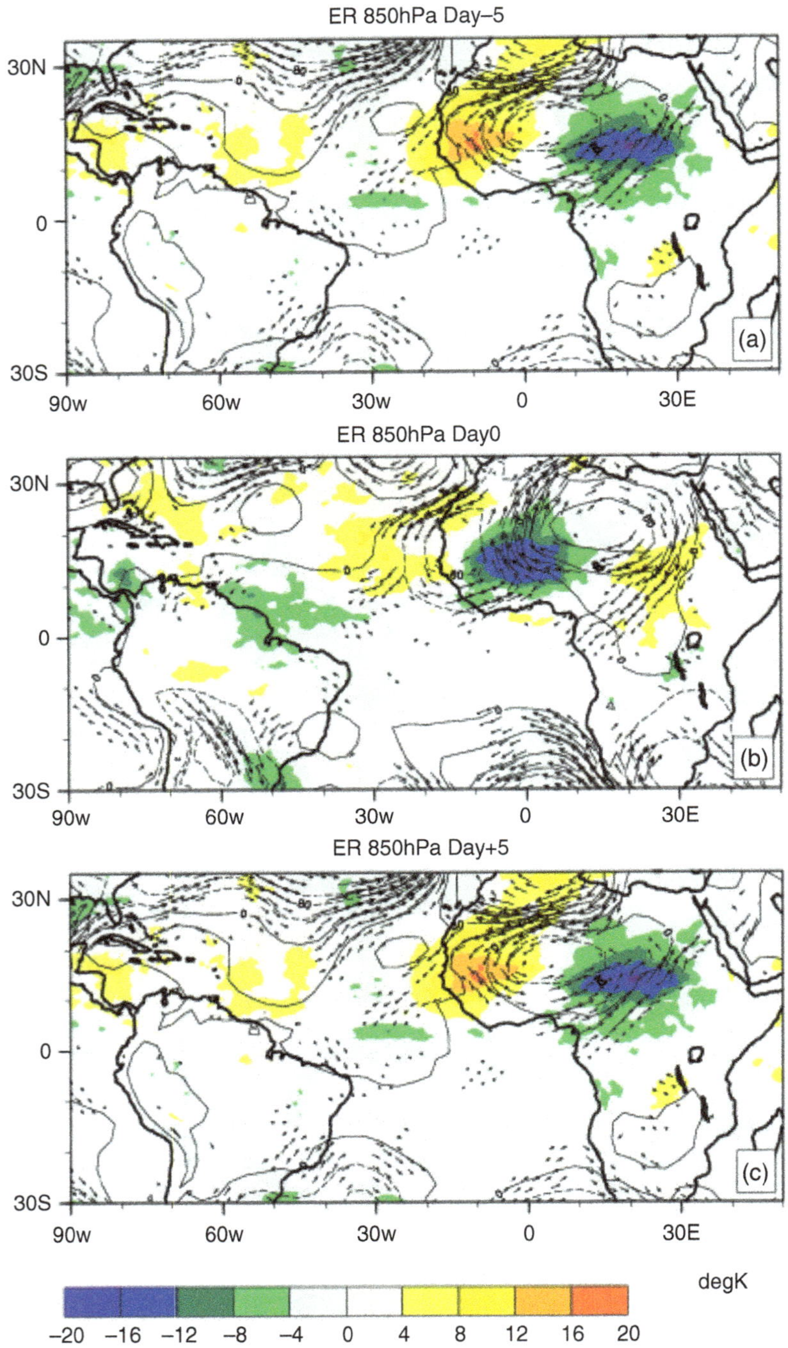


**Figure 7.8** Composite maps of May to October anomalous brightness temperature $T_b$ (shading), geopotential height (contours) and wind (vectors) associated with a −7.5 K perturbation in $n = 1$ ER wave $T_b$ at the base point 15°N, 0°E, for (a) day −5 at 850 hPa, (b) day 0 at 850 hPa, and (c) day +5 at 850 hPa. The contour interval is 40 mgp, with negative contours dashed. Cold shading is for negative and warm shading is for positive $T_b$ perturbations. $T_b$ and wind vectors are locally significant at the 95% level, with the largest vectors around 2 m s$^{-1}$.

signal (bottom). These panels allow one to see how this filtering procedure can extract a high fraction of the raw signal over some sequence of time – for example, during July over the Guinea coast (top left) or in August over the Sahel (top right) – or miss most of the raw signal over other time sequences.

Further analyses have been carried out based on a regional EOF decomposition in order to extract the main modes of variability of the 10-25-day timescales. Two modes have been highlighted: (1) from EOF1, a quasi-stationary zonal dipole of convection within the convergence zone over land (described in Section 7.1.5.1.1); (2) from EOF2, a westward-propagating signal from eastern Africa to the tropical Atlantic, with increased cyclonic vorticity ahead of an area of enhanced convection, referred to as the Sahel mode (Janicot *et al.*, 2010; Section 7.1.5.1.2). Both modes are displayed in Figure 7.11 based on composite anomalies of NOAA OLR over June–September. Dates where daily values of the corresponding PC1 and PC2 standardised time series are maximum and greater than one standard deviation are retained to identify strong

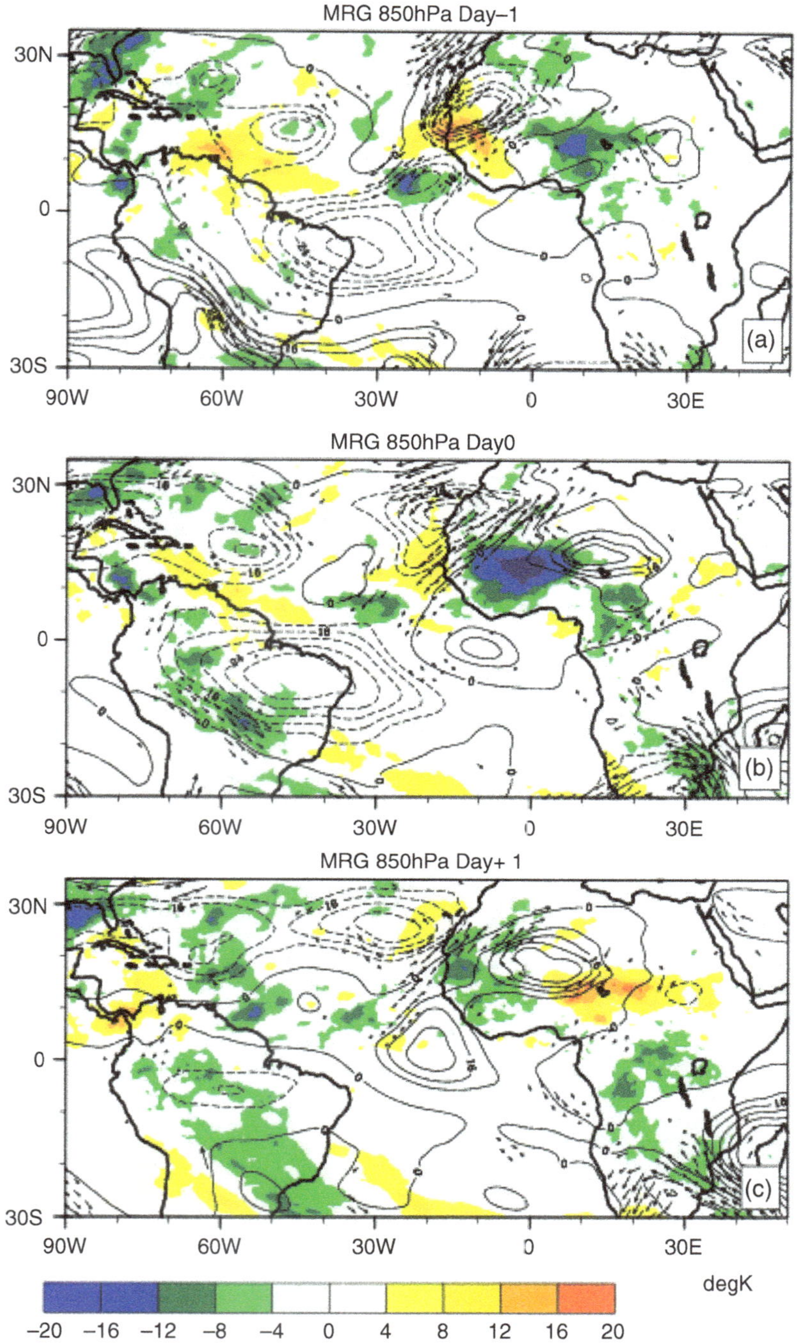


**Figure 7.9** Composite maps of May to October anomalous brightness temperature $T_b$ (shading), stream function (contours), and wind (vectors) associated with a –7.5 K perturbation in MRG wave $T_b$ at the base point 15°N, 0°E, for (a) day –1 at 850 hPa, (b) day 0 at 850 hPa, and (c) day +1 at 850 hPa. The contour interval is $4 \times 10^5$ m$^2$ s$^{-1}$, with negative contours dashed. Cold shading is for negative and warm shading is for positive $T_b$ perturbations. $T_b$ and wind vectors are locally significant at the 95% level, with the largest vectors around 2 m s$^{-1}$.

convective phases of the EOF1 and EOF2 modes respectively. Dates where these daily values are negative and lower than minus one standard deviation correspond to weak convective phases. The composites of strong and weak phases are computed by averaging unfiltered OLR fields for the selected dates of PC1 and PC2. The differences between the strong and weak composite phases are displayed for the EOF1 mode (Figure 7.11 top) and the EOF2 mode (Figure 7.11 middle).

##### *7.1.5.1.1 The quasi-biweekly Zonal Dipole Mode*

Mounier *et al.* (2008) explored the nature of the 10–25-day variability. They proposed the existence of a QBZD (Figure 7.11 top) structure. This mode is the leading one over West Africa. The 'dipole' aspect of this refers to a nearly standing oscillation with an out-of-phase relationship that African rainfall has with rainfall in the western Atlantic and Americas. It is controlled both by equatorial atmospheric dynamics and by land surface processes over Africa, inducing combined fluctuations

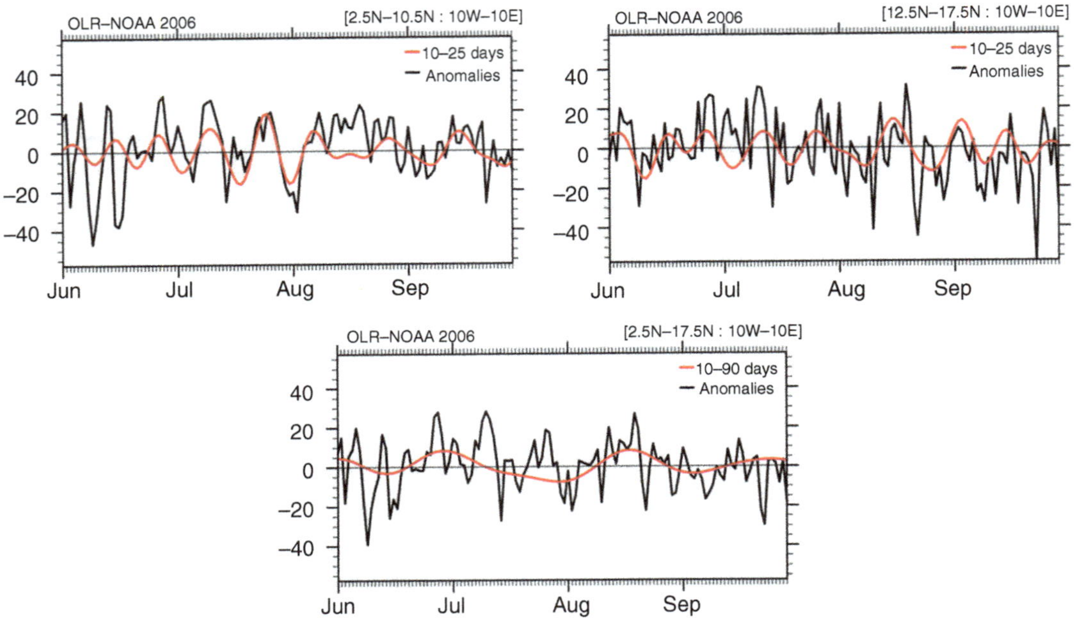


**Figure 7.10** Time series of deseasonalised (black) and filtered (red) OLR indices (W m$^{-2}$) over 10°W–10°E in summer 2006. Top right : index averaged over 12.5–17.5°N and 10–25-day filtered values. Top left: index averaged over 2.5–10°N and 10–25-day filtered values. Bottom: index averaged over 2.5–17.5°N and 25–90-day filtered values.

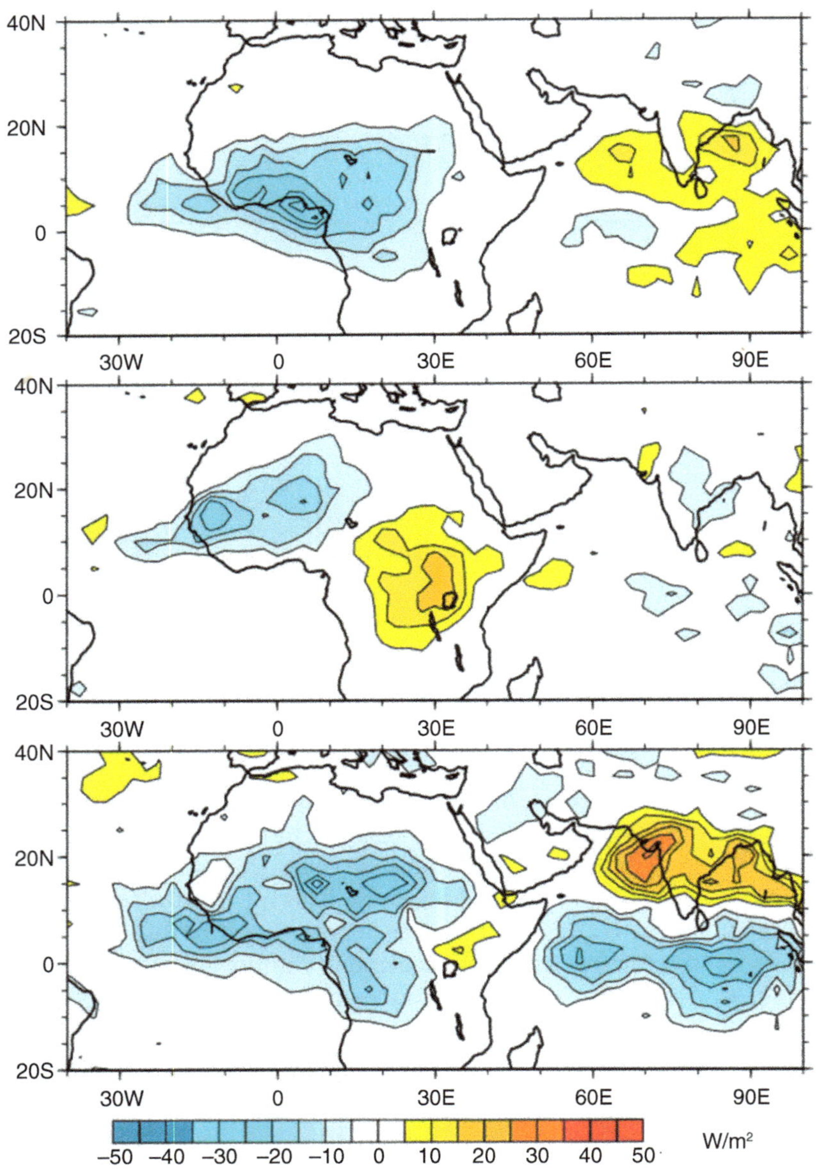


**Figure 7.11** Composite pattern of the subseasonal modes in terms of non-filtered OLR modulation (W m$^{-2}$): (top) quasi-biweekly zonal dipole (QBZD) mode; (middle) 'Sahel mode' and (bottom) 'MJO mode'. Adapted from Sultan *et al.* (2003).

in surface temperature, surface pressure and low-level zonal winds off the coast of West Africa. When convection is at a minimum over West and Central Africa, net shortwave flux at the surface is high, which increases surface temperatures and lowers surface pressures. This creates an east–west pressure gradient at the latitude of both the ITCZ and the Saharan heat low (SHL), leading to an increase in eastward moisture advection inland.

The arrival from the Atlantic of the positive pressure signal, associated with a Kelvin-wave-type pattern, amplifies the low-level westerly wind component and the moisture advection inland leading to an increase in convective activity over West and Central Africa. Then the opposite phase of the dipole develops. When filtering out the Kelvin eastward-propagating signal, the stationary character of the QBZD is highlighted (Mounier *et al.*, 2008). This mode is also present in spring.

#### *7.1.5.1.2 The Sahel Mode*

The second EOF mode of 10–25-day variability (the 'Sahel mode'; Figure 7.11 middle) has been detected by Sultan *et al.* (2003) and Mounier and Janicot (2004). The convection increase in the African ITCZ is associated with a propagating mode appearing first over Central Africa, moving to the north toward the Sahelian latitudes, and then propagating westward toward the eastern tropical Atlantic. This pattern is associated with a cyclonic circulation positioned to the northwest of the enhanced convective pole (the enhanced convective zone in the dipole structure), increasing the advection of moisture toward this pole.

Taylor (2008) showed that land–atmosphere and radiation–atmosphere interaction processes contribute significantly to the maintenance and the westward propagation of the convective pole. Extensive areas of wet soil associated with the positive phase of the Sahel mode induce weak surface heat fluxes and low-level anticyclonic circulation, bringing moisture to the west of the convective zone and helping its westward propagation.

Model simulations with and without soil moisture coupling at intraseasonal timescales suggest that this westward-propagating mode is also controlled by a purely internal atmospheric mode (Lavender *et al.*, 2010). In fact it appears to be partly associated with the occurrence of a convectively coupled ER wave propagating westward at a similar periodicity range (Janicot *et al.*, 2010).

Another mechanism could be governed by interactions with the mid-latitudes. Vizy and Cook (2009) showed that strong cold air surges coming from the Mediterranean area can lead to African monsoon breaks, through a southwestward propagation. This investigation has been widened thanks to the analysis of the SHL dynamics. Chauvin *et al.* (2010), Lavaysse *et al.* (2010) and Roehrig *et al.* (2011) showed that it corresponds to a robust mode over North Africa and the Mediterranean. This mode consists of two phases: a west phase and an east phase. The west phase corresponds to a maximum temperature over the coast of Morocco–Mauritania, propagating southwestward, and a minimum between Libya and Sicily, propagating southeastward. The east phase corresponds to the opposite temperature structure, which propagates as in the west phase. This SHL mode was preceded by large-scale, mid-latitude, intraseasonal fluctuations of the atmosphere.

In essence, the 10–25-day intraseasonal variability of convection over the Sahel can be partly explained by mid-latitude intraseasonal variability, through a major role played by the SHL (about one-third of the cases selected in Figure 7.11 (middle)) and by ER wave dynamics (about one-third of the cases selected in Figure 7.11 (middle)). These two mechanisms are likely to interfere together in a constructive or destructive way, leading to high temporal and spatial variability of the Sahelian dry and wet spells. The monitoring of both the SHL intraseasonal activity and that of the ER wave should then provide relevant information to forecast at least two-thirds of such high-impact events.

The monitoring of these intraseasonal modes (QBZD mode, Sahel mode and SHL mode) was established thanks to the setting up of the Service de Données (SEDOO) platform by l'Observatoire Midi-Pyrenées (OMP) at Toulouse (France). These indices were first used during the Fennec field campaign in 2011 and are now updated during each summer season through the Real-Time Monitoring and Forecasting of Intraseasonal Variability over Africa project (see these indices and other diagnostics from the ECMWF forecast system on the website http://isv.sedoo.fr).

### 7.1.5.2 Periodicities Between 25 and 90 Days and the Madden–Julian Oscillation

Recent advances in the understanding of the MJO and its impacts on tropical climate and advances in the use of NWP products have enabled the development of tools for subseasonal forecasting. In the following we describe the real-time monitoring and prediction tools that enable the operations of subseasonal forecasting.

#### *7.1.5.2.1 Definition*

The MJO is a planetary wave number 1 to number 3 pattern that propagates from west to east with an area of enhanced rainfall activity associated with upward motion and divergence in the upper troposphere, followed by an area of suppressed rainfall associated with descending motion. It is known to be the dominant mode of tropical variability on the intraseasonal timescale.

The MJO is episodic, but when regular has a period of around 30–60 days. Although it is centred along the

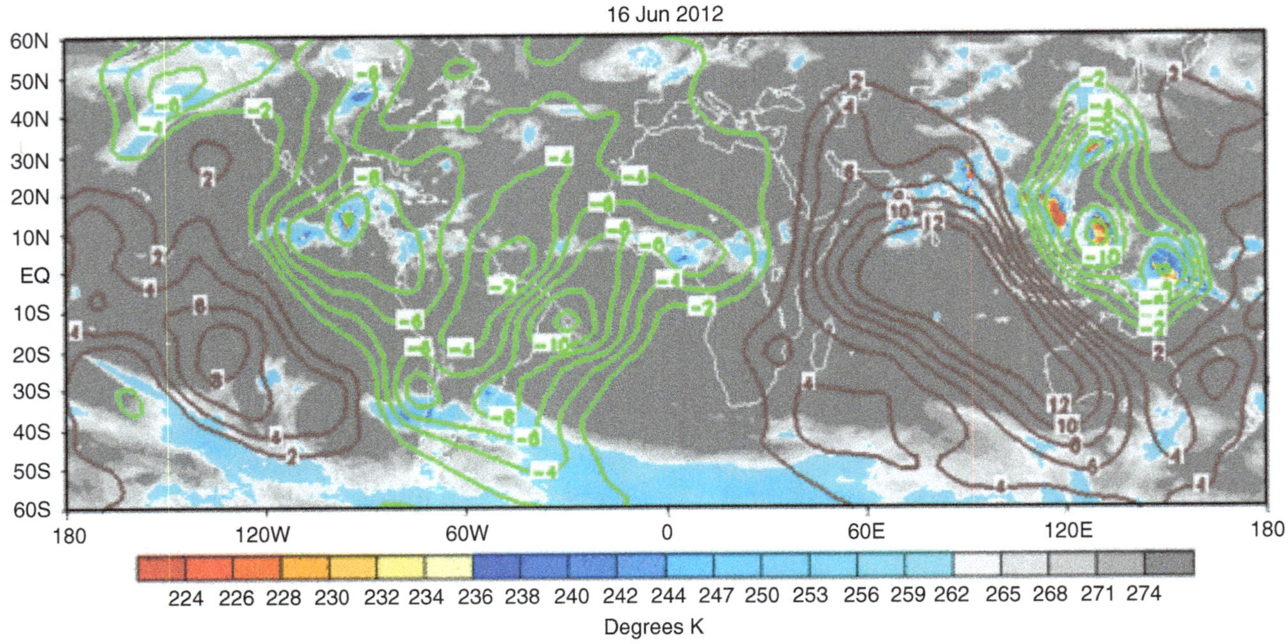


**Figure 7.12** OLR (NOAA 18 AVHRR IR window channel measurements by NESDIS/ORA) for 16 June 2012, shading in kelvin, overlaid with 200 hPa velocity potential contours. Contour interval is $10^6$ $m^2$ s. Low values of OLR are indicative of convection. Green (brown) contours are associated with upper level divergence (convergence). *Source*: NOAA's Climate Prediction Center MJO web portal: http://www.cpc.ncep.noaa.gov/products/precip/CWlink/MJO/mjo.shtml.

Equator, it straddles both hemispheres. An example of an MJO event for 16 June 2012, as depicted by OLR and 200 hPa streamfunction, is shown in Figure 7.12. This event resembles that of a wave number 2 pattern. The primary area of large-scale upper level divergence is centred over the Western Hemisphere and extends from the eastern Pacific to central Africa. Convection, as portrayed by low-level values of OLR, is observed in this region. A secondary area of convection is observed over the western Pacific. In between the two convective areas, large-scale subsidence prevails over the Indian Ocean and the central Pacific. When present, the MJO can have an impact on weather and climate patterns in many parts of the globe, including West Africa. The MJO, and associated rainfall, is often more evident over the Indian Ocean, and remains apparent as it propagates eastward over the very warm waters of the western and central tropical Pacific. Over the cooler ocean waters of the eastern Pacific, the pattern of tropical rainfall generally becomes less evident, but often reappears over the tropical Atlantic and Africa. Along with these variations in tropical rainfall, there are distinct patterns of lower- and upper-level atmospheric circulation anomalies in the tropics and subtropics. These features extend around the globe and are not confined to the Eastern Hemisphere. Thus, they provide important information regarding the regions of ascending and descending motion associated with particular phases of the oscillation.

#### *7.1.5.2.2 Madden–Julian Oscillation Detection and Diagnostics*

Global operational monitoring centres of weather and climate information monitor, assess and predict the MJO for two important reasons. (1) The MJO alters the pattern of tropical convection substantially and can lead to extended periods of above- or below-average rainfall that can lead to flooding or exacerbate drought conditions. Knowledge of circulation changes linked to the MJO increase predictability of subseasonal forecasts such as the week-1 and week-2 forecasts. (2) The MJO can affect the onset, decay and evolution of the ENSO through its modulation of the low-level wind field that can produce important oceanic surface and subsurface changes (Kessler and Kleeman, 2000; Zhang and Gottschalck, 2002).

Several parameters are used to detect and diagnose the strength of the MJO and to monitor and predict its evolution. These include OLR, low- and upper-level wind fields and other derived variables from the zonal and meridional components of the wind, such as VP (Figure 7.12); also see Ventrice *et al.* (2013).

The Wheeler–Hendon diagram (Wheeler and Hendon, 2004) is a seasonally independent index that is used for operational real-time monitoring of the state of the MJO. It is based on a pair of EOFs of the combined fields of near-equatorially averaged 850 hPa and 200 hPa zonal wind, and OLR. Projection of the daily observed data

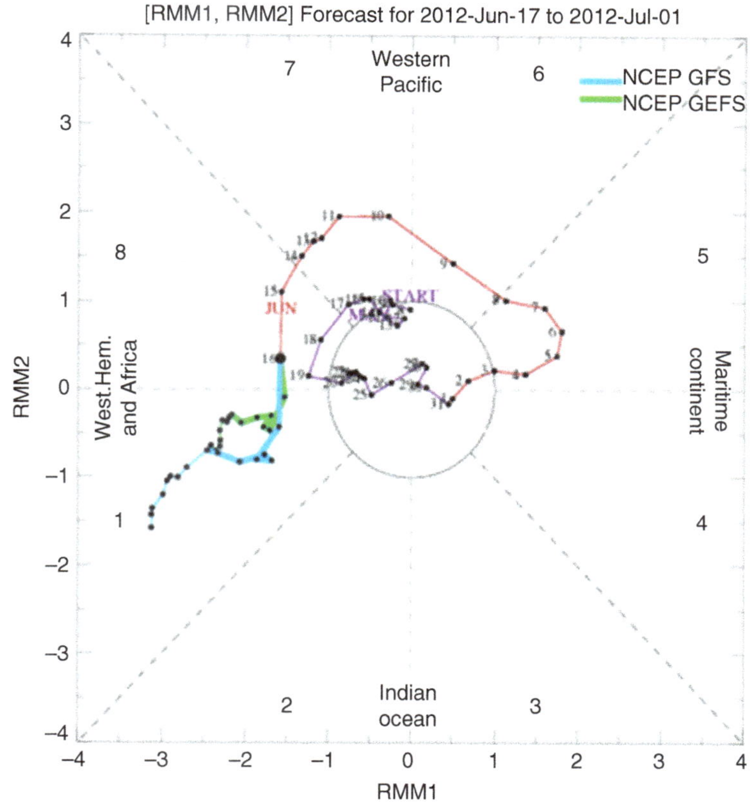


Figure 7.13 Phase-space representation of the two-component MJO index, for the period 1 May–16 June 2012 (thin lines) and the forecast period valid 17 June–1 July 2012. Each dot represents the value of the index on a particular day, with the starting and ending days labelled. Also shown are the eight defined phases of the MJO and the region used to signify weak MJO activity. Also label ed (with words) are the approximate locations of the near-equatorial enhanced convective signal of the MJO for each quadrant of the phase space (e.g. the 'Indian Ocean' for phases 2 and 3). After Wheeler and Hendon (2004). *Source*: NOAA's Climate Prediction Center MJO web portal: http://www.cpc.ncep.noaa.gov/products/precip/CWlink/MJO/mjo.shtml.

onto the multiple-variable EOFs, with the annual cycle and components of interannual variability removed, yields PC time series that vary mostly on the intraseasonal timescale of the MJO only. This projection thus serves as an effective filter for the MJO without the need for conventional time filtering, making the PC time series an effective index for real-time use. Wheeler and Hendon (2004) call the pair of PC time series that form the index the real-time multivariate MJO series 1 (RMM1) and 2 (RMM2).

RMM1 and RMM2 are dominated by the circulation components (Straub, 2013), but nevertheless correlate highly with the MJO convective signal. Therefore, an index formed by RMM1 and RMM2 can be used to describe the evolution of the MJO along the Equator, independent of season. Figure 7.13 shows the phase-space diagram (RMM1, RMM2) of the Wheeler–Hendon MJO index for the MJO event of 12–16 June 2012 and the NCEP Global Forecast System (GFS) and Global Ensemble Forecast System (GEFS) forecasts valid 17–23 June and 24–30 June 2012. The phase space is divided into eight MJO phases, plus a central circle signifying weak MJO activity. Since the RMM1 and RMM2 patterns include spatial structure along the Equator, their values are also associated with the geographic location of the peak precipitation along the Equator, as indicated by the labels in the diagram. For instance, both negative and positive values of RMM2 and negative values of RMM1 are associated with enhanced convection in the Western Hemisphere and Africa. The MJO signal propagates eastward or counterclockwise around the diagram. The distance from the origin is an indicator of the strength of the MJO signal: the further the index is from the circle the stronger the MJO. The MJO becomes weak or incoherent near or inside the circle. The diagram shown in Figure 7.13 reveals that in late June the MJO signal was quite strong over Africa.

#### 7.1.5.2.3 Madden–Julian Oscillation Signal over West Africa

As shown earlier for the 10–25-day filtered signal, the 25–90-day filtered signal over West Africa (Figure 7.10 bottom) for the summer of 2006 features weaker convective activity at the beginning of July (at the time of the monsoon onset) and around mid-August, and higher convective activity at the end of July. Further analyses have been carried out based on the first EOF decomposition of the 25–90-day timescales (Figure 7.11 bottom) using the same composite approach as for the 10–25-day timescale.

In this range of variability the dominant mode is the 'African MJO' (AMJO) mode with an apparent 40-day periodicity (Matthews, 2004; Janicot *et al.*, 2008, 2009; Lavender and Matthews, 2009). AMJOs proceed from interactions between equatorial atmospheric dynamics and convection and therefore have impact mainly when the ITCZ is closer to the Equator in northern spring, but can also have significant impacts on West Africa in summer. For this reason their occurrences could explain part of the variability of African monsoon onset dates, as in 2006 during the AMMA experiment (Janicot *et al.*, 2008).

The AMJO mode was first associated with the propagation along the Equator of a dry Kelvin wave moving eastward and a dry ER wave induced by a decrease of convection over the equatorial warm pool and travelling westward. When these atmospheric waves meet over African longitudes about 20 days later, they favour deep moist convection (Matthews, 2004). Over West Africa, the dry Kelvin wave component contains lower-tropospheric westerly anomalies that act to increase the boundary-layer monsoon flow and moisture supply. The westerly anomalies also increase the cyclonic shear on the equatorward flank of the African easterly jet (AEJ), leading to enhanced AEW and transient convective activity. However, the westward migration of convective anomalies observed over Africa suggests more the action of an ER wave coupled with convection and modulated by the active and break phases of the Indian monsoon (Janicot *et al.*, 2009; Mohino *et al.*, 2012). During an active phase of the Indian monsoon, more convection occurs over the Indian subcontinent than near the Equator.

This establishes a dipole of convection along the meridional plane and favours the development of low-pressure cells with cyclonic circulation over the northern part of the dipole. These cells migrate westward and reach West Africa where they enhance moisture advection inland. As a result the convection grows northeast of Lake Chad, then propagates westward before dissipating over the western part of Africa in conjunction with the northward migration of the dipolar structure between the Equator and the Indian subcontinent. These results highlight the use of MJO events to potentially predict regional-scale anomalous convection and rainfall spells over West Africa with a time lag of approximately 15–20 days. See also Leroux *et al.* (2010), who relate changes in transient wave activity over the WAM region to the MJO.

#### *7.1.5.2.4 Monitoring the Impact of the Madden–Julian Oscillation on West Africa*

Most studies on the impact of the MJO on West African summer rainfall and the WAM intraseasonal variability include filtering of precipitation, OLR or wind data, to retain the 30–60-day periodicity of the MJO timescale. However, these techniques are often not applicable to real-time operational monitoring and forecasting of the WAM on the subseasonal timescale. Despite this shortcoming, MJO composites of monthly or seasonal winds and precipitation can readily be used for guidance in subseasonal forecasting, especially when the MJO is active. Composites of 850 hPa wind anomalies and probabilities of weekly rainfall occurring in the upper decile for each MJO phase during May–July are shown in Figure 7.14. Rainfall probabilities refer to the chance of weekly rainfall exceeding the upper decile (90th percentile) expressed as a ratio with the mean probability (nominally 33%). These maps were constructed following the methodology of Wheeler *et al.* (2009). Fluctuations in rainfall probability are evident, especially during phases 1 and 5 of the MJO with scaled rainfall probabilities respectively exceeding 1.6 and falling lower than 0.6 in some localities in the western Sahel. This likelihood of enhanced rainfall during phase 1 occurs in conjunction with anomalous low-level westerly winds. In contrast, suppressed rainfall in phases 3 to 7 occurs in conjunction with easterly wind anomalies. Rainfall probabilities, expressed as the chance of weekly rainfall exceeding the upper tercile (67th percentile), yield the same patterns as for the 90th percentile, except that the signal is noisier and indicative of the role of mechanisms other than the MJO at this rainfall probability. MJO-based rainfall and wind anomaly composites are used as guidance in subseasonal forecasting, and are discussed further in Section 7.2.1.

### 7.1.6 Mechanisms for Dry- and Wet-spell Frequency

Extratropical/tropical interactions play a prominent role in influencing subseasonal wet/dry spells over West Africa, year round, by affecting the atmospheric moisture content, stability and moisture convergence. Knippertz and Fink (2008, 2009) show that an extratropical trough over northwestern Africa can enhance low-level southerly moisture transport from the Gulf of Guinea into the continental interior several days ahead of intense rainfall events in boreal winter. Knippertz and Fink (2006) and Knippertz and Todd (2012) associate anomalously strong Harmattan flow with Saharan dust outbreaks, and find that these outbreaks are typically associated with strong anticyclogenesis following the passage of a cyclone disturbance (see Chapter 5).

Vizy and Cook (2009, 2014) discuss the impact of synoptic cold air surges from the Mediterranean Sea on northern African rainfall. A cold air surge is a shallow dome of relatively cold, dry air that advances equatorward, accompanied by a ridge of high pressure and strong meridional low-level winds. Such events occur year round. They can have profound cooling and drying effects, and are a notable sink of energy for the tropics (Garreaud, 2001). During the peak of the Sahel rainy season, surge events can last for 2–10 days and travel equatorward at approximately 5.5 m $s^{-1}$. The African cold air surge is preceded by the passage of a shortwave trough and the intensification of the westerly jet streak over the Mediterranean Sea. Over the western Sahel, the passage of the surge is first associated with a brief

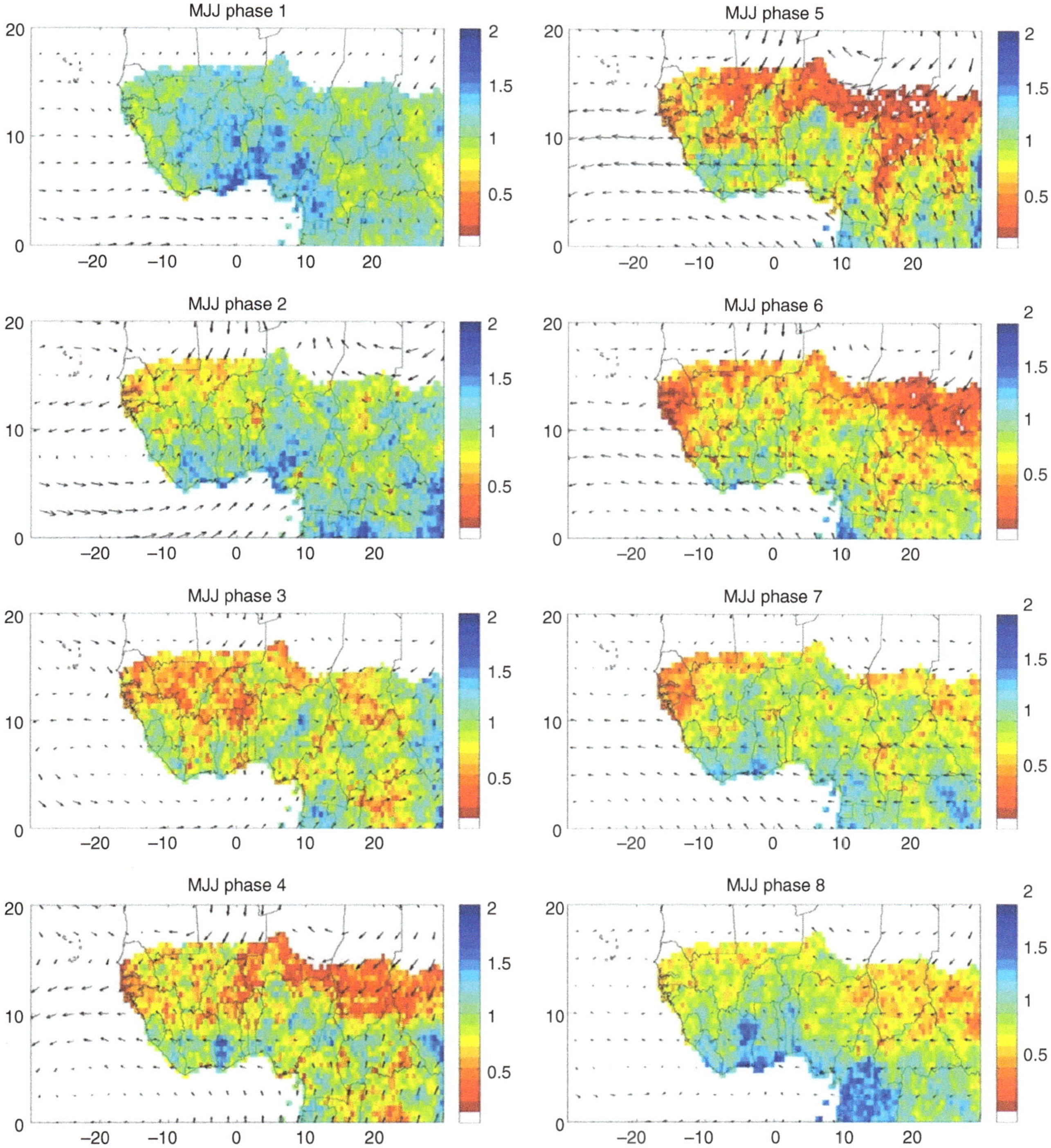


**Figure 7.14** MJO composites of weekly rainfall probabilities (contours and shading) and 850 hPa wind anomalies (vectors) for the MJJ season for phases 1–8. Rainfall probabilities refer to the chance of weekly rainfall exceeding the upper decile, expressed as a ratio with the mean probability (nominally 33%). *Source*: adapted after Wheeler *et al.* (2009).

enhancement of convection 2–5 days before the surge reaches the eastern Sahel, associated with enhanced low-level moisture convergence. Over the central and eastern Sahel, the surge passage is associated with an approximate 6-day break in convective activity. Figure 7.15 illustrates the 850 hPa temperature, wind and rainfall anomalies associated with a typical summer cold air surge event. At Day 0 (Figure 7.15a and b) cold air has begun to move equatorward over northeastern Africa. Significant rainfall anomalies are located over the Mediterranean Sea and across the northernmost Sahel, with the latter being less spatially coherent. By Day +2

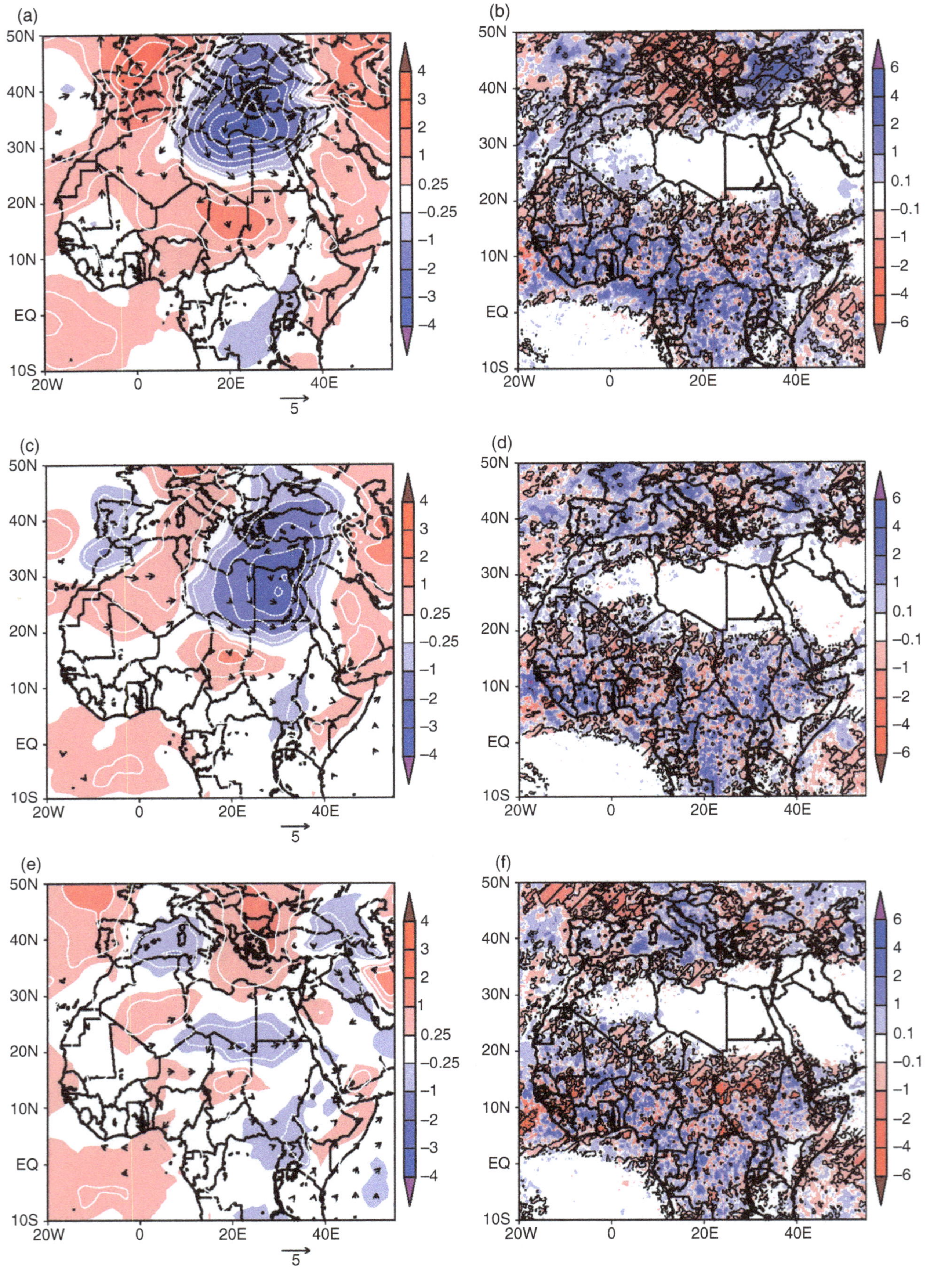


**Figure 7.15** Summer-time (July–September) daily composites of cold air surges. The 850 hPa temperature (K) and wind (m $s^{-1}$) anomalies (left column) and NASA TRMM 3B42V7 precipitation anomalies (mm $day^{-1}$) (right column) at composite Day 0 (a, b), Day +2 (c, d) and Day +5 (e, f). Day 0 is defined as the date when the equatorward-moving 850 hPa cold temperature anomaly averaged between 10°E and 30°E associated with a cold air surge has first crossed 30°N. Hatching denotes rainfall anomalies statistically significant at the 90% level of confidence. Only statistically significant wind vectors at the 95% level of confidence are shown. *Source*: Vizy and Cook (2014). Reproduced with permission of Springer.

(Figure 7.15c) 850 hPa cold air anomalies continue to move equatorward with the strongest cooling greater than 2.5 K positioned over Libya and Egypt. Significant precipitation anomalies extend across the northernmost Sahel at this time (Figure 7.15d). The air mass associated with the cold air surge modifies as it continues to move equatorward, and by Day +5 (Figure 7.15e) cold anomalies of magnitude 0.5–1.25 K are positioned over the central and eastern northern Sahel. A significant reduction in rainfall accompanies the arrival of the cold air surge over Sudan, Chad and Niger (Figure 7.15e). These dry conditions can persist for 4–8 days. Winter surge events are concomitant with stronger cooling and penetrate further equatorward, with breaks in rainfall activity over the Guinea Coast extending eastward into the northern Congo Basin as the associated drier, northerly low-level flow stabilises the atmospheric column over this region (Vizy and Cook, 2014).

As mentioned in Section 3.3.2, variations of the SHL are also related to intraseasonal wet and dry spells. Chauvin *et al.* (2010) and Roehrig *et al.* (2011) identify west and east phases of the SHL and discuss their ramifications for rainfall variability over sub-Saharan northern Africa. During the west phase, the extension of the North Atlantic anticyclone retreats westward from over Europe and the Mediterranean, and is associated with warm temperature anomalies over northwestern Africa and cold anomalies over northeastern Africa. The east phase corresponds to the opposite temperature and circulation patterns. Negative OLR anomalies at the peak of the west phase indicate that convection intensifies over the eastern Sahel at this time and propagates westward at a speed about 600 km day$^{-1}$. This is followed by a 3–6-day period of positive OLR anomalies or reduced convective activity. Circulation characteristics associated with the west phase and cold air surges share many similarities, suggesting that cold air surges are likely to form one aspect of the defined SHL mode (Chauvin *et al.*, 2010).

Another mechanism associated with wet and dry spells is the development and propagation of ER waves (see Section 7.1.4.2). Janicot *et al.* (2010) show that within the 10–30-day periodicity band, the first mode of variability is an ER wave pattern. These waves have an approximate wavelength of 7500 km, a westward phase speed of 6 m s$^{-1}$ and a period of 15 days. Their initiation, at least in terms of OLR signature, occurs over the East African highlands, and they are likely to be forced by deep convective heating, though Leroux *et al.* (2011) show that the remote influence triggered by dynamical precursors coming from the extratropics may be just as influential. Interestingly, the second main mode, termed the Sahel mode (Section 7.1.5.1.2), may also be related to ER waves to some extent. The main difference is that the Sahel mode has a stronger OLR and circulation anomaly response over central and eastern Africa. Janicot *et al.* (2010) estimate that about one-third of the Sahel mode events are related with ER waves, with the rest associated with other factors, including extratropical forcing (Roehrig *et al.*, 2011).

### 7.1.7 Onset of the Monsoon and Rainfall

#### 7.1.7.1 Detection

The onset of the rainfall season is of paramount importance in all monsoon regions, including West Africa. There is a strong demand from society for the predictions of the onset of the rains. Numerous studies have been undertaken to enhance our physical understanding of the onset of the rains and to predict it. Omotosho (1987) used the difference in vertical wind shear at a level below and above the position of the AEJ to define a monsoon index and suggested it was important to the set-up of the onset of the rains at certain locations. He argues that the triggering of the onset of the rains requires that the difference in vertical wind shear between the surface and 700 hPa ($\Delta$UL) and twice the vertical wind shear between 700 and 400 hPa ($\Delta$UM) be positive or equal to zero; i.e. $|\Delta \mathrm{UL}| - 2|\Delta \mathrm{UM}| \geq 0$ (Omotosho, 1992). The rationale behind this analysis was the establishment of both the AEJ and the tropical easterly jet (TEJ), and the poleward retreat of the subtropical jet, which creates a rather sudden wind change from westerlies to easterlies above 400 hPa. It is argued that the onset of the rains occurs 10 weeks after these wind conditions become established at a given location. This approach has been put into practice but suffers from a lack of observed upper air wind data over West Africa. Also, there was no clear distinction between the onset of the rains in the Sahel and that in the Gulf of Guinea region.

Other studies focused on the boundary-layer conditions. Eltahir and Gong (1996) showed that the dynamics of the monsoon are strongly regulated by the meridional gradient of the boundary-layer entropy and the moist static energy (MSE) that accounts for entropy and latent heat. The MSE is widely used to characterise monsoon and boundary-layer variation (Fontaine *et al.*, 1999, 2008; Chakraborty *et al.*, 2006; Peyrillé *et al.*, 2007; Dalu *et al.*, 2008). Thiaw and Kumar (2001) suggested that enhanced MSE in Central Africa and the Gulf of Guinea region during the pre-monsoon season tends to contribute to above-average rainfall in the Sahel. Indeed, it has been documented that the onset could be delayed due to lack of moisture supply (Okumura and Xie, 2004; Coëtlogon *et al.*, 2010).

The role played by the dynamics of the ITF has also been studied. Sultan and Janicot (2003) defined a 'pre-onset' by a first positioning of the ITF at 15°N. Then, the onset stage corresponds to an abrupt latitudinal shift of the monsoon precipitation from 5°N to 10°N with a climatological date around 24 June for the continental Sahel (between 10°W and 10°E).

#### 7.1.7.2 Mechanisms

The mechanisms associated with the onset of the rainfall season have been studied more thoroughly. Hagos and Cook (2007), using a regional climate model, documented the 'monsoon jump' theory and confirmed the important role of the boundary-layer circulation and the supply of moisture, which preconditions the atmosphere. Figure 7.16 shows a Hovmöller diagram of the daily precipitation climatology over West Africa for 1998–2012 from TRMM 3B42 V7 (Huffman *et al.*, 2007). Values are averaged from 12°W to 6°E to avoid the precipitation maximum over the Cameroon Highlands (in the Cameroon Highlands, the rainfall has a seasonal signal that is distinct from that to the west). Because of strong orographic influences, rainfall near the topography is maintained throughout the summer. During boreal winter, the precipitation maximum in the 12°W to 6°E longitude band stays within 3° of latitude of the Equator, and during these months the rainfall can be characterised as part of the marine Atlantic ITCZ without strong continental influence. In early May the rainfall intensifies as solar heating increases, and the precipitation maximum moves further north over the Guinea Coast region.

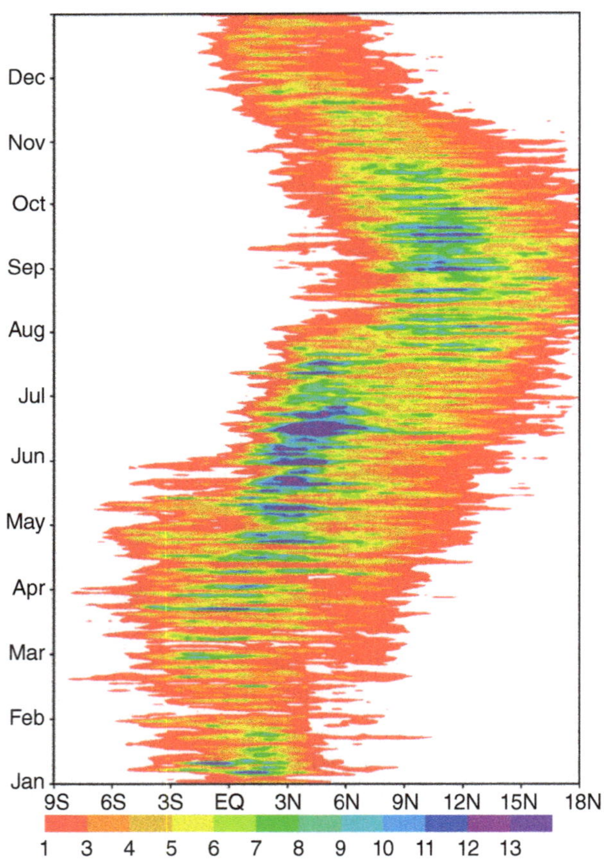


**Figure 7.16** Daily precipitation climatology (mm day$^{-1}$) for 1998–2012 from TRMM 3B42 V7 (Huffman *et al.*, 2007). This may be compared with Figure 1.3b.

Rainfall rates increase along the Guinea Coast (approximately 4–6°N) throughout May, and strong rainfall is maintained in this region through June. In early summer, Guinea Coast rainfall declines and the rainfall maximum re-establishes over the Sahel (about 10–15°N). The sudden movement of the precipitation maximum from the coast into the Sahel is referred to as the WAM jump (Sultan and Janicot, 2003; Okumura and Xie, 2004; Hagos and Cook, 2007). The average date of the jump, defined as the date on which the zonal-mean Sahel rainfall exceeds the Guinea Coast rainfall, is 7 July in the TRMM climatology. Grodsky and Carton (2001) examined high-resolution, satellite-observed winds and reported that the beginning of the rainy season in the Sahel is preceded by a biweekly oscillation of the zonal wind. Hagos and Cook (2007) relate the apparent sudden movement of the precipitation maximum to the development of inertial instability that is associated with the formation of the AEJ. Camberlin *et al.* (2010) and Flaounas *et al.* (2012) suggest that there is also a role played by the Indian monsoon in determining the timing of the monsoon onset. Mounier *et al.* (2008) also provide evidence that the combination of QBZD and Kelvin wave interactions could be related to monsoon onset in some years.

#### 7.1.7.3 Applications

Many definitions of onset are found in the literature, reflecting the considerable interest in the topic. Each has its own relevance for different users, including farmers, hydrologists and pastoralists. In this handbook we focus first on the definition mostly used in agricultural applications. Factors to be considered when defining the onset include the local conditions (first rainfall event at one station), a large-scale signal (simultaneous rain event reported over a large domain) and an actual onset, which usually establishes continuity in the rainfall sequences (onset not followed by a long dry spell). For agriculture, a dry spell that occurs after farmers have already planted can result in loss of crops and reduced productivity, with devastating socio-economic impacts, including food insecurity. In the Sahel, a rainfall event is considered to be significant when amounts reach 20 mm over a 3-day period. This rainfall threshold was found to be critical for early-season runoff and moistening of the soil, taking into account the daily evapotranspiration in the warm, sunny and windy climate of the Sahel (Taylor, 2000; Gash *et al.*, 1991). For example, Davy *et al.* (1976) documented that the planting date for millet was observed to coincide with the first occurrence of 20 mm of rain over a 2-day period. Such a rainfall threshold was consistent with that proposed by Stern *et al.* (1982) for an onset definition over the Sahel, and Sivakumar (1998) based on observations of the establishment of the millet crop at the International Crops Research Institute for the Semi-Arid Tropics Sahelian Center in Niger.

For agricultural purposes and to maximise crop production, it is extremely important to distinguish a true start to the rainfall season from rain events that occur at the beginning of the season but that are followed by prolonged dry spells. These short-lived initial rains are referred to as 'false starts' to the rainfall season. A case study of true and false onsets of the rainfall season over Sénégal is discussed. The onset is defined as the first time after the 1 May when there is a 3-day rainfall accumulation of at least 20 mm with no dry period of seven or more consecutive days occurring during the next 20 days (similar to Marteau *et al.* (2009)). To assess the large-scale signals in the climate system associated with true versus false onset of the rainfall season, the first event and the true onset are calculated for each station in Sénégal. These are then classified into homogeneous zones with respect to the onset date. A spatial mean standardised index for the date of the first and true onset events is calculated to identify years with a high number of dry days following the large-scale onset, and years with a low number of dry days. Composite maps of the difference in number of dry days between the most severe false starts (upper quartile of the dry days number distribution) and the best true starts (lower quartile) over the northern part of Sénégal are shown in Figure 7.17. The striking feature is the persistence of a high negative precipitable water (PWAT) anomaly over and to the west of the region. Well before the onset event date (time *T*) there is a systematic anomaly of low PWAT (blue colour) located in the vicinity of Sénégal and surrounding areas. During the northern region onset, the negative anomaly of PWAT is seen 8 days prior to the onset, when it extends further north and east. It persists, shifting its axis southward and westward, until 2 days before the onset date. This negative anomaly of PWAT is concomitant with a northeasterly wind to the south of Sénégal, indicative of reduction in the monsoon flow, which is southwesterly. Those two signals are internally consistent, since monsoon flow supplies water vapour. During the time of the rainfall onset event, areas of reduced PWAT values are found across western Sénégal into the eastern Atlantic and over equatorial West Africa. Two days after the onset, negative PWAT anomalies become re-established over Sénégal, and a strong northerly/northeasterly wind advecting cold air emerges in the composite. This cold air surge at low levels will contribute to stabilisation of the air and will have an inhibiting effect on the monsoon surge, while killing any convection. Figure 7.17 also shows that the onset over Sénégal is strongly controlled by the large-scale atmospheric system with propagating features prior to, and after, the onset. The composite analysis of the difference in anomaly shows a clear 'danger' signal to watch for. False onset (compared with true onset) can be characterised by pre-event anomalies of PWAT located at the western fringe of Sénégal and the tropical North Atlantic, with strong northeasterly wind located north of the region (Mauritania to western Europe) advecting cold air. This low-level cold air surge can act to kill any convection. For the northern region, the PWAT anomalies may be established over the tropical Atlantic and Sénégal several days ahead of the 'false' onset event.

#### 7.1.7.4 Summary and Outlook

Unlike short- and medium-range weather forecasting and seasonal to interannual climate forecasting, subseasonal forecasting is at its infancy and remains challenging. This is primarily due to the fact that boundary conditions are not sustained long enough at this timescale to counterbalance the chaotic nature of the atmosphere. Nonetheless, the importance of this timescale in decision-making in various socio-economic sectors, including agriculture and food security, has enticed scientists to better understand the mechanism associated with intraseasonal variability in order to improve predictions. While the study of intraseasonal variability often requires filtering of the data in order to isolate dominant modes of variability at this timescale, forecasters need to work with total fields and anomalies to issue forecasts on time. This represents a huge challenge, as many of the intraseasonal variability modes are hard to diagnose based solely on time-averaged weather charts. However, recent progress in the understanding of the MJO and its relative influence on the WAM variability has enabled the beginning of science-based operational subseasonal forecasting for West Africa and the global tropics. This effort is sustained by recent advances in NWP and reanalysis, combined with forecasters' expertise in the analysis and interpretation of the NWP outputs. In this chapter, the proposed approach to operational subseasonal forecasting takes into account the MJO, whether in its active phase or not, and the interpretation of NWP outputs. Future generations of NWP at even higher resolution and improved physics will lead to improved subseasonal forecasting over the next few years.

## 7.2 Operational Methods

One of the challenges of operational subseasonal forecasting is that it involves some degree of subjectivity; it also heavily draws on the forecaster's experience. At present there is a gap in transitioning recent advances in intraseasonal variability studies to real-time operational subseasonal forecasting. This is due in part to the fact that the forecaster has to issue an outlook at a given day and time, which makes it almost impractical to derive forecasts from filtered data to isolate intraseasonal variations. For this reason, this section does not address the AMJO as described in Section 3.3.2. It is good practice, however, for forecasters to be aware of the AMJO, and to allow forecast decisions, as far as possible, to be guided

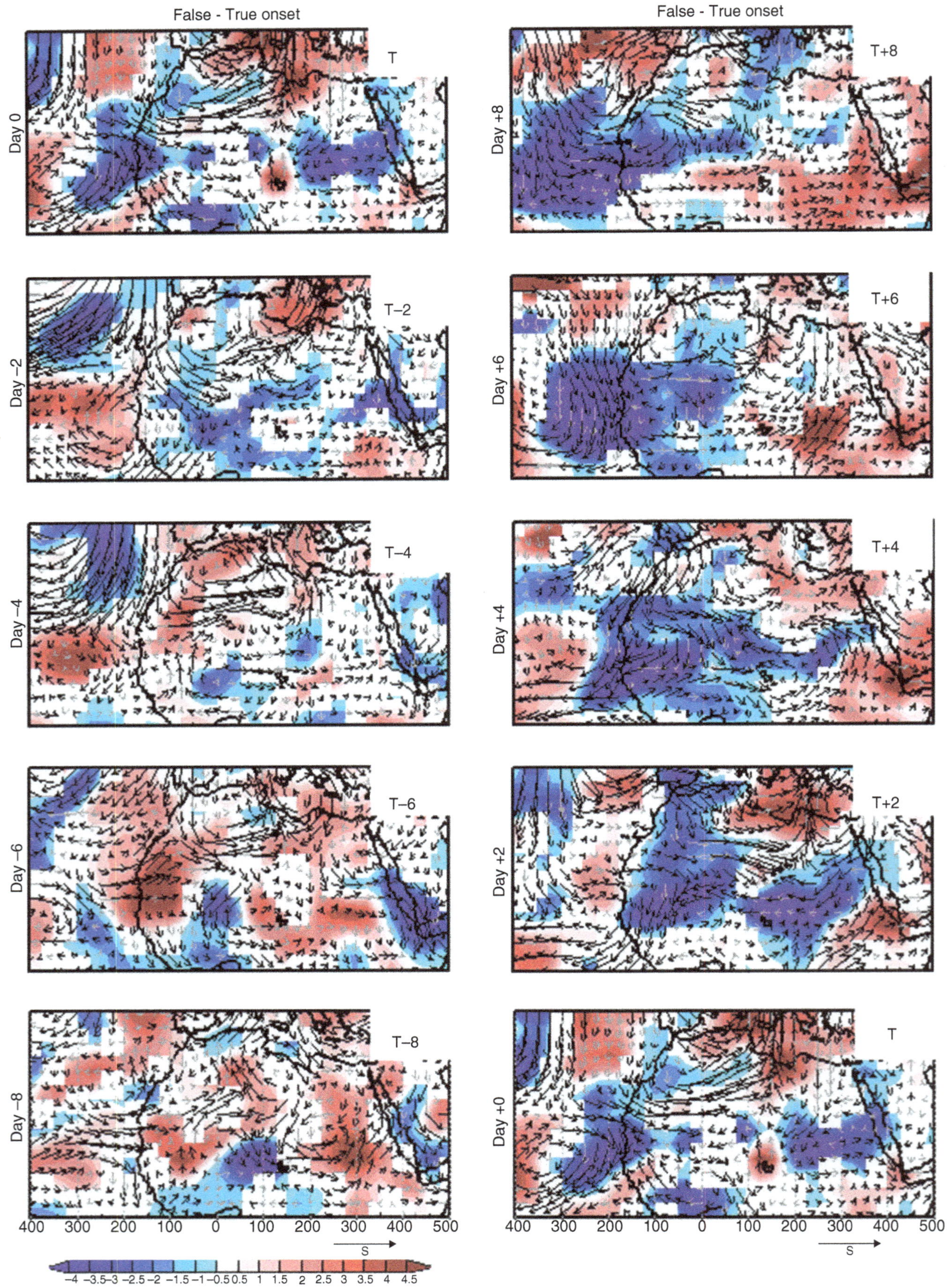


**Figure 7.17** Sequences of composite fields of PWAT difference (kg m$^{-2}$, shaded) and 925 hPa wind difference (vector, m s$^{-1}$) between a false onset and a true onset for days $T-8$ through $T+8$, where $T$ represents the date of the first large-scale rainfall event index over northern Sénégal region. Note that blue colours indicate lower PWAT values in false-onset cases. *Source*: Adapted from Ndiaye (2010).

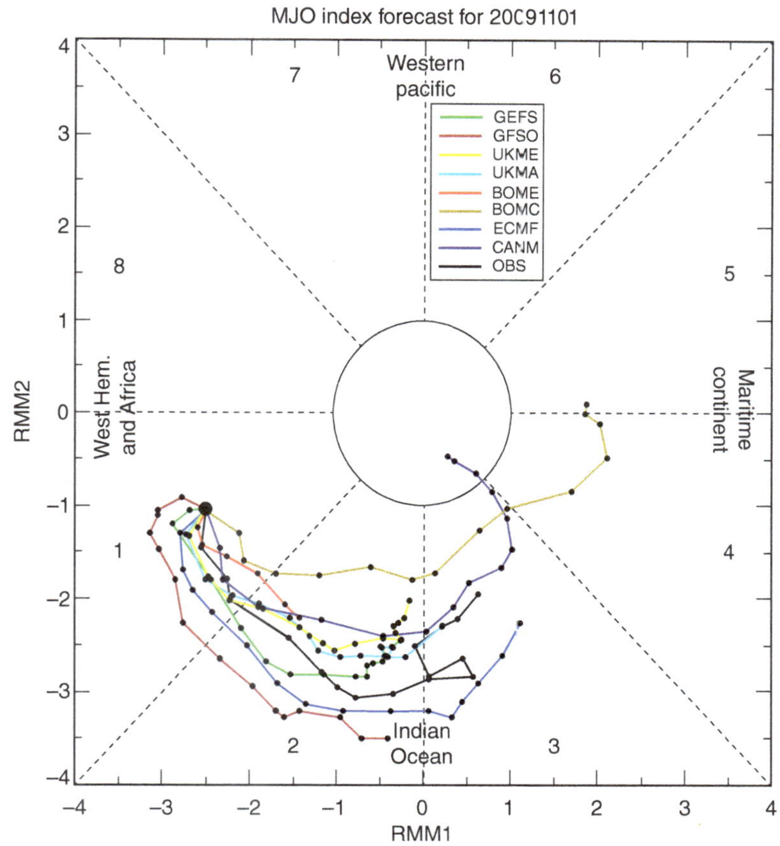


**Figure 7.18** Two-component MJO index (as in Figure 7.13), for projections of the MJO index from dynamical coupled ocean–atmosphere models, 1 November 2009 initial conditions, through week-2 forecasts. *Source*: NOAA's Climate Prediction Center MJO web portal: http://www.cpc.ncep.noaa.gov/products/precip/CWlink/MJO/mjo.shtml.

by their own intuition with respect to such features. Work is being carried out to develop more objective techniques for subseasonal forecasting similar to seasonal climate predictions.

Despite the challenges, a better understanding of the MJO activity and its representation in numerical models has provided a physical basis for current operational subseasonal forecasting. In this section, we document one approach, based on the MJO and NWP guidance. The outlooks can be produced at any operational meteorological centre with access to the MJO and NWP tools. At present, it is realistic to provide forecasts of enhanced and suppressed rainfall during the week-1 and week-2 time periods. Forecasts for week-3 and week-4 are also being considered.

### 7.2.1 Prediction Tools

#### 7.2.1.1 Madden–Julian Oscillation Predictions

Tremendous progress has been made over the past decades in understanding the physics of the tropical climate. This improved understanding is confirmed by the performance of dynamical climate models especially in the prediction of the MJO. Currently, dynamical model forecasting skill for the MJO is higher than persistence and statistical models (Rashid *et al.*, 2010). To translate this advancement of the science into operational forecasts, a subgroup of the US CLIVAR program MJO working group, now reformulated as the World Climate Research Programme/World Weather Research Programme-THORPEX Year of Tropical Convection Task Force, has been tasked with the development of an MJO forecast metrics for the comparison of dynamical models from operational global prediction systems. The NCEP Climate Prediction Center hosts the application and provides displays and evaluations of the MJO model forecasts through a website.[1] Participating global prediction centres include NCEP, ECMWF, UK Met Office, and the weather services of Australia, Brazil, Canada, India, Japan and Taiwan.

Figure 7.18 shows predictions of the MJO index from the ensemble mean of eight models, from 1 November 2009 initial conditions. The observed MJO index is also displayed. It can be seen that all models depicted the eastward propagation of the MJO reasonably well, despite a considerable spread in the models with respect to the amplitude of the index. The consistency of the models in the east–west propagation of the MJO suggests that the RMM MJO index can be used for subseasonal forecasting in the tropics, especially for areas where the teleconnection is very strong.

#### 7.2.1.2 Prediction of Regional Climate Anomalies

Although the MJO is a powerful tool for subseasonal forecasting, its influence on the WAM is not as prominent as on other tropical regions, such as equatorial eastern and central Africa. The summertime rainfall in West Africa is primarily modulated by westward-propagating synoptic-scale events known as AEWs. Thus, it is essential to take into account regional medium-range weather forecasts derived from NWP outputs.

The NCEP African Desk forecasting team examines outputs from the NCEP GFS, the GEFS, the CFS and both the GFS and Climate Forecast System (CFS) bias-corrected precipitation forecasts. Emphasis is placed on forecasts of 7-day precipitation, surface temperature, mean sea-level pressure, wind fields at low, mid and upper levels, both at zero lead for week-1 outlooks and at 7-day lead for week-2. The forecaster looks for a convergence of evidence between the predicted circulation anomalies and the precipitation forecasts.

### 7.2.2 Guidelines for Operational Forecasting

In this section, a step-by-step approach for week-1 and week-2 outlooks is described. The approach provides guidance on how to utilise a combination of subseasonal prediction, along with weather and climate monitoring tools to prepare the week-1 and week-2 forecasts. Assessing the current state of the MJO is the first step to the forecasting process (Thiaw and Kumar, 2015). The next step is to understand the influence of the MJO on African rainfall so that areas of potential impacts can be identified when the MJO is active. Dynamical and statistical MJO forecasts out to 3 weeks, NWP outputs for precipitation, temperature and circulation out to 14 days, and the monitoring of SST and SST tendency are additional critical inputs to the subseasonal forecasts. The outlook period of 19–25 June 2012 is used here to illustrate the subseasonal forecasting process for West Africa as followed at NCEP's African Desk.

#### 7.2.2.1 State of the Madden–Julian Oscillation

Since the MJO is the dominant mode of intraseasonal variability in the tropics, much emphasis is given to its state at the time of the forecasts and its projected phases on the Wheeler–Hendon diagram. Thus, at the outset, the forecaster needs to refer to the MJO monitoring and prediction tools described in Section 7.2.1 to determine whether the MJO is active. Typically, the MJO is suspected of being active when the global OLR, superimposed with the 200 hPa VP anomaly map (Figure 7.12), reveals the presence of a coherent wave number 1, or sometimes wave number 2, pattern with distinct large areas of enhanced convection and areas of suppression. The presence of the MJO is further confirmed by examining additional information, including time–longitude diagrams of low-level wind, OLR and 200 hPa VP, to look for evidence of a relatively slow west–east propagation of the wave. This information is critical, as sometimes fast-moving Kelvin waves can alter the atmospheric circulation such that it resembles that of MJO activity (e.g. Roundy *et al.*, 2009).

These high-frequency waves usually do not last long and can be misleading, although there are occasions when they can be useful tools for the week-1 outlooks. The Wheeler–Hendon diagram (Figure 7.13) is extremely useful in determining the phase and amplitude of the MJO. For an MJO event to be considered active, the MJO index must be outside of the centre circle of the diagram and show evidence of an eastward propagation over the past several days, preferably of at least 1 week's duration.

Once it has been determined that the MJO is active, the forecaster next needs to examine the dynamical and statistical model predictions of the MJO index on the Wheeler–Hendon diagram. Questions to ask include: Do the models predict a coherent MJO pattern? Are they in agreement? What is the projected phase of the MJO during the week-1 and week-2 outlook periods? MJO-based May to July rainfall and 850 hPa wind anomaly composites for West Africa are displayed in Figure 7.14. Enhanced convection is favoured over the Gulf of Guinea region and along the coast during phase 8 with a northward extension into the Sahel during phase 1. These rainfall patterns are associated with westerly wind anomalies consistent with the eastward propagation of the MJO. As the MJO continues to propagate eastward, rainfall is gradually suppressed along the Guinea Coast, while moisture deficits intensify over the Sahel during phases 3 to 6 with a remarkable wind anomaly reversal. Rainfall gradually returns along the Guinea Coast during phase 7, and the northward progression into the Sahel reappears in phase 8. The above analysis suggests that the MJO-based rainfall and wind anomaly composites can contribute significantly to the subseasonal outlooks when the MJO is active. Thus, two scenarios are considered during the forecast preparation:

1) *MJO scenario I: MJO is active and projected to remain active through the outlook period*
   In the scenario of an active MJO throughout the outlook period, the MJO is used as the main forecast guidance. This is illustrated in the case of the forecast period of 19–25 June 2012. The forecast was issued on 18 June 2012. Figure 7.19 shows that convection associated with MJO activity was centred over the Western Hemisphere, approximately over South America, corresponding to phase 8 of the Wheeler–Hendon diagram, and projected to propagate eastward to Africa corresponding to phase 1 and then to

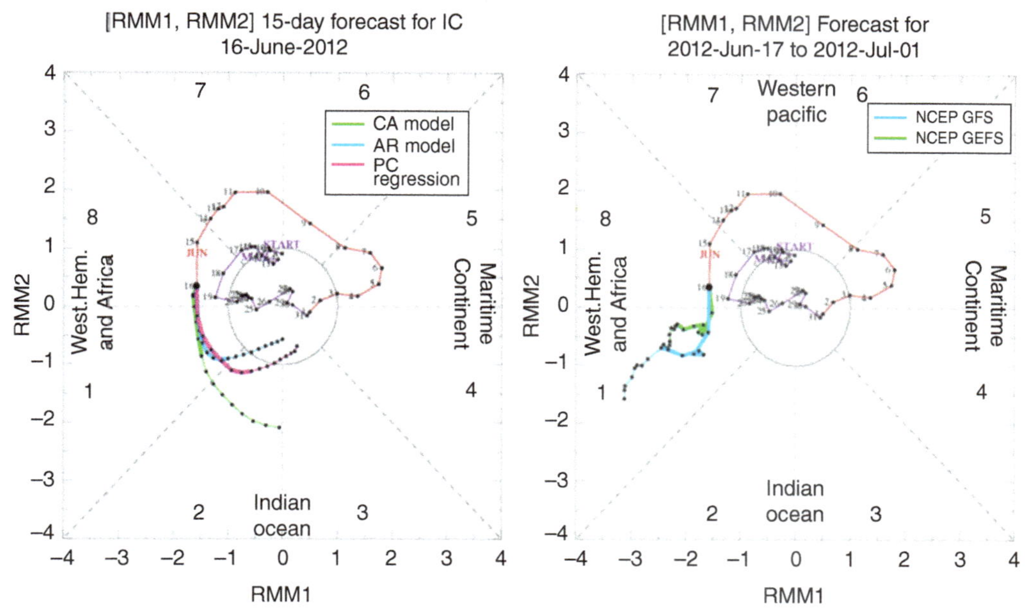


**Figure 7.19** Plot of forecast two-component MJO index. The right panel is the same forecast as shown in Figure 7.13. The left panel includes statistical forecasts of the MJO index: constructed analogues (green); autoregression model (blue); and principal component regression (pink). *Source*: NOAA's Climate Prediction Center MJO web portal: http://www.cpc.ncep.noaa.gov/products/precip/CWlink/MJO/mjo.shtml.

the regions in phase 2 according to the NCEP GFS, GEFS and statistical models. Based on the projected phase of the MJO and guided by the MJO rainfall anomaly composite maps (Figure 7.14), the forecaster may draw polygons (see Figure 7.22 for an example) to highlight areas with higher probabilities of enhanced or suppressed rainfall during the week-1 and/or week-2 outlook periods. However, the forecaster must also refer to the NWP guidance, especially for West Africa, where the MJO signal is not quite robust. Consistency between the MJO rainfall anomaly composite maps and the NWP guidance will yield a higher confidence in the forecasts. The low-level 850 hPa wind forecasts are displayed in Figure 7.20. The outflow from the Saint Helena high-pressure system in the South Atlantic results in winds turning into southwesterlies in the Gulf of Guinea region and westerlies across the western and central Sahel. The GFS and GEFS precipitation forecasts (Figure 7.21) were consistent with the wind forecasts and indicated an enhancement of rainfall along the Guinea Coast. The weekly SST anomaly and SST anomaly tendency for the past and current weeks were also examined to locate the sources of moisture for West Africa at the time of the forecasts. Positive SST anomalies were prominent in the eastern Atlantic near the Equator. Assuming that these SST anomalies persist through the forecast period, they will act to enhance precipitation in the Gulf of Guinea region and contribute to the suppression of rainfall in the Sahel.

An agreement between the MJO projected rainfall anomaly composites and the NWP guidance will enhance the forecaster's confidence in highlighting areas of enhanced or suppressed precipitation.

An example of forecasts with a validity period of 19–25 June 2012 is displayed in Figure 7.22. They indicate an area of rainfall surplus over the Gulf of Guinea region and suppression to the north in the western sector of the Sahel. Sometimes there are discrepancies between the MJO rainfall anomaly composites and the NWP forecasts outputs. In such cases, intuition and knowledge of the West African climate will guide the forecaster in their decision process.

2) *MJO scenario II: the MJO is projected to be present, but an incoherent pattern during the outlook period is expected*
NWP outputs are the primary tools for guiding the week-1 and week-2 forecasts in this case. The NWP forecast guidance most relevant to the week-1 and week-2 rainfall outlooks includes winds, surface temperature, quantitative precipitation forecasts, rainfall probability of exceedance of climatological rainfall derived from the ensemble forecasts and bias-corrected rainfall forecasts. The forecaster will locate areas where NWP models predict high or low rainfall amounts.

The credibility of the NWP guidance is measured by the spread in the model ensembles, the consistency among the different models, and the consistency between precipitation forecasts and predicted circulation anomalies. The latter is critical in guiding the forecaster, since

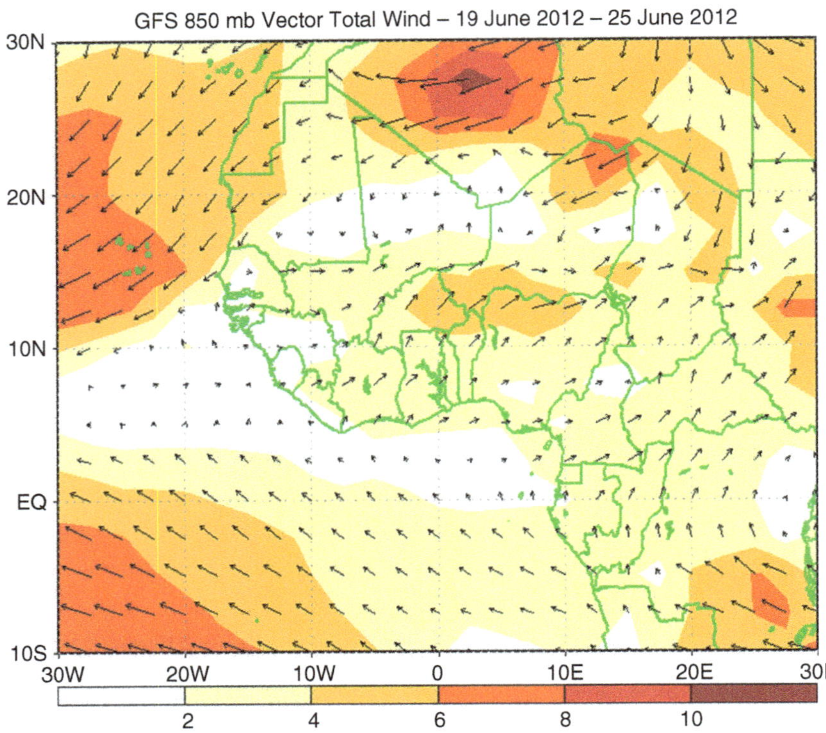


**Figure 7.20** GFS 850 hpa wind forecast (m $s^{-1}$), valid 19–25 June 2012. Arrows indicate wind direction, and the magnitude of the wind is represented by the colour shades.

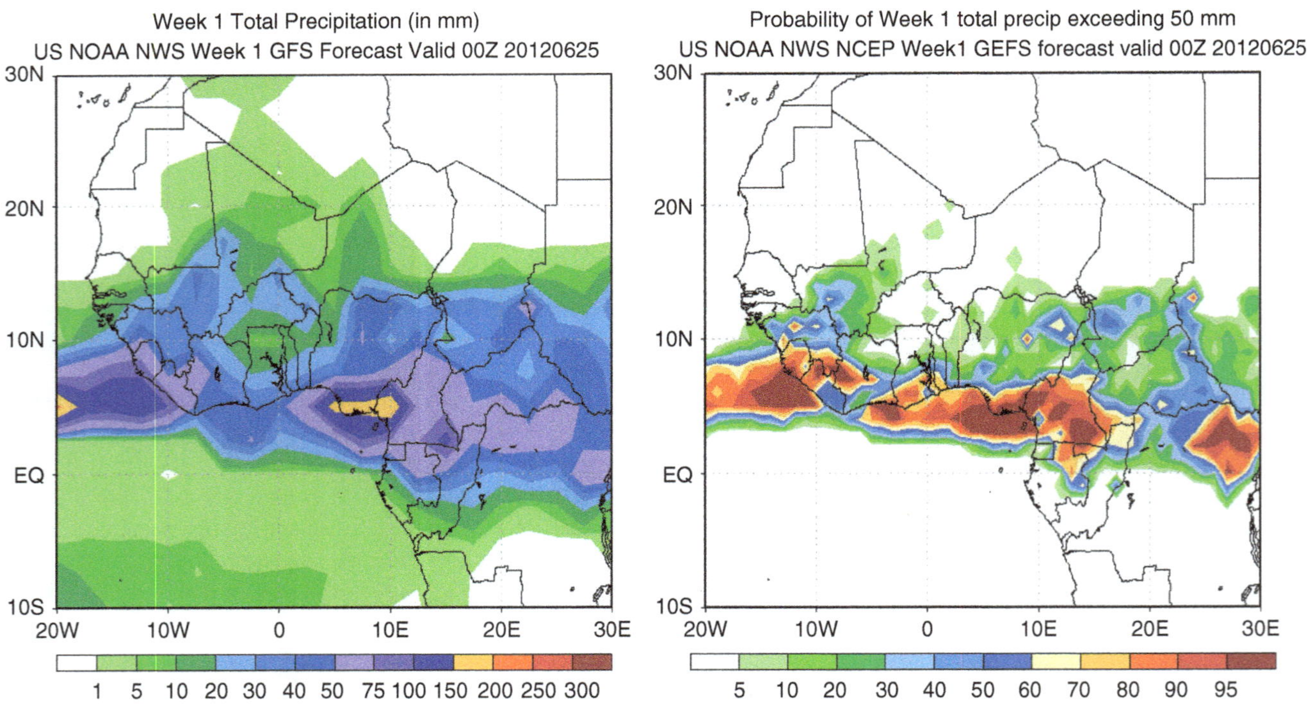


**Figure 7.21** GFS 7-day precipitation forecast, total in millimetres (left) and probability 7-day total precipitation exceeding 50 mm (right).

NWP models have higher skills in predicting winds than precipitation (see Chapter 10). More weight is given to the NWP precipitation forecast tools that are more consistent with the predicted circulation anomalies.

Typically for West Africa, when the MJO is not active, the forecaster will examine the structure of the wind forecasts and anomalies at 850, 600 and 200 hPa to locate areas of predicted low-level convergence and to assess

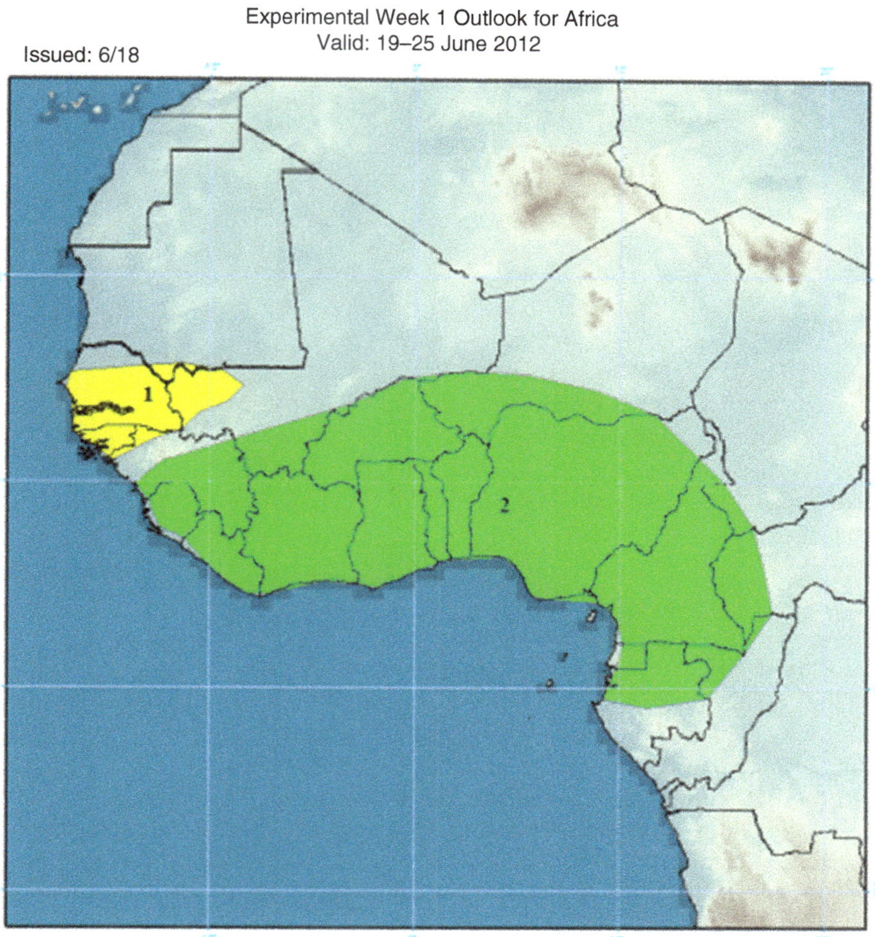


Figure 7.22 Week-1 precipitation outlook, valid 19–25 June 2012. Green (yellow) shades indicate areas with a higher chance for above (below)-average rainfall.

the structure and intensity of the AEJ and the TEJ during the outlook period. Surface temperature and mean sea-level pressure forecasts out to 14 days will provide insights into the location of the heat low, a prominent feature of the WAM and a driver of convection and rainfall in West Africa. As described in Chapter 2, the strength of the heat low will act to intensify the low-level southwesterly and northerly flows, and its upward motion generates an anticyclonic circulation aloft (around 700 hPa), which contributes to the strengthening of the AEJ. In addition to the NWP, the forecaster will examine the patterns of weekly SST, SST anomaly and SST anomaly tendency over the Atlantic. The SST anomaly tendency is defined as the difference between the current week's SST anomaly and the previous week's SST anomaly, so that a positive SST anomaly tendency implies that SST is increasing. Because SST does not change rapidly, the SST anomaly tendency provides a good approximation of SST patterns during the outlook period. For example, an increased SST anomaly tendency in the tropical North Atlantic and a decreased SST anomaly tendency in the Gulf of Guinea will act to enhance the surface temperature gradient between the Guinea region and the Sahel, will accelerate the LLWJ, especially in the middle of the monsoon season, and will favour enhanced rainfall in the Sahel during the summer season. Conversely, a decreased SST anomaly tendency in the tropical North Atlantic with increased SST anomaly tendency in the Gulf of Guinea, will reduce the surface temperature gradient, weaken the LLWJ and suppress rainfall. Thus, a careful examination of the NWP tools available at the global forecast producing centres will help the forecaster make an informed decision about the forecasts through the outlook period. The NWP tools may include winds totals and anomalies, bias-corrected precipitations and temperature forecasts, ensemble probability of exceedance maps, and so on.

#### 7.2.2.2 Consensus Outlooks

Upon making a preliminary forecast, the forecaster should conduct a forecast briefing to review the forecasting tools and to discuss the forecasts before they are made final. The briefing is very important, especially when there are inconsistencies among the prediction tools. The intuition and experience of different forecasters will help the arrival at a consensus outlook.

A geographic information system tool is then used to draw polygons that indicate the chances of above- or below-average precipitation, as displayed in Figure 7.22. The outlooks are disseminated via a distribution list and posted on the web site.

#### 7.2.2.3 Forecast Verifications

Forecast verifications are important as they help determine the forecast accuracy over time, so that the user can make informed decisions based on the forecasts. The week-1 and week-2 forecasts call for areas of enhanced or suppressed precipitation. Preferably, gauge data should be used to verify the forecasts for the wet and dry categories. However, satellite rainfall estimates are the basis for the verifications here. Two verification approaches are discussed: subjective and objective.

##### *7.2.2.3.1 Subjective Verifications*

Contours of the week-1 or week-2 wet or dry polygon forecasts are drawn on to the observed rainfall anomaly maps, as shown in Figure 7.23. The red contour indicates a forecast for below-average rainfall, and the blue contour for above-average rainfall. One can see that this forecast was partially successful; most areas in the red polygon in this case received below-average rainfall, except for the eastern sector of Sénégal. Similarly, heavy rains pounded many areas within the blue contour line, although some dry areas were noticeable, especially over the western part of the Gulf of Guinea region and along the boundary between Nigeria and Cameroon.

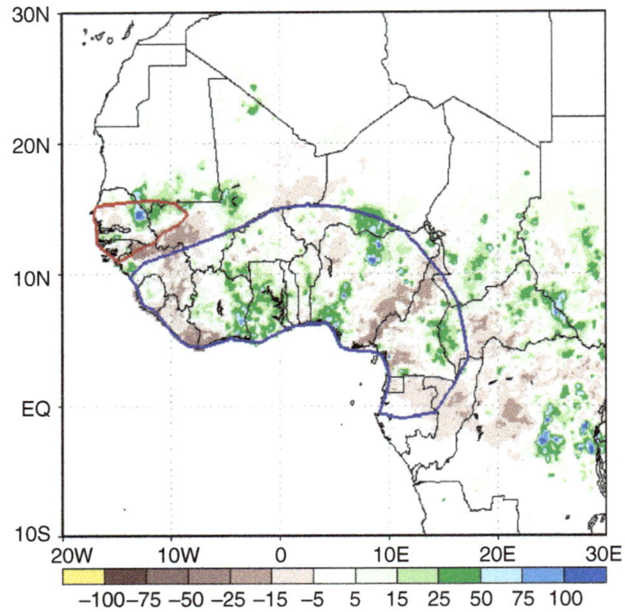


Figure 7.23 Satellite rainfall estimate (ARC2) anomaly (mm) for 19–25 June 2012 used in the verification of the week-1 outlook valid 19–25 June 2012.

##### *7.2.2.3.2 Objective Approach*

Standard verification statistical tools may be used to measure the skill of week-1 and week-2 forecasts. The Heidke skill score (HSS) is the approach employed here. Forecast polygon areas are converted into the same grid as the satellite rainfall estimates and values of +1 or −1 assigned to each grid cell that receives above- or below-average rainfall respectively. The Cohen's $K$ HSS:

$$\text{HSS} = \frac{(\text{hits} + \text{correct negatives}) - (\text{expected correct})_{\text{random}}}{N - (\text{expected correct})_{\text{random}}}$$

where

$$(\text{expected correct})_{\text{random}} = \frac{1}{N}\Big[(\text{hits} + \text{misses})(\text{hits} + \text{false alarms}) + (\text{correct negatives} + \text{misses})(\text{correct negatives} + \text{false alarms})\Big]$$

is then used to compute the forecast skill scores for each grid point. HSS is a measure of how well a forecast performed relative to a randomly selected forecast. A score of zero means that the forecast did not perform any better than a random forecast. A score of 1 depicts a 'perfect' forecast and a score of −0.5 depicts the 'worst possible' forecast. The performance of the week-1 outlook for West Africa issued at the NCEP's African Desk from 14 April 2008 to 31 December 2012 is displayed in Figure 7.24 and reveals moderate skill (better than random chances) over portions of West Africa, including southeastern Sénégal, Mali, Burkina Faso and western Niger. Skill was much lower along the Guinea Coast.

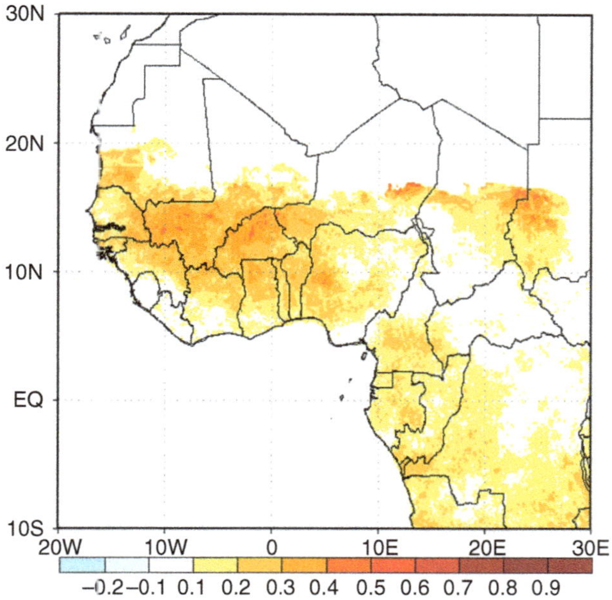


**Figure 7.24** HSS for week-1 rainfall outlooks issued 14 April 2008 through 31 December 2012. *Source*: Thiaw and Kumar (2015). © American Meteorological Society. Used with permission.

## Acknowledgements

We would like to thank George N. Kiladis for his careful and constructive review of the manuscript.

## Appendix: Acronyms

| | |
|---|---|
| AEJ | African easterly jet |
| AEW | African easterly wave |
| AMJO | African Madden–Julian oscillation |
| AMSU | Advanced Microwave Sounding Unit |
| AVHRR | Advanced Very High Resolution Radiometer |
| CCEW | convectively coupled equatorial wave |
| CDAS | Climate Data Assimilation System |
| CFS | Climate Forecast System |
| CLAUS | Cloud Archive User Service |
| CLIVAR | Climate and Ocean: Variability, Predictability and Change |
| CPC | Climate Prediction Center |
| ECMWF | European Centre for Medium Range Weather Forecasting |
| ENSO | El Niño Southern Oscillation |
| EOF | empirical orthogonal function |
| ER | equatorial Rossby |
| GEFS | Global Ensemble Forecast System |
| GFS | Global Forecast System |
| HSS | Heidke skill score |
| IR | infrared |
| ISCCP | International Satellite Cloud Climatology Project |
| ITCZ | intertropical convergence zone |
| ITF | intertropical front |
| LLWJ | low-level westerly jet |
| MJO | Madden–Julian oscillation |
| MRG | mixed Rossby–gravity |
| MSE | moist static energy |
| NCEP | National Centers for Environmental Prediction |
| NOAA | National Oceanic and Atmospheric Administration |
| NWP | numerical weather prediction |
| OI | optimum interpolation |
| OLR | outgoing longwave radiation |
| OMP | Observatoire Midi-Pyrénées |
| PC | principal component |
| PWAT | precipitable water |
| QBZD | quasi-biweekly zonal dipole |
| RFE2 | Rainfall Estimate Version 2 |
| RMM | Real-time multivariate MJO series |
| SEDOO | Service de DOnnée |
| SHL | Saharan heat low |
| SSM/I | Special Sensor Microwave/Imager |
| SST | sea surface temperature |
| TB | brightness temperature |
| TD | tropical disturbance |
| TEJ | tropical easterly jet |
| THORPEX | The Observing-System Research and Predictability Experiment |
| TOA | top-of-the-atmosphere |
| TRMM | Tropical Rainfall Measurement Mission |
| VP | velocity potential |
| WAHL | West African heat low |
| WAM | West African monsoon |
| WAWJ | West Africa westerly jet |

## Note

1 http://www.cpc.ncep.noaa.gov/products/precip/CWlink/MJO/CLIVAR/clivar_wh.shtml#for.

## References


Arkin P. 1983. Tropical circulation anomalies associated with the 1982–83 ENSO event. In: *Proceedings of the 8th Climate Diagnostics Workshop*, Toronto, 17–21 October.

Camberlin B, Fontaine B, Louvet S, *et al.* 2010. Climate adjustments over West Africa accompanying the Indian monsoon onset. *J. Climate* **23**: 2047–2064.

Chakraborty A, Nanjundiah RS, Srinivasan J. 2006. Theoretical aspects of the onset of Indian summer monsoon from perturbed orography simulations in a GCM. *Ann. Geophys.* **24**: 2075–2089.

Chauvin F, Roehrig R, Lafore J-P. 2010. Intraseasonal variability of the Saharan heat low and its link with midlatitudes. *J. Climate* **23**: 2544–2561.

Chelliah M, Arkin P. 1992. Large-scale interannual variability of monthly outgoing longwave radiation anomalies over the global tropics. *J. Climate* **5**: 371–389.

Christoph M, Ulbrich U, Haak U. 1995. Faster determination of the intraseasonal variability of storm tracks using Murakami's recursive filter. *Mon. Weather Rev.* **123**: 578–581.

Coëtlogon G, Janicot S, Lazar A. 2010. Intraseasonal variability of the ocean atmosphere coupling in the Gulf of Guinea during boreal spring and summer. *Q. J. R. Meteorol. Soc.* **136**(s1): 426–441.

Dalu GA, Gaetani M, Baldi M. 2008. A hydrological onset and withdrawal index for the West African monsoon. *Theor. Appl. Climatol.* **96**: 179–189.

Davy E, Mattei GF, Solomon SL. 1976. An evaluation of the climate of water resources for development of agriculture in the Sudano-Sahelian zone of West Africa. Special Environmental Report 9. WMO-No. 459. World Meteorological Organization: Geneva.

Dee DP, Uppala SM, Simmons AJ, *et al.* 2011. The ERA-Interim reanalysis: configuration and performance of the data assimilation system. *Q. J. R. Meteorol. Soc.* **137**: 553–597.

Eltahir EAB, Gong C. 1996. Dynamics of wet and dry years in West-Africa. *J. Climate* **9**: 1030–1042.

Flaounas E, Janicot S, Roca R, Mohino E. 2012. The role of the Indian monsoon onset in the West African monsoon onset: observations and AGCM nudged simulations. *Clim. Dyn.* **38**: 965–983.

Fontaine B, Philippon N, Camberlin P. 1999. An improvement of June–September rainfall forecasting in the Sahel based upon region April–May moist static energy content (1968–1997). *Geophys. Res. Lett.* **26**(14): 2041–2044.

Fontaine B, Louvet S, Roucou P. 2008. Definition and predictability of an OLR-based West African monsoon onset. *Int. J. Climatol.* **28**: 1787–1798.

Garreaud R. 2001. Subtropical cold surges: regional aspects and global distribution. *Int. J. Climatol.* **21**: 1181–1197.

Gash JHC, Wallace JS, Lloyd CR, *et al.* 1991. Measurements of evaporation from fallow Sahelian savannah at the start of the dry season. *Q. J. R. Meteorol. Soc.* **117**: 749–760.

Grodsky SA, Carton JA. 2001. Coupled land/atmosphere interactions in the West African monsoon. *Geophys. Res. Lett.* **28**: 1503–1506.

Hagos SM, Cook KH. 2007. Dynamics of the West African monsoon jump. *J. Climate* **20**: 5264–5284.

Hodges KI, Chappell DW, Robinson GJ, Yang G. 2000. An improved algorithm for generating global window brightness temperatures from multiple satellite infra-red imagery. *J. Atmos. Oceanic Technol.* **17**: 1296–1312.

Huffman GJ, Adler RF, Bolvin DT, *et al.* 2007. The TRMM Multi-satellite Precipitation Analysis: quasi-global, multi-year, combined-sensor precipitation estimates at fine scale. *J. Hydrometeorol.* **8**(1): 38–55.

Janicot S, Sultan B. 2001. Intraseasonal modulation of convection in the West African monsoon. *Geophys. Res. Lett.* **28**: 523–526.

Janicot S, Thorncroft CD, Ali A, *et al.* 2008. Large scale overview of the summer monsoon over the summer monsoon in West Africa during the AMMA field experiment in 2006. *Ann. Geophys.* **26**: 2569–2595.

Janicot S, Mounier F, Hall NM, *et al.* 2009. The dynamics of the West African monsoon. Part IV: analysis of 25–90-day variability of convection and the role of the Indian monsoon. *J. Climate* **22**: 1541–1565.

Janicot S, Mounier F, Gervois S, *et al.* 2010. The dynamics of the West African monsoon. Part V: the detection and role of the dominant modes of convectively coupled equatorial Rossby waves. *J. Climate* **23**: 4005–4024.

Janicot S, Caniaux G, Chauvin F, *et al.* 2011. Intraseasonal variability of the West African monsoon. *Atmos. Sci. Lett.* **12**: 58–66. doi: 10.1002/ASL.280.

Kalnay E, Kanamitsu M, Kistler R, *et al.* 1996. The NCEP/NCAR 40-year reanalysis project. *Bull. Am. Meteorol. Soc.* **77**: 437–471.

Keppenne CL, Ghil M. 1992. Adaptive filtering and prediction of the Southern Oscillation index. *J. Geophys. Res.* **97**: 20449–20454.

Kessler WS, Kleeman R. 2000. Rectification of the Madden–Julian oscillation into the ENSO cycle. *J. Climate* **13**: 3560–3575.

Kiladis GN, Wheeler MC, Haertel PT, *et al.* 2009. Convectively coupled equatorial waves. *Rev. Geophys.* **47**: RG2003. doi: 10.1029/2008RG000266.

Knippertz P, Fink AH. 2006. Synoptic and dynamic aspects of an extreme springtime Saharan dust outbreak. *Q. J. R. Meteorol. Soc.* **132**: 1153–1177.

Knippertz P, Fink AH. 2008. Dry-season precipitation in tropical West Africa and its relation to forcing from the extratropics. *Mon. Weather Rev.* **136**: 3579–3596.

Knippertz P, Fink AH. 2009. Prediction of dry-season precipitation in tropical West Africa and its relation to forcing from the extratropics. *Weather Forecast.* **24**: 1064–1084.

Knippertz P, Todd MC. 2012. Mineral dust aerosols over the Sahara: meteorological controls on emission and transport and implications for modeling. *Rev. Geophys.* **50**: RG1007. doi: 10.1029/2011RG000362.

Laing AG, Carbone RE, Levizzani V. 2011. Cycles and propagation of deep convection over equatorial Africa. *Mon. Weather Rev.* **139**: 2832–2853.

Lamb PJ. 1978. Case studies of tropical Atlantic surface circulation patterns during recent sub-Saharan weather anomalies: 1967 and 1968. *Mon. Weather Rev.* **106**: 482–491.

Lavaysse C, Flamant S, Janicot S, *et al.* 2009. Seasonal evolution of the West African heat low: a climatological perspective. *Clim. Dyn.* **33**: 313–330.

Lavaysse C, Flamant C, Janicot S, Knippertz P. 2010. Links between African easterly waves, midlatitude circulation and intraseasonal pulsations of the West African heat low. *Q.J. R. Meteorol. Soc.* **136**: 141–158.

Lavender SL, Matthews AJ. 2009. Response of the West African monsoon to the Madden–Julian oscillation. *J. Climate* **22**: 4097–4116. doi: 10.1175/2009JCLI2773.1.

Lavender SL, Taylor CM, Matthews AJ. 2010. Coupled land–atmosphere intraseasonal variability of the West African monsoon in a GCM. *J. Climate* **23**: 5557–5571.

Leroux S, Hall NMJ, Kiladis GN. 2010. A climatological study of transient–mean-flow interactions over West Africa. *Q. J. R. Meteorol. Soc.* **136**: 397–410.

Leroux S, Hall NMJ, Kiladis GN. 2011. Intermittent African easterly wave activity in a dry atmospheric model: influence of the extratropics. *J. Climate* **24**: 5378–5396.

Marteau R, Moron V, Philippon N. 2009. Spatial coherence of monsoon onset over western and central Sahel (1950–2000). *J. Climate* **22**: 1313–1324. doi: http://dx.doi.org/10.1175/2008JCLI2383.1.

Matsuno T. 1966. Quasi-geostrophic motions in the equatorial area. *J. Meteorol. Soc. Jpn.* **44**: 25–43.

Matthews AJ. 2004. Intraseasonal variability over tropical Africa during northern summer. *J. Climate* **17**: 2427–2440.

Mekonnen A, Thorncroft CD, Aiyyer AR. 2006. Analysis of convection and its association with African easterly waves. *J. Climate* **19**: 5405–5421.

Mekonnen A, Thorncroft CD, Aiyyer AR, Kiladis GN. 2008. Convectively coupled Kelvin waves over tropical Africa during the boreal summer: structure and variability. *J. Climate* **21**: 6649–6667.

Mohino E, Janicot S, Douville H, Li LZX. 2012. Impact of the Indian part of the summer MJO on West Africa using nudged climate simulation. *Clim. Dyn.* **38**: 2319–2334.

Mounier F, Janicot S. 2004. Evidence of two independent modes of convection at intraseasonal time scale in the West African summer monsoon. *Geophys. Res. Lett.* **31**: L16116. doi: 10.1029/2004GL020665.

Mounier F, Kiladis GN, Janicot S. 2007. Analysis of the dominant mode of convectively coupled Kelvin waves in the West African monsoon. *J. Climate* **20**: 1487–1503.

Mounier F, Janicot S, Kiladis GN. 2008. The West African monsoon dynamics. Part III: the quasi-biweekly zonal dipole. *J. Climate* **21**: 1911–1928.

Murakami M. 1979. Large-scale aspects of deep convective activity over the GATE area. *Mon. Weather Rev.* **107**: 994–1013.

Ndiaye O. 2010. The predictability of the Sahelian climate: seasonal Sahel rainfall and onset over Sénégal. PhD thesis, Columbia University.

Nguyen H, Thorncroft CD, Zhang CD. 2011. Guinean coastal rainfall of the West African monsoon. *Q. J. R. Meteorol. Soc.* **137**: 1828–1840.

Novella NS, Thiaw WM. 2013. African Rainfall Climatology Version 2 for famine early warning systems. *J. Appl. Meteor. Climatol.* **52**: 588–606.

Okumura Y, Xie SP. 2004. Interaction of the Atlantic equatorial cold tongue and the African monsoon. *J. Climate* **17**: 3589–3602.

Omotosho JB. 1987. Richardson number, vertical wind shear and storm occurrences at Kano, Nigeria. *Atmos. Res.* **21**: 123–137.

Omotosho JB. 1992. Long-range prediction of the onset and end of the rainy season in the West African Sahel. *Int. J. Climatol.* **12**: 369–382.

Peyrillé P, Lafore JP, Redelsperger JL. 2007. An idealized two-dimensional framework to study the West African monsoon. Part I: validation and key controlling factors. *J. Atmos. Sci.* **64**(8): 2765–2782.

Pu B, Cook KH. 2010. Dynamics of the West African westerly jet. *J. Climate* **23**: 6263–6276.

Pu B, Cook KH. 2012. Role of the West African westerly jet in Sahel rainfall variations. *J. Climate* **25**: 2880–2896.

Rashid HA, Hendon HH, Wheeler MC, Alves O. 2010. Prediction of the Madden–Julian oscillation with the POAMA dynamical prediction system. *Clim. Dyn.* **36**: 649–661.

Reynolds RW. 1988. A real-time global sea surface temperature analysis. *J. Climate* **1**: 75–86.

Reynolds RW, Marsico DC. 1993. An improved real-time global sea surface temperature analysis. *J. Climate* **6**: 114–119.

Roehrig R, Chauvin F, Lafore J-P. 2011. 10–25 day intraseasonal variability of convection over the Sahel: a role of the Saharan heat low and midlatitudes. *J. Climate* **24**: 5863–5878.

Roundy PE, Frank WM. 2004. A climatology of waves in equatorial region. *J. Atmos. Sci.* **61**: 2105–2132.

Roundy PE, Schreck CJ, III, Janiga MA. 2009. Contributions of convectively coupled equatorial Rossby waves and Kelvin waves to the real-time multivariate MJO indices. *Mon. Weather Rev.* **137**: 469–478.

Salby ML. 2012. *Fundamental of Atmospheric Physics*, 2nd edition. Cambridge University Press; 647 pp.

Sivakumar MVK (ed.). 1998. Climate variability prediction, water resources and agricultural productivity: food security issues in tropical sub-Saharan Africa. Report of a Joint START/WCRP/SCOWAR Workshop, 22–25 July 1997, Cotonou, Benin. International START Secretariat: Washington DC.

Stern RD, Dennett MD, Dale IC. 1982. Analysing daily rainfall measurements to give agronomically useful results. II. A modelling approach. *Exp. Agr.* **18**(3): 237–253.

Straub KH. 2013. MJO initiation in the real-time multivariate MJO index. *J. Climate* **26**: 1130–1151.

Sultan B, Janicot S. 2003. The West African monsoon dynamics. Part II: the 'preonset' and 'onset' of the summer monsoon. *J. Climate* **16**: 3407–3427.

Sultan B, Janicot S, Diedhiou A. 2003. The West African monsoon dynamics. Part I: documentation of intraseasonal variability. *J. Climate* **16**: 3389–3406.

Takayabu YN. 1994a. Large-scale cloud disturbances associated with equatorial waves. Part I: spectral features of the cloud disturbances. *J. Meteorol. Soc. Jpn.* **72**: 433–448.

Takayabu YN. 1994b. Large-scale cloud disturbances associated with equatorial waves. Part II: westward propagating inertio-gravity waves. *J. Meteorol. Soc. Jpn.* **72**: 451–465.

Takayabu YN, Nitta TS. 1993. 3–5 day-period disturbances coupled with convection over the tropical Pacific Ocean. *J. Meteorol. Soc. Jpn.* **71**: 221–246.

Taylor CM. 2000. The influence of antecedent rainfall on Sahelian surface evaporation. *Hydrol. Process.* **14**(7): 1245–1259.

Taylor CM. 2008: Intraseasonal land–atmosphere coupling in the West African monsoon. *J. Climate* **21**: 6636–6648.

Thiaw WM, Kumar VB. 2001. Effects of tropical convection on the predictability of the Sahel summer rainfall interannual variability. *Geophys. Res. Lett.* **28**(24): 4627–4630.

Thiaw WM, Kumar VB. 2015. NOAA's African desk: twenty years of building capacity in weather and climate forecasting in Africa. *Bull. Am. Meteorol. Soc.* **96**: 737–753.

Ventrice MJ, Thorncroft CD, Janiga MA. 2012. Atlantic tropical cyclogenesis: a three-way interaction between an African easterly wave, diurnally varying convection, and a convectively coupled atmospheric Kelvin wave. *Mon. Weather Rev.* **140**: 1108–1124.

Ventrice MJ, Wheeler MC, Hendon HH, *et al.* 2013. A modified multivariate Madden Julian oscillation index using velocity potential. *Mon. Weather Rev.* **139**: 2704–2722.

Vizy EK, Cook KH. 2009. A mechanism for African monsoon breaks: Mediterranean cold air surges. *J. Geophys. Res.* **114**: D01104. doi: 10.1029/2008JD010654.

Vizy EK, Cook KH. 2014. Impact of cold air surges on rainfall variability in the Sahel and wet African tropics: a multi-scale analysis. *Clim. Dyn.* **43**: 1057–1081. doi: 10.1007/s00382-013-1953-z.

Wheeler MC, Hendon HH. 2004. An all-season real-time multivariate MJO index: development of an index for monitoring and prediction. *Mon. Weather Rev.* **132**: 1917–1932.

Wheeler M, Kiladis GN. 1999. Convectively coupled equatorial waves: analysis of clouds and temperature in the wave number-frequency domain. *J. Atmos. Sci.* **56**: 374–399.

Wheeler MC, Hendon HH, Cleland S, *et al.* 2009. Impacts of the Madden–Julian oscillation on Australian rainfall and circulation. *J. Climate* **22**: 1482–1498.

Yang G-Y, Slingo J. 2001. The diurnal cycle in the tropics. *Mon. Weather Rev.* **129**: 784–801.

Zhang C, Gottschalck J. 2002. SST anomalies of ENSO and the Madden–Julian oscillation in the equatorial Pacific. *J. Climate* **15**: 2429–2445.

# 8

# Seasonal Forecasting

**Lead author:** Andrew Colman[a]
**Contributing authors:** David Rowell[a], Andre Kamga Foamouhoue[b], Ousmane Ndiaye[c], Belén Rodríguez-Fonseca[d], Roberto Suarez[d] and Pascal Yaka[e]
**Reviewers:** Douglas J. Parker, (University of Leeds, Leeds, UK) and Mariane Diop-Kane (Agence Nationale de l'Aviation Civile et de la Météorologie du Sénégal–ANACIM, Sénégal)

[a] Met Office Hadley Centre, FitzRoy Road, Exeter, Devon, EX1 3PB, UK
[b] African Center for Meteorological Applications for Development, (ACMAD), Niamey, Niger
[c] Agence Nationale de l'Aviation Civile et de la Météorologie (ANACIM), Dakar, Sénégal
[d] Universidad Complutense de Madrid, Madrid, Spain
[e] Direction Generale de la Meteorologie, Ouagadougou, Burkina Faso

## 8.1 Scientific Background

### 8.1.1 Introduction

Seasonal forecasting plugs the gap between weather forecasting, which predicts events out to 2 weeks ahead, and climate change projection, in which one is trying to predict the statistics of climate years or centuries ahead. This is one of the most important timescales for agricultural planning in particular (Figure 8.1).

Seasonal forecasting involves the prediction of weather and climate events that occur on timescales of 1 month to 1 year. An example of such an event could be an anomalously dry season. This could be defined as a season where the total or average rainfall was substantially below what normally occurs during the season. Such a season may also be referred to as a season with a 'significantly negative' anomaly in mean rainfall, or it could be called a 'drought'. In a similar way, one can predict seasons where the average rainfall was above what normally occurs, or a season where the average temperature was below or above what is normally expected. It could be an anomaly of some other variable, such as wind speed or humidity or cloud cover. All these events are similar, in that they are long-period averages. Seasonal forecasting does not have to be restricted to seasonal averages: one can also try to predict the likelihood or frequency of storms for example, or the onset of a rainy season; Chapter 7 of this book, which is concerned with 'sub-seasonal' forecasting, describes the prediction of such events. The main purpose of this chapter is to describe the prediction of seasonal timescale events such as seasonal means; its focus will be forecasts of 1–6 months ahead.

We define West Africa to be the area between the Sahara and the tropical Atlantic coast to the south. It can here be divided into three regions (Figure 8.2) that are climatically homogeneous during the wet season (July–September (JAS)). Seasonal predictability tends to be evident on larger scales (Arribas *et al.*, 2011); therefore, much of this chapter discusses predictability for these three regions. However, predictability at smaller scales is also discussed when appropriate.

Nowhere in the world have such seasonal climate anomalies had a greater impact than in West Africa. The vast majority of people who live in this region depend on rain-fed agriculture for their livelihood and lives. For example, a drier than average wet season in 1984 worsened a pre-existing drought and led to a consequent famine, which inspired the Band Aid and Live Aid concerts and an international response.

So what is the cause of events like the 1984 drought? Can such events be predicted in any way, or is it just down to 'chaos'. For a long time, many regarded seasonal forecasting to be impossible because of the 'Butterfly effect' (Lorenz, 1963). The theory suggests that a microscopic event such as someone waving an arm could shift the balance of the local atmosphere, causing a bigger event such as formation of a bubble of warm air that could rise and form a cloud, which in turn could shift the balance of atmosphere on a larger scale and deviate winds, ultimately leading to a drought. However, particularly in the last 30 years, evidence has been gathered

*Meteorology of Tropical West Africa: The Forecasters' Handbook*, First Edition. Edited by Douglas J. Parker and Mariane Diop-Kane.

Figure 8.1 Scientists in Sénégal working with local farming groups to communicate seasonal forecasts. Photo (c) Dorian Speakman.

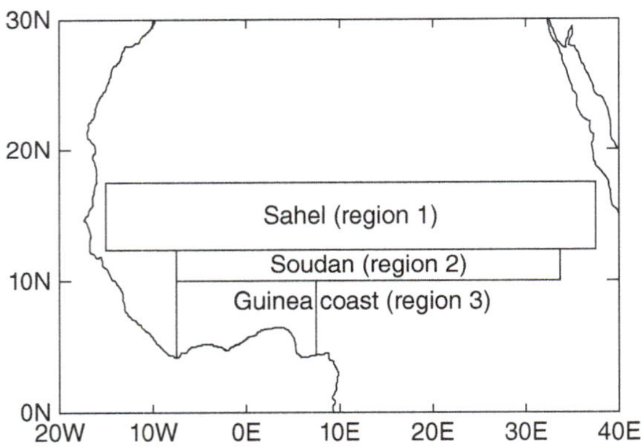


Figure 8.2 West Africa climate zones. *Source*: Met Office. © Crown Copyright.

suggesting that while chaos plays a part and cannot be discounted, there is some predictability of the statistics of weather and climate on the seasonal timescale. In order to determine climate phenomena that might influence and or cause seasonal anomalies, like the 1984 drought observed in West Africa, we can specify some criteria that these phenomena will need to satisfy:

1) Any phenomenon that is considered to be associated with or perhaps a cause of droughts or other seasonal timescale anomalies would need to be active on similar timescales.
2) The phenomenon should involve the transfer of sufficient energy to physically cause or respond to atmospheric changes related to droughts.
3) The phenomenon should be located near enough to the affected anomaly in order to have a credible physical impact.

A phenomenon that fits all these criteria is the pattern of large-scale anomalies in sea surface temperature (SST). SST anomalies tend to last for months, unlike most atmospheric phenomena, which typically do not last for more than a few days. SST anomalies involve transfer of very large amounts of energy from the ocean to atmosphere and vice versa; this is easily on the same scale as energy changes associated with drought season or wet seasons in West Africa (Bjerknes, 1969; Namias, 1972). Some of the most energetic SST anomalies are located in the tropics, not too far from West Africa. Last but not least, SSTs are known to modulate climate around the whole globe, particularly in the tropics through the El Niño Southern Oscillation (ENSO) phenomenon (Ropelewski and Halpert, 1987, 1989).

Many authors have investigated the link between West African seasonal anomalies and SST variability (e.g. Lough, 1986; Folland *et al.*, 1991; Ward, 1993; Ward *et al.*, 1993; Fontaine and Janicot, 1996). Whilst some have used statistical analyses of historical SST to develop their theories (e.g. Ward, 1998), others (e.g. Folland *et al.*, 1986; Rowell *et al.*, 1995; Janicot *et al.*, 1998) have simulated the atmosphere and ocean using general circulation models (GCMs) to detect links between SST and seasonal anomalies.

SST is not the only variable that varies on seasonal timescales in a way that could explain droughts. Sea ice cover and snow cover over land also vary on seasonal timescales and involve considerable amounts of energy. However, these phenomena are located in the extratropics, a long way from West Africa. The role of land surface changes is arguably weaker, as there is less moisture involved and consequently less energy. Soil moisture levels can evolve on a seasonal timescale and effect vegetation and albedo. The role of soil moisture deficits in prolonging or enhancing droughts is discussed by Nicholson (2000), Thiaw and Mo (2005) and others. Charney (1975) argued that the effect of soil moisture deficits could be enhanced through a positive feedback of increasing albedo and consequent reduction in evaporation and this may have been the cause of the series of Sahel droughts in the early 1970s. Fontaine *et al.* (1999) argued that adding information about pre-season moisture content and temperature in the West African region could improve on predictions made using pre-season SST alone.

Most of the literature on seasonal forecasting discusses prediction of seasonal rainfall during the West African wet season, which lasts roughly from July to September. This is understandable, as it is water shortages caused by below-average seasonal rainfall that have most impact on the indigenous population. Other variables, such as wind, temperature and humidity, also tend to vary with rainfall. Thus, if one can predict rainfall, one will have a very good idea of what temperature, humidity and wind will be like. For these reasons, this chapter focuses on discussing predictions of seasonal rainfall totals for the main wet season (JAS).

For all the methods discussed, we try to make an estimate of performance over 1996–2009. This testing period is chosen as there are forecasts available from most forecasting systems (statistical, dynamical and combined). Owing to the lead author's affiliation, a number of examples are taken from the Met Office GloSea system, but the results described here are of more general applicability.

### 8.1.2 Sea Surface Temperature Teleconnections

Year-to-year variations in SST patterns are known to cause variations in seasonal rainfall totals across Africa, through so-called teleconnections. Each of these is essentially an atmospheric bridge between an ocean region and the land region of interest. Since SST anomalies evolve quite slowly (on timescales from months to years), they provide a particularly useful source of seasonal predictability. This slow evolution has two consequences. First, the SST anomalies themselves can often be predicted on seasonal timescales. Second, their impact on the atmosphere usually persists throughout a season, leading to significant changes in seasonal rainfall totals – the primary focus of this chapter – and sometimes also other seasonal characteristics, such as the number of extreme wet events or the typical length of dry spells.

However, the impact of SST anomalies on West Africa varies from one part of this region to another. In particular, impacts differ substantially between the Sahel and 'Guinea Coast' (regions shown in Figure 8.2). The transition zone between these two regions, the 'Soudan', experiences a blend of the Sahel and Guinea Coast responses to SSTs, and will not be discussed further in this section.

In terms of its variety of teleconnections from SSTs, the Sahel represents one of the most complex regions of the globe, being influenced by all three tropical oceans, and additionally by the Mediterranean. This is illustrated in the schematic of Figure 8.3. Also, a significant portion of the Sahel's seasonal rainfall anomalies often persist from year to year, resulting in extended periods of drought or abundance. This, too, derives from the character of SST influences on the Sahel.

Figure 8.4a maps the correlations between observed time series of JAS mean Sahel rainfall and JAS mean grid point SSTs around the globe. The areas of largest correlations indicate the ocean sources of the strongest teleconnections to Sahel rainfall. First, the influence of the Atlantic is clear, with cooler than average SSTs in the equatorial and South Atlantic tending to lead to an abundant wet season, and conversely warmer than average SSTs often leading to Sahel drought. There is also an opposite, but weaker, impact of SST variations in the north tropical Atlantic. These links were first discovered by Lamb (1978), and have since been substantiated by numerous studies. In order to use such SST–rainfall teleconnections as part of a seasonal forecasting system, we must have confidence that they do not arise merely from a chance statistical association, but rather that they are founded on an understanding of their physical mechanisms. A popular approach to uncover the chain of mechanisms that causes an SST-to-rainfall teleconnection involves running a GCM of the atmosphere for several years forced by the global climatological pattern of SSTs. This is then repeated by adding realistic SST anomalies in the ocean region of interest, with one experiment incorporating a positive anomaly and another a negative anomaly. For example, Losada *et al.* (2010) describe this kind of approach for the tropical Atlantic. By analysing the differences between the positive and negative 'sensitivity experiments', they were able to suggest that warm SSTs in the equatorial and South Atlantic lead to a decrease in the surface temperature gradient to the Sahel, weakening the West African monsoon flow and so weakening the surface moisture convergence over the Sahel.

An influence of Mediterranean surface temperature anomalies on the Sahel is also evident from Figure 8.4a, whereby warm anomalies often lead to an abundant wet

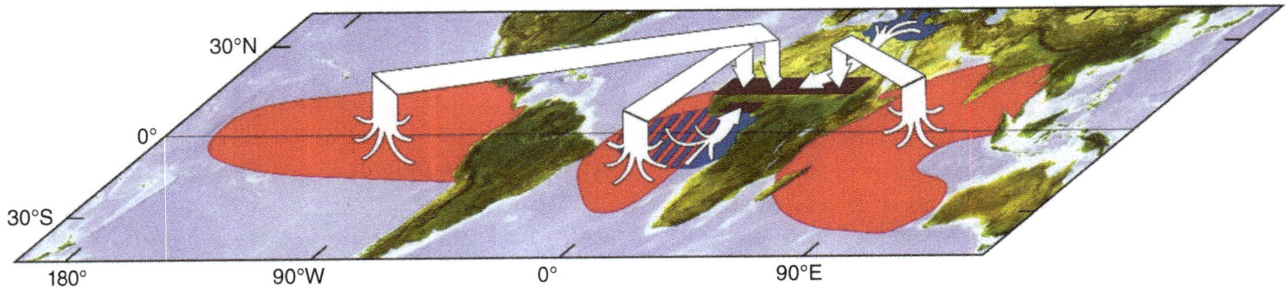


Figure 8.3 Schematic illustration of SST teleconnections that often cause drier than average July-to-September rainfall totals over the Sahel and 'Guinea Coast'. Red ocean areas show where warmer than average SSTs are linked to drier than average seasons, and blue ocean areas show where cooler than average SSTs are linked to drier seasons. Note that the striped Atlantic region shows an overlap of areas that affect either the Sahel or Guinea Coast. *Source*: Met Office. © Crown Copyright.

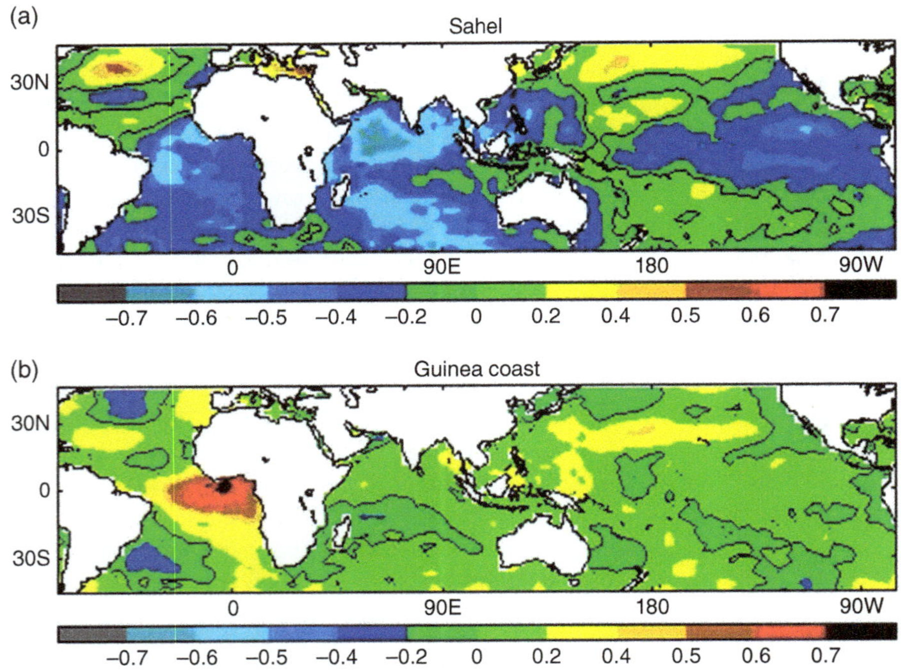


Figure 8.4 Correlations of observed rainfall averaged over (a) the Sahel or (b) the Guinea Coast, against observed SSTs on a 2.5° × 3.75° grid. Data are July–September averages for the period 1922 to 1994. *Source*: Met Office. © Crown Copyright.

season and cool anomalies often lead to drought (Ward, 1994). Again, this has been substantiated and elaborated using GCM sensitivity experiments. The mechanism suggested by Rowell (2003) is that when the Mediterranean is warmer than usual its evaporation is enhanced and the additional moisture is advected across the Sahara by the mean flow, where it enhances rainfall over the Sahel. This effect is then amplified by a variety of positive-feedback mechanisms.

Besides these two regional SST influences, there are also remote influences on the Sahel from the Indian and Pacific Oceans, originally uncovered by Folland *et al.* (1986) and Palmer (1986). Figures 8.3 and 8.4a illustrate these links, whereby abundant rains over the Sahel are usually associated with cooling in the Indian Ocean and the east and central equatorial Pacific (i.e. La Niña events) or, conversely, drought can result from Indian Ocean warming and/or El Niño events. Rowell (2001) examined these relationships in detail, and also interpreted them, in part, as a response of the Sahel to the large-scale gradient of SSTs from the West Pacific to the eastern Indian Ocean. He, too, used GCM experiments to describe the teleconnection mechanisms, suggesting that a stationary Kelvin wave provides an eastward link from the East Pacific to the Sahel, and a Rossby wave provides a westward link from the West Pacific–Indian Ocean SST gradient. The dynamics of these Kelvin and Rossby waves is described in more detail in Section 7.1.4.

The variations of the Atlantic, Indian and Mediterranean SSTs contain a sizable decadal component, and it is this that drives the large decadal variation – persistent wet or dry periods – found in Sahel rainfall. These regional SST variations are often combined into a single pattern on decadal timescales, known as the interhemispheric thermal contrast (IHC; e.g. Folland *et al.*, 1999). For example, from the 1950s to 1980s the Southern Hemisphere oceans warmed more than those of the Northern Hemisphere, leading to a transition from abundant rainfall to drought over the Sahel. At present, the origin of the IHC is unclear. It may simply be an artefact of separate regional SST modes; it may result from separate regional responses to a common forcing (e.g. anthropogenic or aerosol emissions); or it may be an inherent natural mode of the coupled ocean–atmosphere system with global-scale mechanisms linking the different ocean basins. Despite these mechanistic uncertainties, however, we will see in the following sections that it proves valuable as one of the predictors of seasonal Sahel rainfall anomalies.

For the Guinea Coast, the role of SSTs in year-to-year rainfall variations is much more straightforward. Before July, and after September, SST fluctuations have little or no impact on this region; or, in other words, rainfall anomalies are primarily governed by chaotic variability. However, during JAS, rainfall totals exhibit a strong relationship with SST anomalies over the neighbouring equatorial East Atlantic; see Figures 8.3 and 8.4b. Although these months partly coincide with a break in the rains over the Guinea Coast, prediction of the intensity of this break may be useful. A likely mechanism for this teleconnection is that a Rossby wave response to warmer-than-average SSTs produces anomalous ascent along the Guinean Coast, so increasing rainfall here (Vizy and Cook, 2001). Conversely, cool SSTs weaken the monsoon flow, leading to below-normal rainfall.

A final point, critical to the process of prediction, is that we should consider whether an SST-to-rainfall teleconnection has been stationary (i.e. the sign and magnitude of relationship has been maintained for many decades), and will remain stationary into the future. This has been investigated for a number of African regions (Rowell, 2013), including the Sahel and Guinea Coast, with the result that what appear to be fluctuations in the strength of SST–rainfall teleconnections are in fact more likely to have arisen from sampling effects rather than from genuine multidecadal variations of the climate system. In other words, the predictive relationship between SST and rainfall has probably been more or less stationary. Nevertheless, these relationships may change at some point in the future, as the magnitude of climate change increases due to anthropogenic emissions.

### 8.1.3 Statistical Forecasts

Statistical forecasts involve studying past historical data in order to deduce a forecast for the future. Statistical forecasts are widely used in the fields of medicine and economics and are also used in the fields of weather and climate. Up until a few years ago when powerful computers opened up the opportunity of making forecasts with GCMs, nearly all seasonal forecasts were produced using statistical methods.

#### 8.1.3.1 Methods

There is a wide range of statistical methods available and used for seasonal prediction from the simple to the advanced and complex. The simplest statistical technique used is analogues. This technique was widely used by the Met Office, National Centers for Environmental Prediction (NCEP) and other forecasting services in the 1960s and 1970s, the early days of seasonal forecasting.

##### *8.1.3.1.1 Analogues*

The analogue technique works as illustrated in the following example. We want to make a prediction for June 2012. We note that the preceding May was very wet and the preceding April very warm. We find three years in the historical record when May was wet and April was warm. During these three years, there were two warm Junes and one cold June. The forecast for June 2012, therefore, is warm with a probability of 2 out 3 (0.67).

To describe in more general terms, the forecaster has a predicted event that they are trying to predict based on one or more precursor events. One wants to know what will happen in the 'current' case given observed precursor events. To deduce a forecast, the forecaster scans the historical record for cases when similar precursor events occurred and assesses what happens to the predicted event in these 'analogue' cases. The forecaster then assumes the same will happen in the 'current' case, the event one is trying to predict.

Whilst the analogue technique is simple, all statistical forecasts work on the same basic principle. For statistical forecasts to work, a number of assumptions have to be made:

1) The relationship between the precursor event and the event being predicted does not change with time (in other words, stationarity of the relationship).
2) There are one or more precursor cases that are sufficiently similar to the current case for a forecast to be deduced.
3) There are a sufficient number of accurate observations of the predictor events and the predicted events to make a deduction.

Statistical forecasts do have the following advantages:

1) Significantly less computer power is generally needed to produce statistical forecasts than the dynamical alternative.
2) Statistical forecasts are free of many of the biases complex dynamical models can bring.

#### 8.1.3.1.2 Linear Regression

Analogues are amongst the simplest statistical methods used for seasonal forecasting; linear regression is slightly more advanced. Linear regression has the advantage that use is made of all the historical data available rather than a few analogue years. However, the constraint of linearity is introduced. Regression is used to predict the value of a single 'dependent' variable from one or more independent variables. A number of forecasters have used linear regression to predict seasonal rainfall in West Africa. Linear regression works by constructing a mathematical model of the form

$$Y = a_1X_1 + a_2X_2 + \cdots + a_nX_n + C \tag{8.1}$$

where $Y$ is the predicted value (forecast), $X_1, X_2, \ldots, X_n$ are the predictor values (e.g. observations), $a_1, a_2, \ldots, a_n$ are regression coefficients and $C$ is a constant. This equation is also referred to as a regression model. The regression model is constructed using historical data, also referred to as training data. In order to make a forecast, the appropriate recent predictor values are substituted into the equation.

Canonical correlation analysis (CCA) and discriminant analysis are variants of linear regression used by the Met Office Hadley Centre, NCEP and the International Research Institute for Climate and Society (IRI) in the past to produce forecasts. CCA and discriminant analysis are similar to regression, in that the forecaster is also trying to predict an independent variable from independent variables and a model (equation) is produced into which recent predictor values are substituted to produce a forecast.

#### 8.1.3.1.3 Canonical Correlation Analysis

In CCA, the variables are 'fields' containing many values. CCA identifies pairs of patterns in a predicted and a predictor field that are highly correlated in time. The logic is that if the predictor field resembles a CCA field, then the corresponding predicted field will resemble its partner field. Barnett and Preisendorfer (1987), describe the mathematics of CCA in detail and developed a particular form of CCA for seasonal forecasting. Barnett and Preisendorfer (1987) showed that for climate and seasonal forecasting applications there was often a requirement of CCA to make predictions of fields containing many (hundreds or thousands) of values. However, one may only have 10–100 cases of data to work with. In this situation, CCA tends to cause data fitting and spurious results. Barnett and Preisendorfer's (1987) answer was to reduce the number of variables (dimensionality) by calculating empirical orthogonal functions (EOFs) of both the predictor and predicted fields. This brings down the number of variables (dimension) from hundreds to no more than around 10. NCEP developed and tested the Barnett and Preisendorfer (1987) CCA seasonal forecasts for many parts of the world, including West Africa (Barnston *et al.*, 1996; Korecha and Barnston, 2007).

A pair of CCA fields like those used by Barnston *et al.* is presented in Figure 8.5 as an example. This is the first pair of fields from a CCA analysis of March mean SST and subsequent JAS West African rainfall. The '$X$ spatial loadings' (Figure 8.5 upper) is the SST component and the '$Y$ spatial loadings' the rainfall component. The CCA analysis has determined that out of all the possible pairs of patterns that could be constructed from EOFs of SST and rainfall, this is the pair of patterns that explains the maximum amount of common variance over the analysis period, 1948–2010. The SST field shows an interhemispheric contrast in anomaly, with predominantly negative weights in the Southern Hemisphere and Indian Ocean and predominantly positive weights in the north. The rainfall field (Figure 8.5 lower) has predominantly positive weights except near the Guinea Coast and in the far southeast. This CCA mode resembles the teleconnection link between interhemispheric contrast in SST and seasonal rainfall discussed in Section 8.2 and illustrated in Figure 8.4a. The temporal scores plot (Figure 8.5 middle) illustrates the common variance. The red line is the index calculated by projecting the SST ($X$) component of the CCA pair onto the 63 March SST anomaly fields during the analysis period. Similarly, the green line is the index calculated by projecting the rainfall ($Y$) component of the CCA pair onto the 63 JAS rainfall fields.

The first CCA pair explains the highest common variance, but not all the common variance. These patterns can be called pair 1. CCA also detects additional pairs of patterns with the constraint that all the $X$ components are orthogonal to each other and all the $Y$ components are orthogonal to each other. The second pair, pair 2, thus is the pair of patterns that explain the highest common variance whilst being orthogonal to pair 1. The third pair, pair 3, is the pair of patterns that explains the most common variance whilst being orthogonal to pairs 1 and 2, and so on.

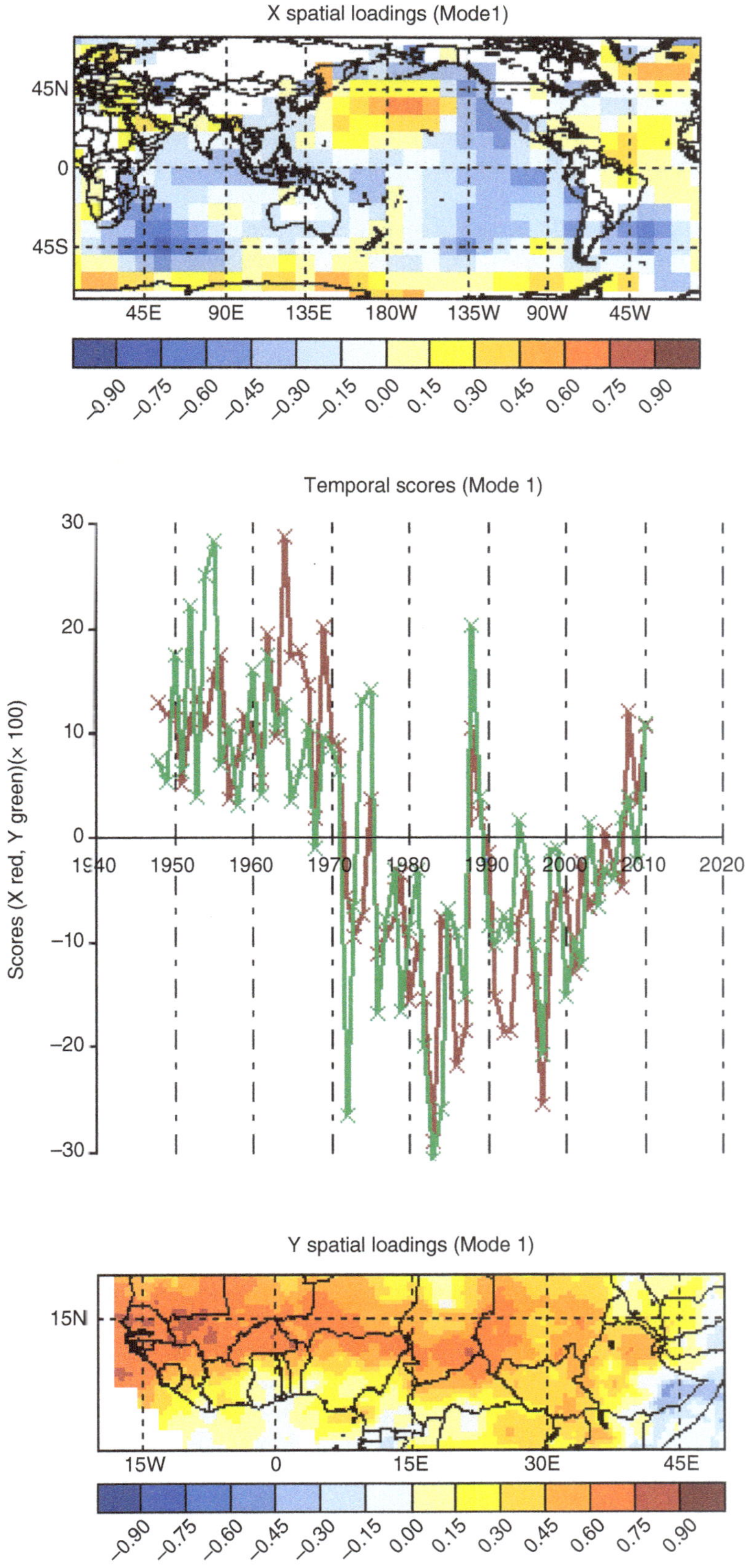


**Figure 8.5** Example pair of CCA patterns and temporal scores representing the association of West African rainfall and pre-season SST. See text for more details. *Source*: Met Office. © Crown Copyright.

#### *8.1.3.1.4 Principal Component Regression*

Principal component regression (PCR) is a variant of linear regression where the independent variables are principal components (PCs). It is a 'halfway house' between linear regression and CCA where the predictors are 'fields' represented by PCs or EOFs, and predictands are a single 'dependent' variable. EOFs or PCs represent patterns of variability in the data and are discussed further in Appendix B. As in the CCA algorithm of Barnett and Preisendorfer (1987), EOFs are used to reduce dimensionality in the predictor field, reducing many independent variables to a more manageable number of representative EOFs and associated PCs. PCR has the added advantage that PC predictors are by definition 'orthogonal', removing the problem of collinearity in predictors that can adversely affect linear regression. The Met Office statistical forecasts discussed in Section 8.1.3.2 and assessed in 8.1.3.4 are produced using a variant of PCR where a selected subset of EOFs/PCs is used as predictors. PCR is one of the methods offered by the Climate Predictability Tool (CPT, see Section 8.2.3).

#### *8.1.3.1.5 Discriminant Analysis*

The product of discriminant analysis is not a point value as in regression, nor a field of values, as with CCA, but, rather, probabilities of two or more events. Probabilities are predicted for a fixed number of events that are mutually exclusive and cover the full range of possibility. An example could be categorising a season's rainfall into three equiprobable tercile categories. Discriminant analysis determines the 'true' probability of an event, given a forecast of that event, by reference to the retrospective performance of the forecast system. The method is based on finding a generalised squared distance $D_t^2$ for each predicted category $t$ that measures the difference in real-time values of a predictor index from its historical average value for the occasions when category $t$ is subsequently observed. Typically for seasonal forecasts, $t=1,2,3$ (tercile categories). The $t$ categories may be stratifications of any meteorological variable (e.g. temperature or precipitation), or may be an 'impact' variable such as river flow. The smaller the generalised distance for a given category, relative to the distance for the other two categories, the higher the predicted probability of that category. The generalised squared distance for a vector of predictor variables $\mathbf{x}$ may be written as

$$D_t^2(\mathbf{x})=(\mathbf{x}-\mathbf{m}_t)' S^{-1}(\mathbf{x}-\mathbf{m}_t)-2\log_e(q_t) \quad (8.2)$$

where $\mathbf{m}_t$ is the vector of corresponding historical mean values of the predictors when category $t$ was observed, and $S$ is the covariance matrix of the different predictors (defined from hindcasts). The variable $q_t$ is the prior probability of category $t$. The prior probability is the chance or default probability of category $t$ and the expected forecast probability if random data were used to make forecasts. The final term in Equation 8.2 can be ignored if all the prior probabilities are equal; for example, if three tercile categories all had prior probabilities of 1/3. This is often the case, but not exclusively so. For example, the background climate may be changing due to global warming.

The probability of category $t$ determined by the combined predictors $\mathbf{x}$ is then given by

$$p_t(\mathbf{x})=\frac{\exp\left[-0.5D_t^2(\mathbf{x})\right]}{\sum_{u=1,3}\exp\left[-0.5D_u^2(\mathbf{x})\right]} \quad (8.3)$$

The formulation derives from Bayes' thereom (Bayes and Price, 1763).

#### *8.1.3.1.6 Non-stationary Methods*

A problem with statistical prediction is non-stationarity in the system. Teleconnections as described in Section 8.1.2 can emerge and disappear with time. Consequently, different predictors appear, depending on the time period analysed.

In a first approximation, the main co-variability mode between SST and precipitation anomaly fields may be studied using maximum covariance analysis (MCA). The MCA method identifies pairs of patterns that are projected onto observed anomalies to calculate pairs of 'expansion coefficients', which represent the strength and sign of the MCA pattern in the observed anomaly pattern. The selected patterns are chosen to maximise the covariance between the pairs of expansion coefficients. Thus, MCA is similar to CCA, but covariance is maximised rather than correlation. As an example, the S4CAST model (SST-based Statistical Seasonal ForeCAST model; Suárez-Moreno and Rodríguez-Fonseca, 2015) has been applied to study the predictability of rainfall anomalies in the Sahelian region (18–10°W; 12–18°N) during the monsoon season (JAS) taking anomalous SST in the tropical Atlantic (60°W–20°E; 20°S–5°N) as a predictor field. MCA is carried out over the whole analysis period (1901–2013) and pairs of patterns calculated along with their expansion coefficients. These expansion coefficients representing the SST and rainfall are then correlated over all 20-year sub-periods (moving windows) within the 113-year analysis period

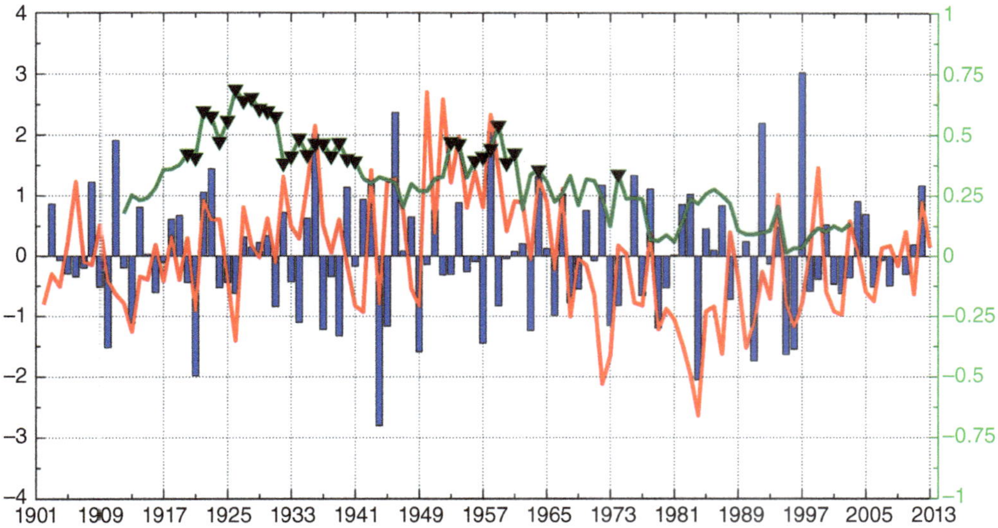


**Figure 8.6** Twenty-one-year centred moving correlation windows (green line, right *y*-axis) between the expansion coefficients *U* corresponding to predictor field (blue bars, left *y*-axis) and *V* corresponding to predictand field (red line, left *y*-axis) obtained from the leading mode of co-variability from MCA between both anomalous fields considering monthly lag 3 (April–June) for predictor. Coloured triangles indicate significant correlation under a Monte Carlo test at 90%. *Source*: B Rodríguez Fonseca and R Suarez (Universidad Complutense de Madrid, Spain).

(Figure 8.6). The non-stationary influence periods are determined from the significant correlation between the expansion coefficients of the SST predictor and rainfall predictand fields. For this analysis, the rainfall predictands for JAS lag the SST predictors for April–June by 3 months.

For evaluation, three sets of MCA hindcast predictions are made using cross-validation. First, non-stationary predictions are based on the model calculated over the 'statistically significant correlated' (SC) and 'no significant correlated' (NSC) periods as indicated in Figure 8.6; second, stationary predictions are made using the entire period (EP). The correlation skill-score for the three sets of predictions along with the predictor and predictand leading modes of covariability used to generate the hindcasts is illustrated in Figure 8.7. It is apparent that when assuming non-stationary predictability for significant correlation periods (SC; (b) panels), the skill-score extends to a larger part of the study region than if a stationary method (EP; (a) panels) is used. For the no-significant correlation period (NSC; (c) panels), no skill is observed.

Whatever the predictor and predictand fields, these results highlight the need for study of the stationarity of the relationship between them. To better apply non-stationary models such as S4CAST, there is a need to seek and understand the physical mechanisms behind the modulations on different timescales (e.g. interannual, multidecadal).

#### 8.1.3.2 Representation of Predictors: Sea Surface Temperature Empirical Orthogonal Function Predictors

A number of physically realistic teleconnections between SST and seasonal rainfall have been identified and were discussed in Section 8.1.2. An obvious application of these teleconnections is to develop statistical seasonal forecasts based on the identified teleconnection relationship.

In order to make use of the teleconnections in a statistical forecast system such as linear regression or discriminant analysis, one needs to calculate time indices of the teleconnection patterns. This is not straightforward. Any correlation map evaluated between SST and rainfall, like Figure 8.4, will contain an element of noise as well as signal. One could use the correlation map to define the pattern and calculate a time series, but unless the correlation is perfect there will be a noise component in the series.

An alternative way of representing teleconnections is to define rectangular box areas where correlation is highest; for example, SST averages for the Nino3 box (5°N–5°S, 90–150°W) is commonly used to represent ENSO (e.g. http://www.cpc.ncep.noaa.gov/data/indices/). Rectangular box area SST averages are widely used in the Regional Climate Outlook Fora (RCOFs) capacity-building process (Mason and Chidzambwa, 2009). Small-scale detail, which has a high chance of being noise, is thus removed. However, this form of averaging does not eliminate the noise problem fully, and it assumes the teleconnection patterns fit nicely into latitude–longitude rectangles, which in practice is very unlikely.

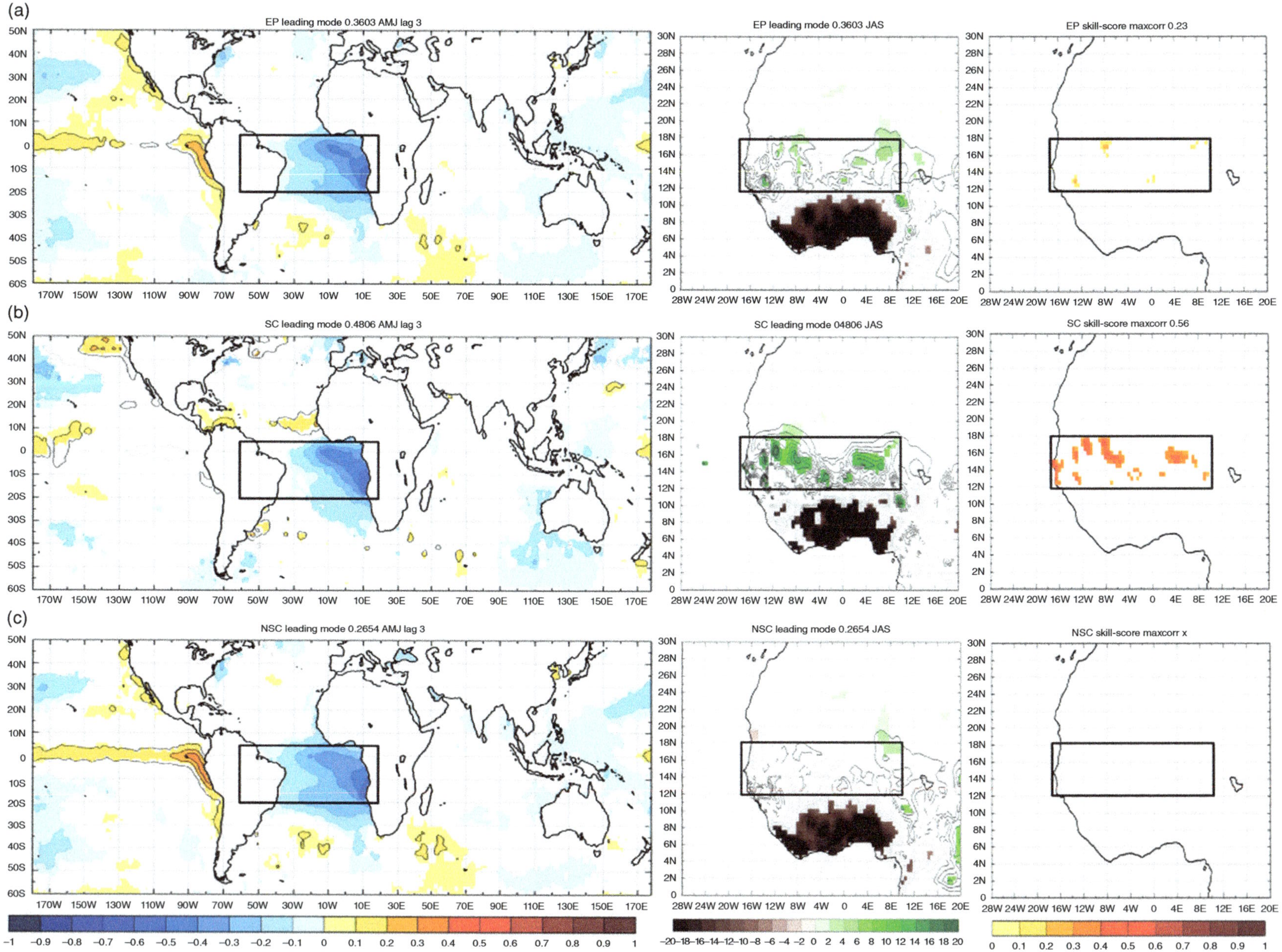


**Figure 8.7** Maps corresponding to the leading mode of co-variability and monthly lag 3 (April–June) for the predictor. (a) Panels refer to the whole period; (b) panels refer to the SC period; (c) panels refer to no-significant correlation period. Left panels show the leading mode for the predictor field (squared region) obtained from projecting the expansion coefficient *U* into the global domain (°C per standard deviation). Central panels show the leading mode for the predictand field (squared region) obtained from projecting the expansion coefficient *U* into the local domain (mm day$^{-1}$ per standard deviation). Right panels show correlation skill-score obtained by Pearson correlation coefficients between observed and simulated maps showing a maximum correlation value of 0.23 (EP), 0.56 (SC) and no skill (NSC). Coloured regions indicate significant values under a Monte-Carlo test at 90%. *Source*: B Rodríguez Fonseca and R Suarez (Universidad Complutense de Madrid, Spain).

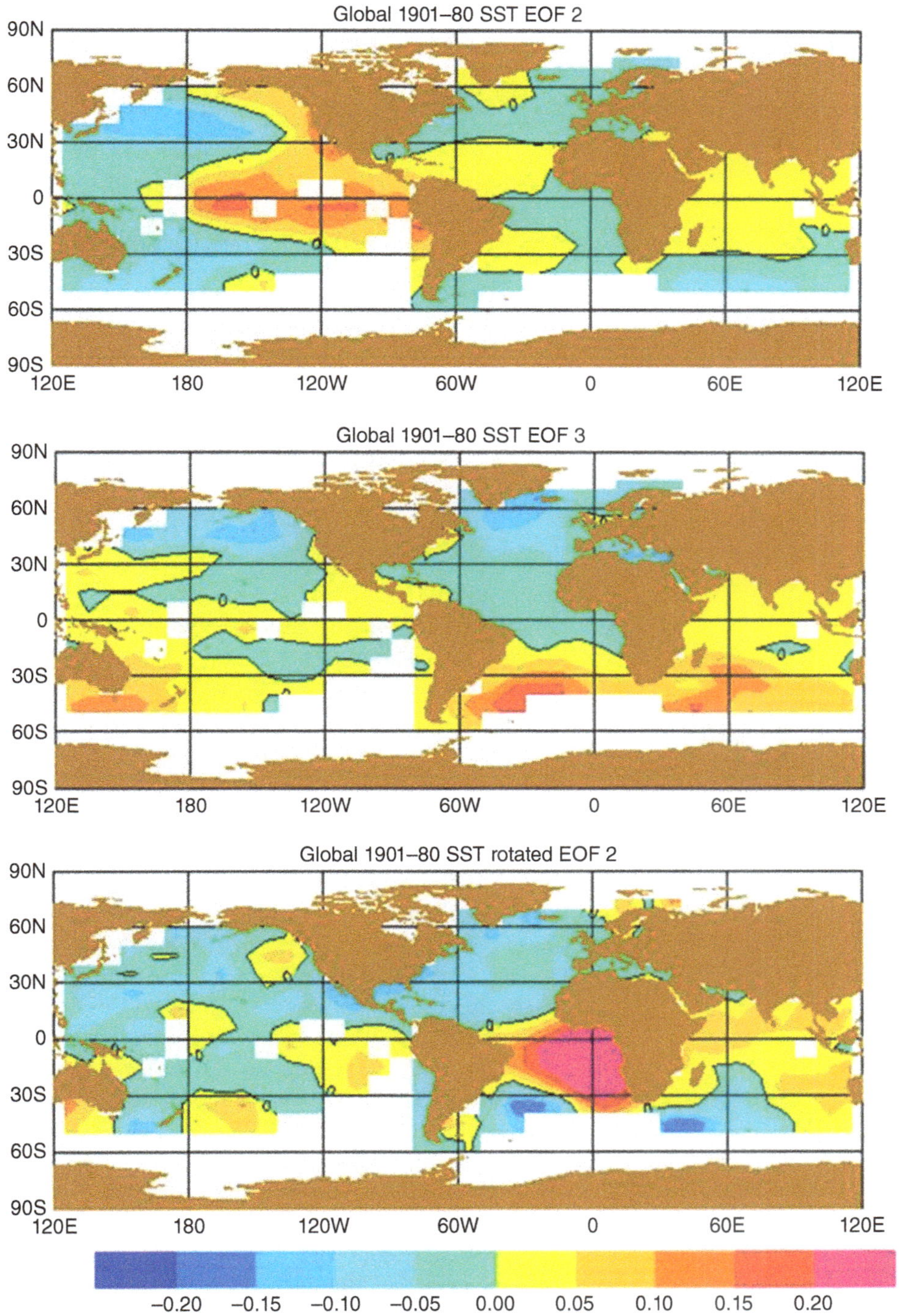


Figure 8.3 Three global SST EOF patterns used for forecasting West African rainfall. The final panel is EOF2 of the 'varimax rotated' EOFs; see text. *Source*: Met Office. © Crown Copyright.

Another approach (e.g. Folland *et al.*, 1991; Ward *et al.*, 1993) is to use EOFs. EOFs represent patterns of variance in the data, and thus indicate how SST anomalies evolve independent of any teleconnection (Appendix B). Folland *et al.* (1991) identified three regional–global-scale SST EOF patterns, which have been used for seasonal forecasting and continue to be used today (Figure 8.8; see also Folland *et al.* (1991)).

The first EOF (EOF1, not shown) has weights of the same sign over most of the globe and represents global mean SST anomaly. It is not related to West African rainfall, at least over the historical record, which goes back to 1900. EOF2 is a representation of the ENSO cycle that has some influence on West African rainfall. The most important pattern for African rainfall forecasting is EOF3. This shows a contrast in sign between Northern and Southern Hemispheres and is considered to represent the IHC pattern discussed in Section 8.2. The identification of EOF3 was perhaps the most important result of the Met Office research to develop seasonal predictions for West Africa in the 1980s and 1990s. It is still widely used as the key predictor of West African rainfall by the African Center of Meteorological Application for Development (ACMAD), PREvision Saisonairre en Afrique del l'Ouest (PRESAO) and others.

The remaining lower order EOFs were not thought to represent any physical pattern. However, when a 'varimax rotation' is applied to EOF4–EOF13, the result was a set of more localised patterns, including EOF2 (shown here in the third panel of Figure 8.8), which is also a good predictor of African rainfall.

#### 8.1.3.3 Representation of Predictands

Uncertainty in seasonal forecasting systems (whether statistical or dynamic) that answer the question 'What will next season be like?' is well appreciated. Often not so well appreciated is uncertainty in observations, when one asks the question 'What was last season like?'

A forecast system requires a variable or set of variables to predict. Station data have traditionally been the fundamental building block of seasonal rainfall records used for this purpose. Averaging data for a group of stations in a climatologically homogeneous zone can deliver the following benefits.

1) Individual station data can be adversely impacted by local topography, the proximity of nearby buildings, trees, rivers and other features. It can be subject to human observer error.
2) Storms may miss a station or a station may capture a storm not observed elsewhere in the vicinity.
3) Seasonal climate tends to be predictable on large scales, so a signal may be picked up over a large area that would not be identifiable at smaller scales.

Nicholson (1986) and others developed seasonal rainfall indices for large climatically homogeneous regions in West Africa, including for example the Sahel, arguing that African rainfall behaves in a homogeneous way on large space scales. Hulme (1994) created a gridded dataset using the same grid as the Met Office climate model to enable evaluation of this model. The gridded data are evaluated by weighting station values according to their distance from the grid point or grid box.

The big problem with station data, however, is lack of coverage, and this problem is particularly acute in West Africa, where for some locations the nearest station can be hundreds of miles away, in a different climate zone (Chapter 1, Figure 1.2).

Two ways of remedying this problem are being developed.

1) Use of satellite observations (Xie and Arkin 1998; Adler *et al.*, 2012; GPCP, http://www1.ncdc.noaa.gov/pub/data/gpcp/monthly-v2.2/documentation/V2.2_doc.pdf).
2) Use of reanalyses created by GCMs.

Both these systems have the advantage that they can offer complete coverage on resolutions down to a few kilometres. More details about satellite observations can be found in Chapter 9. Satellite observations have the following disadvantages:

1) The ground can be obscured, and deductions often have to be made based on radiation data from cloud tops.
2) Satellites have a lifetime of a few years, after which a replacement satellite may have different instruments with differing measurement biases.
3) Biases can arise due to tracking times in the case of non-geostationary satellites.

Reanalysis data also have limitations:

1) Model error. The GCMs used to create reanalyses do not perfectly represent atmospheric processes, particularly at small scales.
2) Initialisation error. The GCMs may be initialised with inaccurate data.
3) Boundary forcing error. The GCMs may be forced with incorrect boundary conditions (e.g. soil moisture or vegetation type), particularly on small scales.

A number of databases have now been created using satellite data alone or satellite data blended with observations. However, there are still large uncertainties in observed data. As observations are required to produce statistical forecasts, observational uncertainty can adversely affect statistical forecast systems.

To illustrate this uncertainty four example maps of observed precipitation anomalies for the same period, June–August (JJA) 2011, are presented in Figure 8.9.

The messages from the four maps for West Africa are quite different: the satellite-based TRMM (http://trmm.gsfc.nasa.gov/) plot suggested it was mostly wet except near the coast; CPC-FEWS ARC2, a blended satellite/gauge dataset (Novella and Thiaw, 2013), has a near equal coverage of wet and dry; and the CAMS-OPI map (another satellite outgoing longwave radiation-based database; Janowiak and Xie, 1999) suggests it was mostly dry. The fourth map from NCDC (http://www.ncdc.noaa.gov), based on gauge data, does not show a strong signal either way.[1]

##### *8.1.3.3.1 Representation of Rainfall Using Box Zones*

Box zones have been used to define regions of homogeneous climate that can be predicted as one variable. The three regions in Figure 8.2 are an example for JAS rainfall. For example, having considered each of these regions to be climatically homogeneous, the Met Office has

(a)

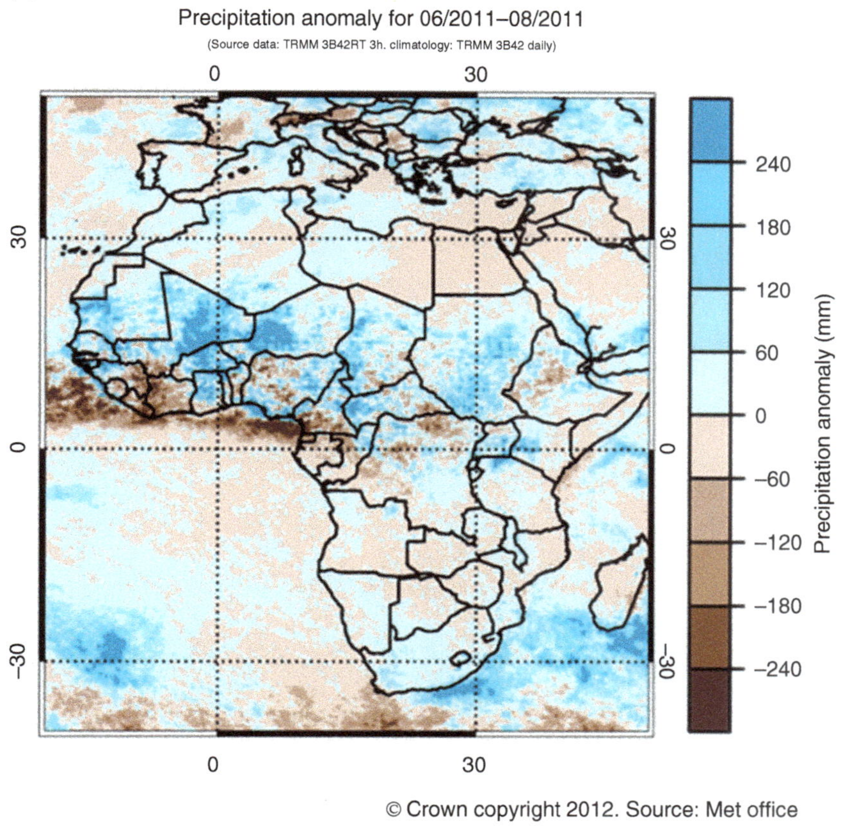


(b)

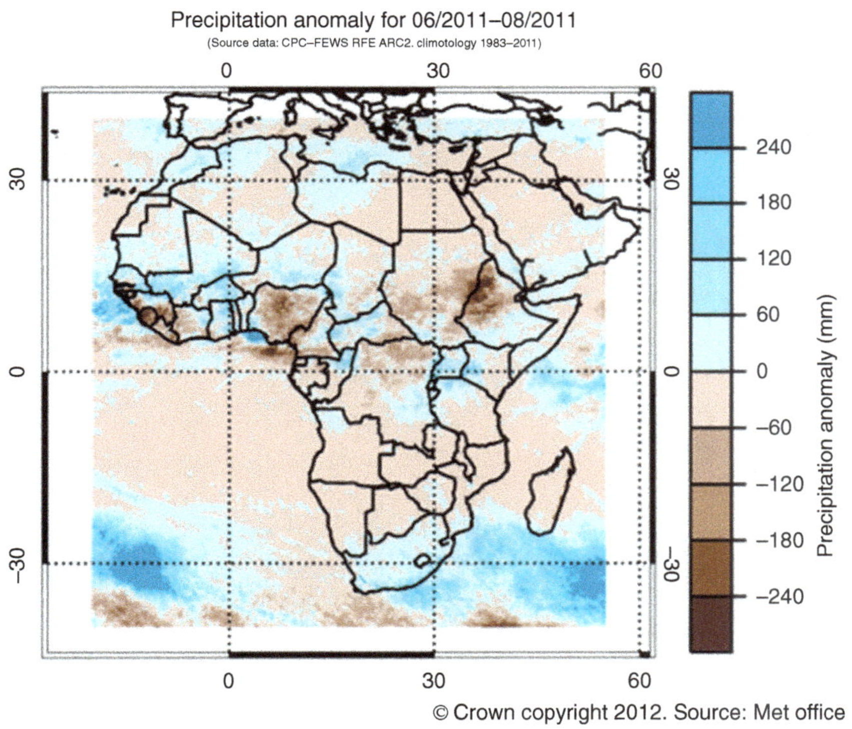


**Figure 8.9** Four estimates of rainfall over Africa for JJA 2011. (a) TRMM anomaly for JJA 2011 (relative to 1998–2011 climatology). *Source*: Lizzy Good (Crown Copyright Met Office). (b) FEWS ARC2 anomaly for JJA 2011 (relative to 1983–2011 climatology). *Source*: Lizzy Good, (Crown Copyright Met Office). (c) CAMS-OPI JJA 2011 anomaly from 1979–2000. *Source*: Reproduced with permission of International Research Institute for Climate and Society, (iridl.ideo.columbia.edu). (d) NCDC precipitation anomalies for JJA 2011 as percentage 1961–1990 climatology (from National Oceanic and Atmospheric Administration (NOAA) National Centers for Environmental Information, State of the Climate: Global Analysis for August 2011, published online September 2011). *Source*: Reproduced with permission of National Centres for Environmental Research (NCEI).

(c)

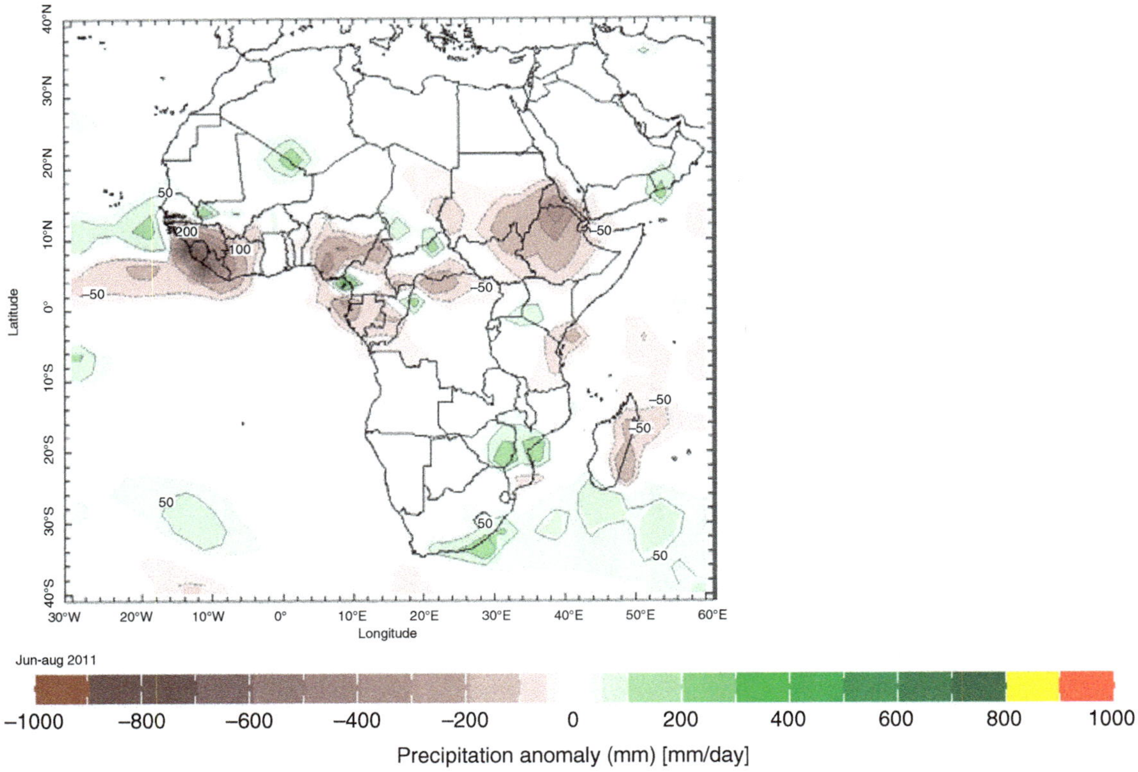


(d)

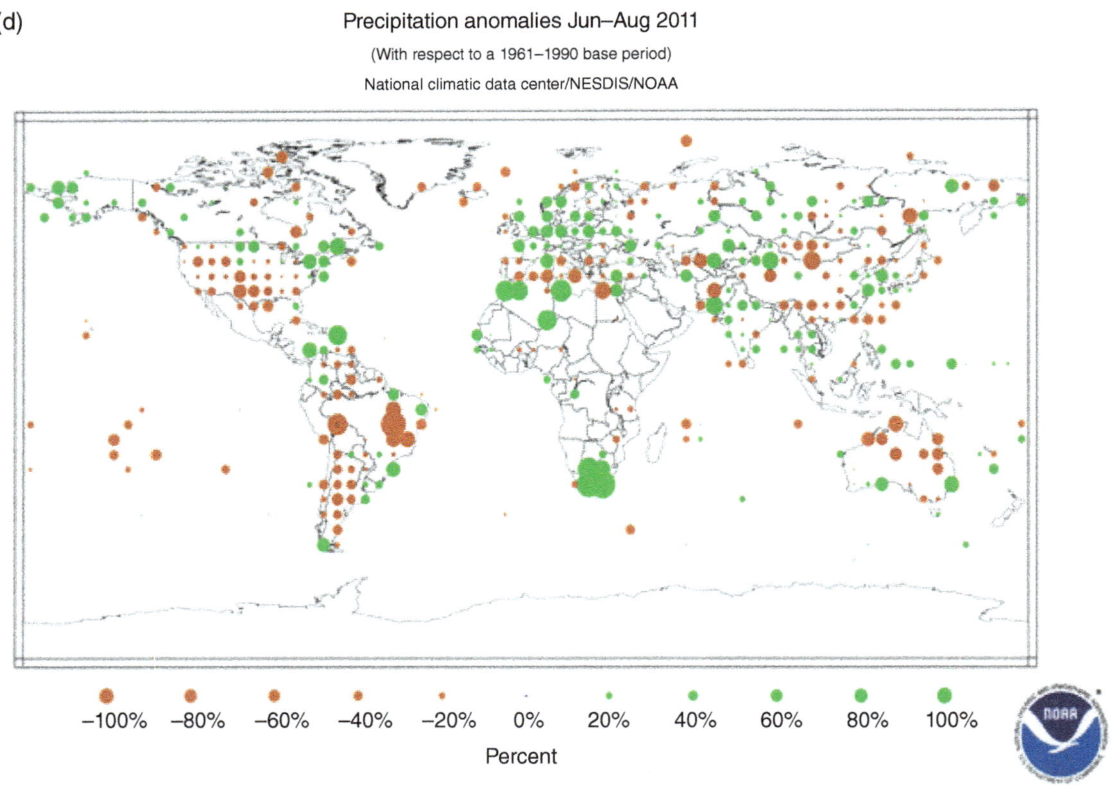


**Figure 8.9 (Cont'd)**

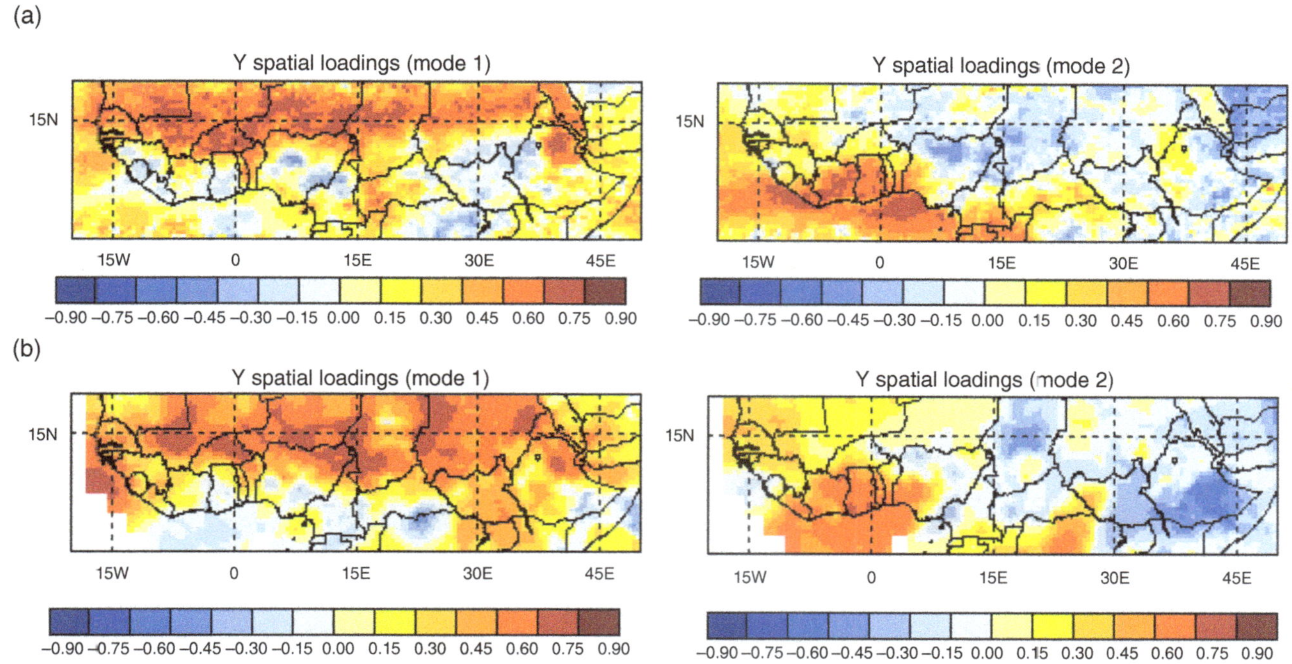


**Figure 8.10** First two EOFs of (a) FEWS ARC2 for JAS (1983–2012) and (b) NCEP PREC/L for JAS (1983–2012). *Source:* Met Office. © Crown Copyright.

regularly produced seasonal forecasts of mean rainfall for JAS for the three regions since 1992.

The homogeneity of the regions shown in Figure 8.2 can be seen by viewing the leading eigenvectors/EOF of gridded rainfall datasets (Figure 8.10). The first EOF of both datasets shows a bimodal structure with a divide at about 10°N. The area to the north of 10°N coincides quite well with the Sahel and Soudan regions. The second modes both have strong positive weights for a region between about 7.5°W and 7.5°E and south of 10°N, and opposite sign elsewhere. The region with strong weights coincides quite well with the Guinea Coast region. The locations of the regions are determined by the position of the zone of maximum rainfall in JAS. These homogeneous regions are season dependent; however, if one were making predictions for another time of year, the choice of boundaries may well be different, as rain bands (particularly the maximum rainfall zone) shift with the season.

#### 8.1.3.4 Assessment of Statistical Methods

Assessing statistical methods can be problematic, as statistical forecast systems represent apparent relationships in the historical data. If one tests a statistical system on the same data that were used to produce it, the result will suggest there is skill. A solution to this is to test a statistical forecast system on independent data; that is, data that were not used to produce the forecasts. This presents a problem for the following reasons:

1) Data are often in short supply. Reliable records often only go back a few years.
2) If there are good spare data, would it not be better to use it in producing the statistical model?

Cross-validation or jack-knifing is a technique to make best use of a restricted amount of data. In its simplest form, if one has $N$ years of data to create a forecasting system, one removes the first year from the analysis and makes a prediction equation using the $(N-1)$ remaining years. That prediction equation is then tested on the year that was excluded (the first year). The process is then repeated in turn for all $N$ years. Thus, $N$ equations are produced, each equation created using a slightly different set of $(N-1)$ years and used to make a prediction for each of the $N$ years.

Variants of this process exist. For example data from extra years can be removed from either side of the predicted year. The number of years with data removed is referred to as the cross-validation window. If just one year's data is removed at a time, the cross-validation window has a value of 1. If data for a year either side is removed as well, then data for three years are removed in total and the cross-validation window has a value of 3. Using cross-validation windows >1 is recommended when the data are highly autocorrelated (meaning that there is 'memory' in the data from year to year), or when skill is low and removal of 'spike' years is likely to have an excessive influence on the prediction equation.

Cross-validated assessments of predictions of rainfall using (a) EOF3 and (b) EOF2, EOF3 and rotated EOF2 of SST are presented in Figure 8.11a and b respectively. Correlations evaluated over 1901–2011 are comparable to

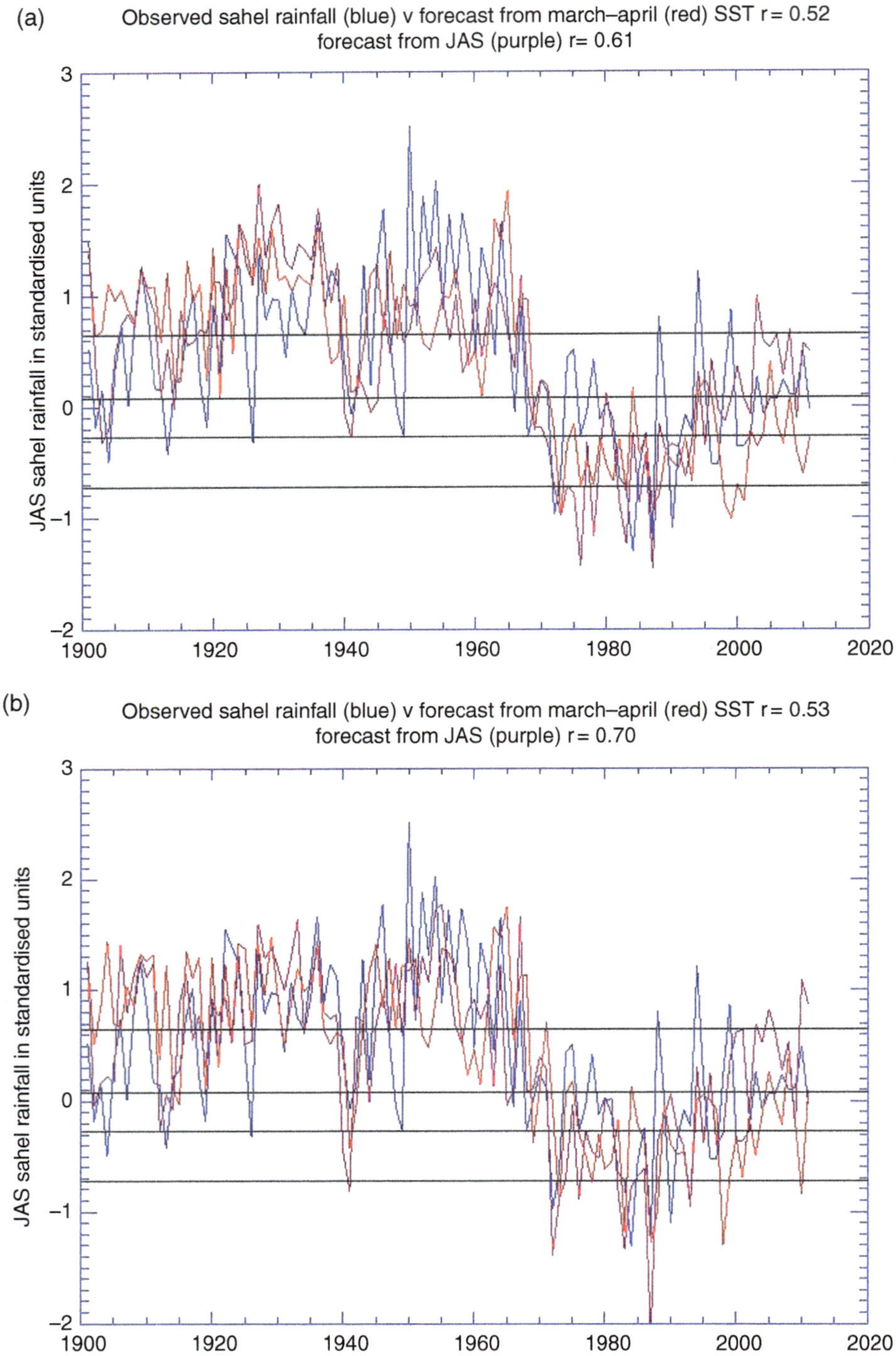


**Figure 8.11** Cross-validated forecasts of Sahel rainfall plotted against observations. *Source:* Met Office. © Crown Copyright.

correlations evaluated over 1901 to around 1990 in Ward *et al.* (1993) and Folland *et al.* (1991), though we now have 20 years of extra data since these publications. The correlation between cross-validated forecasts of JAS Sahel rainfall using March–April EOF3 and observations is $r = 0.52$. The correlation rises to $r = 0.61$ if we use concurrent (JAS) SST to predict rainfall. Forecasts from the three predictors (Figure 8.11b) are a slight improvement on using EOF2 alone, if concurrent JAS SST is used ( $r = 0.70$ ), but the skill of the three predictors using March–April SST ( $r = 0.53$ ) is similar to using EOF3 alone ( $r = 0.52$ ).

#### 8.1.3.5 Performance over 1996–2009

Whilst the EOF-based predictors show skill over the 111-year period (Figure 8.11), unfortunately it is a different story when forecasts are assessed over the 14-year period 1996–2009 (Figure 8.12). Assessments are made against observations from five different datasets in

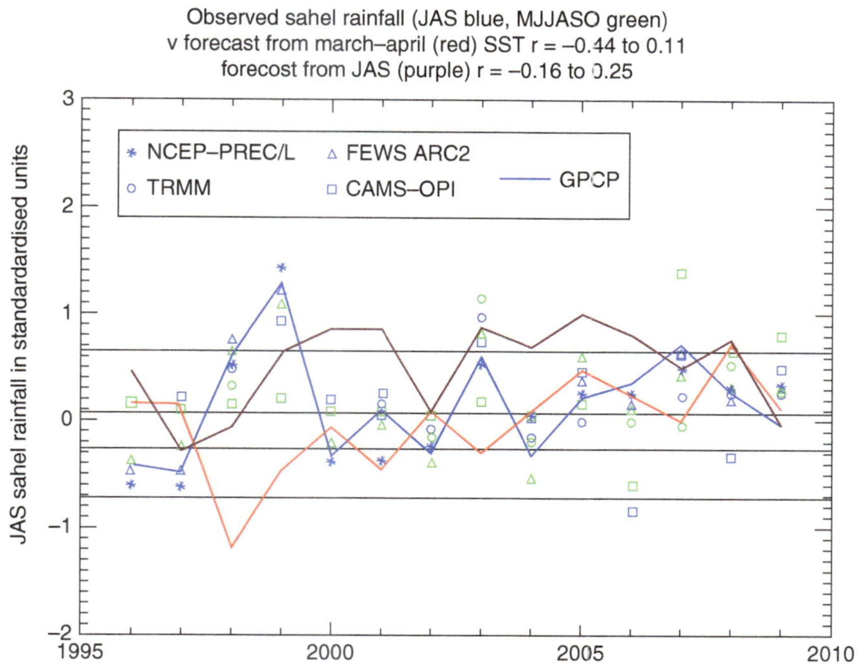


**Figure 8.12** Sahel forecasts for JAS 1996–2009 verified using five different datasets. *Source:* Met Office. © Crown Copyright.

order to take account of observed uncertainty as discussed in Section 8.1.3.3. Skill from March–April is reduced to near or below zero and skill from JAS from all three predictors is 0.25 or lower, well short of the 5% significance level. In particular, the March–April forecasts failed completely to predict the relatively extreme dry and wet years between 1996 and 1999, but a weak upward trend from 2000 to 2009 has been correctly predicted by the March–April forecast. However, this trend is not predicted by JAS forecasts, suggesting that the trend prediction from March–April is likely to be accidental.

### 8.1.4 Dynamical Forecasts

Dynamical models simulate the atmosphere by replicating the physics (see Chapter 10). Dynamical models have been used for weather forecasting for over half a century, but seasonal dynamical models have only been widely available for about 15 years, since sufficiently powerful computers have become available. Seasonal forecasting for West Africa has been a relative success story for dynamical models when compared with seasonal forecasts for some other regions. Folland *et al.* (1986, 1991) were able to successfully simulate droughts in the Sahel using the currently available climate model. In this regard, climate models are helped by the fact that much of West Africa is quite flat and climate zones regularly shaped. This is in contrast to East Africa, where there are several mountain ranges and large lakes influencing climate on a local level.

Dynamical models are generally run in groups of repeated runs called ensembles (see Section 10.1.3). Ensembles are run to allow for uncertainty in initial conditions and sometimes to allow for uncertainty in model formulation. Ensembles can also be used to produce probability forecasts. This is done by classifying the forecasts of some variable (e.g. seasonal mean precipitation forecast for the Sahel) from each of the ensemble members into a set of categories (e.g. quintile categories, which are five equiprobable categories). Probabilities are calculated in direct proportion to frequency of ensemble members in each category. These probabilities are called 'raw ensemble' probabilities to distinguish them from calibrated probabilities discussed in Section 8.1.4.2, where further corrections for model error are made.

The formulation of raw ensemble probabilities is illustrated in Figure 8.13. The purple crosses (+) in the top represent 70 forecasts of Sahel rainfall for the JAS 2012 season. The four horizontal lines represent the four quintiles dividing five quintile categories. Raw ensemble probabilities are based on the number of purple crosses in each of the five categories separated by the four horizontal lines. The five categories (lower to upper) contain 7, 39, 20, 4 and 0 purple crosses, so the raw ensemble probabilities for the five categories are 0.10, 0.56, 0.29, 0.06 and 0.00 respectively. Maps of raw ensemble probabilities are presented on the Met Office and European Centre for Medium-Range Weather Forecasts (ECMWF) seasonal forecast web sites.

#### 8.1.4.1 Performance over 1996–2009

The performance of the Met Office GloSea4 model in predicting JAS rainfall for the three regions displayed in Figure 8.2 is also shown in Figure 8.13. Hindcasts for the 1996–2009 seasons are plotted as red crosses against corresponding observations represented by X symbols. The correlation score in the title provides a measure of performance. The correlations of 0.696 for Sahel forecasts

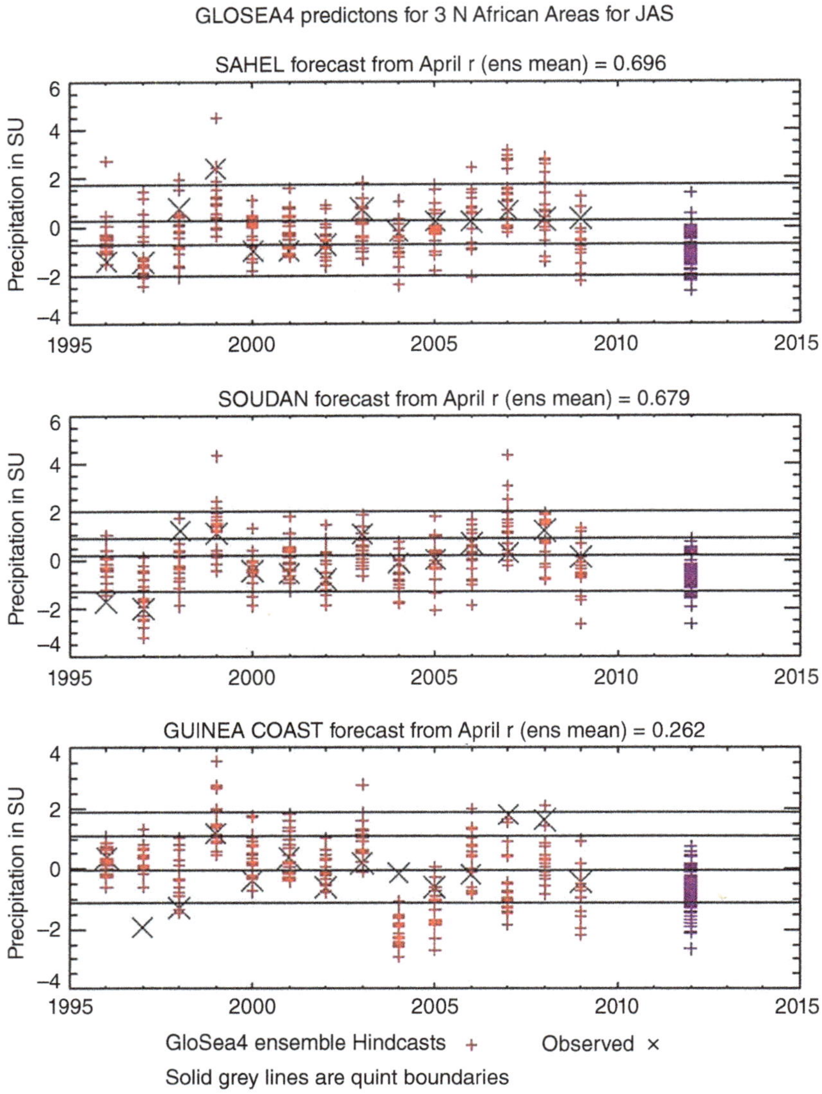


**Figure 8.13** Dynamical predictions from the Met Office GloSea4 system for the three regions shown in Figure 8.2. *Source:* Met Office. © Crown Copyright.

and 0.679 for the Soudan are good for seasonal forecasts. The Guinea Coast correlation is somewhat lower at 0.282 and disappointing. A possible explanation for this low skill, related to poor model performance in predicting tropical Atlantic SST, is discussed in Section 8.1.4.3.

#### 8.1.4.2 Correction of Systematic Errors in Models

Although GCMs are our best way of representing processes in the atmosphere, they do have some limitations. Their limitations come from various sources: our understanding and representation of some processes in the climate system, including coarse resolution of the climate model to take into consideration small-scale processes (turbulence, convection), lack of knowledge of the initial state of the forecasting system, truncation error, and computational limitations (storage or processors) (see Chapter 10). Each limitation will have an impact on the outputs and could generate errors that grow as the model steps forward through time.

There are different kinds of model errors. Forecast errors from chaotic evolution of the atmosphere might not be corrected, but taken into consideration by ensemble representation and/or probabilistic aspects of the forecast. There are some types of error that are not represented by ensembles, such as error due to incorrect representations of dynamical and/or physical processes (ENSO, monsoon dynamics). A model might capture actual physical process but with some errors due to, for example, coarse grid-scale resolution. In such a case, the model might capture the large-scale circulation patterns in the winds or in SST, but it may fail to simulate the processes linking circulation to rainfall and thus incorrectly predict the rainfall. Such an error may be quantifiable and ultimately correctable through the approach of bias correction.

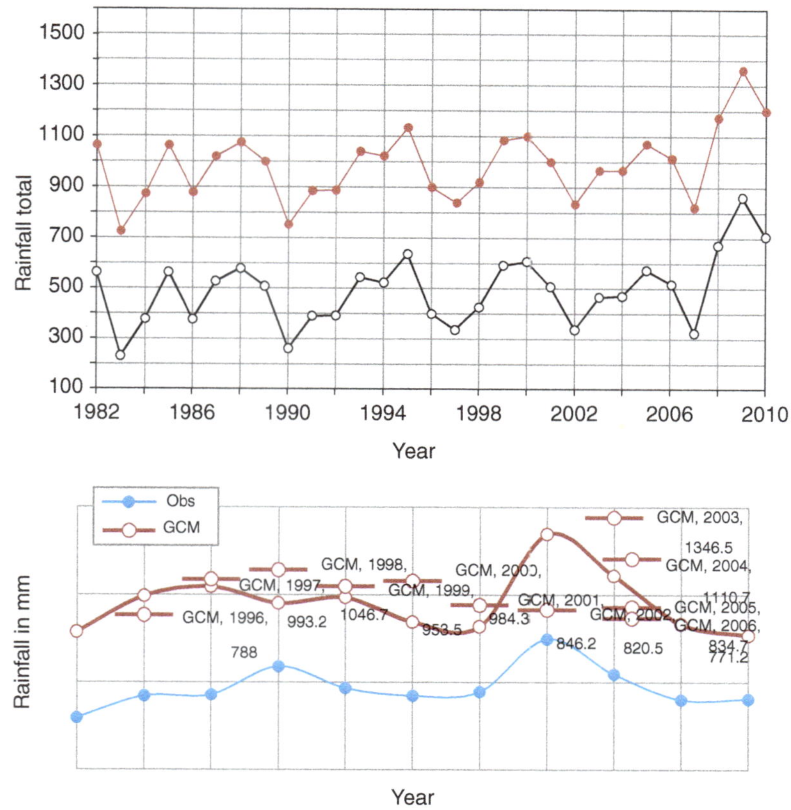


**Figure 8.14** Example of systematic mean bias error between model and observation. *Source:* O Ndiaye (Sénégal Met Service).

Many studies have been conducted to correct for systematic GCM error (Ward and Navarra, 1997; Feddersen *et al.*, 1999; Goddard and Mason, 2002). Most of this kind of model correction has been focused on rainfall because of its large social implications for many parts of the world.

For some models, systematic errors are obvious and easy to correct. Examples of such are errors in the mean or in the variance, including the example in Figure 8.14. It is clear in Figure 8.14 that the GCM is able to capture well the rainfall variability but present systematic positive bias on the mean.

A simple comparison between model and observation statistics at a specific location allows us to identify the systematic difference between the two, and then to implement a correction. Correction can be applied either in the mean or in the variance, or in both, as in the equation;

$$X_{\text{corr}} = \frac{X}{\sigma_X}\sigma_{X_{\text{obs}}} + \left(\bar{X}_{\text{obs}} - \bar{X}\right) \tag{8.4}$$

where $X$ is the raw value, $\bar{X}$ is the model forecast mean, $\bar{X}_{\text{obs}}$ is the corresponding observed mean, $\sigma X_{\text{obs}}$ the observed standard deviation and $\sigma_X$ the standard deviation of the model forecasts.

Other biases might not be as obvious as the systematic difference because model and observation are not collocated at the same grid point, when there is a shift or misrepresentation in space. In such cases, decomposition or filtering of the GCM outputs into modes of variability is necessary to 'see' the captured signal. Usually, EOF analysis is used. Tippet and Giannini (2006), using signal-to-noise analysis, identified the most reproducible GCM patterns of African summer precipitation related to the SST forcing. They used EOF analysis, rather than the ensemble mean, to identify the most predictable components of the GCM and they were able to increase GCM skill over the Sahel.

Ndiaye *et al.* (2009, 2011) identified the low-level wind over the Atlantic basin as a good predictor of Sahelian rainfall. They used the first EOF (Figure 8.15a), which captures 33% of the total wind variance, to correct poor GCM rainfall skill. They showed that GCMs in general were able to represent wind better than rainfall over this region. They were able to recover most of the rainfall variability (Figure 8.15b) through a model output statistics (MOS) approach (see Section 10.1.2.5).

There might also be consistent spatial 'distortion' in the way a model represents a given mode of variability. In other words, a GCM can dislocate an observed pattern to a different place or distort it into a different shape. A comparison between decomposed modes from observations and GCMs, or linear combinations of these, is then necessary in order to identify modes that behave consistently in time and then correct GCM modes back to the right structure and location. CCA (discussed in Section 8.1.3.1) is generally used for this purpose. CCA maximizes the correlation between the two fields using

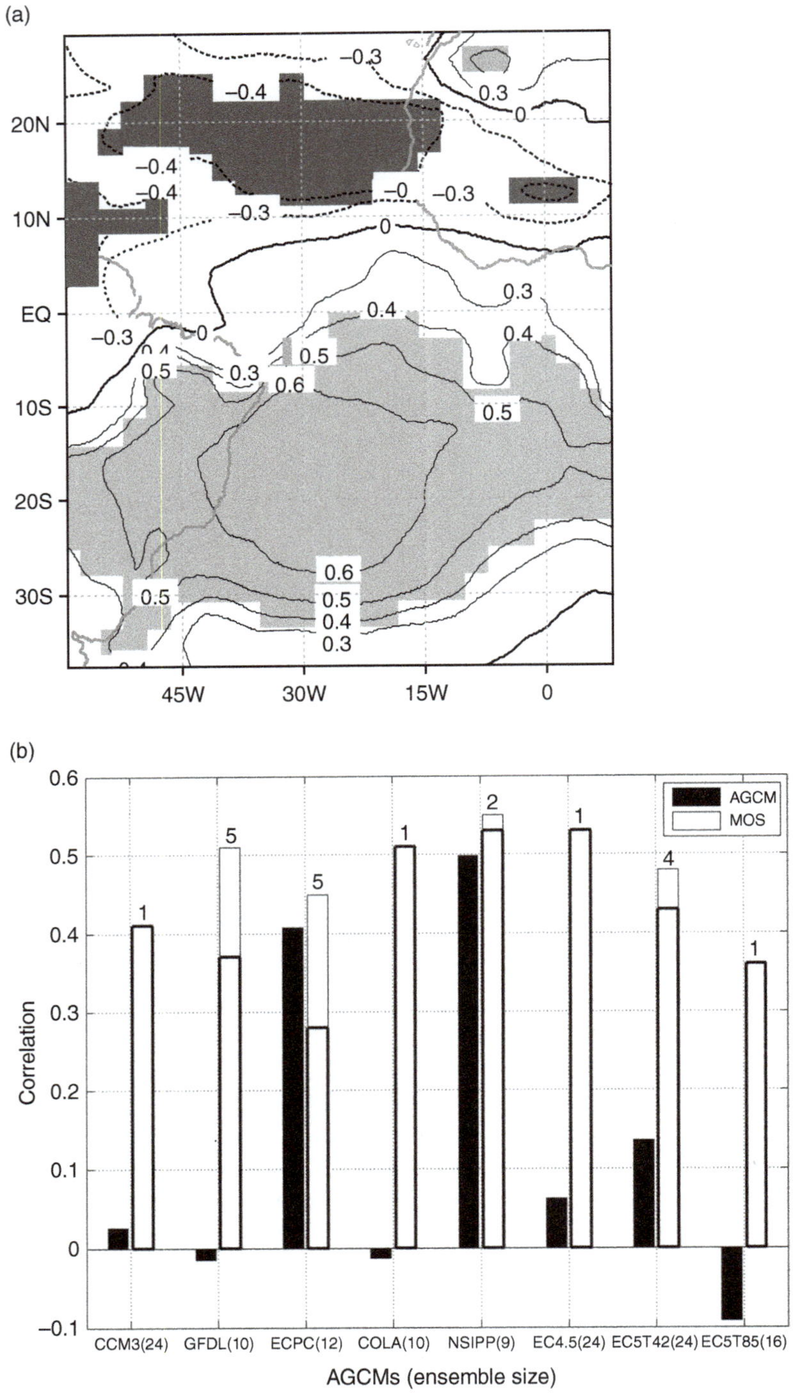


**Figure 8.15** (a) Spatial pattern of EOF1 of the 925 hPa GCM zonal wind used for the MOS and (b) Sahel rainfall skill from various AGCMs forced with observed SST, for rainfall (shaded bars) and after applying MOS (open bars) over 1968–2001. *Source:* (b) Ndiaye *et al.* (2011). © American Meteorological Society. Used with permission.

weighted combinations of the few first EOFs from each of the two fields; for example, SST and precipitation – see Ward (1998) and Barnston *et al.* (1996).

We have to recognize some limitations of the bias-correction approach. The first is the quality of the observed data that we take as the truth: any error in the observed data will affect the bias correction. The bias correction assumes that the correcting factor or transfer function is time independent and could be applied in the future, which might not hold. Fluctuations, due to different physical mechanisms acting on various timescales, are mixed together when applying a single bias correction. Most bias corrections are for spatial shift and not a time shift; for example, error related to the exact timing of some major phenomenon, such as the onset date in a monsoon system, cannot be corrected. To overcome this problem,

Haerter *et al.* (2011) proposed a 'cascade bias correction' approach that generates a cascade of bias correction functions, each operating on a different timescale. Errors can occur in the correction factor, particularly if there are insufficient data to provide a statistically robust estimate; this tends to lead to overcorrection, as random noise is included in the correction factor, increasing its magnitude. A remedy to inaccurate bias correction is to mix back some of the original signal into the bias-corrected forecast. This is discussed further in Section 8.1.5

So, owing to poor spatial representation of some aspects of the climate system (such as vegetation, elevation, lakes and rivers) or small-scale processes, GCMs can make systematic errors in some parameters affected by spatial resolution, like rainfall. Comparison over a long time period of actual or decomposed modes between observed and GCM parameters can allow for correction of the time-independent error. Bias correction is suitable when GCM output is used to force another system (crop, hydrological or impact assessment models or a regional dynamic climate model). Incorrect GCM distributions will have a strong impact on the output of the other system (dry and extreme events). For example, model misrepresentation of the likelihood of the occurrence of a given magnitude of daily precipitation or temperature will affect the adequacy of representation of the risk of flood and drought on streamflow.

#### 8.1.4.3 Future Models: Potential for Improving Seasonal Forecasts

The assessments of model performance discussed above suggest that there is still substantial potential for improvement. One example of where models can be improved is now presented.

**Example: Need to Improve Simulation of South Atlantic SST**

The performance of the seasonal forecast for the Guinea Coast region is substantially worse than that of the forecasts for the Sahel and Soudan zones. One reason for this is that models poorly represent SST in the Atlantic. The Guinea Coast has a much stronger association with the tropical South Atlantic SST than the Soudan and Sahel regions.

One of the worst failures of GloSea4 in predicting Guinea Coast rainfall was 2004, in which all the forecast members were too dry. A clue to the dryness of the forecast is seen in Figure 8.16. One of the clearest features of this SST anomaly map is the colder than average SST in the tropical South Atlantic, particularly in the Gulf of Guinea. Such cold anomalies are associated with drier than average conditions near the Guinea Coast. However, come the height of the wet season in August, the anomalies in the South Atlantic are very different: the colder than

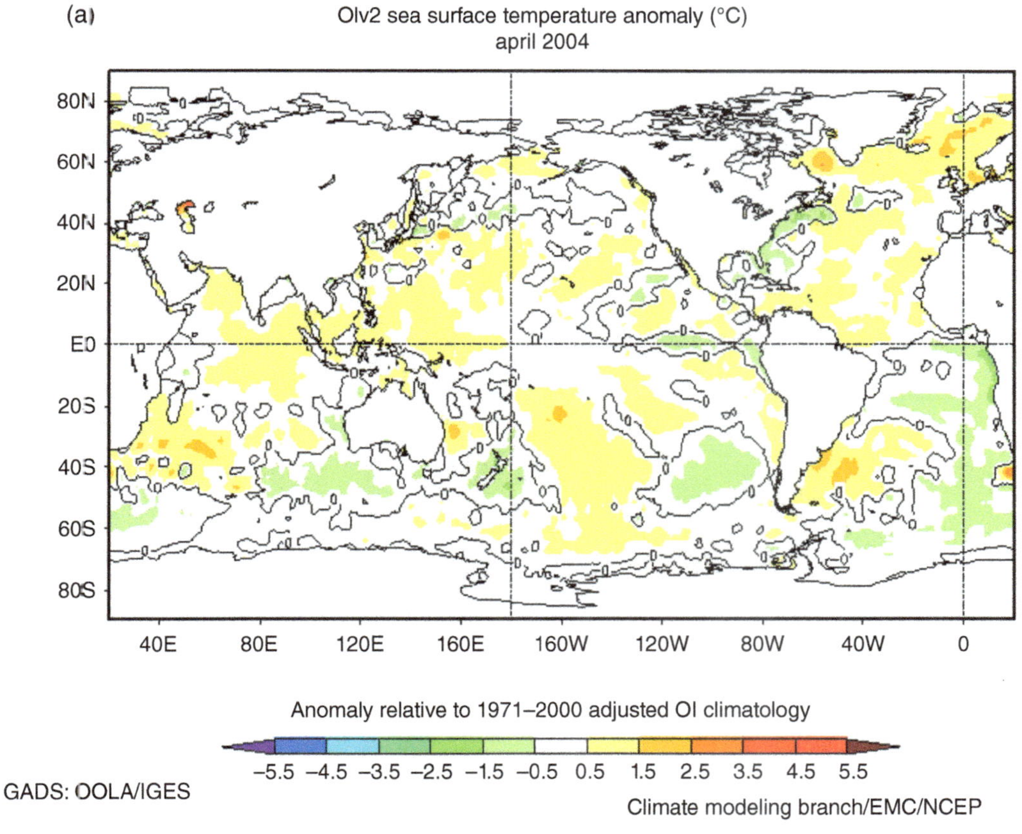


Figure 8.16 Observed SST anomalies in the South Atlantic from (a) April 2004 and (b) August 2004. *Source:* NCEP, http://www.emc.ncep.noaa.gov/research/cmb/sst_analysis/#_sstplots/. (c) Forecast SST anomalies from GloSea4 model for August 2004. *Source:* Met Office. © Crown Copyright.

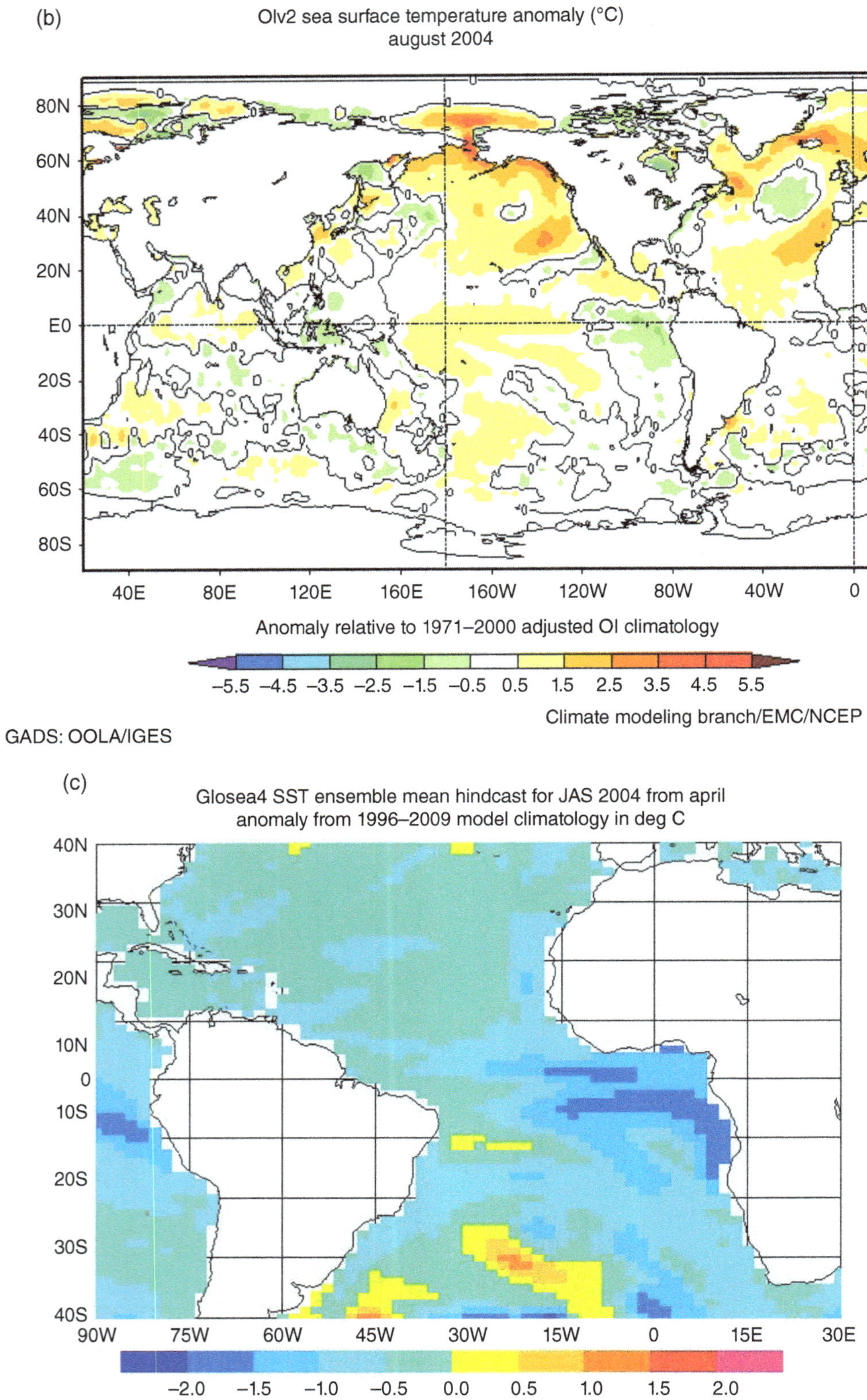


**Figure 8.16 (Cont'd)**

average anomalies have all but disappeared and the majority of the tropical South Atlantic is now warm. These anomalies are consistent with the near-average season that was observed on the Guinea Coast. In contrast, the GloSea4 model predicted colder than average SST anomalies to persist in the Gulf of Guinea (Figure 8.16c).

An underlying cause of this problem may be in the process of developing dynamical models for seasonal forecasting. A key metric used in development of the models is how well they simulate ENSO events in the Pacific. In contrast, less attention is paid to the Atlantic. However, results like this show a need to correctly model the Atlantic too; a challenge for developers of future models.

#### 8.1.4.4 Long-lead Predictability

Predictability at long lead times is nearly always likely to be harder than at shorter lead times. SST anomalies evolve slowly enough to make short-lead seasonal forecasts viable, but over longer periods the SSTs may change significantly, and thus it may be harder to detect a predictive signal.

As an example of long-lead predictability, Figure 8.17 shows predictions of rainfall for the three African zones (Sahel, Soudan and Guinea Coast) for JAS from the Met Office DEPRESYS model (http://www.metoffice.gov.uk/research/modelling-systems/unified-model/climate-models/depresys) and the Met Office GLOSEA4 model, initialised at the beginning of the previous November. GLOSEA4 and DEPRESYS are two predictions systems using the HadGEM3 model. The main difference is the data assimilation system: DEPRESYS works in anomaly space, whilst GLOSEA works with actual values.

Whilst correlation skill is mostly positive, skill is much lower compared with the shorter lead dynamical forecasts in Figure 8.13. In particular, the models failed to pick up the relatively wet years in the 1960s in the Sahel, while in recent years the forecasts have generally been too wet, showing an excessive positive trend. The use of CCA calibration to enhance the skill was considered; however, no extra signal could be found from using CCA or proxy forecasts.

### 8.1.5 Combined Forecasts

Many forecasting techniques have been discussed so far, but most users just want one forecast indicating what the forthcoming season is going to be like. Combining forecasts of different types has the added advantage in that errors can be minimised. Hagedorn *et al.* (2005) and Weisheimer *et al.* (2009) discuss the merits of combining seasonal forecasts from several different models.

The logic behind the use of multiple models is related to the uncertainty in their representation of the ocean and atmosphere. A forecast system can be represented by the following equation containing three vectors:

$$\mathbf{O} = \mathbf{F}_A + \mathbf{E}_A \tag{8.5}$$

where $\mathbf{O}$ is the observation vector, $\mathbf{F}_A$ is a forecast vector from model A and $\mathbf{E}_A$ is the model error vector. Considering now the case of more than one model, then the forecasts from these models would be $\mathbf{F}_B$ and $\mathbf{F}_C$ and the errors $\mathbf{E}_B$ and $\mathbf{E}_C$. If one uses the mean of $N = 3$ models A, B and C then the error $\mathbf{E}_M$ will be $(\mathbf{E}_A + \mathbf{E}_B + \mathbf{E}_C)/N$.

Unless $\mathbf{E}_A$, $\mathbf{E}_B$ and $\mathbf{E}_C$ are perfectly correlated, the magnitude of $\mathbf{E}_M$, $|\mathbf{E}_M|$ will always be less than the average magnitude $(|\mathbf{E}_A|+|\mathbf{E}_B|+|\mathbf{E}_C|)/N$. If the errors are random then $\mathbf{E}_M$ will tend towards zero as $N$ becomes large. Some errors, however, are common to all models;, for example, errors related to model resolution or to the lack of available observations for initialisation. These errors will remain after models have been combined.

Therefore, it is a good idea to have a collection of models, each with different errors that will ultimately cancel each other out. It is important, however, that all the models used have good skill and pick up seasonal predictability, otherwise the forecasts from poor models will just generate noise and consequently forecasts will always be close to climatology and of little practical use.

Dynamical forecasts can be divided into three components:

$$\text{Forecast} = \text{direct signal} + \text{indirect signal} + \text{noise}$$

Direct signal is when the model directly predicts the event; for example, the mean of precipitation output from the model predicts an anomalously wet or dry season. Indirect signal is when the model fails to predict the event, but does predict climate phenomena related to the event. For example, the model may fail to predict rainfall in a particular location, but it may predict El Niño events that are known to be related to the rainfall. The remaining component is noise. This includes atmospheric variability that is unpredictable due to chaos or due to the model not being good enough.

In a forecast system, raw ensemble probabilities represent the direct signal and CCA-calibrated forecasts represent the indirect signal, while all forecasts contain an element of noise. In order to produce a combined forecast one needs to know the ratio of direct signal to indirect signal, in order to determine how much weight to give each forecast. This is not a straightforward question and can only be estimated using a relatively small number of hindcasts. We have done this, however, for West Africa seasonal precipitation forecasts using the relative operating characteristic (ROC) skill measure (Stanski *et al.*, 1989; Broecker, 2012). ROC scores are also discussed on the Met Office web site (http://www.metoffice.gov.uk/research/areas/seasonal-to-decadal/gpc-outlooks/user-guide/interpret-roc).

**Example: Met Office Forecast for PRESAO15**

For producing a forecast of West Africa seasonal rainfall to contribute to PRESAO, we had available data from five Global Producing Centres (GPCs) models. For each model we have two types of forecast: we had a raw ensemble probability forecast and a calibrated forecast evaluated using CCA, making a total of 10 forecasts. Since the ratio of direct to indirect signal is unknown, the forecast system was tested giving various weights to the CCA-calibrated forecasts (representing

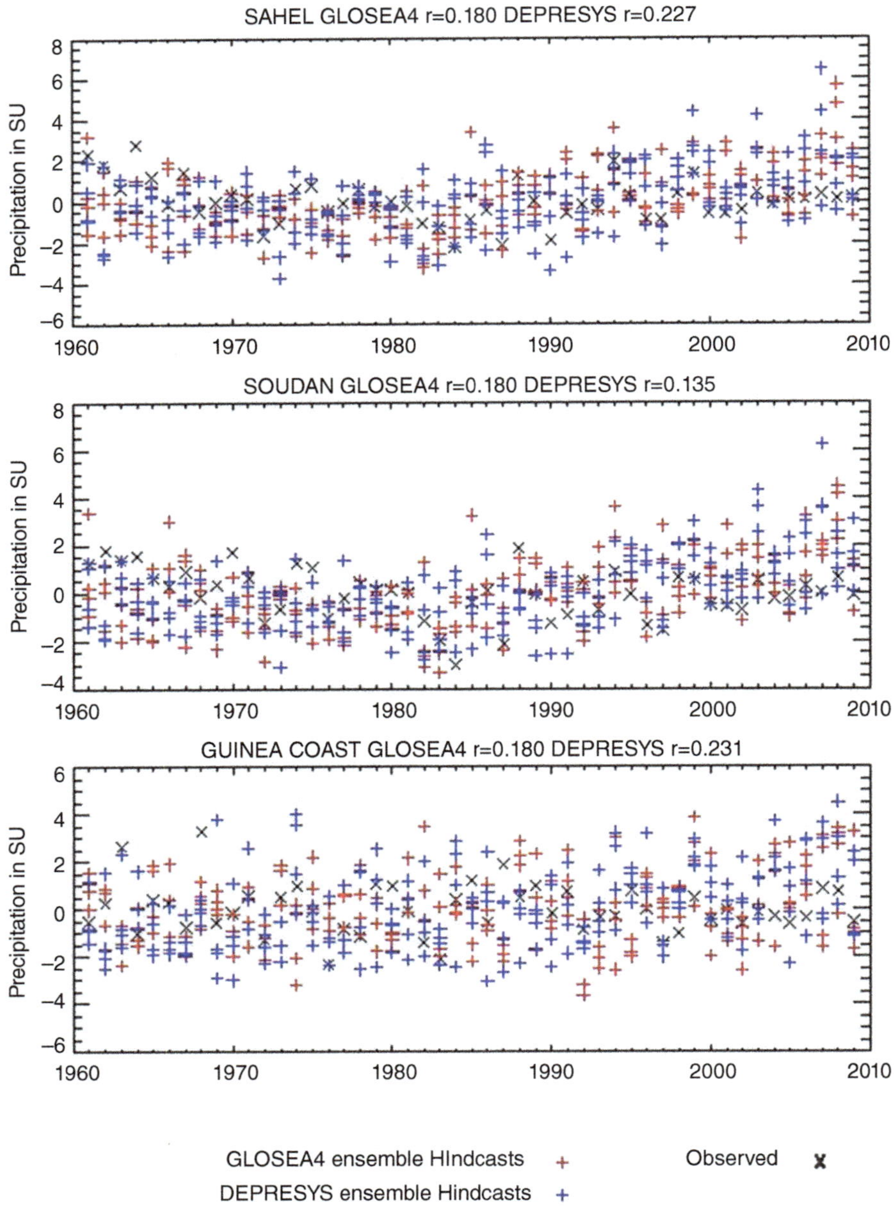


**Figure 8.17** Long-lead predictions of West Africa climate zone rainfall anomalies for JAS from preceding November using two models. *Source:* Met Office. © Crown Copyright.

the indirect signal) and the raw ensemble forecasts. Combinations using 0, 50, 70 and 100% weights to the CCA-calibrated predictions were tested and the ROC skill is presented in Figure 8.18.

#### 8.1.5.1 Performance over 1996–2009

Chance ROC skill is 50%, so blue in Figure 8.18 represents areas where there is no skill. Highest skill is indicated by orange and red. For the dry and wet terciles, patterns of skill from the different weightings are quite similar. For the dry tercile, skill appears highest in the centre-west, patchy in the east and absent in the central-south. Wet tercile skill is less widespread and more restricted to the centre and west of the region. Differences in skill between the weightings are relatively small. The 100% CCA-calibrated forecast has the highest average skill (56.7%) for the wet tercile category, whilst the 50% weighting has the highest skill on average (61.2%) for dry. Overall, the 70% weighting has the highest skill on average ((60.2 + 56.6) / 2 = 58.4%).

**Example of a Combined Forecast (for JAS, 2012)**

Forecasts from the different combinations for JAS 2012 are shown in Figure 8.19. The 70% weighting is regarded as the best guess, as this combination had the highest average ROC skill overall.

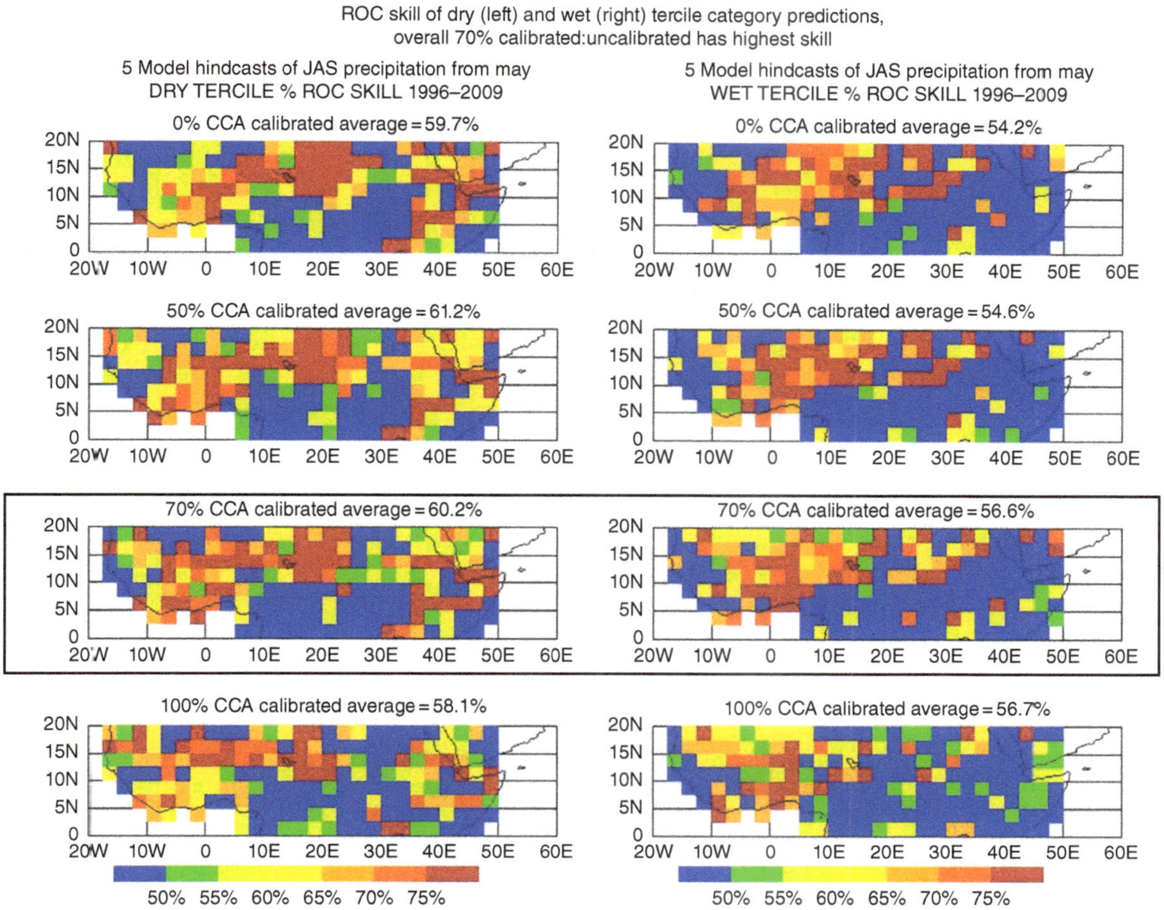


Figure 8.18 Assessments of combined forecasts from five GPC models giving various weights to the calibrated and uncalibrated components (see text). *Source:* Met Office. © Crown Copyright.

## 8.1.6 Other Variables and Seasons

### 8.1.6.1 Prediction of Outflow from Lake Volta

The Volta Basin covers Ghana and extends north covering more than half of Burkino Faso and also parts of Benin, Mali, Togo and Ivory Coast (Figure 8.20a); thus, it bisects the Guinea Coast, Soudan and Sahel zones of Figure 8.2. Lake Volta outflow occurs predominantly between June and November and peaks in September each year, following the West African wet season (Figure 8.20b).

Given that wet-season rainfall had been found to be predictable, it was reasonable to assume that Lake Volta outflow would also be predictable as it is strongly linked to rainfall. Therefore, a system to predict outflow was developed. Initially, statistical methods using precursor SST pattern indices were used to produce the forecast, but when it became evident that these statistical forecasts were not contributing to skill they were discontinued and replaced with a linear regression model that uses three predictors:

1) Forecast rainfall from the GloSea model for the Volta basin region.
2) Observed rainfall in the Volta basin 1–3 months prior to the forecast date.
3) Observed upstream flow in two tributaries flowing into Lake Volta (Black Volta and Saboba) for 1–3 months prior to forecast date.

Six forecasts are produced annually: at the end of April, a forecast for June–November; at the end of May, a forecast for June–November; at the end of June, a forecast for July–November; at the end of July, a forecast for August–November; at the end of August, a forecast for September-–November; and finally at the end of September, a forecast for October–November.

The performance of forecasts from 1 May is presented in Figure 8.20c. Skill is comparable to the performance of climate zone rainfall forecasts shown in Figure 8.13.

### 8.1.6.2 Prediction of Meningococcal Meningitis in the West Africa Dry Season

Not all West Africa's problems are linked to the wet season. Meningococcal meningitis (MCM) epidemics, occurring during the dry and hot season, have been associated with wind meridional component (Yaka *et al.*, 2008). MCM is a contagious, infectious disease caused by the bacteria *Neisseria meningitis*. MCM

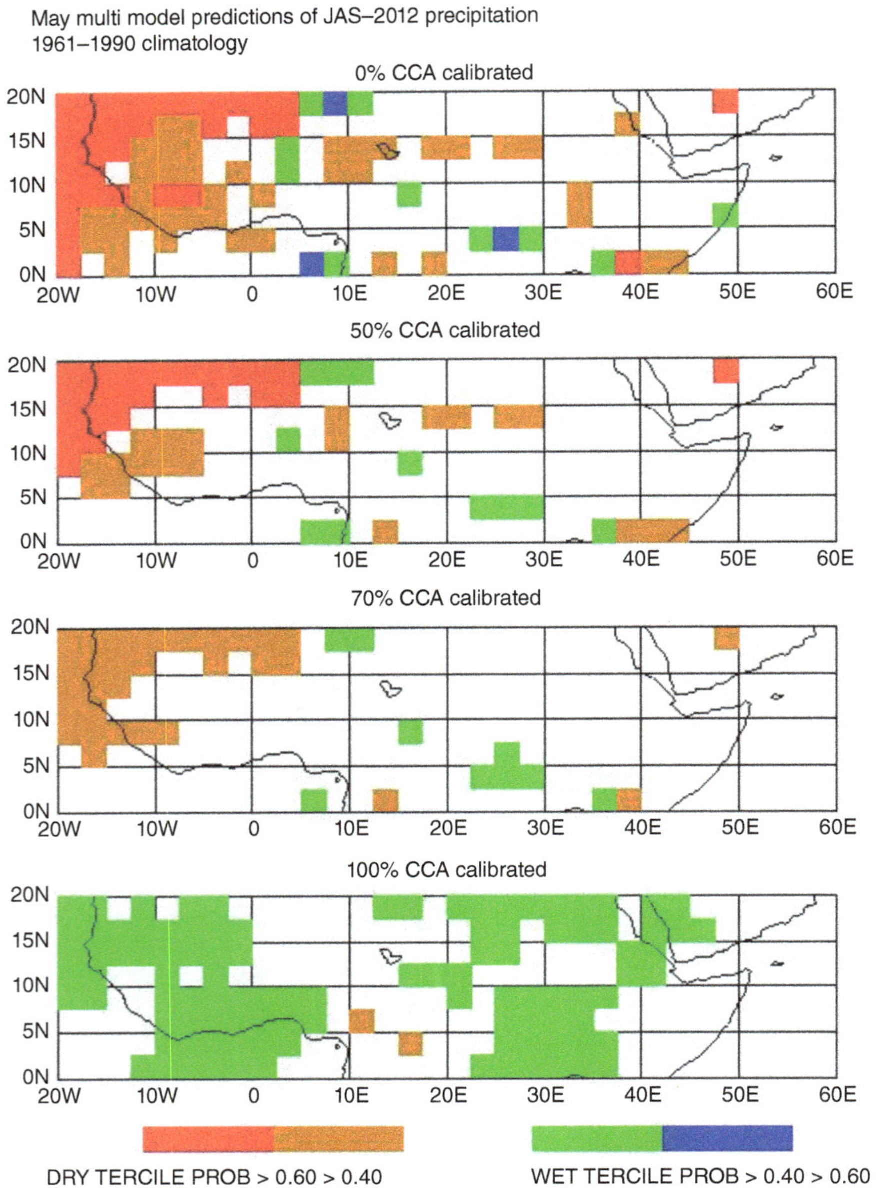


**Figure 8.19** Combined forecasts from five GPC models giving various weights to the calibrated and uncalibrated components. *Source:* Met Office. © Crown Copyright.

epidemics occur worldwide, but the highest incidence is observed in the 'meningitis belt' of sub-Saharan Africa, stretching from Sénégal to Ethiopia. Epidemics tend to occur during the latter part of the dry season (February–April) and usually end with the onset of rains. The severity of the MCM epidemics varies from year to year, with the number of cases in the region varying from 25 000 to 250 000. Children under 15 are particularly affected. Mortality rates average around 10%, with 10–20% of survivors affected by serious neurological repercussions.

The factors behind these epidemics are not fully understood but are known to involve a complex interplay of social interactions, new strains, susceptibility of populations, asymptomatic carriage and environmental conditions. Climate conditions are also considered to be important in influencing resurgence and dispersion of the disease. Evidence for this climatic link is that the spatial extent of this disease coincides with the climatically homogeneous Sahel zone extending from Sénégal to Ethiopia.

Data from synoptic stations and NCEP reanalysis data have been investigated to identify links between the severity of MCM epidemics and climate. In particular, relationships where there is a time lag between the climate anomaly and subsequent epidemics have been sought, and such relationships open up the opportunity of making forecasts of MCM epidemics. These studies have found that enhanced easterly (Harmattan) winds at the beginning of the dry season (particularly the meridional wind component of October in Burkina Faso, or the

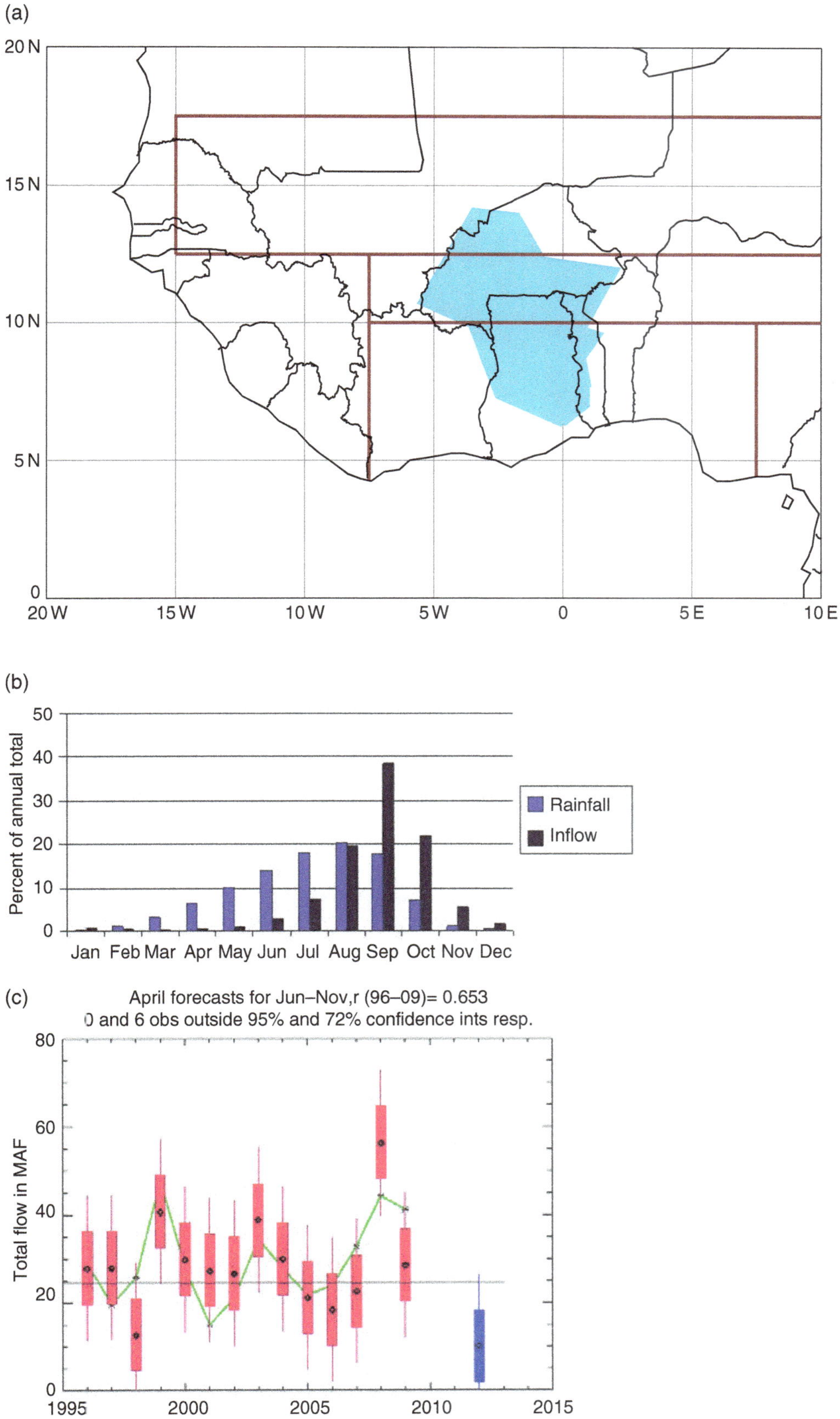


**Figure 8.20** (a) Volta Basin region (blue) relative to Sahel, Soudan and Guinea coast regions (outlined in red). (b) Annual cycle of rainfall and inflow. (c) Hindcasts for 1996–2009 and forecast for 2012. *Source:* Met Office. © Crown Copyright.

November and December meridional wind component in Niger) have been associated with MCM outbreaks and epidemics during the following year.

Yaka *et al.* (2008) have made a statistical analysis of the relationship between MCM and observed Harmattan winds in the winter season, finding that the winter climate can explain 25% of the interannual MGM variance in Niger (albeit with lower skill in Burkina Faso). Overall, the skill of the statistical models is encouraging, and development of similar forecasts for other parts of the Sahel is recommended.

#### 8.1.6.3 Applying Seasonal Forecasts to Assist Agriculture in Sénégal

In the Sahel region, more than 80% of the population live off agriculture and pastoralism, both of which depend on rainfall. Particularly when resources are scarce, this dependency will have a direct impact on food security, and anticipation is needed.

Therefore, it is very important that climate information (weather to seasonal climate forecasts) is communicated in accessible and meaningful ways.

To this end, a pilot project was set up by the Sénégal National Meteorological Agency to communicate climate information to farmers in the Kaffrine district of Sénégal. The project involved building a bridge of trust between the climate scientists producing the forecasts and the farmers as users; this included understanding how the farmers currently use indigenous knowledge to predict climate and weather events related to their needs, getting the farmers to share their needs for climate information and to identify what new information could be useful to them, and investigating ways of communicating forecast information to the farmers.

Initially, 33 farmers were involved in the project, but it has been recently expanded to involve up to 1000 farmers. It was found that farmers use indigenous knowledge in an analogous way to users of multimodel forecasts. They get information from several sources and have various levels of confidence in the different information. When they discuss their indigenous forecast there is no 100% consensus amongst farmers on the signal – some will say they observed it clearly, others will say not very clearly, and this is a kind of uncertainty in our system. They also clearly use different indicators (birds, trees, wind, stars, etc.), which is similar to our multimodelling approach. More interestingly, farmers can say 'signals are not there yet, for example the baobabs are not flowering,' so they cannot say for sure now, but might have more confidence in a month's time; this is similar to our updating approach of seasonal forecast, which gets more accurate as the targeted season approaches. When farmers understand that the climate scientist's approach is similar, it is likely to build more trust.

Farmers were divided into groups, some gender specific, to say what climate information they needed and answer the question 'If you knew the future climate, what would you do differently?' Foreknowledge of the rainy season onset date was considered important as it guided what and when to plant. There was also interest in the length of the wet season and the onset data for next year's season.

Ways of communicating were discussed. Use of SMS texting and radio broadcasts was discussed, but it was pointed out that electricity supplies are often unreliable. Personal contact was considered better. Women identified social gatherings (e.g. naming ceremony) as a good medium, as people would be in attendance for other (cultural or social) reasons. In Kaffrine it appeared that communicating through an 'extension person' on location was probably the most efficient route.

Communicating forecasts and the results of this pilot were discussed with other agricultural experts from Sénégal. It was suggested that examples from the historical record should be used to convey the message. For example, if the forecast is 'wet,' the growing seasons in other 'wet' years should be used as examples of what is expected.

Finally, this process was considered to be dynamic and ongoing. Communication between farmers and forecast-producing climate scientists should be an ongoing two-way process. To this end, some farmers were supplied with rain gauges so they could provide observations and contribute to the forecast production process.

## 8.2 Operational Methods

Before endeavouring to produce a seasonal forecast, readers are advised to search through the many seasonal forecast products available on the web, most of which are free. International sites, including a brief description of each site, are listed in Section 8.2.1. These sites show forecast products for the whole region. If one is interested in a forecast for a particular country, investigating the appropriate National Met Service web site is recommended; these are listed in Section 8.2.2. In Section 8.2.3 the production of consensus forecasts is discussed, including the PRESAO process. Finally, in Section 8.2.4 some useful tools are listed for producing your own forecast.

### 8.2.1 International Seasonal Forecast web sites: Sources of Information

The following web sites contain seasonal forecasts and verifications of seasonal forecasts for West Africa and other parts of the world.

#### 8.2.1.1 World Meteorological Organization Lead Centre Multimodel Ensemble web site

https://www.wmolc.org/

This web site was set up as a single location where seasonal forecasts from all reliable sources could be viewed together in a common format and, therefore, compared and combined. The seasonal forecasts presented here are all produced by World Meteorological Organization (WMO)-approved lead centres. There are currently 12 lead centres, shown in Figure 8.21. These lead centres are also known as GPCs. Forecasts are presented of means of precipitation, surface temperature, surface pressure and 500 mb height for the next three calendar months. Forecasts are updated around the 20th of the month. A UserID and password are required to view the forecasts, which can be obtained by accessing the site and clicking on 'sign up'.

#### 8.2.1.2 Long-range Forecast Verifications

http://www.bom.gov.au/wmo/lrfvs/

This site contains skill assessments of the lead centre models presented in Section 8.2.1.1 using the ROC and root-mean-square error skill measures. Checking skill is recommended before making use of model forecasts from www.wmolc.org. However, as with interpretation of seasonal forecasts, skill maps should be interpreted with caution. Assessments are made over a relatively small number of years and may not provide an accurate assessment of the model. Beware of local detail, as this is likely to be due to unpredictable noise, and perhaps observational error.

#### 8.2.1.3 UK Met Office web site

Met Office forecasts produced using the GloSea system of 3-month mean precipitation, surface temperature, surface pressure and 500 mb height out to 6 months are available here for free public access. They include global maps and maps for specific regions, including West Africa. Also available are ENSO 'plumes', forecast time series of ENSO-related SST time series, including the Nino3, Nino 3.4 and Nino 4 regions.

#### 8.2.1.4 ECMWF/EUROSIP web site

http://www.ecmwf.int/en/forecasts/documentation-and-support/long-range/seasonal-forecast-documentation/eurosip-user-guide/multi-model

This site contains forecast output from the EUROSIP models (Met Office GloSea4, Meteo-France ARPEGE, NCEP Climate Forecast System (CFS) as well as ECMWF). Only unweighted averages of the four models are available; access to individual model output is restricted to a small number of ECMWF member state users. Included are forecasts of surface temperature, surface temperature, precipitation, 500 mb height and 850 mb temperature for 3-month periods out to 6 months ahead. Forecasts for the tropics (30°N–30°S including West Africa) are publicly available. Also available are the 'ENSO plumes', as on the UK Met Office site but produced using all four EUROSIP models.

#### 8.2.1.5 NOAA/NCEP web site

http://www.cpc.ncep.noaa.gov/products/people/wwang/cfsv2fcst

Forecasts of 1 and 3-month averages of surface temperature, precipitation, 200 mb height and 700 mb height from the CFS model out to 9 months ahead are publicly available here in the form of global maps. ENSO plumes are also shown, as on the ECMWF and UK Met Office site.

#### 8.2.1.6 International Research Institute for Climate and Society Seasonal Forecast web site

http://iri.columbia.edu/climate/forecast/net_asmt/

This site provides forecasts of 3-month mean temperature and precipitation out to 6 months ahead, which are presented as maps for regions, including Africa and for the globe. The IRI products are based on a combination of dynamical and statistical models and differ from those

Figure 8.21 The 12 lead centres. *Source:* www.wmolc.org/.

discussed in Sections 8.2.1.1, 8.2.1.3, 8.2.1.4 and 8.2.1.5 in that human forecaster intervention is involved in producing the forecasts. There is also a discussion section that gives details of how the forecasts are produced and which models are used.

#### 8.2.1.7 African Center of Meteorological Application for Development web site

http://www.acmad.net/

ACMAD is in a demonstration phase, preparing to become a WMO designated Regional Climate Centre for Africa with a long-range forecasting mandate for West Africa. Publications from recent PRESAO events, including consensus forecasts, are posted on the web site alongside shorter range (1–10-day) predictions and announcements of upcoming climate outlook forums, including PRESAO events and relevant training workshops.

### 8.2.2 Consensus Forecasts

Section 8.2.1 describes several sources of seasonal forecast information for West Africa. How should one use this information? After all, most users require just one forecast. The RCOF process was set up in the late 1990s to help resolve this problem. RCOFs are events where forecasters can get together to produce a consensus forecast and disseminate it to users.

#### 8.2.2.1 The PRESAO Process

Follow links to ACMAD web site: http://acmad.net/rcc/and http://www.acmad.net/new/.

For the purpose of organising RCOFs, Africa and other parts of the world are subdivided into climatically homogeneous regions. West Africa is covered by PRESAO. PRESAO is one of five RCOFs covering mainland Africa. The others are PRESAC (covers Central Africa), SARCOF (covers Southern Africa), GHACOF (covers East Africa) and PRESANOR, which covers North Africa (north of the Sahara).

The consensus forecast is produced by a conference of experts from (1) West African National Met Services, (2) ACMAD, (3) overseas institutes such as the UK Met Office and Meteo-France. Figure 8.22 is an example consensus forecast produced by PRESAO for the 2012 season.

To produce the consensus forecast the following information is used.

1) Calibrated (systematic errors corrected) and downscaled GPC model dynamical forecasts for climatically homogeneous regions in individual countries (method described in Section 8.1.4.2).
2) Forecasts for these same zones using pre-season observed SST produced using PCR and simple multiple linear regression (method described in Section 8.1.3.1).
3) SSTs and precipitation forecast output from the 12 GPC models as posted on the lead centre website (www.wmolc.org) and individual GPC websites (e.g. www.ecmwf.int).
4) The IRI seasonal forecast from http://iri.columbia.edu/climate/forecast/net_asmt/.
5) ACMAD's Regional Climate Centre Demonstration phase's consolidated long-range forecast product for Africa. These forecasts are produced by a subjective analysis of observed sea temperatures and trends. Rainfall predictions are derived from the sea temperatures based on known teleconnections (like those discussed in Section 8.1.2) and from investigating 'analogue' years with similar observed sea temperature patterns and trends to the current year, particularly in the tropics. The forecasts are cross-checked against the GPC forecasts referred to in Section 8.2.1.2.
6) Expert interpretation of recent SST and subsurface sea temperature anomalies and their relation to teleconnections (including recent values of Global EOF3; see Section 8.1.3.2).

A forecast map like the one in Figure 8.22 is drawn by hand based on the aforementioned information. The forecast is issued in the form of probabilities of tercile categories.

The three tercile categories are 'above normal', which represents rainfall above the upper tercile, 'below normal', which represents rainfall less than the lower tercile, and 'near normal', which represents rainfall between the upper and lower terciles. The three categories are climatologically equiprobable, with a probability of 33%.

Effort is made to generate additional climate outlooks for each participating country ensuring consistency with neighbouring countries and the regional consensus outlook (Figure 8.23 for Burkina Faso).

West Africa is divided into regions where the forecast is similar, and the probabilities for the three categories are printed on the map.

#### 8.2.2.2 Assessment of PRESAO Forecasts 1998–2007

An assessment of the first 10 PRESAO forecasts issued between 1998 and 2007 is included in Mason and Chidzambwa (2009); see http://iri.columbia.edu/docs/publications/TechRpt09-02%20-RCOF.pdf. When verified against a blend of CAMS-OPI (Janowiak and Xie, 1999) and CRU TS2.1 (Mitchell and Jones, 2005) observed gridded rainfall data, PRESAO forecasts for all three tercile categories exhibit modest skill, with ROC scores around 55–60%, a similar level to the average skill of combined forecasts illustrated in Figure 8.18.

### 8.2.3 Software and Tools

To develop your own forecasting system, the following tools are available.

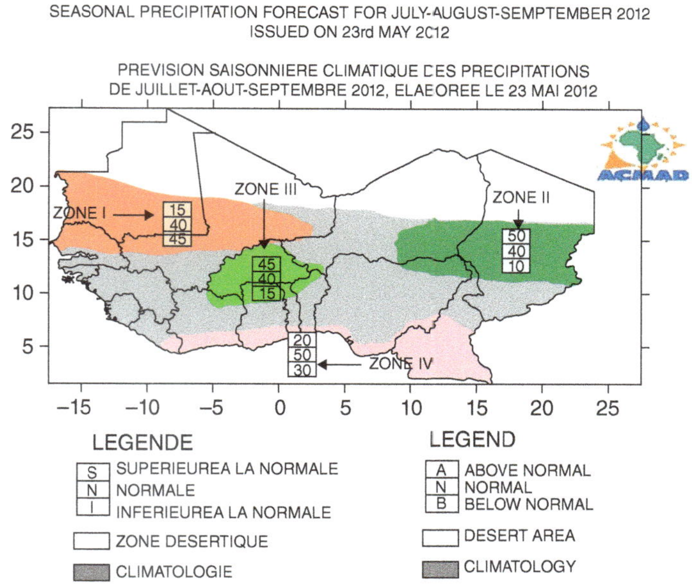


**Figure 8.22** Consensus forecast for 2012 season issued by PRESAO15. *Source:* Andre Kamga, ACMAD (http://www.acmad.net/new/)

#### 8.2.3.1 The Climate Predictability Tool

The CPT (http://iri.columbia.edu/our-expertise/climate/tools/cpt/) is a powerful, purpose-built and easy to use facility for developing seasonal forecast models and forecasts. It is free and easily downloadable from the link given. A set of help pages is included with the product, and more help can be found on the web site, including a CPT tutorial.

In order to use CPT you require two datasets:

1) *X* variable. A dataset of precursor observations (e.g. SST indices) or of forecasts from some dynamical model. The data need to be for a number of historical years and stored in CPT format.
2) *Y* variable. A data set of predictand observed variables; for example, West African station rainfall. There must be data for at least some of the years for which there is *X* variable data.

Details of CPT formats can be found in the CPT tutorial linked to on the web site. Data in CPT format can be directly downloaded from the IRI web site. Statistical models can then be created using multiple linear regression, PCR or CCA (discussed in Section 8.1.3).

CPT automatically produces a variety of skill measures, including Pearson correlation, ROC, linear error in probability space, GERRITY and others. Output also includes skill maps and time-series plots.

Finally, one can produce forecasts by adding an extra year or set of years to the *X* variable. CPT can convert the forecasts to probabilities by assuming regression forecasts have a normal distribution with standard deviation equal to the standard error of the forecasting system. CPT presents the forecasts on forecast versus observation time-series plots, and as probability maps.

#### 8.2.3.2 International Research Institute for Climate and Society Data Library

http://iridl.ldeo.columbia.edu/

A large selection of data is available here that can be used directly in CPT.

#### 8.2.3.3 KNMI Climate Explorer

http://climexp.knmi.nl/start.cgi?id=someone@somewhere

The KNMI Climate Explorer is a web application to analyse climate data statistically. It started in late 1999 as a simple web page to analyse ENSO teleconnections and has grown over the years to more than 1 TB of climate data and dozens of analysis tools. It is now part of the WMO Regional Climate Centre at KNMI. Its main purpose is to validate model runs and study teleconnections. It does not apply the statistical post-processing used by CPT (regression, CCA, cross-validation); it simply compares selected pairs of data.

#### 8.2.3.4 NOAA Earth System Research Laboratory Interactive web site

http://www.esrl.noaa.gov/psd/data/correlation/

### 8.2.4 Other web sites Including Forecast Applications

#### 8.2.4.1 FEWS NET

www.fews.net

FEWS NET produce high-resolution maps showing food security risk based on available seasonal forecasting information and precursor conditions. Maps are usually for the next 3 months.

#### 8.2.4.2 AGRHYMET, Niger

www.agrhymet.ne

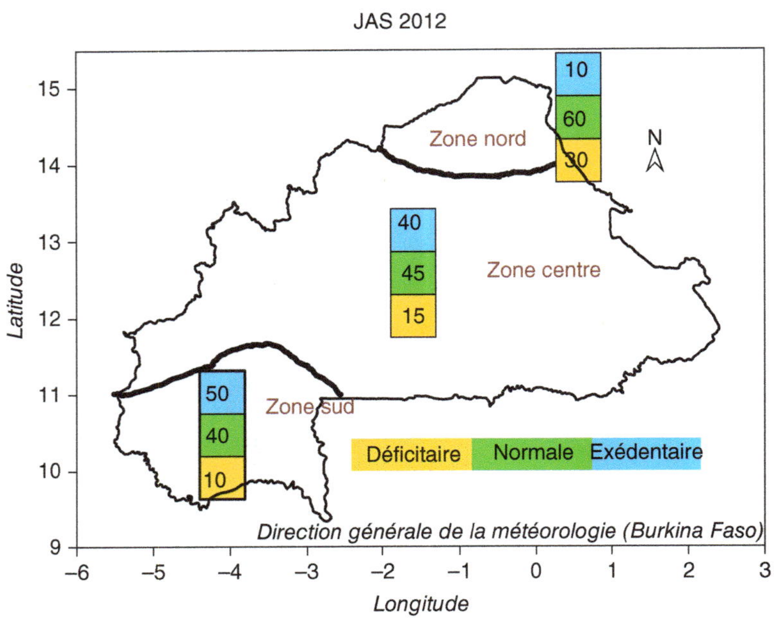


**Figure 8.23** JAS 2012 forecast for Burkina Faso consistent with neighbouring countries and the regional consensus of May 2012. *Source:* Andre Kamga, ACMAD (http://www.acmad.net/new/).

### 8.2.5 Hints and Recommendations

Statistical forecasts are becoming obsolete for most seasonal forecast applications in West Africa. This is due to the weakness of statistical methods, a consequence of poor-quality observations, climate change and changing land use, and to the relative success of dynamical methods.

However, there may be some localised applications where statistical methods may still be of use. This may correspond to applications where dynamical model resolution is too coarse to provide a reasonable prediction. For example, for years during which the Mediterranean SSTs and/or northern Africa near-surface land temperature influence significantly Sahel rainfall (e.g. 2012), statistical tools are more useful because very few dynamical models have Mediterranean SST/Sahel rainfall mechanisms well represented.

## Appendix A: Acronyms

| | |
|---|---|
| ACMAD | African Center of Meteorological Application for Development |
| CCA | canonical correlation analysis |
| ENSO | El Niño Southern Oscillation |
| EOFs | empirical orthogonal functions |
| GCM | general circulation model |
| IHC | interhemispheric thermal contrast |
| IRI | International Research Institute for Climate and Society |
| JAS | July–September |
| MCA | maximum covariance analysis |
| MOS | model output statistics |
| NCEP | National Centers for Environmental Prediction |
| PRESAO | PREvision Saisonairre en Afrique del l'Ouest |
| RCOF | Regional Climate Outlook Forum |
| ROC | relative operating characteristic |
| S4CAST | Sea Surface Temperature-based Statistical Seasonal Forecast Model |
| SST | sea surface temperature |

## Appendix B: Empirical Orthogonal Functions

EOFs represent patterns in a dataset that explain a maximum amount of variance, and are orthogonal to each other. Usually, in climate data analysis, the dataset is composed of spatial data that varies in time. The covariances of the time series of data from every pair of points (spatial locations), $i$ and $j$, then form a covariance matrix $C_{ij}$. The EOF is computed by finding the eigenvectors and eigenvalues of the matrix $C$. Therefore, EOFs are also referred to as 'eigenvectors'.

The first EOF or eigenvector is the pattern that explains the most variance (in other words, it has the highest eigenvalue) in the data studied. The second EOF is the pattern that explains the most of the remaining variance (next-highest eigenvalue) and is orthogonal to the first EOF. The third EOF is the pattern that explains the most of the remaining variance (not explained by the first or second EOF) and is orthogonal to the first and second EOF. Given that the covariance matrix is of dimension $N \times N$ (where $N$ is the number of spatial points in the dataset), the method leads to $N$ EOFs being identified. The first, second and third EOF, and so on, can be referred to as EOF1, EOF2, EOF3, …, EOF$N$. Use of EOFs beyond the first and second should be undertaken with great caution, especially when they explain only a small fraction of the variance.

Once the EOFs have been computed, the original dataset can be decomposed uniquely into a linear combination of these functions, each varying in time. The 'principal component' is then the time series of amplitude of that EOF in the original dataset. EOF analysis is sometimes alternatively known as 'principal component analysis'.

Further information on EOFs can be found at https://en.wikipedia.org/wiki/Empirical_orthogonal_functions, https://en.wikipedia.org/wiki/Principal_components_analysis, https://en.wikipedia.org/wiki/Eigenvalues_and_eigenvectors.

## Note

1 A possible explanation for the differences is the use of different climatology periods. CAMS-OPI shows anomalies from 1979–2000, whilst TRMM shows anomalies relative to 1998–2011. If the CAMS-OPI climatology is wetter than the TRMM climatology then the CAMS-OPI maps are likely to look drier. However, most datasets suggest that the 1998–2011 climatology period was wetter than the 1979–2000 climatology period (the driest seasons on record were in the 1980s). Therefore, it is unlikely that climatology period differences can explain the differences between the four maps.

## References


Adler RF, Gu G, Huffman GJ. 2012. Estimating climatological bias errors for the Global Precipitation Climatology Project (GPCP). *J. Appl. Meteorol. Climatol.* 51(1). doi: 10.1175/JAMC-D-11-052.1, 84-99.

Arribas A, Glover M, Maidens A, *et al.* 2011: The GloSea4 ensemble prediction system for seasonal forecasting. *Mon. Weather Rev.* 139(6): 1891–1910.

Barnett TP, Preisendorfer R. 1987. Origins and levels of monthly and seasonal forecast skill for the United States surface air temperatures determined by canonical correlation analysis. *Mon. Weather Rev.* 115: 1825–1850.

Barnston AG, Thiao W, Kumar V. 1996. Long-lead forecasts of seasonal precipitation in Africa using CCA. *Weather Forecast.* 11: 506–520.

Bayes T, Price R. 1763. An essay towards solving a problem in the doctrine of chance. By the late Rev. Mr. Bayes, communicated by Mr. Price, in a letter to John Canton, A. M. F. R. S. *Phil. Trans. R. Soc. London* 53: 370–418. doi: 10.1098/rstl.1763.0053.

Bjerknes J. 1969. Atmospheric teleconnections from the equatorial Pacific. *Mon. Weather Rev.* 97: 163–172.

Broecker J. 2012. Probability forecasts. In: *Forecast Verification: A Practitioners Guide in Atmospheric Science*, 2nd edition, IT Jolliffe, DB Stephenson (eds). John Wiley & Sons, Ltd: Chichester; pp. 119–140.

Charney JG. 1975. Dynamics of deserts and drought in the Sahel. *Q. J. R. Meteorol. Soc.* 101: 193–202.

Feddersen H, Navarra A, Ward MN. 1999. Reduction of model systematic error by statistical correction for dynamical seasonal prediction. *J. Climate* 12: 1974–1989.

Folland CK, Parker DE, Palmer TN. 1986. Sahel rainfall and worldwide sea temperatures 1901–85. *Nature* 320: 602–607.

Folland CK, Owen J, Ward MN, Colman AW. 1991. Prediction of seasonal rainfall in the Sahel region using empirical and dynamical methods. *J. Forecast.* 10: 21–56.

Folland CK, Parker DE, Colman AW, Washington R. 1999. Large scale modes of ocean surface temperature since the late nineteenth century. In: *Beyond El Niño: Decadal and Interdecadal Climate Variability*, A Navarra (ed.). Springer-Verlag: Berlin; pp. 73–102.

Fontaine B, Janicot S. 1996. Sea surface temperature fields associated with West African rainfall anomaly types. *J. Climate* 9: 2935–2940.

Fontaine B, Phillipon N, Camberlin P. 1999. An improvement of June–September rainfall forecasting in the Sahel based upon region April–May moist static energy content. *Geophys. Res. Lett.* 26: 2041–2044.

Goddard L, Mason SJ. 2002. Sensitivity of seasonal climate forecasts to persisted SST anomalies. *Clim. Dyn.* 19, 619–631.

Haerter JO, Hagemann S, Moseley C, Piani C. 2011. Climate model bias correction and the role of timescales. *Hydrol. Earth Syst. Sci.* 15: 1065–1079.

Hagedorn R, Doblas-Reyes FJ, Palmer TN. 2005. The rationale behind the success of multi-model ensembles in seasonal forecasting – I. Basic concept. *Tellus A*, 57: 219–233.

Hulme M. 1994. Validation of large-scale precipitation fields in general circulation models. In: *Global Precipitation and Climate Change*, M Desbois, F Desalmand (eds). NATO ASI Series, vol. 126. Springer-Verlag: Berlin.

Janicot S, Harzallah S, Fontaine B, Moron V. 1998. West African Monsoon dynamics and eastern equatorial Atlantic and Pacific SST anomalies (1970–88). *J. Climate* 11: 1874–1882.

Janowiak JE, Xie P. 1999. CAMS_OPI: a global satellite-rain gauge merged product for real-time precipitation monitoring applications. *J. Climate* 12: 3335–3342.

Korecha D, Barnston AG. 2007. Predictability of June–September rainfall in Ethiopia. *Mon. Weather Rev.* 135: 628–650. doi: 10.1175/MWR3304.1.

Lamb PJ. 1978. Large scale surface circulation patterns associated with Subsaharan weather anomalies. *Tellus* 30: 240–251.

Lorenz EN. 1963. Deterministic non-periodic flow. *J.Atmos. Sci.* 20: 130–141.

Losada T, Rodriguez-Fonseca B, Janicot S, *et al.* 2010. A multi-model approach to the Atlantic equatorial mode: impact on the West African monsoon. *Clim. Dyn.* 35: 29–43.

Lough JM. 1986. Tropical Atlantic sea surface temperatures and rainfall variations in sub-Saharan Africa. *Mon. Weather Rev.* 114: 561–570.

Mason S, Chidzambwa S. 2009. Position Paper: Verification of African RCOF forecasts. *IRI Technical Report 09-02*. International Research Institute for Climate and Society: Palisades, NY.

Mitchell TD, Jones PD. 2005. An improved method of constructing a database of monthly climate observations and associated high resolution grids. *Int. J. Climatol.* 25: 693–712.

Namias J. 1972. Space scales of sea-surface temperature patterns and their causes. *Fish. Bull.* 70: 611–617.

Ndiaye O, Goddard L, Ward MN. 2009. Using regional wind fields to improve general circulation model forecasts of July–September Sahel rainfall. *Int. J. Climatol.* 29: 1262–1275.

Ndiaye O, Ward MN, Thiaw W. 2011. Predictability of seasonal Sahel rainfall using GCMs and lead-time improvements through the use of a coupled model. *J. Climate* 24: 1931–1949.

Nicholson SE. 1986. The spatial coherence of African rainfall anomalies: interhemispheric teleconnections. *J. Clim. Appl. Meteorol.* 25: 1365–1381.

Nicholson SE. 2000. Land surface processes and Sahel climate. *Rev. Geophys.* 38: 117–140.

Novella NS, Thiaw WM. 2013. African Rainfall Climatology Version 2 for famine early warning systems. *J. Appl. Meteorol. Climatol.* 52: 588–606.

Palmer TN. 1986. Influence of Atlantic, Pacific and Indian Oceans on Sahel rainfall. *Nature* 322: 251–253.

Ropelewski CF, Halpert MS. 1987. Global and regional scale precipitation patterns associated with the El Nino/ Southern Oscillation. *Mon. Weather Rev.* 115: 1606–1628.

Ropelewski CF, Halpert MS. 1989. Precipitation patterns associated with the high index phase of the Southern Oscillation. *J. Climate* 2: 268–284.

Rowell DP. 2001. Teleconnections between the tropical Pacific and the Sahel. *Q. J. R. Meteorol. Soc.* 127(125): 1683–1706.

Rowell DP. 2003. The impact of Mediterranean SSTs on the Sahelian rainfall season. *J. Climate* 16: 849–862.

Rowell DP. 2013. Simulating large-scale teleconnections to Africa: what is the state of the art? *J. Climate* 26(15): 5397–5418.

Rowell DP, Folland CK, Maskell K, Ward MN. 1995. Variability of summer rainfall over tropical North Africa (1906–92): observations and modelling. *Q. J. R. Meteorol. Soc.* 121: 669–704.

Stanski HR, Wilson LJ, Burrows WR. 1989. Survey of common verification methods in meteorology. World Weather Watch Technical Report No. WMO/TD 358. WMO: Geneva.

Suárez-Moreno R, Rodríguez-Fonseca B. 2015. S4CAST v2.0: sea surface temperature based statistical seasonal forecast model. *Geosci. Model Dev. Discuss.* 8: 3971–4018. doi: 10.5194/gmdd-8-3971-2015.

Thiaw WM, Mo KC. 2005. Impact of sea surface temperature and soil moisture on seasonal rainfall prediction over the Sahel. *J. Climate* 18: 5330–5343.

Tippett MK, Giannini A. 2006. Potentially predictable components of African summer rainfall in a SST-forced GCM simulation. *J. Climate* 19: 3133–3144.

Vizy EK, Cook KH. 2001. Mechanisms by which Gulf of Guinea and eastern North Atlantic sea surface temperature anomalies can influence African rainfall. *J. Climate* 14: 795–821.

Ward MN. 1993. Tropical North African rainfall and worldwide monthly to multi-decadal climate variations. Directed towards the development of a corrected ship win dataset and improved diagnosis, understanding and prediction of North African rainfall. PhD thesis, sponsored by Reading University and the Met Office.

Ward MN. 1994. Tropical North-African rainfall and worldwide monthly to multi-decadal climate variations. PhD thesis, University of Reading; 313 pp.

Ward MN. 1998. Diagnosis and short-lead time prediction of summer rainfall in tropical North Africa at interannual and multidecadal timescales. *J. Climate* 11: 3167–3191.

Ward MN, Navarra A. 1997. Pattern analysis of SST-forced variability in ensemble GCM simulations: examples over Europe and the tropical Pacific. *J. Climate* 10: 2210–2220.

Ward MN, Folland CK, Maskell K, *et al.* 1993. Experimental seasonal forecasting of tropical rainfall at the Meteorological Office. In: *Prediction of Interannual Climate Variations*, J Shukla (ed.). NATO ASI Series, vol. 16. Springer; pp. 197–216.

Weisheimer A, Doblas-Reyes FJ, Palmer TN, *et al.* 2009. ENSEMBLES: a new multi-model ensemble for seasonal-to-annual predictions – skill and progress beyond DEMETER in forecasting tropical Pacific SSTs. *Geophys. Res. Lett.* 36: L21711.

Xie P, Arkin PA. 1998. Global monthly precipitation estimates from satellite-observed outgoing longwave radiation. *J. Climate* 11: 137–164.

Yaka P, Sultan B, Brotin H, *et al.* 2008. Relationships between climate and year-to-year variability in meningitis outbreaks: a case study in Burkina Faso and Niger. *Int. J. Health Geogr.* 7: 34. doi: 10.1186/1476-072X-7-34.

# 9

# Remote Sensing

**Lead author:** *Arlene Laing[a]*

**Contributing author:** *Hama Hamidou[b]*

**Reviewers:** *Mariane Diop-Kane (Agence Nationale de l'Aviation Civile et de la Météorologie du Sénégal–ANACIM, Sénégal) and Tony Wardle (UK Met Office)*

[a] Cooperative Institute for Research in the Atmosphere (CIRA), Colorado State University, Fort Collins, CO, USA
[b] L'Ecole Africaine de la Météorolgie et de l'Aviation Civile (EAMAC), Niamey, Niger

## 9.1 Scientific Background

Remote sensing offers remarkable possibilities for the analysis of weather patterns over tropical Africa. On a continent where the ground-based observing networks have substantial problems, the spatial and temporal coverage provided by satellite data has, over recent decades, transformed the level of information available to a forecaster.

While in other parts of the world ground-based radar networks have become the cornerstone of real-time analysis of storms and rainfall, efforts to implement operational radar networks in West Africa have generally not met with much success. However, given the desire among the meteorological community in Africa to develop rainfall radar networks in future, as well as the existence of useful space-borne weather radar data, there is a need for forecasters to understand the basic principles of meteorological radar observation.

New satellites and sensors are constantly in development, and the range of datasets available can be bewildering. Here we summarise the important datasets available, and conversely we describe those remote sensing data that are best applied to each of the key weather phenomena for West Africa, along with examples of their application to real case studies from West Africa. Through training based on the materials referred to in this chapter, learners should be able to:

- Explain the basic principles of remote sensing by environmental satellites and radar.
- Describe the different platforms, sensors and techniques used to retrieve information about meteorological parameters useful for forecasting in West Africa.
- Describe how single-channel and multispectral products are used in analysing and forecasting key meteorological phenomena in West Africa.

### 9.1.1 The Global Satellite Observing System

The global satellite observing system (Figure 9.1) is vital to forecasting in Africa, particularly because of the scarcity of observations from the surface and upper-air station networks. The network has been very valuable for numerical weather prediction. Satellite measurements are commonly used in operational meteorology to monitor and analyse the atmosphere and surface, including:

- surface temperature and reflectance
- water vapour
- clouds and precipitation
- vertical profiles of temperature and humidity
- wind velocity (atmosphere and ocean surface)
- dust, smoke and other aerosols
- lightning.

#### 9.1.1.1 Orbital Geometry

Meteorological satellites are categorised by their orbits (Figure 9.2):

- Geostationary or geosynchronous, which orbit at about 35 800 km above the Equator and move at the velocity of the Earth's rotation, therefore observing above a fixed point on the Earth itself.
- Low-Earth-orbit (LEO) satellites, which are:
  - Polar-orbiting Environmental Satellites (POES) that orbit around the poles at approximately 850 km above the surface; most are Sun-synchronous.

*Meteorology of Tropical West Africa: The Forecasters' Handbook*, First Edition. Edited by Douglas J. Parker and Mariane Diop-Kane.

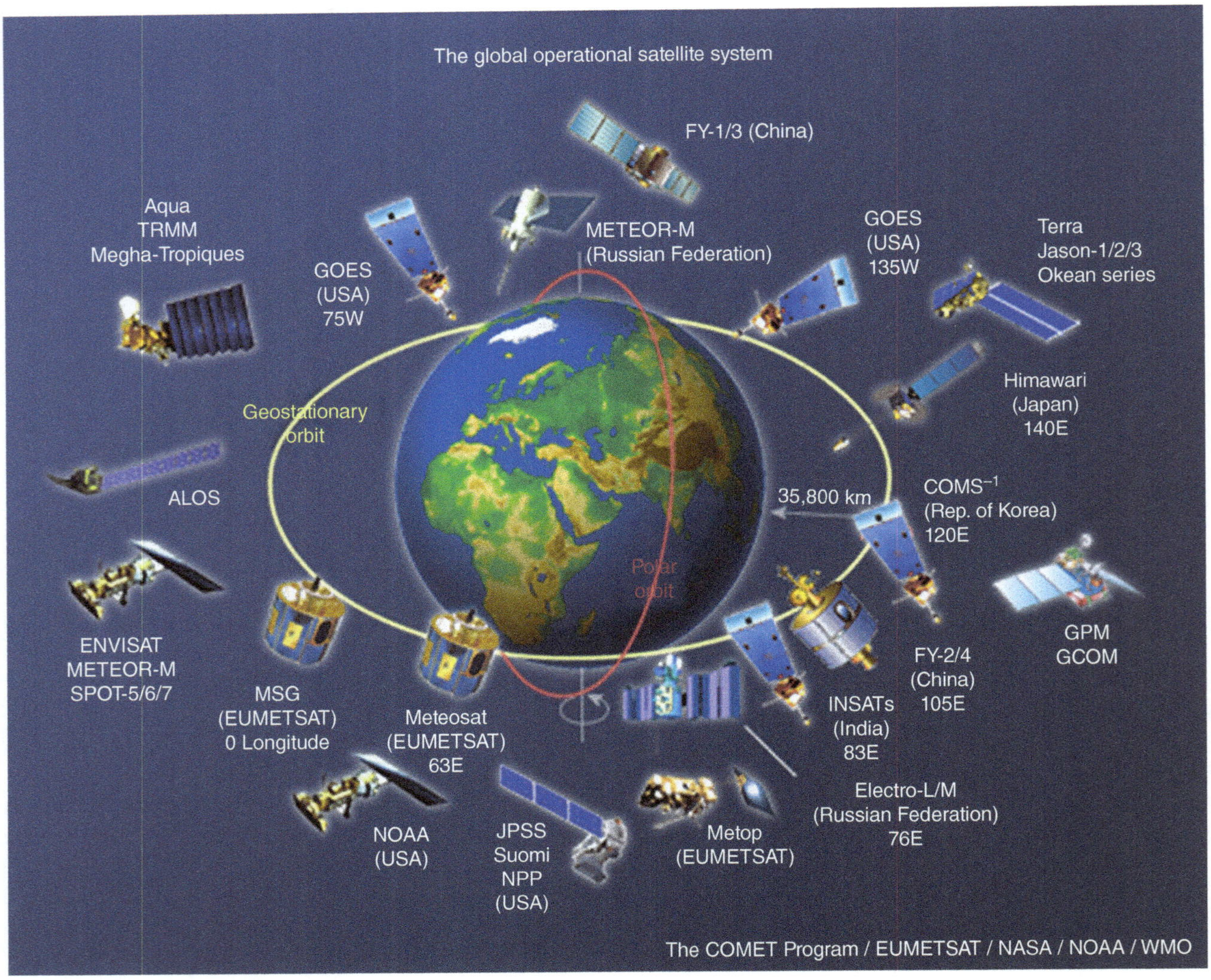


**Figure 9.1** The global operational satellite observing system (World Meteorological Organization, 2005). *Source*: reproduced with permission of the COMET Program.

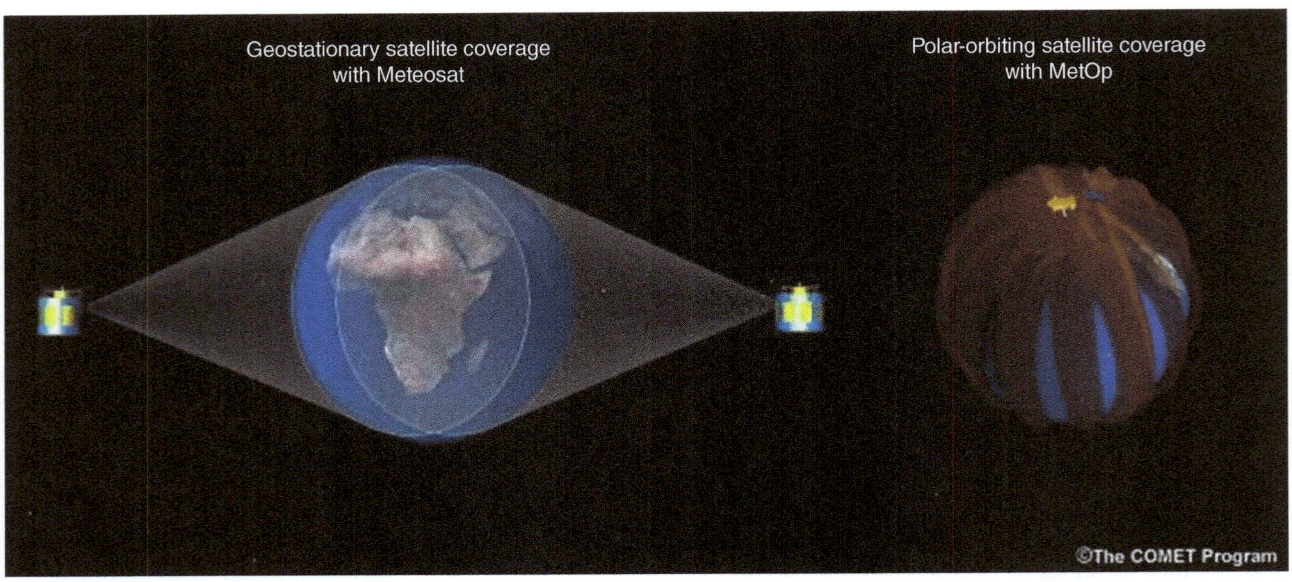


**Figure 9.2** EUMETSAT geostationary Meteosat coverage and EUMETSAT Polar System (EPS) orbiting coverage with MetOp. *Source*: reproduced with permission of the COMET Program.

– Research and development satellites that orbit between certain latitudes a few hundred kilometres above the surface. For example, the Tropical Rainfall Measurement Mission (TRMM; Simpson *et al.*, 1996) satellite orbit is circular and non-Sun-synchronous. It orbits between 35°N and 35°S at 402.5 km above the surface. The Megha-Tropiques satellite, launched in 2011, orbits between 20°N and 20°S at 866 km and the Global Precipitation Mission (GPM) Core, launched in 2014, orbits between 65°N and 65°S. Some research satellites, like the TRMM, have been used for weather forecasting.

#### 9.1.1.2 Advantages and Limitations of Geosynchronous and Low-earth-orbit Satellites

Geostationary satellites, such as Meteosat, have a wide field of view and lower spatial resolution but have the advantage of high temporal resolution and continuous observations of the same location, and so can be used to track weather systems over great distances.

LEO satellites, such as Terra, MetOp, GPM Core and TRMM, have the advantage of a high spatial resolution over a narrow field of view but poor temporal resolution as their orbits rarely pass over the same location more than once per day. While polar orbiting satellites provide good coverage of the poles, individual POES orbits leave wide gaps in the tropics (e.g, MetOp orbit in Figure 9.2). Having a constellation of satellites with similar instruments diminishes those gaps. In a given day, the combination of geosynchronous and LEO satellites provides global coverage of significant meteorological phenomena.

#### 9.1.1.3 Scanning and Viewing Geometry

As a satellite instrument scans, its field of view or footprint on the Earth's surface varies. The nadir view, at the sub-satellite point, has the smallest footprint and maximum resolution, while the limb view, at its maximum viewing angle, has the largest footprint and minimum resolution (e.g. Figure 9.3). Satellite instruments commonly scan in three ways: across their orbit track (cross-track, Figure 9.3), along an arc (conical) or like a 'push broom'.

### 9.1.2 Scientific Basis of Remote Sensing

The transfer of electromagnetic (EM) radiation by the Earth–atmosphere system is the basis for remote sensing applications. This section is a summary description of EM radiation, how it interacts with the Earth and atmosphere, and how the radiation measured by remote sensors is used to infer information about the atmosphere and the surface.

#### 9.1.2.1 Electromagnetic Spectrum

EM waves are described by the following fundamental quantities:

- Wavelength $\lambda$, the distance between crests of an electric or magnetic field (SI unit: metre; common unit,: micrometre).
- Wavenumber $\kappa$, the reciprocal of the wavelength, $\kappa = 1/\lambda$ (SI unit: 1/metre; common unit: 1/centimetre).
- Frequency $\nu$, the rate of oscillation of radiation at a point, $\nu = c/\lambda$ (SI unit: hertz), where $c$ is the speed of light in a vacuum ($2.99792458 \times 10^8$ m s$^{-1}$).

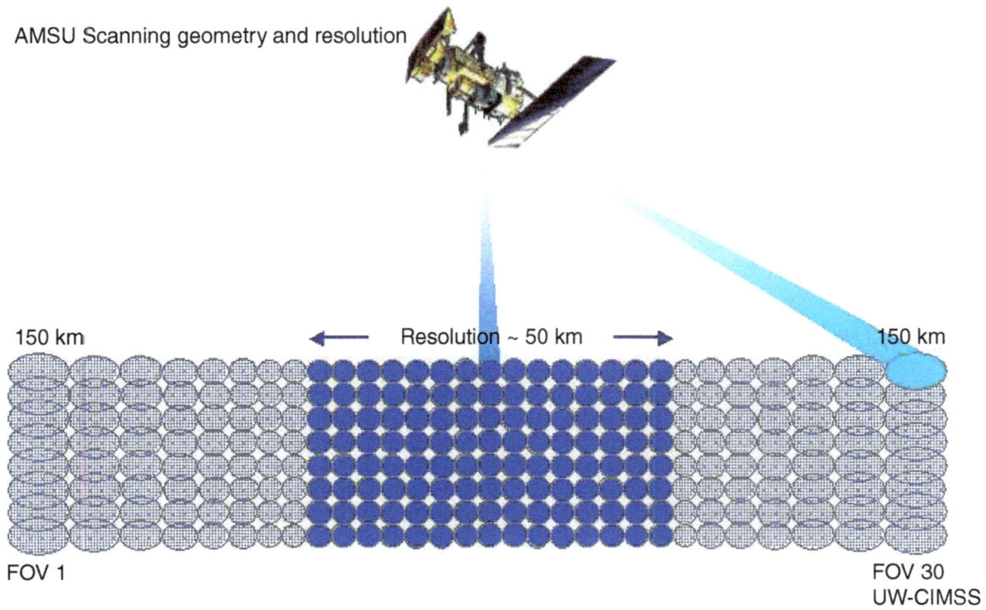


**Figure 9.3** Advanced Microwave Sounding Unit (AMSU) scanning geometry and resolution. As the instrument scans away from nadir, its field of view (FOV) is larger and more skewed. *Source*: courtesy of Cooperative Institute for Meteorological Satellite Studies CIMSS, University of Wisconsin–Madison.

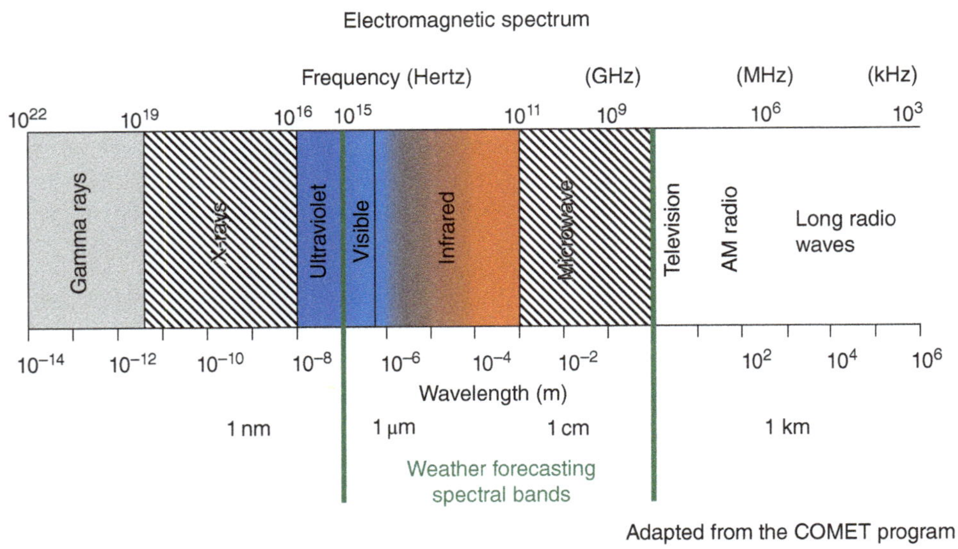


**Figure 9.4** EM spectrum from gamma rays to radio waves showing scales in wavelength and frequency. *Source*: reproduced with permission of the COMET Program.

Wavelength increases from left to right on the EM spectrum (Figure 9.4). Forecast applications mainly utilise the visible to microwave part of the spectrum.

#### 9.1.2.2 Radiance and Blackbody Radiation

Radiation can be described in terms of particles of radiant energy, called photons. The energy per photon emitted is

$$E = h\nu \tag{9.1}$$

where $h$ is Planck's constant ($6.625 \times 10^{-34}$ J s) and $\nu$ (Hz) is the frequency, which is inversely proportional to the wavelength.

Radiant flux or power measures the rate of transfer of radiant energy (units of joules per second, or watts), which when normalised by area gives the radiant flux density or irradiance (units of watts per square metre).

A satellite telescope has a field of view that is most naturally measured by a solid angle (Figure 9.5). Imagine a beam of energy travelling in a given direction $\Omega$, and intersecting an infinitesimal cross-sectional area perpendicular to the direction of the beam. The *radiance* or *specific intensity* is the energy crossing that unit area, per unit time, per unit solid angle in the given direction, per unit wavelength (or frequency or wavenumber), expressed as

$$dI_\lambda = \frac{dE_\lambda}{\cos\theta \ d\Omega \ dt \ dA \ d\lambda} \tag{9.2}$$

where $E$ is radiant energy, $A$ is cross-section area, $t$ is time, solid angle is $\Omega$ in steradians and $\lambda$ is wavelength. Since radiance measures radiation perpendicular to the beam, $\cos\theta$ weights the energy from other directions. Radiance has SI units of W m$^{-2}$ sr$^{-1}$ µm$^{-1}$.

*Monochromatic* or spectral radiance is the most fundamental unit measured by satellites, where monochromatic refers to a very narrow band $d\lambda$ centred on a given $\lambda$. Note that, as a satellite instrument scans, its solid angle remains constant but its field of view varies (Figure 9.3).

##### 9.1.2.2.1 *Blackbody Radiation*

A useful concept in radiation is a blackbody, an object that absorbs all incident radiation and emits the maximum amount of energy at all wavelengths and emits equally in all directions (isotropically). *Planck's law* expresses the energy emitted by a blackbody as a function of temperature and wavelength:

$$B_\lambda(T) = \frac{c_1\lambda^{-5}}{\exp(c_2/\lambda T) - 1} \tag{9.3}$$

where $\lambda$ is wavelength, $T$ (K) is absolute temperature, $c_1 = 1.1910439 \times 10^{-16}$ W m$^{-2}$ sr$^{-1}$ and $c_2 = 1.438769 \times 10^{-2}$ mK.

Integrating the Planck function (Equation 9.3) over all wavelengths gives us the blackbody flux density or irradiance, which is represented by the area under the blackbody curves in Figure 9.6. The energy emitted per unit area (from all wavelengths) is related to the absolute temperature through the Stefan–Boltzmann law:

$$E = \sigma T^4 \tag{9.4}$$

where the Stefan–Boltzmann constant $\sigma = 5.67 \times 10^{-8}$ W m$^{-2}$ K$^{-4}$.

The wavelength of maximum blackbody emission is inversely proportional to its absolute temperature; that is, the hotter the object the shorter the peak wavelength at which it emits most of its energy, expressed as Wien's displacement law:

$$\lambda_{max} = \frac{2897.9}{T} \tag{9.5}$$

Figure 9.5 Conceptual diagram of the spectral radiance or specific intensity of radiation.

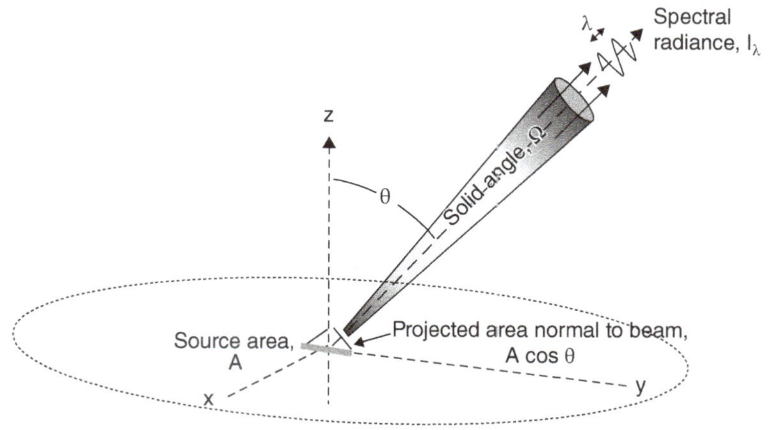


Figure 9.6 The radiance (W $m^{-2}$ $sr^{-1}$ $\mu m^{-1}$) normalised per wavelength on a logarithmic scale. The areas under the yellow and red lines represent solar and terrestrial radiance respectively. *Source*: reproduced with permission of the COMET Program.

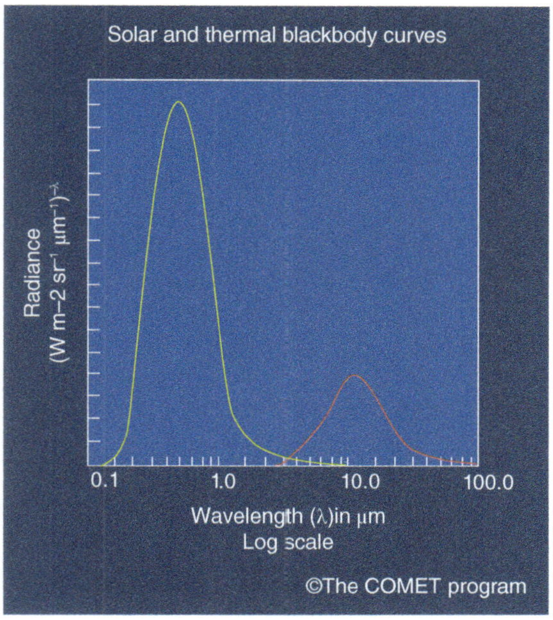


Therefore, visible wavelengths (shortwave) have much higher peak energy than infrared (IR) wavelengths (longwave) (Figure 9.6). As wavelength decreases, the radiance increases more rapidly with temperature (Figure 9.7). This characteristic is the basis for detecting fires and other hot areas using the shortwave IR channel, centred at 3.9 µm. Its radiance response to temperature increase is an order of magnitude greater than the response at longer IR wavelengths. It is also the peak wavelength of a blackbody at 743 K (Figure 9.7), which approximates the temperature of a forest fire.

#### 9.1.2.2.2 *Brightness Temperature*

The brightness temperature *T* is determined by inverting the Planck function. At microwave wavelengths (millimetres, centimetres) and for the Earth and atmospheric temperature range, radiance to temperature is a simple proportion. While Planck's law of blackbody radiation allows us to equate spectral radiance with the brightness temperature of an emitting object as if it were a blackbody, most objects are not blackbodies. At the IR window near 10.8 µm, the brightness temperature is close to the physical temperature. However, at microwave wavelengths, the brightness temperature is much less than the physical temperature of the emitting object. The satellite viewing angle affects the observed brightness temperature. As the instrument scans further away from nadir, it senses a layer that is thinner, higher in the troposphere, and has a longer path length. Therefore, the near-limb areas of IR images appear cooler, a phenomenon known as 'limb darkening' or 'limb cooling'.

### 9.1.2.3 Radiative Transfer

A beam of radiation passing through the atmosphere can be changed by

- absorption
- emission
- scattering out of the beam, and
- scattering into the beam of radiation.

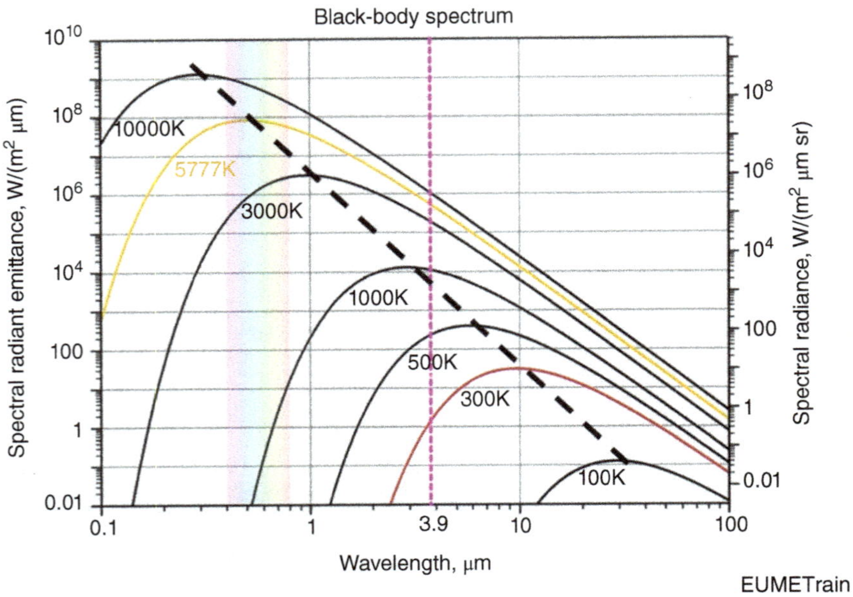


**Figure 9.7** Log-scale plot of wavelength and spectral radiance and emittance for various blackbody temperatures. Coloured annotations mark the emissions of the Sun (5777 K, centred on 0.5 µm, visible spectrum), the Earth's surface (300 K, centred on 10 µm) and forest fires (600–1000 K, centred on 3.9 µm). The dashed diagonal line runs through the wavelengths of peak emission, illustrating Wien's displacement law. *Source*: reproduced with permission of EUMETrain.

For example, water vapour is a major absorber of long-wave radiation in the atmosphere, snow scatters visible light, and ice particles scatter microwave radiation.

The atmosphere is largely transparent to visible and microwave radiation and opaque to IR radiation except for a small window close to 10 µm (Figure 9.8). The optical depth or optical thickness of a medium is a measure of the amount of radiation removed from a beam of radiation by absorption or scattering along its path through that medium. The atmosphere is said to be optically thick where little radiation is transmitted and optically thin when most of the radiation is transmitted.

Since gases do not behave exactly as blackbodies, we define measures such as the (monochromatic) emissivity $\varepsilon_\lambda$, the ratio of the emitted intensity to the corresponding blackbody radiation at a given wavelength:

$$\varepsilon_\lambda = \frac{I_\lambda}{B_\lambda(T)} \tag{9.6}$$

The absorptivity $\alpha_\lambda$, transmissivity $\tau_\lambda$ and reflectivity $r_\lambda$ measure the fraction of incident radiation that is absorbed, transmitted and reflected respectively. According to Kirchhoff's law, for a given wavelength, emissivity is equal to absorptivity.

The rate of change of radiance for a given wavelength $I_\lambda$ over distance $s$ is a sum of the extinction (absorption and/or scattering) and sources (emission and/or scattering into the beam).

This rate of change of radiance may be written as the *radiative transfer equation*:

$$\frac{dI_\lambda}{ds} = -k_\lambda \rho r I_\lambda + j_\lambda \rho \tag{9.7}$$

Here, $k_\lambda$ is the extinction cross-section (in units of area per mass) for wavelength $\lambda$, $j_\lambda$ is the source function coefficient, $\rho$ is the density of the medium and $r$ is the mass of the absorbing gas per unit mass of air. The coefficients measure the efficiency of molecules in absorbing or scattering radiation at wavelength $\lambda$. Figure 9.9 is a simple schematic of the change of radiance over a small distance.

Satellite instruments sense radiation from the surface, and from each atmospheric layer that is transmitted to the layer above. At IR wavelengths we focus on absorption and emission only for satellite sensing of the Earth and atmosphere because scattering is negligible. Then the rate of change of radiation along the path to a satellite sensor is

$$dI_\lambda(\text{absorption}) - dI_\lambda(\text{emission})$$

$$dI_\lambda = -k_\lambda \rho r I_\lambda \, ds - B_\lambda(T)\varepsilon_\lambda$$

This is known as Schwarzschild's equation, written simpler as:

$$dI_\lambda = -\left(I_\lambda - B_\lambda(T)\right)k_\lambda \rho r ds \tag{9.8}$$

When integrated, Equation 9.8 gives the intensity of radiation reaching a given atmospheric layer or satellite orbit, as illustrated schematically in Figure 9.9:

$$I_\lambda(s_i) = I_\lambda e^{-\tau_\lambda(s_i,0)} + \int_0^{s_i} k_\lambda \rho r B_\lambda\left[T(s)\right]e^{-\tau_\lambda(s_i,s)} ds \tag{9.9}$$

where $\tau_\lambda$ represents the transmissivity of the radiation along different sections of the path.

Remote sensors measure radiation transmitted through and by the atmosphere. A detector, usually a radiometer, records an EM signal after it interacts with target molecules, particulates or surfaces. If $T$ and $S$ denote the target and signal, we can represent their relationship by

$$S = F(T) \quad \text{and its inverse} \quad T = F^{-1}(S) \tag{9.10}$$

where $F$ is a function that governs radiative transfer.

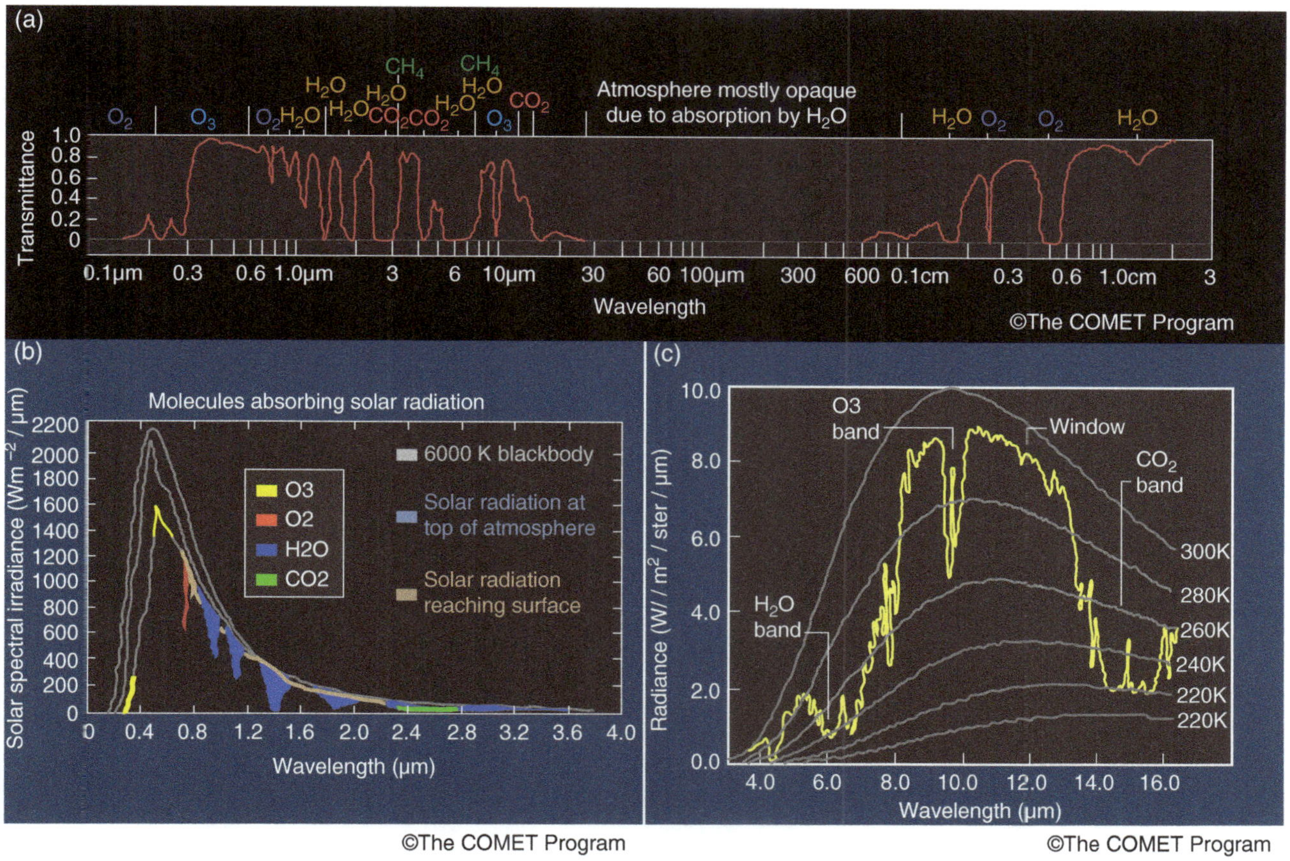


**Figure 9.8** (a) Transmittance, a measure of the fraction of radiation passing through the atmosphere, from ultraviolet to microwave wavelengths. The higher the red line the greater the amount of radiation transmitted through the atmosphere. Wavelengths with transmittance at or near 1.0 are referred to as atmospheric windows. (b) Solar blackbody irradiance, amount at the top of the atmosphere, and amount at the surface after atmospheric absorption. (c) Terrestrial blackbody radiance for different temperatures and the reduced surface radiance transmitted because of atmospheric absorption. *Source*: reproduced with permission of the COMET Program.

**Figure 9.9** (a) Schematic diagram of the change of radiance (for a given wavelength) over a distance *s* through the atmosphere. (b) Transmittance and emission from a layer in a plane-parallel atmosphere.

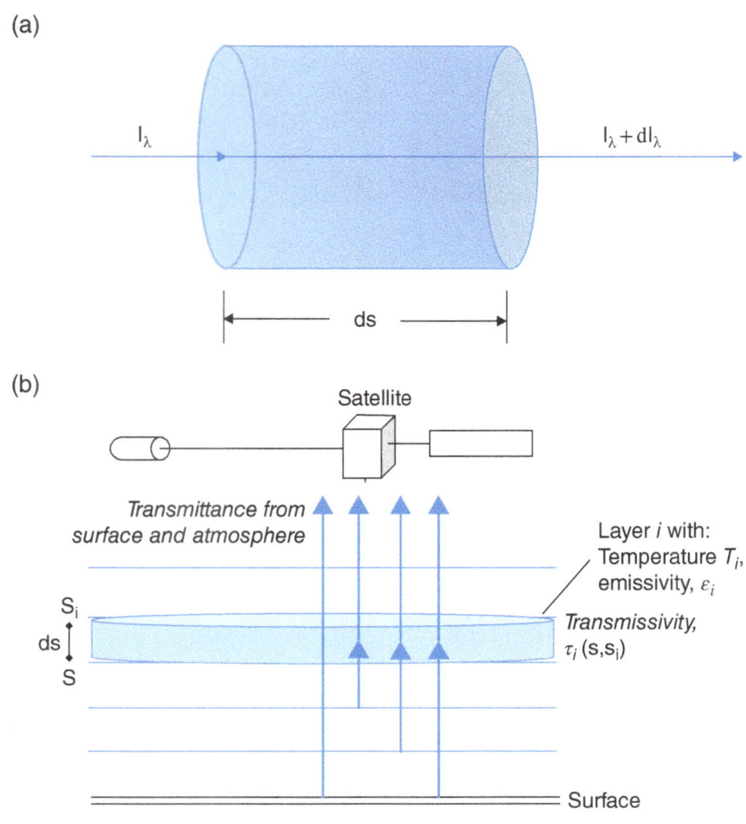

While radiative transfer is characterised by a specific wavelength and its sensitivity to some physical aspect of the transmitting medium, the inversion solution may not be unique. A combination of unknown parameters can lead to the same radiance measurement. This *non-uniqueness creates an inherent uncertainty in remotely sensed measurements.* Errors also occur because the radiation transfer function is not necessarily linear and may not exist in a functional form, and in the mathematical solutions used. For further information about radiative transfer and satellite remote sensing, see Kidder and Vonder Haar (1995) and Liou (2002).

#### 9.1.2.4 Reflectance

Reflected radiation, especially reflected solar (Figure 9.10a) and microwave radiation, is a critical measurement in satellite meteorology. Reflected radiance is used mainly for the daytime identification of cloud, especially lower

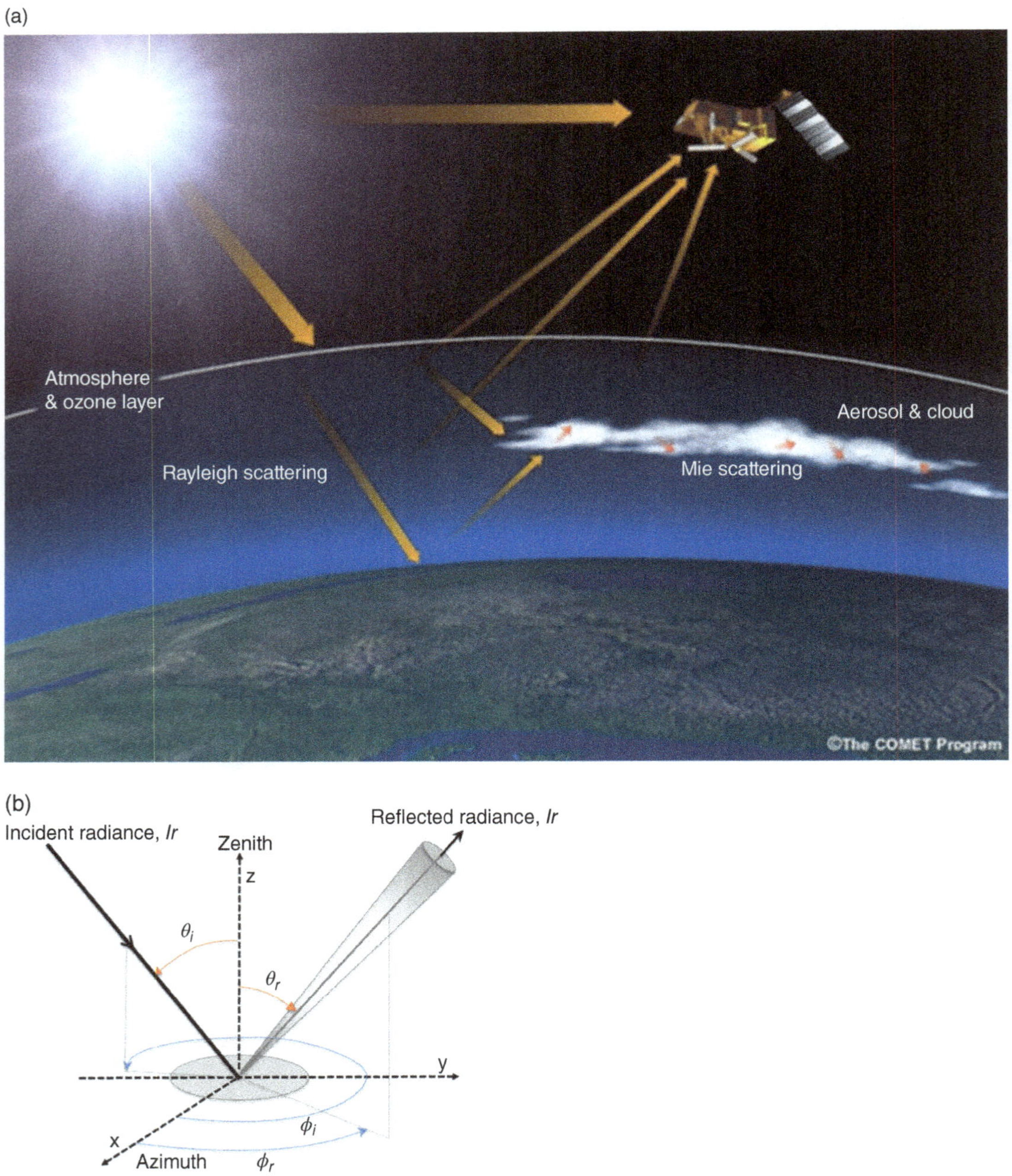


**Figure 9.10** (a) Schematic of satellite sensing direct sunlight and light scattered by the atmosphere and surface. *Source*: Reproduced with permission of the COMET Program. (b) Conceptual diagram of the incident radiance $I_i$ and reflected radiance $I_r$. Both the incident and reflected quantities are dependent on the zenith angle $\theta$ and the azimuth angle $\phi$. The fraction of the incident radiation that is reflected is termed the *reflectance* and varies according to the medium on which the radiation is incident.

and middle tropopause, but may be used in combination with other remotely sensed parameters to distinguish surface features, fires, smoke, dust, fog and snow.

Radiance reflected from a small unit of surface, shown conceptually in Figure 9.10b, can be expressed as

$$I_r(\theta_r,\phi_r)=\int_0^{2\pi}\int_0^{\pi/2} I_i(\theta_i,\phi_i)\gamma_r(\theta_r,\phi_r;\theta_i,\phi_i)\cos\theta_i \sin\theta_i \, d\theta_i \, d\phi_i \quad (9.11)$$

where $\theta$ is the zenith angle $\phi$ is the azimuth angle. The incident radiance is $I_i$ (zenith/azimuth), and the reflected radiance is $I_r$ (zenith/azimuth). The bidirectional reflectance $\gamma$ is related to the fraction of incident radiation that is reflected in the zenith and azimuthal directions.

The reflectance is used to identify various features and atmospheric phenomena (Figure 9.11). Notice the high reflectance of clouds at visible wavelengths and the high reflectance of fire and hot areas at near-IR wavelengths, properties that helps to distinguish among them.

#### 9.1.2.5 Passive and Active Remote Sensing

Passive sensors measure radiation that is emitted or reflected naturally; for example, the Meteosat Spinning Enhanced Visible and Infrared Imager (SEVIRI). Active sensors emit pulses of energy (usually microwave) and measure backscattered energy; for example, CloudSat cloud radar and the TRMM precipitation radar. Radar will be covered Section 9.1.6.

### 9.1.3 Spectral Bands or Channels

#### 9.1.3.1 Visible and Infrared Channels

Satellite instruments measure radiance from particular wavelengths, referred to as spectral bands or 'channels'.

Channels are chosen based on the interactions of particular wavelengths with atmospheric and surface constituents; for example, absorption, emission and scattering (Figure 9.12). At 'window' channels, radiation passes through the atmosphere with little or no absorption.

Channels are usually numbered in increasing order from shortwave to longwave bands (e.g. Figure 9.13). Channels 1, 2, 3 and 12 rely on reflectance (solar radiation dominates), as shown in Figure 9.12. Channel 12, the high-resolution visible (HRV) is a broadband visible channel with a 1 km sampling interval to better monitor small-scale features. Channel 3, the near-IR band, also relies on absorption (e.g. differences between water and ice absorption). Channel 4, 3.9 μm, relies on both solar reflectance and terrestrial emission/absorption. Channels 4–11 rely on emission/absorption (terrestrial radiation dominates).

Each of the channels represents particular physical properties (Table 9.1) that are useful for differentiating among phenomena; for example, differences in reflectance for channels 1–3 help to distinguish between vegetation and soil (Figure 9.14).

#### 9.1.3.2 Microwave Channels

Microwave atmospheric window regions allow sensing of the surface through clouds and are used to distinguish land, ocean, clouds and precipitation. The high absorption regions are sensitive to gases such as water vapour and oxygen (Figure 9.15) and are used mainly for soundings. Both window and high-absorption regions are used to derive various products for identifying surface and cloud properties. Note that the energy per unit area in the microwave region is many orders of magnitude less

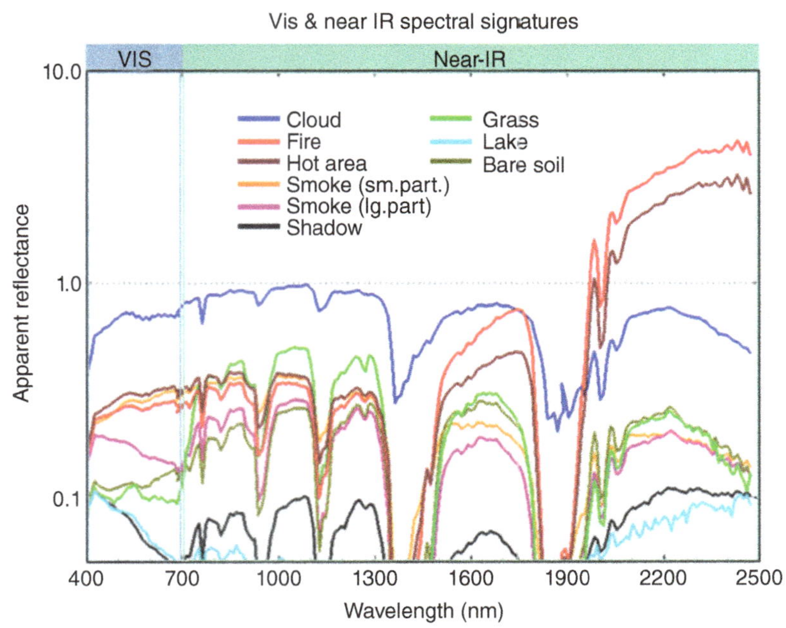


Figure 9.11 Apparent reflectance from various atmospheric phenomena and surface cover. Source: Griffin *et al.* (2000).

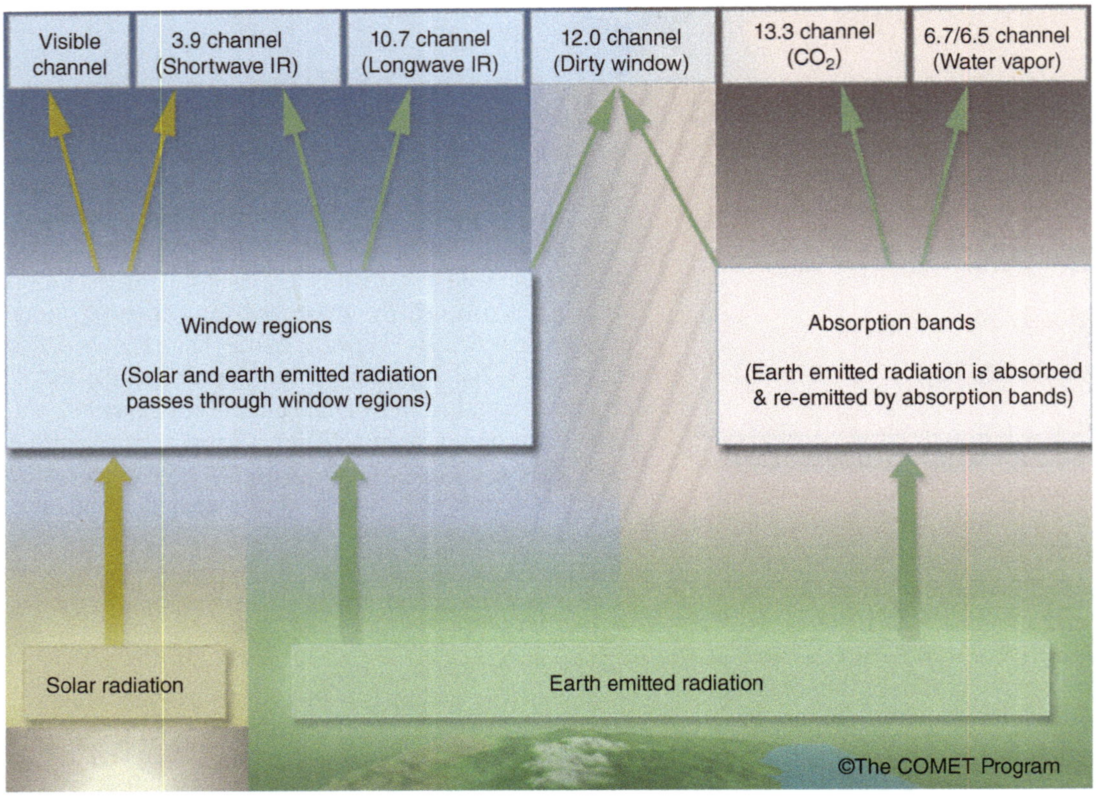


Figure 9.12 Schematic of channels that represent window regions and absorption bands on the Meteosat SEVIRI. *Source*: reproduced with permission of the COMET Program.

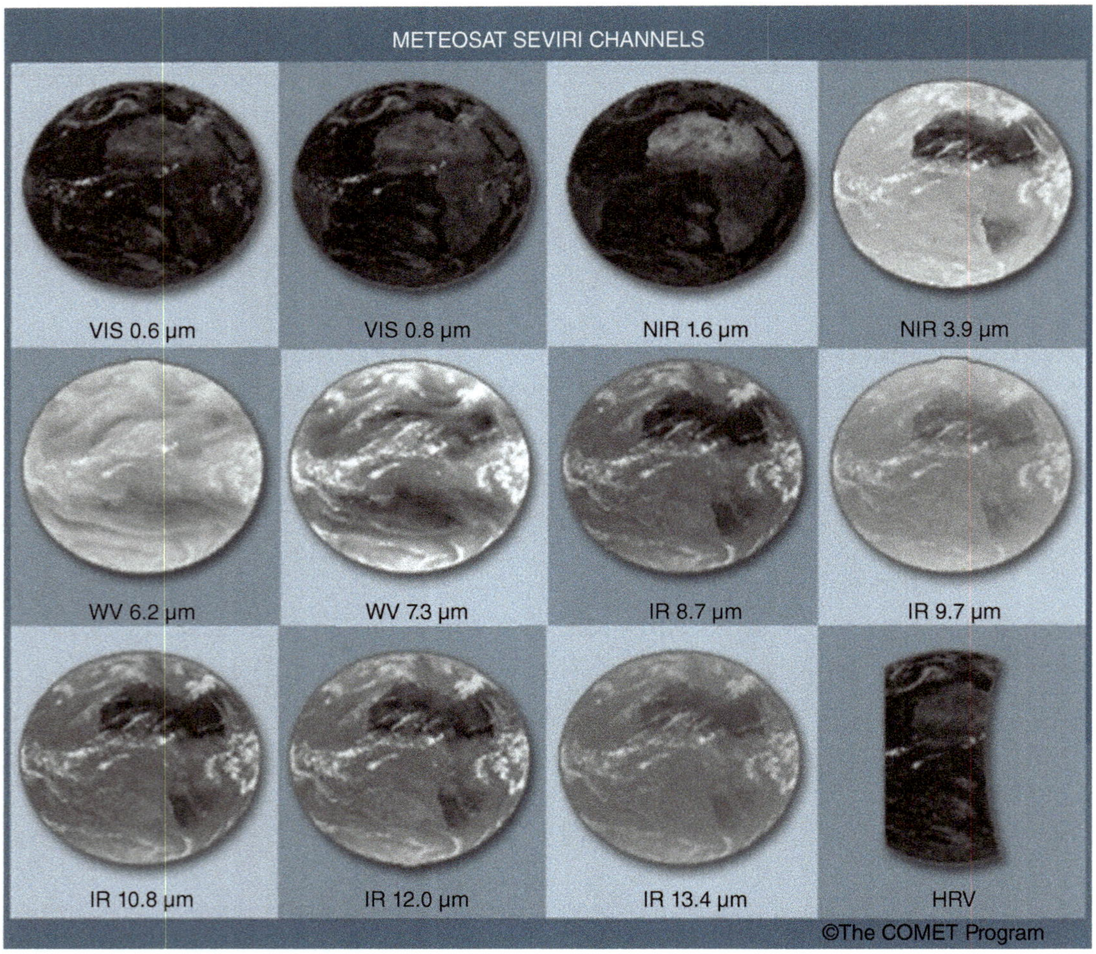


Figure 9.13 The channels on the Meteosat Second Generation (MSG) SEVIRI instrument. *Source*: reproduced with permission of EUMETSAT/The COMET Program.

Table 9.1 Physical properties represented by the Meteosat SEVIRI channels (EUMETSAT).

| Channels | Represented physical properties |
|---|---|
| VIS0.6 | Optical thickness and amount of cloud water and ice |
| VIS0.8 | Optical thickness and amount of cloud water and ice, 'greenness' of vegetation |
| NIR1.6, IR3.9 | Particle size and phase of water |
| WV6.2, WV7.3 | Mid- and upper-level moisture |
| IR8.7, IR10.8, IR12.0 | Cloud top temperature |
| IR8.7–IR10.8 | Phase of water and optical thickness |
| IR12.0–IR10.8 | Optical thickness |
| IR3.9–IR10.8 | Optical thickness, phase of water, particle size |
| IR13.4–IR10.8 | Cloud top height |
| WV6.2–IR10.8 | Cloud top height, overshooting tops |

than in the IR and visible range. This means that sensors of microwave radiation need large fields of view, greater than 10 km, to collect enough energy for a measurement. Therefore, microwave sensors are carried by LEO satellites. Microwave satellite channels are usually referred to by their frequency, with the most commonly used being the 85–89 GHz and 37 GHz channels (window regions).

### 9.1.4 Multispectral Analysis: Visualised Products

The number of spectral channels on geostationary and polar orbiting sensors continues to expand. The increasing amounts of data present an opportunity to expand feature identification, but also increase the challenge of distilling the information so that it is easy for forecasters to interpret. Features are identified from multiple channels by:

- classification of individual channel brightness temperature and/or reflectance according to an appropriate colour scale or shades of grey;

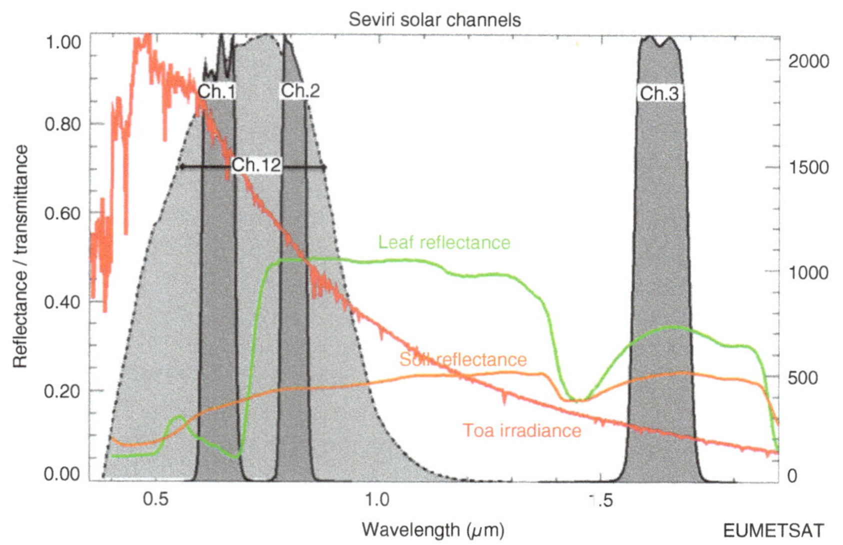


Figure 9.14 Reflectance from leaves (green curve) and soil (orange curve) in the visible and near-IR region of the spectrum, as sensed by four SEVIRI channels: 0.6, 0.8 and 1.6 μm (dark grey) and HRV (light grey). The red line is the irradiance (W m$^{-2}$) at the top of the atmosphere. Notice that vegetation is strongly reflective at longer visible wavelengths, while bare soil has high reflectance near the centre of the visible spectrum. *Source*: reproduced with permission of EUMETSAT.

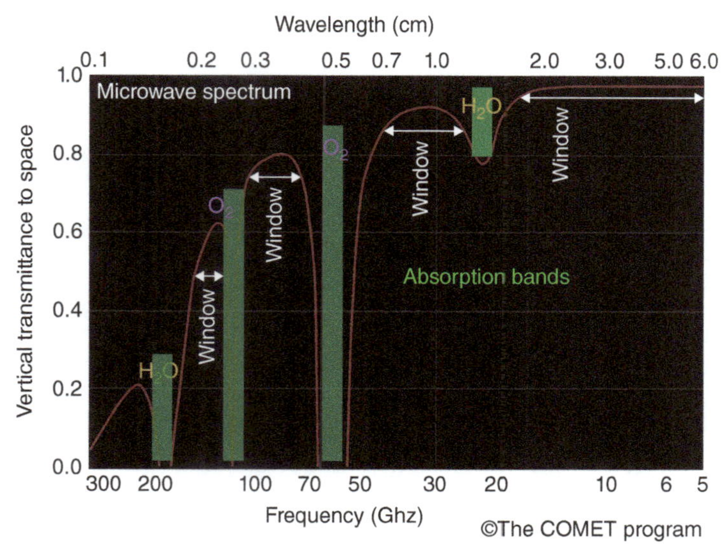


Figure 9.15 The window and absorption bands at microwave wavelengths. *Source*: reproduced with permission of the COMET Program.

- calculating differences or ratios of two or more channels;
- quantitative feature extraction via algorithms and a lookup table of discrete values that use thresholds or pattern recognition to identify specific features from multispectral images, and create a single-channel image;
- combining channels to create colour composite images based on the primary colours of red, green, and blue (RGB);
- creating blended or 'sandwich' products by combining two input images into one output image by a method of 'layer blending', an advanced function in some graphics software.

These methods are elaborated in the following subsections.

#### 9.1.4.1 Classification Using Individual Channels

With this method, brightness temperatures or reflected radiance are mapped to a lookup table of colours or grey scale with suitable brightness or contrast for individual channels. For instance, light shades are used for high reflectivity and IR images are usually inverted so that low brightness temperatures are in lighter shades, providing a more intuitive view of clouds. Typically, a given image will have 256 colours with ranges or values assigned to each colour or shade. Specific features are detected using colour or grey-scale enhancements and displays of individual channels. Features with similar reflectance in the visible are then differentiated using brightness temperature in the IR and water vapour images. However, not all distinctions can easily be made by this method; for example, dense cirrus has the same cloud top temperature as cumulonimbus; so, in differentiating between cloud types, the cloud form and movement have to be considered. This method is appropriate for imagers with only a few channels.

#### 9.1.4.2 Differencing or Ratio of Two Channels

The use of channel differences or ratios helps to enhance particular features of interest and is known as a 'split-window' or 'reverse absorption' technique. Channel differencing is used to detect phenomena such as snow, fog, fire and smoke, vegetation, volcanic ash and cloud types. Brightness temperature difference (BTD) is commonly used; for example, the BTD between 3.9 μm and 10.8 μm helps to identify fog and low stratus, except at dawn and dusk when the difference is not visible. BTD between 12.0 μm and 10.8 μm also provides a clear method of delineating between cirrus and dust, since cirrus absorbs more radiation at 12.0 μm, while dust absorbs more at 10.8 μm (Figure 9.16).

#### 9.1.4.3 Quantitative Feature Extraction

Quantitative extraction of satellite image features is done via algorithms and/or a lookup table of discrete values that use thresholds or pattern recognition to detect specific features from multispectral images, and create a single channel image (NWCSAF, 2012). For example, the Satellite Application Facilities Support to Nowcasting and Very Short Range Forecasting (SAFNWC) Cloud Mask algorithm identifies pixels that are cloud free (Figure 9.17), where other meteorological parameters can be retrieved (e.g. precipitable water, land or sea surface temperatures, dust, volcanic ash, snow/ice). It uses a multispectral threshold method to delineate between brightness temperatures of cloud-free pixels and pixels with clouds or aerosols. Thresholds are tuned according to the type of phenomena, atmospheric conditions, and the positions of the Sun and satellite. The Cloud Top Temperature and Height algorithm is used in nowcasting applications, especially aviation forecasting and early warning of thunderstorm development. This method uses the brightness temperature from various channels to compute the physical temperature of *opaque* cloud tops and retrieve the corresponding height from numerical weather prediction. This type of quantitative processing is also used for ground validation of satellite image products.

#### 9.1.4.4 RGB Products

RGB compositing is a commonly used method for integrating multispectral channels into a single image. Different channels are combined to highlight features in

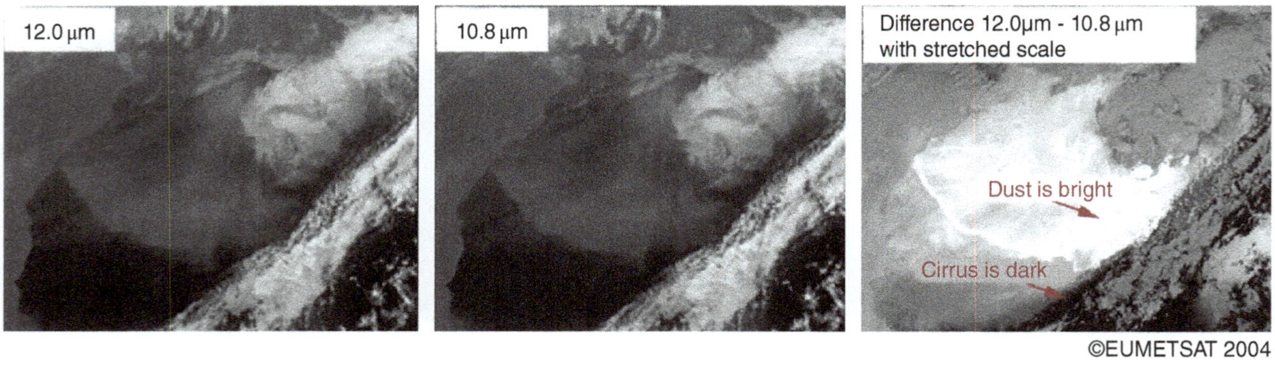


**Figure 9.16** Images of Meteosat SEVIRI IR12.0, IR10.8 and IR12.0–IR10.8 difference. *Source*: reproduced with permission of EUMETSAT and the COMET Program.

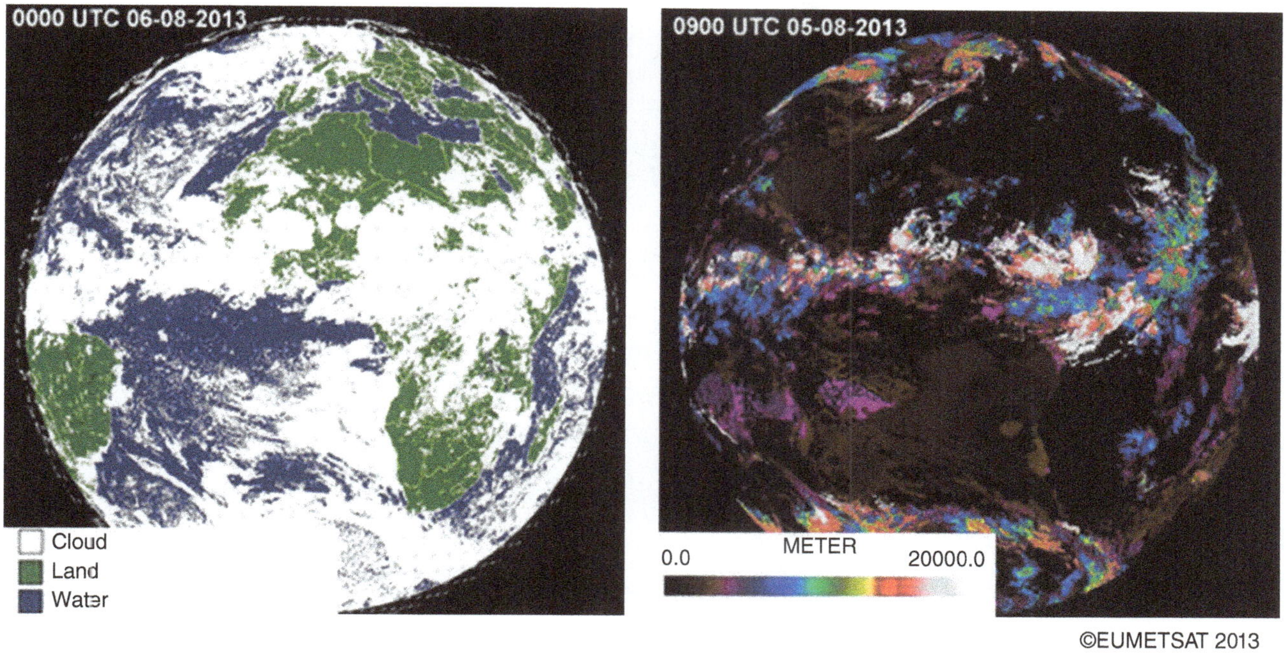


Figure 9.17 Cloud Mask (left) and Cloud Top Height (right) products derived from quantitative extraction algorithms developed by the EUMETSAT SAFNWC. *Source*: reproduced with permission of EUMETSAT.

the atmosphere or on the surface that are difficult to identify with single-channel images only. RGB processing combines three or more spectral channels, where each channel or channel difference is assigned to a primary colour (red, green or blue). The resulting images are usually realistic-looking in terms of texture and patterns but with significantly greater ability to distinguish specific features or phenomena.

Each primary colour is related to a physical property in the composite image. For example, in Figure 9.18, RGB processing is used to create a 'true colour', or a human eye view, from the VIIRS sensor by assigning the visible channels as red 0.671 μm, green 0.551 μm and blue 0.443 μm.

Channels selected to create an RGB composite should be based on the physical properties represented by the channels. It is best to combine channels that represent different physical properties (e.g. IR10.8 provides cloud top temperature and VIS0.6 provides cloud optical thickness). The following sections highlight the Meteosat RGBs that are most commonly used in West Africa.

#### *9.1.4.4.1 Air Mass RGB*

The air mass RGB is designed to detect cyclones, particularly rapid cyclogenesis, jet streaks and potential vorticity anomalies. Because it is mainly composed of IR channels in the water vapour and ozone absorption bands, it provides information mostly about the middle and upper troposphere. It distinguishes between ozone-poor tropical air masses and ozone-rich polar air masses, clouds at different levels, and dry air masses related to jet streaks, deformation zones and subtropical high-pressure systems (Figure 9.19).

**Meteosat SEVIRI bands and physical interpretation**

- Red = difference WV6.2 μm – WV7.3 μm, water vapour content at about 700–400 hPa and 500–200 hPa levels.
- Green = difference IR9.7 μm – IR10.8 μm, total ozone concentration (tropopause height), responds to strong diurnal and seasonal cycle in surface temperature (high BTD in cloud free regions over warm surface).
- Blue = WV6.2 μm, upper tropospheric humidity.

**Time**

- Day and night.

#### *9.1.4.4.2 Natural Colour RGB*

The natural colour RGB is derived from SEVIRI channels 1, 2 and 3. It is similar to the true colour RGB, except that it uses the phase difference of water and ice to differentiate between water clouds (whitish or pinkish) and ice clouds and snow (cyan), as shown in Figure 9.20. It is difficult to distinguish ice clouds from snow because they have similar reflectance and, in the case of ice clouds, similar brightness temperature and phase. Ice clouds can be distinguished from snow by their more intense colour and often clumpy appearance and with the use of animation or climatological knowledge of the area. Thin cirrus is difficult to identify with this composite.

**Meteosat SEVIRI bands and physical interpretation**

- Red = VIS0.6 μm, optical thickness and amount of cloud water and ice.
- Green = VIS0.8 μm, optical thickness, amount of cloud water and ice, greenness of vegetation.

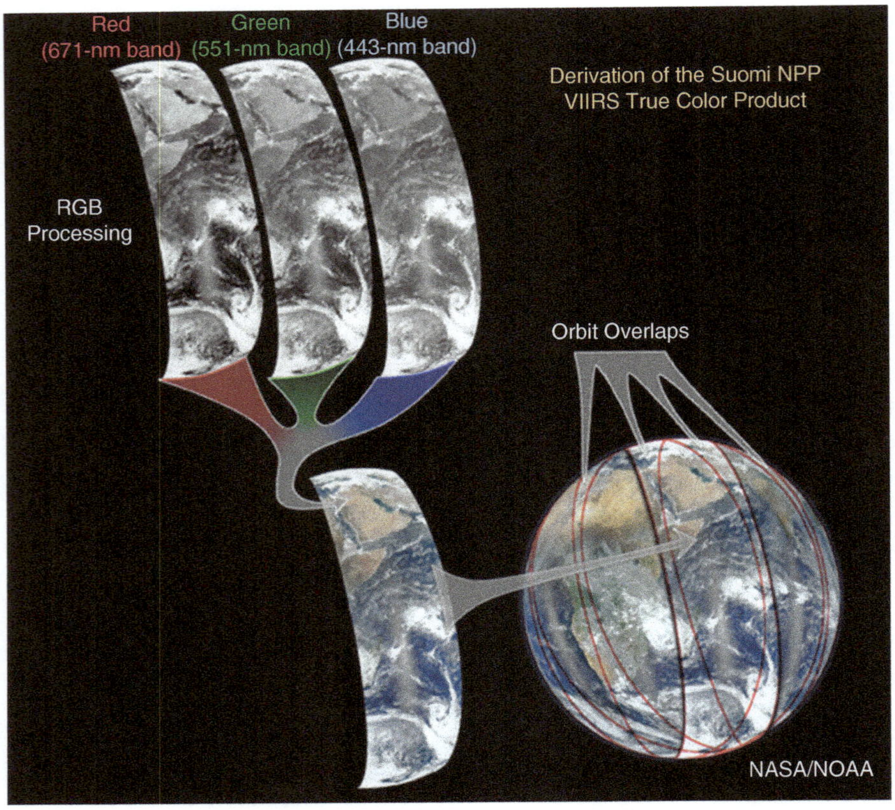


**Figure 9.18** True-colour image produced from RGB processing of data from the Suomi National Polar-orbiting Partnership (NPP) Visible Infrared Imaging Radiometer Suite (VIIRS).

(a)

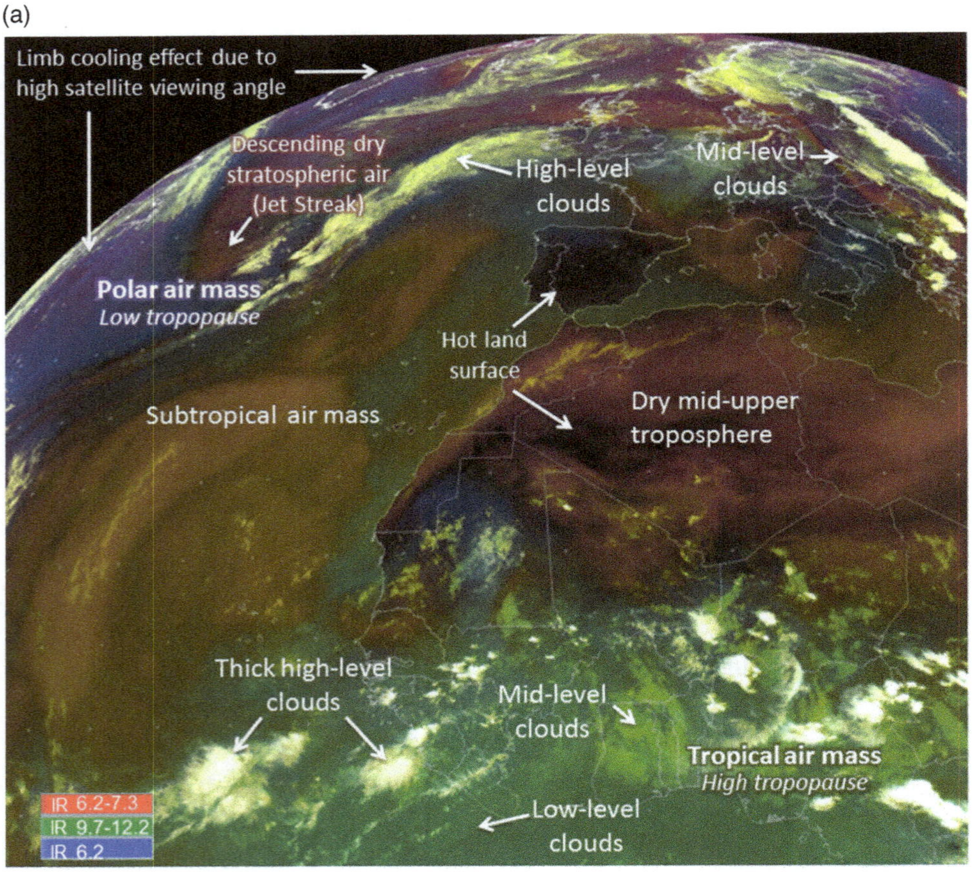


**Figure 9.19** (a) Air mass RGB image with interpretation of air masses and cloud levels. *Source*: reproduced with permission of EUMETSAT.

(b)

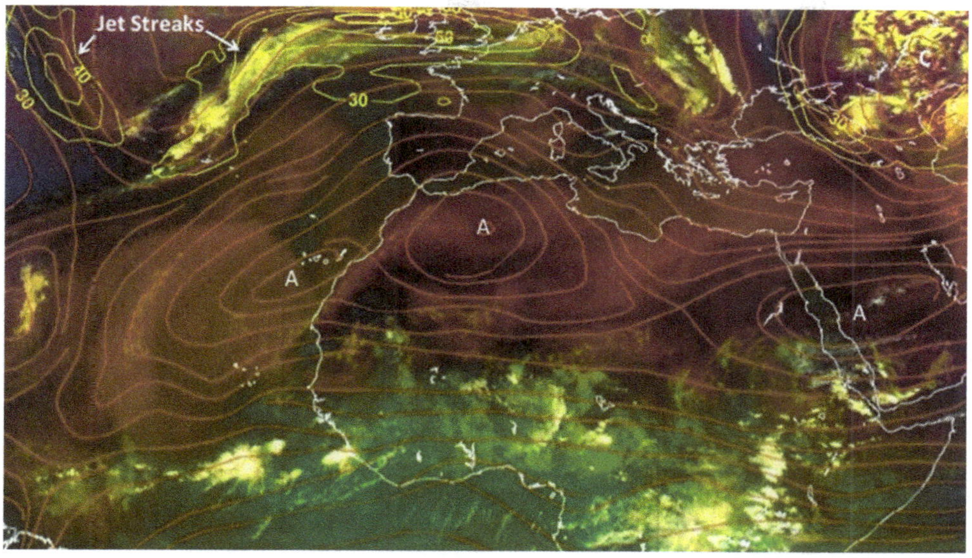


©EUMETSAT 2013/ECMWF/EuMeTrain

Figure 9.19 (Cont'd) (b) Air mass RGB image overlaid with streamlines (orange) and isotachs (yellow) at 300 hPa. *Source*: reproduced with permission of EUMETSAT/ECMWF/EuMeTrain.

- Blue = NIR1.6 μm, distinguishes particle size and phase of water.

**Time**

- Day (needs solar reflectance).

*9.1.4.4.3 Dust RGB*

The dust RGB utilises the difference in emissivity of dust and desert surfaces in the IR bands. Bright magenta indicates low-level dust during the day and dark magenta shows low-level dust at night (Figure 9.21). During daytime, it is easier to detect low-level dust because of the larger temperature difference between the hot desert surface and the cooler dust cloud. At night, the smaller thermal contrast makes it difficult to detect low-level dust. Higher-level dust absorbs more at 10.8 μm than cirrus clouds, which are more sensitive to energy at 12.0 μm, thereby producing a negative difference between the channels and a clear delineation between dust and higher clouds.

**Meteosat SEVIRI bands and physical interpretation**

- Red = difference IR12.0 μm − IR10.8 μm, emissivity of dust, desert surfaces, and clouds; dust absorbs more at 10.8 μm, cirrus more sensitive to energy at 12.0 μm.
- Green = difference IR10.8 μm − IR8.7 μm, BTD between the hot desert surface and the cooler dust cloud, reduced at night.
- Blue = IR10.8 μm.

**Time**

- Day and night.

Visible channels, especially the HRV channel, are best for tracking dust over the ocean during daytime; for example, dust is more distinct relative to the underlying ocean with the natural colour RGB than with the dust RGB (Figure 9.22). The dust RGB also helps with low-level humidity analysis; the dark blue areas are humid air masses, and lighter blue areas are drier. It also very useful for tracking deep convective clouds (dark red). Further examples of the use of satellite and other data for the monitoring of dust are provided in Chapter 5.

*9.1.4.4.4 Convection RGB*

The convection RGB provides information about the evolution of convective systems and high-level clouds (Figure 9.23). The composite is based on the differences in the radiative properties of ice and liquid water (phase and optical thickness) and the effect of updraft strength and cloud depth on ice particle size (Lensky and Rosenfeld, 2008). For example, because of the sensitivity of the reflectance at NIR1.6 and IR3.9, clouds with small water droplets are darker than ice clouds during daytime. The yellow colour on this RGB indicates deep penetrating convection that causes severe weather or thick cirrus with small ice particles generated by mountain waves. Small ice particles also dominate polluted clouds, which often means dust for West Africa.

**Meteosat SEVIRI bands and physical interpretation**

- Red = difference WV6.2 μm − WV7.3 μm brightness temperature, near zero difference in overshooting cumulonimbus.

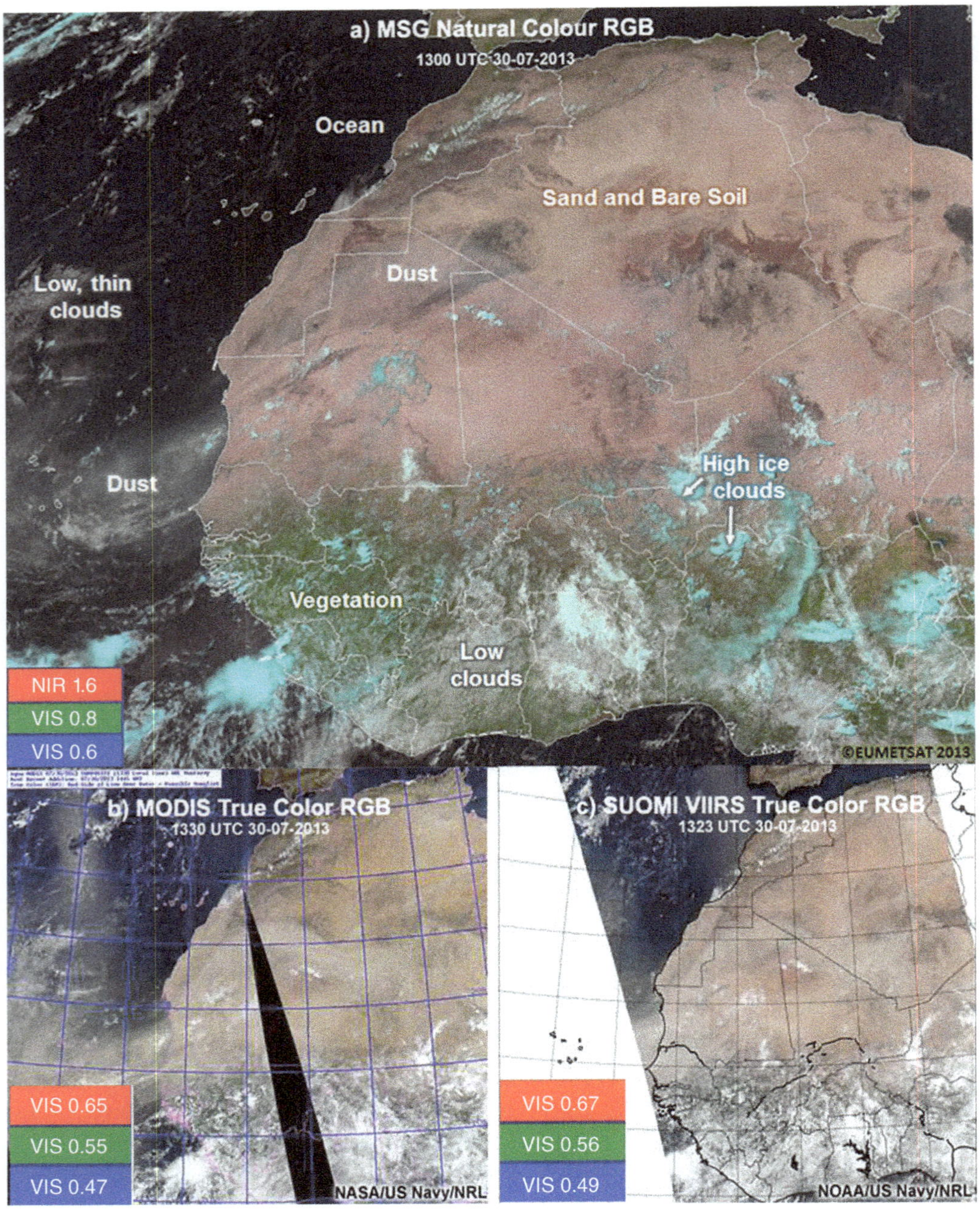


**Figure 9.20** Natural colour RGB image from the MSG SEVIRI (upper) and true colour RGB images from Moderate-resolution Imaging Spectroradiometer (MODIS) Aqua (lower left) and Suomi-NPP VIIRS (lower right). *Source*: (upper) reproduced with permission of EUMETSAT.

- Green = difference IR3.9 μm − IR10.8 μm brightness temperature, optical thickness of small ice particles that form from nearly homogeneous freezing in severe thunderstorm updrafts.
- Blue = difference in NIR1.6 μm − VIS0.6 μm reflectance, sensitivity to particle size and phase of water. NIR1.6 has higher reflection from water droplets than from ice.

**Time**

- Day.

#### *9.1.4.4.5 Day microphysics RGB*

The day microphysics RGB is used for cloud analysis (Figure 9.24) and monitoring convection, fog, snow and fires. The cloud optical depth and amount of liquid water in clouds are indicated by the VIS0.8 reflectance. Non-precipitating water clouds, with small cloud drops, appear white, while precipitating clouds with large drops appear pink. The IR3.9 channel (green) measures the phase and particle size, while the IR10.8 brightness temperature modulates the blue.

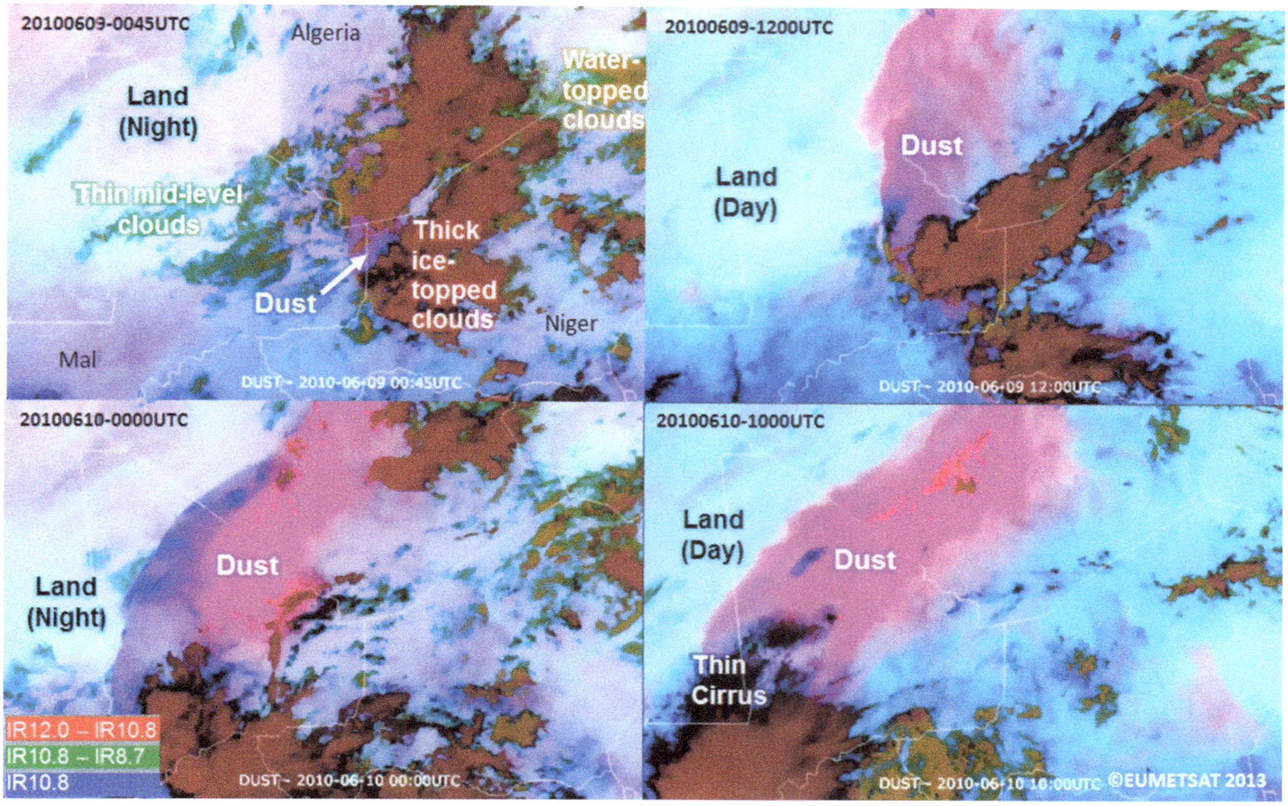


**Figure 9.21** Dust RGB images associated with a haboob (dust storm), associated with a strong thunderstorm outflow, which moved across Niger, Mali and Algeria on 9–10 June 2010. *Source*: reproduced with permission of EUMETSAT.

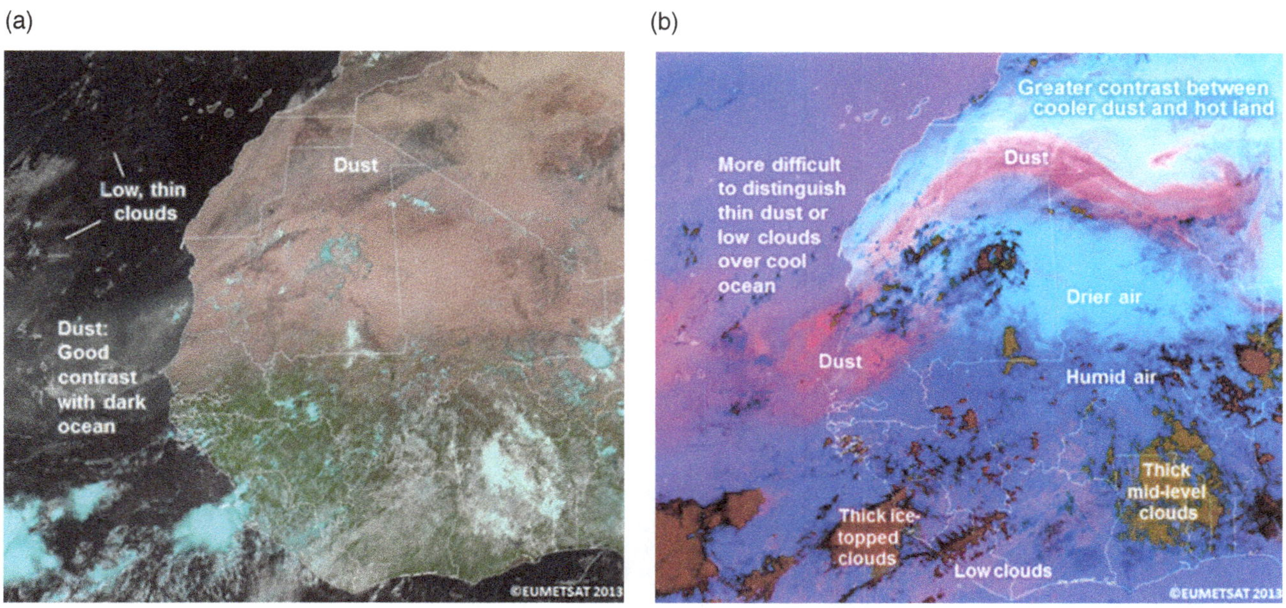


**Figure 9.22** Dust over West Africa detected by (a) natural colour RGB and (b) dust RGB. *Source*: reproduced with permission of EUMETSAT.

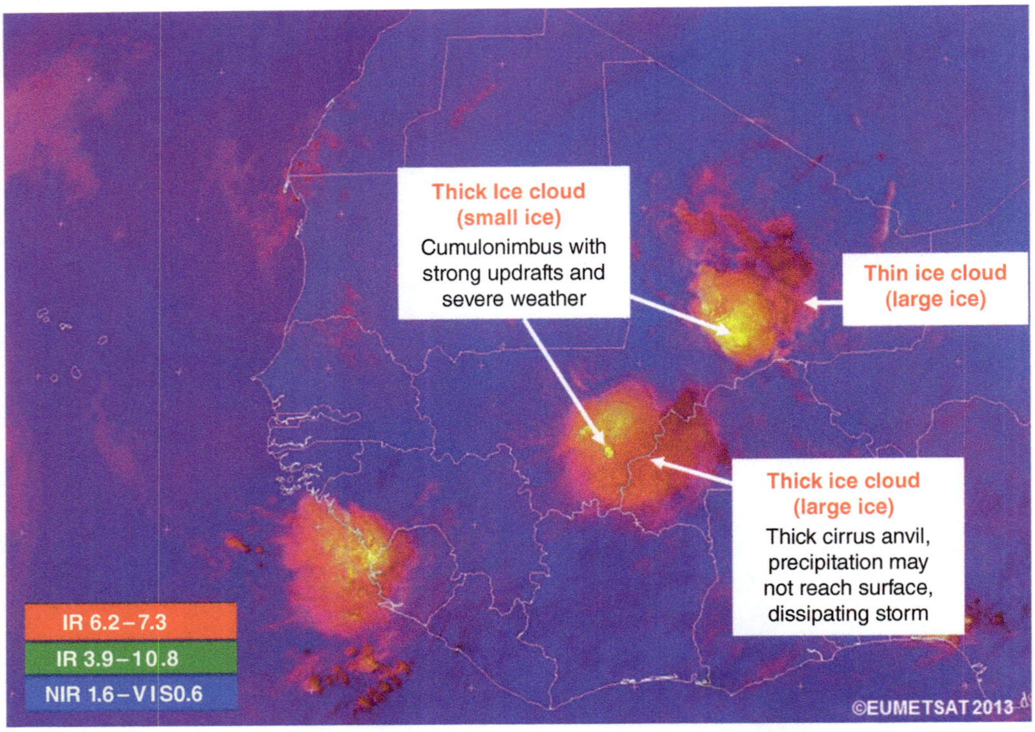


Figure 9.23 Convective RGB image of mesoscale convective systems over West Africa. *Source*: reproduced with permission of EUMETSAT.

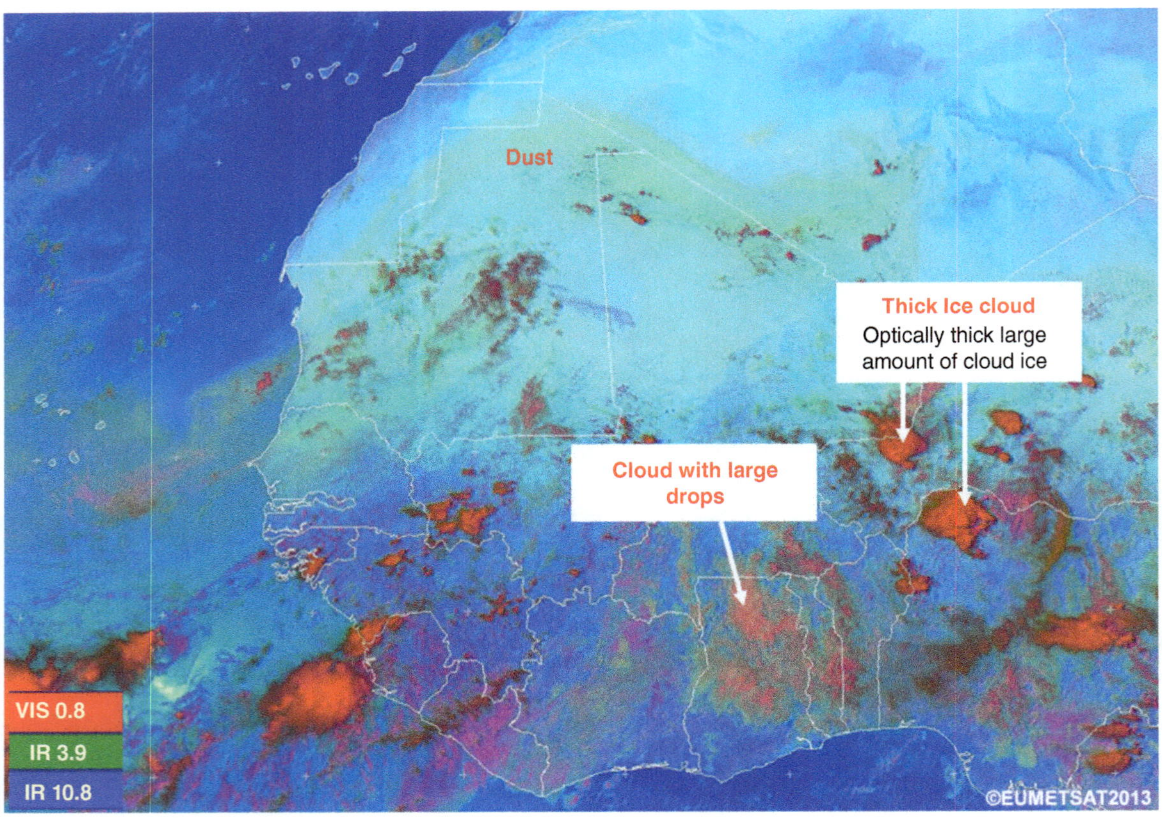


Figure 9.24 Day microphysics RGB image. *Source*: reproduced with permission of EUMETSAT.

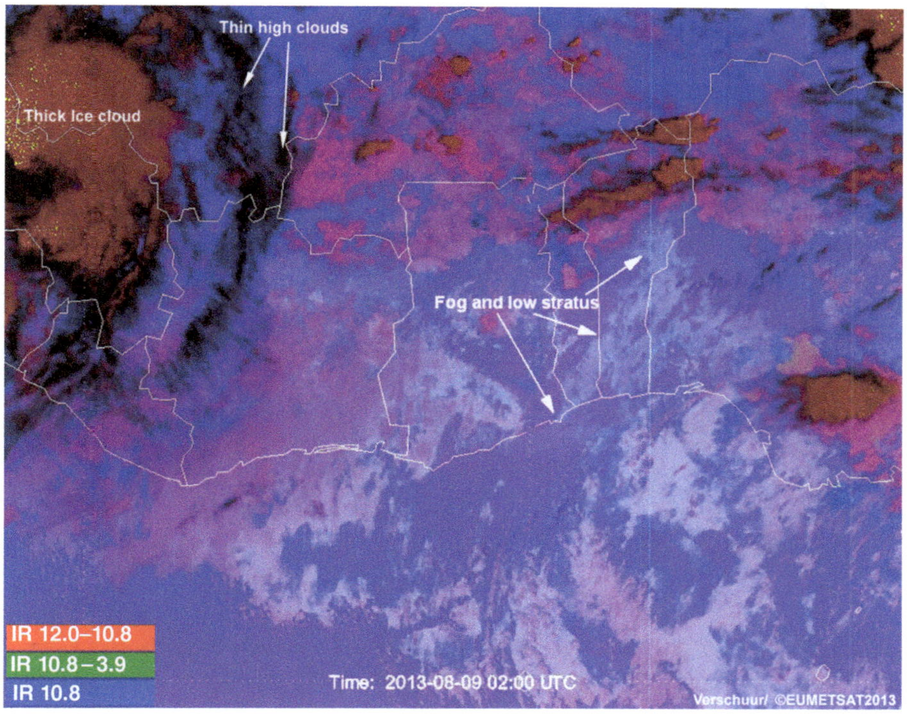


Figure 9.25 Fog and low stratus near the coast of the Gulf of Guinea. Thunderstorms near the west coast of Africa are bordered by thin cirrus outflow. *Source*: reproduced with permission of Henk Verschuur and EUMETSAT.

**Meteosat SEVIRI bands and physical interpretation**

- Red = VIS0.8 μm, optical depth and amount of cloud water and ice.
- Green = IR3.9 μm reflectance and difference in phase and size of particles.
- Blue = IR10.8 μm brightness temperature.

**Time**

- Day.

#### 9.1.4.4.6 *Fog/Low Stratus or Night Microphysics RGB*

This RGB monitors the evolution of night-time fog/low stratus (Figure 9.25), and secondarily the detection of fires, low-level moisture boundaries, thin cirrus clouds, snow and general cloud classification. It is tuned for night-time conditions and is, therefore, not recommended for use during daytime. Owing to low surface emissivity, this RGB is not recommended for fog detection over desert areas.

At night, fog/low stratus appears light green; the smaller the droplets the stronger the input from green. By comparison, the cloud-free land appears a pinkish colour, the thick high-level clouds have red tones or appear with a red–yellow speckled colour (an indicator of very cold clouds) and the thin high-level clouds appear dark blue. The ocean appears pale violet and low-level clouds appear pale blue. Thick mid-level clouds appear brown, and thin mid-level clouds appear green.

**Meteosat SEVIRI bands and physical interpretation**

- Red = difference IR12.0 μm – IR10.8 μm, cirrus is more sensitive to energy at 12.0 μm than the near-surface clouds.
- Green = difference IR10.8 μm – IR3.9 μm, BTD is about 6–10 K, emissivity of water cloud at 3.9 μm is less than at 10.8 μm. The IR3.9 shows more reflection of cold atmosphere above. Indicates particle size difference, but only for warm clouds at night.
- Blue = IR10.8 μm, brightness temperature distinguishes the surface from clouds.

**Time**

- Night.

#### 9.1.4.4.7 *Volcanic ash RGB*

The ash RGB is designed to identify volcanic ash and sulphur dioxide ($SO_2$) clouds from volcanic eruptions. It is similar to the dust RGB except the temperature ranges assigned to certain colours allow the distinction among ash, $SO_2$ and ice crystals (hazards to aviation and air quality). The IR8.7 μm channel detects volcanic ash aerosols, and the difference between that channel and the IR10.8 μm channel is usually positive. For ice clouds the BTD will be negative. The ash RGB colour varies by temperature, height and particle size, with red indicating when ash is cold, magenta when it is warm and yellow when it is mostly small ash particles. Different colours

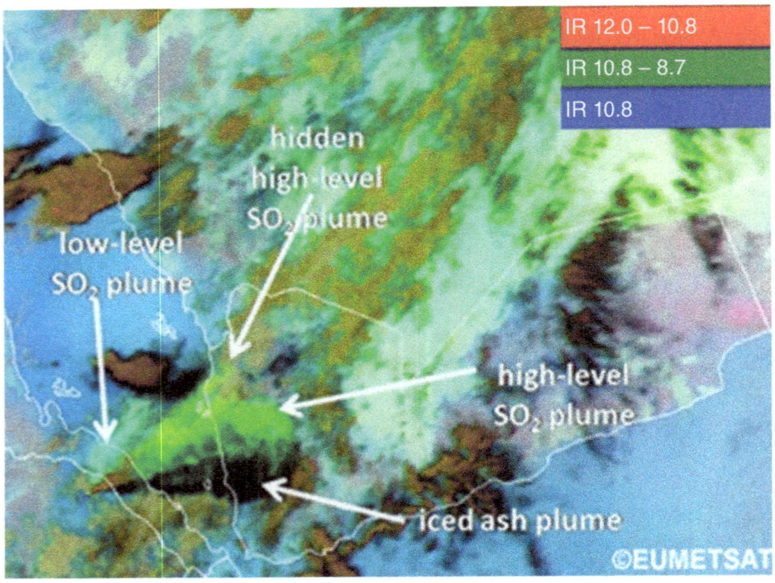


**Figure 9.26** Example of Ash RGB image showing $SO_2$ plumes at different levels and an iced ash plume from volcanic eruption in Ethiopia. *Source*: reproduced with permission of EUMETSAT.

also indicate temperature and height of $SO_2$ (Figure 9.26), which is more transparent at 10.8 μm. Note that ash is only detectable at a certain concentration and it is difficult to identify $SO_2$ and ash embedded in ice clouds.

**Meteosat SEVIRI bands and physical interpretation**

- Red = IR12.0 μm – IR10.8 μm, helps to distinguish between cirrus and aerosol absorption.
- Green = IR10.8 μm – IR8.7 μm, the IR8.7 μm detects the lower emissivity of the aerosols. The BTD measures the aerosols and trace gases absorption relative to the IR window channel.
- Blue = IR10.8 μm, brightness temperature, distinguishes among surface and cloud temperatures.

**Time**

- Day and night.

The specialised SEVIRI RGB products are summarised in Table 9.2. RGB products will become more common as the next generation satellites, such as the United States GOES-R and Meteosat Third Generation (MTG), will have imagers with more spectral bands.

**Table 9.2** Meteosat SEVIRI RGB composite products, composite channels and applications.

| RGB composite & channels | Applications | Time |
|---|---|---|
| Dust and ash RGB 10–09, 09–07, 09 | Dust, clouds (thickness, phase), contrails, fog, ash, $SO_2$, low-level humidity | Day & night |
| Air mass RGB 05–06,08–09, 05 | Severe cyclones, jets, potential vorticity analysis | Day & night |
| Fog/low stratus RGB 10–09, 09–04, 09 | Clouds, fog, contrails, fires | Night |
| Day microphysics RGB 02, 04r, 09 | Clouds, convection, snow, fog, fires | Day |
| Convection RGB 05–06, 04–09, 03–01 | Severe convection | Day |
| Cloud over snow RGB 02, 03, 04r | Snow, fog | Day |
| Natural colour RGB 03, 02, 01 | Vegetation, snow, smoke, dust, fog | Day |

#### 9.1.4.5 Blended or 'Sandwich' Image Products

Blended or sandwich products are used to collocate cloud top features (such as overshooting tops, cloud plumes and gravity waves) with the related cold or warm features in the brightness temperature field. The output image is created by combining information from one (typically) black-and-white image and one colour image (either colour-enhanced single-band image or RGB product). Using a graphics editor, the images are layered to produce an image that emphasises the cloud top features in both input images. For example, Figure 9.27 shows the location of the overshooting tops from the texture and shadows of the visible image and the cold (red) areas in the brightness temperature image. Brightness temperature scales are optimally shifted or stretched depending on location and season; generally colder temperatures are used for the tropics and for the warm season in mid-latitudes.

### 9.1.5 Satellite Retrieval of Meteorological Parameters

As noted earlier, satellite observations of particular features depend on the chosen spectral intervals (window and absorption band). Observation capability

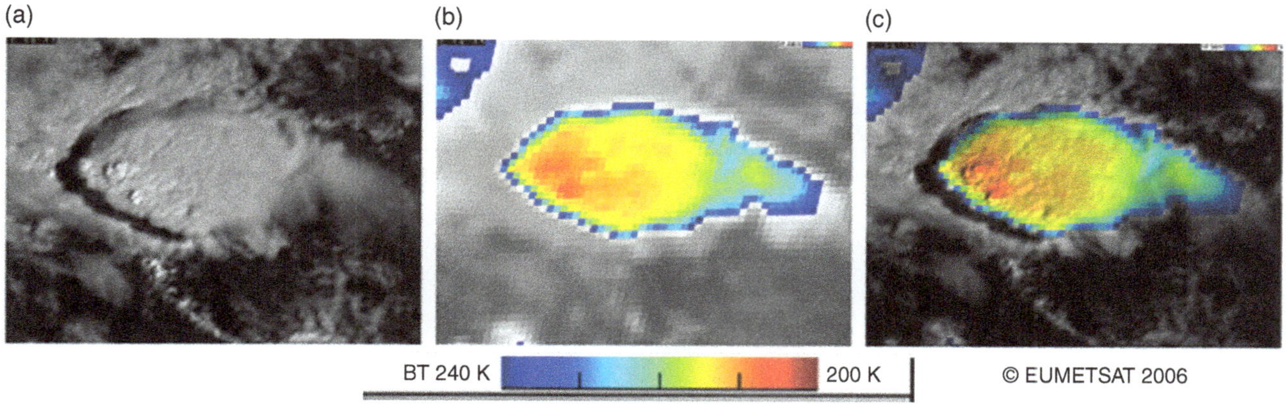


**Figure 9.27** MSG-3 (a) visible, (b) colour-enhanced IR10.8 μm brightness temperature and (c) sandwich product of the visible and IR images highlighting the overshooting tops of a supercell thunderstorm over Italy. *Source*: reproduced with permission of EUMETSAT.

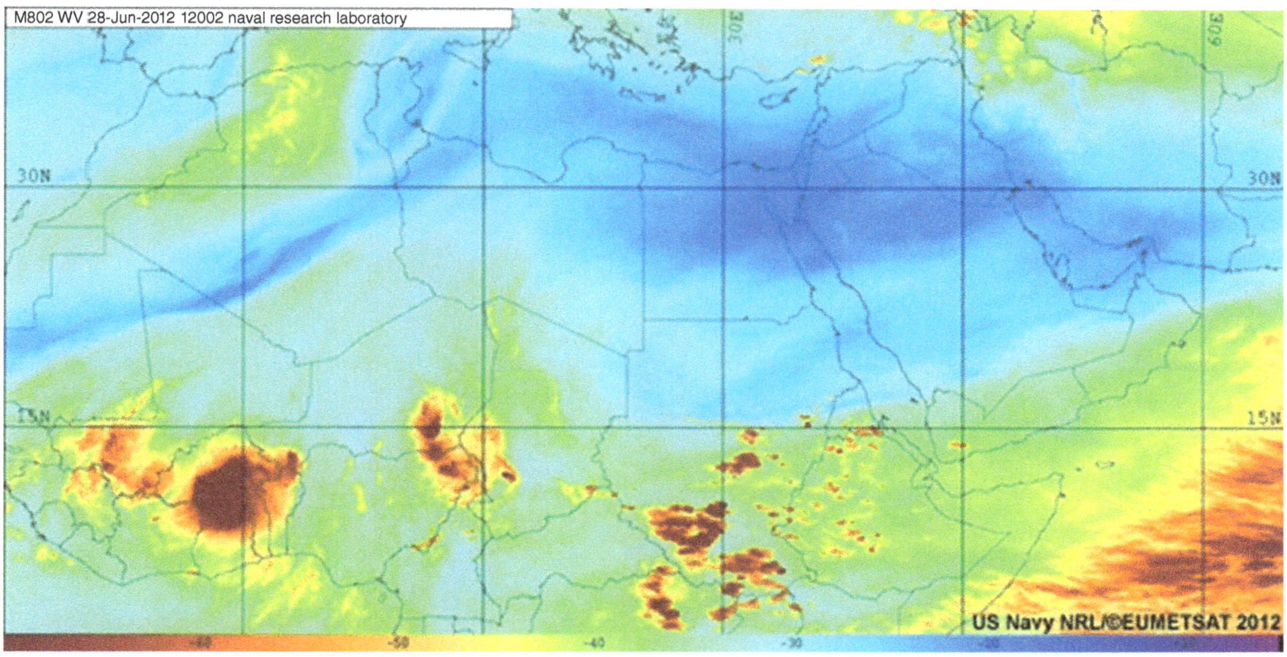


**Figure 9.28** Example of colour-enhanced IR water vapour image for northern Africa; dark red areas indicate high water vapour content in the mid-upper troposphere and blue to violet areas show the dry mid-upper troposphere. *Source*: reproduced with permission of US Navy NRL/EUMETSAT.

also depends on the resolving power of the sensor telescope, time of day (night or day), atmospheric state (temperature, humidity and clouds) and the underlying surface (temperature, ocean, land, vegetation). This section explores the retrieval of parameters that are critical to weather forecasting and seasonal outlooks in West Africa.

#### 9.1.5.1 Water Vapour

##### *9.1.5.1.1 Infrared Water Vapour*

Wavelengths between 6 and 7 μm are known as the 'water vapour channels' because water vapour absorbs and emits strongly in that band. Geostationary satellite images of IR water vapour emission are used to follow motion in the mid-upper troposphere. They are commonly used to detect upper-level short waves, rapid cyclone genesis, jet streams, potential vorticity anomalies and thunderstorm tops. Feature identification is made easier using colour enhancement based on the brightness temperature. For example, in Figure 9.28, dark red areas show where the satellite senses water vapour high in the troposphere where it is cold. Blue indicates dry areas in the mid-upper troposphere, which allows the sensor to 'see' lower down into the troposphere where it is warmer. Isolated dark red areas over West Africa can be interpreted as convective systems, while the striated red areas over the Indian Ocean are likely to be cirrus clouds associated with the tropical

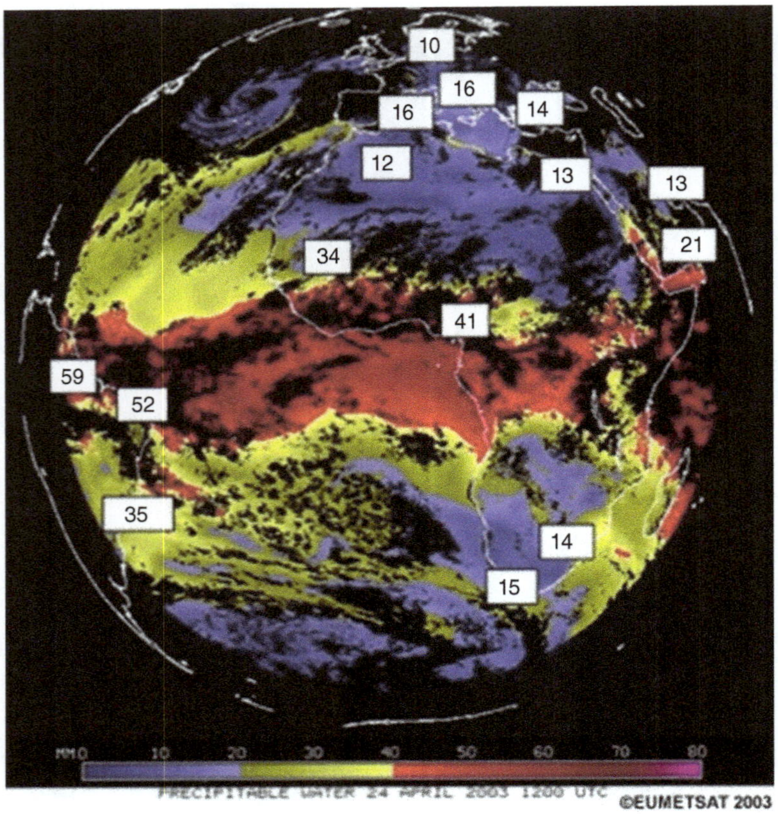


**Figure 9.29** Precipitable water product derived from MSG SEVIRI. Boxed values are precipitable water from soundings. *Source*: reproduced with permission of EUMETSAT.

easterly jet, which has its maximum in the upper troposphere near 150 hPa (see Figure 1.30c). Blue areas wrapping around an upper-level vortex over Tunisia indicate dry air intruding from the extratropics.

Estimates of precipitable water are derived using the water vapour channels (e.g. Figure 9.29). The figure shows radiosonde measurements and satellite-estimated precipitable water.

Low-level humidity can be determined using the BTD between IR10.8 and IR12.0. The distinction is possible because more water vapour is absorbed at IR12.0 than at IR 10.8 and the difference is mostly due to humidity concentrated in the low troposphere. Therefore, assuming a normal temperature profile (no low-level inversion), then the greater the BTD, the higher the low-level absolute humidity. Since the dust RGB uses the IR10.8–IR12.0 BTD, it can used to monitor low-level humidity. The shift from dry to moist air masses is indicated by the change from light blue to darker blue; for example, across the east–west boundary close to 16°N, in the middle of Figure 9.22b.

#### 9.1.5.1.2 *Microwave Estimates of Water Vapour*

Estimates of water vapour from microwave emissions are measured from LEO satellite sensors. While each satellite has gaps in its coverage, the combination of several orbits makes it possible to monitor the total precipitable water operationally across the globe (Figure 9.30). Of particular concern for forecasters are regions of high precipitable water, which have the potential for heavy rainfall and flooding (see Sections 2.1.3.2 and 3.1.1). These images are also useful for tracking what are termed 'atmospheric rivers', the narrow plumes of high precipitable water that flow from the tropics and lead to heavy precipitation in the subtropics and midlatitudes.

### 9.1.5.2 Vertical Structure of Temperature, Humidity and Winds

#### 9.1.5.2.1 *Satellite Sounders*

Soundings derived from satellite sensors have significantly improved weather forecasting, primarily through the assimilation of sounder radiances into operational numerical weather prediction models. Satellite sounders provide vertical measurements of temperature and moisture that compare favourably with radiosonde profiles in validation experiments (Figure 9.31). For example, the IASI has temperature accuracy of ~1 °C for 1 km and 10% for humidity.

Satellite sounders can also be applied to synoptic weather analysis, especially to severe weather nowcasting, in diagnosis of the pre-convective environment. Hyperspectral sounder data (especially from geostationary satellites) can enhance the microwave-based precipitable water products, making it easier to detect and track moist or dry features. Geostationary sounders, currently on the GOES satellite and planned for the MTG satellites, provide observations at hourly or higher temporal resolution, which is a distinct

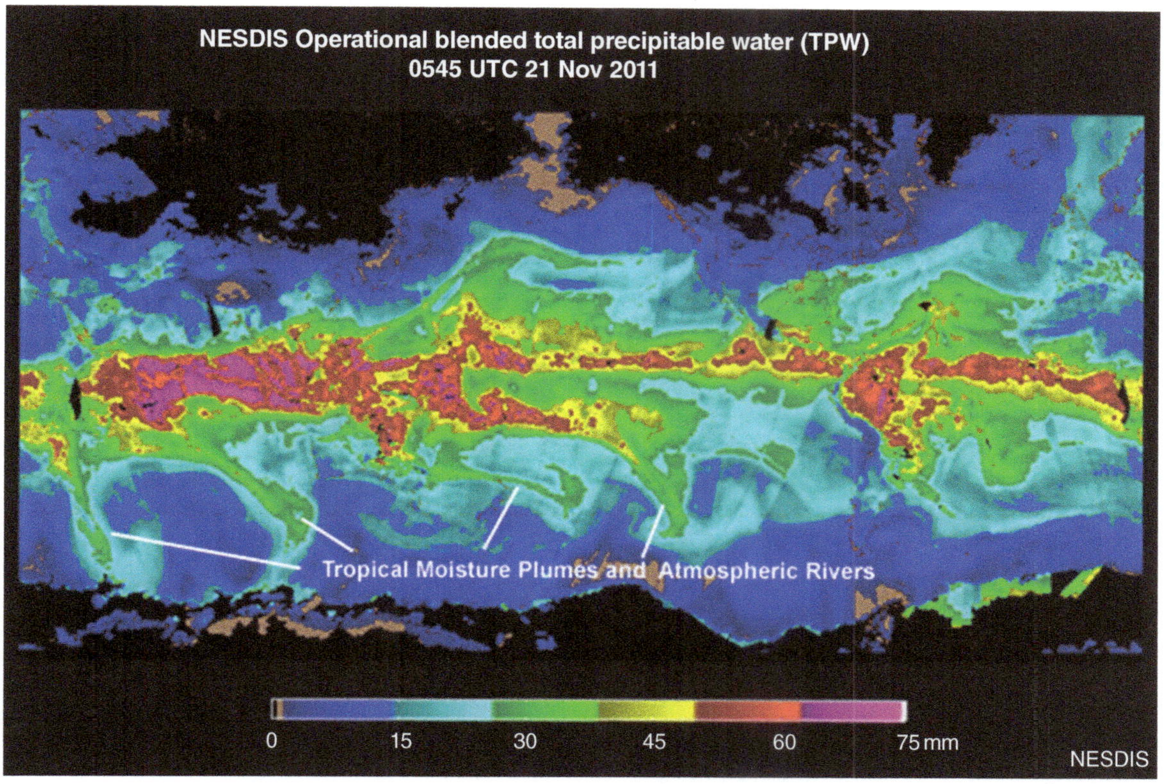


Figure 9.30 Total precipitable water vapour product derived from passive microwave satellite sensors. Examples of tropical moisture plumes and atmospheric rivers are marked.

Figure 9.31 (a) Example of satellite sounder temperature retrieval from the Infrared Atmospheric Sounding Interferometer (IASI) sounder on the MetOp satellites. *Source*: reproduced with permission of EUMETSAT. (b) Skew-*T* log *P* plot of temperature and dew point from the NASA Atmospheric IR Sounder (AIRS), radiosonde and ECMWF analysis for clear sky over Zadar, Croatia. *Source*: reproduced with permission of EUMETSAT.

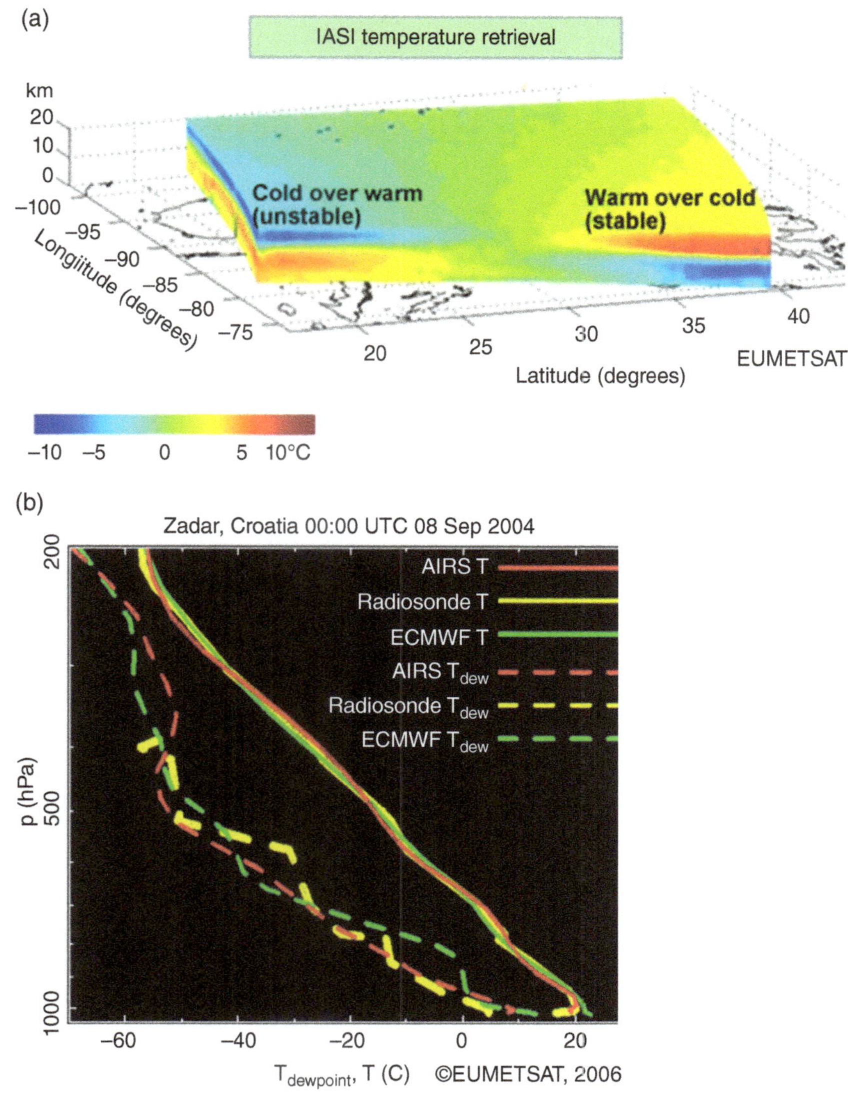

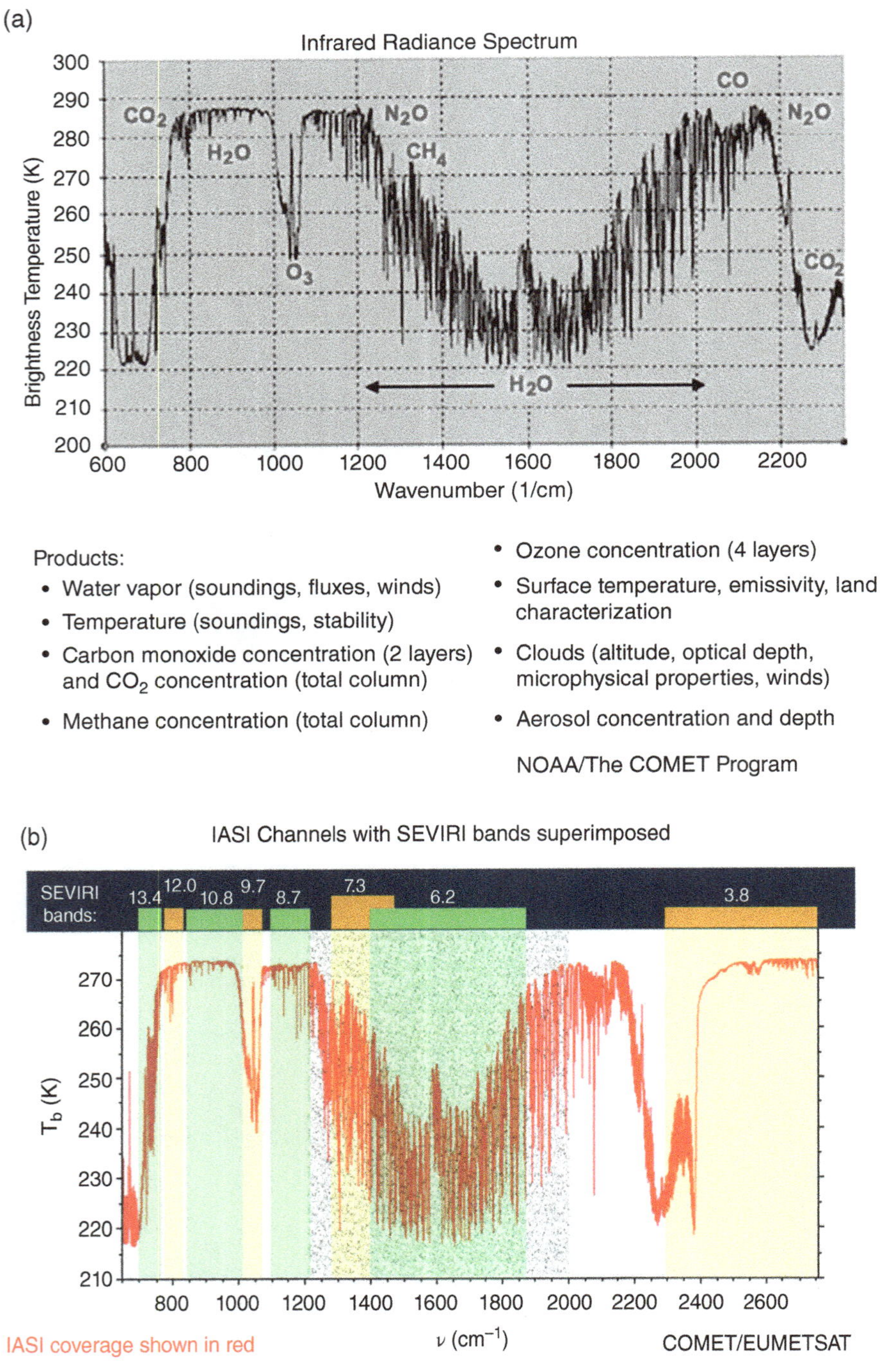


**Figure 9.32** (a) IR radiance spectrum, bands that interact with specific gases, and products retrieved by satellite sounders to monitor the surface and the atmospheric column above. (b) Discrete channels of the hyperspectral IASI sounder superimposed with the MSG SEVIRI channels. *Source*: reproduced with permission of EUMETSAT and the COMET Program.

advantage compared with radiosonde profiles that are rarely available more than twice a day.

Atmospheric emission and absorption occur at discrete wavelengths related to the molecular structure of gases or aerosols (Figure 9.32). Attenuation by gases is strongest near the centre of the absorption region of the spectral band or channel. Sounder channels are sensitive to temperature variations related to carbon dioxide ($CO_2$), ozone and water vapour at different atmospheric levels (compare the hyperspectral IASI with the MSG SEVIRI channels in Figure 9.32). Each band is converted into an image with assigned colours to identify thermal differences (Figure 9.31a).

By selecting different sounding wavelengths, observed radiances are used to infer temperature and humidity profiles, and cloud top pressures. Gases that are relatively abundant and uniformly distributed, such as $CO_2$, are good emission sources on which to base sounding retrievals.

From Schwarzschild's Equation 9.8, we can determine the infrared radiation emitted and transmitted from the surface and different layers of the atmosphere (Figure 9.9).

The upwelling radiation for a satellite channel can also be written in terms of the radiance and a weighting function:

$$\overbrace{I_\lambda}^{\text{Radiance}} = \overbrace{\varepsilon_\lambda B_\lambda T(\text{surface})}^{\text{Planck surface emission}} \overbrace{\tau_\lambda(p_{\text{surface}})}^{\text{Transmission from surface}} + \int_0^\infty \overbrace{B_\lambda[T(p)]}^{\text{Planck temperature profile}} \overbrace{\frac{d\tau_\lambda(p)}{dp}}^{\text{Weighting function}} \overbrace{dp}^{\text{Change of pressure}} \tag{9.12}$$

The weighting function determines how much a layer of unit thickness contributes to the observed outgoing radiation, and hence the atmospheric layer from which radiance measured by a particular channel was transmitted. Weighting functions are applied according to the emission characteristics of gases at particular wavelengths and where bands peak depends on the vertical profile and the satellite-viewing angle. Discrete sounding channels are usually chosen to become more opaque with height. As opacity increases, the sensed signal comes from higher and higher in the atmosphere.

For instance, the weighting function for the GOES 13.3 μm spectral band, a $CO_2$ band, has its peak near the surface (Figure 9.33). With the peak near the surface, clouds and surface features are still visible. However, as observed brightness temperature due to $CO_2$ becomes more dominant lower in the atmosphere, lower clouds and surface features appear less distinct. Note that it is sometimes difficult to determine temperature and water vapour profiles with thick, low-lying clouds and over stretches of land.

#### 9.1.5.2.2 *Convective Indices*

Convective instability indices are derived for cloud-free areas using six Meteosat SEVIRI channels and output from the ECMWF model.

The global instability indices (GII), which describe the instability of the troposphere, include the *K*-index, lifted index, and total precipitable water (Figure 9.34; see also Section 3.2.3.1.2 (**1**)) – note, following links require sign-in to COMET MetEd:

- *Lifted Index* (http://www.meted.ucar.edu/mesoprim/skewt/li.htm) – measures stability as the difference between the parcel and environmental temperature at 500 hPa; greater difference, more unstable.
- *K Index* (http://www.meted.ucar.edu/mesoprim/skewt/ki.htm) – high values indicate instability and heavy rainfall potential; measuring a strong lapse rate and abundant moisture through the mid-troposphere.
- *Total Totals Index* (http://www.meted.ucar.edu/mesoprim/skewt/tt.htm) – increasing values correspond to increasing potential for thunderstorms.
- *Precipitable water content* – high values are associated with heavy precipitation and high precipitation efficiency.

#### 9.1.5.2.3 *Atmospheric Motion Vectors*

Atmospheric motion vectors are generated for layers in the lower, mid and upper troposphere from geostationary imagery (Figure 9.35). Vectors are determined through automatic tracking of water vapour image features in the middle to upper troposphere and cloud elements in the lower troposphere using visible and IR images (Velden *et al.*, 2005). The assimilation of atmospheric motion vectors has contributed significantly to improvements in numerical weather prediction, particularly in the tropics, where radiosonde data are sparse.

#### 9.1.5.2.4 *Soundings from Global Positioning System Radio Occultation*

Radio occultation is a technique by which satellite receivers intercept signals from Global Positioning System (GPS) satellites and infer deviations in the signal's path caused by temperature and moisture gradients (Figure 9.36; Bevis *et al.*, 1992). The refractive index and the temperature and humidity gradient relationships are then used to create soundings (Figure 9.36 inset).

#### 9.1.5.2.5 *Surface-based Global Positioning System Meteorology Applications*

Integrated water vapour is derived from retrievals of refractive indices by surface GPS receivers. Real-time observations (e.g. Figure 9.37) are available from sites around the world such as those listed at http://www.suominet.ucar.edu/indexGlobal.html (Ware *et al.*, 2000). Several GPS stations were installed in West Africa during the African Monsoon Multidisciplinary Analyses (AMMA) field programmes (Lebel *et al.*, 2010).

### 9.1.5.3 Clouds

Cloud optical thickness is detected by reflection/scattering at visible and microwave wavelength and by emission at IR and microwave wavelengths. At IR wavelengths clouds behave almost like blackbodies as they are composed of water and water vapour; that is, good absorbers/emitters. At visible wavelengths, clouds are good scatterers because of Mie scattering by large particles; all visible wavelengths are scattered equally, so thicker clouds appear whiter and brighter and thinner clouds appear grey and duller. For example, in Figure 9.38, the thin cirrus outflow, which allows partial view of underlying features, is distinguished from the whiter, brighter towering cumulonimbus. The gust front or outflow boundary is a kind of density or gravity current that

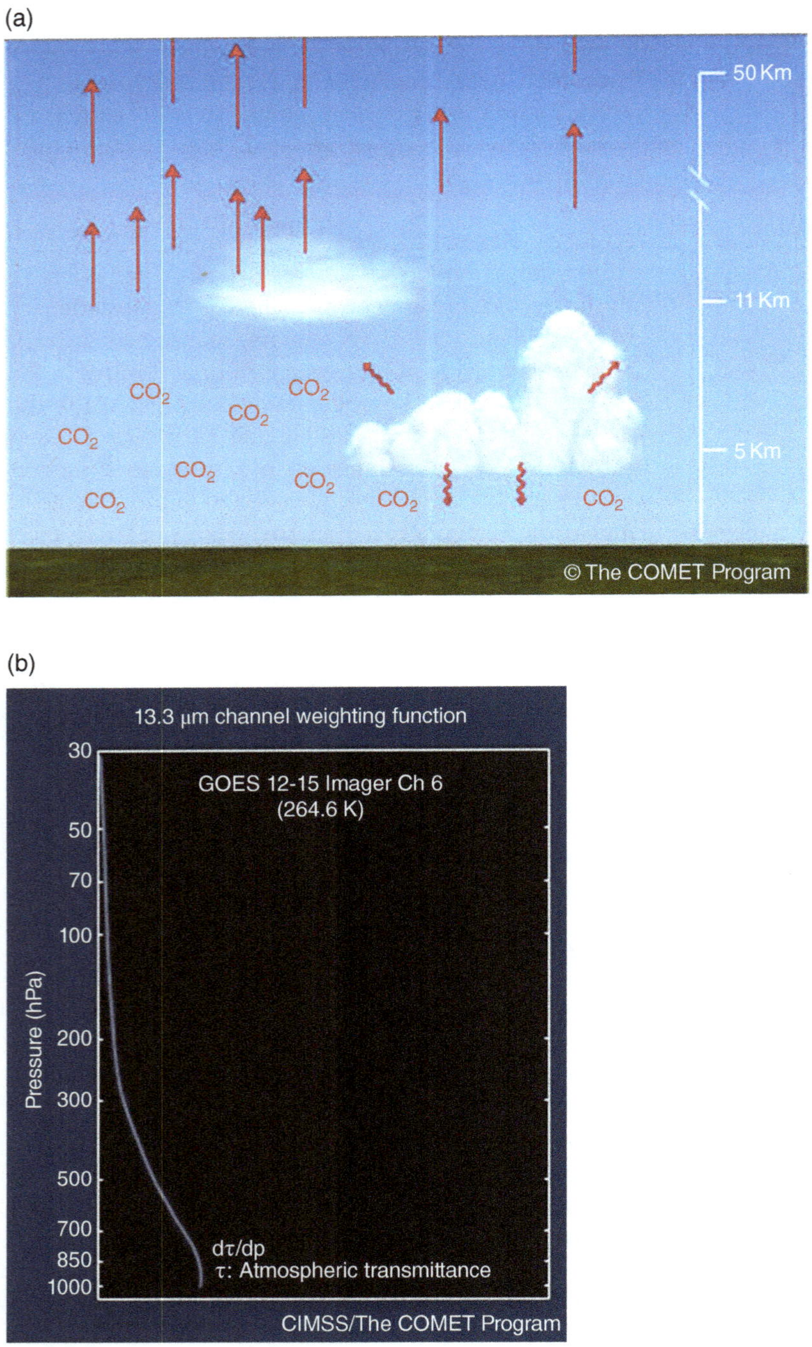


**Figure 9.33** (a) Schematic of emission of $CO_2$ at 13.3 μm wavelength, which is more dominant near the surface; (b) the corresponding weight function for that spectral band. *Source*: reproduced with permission of the COMET Program.

forms when thunderstorm downdrafts spread out at the surface (see Section 3.1.2.2). The cooler air moves away from the parent storm and typically lifts air along its edge, forming an arc-shaped cloud line, a familiar feature in satellite and radar imagery. In the Sahel, gust fronts loft dust, which means that their boundary is recognised as an arc of dust; for example, the grey outflow area in Figure 9.38.

Clouds are also identified by form, texture and duration. Stratiform cloud areas generally have even cloud top heights and fairly large extent. Cumuliform clouds can be isolated cloud clusters with cellular structure and fairly distinct edges. Taller cloud tops cast shadows (e.g. Figure 9.38), which are more pronounced near sunrise and sunset. Cirrus appears streaked and feathered, and anvil cirrus has frayed edges due to strong upper-level winds. Small clouds lines perpendicular to the flow, called transverse lines, occur in cirrus moving in jet streams or blow out of strong convection. Generally, stratiform cloud top heights and shape change very slowly, while convective clouds change height and shape over a short time. Cirrus that is unrelated to thunderstorms generally changes more slowly than anvil cirrus.

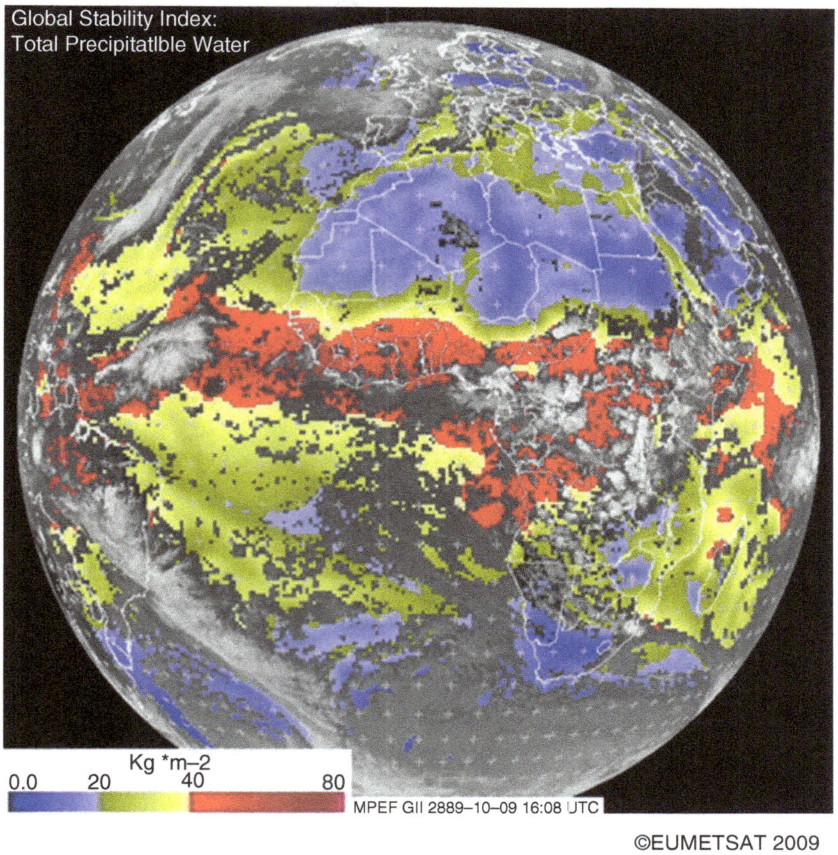


**Figure 9.34** Meteosat total precipitable water, a global stability index derived empirically for cloud-free areas and used to identify convective potential. *Source*: reproduced with permission of EUMETSAT.

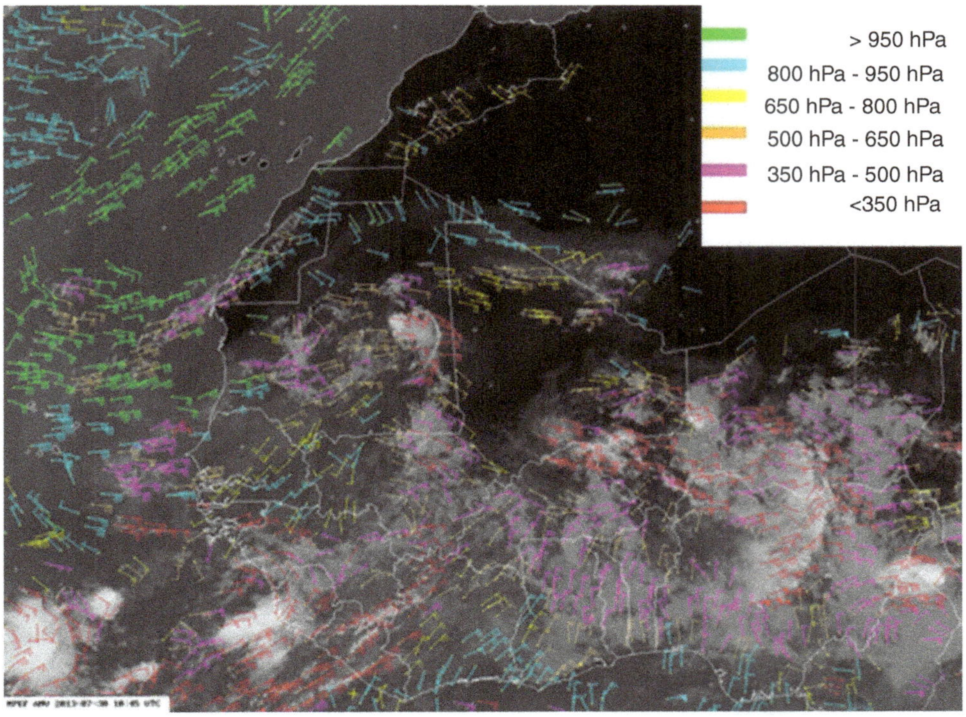


**Figure 9.35** Atmospheric motion vectors for various layers in the atmosphere derived from geostationary imagery. *Source*: reproduced with permission of EUMETSAT.

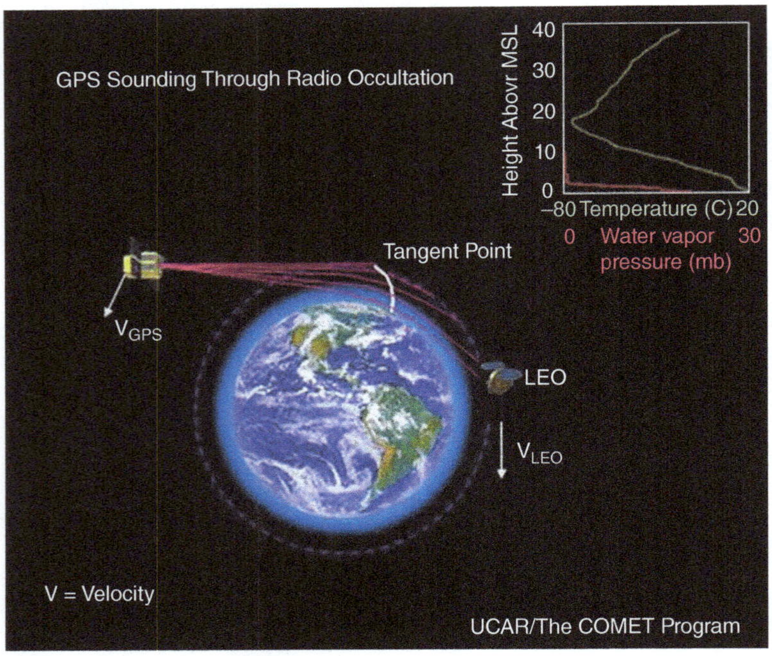


**Figure 9.36** Schematic showing bending of GPS satellite signals due to variations in atmospheric density, a function of temperature and relative humidity. *Source*: reproduced with permission of UCAR/The COMET Program. *Inset:* profiles of temperature and relative humidity (inset) derived from the retrieved refractions.

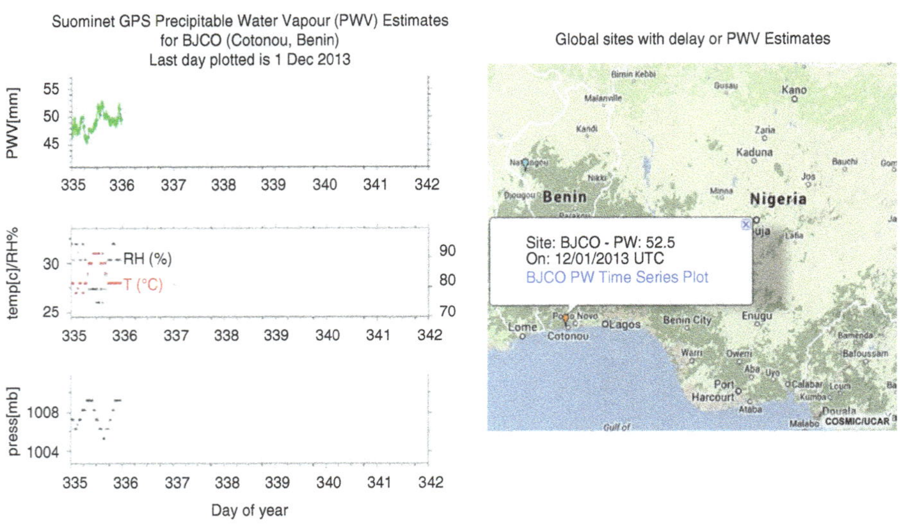


**Figure 9.37** Real-time observations of precipitable water vapour, temperature, relative humidity and pressure from surface GPS instruments at Cotonou, Benin. *Source*: reproduced with permission of COSMIC/UCAR.

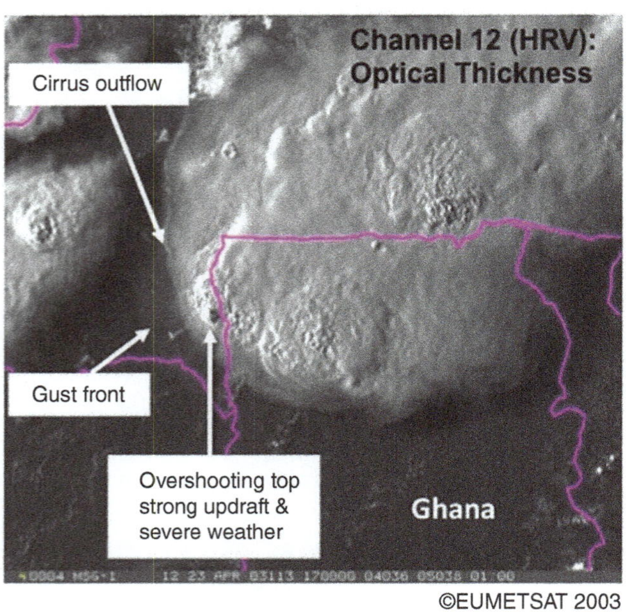


**Figure 9.38** HRV image of convection over Ghana and Burkina Faso. *Source*: reproduced with permission of EUMETSAT.

(a)

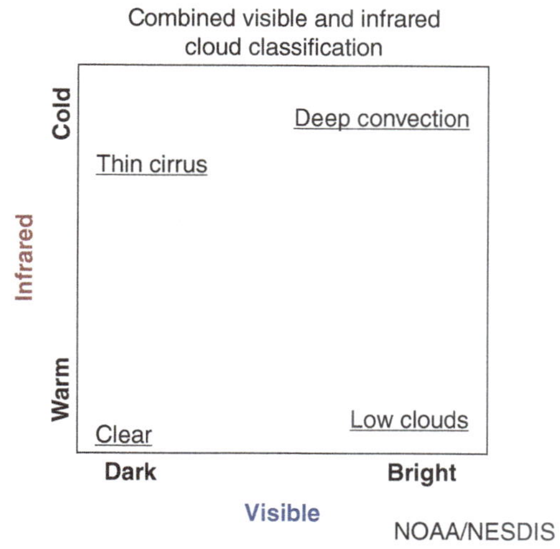


(b)

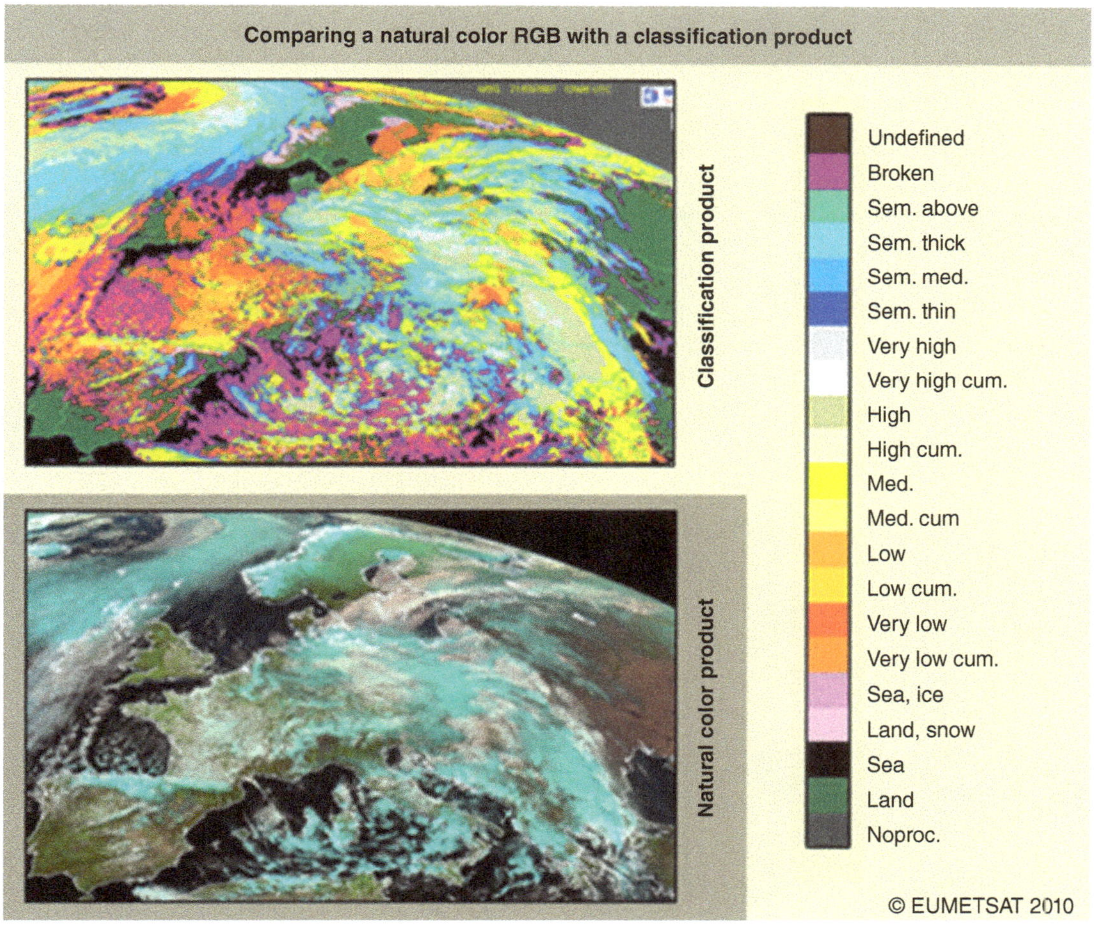


Figure 9.39 (a) Cloud type identification using a combination of visible and IR imagery. (b) Clouds identified by cloud classification compared with natural colour RGB. Main cloud classifications are low, medium and high cumulus and semi-transparent cirrus ('sem.' in the legend of lower panel). Cirrus is also categorized as thin, medium, thick or above other clouds. For example, 'sem. thin' means semi-transparent thin cirrus. *Source*: Reproduced with permission of EUMETSAT.

Cloud types can be identified from their brightness in the visible and emission temperature at IR wavelengths (Figure 9.39a). This technique has weaknesses because of overlap among cloud types. The EUMETSAT Satellite Application Facilities (SAF) Cloud Type image product has several categories based on radiative transfer algorithms and a database of training targets (NWCSAF, 2012). The aim is to differentiate between thin and opaque clouds, water and ice clouds, and roughly estimate the cloud top height. The product identifies major cloud categories, low, medium and high clouds, as well as semi-transparent cirrus. The result is a complex classification image product that, in comparison with the natural texture of the clouds in the RGB product, appears unnatural (Figure 9.39b).

##### 9.1.5.3.1 Sun Glint

When satellite instruments observe sunlight directly reflected by smooth water surfaces, the surface appears brighter than expected (Figure 9.40). This phenomenon, known as sun glint, is often mistakenly classified as cloud cover.

#### 9.1.5.4 Precipitation

##### 9.1.5.4.1 Infrared Estimates of Precipitation

IR emissions are used to estimate precipitation from convective clouds because higher, and thus colder, cloud tops correlate with higher precipitation rates. The advantage of this method is high temporal frequency of images, up to 15 min for geostationary satellites, which is good for studying propagation, thunderstorm development and other nowcasting applications. However, this method cannot detect convective-scale structure, rain from warm clouds, and can struggle to distinguish between thick cirrus and cumulonimbus clouds, which may appear similar.

The Meteosat Multisensor Precipitation Estimate (MPE) product is a combination of high temporal/spatial resolution IR with more accurate rain rate retrieval from microwave sensors. This technique is most useful for mesoscale to large-scale convective rainfall events (e.g. Figure 9.41). For cases of localised precipitation, an algorithm is used to 'smear' the IR and microwave precipitation measurements. While these combination algorithms can detect warm front and orographically induced rainfall, cases are often mislocated by up to 100 km.

##### 9.1.5.4.2 Microwave Observations of Clouds and Precipitation

Microwave satellite sensors use several channels, usually described by their frequency in gigahertz. Channels in atmospheric window regions (Figure 9.15) allow sensing of the surface through clouds. Both window and high-absorption regions are used to derive various products for identifying surface and cloud properties.

Microwave instruments provide better estimates of precipitation than IR because they sense hydrometeors in clouds (Figure 9.42) (Wilheit, 1986; Kummerow *et al.*, 1996). Sensors measure microwave scattering and emission signatures of liquid water or ice particles. Microwave channels have a good atmospheric signal over ocean, but high emission from the land surface diminishes water vapour and liquid water signals. Precipitation is inferred from emission at frequencies <50 GHz, which have the

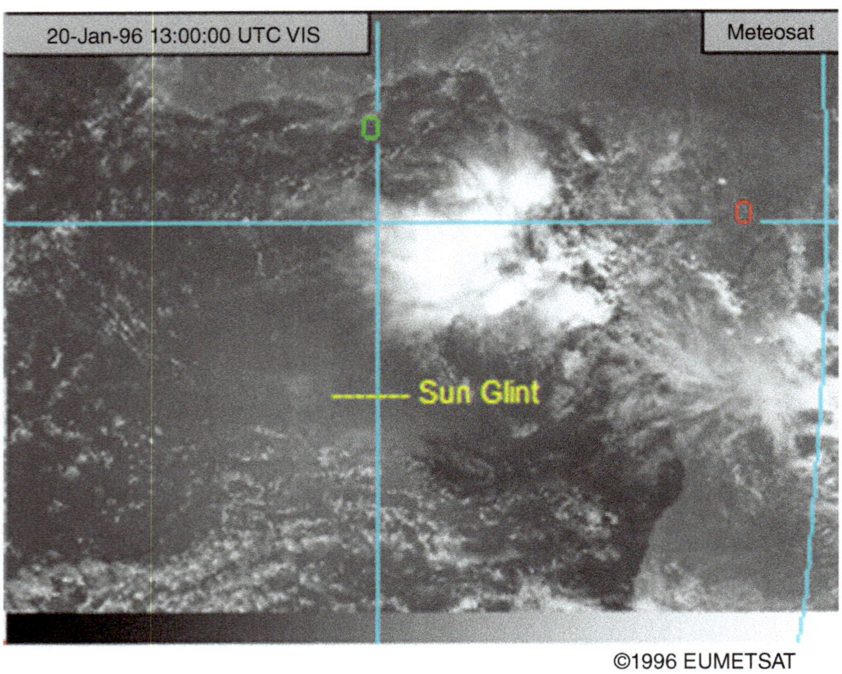


**Figure 9.40** Sun glint over the Gulf of Guinea; note the brightness of the sun glint relative to the rest of the ocean surface. *Source*: Reproduced with permission of EUMETSAT.

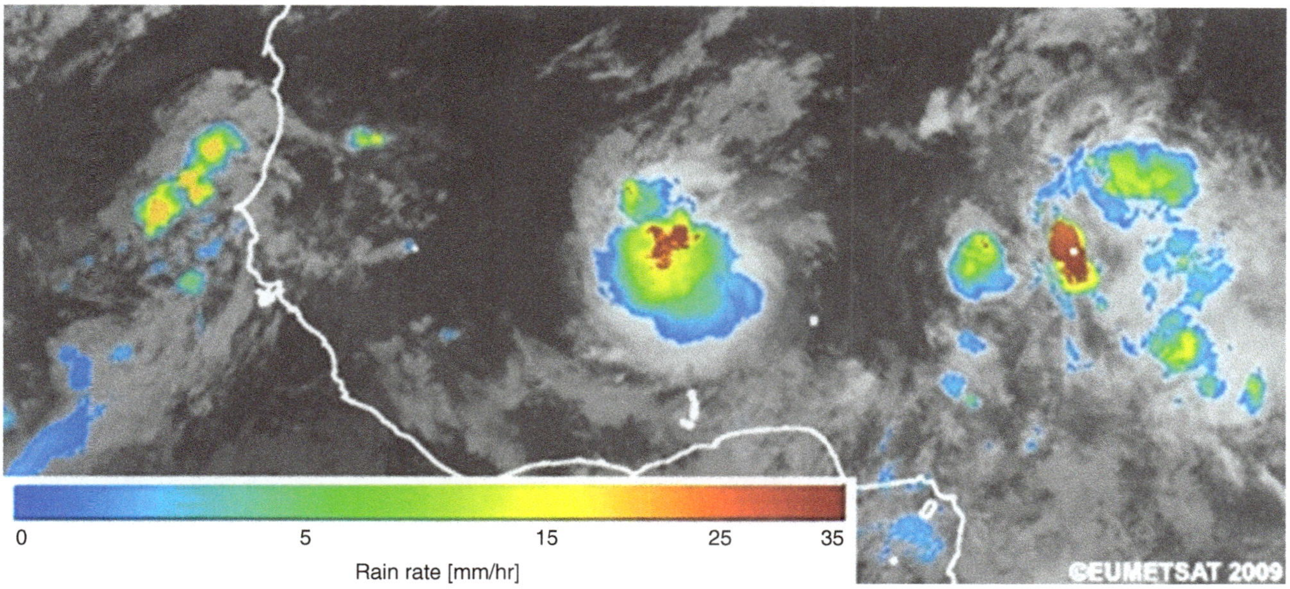


Figure 9.41 Meteosat MPE image showing high rain rates produced by an intensive mesoscale convective system (MCS) that caused record flooding in Ouagadougou on 1 September 2009 (see case study CS04 http://www.cnrm.meteo.fr/waf_handbook_casestudies/). *Source*: reproduced with permission of EUMETSAT.

most direct link to surface precipitation over cold water only. From scattering at frequencies >50 GHz over all surfaces, there is an indirect link to surface precipitation. Specific bands are used for observing clouds and precipitation:

- *The 85–91 GHz band.* Deep convection appears relatively cold. Water clouds and moist air masses have warm brightness temperatures over water surfaces. This band can reveal internal storm structure through thin cirrus anvils. However, it cannot always see circulations associated mainly with low-level water clouds. Its spatial resolution is higher than lower frequency microwave bands.
- *The 37 GHz band.* Shows low-level clouds and rain (Figure 9.42). Water clouds and precipitating clouds appear warm against relatively cold ocean. Displays low-level cloud features because upwelling radiation is largely unaffected by ice particles. It resolves details that are missed by the 85–91 GHz bands.
- *The 23/31 GHz bands.* This combination is used to infer water vapour and liquid water paths.

While microwave alone or combined with IR improves on IR-only estimates, biases remain (Smith *et al.*, 2006).

#### 9.1.5.5 Dust

The Sahara Desert is the primary source of global mineral dust. West Africa is vulnerable to dust storms, which are accompanied by strong winds and reduced visibility, causing hazards to aviation and human health.

Dust is detected by using the differencing of IR channels to distinguish dust from cloud as illustrated by the Meteosat RGB dust product (Figure 9.43). Earlier examples show dust observed by Meteosat natural colour RGB and VIIRS true colour RGB (Figures 9.21 and 9.22).

Dust also influences the formation and intensification of tropical cyclones. Figure 9.44 shows the Sahara air layer dust product created from Meteosat SEVIRI. In this case, a multi-day intense outbreak of dry, dusty air, from West Africa to the Caribbean, inhibits convective weather. Note that dust is an aerosol, like smoke and volcanic ash, but is covered separately because of its chronic effect on weather in West Africa.

#### 9.1.5.6 Fire, Smoke and Haze

As with other aerosols, such as dust and volcanic ash, forecasters need to monitor fires, smoke and haze because they affect air traffic, air quality/health and visibility. A fire is identifiable as a 'high-temperature event' in satellite imagery. However, fire can be indistinguishable from land that appears hot and very bright during daytime.

The near-shortwave IR channels (Figure 9.45) are most useful for identifying fires because of their peak emission at those wavelengths (Figure 9.7), their temperature sensitivity relative to the longwave IR and their sub-pixel response. The difference between the 3.9 μm and 10.8 μm channels is used for detection, as the radiance temperature at 3.9 μm is much greater than the response at 10.8 μm. The sub-pixel response at 3.9 μm is also used to detect fires. As the fractional fire coverage increases, the radiance temperature at 3.9 μm increases relative to 10.8 μm due to the shorter wavelength response to the warm part of the pixel.

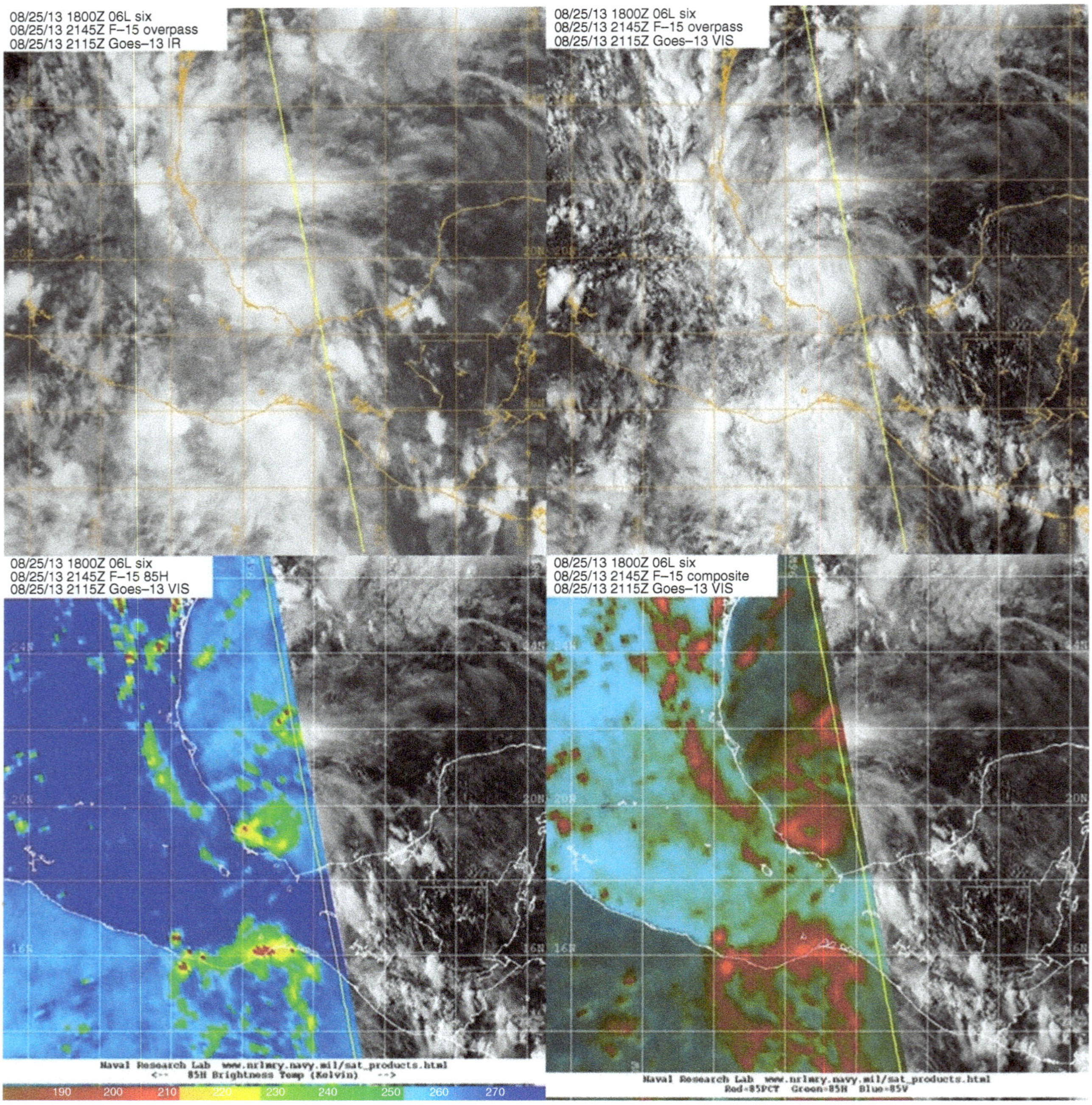


**Figure 9.42** Tropical storm over the Gulf of Mexico observed by (a) geostationary IR and (b) geostationary visible sensors, (c) polar-orbiting microwave 85 GHz polarisation-corrected temperature (PCT) and (d) RGB composite of the 85 GHz PCT and 85 GHz horizontal and vertical polarisation. The RGB shows deep convection, low-level water vapour and precipitation. PCT <255 K almost always indicates precipitation. The eye and eyewall are apparent in the centre of the image in the microwave, but obscured in the IR and visible © NOAA and US Navy/NRL.

Smoke can be detected by combining the 3.9 μm with the HRV, but even then detection is limited over land because there is less contrast over water. With MODIS true colour RGB, smoke is a brighter white than dust (Figure 9.46).

Aerosol indices are based on the reflectance and absorption properties of aerosols (Hyer *et al.*, 2011). Aerosols are best detected using wavelengths smaller than 0.7 μm because there is less reflectance from the surface (forest, types of soil, etc.) to interfere in the detection (Figure 9.11). Products such as the MODIS true colour RGB are used to distinguish between green and charred vegetation. The Global Ozone Monitoring Instrument 2 (GOME-2) on MetOp has four channels for observing ozone, $NO_2$, $SO_3$, and other trace gases associated with natural and industrial combustion. Daily maps of aerosol indices are generated twice per day to help track aerosols in the troposphere. For West Africa,

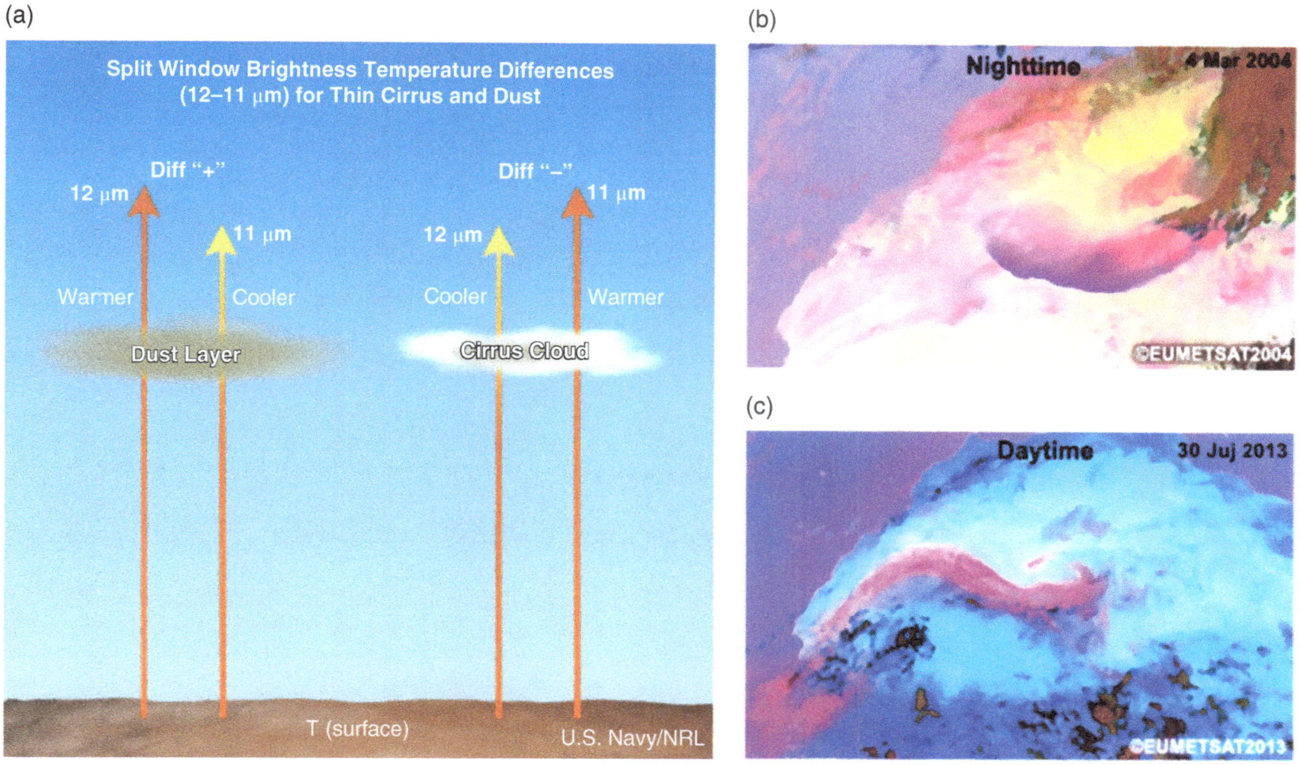


**Figure 9.43** (a) Conceptual model of the split window used to differentiate between dust and cirrus. Dust RGB (b) during night-time 4 March 2004 and (c) daytime 30 July 2013 (same as Figure 9.22b). *Source*: (b, c) reproduced with permission of EUMETSAT.

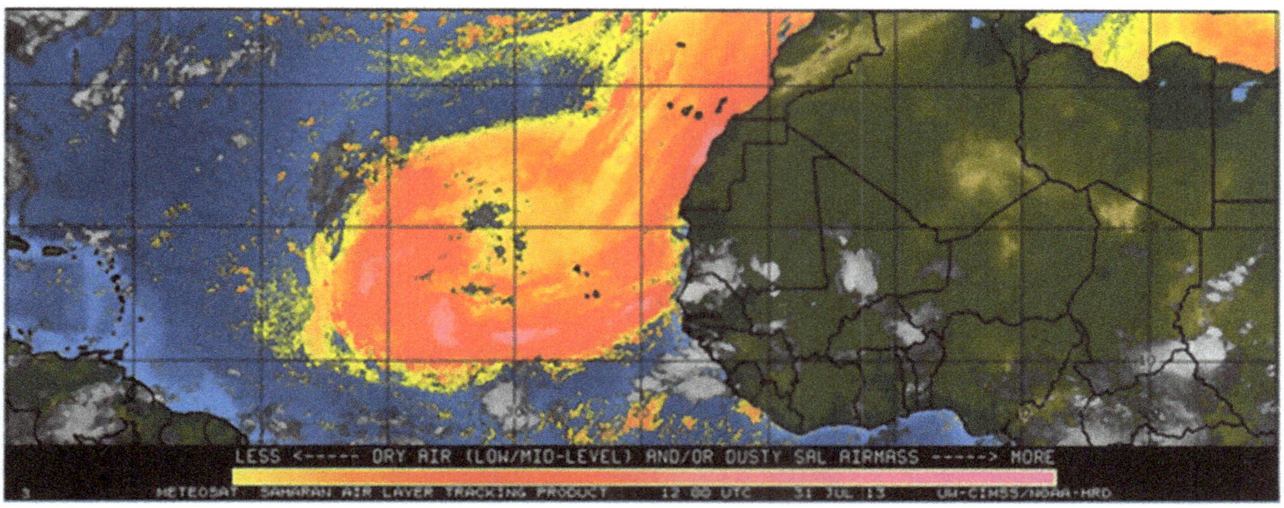


**Figure 9.44** A major dust storm, 31 July 2013, is monitored using the Sahara air layer tracking product derived from Meteosat channels. *Source*: reproduced with permission of University of Wisconsin–CIMSS and NOAA Hurricane Research Division.

the coarser dust aerosols typically dominate in the north, while finer smoke aerosols from fires are further south. The Cloud Aerosol Lidar and Infrared Pathfinder Satellite Observer provides profiles of aerosols at high resolution but is mainly used for climate studies.

#### 9.1.5.7 Volcanic Ash

Volcanic eruptions cause several natural hazards, with volcanic ash posing the greatest threat to larger areas than other volcano hazards, particularly to aviation. Ash also aggregates into large particles that fall quickly and

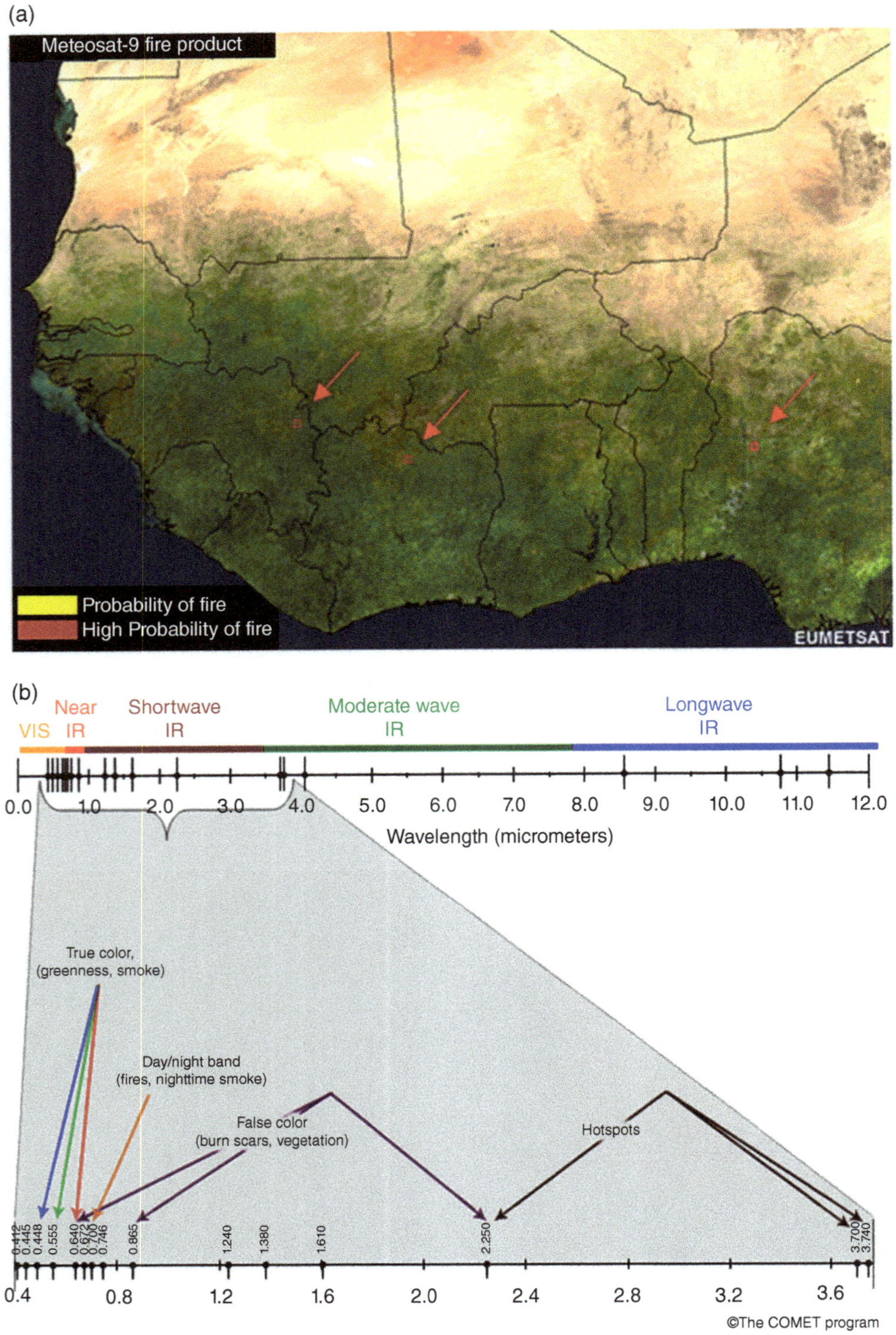


**Figure 9.45** (a) Meteosat-9 Fire Product example. (b) Visible and shortwave IR bands used to monitor fires, smoke and other aerosols. Labels highlight RGBs created from VIIRS measurements. *Source*: reproduced with permission of EUMETSAT and the COMET Program.

causes grave harm to life and property. Volcanic ash is tracked using special satellite products that help to distinguish between ash and other atmospheric constituents, and between ash and surface features.

Volcanic ash has an abnormal negative temperature difference between the 11 and 12 μm IR channels ('split-window' or 'reverse absorption' technique; Section 9.1.4.2). In the resulting image, volcanic ash is differentiated from other clouds. Some false alarms can be produced in the tropics because ice-topped clouds have a similar IR temperature difference. These can usually be distinguished with knowledge of the

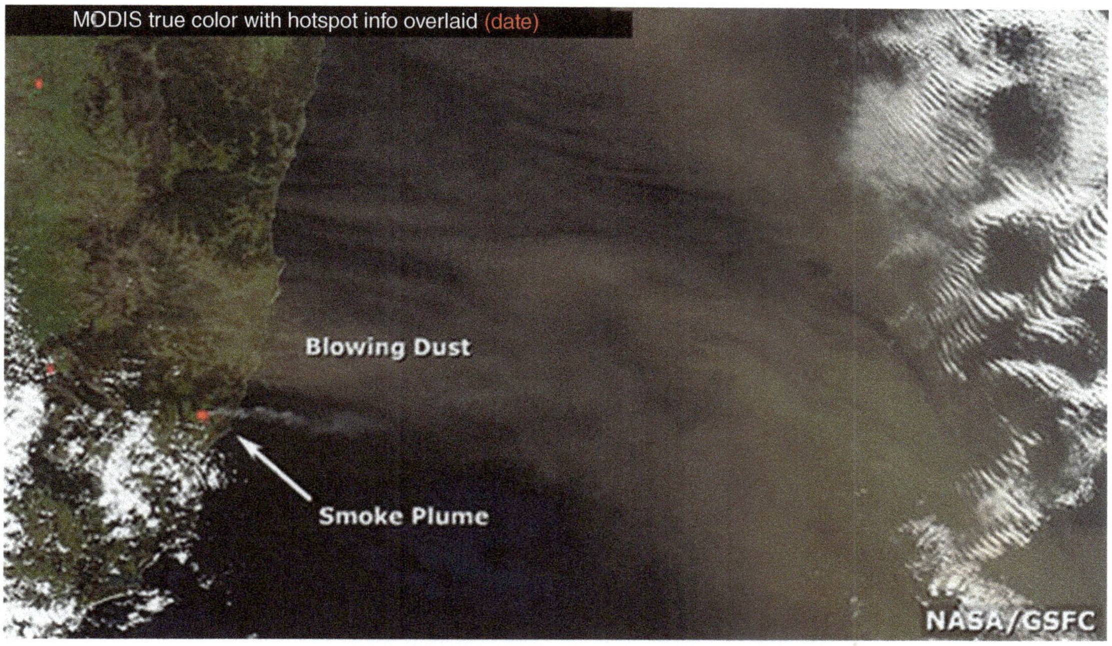


Figure 9.46 MODIS true colour image of smoke and blowing dust off the east coast of Australia. Fires are hot spots in red. Clouds are bright white. Smoke is the grey line extending from a fire spot.

location of active volcanoes, plus associated Sigmets or Volcanic Ash Advisories. The Meteosat Ash product, dust RGB, and the air mass RGB have been applied to volcanic ash monitoring for volcanoes in Africa and Europe (Figure 9.47). The HRV is also used for detecting ash during the morning because of the strong forward scattering of sunlight by ash particles.

#### 9.1.5.8 Lightning

The next generation of geostationary satellites will have lightning sensors (e.g. Figure 9.48). Lightning is a deadly hazard; its variability is an indicator of thunderstorm severity as it is positively correlated with convective available potential energy. The distribution and charge vary with the convective cycle of thunderstorms, so lightning data can be used in nowcasting. Lightning is also a source for the natural generation of nitrogen oxides and tropospheric ozone. The TRMM satellite Lightning Imaging Sensor (LIS) detects lightning with a small, solid-state camera with special filters that admit only the peak optical wavelength emitted by lightning. It separates background emissions from weak lightning signals, thereby maintaining about 90% detection efficiency. In addition to satellite lightning sensors, lightning strikes are also monitored using the UK Arrival Time Difference lightning network, available via EUMETCAST in Africa.

#### 9.1.5.9 Surface Wetness and Vegetation

Surface wetness parameters, such as soil moisture, and vegetation indices can be used to forecast boundary-layer gradients that initiate convection (see Sections 4.1.4.4 and 4.1.5). However, they are most useful for monitoring droughts and for subseasonal and seasonal forecasting.

Soil moisture is typically derived from microwave retrievals and is currently available operationally from a combination of LEO satellite sensors (Figure 9.49). For example, the Microwave Imaging Radiometer using Aperture Synthesis (MIRAS) from the European Space Agency senses emissions from 1.427 GHz from soil, to a depth of about 5 cm, and vegetation cover with water content up to 5 kg m$^{-2}$ (Kerr *et al.*, 2010). MIRAS uses interferometry, which measures the phase difference of incident radiation between waves at two or more receivers that are a known distance apart. Each area along the orbit path is seen from multiple angles, and the origin of the waves can be determined as long as the observed phase difference can be related to waves emitted at the same time.

Common vegetation indices are derived from differences in the visible and near-IR reflectance. The most commonly used vegetation index, the normalised difference vegetation index (NDVI), is based on the principle that dense vegetation has a stronger signal in the near-IR than in the visible spectral bands (Figure 9.11), and is computed as

$$\mathrm{NDVI} = \frac{\mathrm{Ch(NIR)} - \mathrm{Ch(Vis)}}{\mathrm{Ch(NIR)} + \mathrm{Ch(Vis)}} \tag{9.13}$$

Fluorescence – light emitted during photosynthesis – is now being used to monitor changes in vegetation (NASA,

**Figure 9.47** Volcanic ash and gases emanating northwestward from Ethiopia are observed using Meteosat ash RGB (upper) and the air mass RGB (lower). *Source*: reproduced with permission of EUMETSAT.

2013). The GOME-2 on MetOp-A detects this faint fluorescence signal. Measurements are taken from a dark section of the IR part of the solar spectrum, which has little background light, thus allowing detection of the faint fluorescence. Satellite-derived fluorescence, first documented in 2011, can provide information about crop stress and related evapotranspiration feedbacks to weather and climate. Fluorescence measurements, which complement the greenness measures such as NDVI, detect changes in plant productivity more quickly. Fluorescence data from GOME-2 are averaged over areas of 50 $km^2$ about every 10 days.

#### 9.1.5.10 Ocean Surface Winds

Microwave satellite sensors are the primary means of observing the global ocean surface because they are sensitive to small-scale roughness on the ocean surface caused by near-surface winds (e.g. foam and breaking waves) (Gaiser, 2004). Since the sensors are on LEO satellites, individual sensors have gaps in their coverage; for example, the Advanced Scatterometer (ASCAT) satellite provides high-resolution wind direction and speed over the surface ocean but has wide gaps in its orbits (Figure 9.50). However, collectively, satellites in sequence provide three-hourly coverage, which is valuable for coastal and offshore forecasting.

### 9.1.6 Weather Radar

#### 9.1.6.1 Weather Radar Basics

Radar, originally an abbreviation for radio detection and range, rests on the ability of objects to scatter incident radiation. Weather radars transmit high-frequency microwave (millimetre–centimetre scale) pulses to the atmosphere and measure the power returned or 'backscatter' or 'echoes' to the radar receiver (Figure 9.51a).

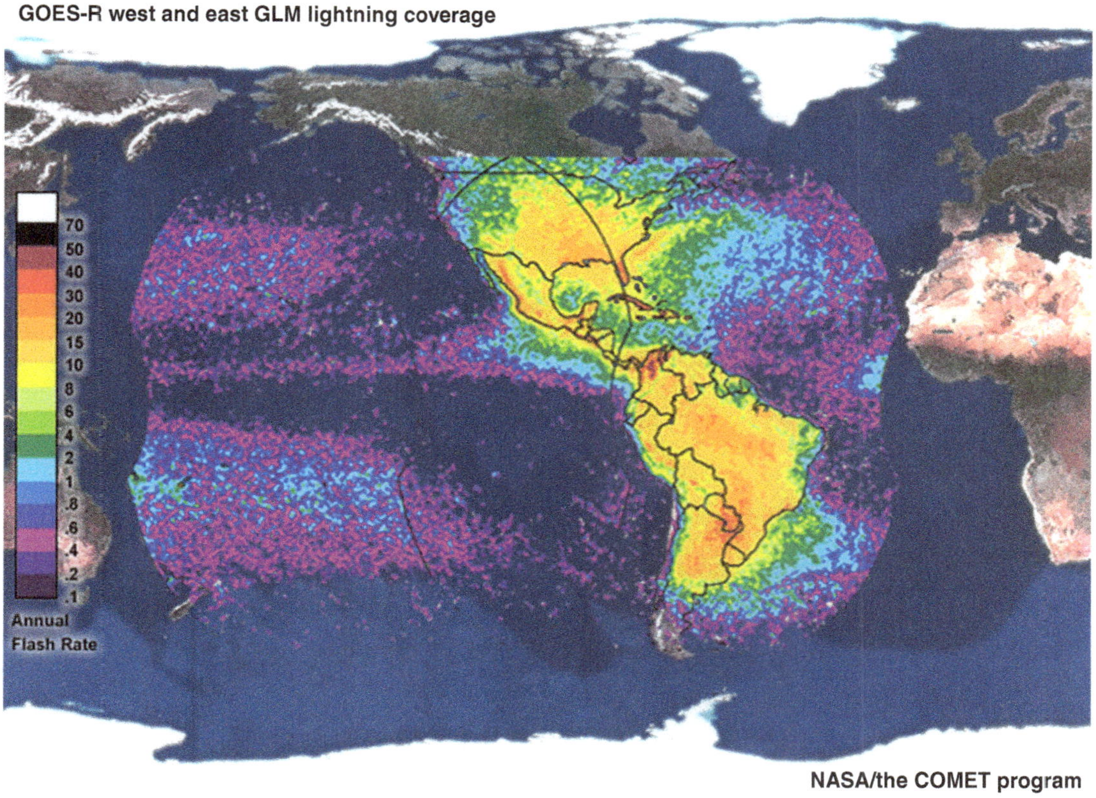


Figure 9.48 Simulation of lightning flash density coverage for the planned geostationary lightning mapper on GOES. *Source*: reproduced with permission of the COMET Program

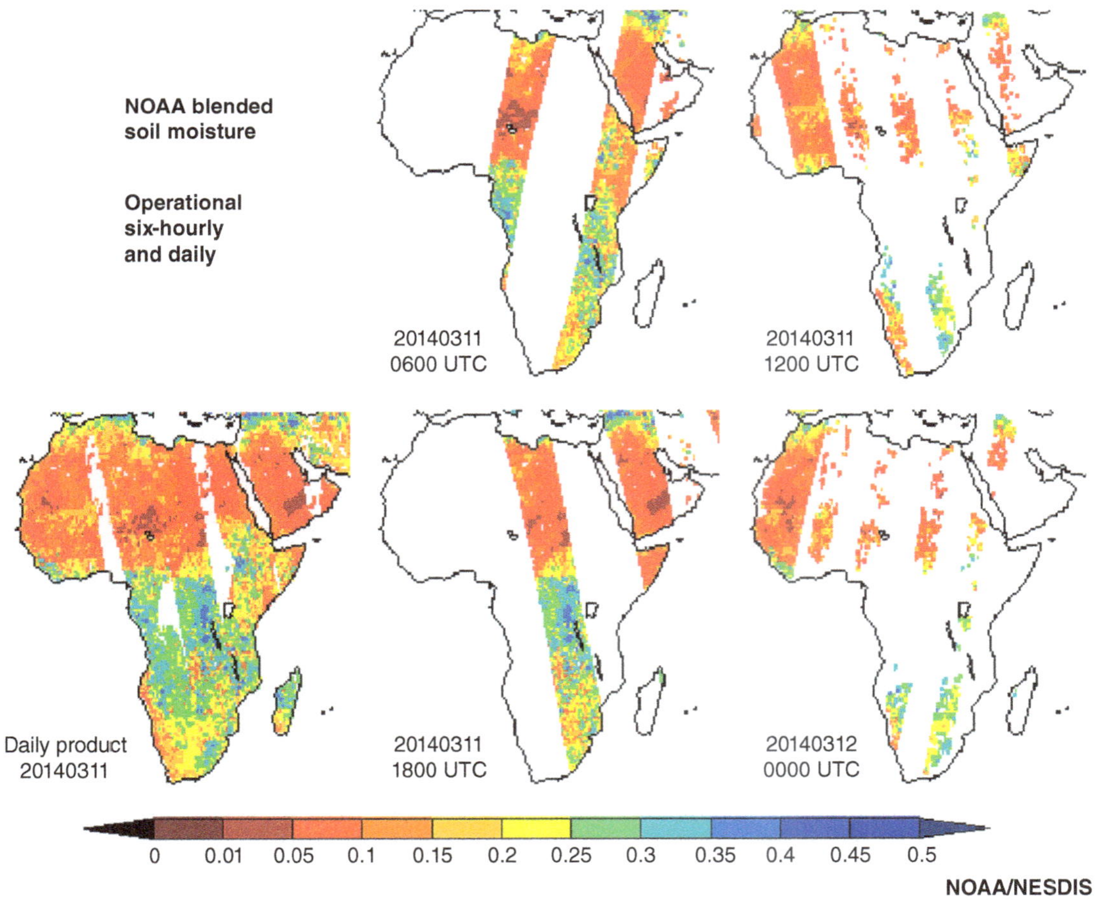


Figure 9.49 Example of blended soil moisture (percentage of water by volume of soil) from satellite microwave sensors available operationally, on a 0.25° × 0.25° grid, at six-hourly and daily intervals.

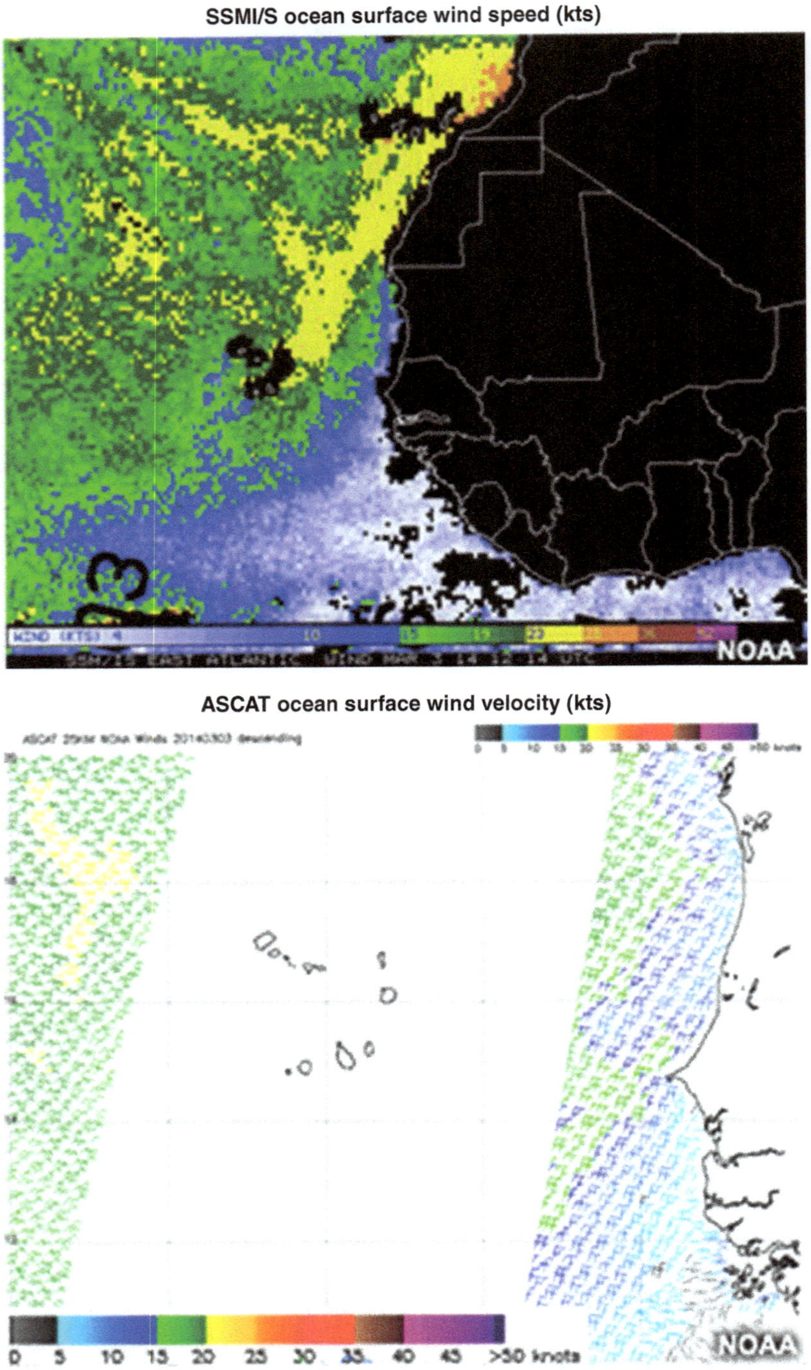


Figure 9.50 Special Sensor Microwave Imager/Sounder (SSMIS) wind speed (left) and Advanced Scatterometer (ASCAT) wind speed and direction (right) on 3 March 2014.

In operational mode, the radar typically scans in circles (plan position indicator, or PPI scans), as the antenna is raised through higher and higher preset angles to provide a three-dimensional view of the atmosphere (Figure 9.51b). The radar receiver amplifies the received power, which is many orders of magnitude smaller than the transmitted radar pulse. This amplitude is used to calculate the radar reflectivity factor $Z$, the most commonly used variable in weather radar analysis. The strength of the returned pulse depends on particle size, the number of particles in the volume sampled by the pulse, their state (e.g. ice or liquid) and their shape. The reflectivity factor $Z$ can be expressed as

$$Z = \int N(D) D^6 \, dD \tag{9.14}$$

where $Z$ is the reflectivity factor, $D$ is the drop diameter and $N(D)$ is the number of drops of a given diameter per cubic metre.

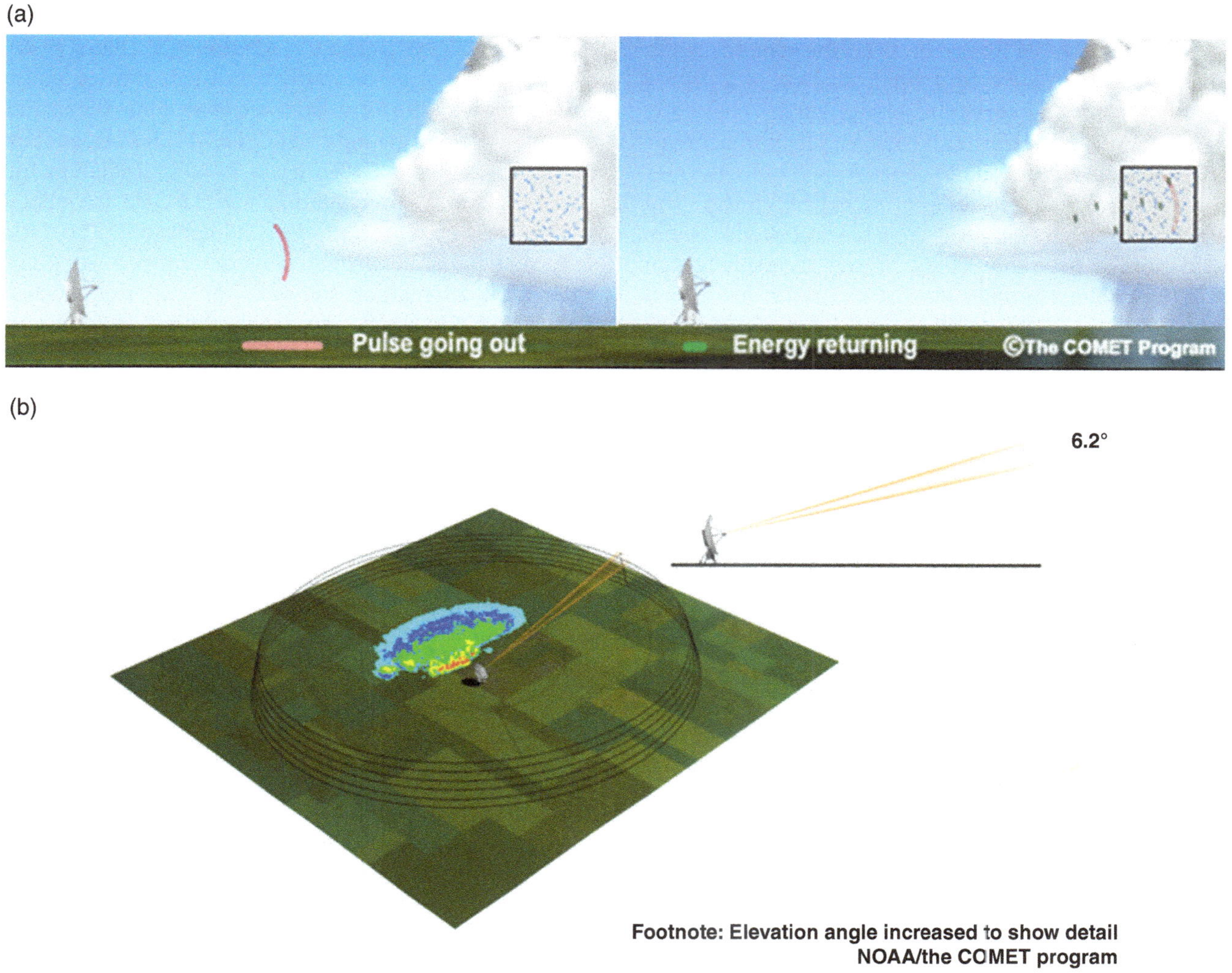


**Figure 9.51** Schematics showing (a) how radar operates to detect hydrometeors in a cloud and (b) how radar fills a volume scan by taking repeated azimuth scans. *Source*: reproduced with permission of NOAA/the COMET Program.

The returned power signal is interpreted to determine estimated intensity of hydrometeors, the evolution of precipitating weather systems, frontal zones and boundaries, dust storms and non-meteorological targets, such as insects and migrating birds. Because atmospheric particle sizes range over many orders of magnitude, so does $Z$. Therefore, reflectivity is usually measured on logarithmic scale, with units of decibels of $Z$ or dBZ.

Generally, we interpret $Z$ as indicating precipitation, and the larger the value of $Z$, the greater the precipitation intensity. However, while Z is proportional to $D^6$, the rain rate is proportional to $D^3$; therefore, radar measurements are biased toward larger diameter hydrometeors. The same value of $Z$ can be due to a few large hydrometeors or many smaller drops, and large hailstones can dominate the reflectivity but contribute little to the rain rate.

From empirical measurements of drop size distributions the rain rate $R$ can be related to $Z$ using a power law:

$$Z = AR^b \tag{9.15}$$

where $A$ and $b$ are constants derived from empirical data and $Z$ units are $mm^6\ mm^{-3}$ and $R$ is in $mm\ h^{-1}$.

The $Z$–$R$ relationship is non-unique. '$Z$–$R$' relationships make assumptions about the raindrop size distribution and size of hydrometeors encountered by the radar beam. Different $Z$–$R$ relationships are used for different atmospheric conditions. For example, a 'tropical' $Z$–$R$ relationship $Z = 250R^{1.2}$ is used with warm moist air and high concentrations of large droplets.

Sampling and other issues should be considered when interpreting radar images. The returned echo weakens as you move further away from the radar because the radar beam broadens with distance and the proportion of the beam that has rain decreases, which reduces the echo power. Increasing distance from the radar also means that the beam points higher up from the ground, as the radar is angled upward by a small fraction of a degree and because of the curvature of the Earth. Therefore, the appearance of a storm can change as it approaches or moves away from the radar. Echoes from buildings, trees and sea spray can create what is called 'clutter,' and mountains can partially or entirely block the radar beam. Refraction of the radar beam by the atmosphere can cause anomalous propagation of the radar beam and display echoes that are displaced from their target locations.

Modern weather radars estimate velocity by using the 'Doppler shift,' the apparent shift in the frequency and wavelength of a wave perceived by an observer moving relative to the source of the wave, as hypothesised by Johann Christian Andreas Doppler. The radial velocity of targets can be determined from the phase shift between the pulse transmitted from the radar and the backscatter received. A positive phase shift indicates movement toward the radar, while a negative shift indicates movement away from the radar. A positive to negative velocity couplet signifies rotation, which helps forecasters to detect mesocyclones and tornadoes.

#### 9.1.6.2 Ground and Satellite Weather Radar Products

Base reflectivity, obtained from the lowest level azimuth scan of the radar, is the most commonly used weather radar measurement. It allows for detection of storm structure and intensity of hydrometeors at or near the surface.

For example, the radar images from Niamey on 11 August 2006 (Figure 9.52) shows the horizontal and vertical structure of an intense squall line. The PPI shows a vigorous leading convective line (reflectivity >45 dBZ), trailing stratiform rain (25–40 dBZ) and a transition zone of weaker reflectivity between them. The RHI of reflectivity shows the leading updraft reaching nearly 15 km, an older weaker updraft behind and a forward anvil. The RHI of radial velocity shows the rearward tilt in the updrafts and a mid-level maximum inflow toward the rear, the 'rear inflow jet' that descends toward the front of the squall line bringing strong winds to the surface (>15 m $s^{-1}$).

The TRMM Precipitation Radar, the first satellite precipitation radar, sends out pulses at 13.6 GHz. Launched in 1997, it has a spatial resolution of 4 km horizontally and 250 m vertically, which provides three-dimensional views of tropical weather systems across the global tropics (e.g. Figure 9.53). Initially developed for a 3-year research mission, the TRMM satellite lasted for almost two decades, becoming a valuable tool for operational forecasting (e.g. Rozante *et al.*, 2010); its service will end in 2016. The precipitation radar swath was increased to 247 km (from 220 km) after TRMM was moved to a higher orbit in 2001. The CloudSat cloud radar, a polar orbiting sensor used primarily for climate studies, emits radio waves at 94 GHz and retrieves fine details of various types of clouds. The GPM Dual-frequency Precipitation Radar (DPR) has a 35 GHz frequency for measuring heavy precipitation and 14 GHz frequency for measuring ice precipitation and light rain. The rain/snow detection is expected to be determined using the differential attenuation between the two frequencies. The GPM DPR has horizontal resolution of 5 km and vertical resolution of 250 m at the 14 GHz band and 250–500 at the 35 GHz band. Its pulse repetition frequency is set to have 0.2 mm $h^{-1}$ sensitivity. These satellites see the same location twice per day at most, so are limited operationally, but they are the only radar observations of most of West Africa.

#### 9.1.6.3 Wind Profilers

Wind profilers are similar to Doppler radars except that they point vertically up from the surface. They make high-resolution measurements of boundary-layer parameters such as wind velocity, temperature and humidity. Forecasters use wind profilers to identify low-level wind shear (an aviation hazard), convergence zones and sharp gradients in temperature and moisture that can initiate severe storms. Some wind profilers have been installed at West African airports and were used in the AMMA field programme (Lebel *et al.*, 2010). Field projects in West Africa have used profilers to monitor low-level jets, the sea-breeze front, convection initiation, fog and low-level clouds. Wind profilers also provide the best measurements of vertical air motion inside convective storms because they can distinguish between the air motion and hydrometeor fall velocities.

## 9.2 Operational Methods

Among the challenges for weather forecasters in West Africa are heavy rainfall, strong winds, dust and sand storms, coastal and radiation fog and stratus, smoke from biomass burning and, in rare cases, volcanic ash. Many of these phenomena can be monitored using satellite and radar products. This section describes the platforms and sensors that are most useful for forecasting; some were introduced in Section 9.1. Next, we review commonly used remote sensing products. Then we recommend appropriate products for phenomena of different scales, starting with large-scale features (e.g. the intertropical discontinuity (ITD)), and then smaller and smaller scales until, finally, local weather phenomena that evolve rapidly (e.g. thunderstorms and cloud lines). Some specific forecast challenges are presented for certain applications (e.g. aviation forecasting).

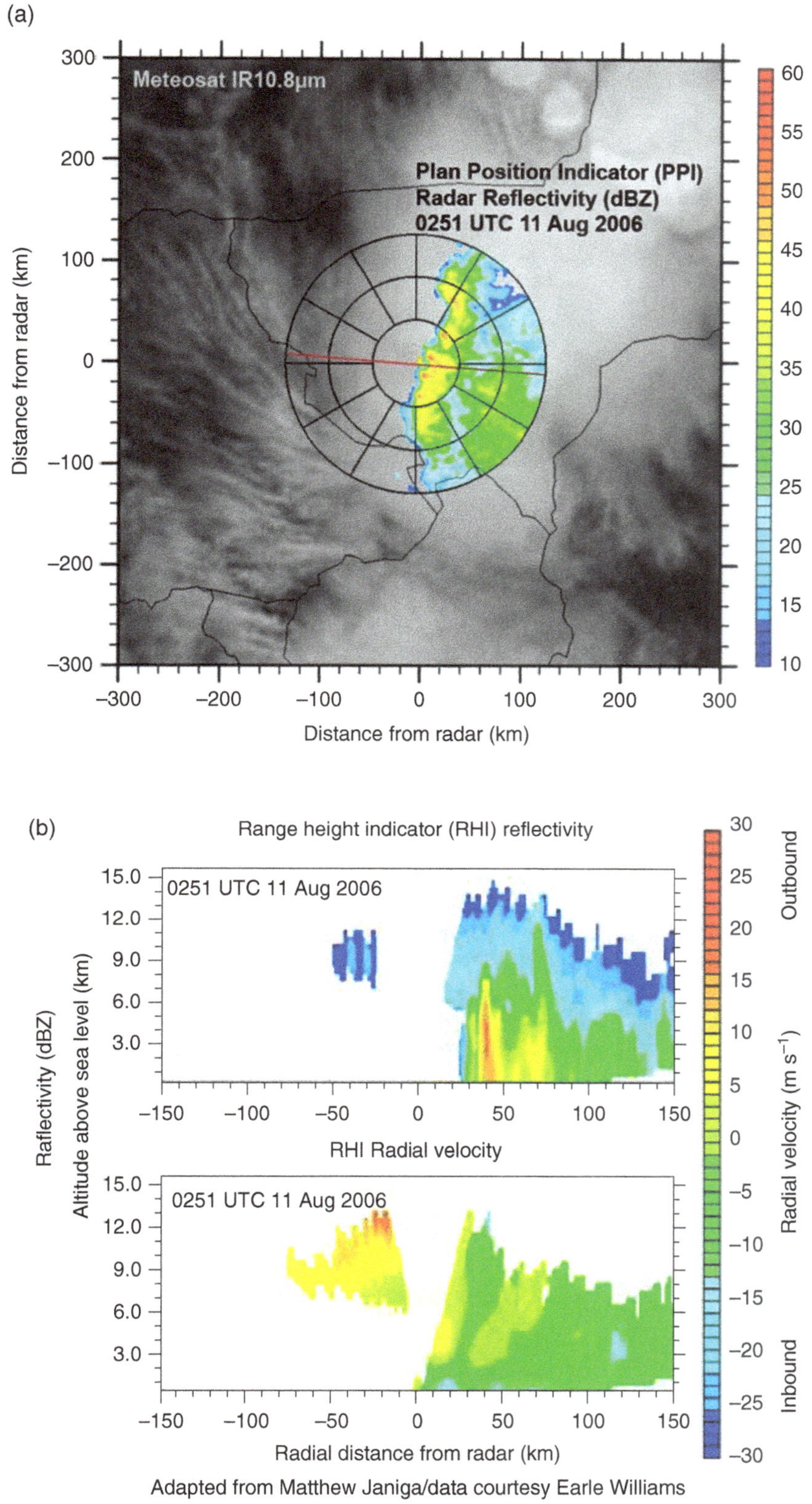


**Figure 9.52** (a) PPI base reflectivity produced by an MCS over Niamey, Niger, at 0251 UTC on 11 August 2006 overlaid on Meteosat IR 10.8 μm. Corresponding range–height indicator (RHI) reflectivity (b) and radial velocity (c) taken along the red line in (a). Notice in (b) and (c) the cone-shaped clear area directly above the radar – an area that is not scanned, and so termed 'the cone of silence'. *Source*: figure courtesy of Matthew Janiga; data courtesy of Earle Williams.

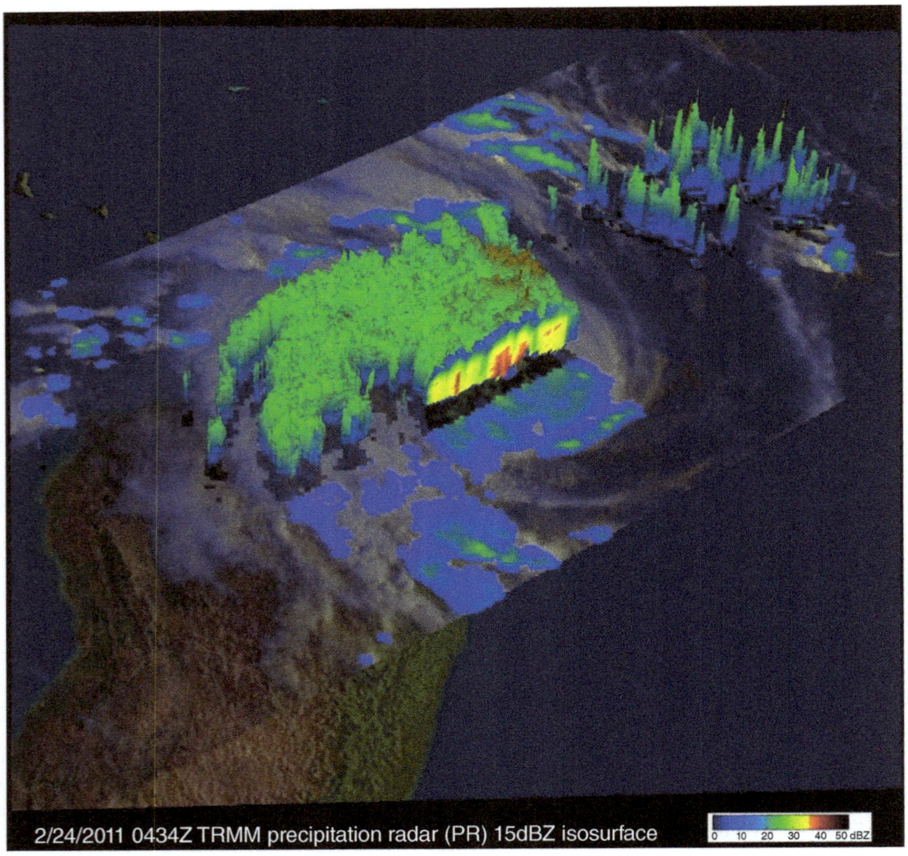


Figure 9.53 Three-dimensional image of radar reflectivity from TRMM Precipitation Radar showing a tropical cyclone making landfall in Madagascar. *Source*: reproduced with permission of NASA/GSFC.

We also discuss standard operational procedure and how to integrate remote sensing products into the analysis and forecasting process. Section 9.2.9 is a list of satellite products that are freely available online.

## 9.2.1 Platforms and Sensors

### 9.2.1.1 Geostationary Satellites and Sensors

The Meteosat SEVIRI is the primary sensor used for operational forecasting in Africa and Europe. Table 9.3 shows the MSG SEVIRI channels and how they are applied to observations. SEVIRI has:

- 12 spectral channels (Figure 9.13)
- images every 15 min
- 3 km horizontal sampling distance at sub-satellite point (0, 0)
- HRV channel with 1 km sampling distance.

The highest resolution Meteosat images are from the HRV, which shows features not seen with the regular visible channel (Figure 9.54). Note that, from 0000 to 1400 UTC the African HRV window only covers the area east of 0° longitudes; in order to optimise the coverage of the Indian Ocean (Figure 9.54c).

The MTG satellites will have 16 channels with higher space and time resolution (2 km IR, 1 km visible available every 10 min).

### 9.2.1.2 Low-earth-orbiting Satellites

Forecasters in West Africa can use information from some LEO satellite sensors for operational forecasting when their orbits are favourable, including MetOp, NOAA, Suomi-NPP, MODIS, TRMM and GPM. For example, the MetOp satellites carry several key instruments for forecasting, including:

- The Advanced Very High Resolution Radiometer, whose six spectral bands provide images of land, water, clouds, SST, ice, snow and vegetation cover.
- The High Resolution Infrared Radiation Sounder and the IASI measure upwelling surface IR radiation to obtain very accurate temperature and humidity profiles. The IASI has 8461 channels between 3.62 μm and 15.5 μm (Figure 9.31).
- The Advanced Microwave Sounding Unit-A (AMSU-A; Ferraro *et al.*, 2005) and Microwave Humidity Sounder (MHS) provide measurements of precipitation as well as temperature and humidity profiles. AMSU has

Table 9.3 MSG SEVIRI channels, spectral bands, and main observational applications.

| Channel no. | Spectral band (μm) | Characteristics of spectral band (μm) Centre | Min. | Max. | Main observational application |
|---|---|---|---|---|---|
| 1 | VIS0.6 | 0.635 | 0.56 | 0.71 | Surface, clouds, wind fields |
| 2 | VIS0.8 | 0.81 | 0.74 | 0.88 | Surface, clouds, wind fields |
| 3 | NIR1.6 | 1.64 | 1.50 | 1.78 | Surface, cloud phase |
| 4 | IR3.9 | 3.90 | 3.48 | 4.36 | Surface, clouds, wind fields |
| 5 | WV6.2 | 6.25 | 5.35 | 7.15 | Water vapour, high-level clouds, atmospheric instability |
| 6 | WV7.3 | 7.35 | 6.85 | 7.85 | Water vapour, atmospheric instability |
| 7 | IR8.7 | 8.70 | 8.30 | 9.1 | Surface, clouds, atmospheric instability |
| 8 | IR9.7 | 9.66 | 9.38 | 9.94 | Ozone |
| 9 | IR10.8 | 10.80 | 9.80 | 11.80 | Surface, clouds, wind fields, atmospheric instability |
| 10 | IR12.0 | 12.00 | 11.00 | 13.00 | Surface, clouds, atmospheric instability |
| 11 | IR13.4 | 13.40 | 12.40 | 14.40 | Cirrus cloud height, atmospheric instability |
| 12 | HRV | Broadband (about 0.4–1.1 μm) | | | Surface, clouds |

*Source*: Reproduced with permission of EUMETSAT.

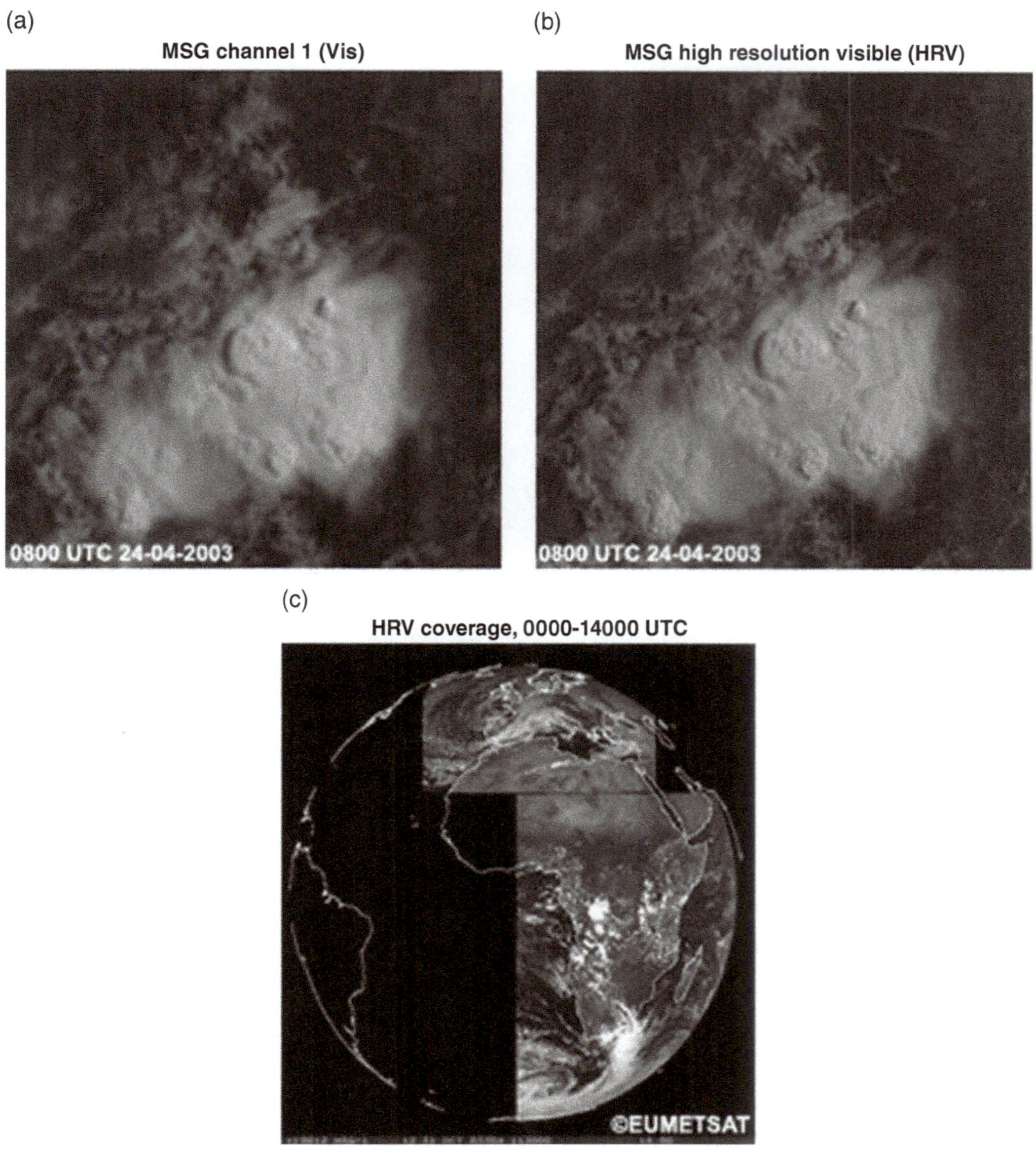


Figure 9.54 Cumulonimbus clouds over Nigeria as seen by (a) channel 1 VIS and (b) channel 12 HRV at 0800 UTC 24 April 2003. (c) Coverage of the HRV from 0000 to 1400 UTC. *Source*: reproduced with permission of EUMETSAT.

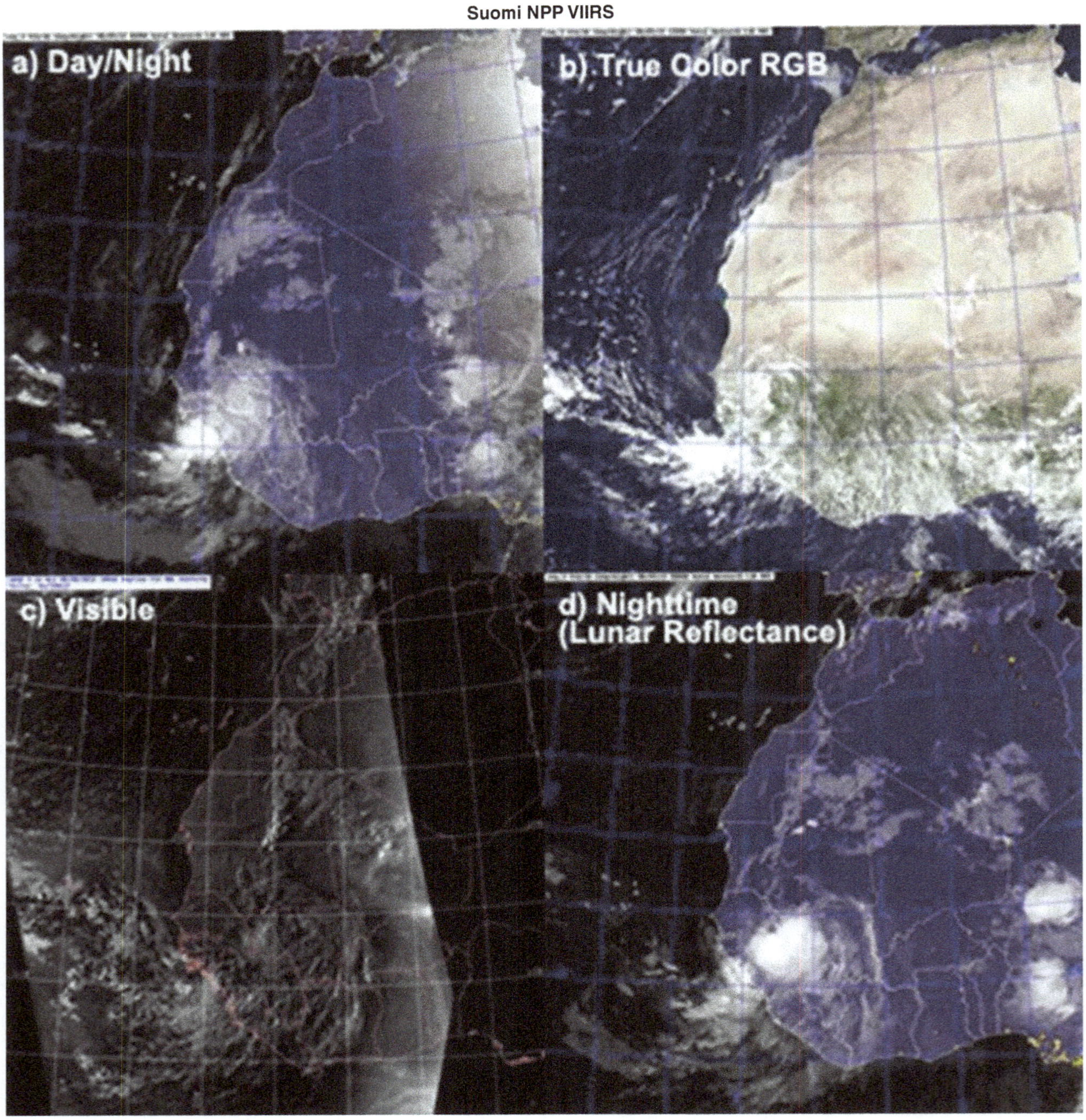


**Figure 9.55** Examples of VIIRS: clockwise from upper left, day/night, true colour RGB, visible, night-time (lunar reflectance) products. *Source*: reproduced with permission of US Navy NRL.

15 discrete channels (23–90 GHz) and MHS has five (89–190 GHz).

- GOME-2, a spectrometer that gives detailed profiles of ozone, nitrogen dioxide, water vapour, oxygen and other gases. Useful for monitoring air quality and volcanic emissions.
- Global Navigation Satellite System Receiver for Atmospheric Sounding – GPS receiver that uses occultation of GPS signals to derive soundings of temperature and humidity (Figure 9.35).
- ASCAT is an active microwave sensor that measures ocean surface wind speed and direction (Figure 9.49).

Suomi-NPP, the first of a new generation of satellites, launched in October 2011, has some instruments that are useful to forecasting:

- VIIRS, a scanning radiometer provides visible and IR imagery, including a night-time visible product (Figure 9.55). It monitors fires, clouds and aerosols properties. Other retrievals are for climate.

- Advanced Technology Microwave Sounder, a 22-channel sounder that provides profiles of atmospheric temperature and humidity.
- Cross-track Infrared Sounder, a spectrometer with 1305 channels, provides high-resolution, three-dimensional temperature, pressure and moisture profiles used in weather forecast models.

Its other two instruments, the Ozone Mapping Profiler Suite, aimed at maintaining stratospheric ozone records, but can track air pollution in the upper troposphere, and the Clouds and the Earth's Radiant Energy System, are used for mainly for climate.

#### 9.2.1.3 Satellite Operational Analysis Tools and Products

- HRV, shortwave IR, thermal IR, water vapour IR imagery (Meteosat, NOAA, MetOp).
- Meteosat RGBs (air mass, dust, convection/severe storm, day microphysics, natural colour, night microphysics or fog/stratus, cloud over snow, volcanic ash).
- Blended or sandwich image products (Meteosat, MetOp).
- Meteosat cloud products (e.g. GII).
- Satellite precipitation estimates (Meteosat MPE, SSMI, AMSU, TRMM).
- Total precipitable water (AMSU, SSMI).
- Atmospheric cloud motion vectors (Meteosat).
- Temperature and humidity soundings from NOAA POES.
- Atmospheric water vapour and surface skin temperature and profiles of temperature and humidity (IASI, AIRS, AMSU).
- Atmospheric profiles of temperature and humidity (real-time skew-*T* plots from AIRS).
- Satellite-derived surface winds (SSMIS, WindSat, ASCAT).
- Soil moisture (combination of WindSat, ASCAT, etc.).
- Precipitable water vapour (surface GPS receivers).

#### 9.2.1.4 Radar Operational Analysis Tools

- Reflectivity (PPI, RHI).
- Velocity (PPI, RHI).
- Velocity altitude display, which shows radial velocity at different altitude with time.
- Precipitation estimates (which aids in forecasting flash floods).

### 9.2.2 Large-scale Feature Identification

#### 9.2.2.1 Intertropical Discontinuity and Intertropical Convergence Zone

- Geostationary imagery (e.g., IR 10.8 μm, dust RGB, natural colour RGB) and LEO products (e.g. precipitable water) to detect the ITD and, additionally, the air mass RGB to identify the ITCZ.
- Meteosat GII also help to delineate the ITD, which typically has maximum instability to its south.
- GPS-derived precipitable water vapour, time series of precipitable water vapour at individual sites can be used to monitor the position of ITD.

Refer to Chapters 1, 2 and 11 for more information on the ITD and ITCZ.

#### 9.2.2.2 Subseasonal Circulations: Madden–Julian Oscillation, Equatorial Waves

- Anomalies of outgoing longwave radiation (OLR) from twice-daily observations are used to identify the Madden–Julian oscillation (MJO).
- Geostationary IR10.8 μm, to identify convectively coupled prominent equatorial Rossby waves (symmetric vortices about the Equator) and mixed Rossby gravity waves (antisymmetric vortices about the Equator; Section 7.1.4).
- Filtered OLR anomalies and forecasts of the MJO and equatorial waves from Australian Bureau of Meteorology and US Climate Prediction Center.

Refer to Chapter 7 for more information on the MJO and equatorial waves.

#### 9.2.2.3 Synoptic Systems

- African easterly waves, jet streams, tropical plumes, upper-tropospheric cyclones, and extratropical intrusions can be monitored using geostationary and LEO imagery (air mass RGB, IR10.8 μm, water vapour, visible).
- African easterly waves are often tracked using longitude–time (Hovmöller) plots of IR10.8 μm (Figure 9.56), available from the US National Weather Service Africa Desk.
- Upper-level systems and extratropical intrusions are usually found using the air mass RGB, water vapour IR, IR10.8 with colour enhancements to identify thunderstorms that form in advance of cold fronts and cold clouds that mark the jet stream.
- Atmospheric motion vectors for the upper troposphere are used to track jet streams.
- Total precipitable water products from LEO sensors show movement of moisture between the tropics and extratropics and atmospheric rivers.
- Sounder products are useful for detecting dry air intrusions from the extratropics.

Refer to Chapter 2 for more information.

### 9.2.3 Mesoscale and Local-scale Analysis

#### 9.2.3.1 Mesoscale Circulations

- Mesoscale circulations are best identified using visible images, which show cloud lines and low-level boundaries such as gust fronts, shear lines or sea-breeze fronts in detail.

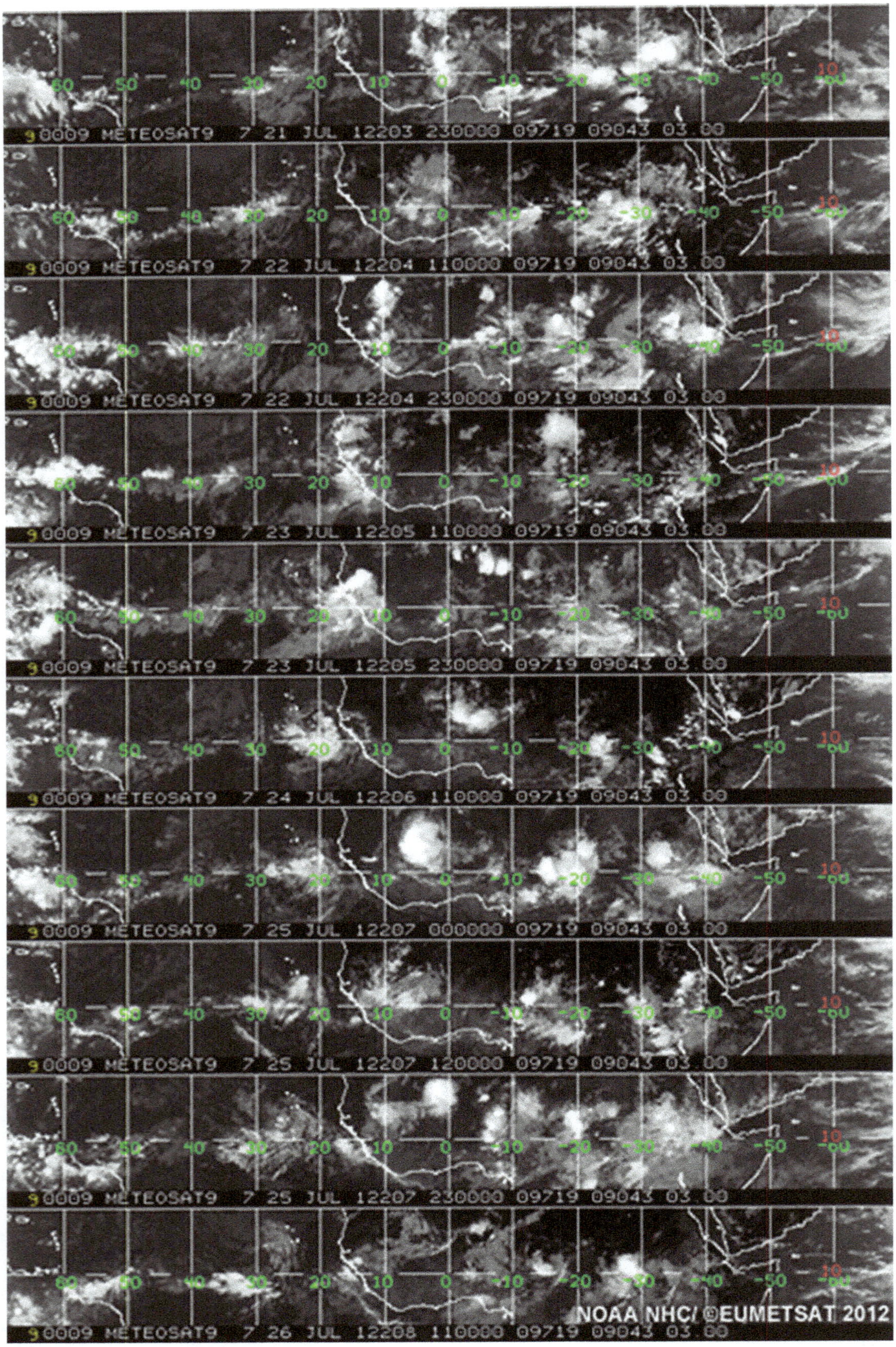


**Figure 9.56** Hovmöller diagram of IR10.8 μm satellite images used to track the movement of synoptic weather systems over Africa and the tropical Atlantic over 5 days. *Source*: reproduced with permission of EUMETSAT and NOAA.

- IR images are useful to differentiate between circulations at different levels; for example, mesoscale comma cloud and upper-level cyclone centre.
- Atmospheric motion vector products are useful for identifying mesoscale shifts in wind speed and/or direction.
- Soil moisture products are also useful for identifying where boundary-layer circulations are likely.
- The air mass RGB and water vapour channel can detect mesoscale vortices and short waves in the mid-upper troposphere.
- Radar reflectivity and base velocity are also used to find mesoscale and local boundary-layer circulations.

Refer to Chapters 3, 4 and 6 for more information on mesoscale and local circulations.

#### 9.2.3.2 Mesoscale Convective Systems

- MCSs are recognised by their extensive area of cold clouds and identified using the convection RGB and enhanced IR10.8 μm images. In Figure 9.57, intense MCSs, which caused flooding in Benin and Nigeria, have large areas of vigorous convection, evident as red areas of Figure 9.57a and yellow areas of Figure 9. 57b. The eastern MCS has an apex or bowed shape, which is an indicator of strong inflow from the rear of the storm that is usually associated with strong surface winds. The apex is also where new convection is forming, which is valuable information for nowcasting.
- LEO microwave sensors at 85–89 GHz show the hydrometeors within MCSs.
- Radar reflectivity and velocity measurements show the underlying precipitation structure of MCSs and would

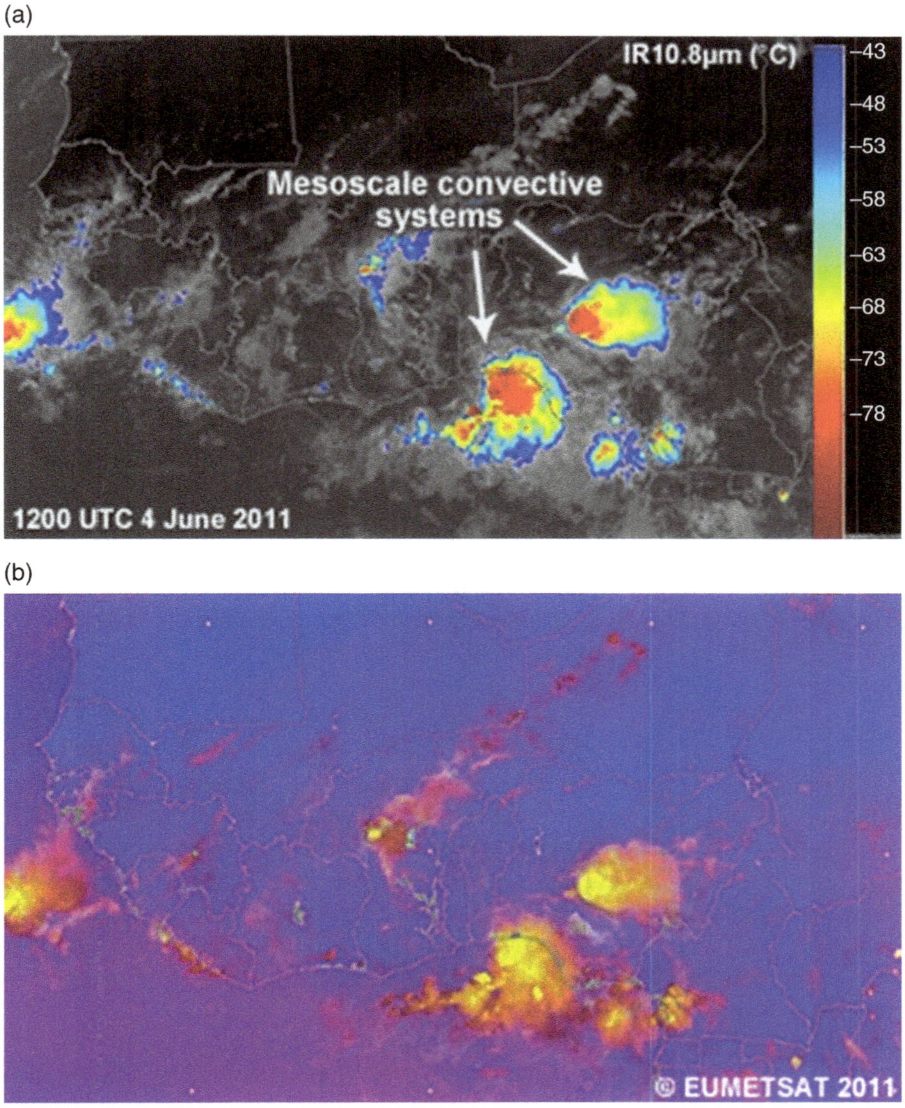


Figure 9.57 Convection satellite products for monitoring mesoscale convective systems: (a) colour-enhanced IR10.8 μm and (b) convection RGB. *Source*: reproduced with permission of EUMETSAT.

confirm whether there is strong rear inflow and descent of strong winds to the surface.

- Visible images are also useful for analysing the evolution of MCSs, which are indicated by contiguous areas of bright white clouds, overshooting tops, cirrus anvils and surface gust fronts.

Refer to Chapters 3 and 6 for more information on MCSs.

#### 9.2.3.3 Dust and Sand Storms

- Dust and sand storms extend across scales but can be treated as mesoscale when linked to mesoscale circulations. They are identified using the dust RGB and natural colour RGB (Figures 9.21 and 9.22) and sometimes the day microphysics RGB (Figure 9.24).
- LEO sensors on the MODIS and Suomi-NPP retrieve aerosol optical thickness used to track dust and other aerosols.
- Radar reflectivity and base velocity are also used to find boundary-layer circulations such as gust fronts or frontal boundaries that will loft dust. Radial velocity images help to identify accelerations in the low-level that can aggravate dust storm hazards.

Refer to Chapter 5 for more information.

#### 9.2.3.4 Convective Cells, Outflow Boundaries, Cloud Lines

- Local-scale circulations are best identified using HRV images, which have resolution of ~1 km and show cellular convection, cloud lines, gust fronts and sea-breeze fronts in detail.
- Soil moisture products, currently derived from LEO microwave sensors, show soil moisture gradients, which help to identify where small-scale boundary-layer circulations are likely to form (e.g. Figure 9.49).
- Radar reflectivity and base velocity are also used to find local boundary-layer circulations such as gust fronts or frontal boundaries that serve to initiate new convection.

Refer to Chapters 4 and 6 for more information on local and mesoscale weather.

### 9.2.4 Fog and Low Stratus

- Fog and low stratus are most easily identified with the Meteosat fog/low stratus RGB (same as night-time microphysics RGB), as shown in Figure 9.25 and described in Section 9.1.4.4.6). However, this product is not useful at dusk or dawn because it uses the difference between 3.9 μm and 10.8 μm, which is not visible at dusk or dawn.
- Visible and natural colour RGB can also help identify fog, which is often aligned with terrain features such as valleys or coastlines.

Refer to Chapter 4 for more information on fog analysis and forecasting.

### 9.2.5 Satellite Soundings

- Sounding profiles of the temperature and humidity in the atmosphere over West Africa are available from LEO satellite sounders, some of which are hyperspectral sounders (i.e. have thousands of discrete channels).
- Real-time plots of temperature and humidity profiles from the AIRS are available on the NASA AIRS website. In the example (Figure 9.58), the satellite sounding nearest to Niamey on 6 March 2014 compares favourably with the radiosonde profile.
- The IASI Atmospheric Temperature Water Vapour and Surface Skin Temperature product, available on EUMETCast for Africa, has vertical profiles of atmospheric temperature and humidity and horizontal sampling of 25 km at nadir.

### 9.2.6 Winds

- Visible and IR channels are useful for analysing winds at cloud levels, and Meteosat Atmospheric Motion Vector products show estimated wind velocity in different tropospheric layers (e.g. Figure 9.35).
- Ocean surface winds, derived from active (e.g. ASCAT) and passive (e.g. WindSAT, SSMIS) microwave sensors are useful for marine forecasting.

### 9.2.7 Aviation Hazards

Specific aviation hazards can be monitored with remotely sensed products:

- *Turbulence.* Clear air turbulence is usually associated with jet streams, which are detected using the air mass RGB, atmospheric motion vectors, water vapour channel, enhanced IR10.8 and the visible channel. Turbulence due to thunderstorms is monitored using convection RGB, enhanced IR, and visible channels to identify intense convection. Near-cloud turbulence is associated with convection-induced gravity waves (Lane *et al.*, 2012), which typically is identified by a series of transverse bands in visible and IR images (e.g. Figure 9.59). Turbulence associated with mountain downslope winds is indicated by cloud lines parallel to the lee of the mountain due to gravity waves. Cloud lines can be observed on visible, HRV, enhanced IR, and natural colour RGB images.
- *Icing.* Day and night-time microphysics RGBs and convection RGB identify ice and water clouds and ice concentration in clouds.
- *Visibility.* Reduction in visibility can be due to phenomena such as dust, heavy precipitation, fog, haze, smoke and ash. Monitor dust using the Meteosat dust RGB, natural colour RGB, and Sahara air layer product. Track heavy precipitation using convection RGB, enhanced

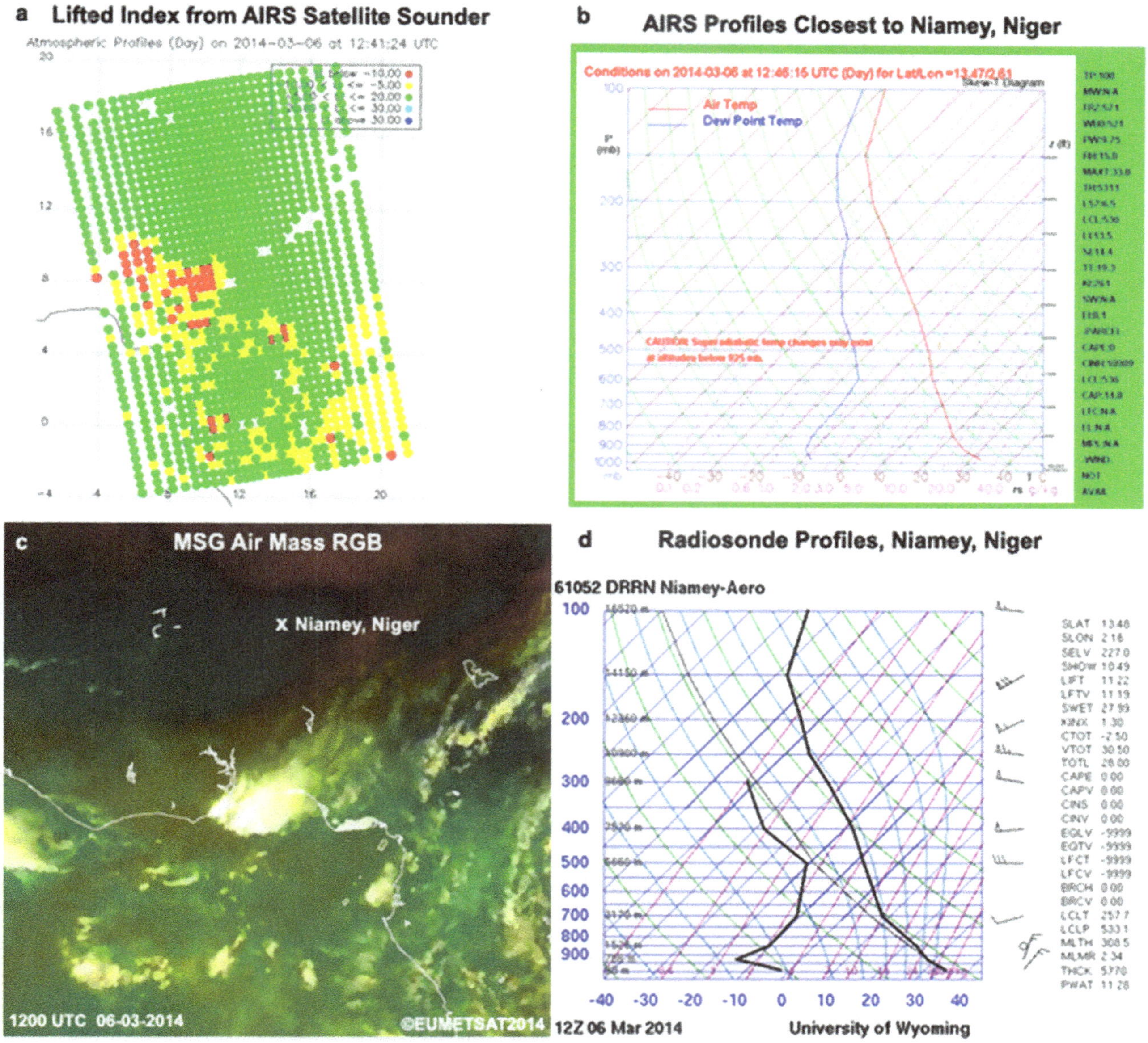


Figure 9.58 Example of real-time products from AIRS, Meteosat and radiosondes on 6 March 2014. (a) Lifted index derived from the sounder channels. *Source*: reproduced with permission of NASA/JPL. (b) skew-*T* log *P* plot at 1246 UTC for location (13.47, 2.61). *Source*: reproduced with permission of NASA/JPL. (c) Air mass RGB image at 1200 UTC. *Source*: reproduced with permission of EUMETSAT. (d) Skew-*T* log *P* plot from radiosonde at 1200 UTC. *Source*: reproduced with permission of University of Wyoming.

IR, and visible channels to identify intense convection. Detect fires and smoke using Meteosat and LEO fire products (e.g. Figure 9.44). Meteosat natural colour RGB, visible and thermal IR. Detect night-time fog with Meteosat fog/low stratus RGB. Identify haze with natural colour RGB. Volcanic ash can be traced with the ash RGB, air mass RGB as well as natural colour RGB, visible and IR10.8 channels.

- *Low ceiling.* Meteosat Cloud Top Temperature and Height algorithm provides early warning of thunderstorm development. Meteosat Fog/low stratus RGB can also be applied at night.
- *Wind shear.* HRV and IR10.8 will help identify boundaries, wind shifts and convergence zones in the cloud pattern. The dust RGB helps to identify low-level moisture boundaries, which are associated with low-level convergence and wind shear. Intense convection that produces strong downdrafts and microbursts can be monitored with Meteosat convection RGB and enhanced IR10.8 channel (warm pixels appearing in the middle of cold cloud tops usually indicates a strong downdraft). Atmospheric motion vectors can indicate shifts in wind velocity between layers and across regions.

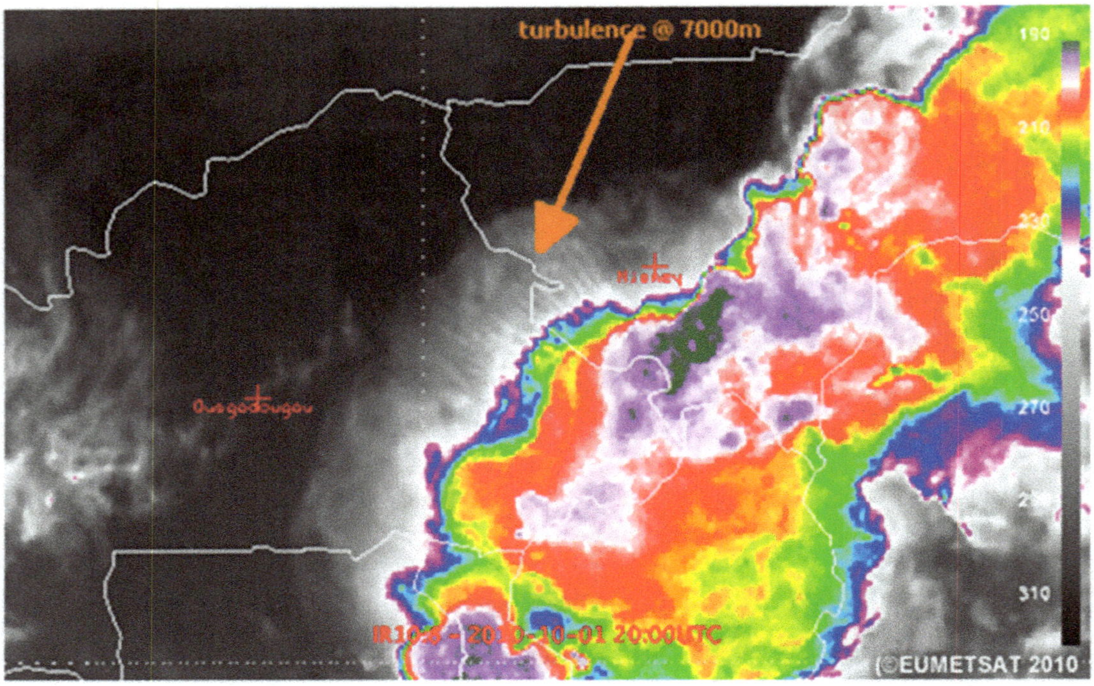


Figure 9.59 Near-cloud turbulence generated by a mesoscale convective system over West Africa. Transverse wave lines mark the area of turbulence. *Source*: reproduced with permission of EUMETSAT.

### 9.2.8 Marine Analysis

- In much the same way as land surfaces, offshore and nearshore boundaries that indicate strong winds and corresponding rough seas are identified using natural colour RGB, visible, IR10.8 and ASCAT winds. Dust RGB can be helpful because dust moving offshore is usually associated with gusty winds.
- Visibility reduction at sea due to Saharan dust can be monitored using the natural colour RGB, Meteosat dust RGB, and the Saharan air layer product.

### 9.2.9 Sources of Satellite Products

- EUMETSAT real-time and archive, topical images, case studies
  http://www.eumetsat.int/
- EUMETRAIN ePort, Atlantic
  http://www.eumetrain.org/eport/archive_atlantic.html?width=1680&height=1050
  Real-time and archives of Meteosat images and ECMWF model analysis at 00, 06, 12, 18 UTC. Images can be overlaid with four fields (e.g. mean sea level pressure).
- Weather satellite products, images and movies, US Naval Research Laboratory
  http://www.nrlmry.navy.mil/sat_products.html
- The Africa Desk, US National Weather Service, NOAA
  http://www.cpc.ncep.noaa.gov/products/african_desk/cpc_intl/africa/africa.shtml.
  Comprehensive source of real-time African weather data and regional model forecasts that includes satellite precipitation estimates, Hovmöller diagrams of synoptic wave evolution, position of the intertropical front, and 24 h loop of enhanced IR images (cold cloud tops).
- Real-time three-hourly tropical rainfall analysis, NASA
  http://precip.gsfc.nasa.gov/rain_pages/3hrly.html.
- Suomi-NPP VIIRS imagery, US Naval Research Laboratory (NRL)
  http://www.nrlmry.navy.mil/VIIRS.html.
  Select West African coast on the global map. Products include HRV and IR images, as well as images created from lunar reflectance and RGB channels. Highlighted colour on each product indicates its age: ≤1 h, ≤12 h, ≤24 h and ≥24 h.
- Regional real-time tropical satellite products, Cooperative Institute for Meteorological Satellite Studies (CIMSS)
  http://tropic.ssec.wisc.edu/
- AIRS temperature and humidity profiles (skew-$T$ plots) from NASA
  http://airsnrt.jpl.nasa.gov/SkewT_index.html
- Operational atmospheric temperature and humidity soundings
  http://www.ospo.noaa.gov/Products/atmosphere/soundings/index.html
- POES temperature and humidity sounding profiles for northern Africa, NOAA
  http://www.osdpd.noaa.gov/pskewt/G33.html

- Dundee satellite receiving station, Dundee University, UK
  http://www.sat.dundee.ac.uk/
  Free images with registration
- Real-time microwave images
  http://weather.msfc.nasa.gov/cgi-bin/sport PublishData.pl?dataset=pmw&product=37H
  Real-time plots of microwave channels, 37 and 89 GHz on AMSR2, SSMI, SSMIS, TMI. Animate through all satellite passes for previous 48 h.
- Ocean surface winds, NOAA NESDIS Center for Satellite Application and Research (STAR)
  http://manati.orbit.nesdis.noaa.gov
- WindSat: description, products and publications, US Naval Research Laboratory
  http://www.nrl.navy.mil/WindSat/
- Metop Program, eoPortal
  http://www.eoportal.org
- Temperature and humidity profiles from GPS Occultation, COSMIC
  http://www.cosmic.ucar.edu
  Free data with registration
- Rain rates from GPM radars
  http://pmm.nasa.gov/GPM/flight-project/DPR

## 9.3 Case Study, Presentations and Other Resources

### 9.3.1 Radar and Satellite Analysis

#### 9.3.1.1 Mesoscale Convective Systems and an African Easterly Wave

Here we highlight a case of an African easterly wave with an easterly wave and an intense squall line MCS that crossed coastal West Africa on 1–2 September 2006. We interpret the satellite and radar products to gather information about the systems and their likely hazards.

The enhanced IR10.8 image shows the synoptic-scale cloud pattern of the easterly wave, clouds arranged in an inverted 'V' pattern (Figure 9.60a and b) with maximum cloudiness near the trough of the wave (Figure 9.60a–d). Areas of rapidly growing cold clouds identify MCSs and their associated hazards such as heavy rain, strong winds and reduced visibility. MCSs are recognised from the colour-shaded cold brightness temperature. The cold cloud shape also gives clues about the heavy rain potential. Notice two adjoining minima in the cloud top temperature at 0130 UTC (Figure 9.60c), a period also observed by radar (Figure 9.61). Notice the apex or 'enhanced-V' shape of the MCS cloud top over the coast of Sénégal at 0000 UTC 2 September (Figure 9.60e), which identifies thunderstorms that are likely producing heavy rain. The coldest temperatures are near the apex along the leading edge of the MCS, where the new, most intense convection develops as older convection is swept back and forms the stratiform cloud area and cirrus outflow. The dust RGB detects the anvil clouds and thin cirrus outflow associated with the updrafts and the low-level outflow boundary (Figure 9.60f).

Radar observations (Figure 9.61) show greater detail of the MCS structure: a leading line of intense convection (reflectivity >45 dBZ), a large area of stratiform rain (reflectivity 20–35 dBZ) and a not well-defined transition zone of weak echoes. The leading line has bow-shaped sections, which usually indicates strong inflow from behind (east of) the line (a rear-inflow jet) that descends toward the front (west) of the system. This is confirmed by the radial velocity image, which shows strong east to west inflow behind the two bow-shaped echoes (>18 m s$^{-1}$). The Dakar sounding showed that easterly winds increased below 850 hPa after the passage of the MCS, probably because of the strong rear inflow of the MCS (Cifelli *et al.*, 2010). The radial velocity image also shows convergence on the western edge of the leading convective line. Low-level convergence enhances lifting and helps sustain the intensity of the MCS over several hours.

Reflectivity cross-sections (Figure 9.62) confirm the satellite brightness temperature interpretation, showing that the leading line remained intense, with reflectivity >50 dBZ, and the MCS maintained a large stratiform rain area. At 0200 UTC, leading cells reached an extremely high 18 km, signifying strong updrafts and downdrafts that lead to turbulence and low cloud ceilings. The extended band of yellow in the stratiform region at 0200 UTC marks the 'bright band', an area of relatively high reflectivity due to melting ice particles. The case shows the combined value of satellite and radar products in providing information about the location of heavy rain, strong winds, intense updrafts and downdrafts, and periods of continuous, steady rain.

#### 9.3.1.2 Tropical Cyclone, Undular Bore and Other Cloud Systems

This case features several types of cloud systems of varying altitudes, including a tropical depression that develops into a tropical storm, MCSs, gust fronts, a gravity wave train (also known as an undular bore; see Section 4.1.4.5), cellular convection and island wake vortices.

The most prominent system is an intensifying tropical depression that became Tropical Storm Bertha (centred south of Cape Verde at 1200 UTC), posing a threat to marine and aviation activities in West Africa (Figure 9.63). Notice convective rain bands along the coast. Thin cirrus in the cyclone outflow allows partial views of the underlying clouds and surface.

Over West Africa there are MCSs with overshooting tops, cirrus anvils and outflow boundaries (associated with the gust fronts, at the leading edge of the convective

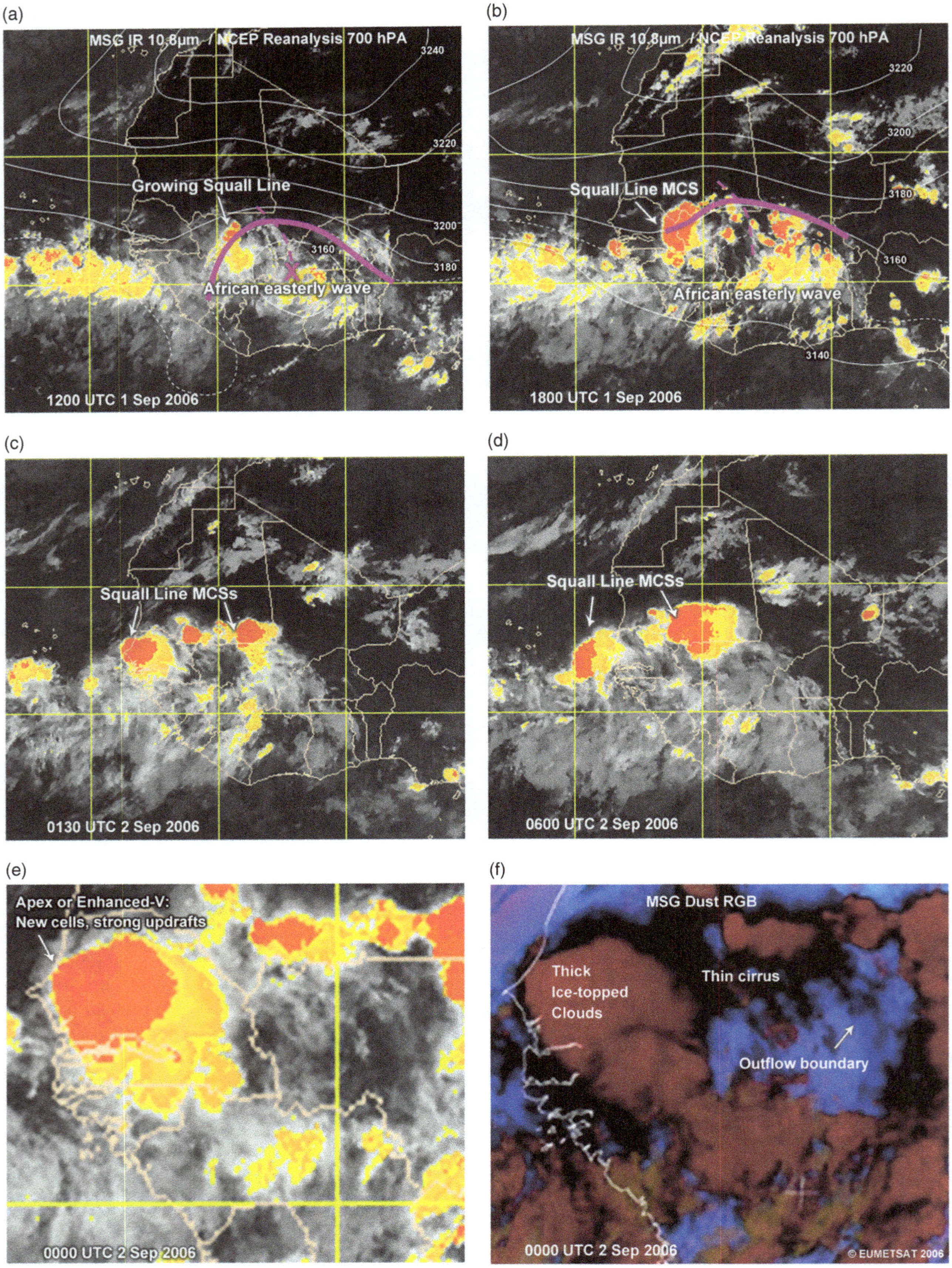


**Figure 9.60** Enhanced Meteosat IR10.8 images of easterly wave and MCSs over West Africa at (a) 1200 UTC, (b) 1800 UTC 1 September, (c) 0130 UTC and (d) 0600 UTC 2 September 2006. (e) Zoomed-in view of MCS at 0000 UTC and (f) dust RGB view at 0000 UTC on 2 September. Orange contour in the IR10.8 images marks cloud tops colder than –65 °C. Panels (a) and (b) include contours of 700 hPa geopotential height with the wave trough and vortex highlighted in magenta. *Source*: reproduced with permission of EUMETSAT.

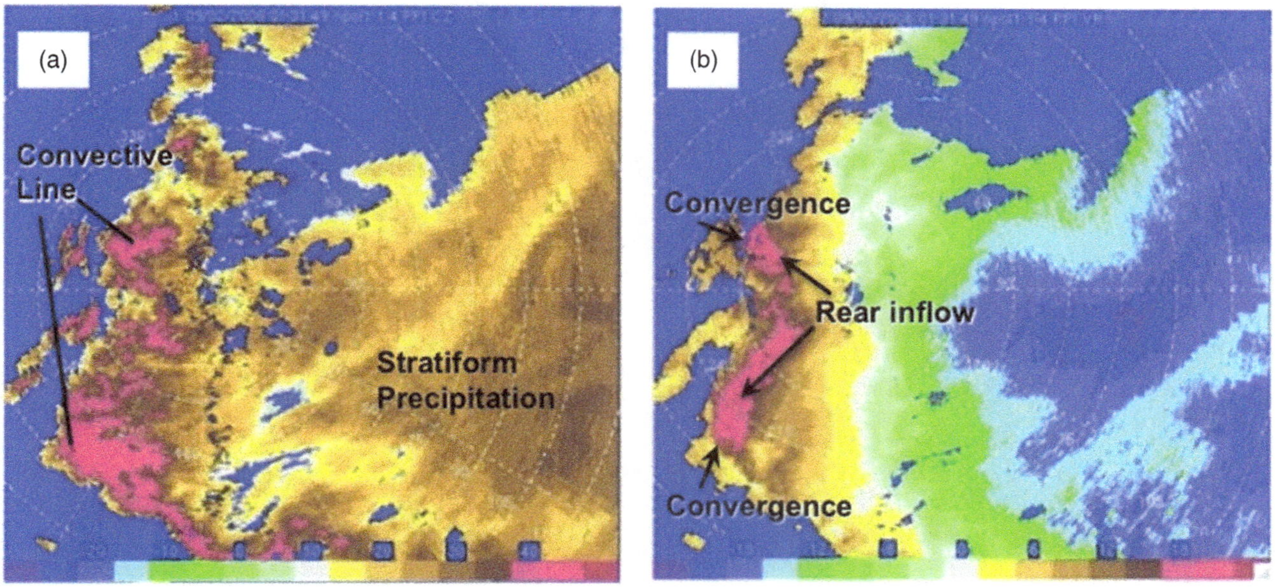


Figure 9.61 Low-level PPI scans of (a) radar reflectivity and (b) radial velocity (m $s^{-1}$) from the NPOL radar, stationed at Dakar, Sénégal, at 0132 UTC 2 September 2006. (b) Here, flow toward the radar is indicated by negative values (blue to green). Flow away from the radar is indicated by positive values (yellow to red). Reproduced from Cifelli et al. (2010). © American Meteorological Society. Used with permission.

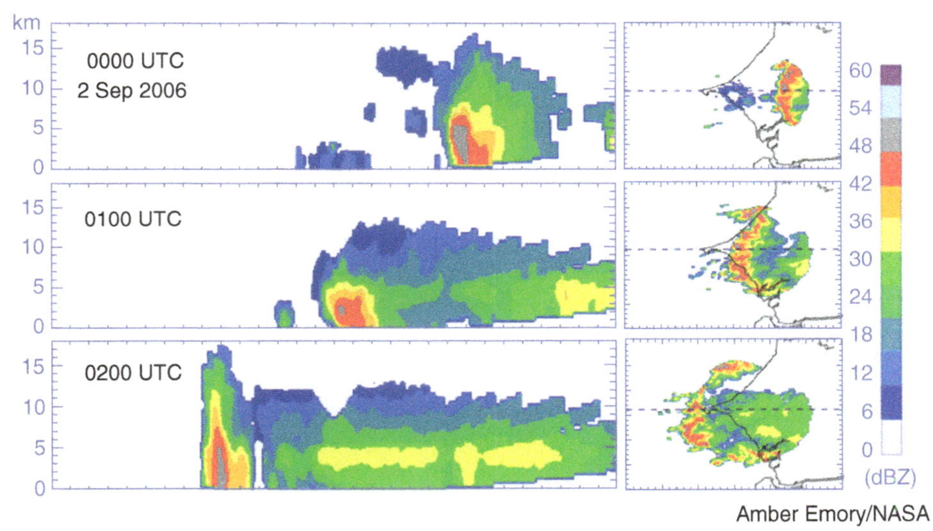


Figure 9.62 Radar reflectivity vertical cross-section and horizontal view of a squall line MCS that moved across the West African coast on 2 September 2006. Cross-sections are taken along the dashed line in the horizontal images. *Source*: reproduced with permission of Amber Emory/NASA.

cold-pool density or gravity currents; Section 3.1.2.2). Convergence and lift ahead of the gravity current produces the arc-shaped clouds spreading out from the MCSs, which are comprised of multiple thunderstorm cells that cast shadows in the morning images.

The outflow from the decaying MCS in the northwest (Figure 9.63 inset) produces a gravity wave train or undular bore over the ocean. The bore contains many narrow, 'roll' or wave clouds (Figure 9.63). The wave clouds become more pronounced over the ocean as the cool, dry air moving from Africa interacts with a stable layer of warm, moist air over the Atlantic Ocean. The temperature inversion above the ocean surface, with a near-neutral Saharan air layer above it, most likely acts as a ducting mechanism for the bore. Notice that the layer of marine stratus that was nearshore dissipates with the passage of the undular bore and that the area east of the wave clouds is almost cloud free. Conceptually, the drier air lifts the moist air, which cools at a peak altitude, sinks and then rises again, but at a lower altitude than the initial peak, thus creating an undulating wave train. Therefore, the most defined cloud is at the front of the group.

The relative brightness and temperatures of the bore clouds show that they are above the marine stratus and stratocumulus layer. However, most are not as bright or as cold as the thunderstorms (Figure 9.64), indicating that they are at or just below the top of the inversion.

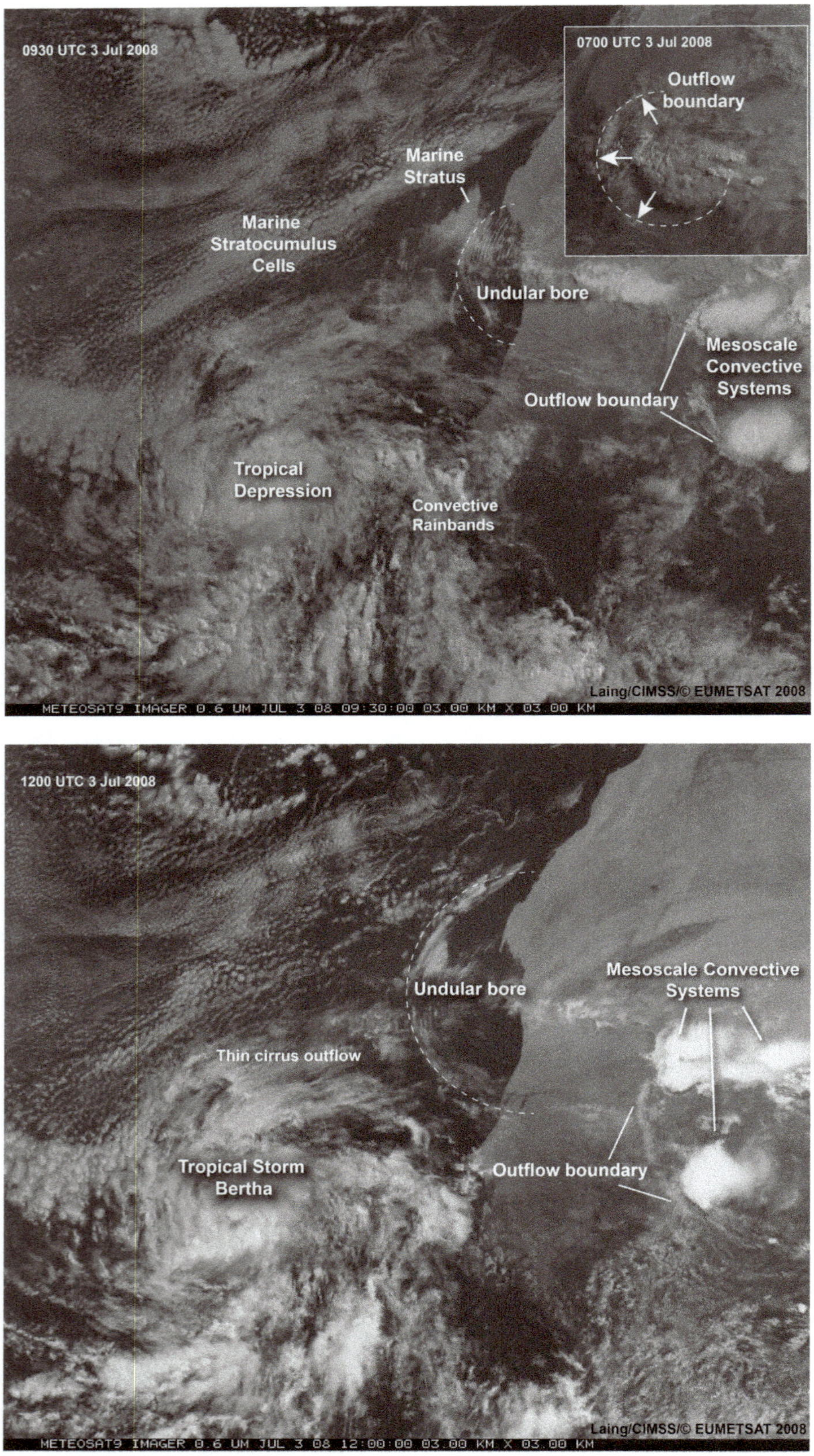


**Figure 9.63** Meteosat visible at 0930 and 1200 UTC, 3 July 2008, with many cloud systems. *Source*: reproduced with permission of EUMETSAT.

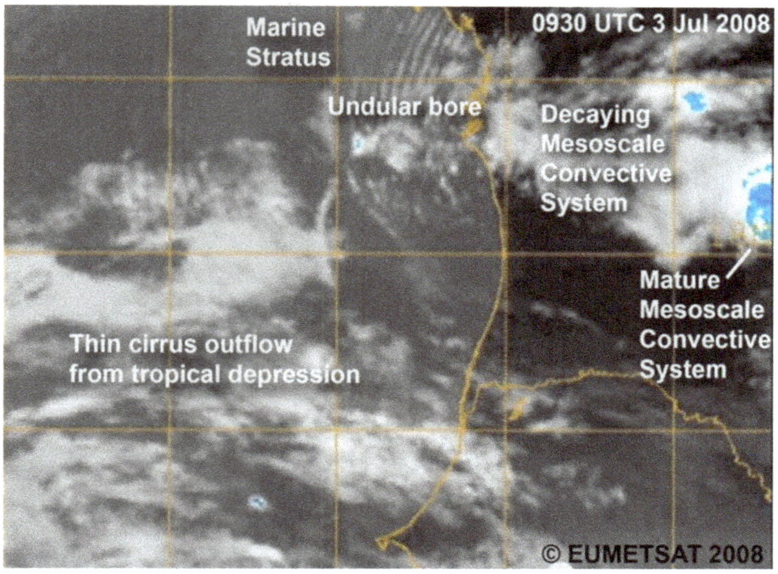


Figure 9.64 Meteosat IR 10.8 at 0930 UTC, 3 July 2008. *Source*: reproduced with permission of EUMETSAT.

Figure 9.65 Meteosat visible at 1415 UTC on 3 July 2008. Notice area of bright ocean due to Sun glint, including small areas of bright white that would be misidentified as cloud. *Source*: reproduced with permission of EUMETSAT.

Curling eddies in the clouds downwind of the Canary Islands, called von Karman vortices, form when the air that splits around or over land comes back together. Another noteworthy feature is the Sun glint along and near the southwest coastal area (Figure 9.65), which is especially bright from 1400 to 1430 UTC, implying a smooth, highly reflective sea surface. Also observed are shadows cast by growing thunderstorms south of the MCSs.

For more information on Hurricane Bertha and the undular bore, see the CIMSS Satellite Blog: http://cimss.ssec.wisc.edu/goes/blog/archives/678.

### 9.3.2 Presentations

- NCAR Colloquium on African Weather and Climate, 2011
  Radar Meteorology: Overview and Applications in Africa by Paul Kucera

http://ral.ucar.edu/csap/events/ISP/presentations/Kucera_Africa_Colloquium_July26_2011.pdf

- Satellite Remote Sensing: A Review by Arlene Laing
  http://ral.ucar.edu/csap/events/ISP/presentations/Laing_2011_NCAR_Africa_colloquium.pdf
- Satellite Products for Weather Analysis in Africa by Estelle De Coning
  http://ral.ucar.edu/csap/events/ISP/presentations/deConing_NCAR_colloquium_2011.pdf
- GPS Applications for Weather Analysis by John Braun
  http://ral.ucar.edu/csap/events/ISP/presentations/Braun_GPS_AfricaWeatherClimate.pdf

### 9.3.3 Education and Training

- African Satellite Meteorology Education and Training (ASMET)
  http://www.meted.ucar.edu/communities/asmet/
  The ASMET project produces online and CD-based learning modules that teach African forecasters how to enhance their forecasts by making better use of meteorological satellite images and products.
- Basics of Satellite Interpretation in the Tropics
  http://rammb.cira.colostate.edu/training/visit/training_sessions/basic_satellite_imagery_interpretation_in_the_tropics/
  An overview of commonly observed features in the tropics, with focus on GOES visible, IR and water vapour imagery, but interpretation is similar for Meteosat imagery.
- Introduction to Tropical Meteorology, UCAR/COMET
  http://www.meted.ucar.edu/tropical/textbook_2nd_edition/
  Chapter 9 of the textbook describes satellite and radar analysis in the tropics at synoptic, mesoscale and local scales.
  Chapter 2 is an overview of remote sensing applications in the tropics.
- The Africa Desk, a presentation by Wassila Thiaw
  http://ral.ucar.edu/csap/events/ISP/presentations/Thiaw_AfricanDesk20110803.pdf
- WMO Virtual Labs for Education and Training in Satellite Meteorology
  http://www.wmo-sat.info/vlab/
  Training and other events, http://www.wmo-sat.info/vlab/calendar-of-events/
- Weather Radar Fundamentals
  https://www.meted.ucar.edu/training_module.php?id=960
  A 2-h COMET module that 'presents the fundamental principles of Doppler weather radar operation and how to interpret common weather phenomena using radar imagery'.
- Conway (1997).

## Appendix: Acronyms

| | |
|---|---|
| AIRS | Atmospheric Infrared Sounder |
| AMSU | Advanced Microwave Sounding Unit |
| ASCAT | Advance SCATterometer |
| BTD | brightness temperature difference |
| EUMETSAT | EUropean organisation for the exploitation of METeorological SATellites |
| GOME-2 | Global Ozone Monitoring Experiment-2 |
| GPM | Global Precipitation Mission |
| GPS | Global Positioning System |
| HRV | high-resolution visible |
| IASI | Infrared Atmospheric Sounding Interferometer |
| IR | infrared |
| LEO | low Earth orbit |
| MODIS | Moderate-resolution Imaging Spectroradiometer |
| MPE | Multisensor Precipitation Estimate |
| MSG | Meteosat Second Generation |
| MTG | Meteosat Third Generation |
| NDVI | normalised difference vegetation index |
| NPP | National Polar-orbiting Partnership |
| POES | Polar-orbiting Environmental Satellite |
| RGB | red, green and blue (primary colours combined for image visualisation) |
| SEVIRI | Spinning Enhanced Visible and Infrared Imager |
| TRMM | Tropical Rainfall Measurement Mission |
| VIIRS | Visible Infrared Imaging Radiometer Suite |

## References


Bevis BG, Bussinger S, Herring TA, *et al.* 1992. GPS meteorology: Remote sensing of atmospheric water vapor using the Global Positioning System. *J. Geophys. Res.* **97**: 15787–15801.

Cifelli R, Lang T, Rutledge SA, *et al.* 2010. Characteristics of an African easterly wave observed during NAMMA. *J. Atmos. Sci.* **67**: 3–25.

Conway E. 1997. *An Introduction to Satellite Image Interpretation.* Johns Hopkins University Press: Baltimore, MD; 242 pp.

Ferraro RR, Weng F, Grody N, *et al.* 2005. NOAA operational hydrological products derived from the Advanced Microwave Sounding Unit (AMSU). *IEEE Trans. Geosci. Remote Sens.* **43**: 1036–1049.

Gaiser PW. 2004. WindSat – remote sensing of ocean surface winds. *NRL Rev.* http://www.nrl.navy.mil/content.php?P=04REVIEW87 (accessed 29 July 2016).

Griffin MK, Hsu SM, Burke HK, Snow JW. 2000. Characterization and delineation of plumes, clouds and fires in an AVIRIS image. In: *Proceedings of the 2000 JPL Airborne Earth Science Workshop*, JPL, Pasadena, CA.

Hyer EJ, Reid JS, Zhang J. 2011. An over-land aerosol optical depth data set for data assimilation by filtering, correction, and aggregation of MODIS Collection 5 optical depth retrievals. *Atmos. Meas. Tech.* **4**: 379–408. doi: 10.5194/amt-4-379-2011.

Kerr YH, Waldteufel P, Wigneron J-P, *et al.* 2010. The SMOS mission: new tool for monitoring key elements of the global water cycle. *Proc. IEEE* **98**(5): 666–687.

Kidder SQ, Vonder Haar TH. 1995. *Satellite Meteorology: An Introduction*. Academic Press; 466 pp.

Kummerow C, Olson WS, Giglio L. 1996. A simplified scheme for obtaining precipitation and vertical hydrometeor profiles from passive microwave sensors. *IEEE Trans. Geosci. Remote Sens.* **34**: 1213–1232.

Lane TP, Sharman RD, Trier SB, *et al.* 2012. Recent advances in the understanding of near-cloud turbulence. *Bull. Am. Meteorol. Soc.* **93**: 499–515.

Lebel T, Parker D, Flamant C, *et al.* 2009. The AMMA field campaigns: Multiscale and multidisciplinary observations in the West African region. *Q. J. R. Meteorol. Soc.* **136**: 8–33. doi: 10.1002/qj.486.

Lensky IM, Rosenfeld D. 2008. Clouds–aerosols–precipitation satellite analysis tool (CAPSAT). *Atmos. Chem. Phys.* **8**: 6739–6753.

Liou K-N. 2002. *An Introduction to Atmospheric Radiation*, 2nd edition. Academic Press; 513 pp.

NASA. 2013. Seeing photosynthesis from space: NASA scientists use satellites to measure plant health. *ScienceDaily.* http://www.sciencedaily.com/releases/2013/07/130724155223.htm (accessed 29 July 2016).

NWCSAF. 2012. Algorithm theoretical basis document for 'Cloud Products' (CMa-PGE01 v3.2, CT-PGE02 v2.2 &CTTH-PGE03 v2.2), SAF/NWC/CDOP/MFL/SCI/ATBD/01.

Rozante JR, Moreira DS, de Goncalves LGG, Vila DA. 2010. Combining TRMM and surface observations of precipitation: technique and validation over South America. *Weather Forecast.* **25**: 885–894.

Simpson J, Kummerow C, Tao WK, Adler RF. 1996. On the Tropical Rainfall Measuring Mission (TRMM). *Meteorol. Atmos. Phys.* **60**: 19–36.

Smith TM, Arkin PA, Bates JJ, Huffman GJ. 2006. Estimating bias of satellite-based precipitation estimates. *J. Hydrometeorol.* **7**: 841–856.

Velden CS, Daniels J, Stettner D, *et al.* 2005. Recent innovations in deriving tropospheric winds from meteorological satellites. *Bull. Am. Meteorol. Soc.* **86**: 205–223.

Ware RH, Fulker DW, Stein SA, *et al.* 2000. Suominet: A real-time national GPS network for atmospheric research and education. *Bull. Am. Meteorol. Soc.* **81**: 677–694.

Wilheit TT. 1986. Some comments on passive microwave measurement of rain. *Bull. Am. Meteorol. Soc.* **67**(10): 1226–1232.

World Meteorological Organization. 2005. The global observing systems. Twenty-second status report on the implementation of the World Weather Watch. WMO-No. 986. http://www.wmo.int/pages/prog/www/StatusReport/22nd_WMO986/Chapter2_GOS.pdf (accessed 28 July 2016).

# 10

# Numerical Weather Prediction over Africa

***Lead author:*** *Sean Milton*[a]

***Contributing authors:*** *Aida Diongue-Niang*[b]*, Benjamin Lamptey*[c]*, Caroline Bain*[a] *and Cathryn Birch*[d]

***Reviewer:*** *Philippe Bougeault (Météo-France/CNRM, Toulouse, France)*

[a] *Met Office, Exeter, UK*
[b] *ANACIM, Dakar, Sénégal*
[c] *ACMAD, Niamey, Niger*
[d] *University of Leeds, Leeds, UK*

## 10.1 Scientific Background

### 10.1.1 Introduction

Operational numerical weather prediction (NWP) grew out of three main strands: (i) a long empirical tradition of observing the weather, (ii) theoretical foundations laid down in fluid dynamics dating back to Newton and applied to extratropical meteorology by Vilhelm Bjerknes, William Napier Shaw, Carl Gustaf Rossby, Jule Charney and many others (see Nebeker (1995)) and (iii) development of high-speed computers allowing numerical forecasts to be produced in real time. The famous numerical forecast carried out by hand by Lewis Fry Richardson in the early 20th century was a landmark for NWP (Richardson, 1922). At the time the forecast was deemed a failure as it suffered from large errors. Later research highlighted the need for methods to provide balanced initial states for such numerical forecasts (see discussion in Lynch (2011)). The first numerical forecast on a computer was performed by Charney, Fjortoft and von Neumann in 1950 building upon the emerging theoretical understanding of barotropic and baroclinic dynamical systems laid down by Rossby (1940), Charney (1947) and Eady (1949). Since then we have seen a rapid improvement in the skill of numerical weather forecasts for the extratropics from all operational centres (Figure 10.1). This has been due to a combination of deployment of more observing systems, particularly remote sensing from satellites (Section 10.1.2.1), improved techniques of data assimilation to derive an optimal initial state from combining observations and model forecasts (Section 10.1.2.2), improved model resolution and the continuing physical and dynamical development of numerical models of the atmosphere (Section 10.1.2.3).

NWP over Africa and the tropics has a similar evolving observational and theoretical heritage. While we have seen some improvements in predicting the large-scale flow over Africa and other tropical regions (Figure 10.2), NWP over the African continent remains a considerable forecasting challenge, incorporating numerous weather and climatic regimes from mid-latitude weather systems at the northern and southern extents, to equatorial rain forests and deserts, and all subject to extreme weather on a range of temporal and spatial scales. The particular geography of Africa leads to some specific weather phenomena. At the synoptic scales are structures such as the African easterly jet (AEJ), African easterly waves (AEWs) and the intertropical front, which play a fundamental role in delivering rain to the Sahel and West Africa (see Chapters 1 and 2). Tropical cyclones are also responsible for extreme weather events in eastern and southern Africa, such as exacerbating the Mozambique floods of 2000, while the AEWs themselves are closely linked to tropical cyclogenesis in the tropical Atlantic. A key difference to mid-latitudes is that many of the weather systems in tropical Africa are strongly coupled to convection across a range of scales, from individual thunderstorms to more organised squall lines and mesoscale convective systems (MCSs). These convective systems, while linked to the large-scale synoptic dynamics, are also closely controlled by the stability of the atmosphere, its diurnal fluctuations and interactions with underlying land surface and topography (see Chapter 3). For NWP, the initialisation and modelling of these convective weather systems remains a major research challenge. The weather

*Meteorology of Tropical West Africa: The Forecasters' Handbook*, First Edition. Edited by Douglas J. Parker and Mariane Diop-Kane.

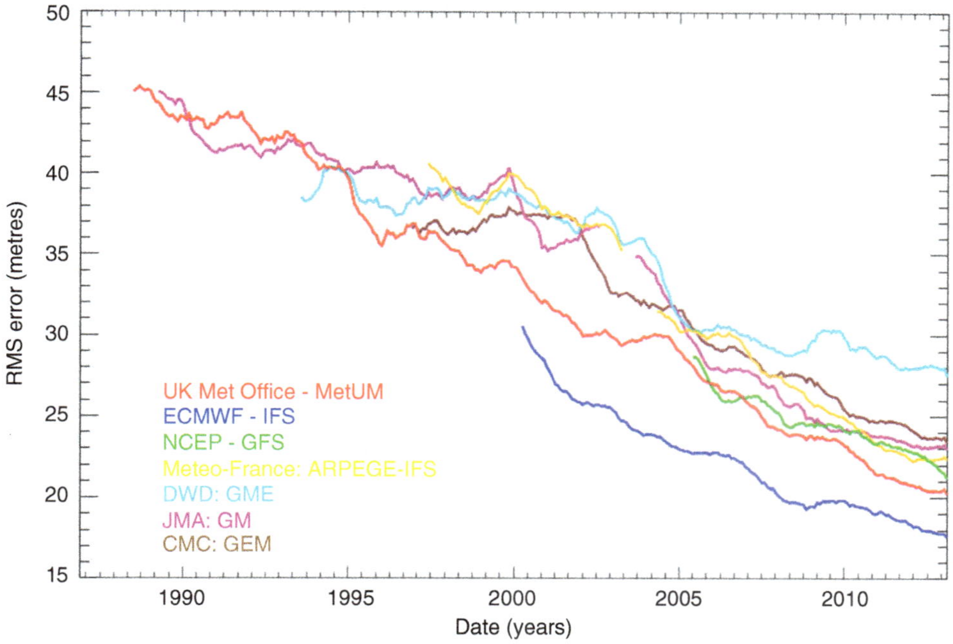


**Figure 10.1** Example of improving NWP forecast performance for the extra-tropics for the period 1987–2013. Time series of monthly averaged Northern Hemisphere (90–20°N) 500 hPa height root-mean-squared (RMS) errors for operational global NWP models. All models show steady improvements in RMS errors of between 2.5% and 5% per year. These verification measures are calculated for model forecasts on a uniform 2.5 degree grid by each centre against it's own operational analysis and exchanged as part of the World Meteorological Organisation (WMO) Commission for Basic Systems (CBS).

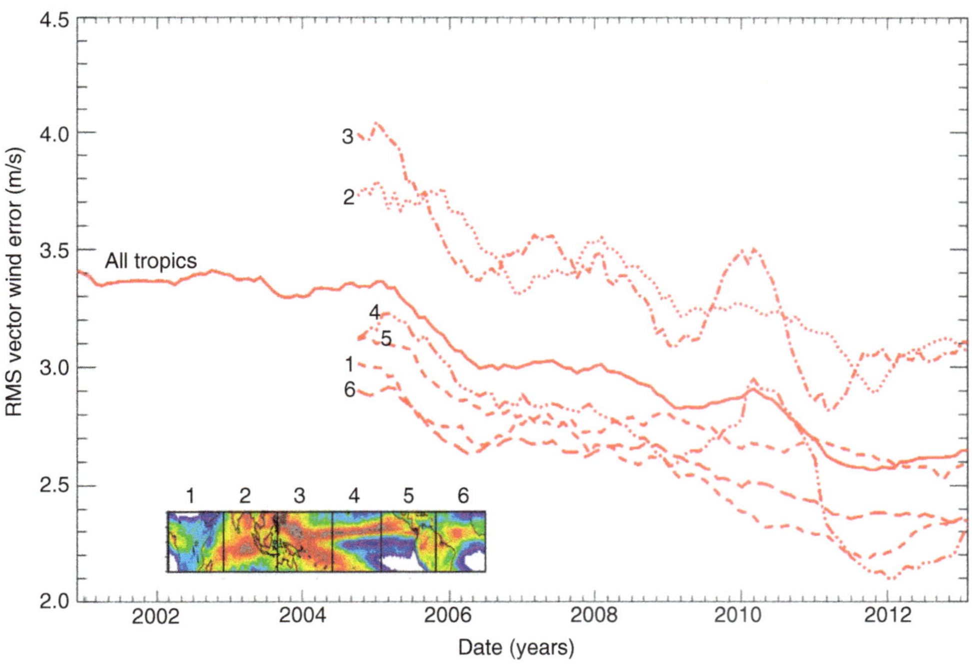


**Figure 10.2** Improving tropical skill in the UK Met Office global model. Time series of 850 hPa RMS vector wind error (RMSVWE) over the entire tropical domain (18.75°N–18.75°S) and over 60° longitude tropical domains (1–6 – see inset). All domains show improving RMS errors. The absolute value of the RMSVWE is partly determined by the natural variability in a given domain (e.g. larger variability in Indian Ocean (domain 2) and tropical Pacific (domain 3) – see inset showing daily mean precipitation standard deviation from GPCP 1DD observations over all days in 2002–2007 as a measure of tropical variability). The degradation in skill, in domains 3 and 4 during winter 2009–2010, may have been associated with this (and other) models' difficulties in predicting an active Madden–Julian oscillation (MJO) episode.

systems over Africa also play a key role in determining the weather and climate of other regions through remote teleconnections, both on intra-seasonal timescales and even after 3 days of forecast (e.g. Faccani *et al.*, 2009). There is also considerable research to show significant variability over Africa on subseasonal timescales of weeks (e.g. equatorial Kelvin waves) to months (e.g. MJO), which can modulate precipitation (Janicot *et al.* 2011; and see Section 7.1.5.2).

Good quality observations over Africa are required for accurate NWP predictions, not only to initialise forecasts, but also to understand the physical and dynamical processes of specific weather phenomena, to study their potential predictability and to help evaluate and improve the numerical models (Tompkins *et al.*, 2005; Faccani *et al.*, 2009; Agustí-Panareda *et al.*, 2010). Alongside the routine meteorological observations, a number of field experiments have shaped our knowledge of the physical and dynamical processes controlling the weather and climate over West Africa. Examples of large international experiments have included, the GARP Atlantic Tropical Experiment (GATE) operating out of West Africa during 1974 and designed to understand the tropical atmosphere and its role in the global circulation (Kuettner, 1974), and more recently the African Monsoon Multidisciplinary Analyses (AMMA) field campaign in 2006 to study the West African monsoon (WAM) physical processes and predictability (Redelsperger *et al.*, 2006; Janicot *et al.*, 2008; Lebel *et al.*, 2010). Smaller scale field experiments have also been carried out by individual NWP and research centres to study specific aspects of African weather systems, such as the Saharan heat low (e.g. Fennec) and the role of dust/aerosols in weather and climate (BodEX, SAMUM, GERBILS). The combination of intensive observations from field experiments and high-resolution numerical modelling is providing a key methodology for improving NWP and climate predictions over Africa (e.g. Garcia-Carreras *et al.*, 2013; Marsham *et al.*, 2013b).

The aims of this chapter are to provide forecasters with a brief review of current NWP techniques used in operational centres to produce daily weather forecasts for Africa and other regions around the globe (Section 10.1). An overview of the current status and capability of operational NWP for Africa is given in Section 10.2. The main focus is on the short-range period of days to a week, but we will also discuss the medium-range (week-2) forecasts and ensemble prediction methods. A large part of Section 10.2 examines the current capability of global (and regional) operational NWP systems and what are the limitations of current NWP products. Section 10.1.4 provides a forward look at how NWP might evolve over the next decade to try and address the challenge of providing accurate NWP forecasts for Africa, and what this may mean for changing forecasting techniques and the enormous role of the forecaster in interpreting NWP products.

### 10.1.2 Deterministic Numerical Weather Prediction Systems

The fundamental elements of any NWP forecast are the observations, used to both initialise and evaluate NWP models; the data assimilation process used to derive the best initial state of atmosphere, land and ocean for starting our NWP forecasts; and the numerical models themselves. Here we discuss these building blocks for a single deterministic NWP forecast. We will see later that ensemble forecasting is also a key technique in providing a risk assessment for severe and hazardous weather, and that ensembles are also playing an increasing role in data assimilation techniques.

#### 10.1.2.1 Observations

The starting point for any numerical weather prediction is an accurate representation of the initial state of the atmosphere (land, ocean and ice) from which to integrate the equations of motion forward in time to provide a forecast. This initial state is usually a measure of the atmospheric temperature, moisture, winds and pressure, and at the surface of the Earth the land surface temperature, moisture content of the soil, vegetation state, snow cover and the sea surface temperature (SST). Clearly, as a starting point we need good observations of these parameters. In general these observations are not regularly distributed in space or time and are a mix of in-situ measurements and remote sensing data from satellites. As an example, Figure 10.3 shows the spatial distribution of some of the quality-controlled observations used to initialise the Met Office global forecast on a single day at 0000 UTC on 27 February 2013. Around 150 000 atmospheric observations are used for this 6 h assimilation period (1800–0000 UTC). All of these data are collected every day and distributed to operational centres via the WMO Global Telecommunications System.

Observing the atmosphere, land and ocean over and around Africa is challenging. For example, the Sahara Desert is probably the least observed region of the globe (Figure 10.3), and we have limited knowledge of the important physical and dynamical processes controlling the meteorology there and influencing the weather in other regions of Africa. However, even in less challenging physical environments, the observational coverage from in-situ measurements is still sparse compared with many other regions of the world.

A number of data types and measurements are used in NWP data assimilation. Surface and marine data (land and ship SYNOPs, buoys, etc.) and radiosondes/pilot balloons[1] (observing profiles of temperature, moisture and winds through the depth of the atmosphere), typically

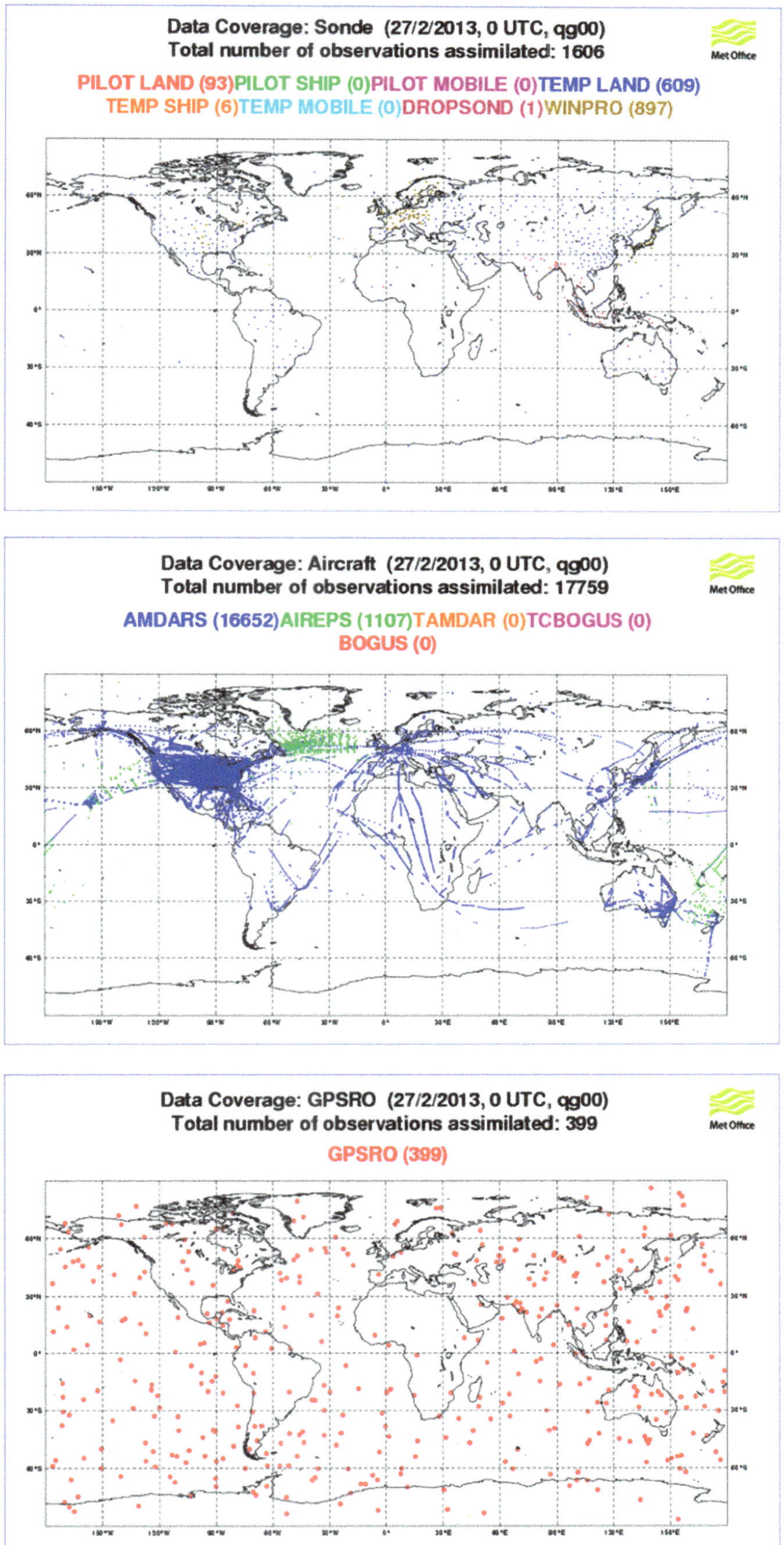


**Figure 10.3** Data coverage of quality-controlled observations used in the Met Office data assimilation to initialise the global NWP forecast at 0000 UTC on 27 February 2013.

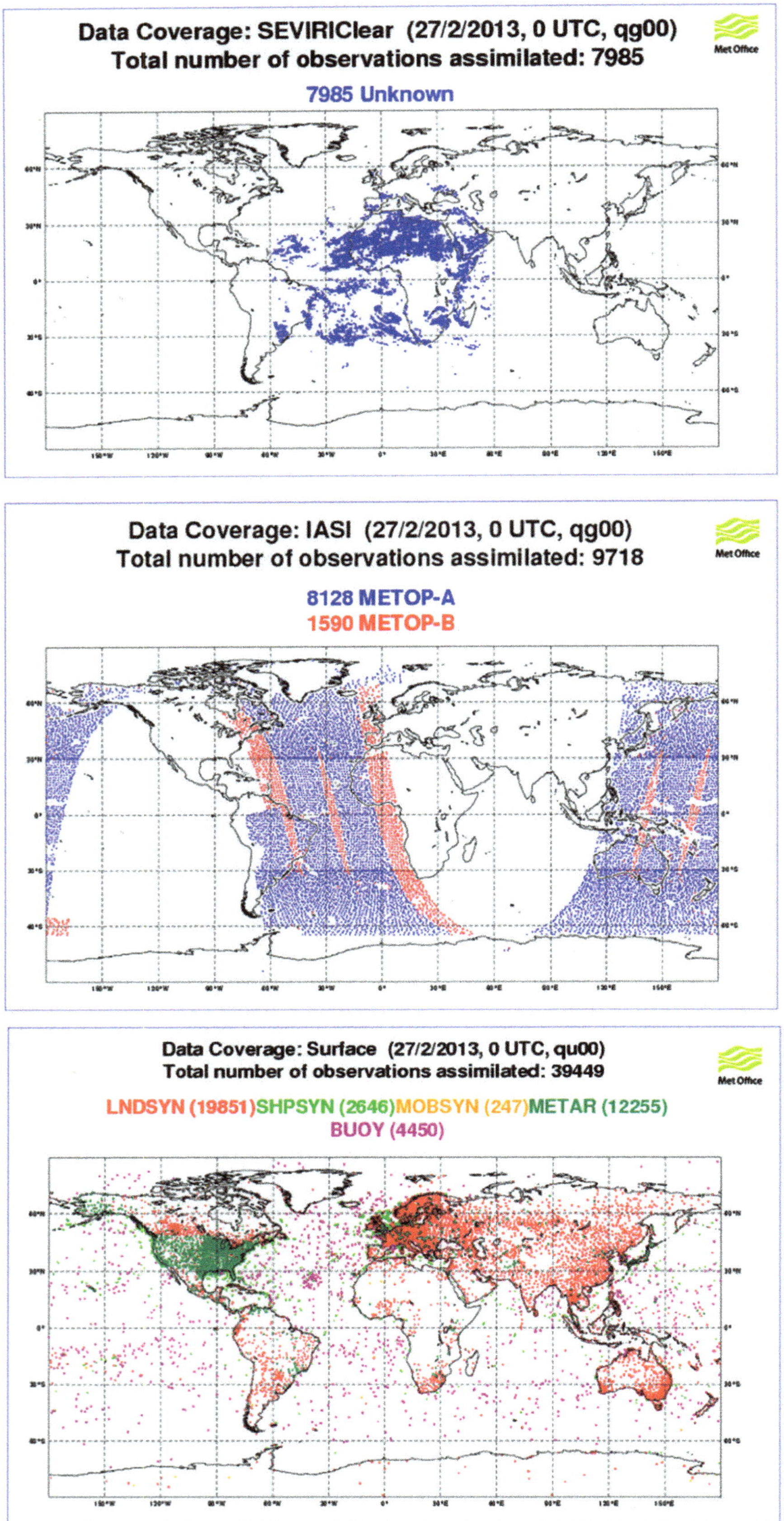


Figure 10.3 (Cont'd)

Figure 10.3 (Cont'd)

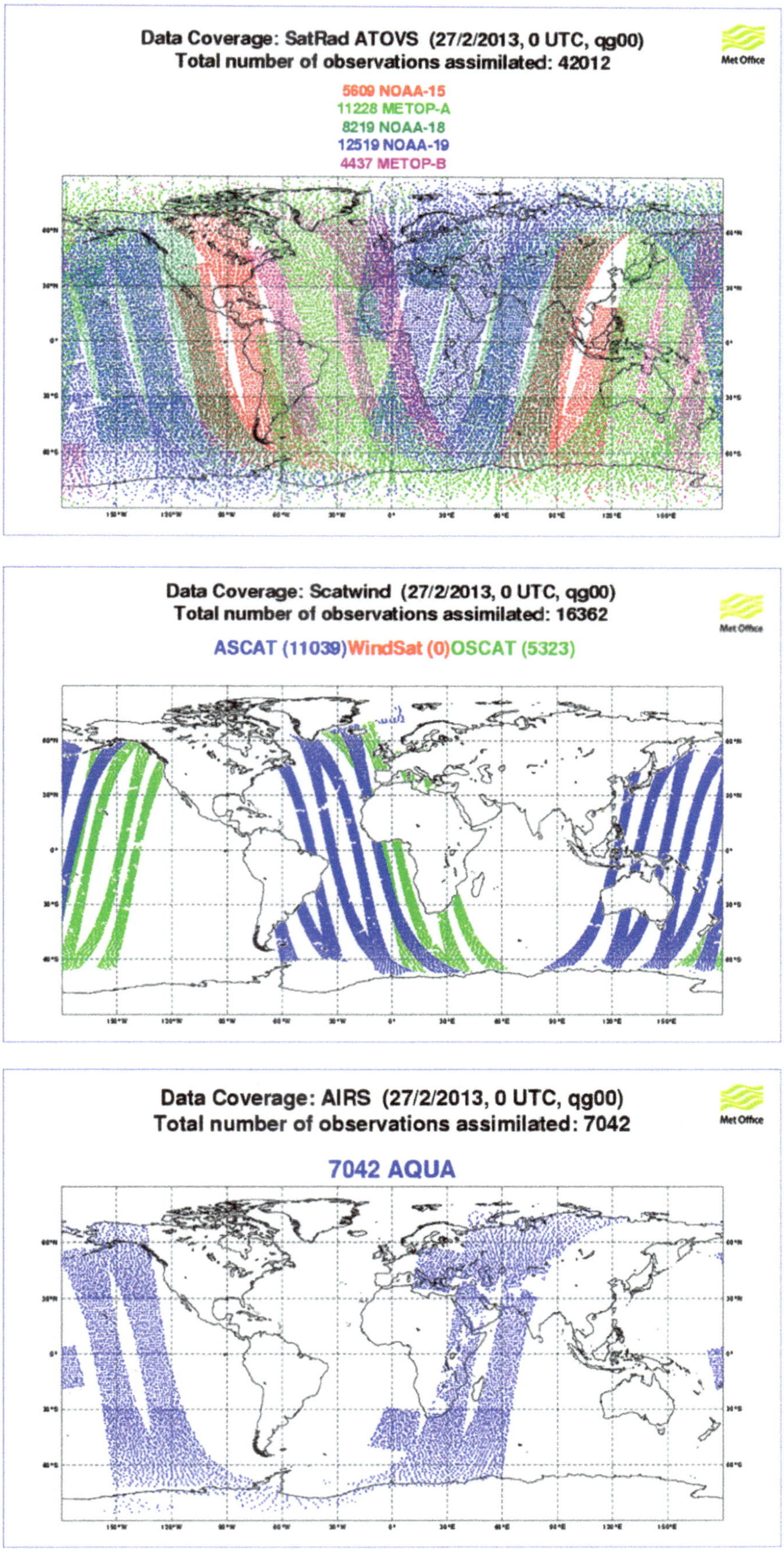

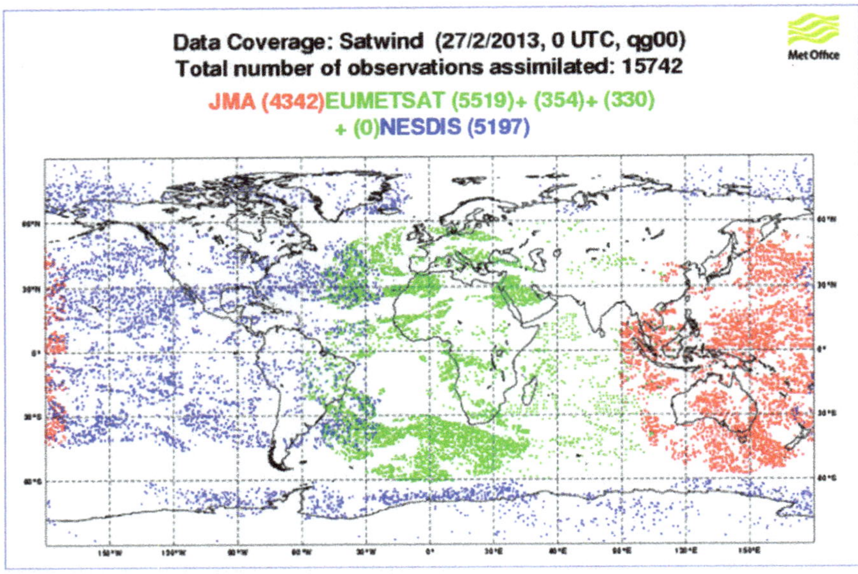


Figure 10.3 (Cont'd)

provide measurements at six-hourly time intervals. Significant data gaps in SYNOPs occur over Africa compared with other regions (Figure 10.3). Similarly, data coming from aircraft are limited over Africa with AMDARS-(Aircraft Meteorological Data Relay) forming the main aircraft data stream. This is a program initiated by the WMO to collect meteorological data worldwide from commercial aircraft. Data are collected from aircraft navigation systems and the onboard standard temperature and static pressure probes, and then preprocessed before linking down to the ground, either via VHF communication (ACARS-Aircraft Communications Addressing and Reporting System) or via satellite link (ASDARS-Aircraft to Satellite Data Relay). Standard measurements are temperature, wind speed/direction and position (latitude, longitude, altitude and time). A few aircraft (mostly over the USA and Europe) also report information on atmospheric moisture, turbulence and icing.

The satellite data are made up of active and passive measurements from remote sensing instruments on board a range of polar orbiting and geostationary satellites. Active remote sensing is where an instrument emits energy in order to scan objects/areas, and a sensor on the satellite then detects the radiation that is reflected or backscattered from the target. An example from current operational NWP is the scatterometer instrument used to measure winds over the ocean surface and soil moisture over land. Passive remote sensing detects natural radiation at the satellite sensor that is emitted or reflected from the atmosphere-ocean-land-ice environment. The ATOVS (Advanced Television and Infrared Observational Satellite (TIROS) Operational Vertical Sounder), AIRS (Atmospheric Infrared sounder) and IASI (Infrared Atmospheric Sounding Interferometer) instruments are passive measurements of radiance from the Earth's surface and atmosphere. In the past the radiance data were converted to temperatures and humidities for use in the data assimilation process, but now the more usual practice is to convert model temperatures and humidities into radiances using a forward model, such as a complex radiative transfer model (e.g. RTTOV-Radiative Transfer for TIROS Operational Vertical Sounder[2]). This presents a model field in the 'observation space' for direct comparison with the observations in a data assimilation scheme (see Section 10.1.2.2). However, satellite retrievals over Africa are still difficult due to poor knowledge of the emissivity of the surface in various observed wavelengths.

In addition to these atmospheric observations there are also the observations used to produce an SST analysis (buoys and remotely sensed data) and snow analysis. Some centres are beginning to exploit microwave observations of soil moisture from instruments such as the Advanced Scatterometer (ASCAT; Dharssi *et al.*, 2011) to provide information on the initial soil moisture state. Using these data in an NWP system is itself challenging, as the microwaves only give information on soil moistures in the top few centimetres of soil and the top model soil layers are typically on the order of 10 cm depth.

Despite the growing use of satellite data in NWP during the last two decades, radiosonde data are still very important for NWP systems, and particularly so over Africa (Fink *et al.*, 2011). Radiosondes provide very high vertical resolution information allowing the important temperature, moisture and wind structures in the boundary layer (and free troposphere) to be represented in the model initial state. These vertical variations in stability and wind shear are important for African weather systems as they influence the formation and growth of convection and also the dynamical evolution of synoptic weather systems. Satellite data have typically had coarser resolution in the vertical, proved difficult to use near the surface due to problems with surface emissivity and have only been available at cloud-free locations, although operational NWP centres are actively working to remove some of these limitations. Information on jet locations and upper-level winds are especially dependent on radiosonde launches as satellites can only deduce winds when cloud is present, using cloud-tracked wind retrievals (SATWINDS – see Figure 10.3). The JET2000 campaign (Thorncroft *et al.*, 2003) demonstrated the value of sonde profiling for NWP through the use of dropsonde data from research aircraft, which better defined the AEJ in model analyses and improved subsequent NWP forecasts (Tompkins *et al.*, 2005; and see discussion in Section 10.2). Unfortunately, the radiosonde network in Africa is even sparser than the ground station network. Radiosonde stations are relatively expensive to maintain and require dedicated staff to run them (Parker *et al.*, 2008). In 2006 as part of AMMA there was an increase in the radiosonde network over West Africa, and several operational NWP centres have used this dataset both to carry out data impact experiments and to evaluate the models (see discussion in Section 10.2).

#### 10.1.2.2 Data Assimilation

The irregular observations (Figure 10.3) cannot be used directly as initial conditions to a numerical prediction. They need to be combined with a 'first guess' or 'background' of the initial state from a previous NWP short-range forecast (usually 6 h), which provides data on the regular model grid. The observations and model background are combined in a statistically optimal way that takes account of known errors in both the model background and observations in a process known as data assimilation (Daley, 1991; Kalnay, 2003). The final outcome is a balanced initial state that is our best estimate of the current temperature, moisture, winds, soil state and so on on a regular grid that can be used in numerical models of the atmosphere, land and ocean to provide a forecast.

Data assimilation techniques used in operational NWP have evolved from simple optimal interpolation techniques used from the 1960s to 1980s, to more complex variational assimilation (three-dimensional variational data assimilation (3D-Var) and four-dimensional variational data assimilation (4D-Var)), Kalman filtering and ensemble techniques. At the heart of many data assimilation systems (e.g. successive correction method, optimum interpolation, 3D-Var, and Kalman filtering) is an analysis equation:

$$x_a = x_b + W\left[y_o - H(x_b)\right] \tag{10.1}$$

Here, $x$ represents a model state and $y$ an observation. The analysis equation states that the model analysis $x_a$ is the background state $x_b$ plus some observation increments or 'innovations', $y_o - H(x_b)$, which measure the difference between the observation $y_o$ and the background (Figure 10.4). The background is converted to the same measurement space as the observation using the forward operator $H$, which could be a radiative transfer model to convert temperature and moisture to radiances. The background also contains very useful *prior* information as it is partly constrained by all previous observations made in earlier assimilation cycles. In modern data assimilation the problem is multidimensional and the $x$ and $y$ in the analysis equation are large vectors, which for the model state $x$ is the total number of model grid points times the number of model variables, and for the observations $y$ is the (irregularly spaced) number of observations (e.g. Figure 10.3). The other term, $W$, is an optimal weight based on statistical estimates of the typical errors (error covariances) of the model and the observations. The determination of the optimal weights is partly what distinguishes the different data assimilation techniques (see Lorenc (1986) for discussion). For example, in 3D-Var and 4D-Var, the analysis $x_a$ is determined from a minimisation of a cost function $J$, which is proportional to the squared distance between the analysis and both observations and background:

$$J = \tfrac{1}{2}\Big[\left(y_o - H(x)\right)^{\mathrm{T}} R^{-1}\left(y_o - H(x)\right) + \left(x - x_b\right)^{\mathrm{T}} B^{-1}\left(x - x_b\right)\Big] \tag{10.2}$$

The first term in $J$ ($J_o$ in Figure 10.4) measures the distance of a model state $x$ from the observations $y_o$ scaled by the observation error covariances $R$, and the second term $J_b$ the distance to model background $x_b$ scaled by background error covariances $B$. So, for example, the larger the background error covariance the smaller weight the model background has relative to observations in determining the analysis. This cost function is minimised in an iterative technique involving the gradient of the cost function, and at the minimum point, $x = x_a$ forms our initial analysis (see red line in Figure 10.4).

Since the late 1990s many operational centres have moved to a four-dimensional data assimilation system in which the time evolution of the observations (and model background) are used to advance the background error covariance (and model state) in time, rather than using a static error covariance, leading to a better estimate of the initial analysis. Extended Kalman filtering (EKF[3]) is considered the 'best' of these techniques for data assimilation (Kalnay, 2003), as it *explicitly* evolves the covariance matrix (and model state) so that at the next analysis cycle both a background model state and an updated error covariance matrix are available. However, the EKF is prohibitively expensive to implement.

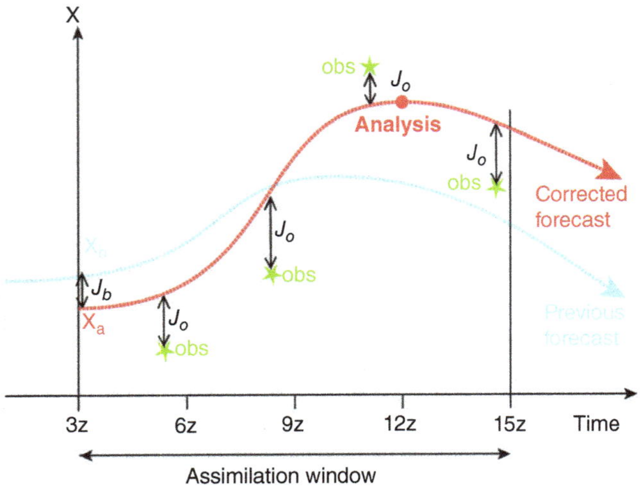


**Figure 10.4** Schematic of the 4D-Var data assimilation process (from Bouttier and Courtier (1999)). This shows a 12-h assimilation window. The analysis cycle begins with the background forecast $x_b$ shown in light blue thatt is gradually corrected in the 4D-Var to the new state ($x_a$ – shown in red) to be more consistent with both the observations (green stars) and the background. $J_o$ and $J_b$ represent the terms in the penalty function (Equation 10.2) that measure the departure of our final optimised 'corrected' forecast from the observations and model background respectively. At some nominal point, shown here at 1200 UTC, an analysis is defined from which to run our NWP forecast forward *Source:* reproduced from Bouttier and Courtier (1999) with the permission of ECMWF.

4D-Var data assimilation has been successfully implemented at a number of operational NWP centres (e.g. Met Office, European Centre for Medium-Range Weather Forecasts (ECMWF), Meteo-France) as a cheaper alternative to the Kalman filter. The major difference is that 4D-Var only propagates the model state with the evolution of the covariance matrix *implicit* in the process but not *explicitly* available to the next analysis cycle (see Figure 10.4 for a schematic of variational assimilation). A further drawback of 4D-Var is that the initial error covariance provided at the start of an analysis cycle is a climatological estimate and there is no knowledge of the particular 'errors of the day'. In recent years research has been active in ensemble filter data assimilation techniques as they evolve flow-dependent forecast error covariances through the use of ensembles of nonlinear forecasts. Many centres have developed ensemble prediction systems (EPSs – see Section 10.1.3) as part of their operational forecast suites and the ensemble perturbed forecast members give information on short-range (6–12 h) 'flow-dependent' error growth that can be used in the data assimilation process. Meteo-France, ECMWF and the UK Met Office all employ these ensemble techniques.

The other important component of any data assimilation system is some form of quality control of observations to ensure that large observation errors do not contaminate the initial conditions. Sources of observation errors include instrument bias and human error in taking measurements (e.g. screen-level temperatures). A good example of the role of observation error and quality control was seen as part of the AMMA campaign where it was found that the different types of radiosondes used as part of the Special Observing Period (SOP) exhibited particular humidity biases when compared with independent data. Correction of these humidity biases had an impact on the subsequent NWP analyses and short-range forecasts (see Section 10.2 for further discussion).

The initialisation of soil moisture and land surface temperatures is also important over Africa in modelling the land–atmosphere interactions. NWP systems have been hampered by lack of observations of soil state variables. One solution was to develop some estimates for soil moisture based on observed screen temperature and moisture variables at SYNOP stations (Mahfouf, 1991). The central idea is that errors in the model soil moisture will cause particular errors in the daytime screen-level temperature and humidity. Knowledge of these screen-level errors (measured from SYNOPs) can be used to slowly correct or nudge the initial soil moisture state toward a state that is hopefully closer to the observed state but which definitely minimises the errors in screen-level temperature and moisture. Soil nudging schemes were originally implemented at ECMWF and Meteo-France. However, there are many sources of error in the screen-level temperature and moisture (e.g. advection, planetary boundary layer (PBL) mixing), so it is important to identify those physical conditions/regimes where the soil moisture deficiencies are contributing to the screen-level errors. For example, the Met Office soil nudging scheme only operates if (i) there is evaporation, (ii) screen temperature and moisture errors are of opposite sign, so the model PBL/surface layer is warm and dry (too little soil moisture) or cold and moist (too much soil moisture), (iii) the lower atmosphere/PBL is unstable and (iv) there is no snow cover.

More recently, NWP centres and the wider research community have been exploring the use of remote sensing to provide observations on soil moisture. Active microwave backscatter measurements from space can be used to extract information about the soil wetness in the topmost 1 cm of soil (Wagner *et al.*, 1999). Other

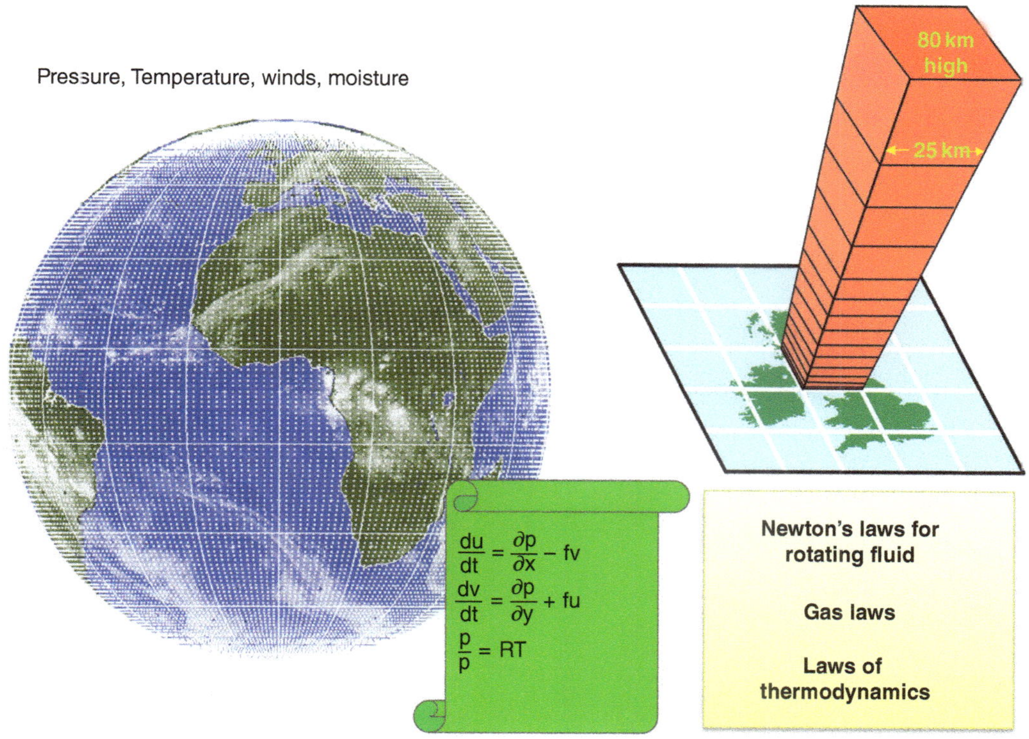


**Figure 10.5** Model horizontal and vertical grids on the sphere used to solve the fluid equations to provide a numerical weather forecast. The dimensions and grid spacings are from the MetUM used for global NWP (circa August 2013).

proposals have been made to use (i) passive microwave measurements, using the fact that microwave emissivity changes with soil wetness, and (ii) the infrared region where the sensitivity of the diurnal cycle of surface temperature to soil moisture can be used to define retrieval methods based on the observed changes of the infrared skin temperature. In October 2006 the ASCAT was launched on the METOP polar orbiting satellite with a daily global coverage of 82%, overpasses at 09:30 and 21:30 local time and delivering products on a 12.5 km grid, which provided a suitable platform for assimilation into NWP models. ASCAT data have been assimilated in the UK Met Office global NWP system since July 2010 (Dharssi *et al.*, 2011) with positive benefits on screen temperature and humidity forecasts in the tropics. Other sources of soil moisture observations include the European Space Agency's current Soil Moisture and Ocean Salinity mission launched on 2 November 2009 and NASA's Soil Moisture Active Passive (SMAP) mission launched on 31 January 2015. SMAP carries an active synthetic aperture radar and passive radiometer, to allow soil moisture to be measured through moderate vegetation and will also give information on the freeze/thaw state of the soil.

These developments in data assimilation techniques and exploitation of new satellite observing platforms have given dramatic improvements in NWP forecasts in the last two decades (Figures 10.1 and 10.2). A good learning resource for data assimilation is the ECMWF training course material available at http://www.ecmwf.int/en/learning/education-material/lecture-notes, and in particular the overview paper by Bouttier and Courtier (1999).

#### 10.1.2.3 Global and Regional Numerical Weather Prediction Models

Weather and climate models mathematically represent the dynamic and thermodynamic processes in the atmosphere on a grid of points (Figure 10.5), using the equations of fluid motion (deriving from Newton's second law), the continuity equation, and the equation of state and the first law of thermodynamics (Kalnay, 2003). Any processes that occur on spatial scales smaller than the model grid spacing must be parametrised. This involves using the resolved temperature, winds, humidity and so on on the model grid together with a sub-model of the physical process to estimate quantities such as heating and cooling rates from clouds and precipitation processes or the vertical transfer of heat and moisture due to turbulent eddies in the boundary layer.

In this section we briefly review the components in a numerical model of the atmosphere; the numerical

**Table 10.1** Details of operational global deterministic NWP models producing forecasts over Africa and ensemble prediction systems used in operations and research (TIGGE – for more details of all OP centres involved in TIGGE, see https://software.ecmwf.int/wiki/display/TIGGE/Models).

| Forecast centre | Model | Resolution | | Forecast range | Assimilation | No. EPS members |
|---|---|---|---|---|---|---|
| | | Horizontal | Vertical | | | |
| *Global deterministic NWP models* | | | | | | |
| Met Office | MetUM – global | N768 – 17 km | 70L (top at 80 km) | 144 h | Hybrid 4D-Var (33 mem) | |
| ECMWF | IFS | T1279 – 16 km | 137L (0.01 hPa top) | 240 h | Ensemble 4D-Var (25 mem) | |
| Météo-France | ARPEGE-IFS | Variable mesh | 70L | 240 h | 4D-Var | |
| JMA | GSM | T959 0.1875° | 100L (top 0.01 hPa) | 216 h | 4D-Var | |
| NCEP | GFS | T1534 T574 | 64L (top 0.3 hPa) | 240 h 240–384 h | Hybrid 3D-Var EnKF | |
| CMC | GM (GEM) | 0.45° long.–0.33° lat. | 80L | | 4D-Var | |
| DWD | GME | 20 km | 60L | 7 days | 3D-Var | |
| *TIGGE EPSs used in this study (research only)* | | | | | | |
| Met Office | MOGREPS-15 (–14 Jul) | N216 (60 km) | 70L (top at 80 km) | 15 days | | 24 |
| | MOGREPS-G (14 Oct–) | N400 (33 km) | 70L (top at 80 km) | 7 days | | 12 |
| ECMWF | IFS-ENS | T639 (32 km) T319 | 91L | 10 days 10–15 days | | 50 |
| NCEP | GEFS-TIGGE | T126 | 28L (top 2.73hPa) | 16 days | | 20 |
| CMC | GEM EPS | ~0.45° | 40L (top at 2 hPa) | 16 days | | 20 |
| JMA | GSM EPS | T479 (0.375°) | 60L (top 0.1 hPa) | 11 days | | 26 |

CMC: Canadian Meteorological Centre; JMA: Japan Meteorological Agency; NCEP: National Centers for Environmental Prediction.

formulation used to model large-scale transports, Coriolis effects and pressure gradients; the physical parametrisations of sub-grid-scale processes; the choice of grid on which to solve the equations, and its grid spacing or resolution in both the horizontal and the vertical. Table 10.1 contrasts the formulation of different operational global NWP models and EPSs at the time of writing this chapter, with the caveat that this information quickly becomes out of date as NWP systems are under constant evolution.

#### *10.1.2.3.1 Model Numerics*

The dynamical core of an NWP model is the component that numerically solves the equations of motion (Equations 10.3, 10.4 and 10.5), continuity (Equation 10.6), equation of state (not shown) and the first law of thermodynamics (Equation 10.7) on a grid of regular points on a sphere (global models) or part of the sphere (regional models).

For illustration we use the primitive equations (Equations 10.3–10.7 expressed here in pressure coordinates), which were the basis for many NWP and climate models in the 1970s and 1980s (Arakawa, 2000b). They are *nonlinear* partial differential equations, which are impossible to solve with analytical methods and require numerical methods to approximate solutions (Haltiner and Williams, 1980; Durran, 1999). These primitive equations are themselves approximations to the full equations of fluid motion and (i) neglect vertical accelerations and friction in the vertical momentum equation (hydrostatic approximation), (ii) neglect some of the Coriolis terms, (iii) approximate the radial distance to the centre of the Earth by a mean radius and (iv) neglect curvature terms in the horizontal and vertical momentum equations. For higher resolution limited-area models (LAMs), where we are trying to model the vertical acceleration in individual mesoscale weather systems, the hydrostatic approximation is not appropriate and those models are fully non-hydrostatic. Indeed, as global models reach the 20–10 km scale in horizontal resolution many are considering a non-hydrostatic formulation: the Met Office global model is fully non-hydrostatic. In addition, the other approximations have already been abandoned in some NWP models to improve accuracy (e.g. see Davies *et al.* (2005) for the Met Office Unified Model (MetUM)).

$$\frac{Du}{Dt}-\left(2\Omega+\frac{u}{a\cos\varphi}\right)v\sin\varphi+\frac{1}{a\cos\varphi}\frac{\partial\Phi}{\partial\lambda}=F_\lambda \quad (10.3)$$

$$\frac{Dv}{Dt}+\left(2\Omega+\frac{u}{a\cos\varphi}\right)u\sin\varphi+\frac{1}{a}\frac{\partial\Phi}{\partial\varphi}=F_\varphi \quad (10.4)$$

$$\frac{\partial\Phi}{\partial p}=-\alpha \quad (10.5)$$

$$\frac{1}{a\cos\varphi}\left[\frac{\partial u}{\partial\lambda}+\frac{\partial(v\cos\varphi)}{\partial\varphi}\right]+\frac{\partial\omega}{\partial p}=0 \quad (10.6)$$

$$C_p\frac{D\theta}{Dt}=\left(\frac{p_0}{p}\right)^{R/C_p}Q \quad (10.7)$$

where $(u, v, \omega)$ are the eastward, northward and vertical velocity (in pressure coordinates) components, $(\lambda, \varphi)$ are longitude and latitude, $\Phi$ is geopotential, $a$ is the radius of the Earth, $\Omega$ is the rate of planetary rotation, $\theta$ is potential temperature, $\alpha$ is the specific volume, and the other terms have a conventional definition, as given for example by Kalnay (2003). Having defined the hydrodynamical equation set, the next challenge is how to discretise these equations so they can be solved numerically on a computer. Some global models and almost all regional models use finite-difference techniques, where the derivatives in the equations are replaced by finite difference approximations at a grid of points covering space and time (Haltiner and Williams, 1980). Other global models and a few regional models use a spectral method for the horizontal spatial discretisation and finite-difference methods in the vertical (see Table 10.1). The spectral method represents the model variables as the sum of a number of global orthogonal mathematical functions (spherical harmonics) with a specified spatial structure – see Haltiner and Williams (1980).

Equations 10.3 to 10.7 provide forecasts by predicting rates of change in temperature, moisture, winds (and pressure) arising from the advection of these quantities and from various sources and sinks including damping due to friction ($F$ on right-hand side of Equations 10.3 and 10.4) and diabatic heating rate due to moist processes ($Q$ on right-hand side of Equation 10.7). The numerical solutions are advanced in increments determined by the time-step of the model.

A popular numerical method for advection adopted by many NWP centres for its accuracy and efficiency is the semi-Lagrangian scheme (Staniforth and Cote, 1991; Durran, 1999). A truly Lagrangian scheme would follow the individual air parcels and their properties throughout the lifetime of a forecast, but this is impractical for weather prediction as the particles may all collect together in one area or leave other areas with very few parcels. The semi-Lagrangian technique avoids this by using the regular grid to solve for the total time derivative ($dx/dt$) for a variable, by finding out where a parcel of air arriving at a particular grid point in a *given time step* came from in the previous time step (the departure point). This is in contrast to Eulerian schemes, where the partial time derivative ($\partial x/\partial t$) is estimated at a grid point requiring numerical estimates of advection of temperature and so on through that point.

The numerical solutions within the finer resolution regional LAMs also require a set of boundary conditions at the perimeter of the limited-area domain. This is both to initialise the forecast and to provide a continuous boundary forcing throughout the period of the forecast. These lateral boundary conditions are supplied by the coarser resolution global NWP prediction systems. The LAMs are typically one-way nested, which means they receive boundary conditions from the global models but do not themselves directly influence the global predictions. Meteo-France (Courtier and Geleyn, 1988) and the CMC (Côté *et al.*, 1998a,b) have both introduced the novel concept of variable-resolution global prediction systems, which have fine resolution over a given target area (e.g. France) and coarser resolution elsewhere on the global domain.

The goal of all numerical schemes is to maintain high accuracy, ensure computational stability of the numerical solutions and to conserve important quantities such as energy, angular momentum and moisture while also being efficient enough to deliver a 5-day forecast in a timely way.

#### 10.1.2.3.2 Model Resolution and Grids

The size of the grid spacing used in both the horizontal and vertical (Figure 10.5) typically depends on computer resources available and the requirement to produce timely forecasts. Horizontal resolution influences the accuracy of topography and land surface details and the accuracy of the model numerics in transporting/advecting important quantities such as moisture, momentum and temperature. The vertical resolution also plays a vital role in the accuracy of large-scale vertical transports in the atmosphere and in the accuracy of modelling the sub-grid-scale parametrised physical processes (see Section 10.1.2.3.3). Examples of the typical horizontal and vertical resolutions of current global operational NWP models (circa 2015), are given in Table 10.1. It is reckoned that a grid spacing of $\Delta x$ can only properly resolve atmospheric features on scales greater than about $7\times\Delta x$ , so that a resolution of, say, 25 km only resolves scales greater than about 175 km with accuracy.

The placement or staggering of the predicted variables (temperature, pressure, etc.) on the horizontal and vertical grids is also an important component of the numerical techniques. Arakawa (2000a) gives a review of the different vertical and horizontal grid staggering and their numerical properties. For global models the traditional latitude–longitude grids on the sphere have the disadvantage that as we move toward the poles the longitude 'meridians' converge and the east–west physical dimensions of a grid box change (this also causes problems for efficient scaling of software codes on massively parallel computers). For example, the Met Office N512 NWP model, operational in 2013, has a 40 km east–west spacing at the equator and 25 km in mid-latitudes. A lot of research is ongoing into new grid structures that remove these limitations.

#### *10.1.2.3.3 Physical Parametrisations*

Many of the physical processes important for weather over Africa are actually occurring on spatial scales that are below the typical grid box of current operational global (17–25 km) or even regional (mesoscale – 10 km and less) NWP models. As a result we need a way to represent the mean effect of these smaller scale features on the larger scales represented by the NWP model grid boxes. This is achieved through a process known as physical parametrisation, where the large-scale variables on the model grid (temperature, moisture, winds, etc.) are used in a sub-grid-scale physically based model of the process (e.g. cumulus convection, large-scale cloud, radiation, boundary-layer turbulence) to derive large-scale heating rates, moisture sinks and sources (precipitation and evaporation) and wind accelerations and mixing due to these small-scale processes.

It is beyond the scope of this book to describe the many different ways to parametrise the sub-grid-scale physical processes employed in different NWP and climate models. In some senses they all follow the same general principles we outline here. However, understanding the performance of these differing physical parametrisations at a detailed process level, evaluating their output (heating rates, surface fluxes) against detailed observations and ultimately improving the parametrisations is at the heart of understanding how these different NWP modelling systems perform on average and on a case-by-case basis over Africa (see Section 10.2). This is the continuing challenge for NWP prediction over Africa. The forecasters can play a vital role in evaluating the performance of the NWP models across a range of situations to help characterise deficiencies related to uncertainty in the model parametrisations.

As a starting point we can express any quantity such as potential temperature $\theta$ as the sum of a grid box average $\bar{\theta}$ and a sub-grid departure from that average $\theta'$. Taking the thermodynamic energy equation (10.7) as an example and plugging in this decomposition, averaging over the grid box and working through the mathematics[4] we can arrive at an expression (Equation 10.8) where all of the 'resolved' terms are on the left-hand side of the equation and all of the sub-grid parametrised values or 'unresolved' processes are on the right and include (i) the radiative heating $Q_R$, (ii) the net condensation $C^*$ due to saturation on the grid scale (handled by parametrisations of the large-scale cloud and precipitation) or cumulus convection on the sub-grid scale, (iii) the vertical flux of temperature due to turbulent transports within the PBL and in convective plumes.

$$\frac{\partial\theta'}{\partial t}+\nabla\cdot\bar{v}\bar{\theta}+\frac{\partial\omega\bar{\theta}}{\partial p}=\left(\frac{p_0}{p}\right)^{R/C_p}\left[Q_R+\frac{L}{C_p}(C^*)\right]-\frac{\partial\overline{\omega'\theta'}}{\partial p} \quad (10.8)$$

The horizontal sub-grid eddy transports ($u'T'$ and $v'T'$) are small and usually neglected for purposes of parametrisation. In addition, the fluxes of heat and moisture (and momentum) at the boundary of land/ocean/ice and atmosphere as a sensible and latent heat flux (and surface frictional stress) are also required.

##### 10.1.2.3.3.1 Planetary Boundary Layer Turbulent Transport

The PBL, which typically varies in depth from hundreds of metres at night to 1 or 2 km during the day, is the lowest part of the troposphere and is directly influenced by the presence of the Earth's surface (see Section 4.1). The PBL responds to surface forcing on timescales of hours and is typically strongly forced by the diurnal cycle of solar radiation. It is critically important for NWP as the interface between the land surface and the atmosphere, and transports the surface fluxes of heat, momentum and moisture into the atmosphere via turbulent eddies. The PBL is not fixed, and its depth is quite variable in space and time, determined by the strength of the surface-generated mixing. During the day the PBL typically grows to 1–2 km (6 km over extreme environments such as the Sahara Desert) under the influence of surface heating and the upward transfer of this heat by turbulence. At night the Earth's surface cools and the heat flux is directed downward toward the surface, suppressing mixing and shrinking the PBL sometimes to within 100 m of the surface. During the day, as heat and moisture are transported upward, the air parcels cool as they rise and expand until the lifting condensation level (LCL) is reached and cloud is formed. The types of cloud usually associated with the PBL are 'fair weather' cumulus, stratocumulus and fog. The other important role is through interactions with the roughness of the surface providing a frictional drag on the overlying atmosphere and slowing the near-surface winds. See Chapter 4 for more discussion of the PBL and local weather associated with it.

In an NWP model we need to parametrise the effects of PBL turbulent mixing and frictional drag on the large-scale budgets of heat, moisture and momentum in both cloud-free and cloudy boundary layers. The vertical eddy fluxes for PBL transports in Equation 10.8 need to be expressed in terms of the resolved variables in order to 'close' the equations into a solvable set. A common approach has been to represent the PBL transports above the surface layer as a 'turbulent diffusion', using the local mean gradients in the so called $K$-theory, in which we have

$$\frac{-\partial\overline{\omega'\theta'}}{\partial p}=\frac{\partial}{\partial p}\left[K\frac{\partial\bar{\theta}}{\partial p}\right] \tag{10.9}$$

where $K$ is an eddy diffusivity dependent on stability of the flow (see also Section 4.1.2.1 and Equation 4.5). However, in the convective PBL the flux is less dependent on the local gradient outside of the near-surface layer and more dependent on transports by large eddies. Many NWP models have adopted a non-local $K$-theory approach (see Holtslag and Boville (1993)) to account for these transports using an eddy diffusivity profile that is a function of the diagnosed height of the PBL and a turbulent velocity scale.

#### 10.1.2.3.3.2 Cumulus Convection

Cumulus parametrisation is a key component for prediction of tropical flows. As discussed by Arakawa (2000b), the cumulus parametrisation is complex and must deliver information on (i) precipitation arriving at the surface due to moist convective instability, (ii) the vertical profiles of heating/cooling and moistening/drying, which impact the large-scale environment and feedback with the large-scale dynamics, (iii) transport of momentum, (iv) producing a cumulus cloud field that can interact with radiation, and (v) interactions with the sub-cloud layer through processes such as precipitation evaporation below cloud base and mass exchange via updrafts and downdrafts. As models become more complex these parametrisations must also deal with transports of liquid water and ice phases as well as atmospheric constituents such as aerosols. A cumulus parametrisation must determine the spatial occurrence of convection and the type (shallow, mid-level or deep), usually through a trigger and diagnosis step. Then a sub-grid cloud model is applied to determine the vertical distribution of heating, moistening and momentum transports. The final element is to determine the overall intensity of convection through some form of 'closure'.

There are three main classes of cumulus parametrisation distinguished by their cloud models and closure: (i) those based on moisture convergence (e.g. Kuo, 1965), (ii) adjustment schemes (Manabe *et al.*, 1965; Betts and Miller, 1986) and (iii) mass flux schemes. The latter are used by most NWP models and owe their heritage to the pioneering multiple plume ensemble of Arakawa and Schubert (1974). A simplification of Arakawa and Schubert's scheme is the single entraining/detraining plume used by Bougeault (1985), Tiedtke (1989), Gregory and Rowntree (1990), Kain and Fritsch (1990), Donner (1993) and Bechtold *et al.* (2001). The vertical eddy transports due to convection (Equation 10.8) can be expressed in terms of a convective mass flux $M_C$ and thermal gradient between cloud and large-scale environment as

$$-\overline{\omega'\theta'}=gM_C\left(\theta_C-\bar{\theta}\right) \tag{10.10}$$

The initial mass flux at cloud base needs to be found from the large-scale variables and a common closure for deep convection relates mass flux to the convective available potential energy (CAPE) in a column using a quasi-equilibrium hypothesis that assumes CAPE is removed over a given timescale, usually a few hours.

It is fair to say that a lot of the uncertainty in predicting tropical flows in NWP models comes from parametrising convection and its evolution through the day, which is particularly challenging over land. Of course, correct modelling of convective processes is not solely due to the convection parametrisation itself but also involves accurate modelling of the surface fluxes, PBL transports and radiative processes, amongst other things.

#### 10.1.2.3.3.3 Land Surface and Subsurface

The land surface and subsurface layers provide important stores of water and energy in the soils, vegetation, snow and land-ice that can be transferred into the atmosphere via surface fluxes of sensible heat and latent heat (moisture flux). The atmosphere replenishes these stores through exchange of moisture in the form of precipitation or dew and in the form of heat from radiative fluxes at the surface and downward mixing of sensible heat at night. The roughness of the surface topography or vegetation is also important in determining the frictional drag on the atmosphere and the magnitude of the fluxes. The vegetation canopy plays an important role in the storage of heat and moisture and its exchange with the atmosphere. For example, the evaporation from plants (evapotranspiration) is partly controlled by the stomata of the plants. During drought-stressed conditions the plant stomata will close down to allow plants to conserve their water, shutting off the evaporation process. The roots of plants also play a role in the evaporative processes. Deep-rooted vegetation can draw on water stores metres below the surface, allowing evapotranspiration to continue in dry spells.

NWP land surface models attempt to capture all of these physical processes impacting the water and energy cycle. They typically have up to four subsurface layers that handle the vertical transfer of water/heat and their storage as soil moisture/temperature. They also account for vegetative canopy processes, including a variety of vegetation types from tundra shrub, to grassland to forested areas. The complexity of these land surface parametrisations continues to increase to take account of other important Earth-system processes, such as the representation of groundwater, the transport of water by rivers, the modelling of wetlands, and the important fluxes of greenhouse gases and modelling of nitrogen and carbon cycles.

#### 10.1.2.3.3.4 Radiative Processes

The radiative heating rate in a given atmospheric layer $Q_R$ is modelled as the vertical divergence in that layer of upward and downward fluxes of shortwave (solar) and longwave (terrestrial) radiation (Peixoto and Oort, 1992). These are determined using a radiative transfer model. The radiation must interact with clouds, water vapour, greenhouse gases ($CO_2$ and methane), ozone and aerosols, and with the land, ocean and ice surfaces. Although complex, the radiative transfer for clear skies is perhaps the best understood of all the parametrisations, but the interactions with clouds and aerosols still require better physical understanding to improve the parametrisations. This is being achieved through improved space-borne observations of clouds from platforms such as CloudSAT and CALIPSO (Stephens *et al.*, 2002) and from ground-based and aircraft observations using cloud radar and lidar, such as those employed at Atmospheric Radiation Measurement (ARM) permanent and mobile sites (see Miller and Slingo (2007) for discussion of measurements at the ARM site stationed at Niamey during AMMA).

#### 10.1.2.3.3.5 Large-Scale Condensation and Cloud

In addition to cloud and precipitation formation through convection processes, all NWP and climate models also have a large-scale cloud and precipitation scheme to model the moist processes due to large-scale ascent. This typically deals with frontal cloud and precipitation mainly in the extratropics and with the stratiform 'anvil' cloud that forms at the outflow of deep convection that is often found in the tropics. More recently, cloud schemes have been developed that try to treat all the sources and sinks of cloud formation processes, including convection. These include the prognostic scheme of Tiedtke (1993) and the PC2 scheme used in the UK Met Office global models (Wilson *et al.*, 2008).

#### 10.1.2.3.3.6 Dust and Other Aerosols

Aerosols provide both a direct effect on the radiation in the atmosphere (and that reaching the surface) through absorption and scattering (see Section 5.1.4), and an indirect effect via changes to the cloud condensation nuclei responsible for cloud formation and lifetime. Many NWP models still use climatological distributions of aerosols, but many are also exploring the use of a fully interactive aerosol on NWP timescales. The role of dust over West Africa is comprehensively explored in Chapter 5, including dust prediction in some current NWP systems.

#### 10.1.2.3.3.7 Gravity Wave Drag and Orographic Boundary Layer Frictional Processes

Airflow over mountains results in atmospheric wave motions across a broad range of scales from planetary-scale stationary waves to mesoscale phenomena. For parametrisation we are interested in the processes operating below the model grid resolution and gravity waves forced by orography fall into this category at typical global model resolutions. They transport momentum upward into the atmosphere until they reach regions where they become unstable and the gravity waves 'break' (rather like ocean waves on the shoreline). The momentum the waves were carrying is transferred to the mean flow. This gravity wave drag (GWD) mechanism acts to slow the winds, and the wave breaking typically occurs in the lower stratosphere or the PBL close to the orography (see Kim *et al.* (2003) for a review of GWD). Another separate drag mechanism parametrised in models is the 'form drag,' which is the mesoscale pressure gradient across the orography caused by turbulent flow (Mason, 1986).

### 10.1.2.4 Convective-scale Models

In the last 5 years significant progress has been made in running regional NWP models at high horizontal resolutions (1–5 km) approaching the size of convective systems, and in implementing such 'convective-permitting' limited-area predictions into operational NWP. The COSMO-DE model developed at the Deutscher Wetterdienst (DWD) became operational in 2007 with a grid spacing of 2.8 km (Baldauf *et al.*, 2011); the AROME model (Seity *et al.*, 2011) was implemented at Meteo-France in 2008 with a grid spacing of 2.5 km; a convective-scale version of the MetUM was fully operational over the UK from 2010 with a grid spacing of 1.5 km. The main advantage of these convective-scale resolutions is that more of the physical processes associated with convection are explicitly represented by the model numerics, rather than being parametrised.

A research example of convective scale modelling for the African domain is the Cascade project (Marsham *et al.*, 2013a). Cascade used the MetUM run with explicit convection and at high resolutions (4–1.5 km) over large domains to investigate the influence of tropical convection over Africa (as well as the Pacific warm pool region). For Africa, two independent LAMs were used at 4 km and 12 km resolutions, along with a global model at approximately 40 km resolution. The 12 km LAM was run in two modes – with and without the parametrisation of convection switched on. The versions with parametrised convection allowed the comparison of models looking at the impact of resolution alone. Switching off the parametrisation meant that convective processes smaller than the grid scale were not represented in the model; however, it also allowed convective processes present at scales larger than the grid to be explicitly represented. In general terms, this led to more mesoscale convective activity and less small-scale sporadic convection and 'drizzle'.

One disadvantage of running a model with explicit convection at scales between 12 km and 4 km is that the strength of the modelled convection tends to be unrealistically strong (see also Beucher *et al.* (2014)). A lack of parametrisation also means that small-scale convective processes such as entrainment and mixing – which act to subdue the strength of convection – are not well represented. In such models, the convection is therefore acting as large-scale 'plumes', with some resemblance to typical Sahelian mesoscale convective systems, but with too strong rain rates. Despite this, one advantage of plume-like explicit model convection is its ability to generate cold pool outflows and represent the diurnal cycle more accurately (see Section 10.2). Cold pools can have a significant impact on local environments by distributing moist, cool air across vast distances (see Chapter 3, and also see Garcia-Carreras *et al.* (2013)). Research is very active in a number of centres on convective-scale (kilometres) and cloud-resolving (~100 m) simulations of convection over West Africa.

#### 10.1.2.5 Model Output Statistics

One way to try to address any systematic biases in operational NWP model forecasts is through model output statistics (MOS) run after the numerical forecasts as part of post-processing. These are statistical models based on the three-dimensional fields from NWP, surface observations and the climatological conditions for specific locations. MOS can correct for local effects that cannot be resolved by the model due to lack of spatial resolution, as well as correct model biases. An example is the correction of model surface temperature for the mismatch between (poorly resolved) model orographic height and true station height at a given point. Forecast parameters used in MOS can include maximum and minimum temperatures, precipitation amount expected, chance for thunderstorms, cloudiness and surface winds.

### 10.1.3 Ensemble Prediction and Atmospheric Predictability

So far we have concentrated on NWP as an initial-value problem involving a single deterministic forecast run from our best estimate of the initial state of the atmosphere. However, Lorenz (1963) demonstrated that the numerical solutions of the nonlinear equations of motion are very sensitive to small perturbations in the initial conditions that grow with time as numerical forecasts progress. Uncertainties in our initial state arise from a number of sources:

1) All observations (including satellite data) suffer from measurement errors.
2) Errors arise in our quality control of observations.
3) Our data assimilation process for optimally combining observations and models has approximations in it.

In order to provide some measure of this initial condition uncertainty the concept of ensemble forecasting (Epstein, 1969; Leith, 1974; Lewis, 2005) has been introduced into the operational NWP process. Ensemble forecasting involves determining multiple perturbations to the initial analysed state of the atmosphere that are consistent with observation and analysis errors. Multiple NWP predictions (typically 20–50 in number) are then run from these perturbed initial states and, as they progress, the perturbations grow and the forecasts diverge and spread (Figure 10.6). The degree of spread of the forecasts is a measure of the uncertainty associated with the predictions. On some days the atmospheric flow may be more stable to perturbation growth and the uncertainty/spread is small and our confidence in the forecast is high. At other times perturbations grow rapidly, uncertainty/spread is large and we have low confidence (see Section 10.2 and Figures 10.19 and 10.20 for examples). This varying spread/uncertainty is itself an inherent property of the atmospheric flow and in theory can be predicted using the ensemble technique. In practice, our estimates of spread may be compromised by our initial perturbations not reflecting the true observation/analysis uncertainty or not projecting onto the fastest modes of perturbation growth in the dynamics, and also by our imperfect models, which may underestimate the true atmospheric variability and cause slower perturbation growth and imperfect spread. The different techniques for generating perturbations are reviewed in Kalnay (2003). These have been more focused on perturbations appropriate for mid-latitude balanced synoptic-scale atmospheric flows rather than the convective scales more relevant for the tropics.

A further source of error growth in NWP arises from the models themselves via the numerical approximations used to solve the equations of motion (including model resolution) and from our lack of knowledge in representing the physical processes (currently parametrised for many important processes such as convection). Some NWP centres have also tried to represent these model uncertainties in ensemble forecasts (Figure 10.6), by perturbing the model physics themselves. Perturbing the model's physics (usually by way of the parametrisations) has proven particularly effective in improving the reliability of ensemble predictions in the tropics. Note that forecasters all over the world commonly use a 'poor man's ensemble' consisting of the forecasts obtained from different modelling centres; this combination of forecasts automatically includes different initial conditions and different model physics.

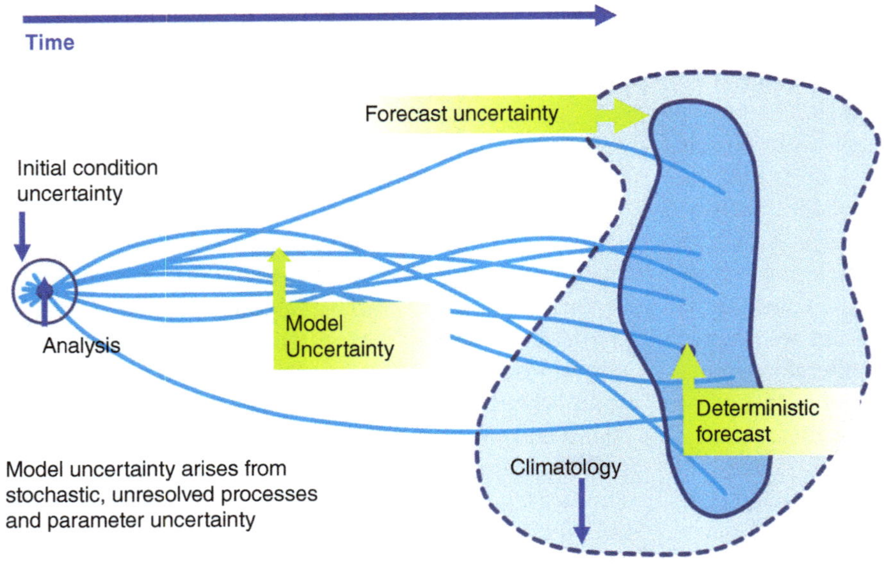


**Figure 10.6** Schematic of the ensemble prediction process. Two types of uncertainty are represented: the initial condition uncertainty given by perturbations to the initial analysis and the model uncertainty. As forecasts progress, both of these sources of uncertainty lead to a divergence of forecast trajectories with time. At some point in the future the forecasts will 'spread' to such a degree that they resemble the climatological distribution of possible model states (shown by region enclosed by dotted line), and all predictability is lost as our ensemble is no more skilful than just predicting climatology. For the case shown, our ensemble has spread but is still less than the climatological distribution, and we still have some predictability and positive information we can use to give a probability of the risk of different weather types.

However, we should also recognise that even if we had a perfect model and perfect observations there is still a limit to predictability. We can never measure all possible time and space scales perfectly, and errors arising from lack of measurement at the very smallest scales represent an error source that can quickly grow as forecasts progress through an upscale transfer of errors from smaller to larger spatial and temporal scales, with the shorter space and time scales (e.g. thunderstorms and MCSs) reaching their predictability limit earlier than the larger synoptic and planetary scales. Statistical techniques, referred to collectively as stochastic physics, have been used to represent or parametrise the energy present in flows below the model grid and to account for their upscale energy transfer, such as the stochastic kinetic energy backscatter scheme (Shutts, 2005).

### 10.1.4 Summary and Future Capability

The purpose of this chapter is to provide forecasters with an appreciation of the various steps in the NWP process, from collecting and processing observations to developing data assimilation methods to provide accurate initial conditions, to the complex numerical modelling of the atmosphere (land and ocean), and the forecasting techniques for predicting the risk of severe weather, such as ensemble prediction and convective-scale modelling.

There continues to be great potential for improving NWP analyses and forecasts over Africa, building on the physical and dynamical knowledge gained from experiments such as AMMA, exploiting new observations, particularly remote sensing platforms, developing improved data assimilation methods to make best use of these observation in data-sparse regions and in biased NWP models, and improving the physical and dynamical formulation of the numerical models themselves. The sparseness of observing systems over Africa remains a challenge, and ultimately these observations are vital to improving the skill of NWP forecasts and to understanding physical and dynamical processes over Africa. Perhaps the increased use of satellite remote sensing data provides a way forward (Karbou *et al.*, 2010a,b), but there will continue to be a requirement for in-situ observations from SYNOPs and radiosondes (Parker *et al.*, 2008) and increasingly a need for mesoscale observations from weather radars to help initialise the next-generation kilometre-scale NWP systems being developed for the tropics. It is to be hoped that improved NWP skill, and better exploitation of that skill by the African meteorological providers, will in future years be a stimulus to the improvement in maintaining the observing (and data communication) networks on which the NWP systems rely.

However, perhaps an equal challenge is to utilise the vast quantity of NWP products already produced at the operational global and regional centres in an effective way for the forecasters on the ground to provide risk-based forecasts of hazardous weather. The WMO Severe

Weather Forecast Demonstration Project (SWFDP – see Section 10.2.2.2) is a good example of bringing together the operational NWP/EPS and local forecasters in Africa.

Looking ahead 5–10 years we might expect to see significant developments in the following areas.

- *Model resolution.* Convective-scale modelling is in its infancy but shows great promise.
- *Parametrisation.* Can parametrisations in coarser scale models be further improved using knowledge gained from the higher resolution models (kilometre scale) and observations from field experiments?
- *Ocean/marine forecasting and coupled ocean–atmosphere NWP systems.* Correct modelling of air–sea interactions may prove crucial for coastal environments and in correctly modelling teleconnections from ocean to land over Africa at longer forecast ranges (15 days to a season).
- *Predictability.* As numerical models, data assimilation and observations improve we should gain improved knowledge of the true limits of predictability of specific phenomena across a range of space and time scales.
- *Improved observations.* More use of remote sensing over the African continent to supplement the sparse in-situ observational network (e.g. Karbou *et al.*, 2010a,b).
- *Improved data assimilation techniques.* For example, ensemble data assimilation. This will allow flow-dependent errors to be taken into account. Research is also active on data assimilation at the convective scale, utilising radar data and so on.
- *Environmental prediction.* Developing NWP capability for composition (chemistry and aerosols) and air quality with direct links to human and animal health.

As the science of NWP over Africa continues to advance, the role of the forecaster is key in delivering timely and accurate advice to end users from a diverse range of complex products. This requires a sophisticated forecasting expertise with detailed knowledge of current NWP practices and of strengths and weaknesses of the various observations, data assimilation techniques, models and ensemble predictions. Evaluation of the NWP products by forecasters also provides a critical feedback loop to model and data-assimilation developers to help continually improve NWP prediction systems over Africa.

## 10.2 Operational Numerical Weather Prediction over Africa

NWP is a valuable tool for forecasters in determining the future day-to-day weather state and the risk of extreme weather. However, as reviewed in Section 10.1, current global and regional NWP systems have their limitations, and there are many sources of error arising from imperfect initial conditions and imperfect numerical models. In this section we review the use of numerical products and highlight some current capabilities and drawbacks of NWP predictions over Africa. African National Met Services (NMSs) have free access to *some* global and LAM products through EUMETCast (e.g. ECMWF, Met Office, Météo-France, NCEP) to facilitate operational weather forecasting. Regional and local modelling are performed in a few NMSs to take more account of their regional and local characteristics and provide diagnostics needed for specific applications. A few national centres are running their own regional modelling systems. For example, Morocco, Algeria and Tunisia (three African members of the ALADIN consortium) and the South Africa Weather Service use a regional version of the MetUM. At present, less than 10 African countries are running operational weather models. The chain of events from collecting observations, running models, interpreting the NWP products by forecasters to issuing final predictions and warnings is shown schematically in Figure 10.7. The forecaster plays a key role in this process, providing the ability to sift through all of the available NWP products, with knowledge of their characteristic strengths and weaknesses, to arrive at the most likely outcome in terms of severe weather forecasts. In the following sections we provide some commentary on the typical strengths and weaknesses of NWP assimilation and models over West Africa.

### 10.2.1 Current Capability of Deterministic Numerical Weather Prediction Forecasts

In this section we briefly review the ability of global NWP prediction systems to capture some of the major seasonal features and weather phenomena discussed in Chapters 1, 2 and 3.

#### 10.2.1.1 Precipitation and Moisture Budget in Numerical Weather Prediction Models and Analyses

Precipitation is a key variable for prediction at weather timescales (e.g. flooding, as in the Ouagadougou event of 2009 presented in Section 2.2.2.2) and for intraseasonal to seasonal timescales (e.g. monsoon breaks and drought) and near-term decadal climate predictions (e.g. interannual variability in long-term Sahel drought).

For a comparison of the capability of numerical models from a number of operational centres we have looked at the NWP control forecasts produced as part of the THORPEX Interactive Grand Global Ensemble (TIGGE) research experiment (Bougeault *et al.*, 2010). The TIGGE dataset consists of control forecasts and ensemble members (between 20 and 50 members) out to 10 or 15 days' forecast lead-time from a number of NWP centres.

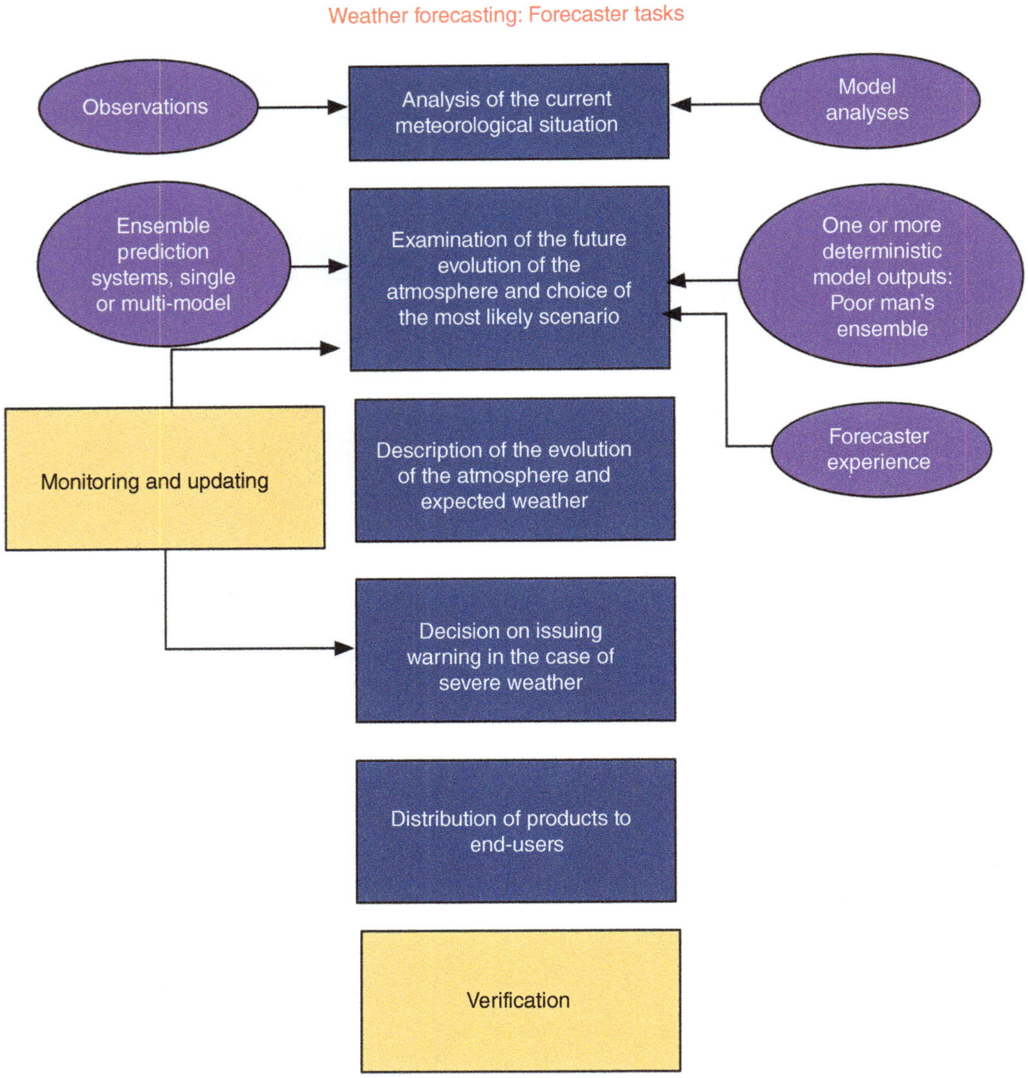


**Figure 10.7** Schematic of typical tasks carried out by the weather forecaster (blue boxes) and the NWP inputs to the process (purple ovals – including forecaster knowledge and experience of NWP characteristics).

The TIGGE models are typically at slightly lower resolution than the deterministic operational NWP global models (see Table 10.1). For this comparison we have chosen forecasts covering June–September 2012 and TIGGE control forecasts from the Met Office, ECMWF, NCEP, CMC, and JMA. All forecasts are initialised at 1200 UTC and model data are plotted from the 1° output grid over West Africa. Details of the TIGGE model formulations are also given in Table 10.1.

#### *10.2.1.1.1 Seasonal Mean Precipitation*

The seasonal mean precipitation is shown in Figure 10.8 for two forecast ranges: 12–36 h and 192–216 h (days 8–9). This is contrasted with the seasonal mean precipitation from the Tropical Rainfall Measuring Mission (TRMM) 3B42, version 7.0, three-hourly precipitation dataset (Huffman *et al.*, 2007). For TRMM we see the familiar features outlined in Section 1.2 for West African precipitation climatology. There is a strong gradient in rainfall, decreasing from south to north as we move from the Soudanian zone (9–12°N) into the Sahel (12–18°N) and then the Sahara. For this year, 2012, the rain reaches quite far north into northern Mali, but is more confined to the southern Sahel in Mauritania and Niger. The 2012 WAM seasonal rainfall anomaly was above normal. Climatological wet regions occur on the southwest coast, Guinean highlands, and west of 0°E across Nigeria, Cameroon, southern Chad and the Central African Republic, and over the Ethiopian Highlands in the east. Drier zones are seen over the Ivory Coast, Ghana, Burkina Faso, Togo and Benin (the Ghana–Dahomey dry zone – Chapter 1) and also further east over the Sudan.

The NWP TIGGE control forecasts at 12–36 h and 192–216 h show reasonably good distributions of seasonal mean precipitation averaged over all forecasts for the June–September 2012 WAM season. All models capture

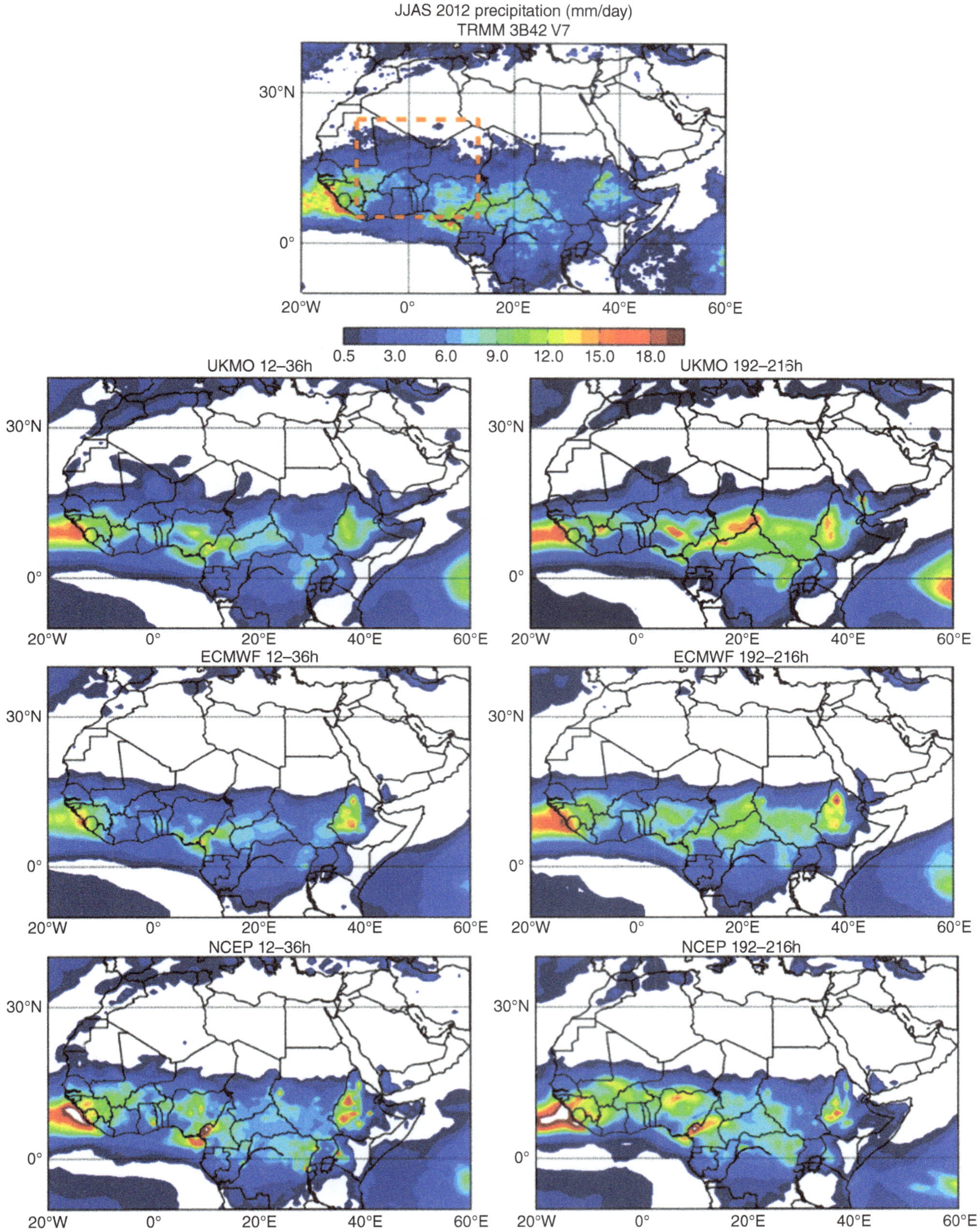


**Figure 10.8** Seasonally averaged (June–September 2012) precipitation (mm day$^{-1}$) for TIGGE control forecasts (on 1° grid) at 12–36 h (left) and 192–216 h (8–9 days – right) compared with the TRMM 3B42 version 7 precipitation (top panel – on 0.25° grid).

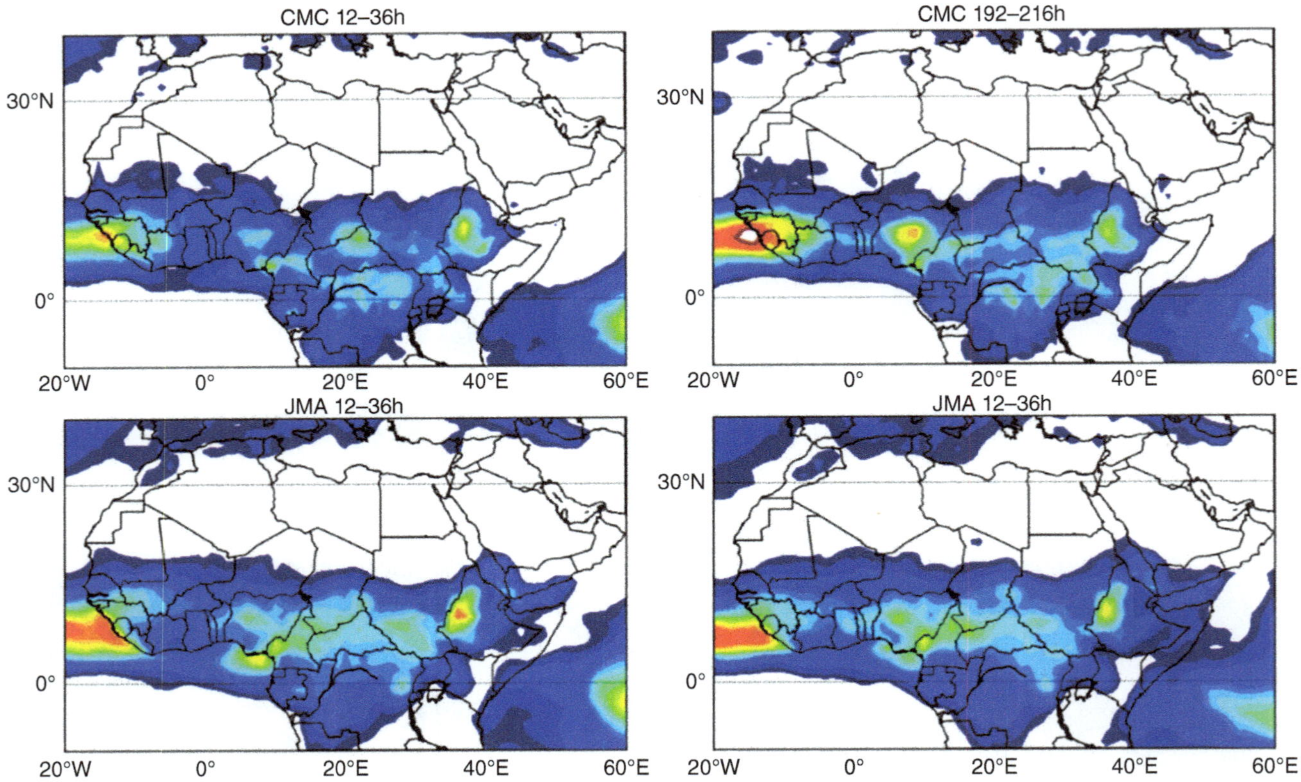


**Figure 10.8 (Cont'd)**

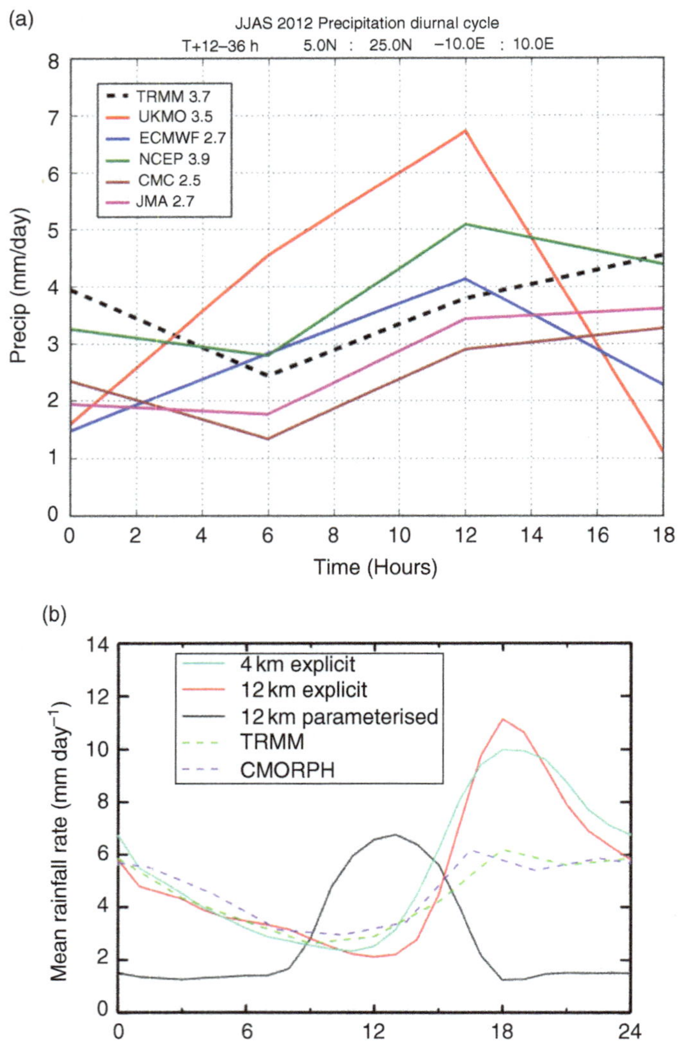


**Figure 10.9** Seasonal mean precipitation (mm day$^{-1}$) as a function of the diurnal cycle for (a) TIGGE control forecasts (6 h sampling) averaged over the domain (5–25°N, 10°W–10°E) shown in Figure 10.8 and (b) diurnal cycles of precipitation (mm day$^{-1}$) over West Africa for observational estimates from TRMM and CMORPH (NOAA CPC Morphing Technique –http://www.cpc.ncep.noaa.gov/products/janowiak/cmorph_description.html) datasets compared with a 12 km simulation of the MetUM with parametrised convection and 12 km and 4 km simulations with explicit convection. *Source:* Marsham *et al.* (2013a). ©John Wiley and Sons. Reproduced with permission.

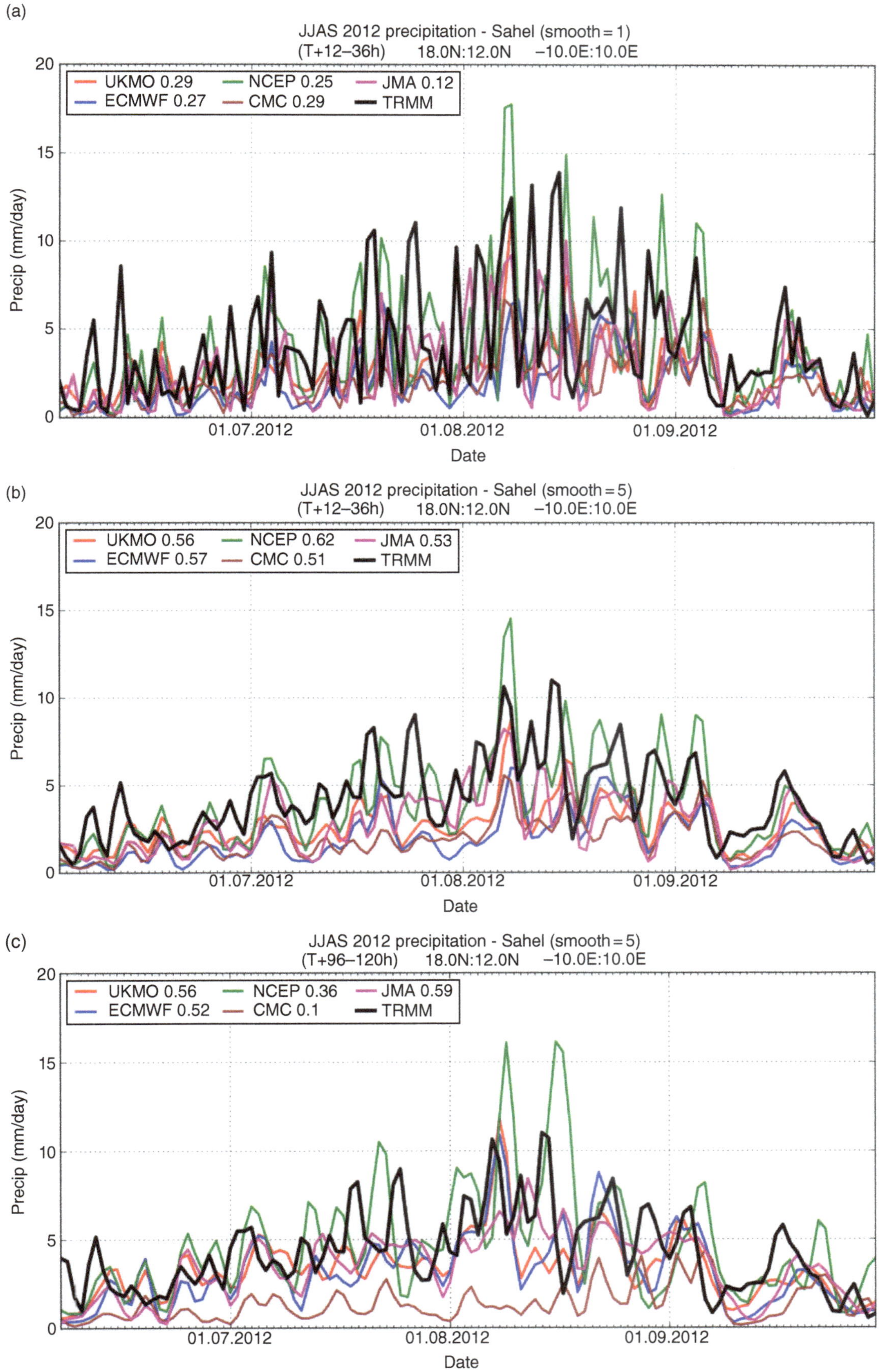


**Figure 10.10** Evolution of daily mean precipitation during the June–September 2012 WWAM (10°W–10°E) season for forecasts at 12–36 h and TRMM over (a) the Sahel region (12–18°N), (b) as (a) but for 5-day smoothed precipitation and (c) as (b) but for $T + 96$–120 h, (d) as (b) but for $T + 192$–216 (days 8–9 of forecasts). The numerical value in the legends is the time series correlation between each forecast and the TRMM data.

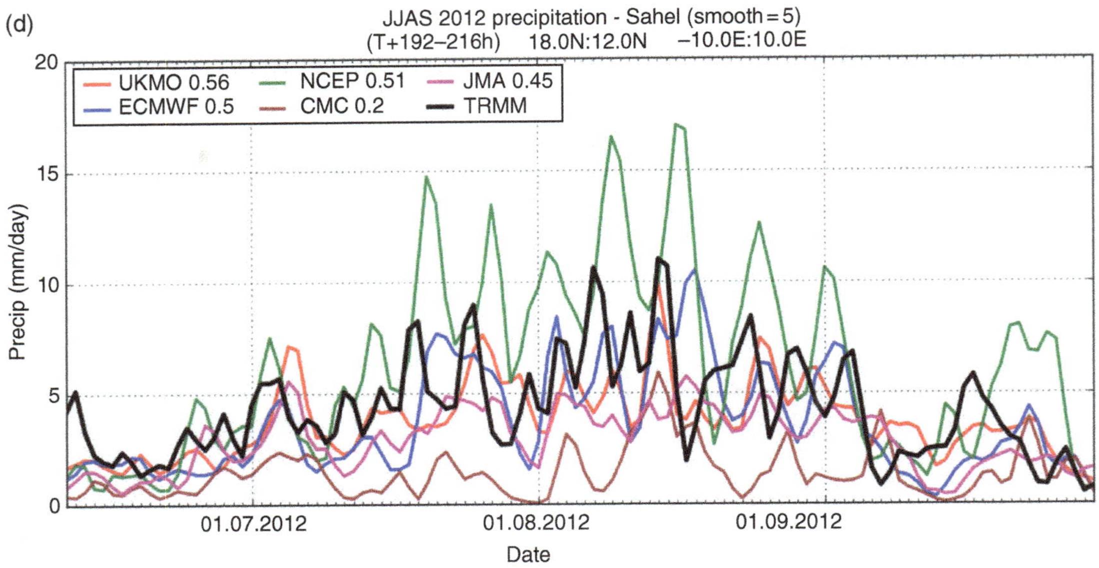


**Figure 10.10** (Cont'd)

the enhanced precipitation off the southwest coast, although it is a little too strong in Met Office, NCEP and JMA forecasts at 12–36 h, and in most models at 192–216 h over both the coast and the oceanic intertropical convergence zone. The CMC and JMA have a good representation of the Ghana–Dahomey dry zone, which is maintained out to 192–216 h, although overall the precipitation is a little low in CMC (see Figures 10.9 and 10.10). The MetUM does a good job of capturing the northward extent of the rainfall and some of the east–west variations, such as the wetter conditions over northern Mali and dry slot over neighbouring northern Niger. These east–west variations may be linked to individual synoptic events associated with MCS and AEWs within the 2012 season. CMC show similar east–west variations, but in some of the other models the precipitation does not push far enough north into the Sahel and is too uniform in the east–west direction. All of the models show an increase or spin-up in precipitation between 12–36 h and 192–216 h. The precipitation at the later forecast ranges is typically larger than the TRMM observations.

This good performance in MetUM TIGGE forecasts for 2012 is in contrast to free-running experiments with the MetUM at various resolutions discussed in Marsham *et al.* (2013a). In these runs for the 2006 AMMA period the MetUM had too little precipitation in the Sahelian region, and longer climate integrations with MetUM also suffer from this dry bias. Possible reasons for the differences could include the soil moisture initialisation. Marsham *et al.* (2013a) uses ECMWF initial data from 2006 based largely on a soil moisture nudging scheme, whereas 2012 MetUM initial conditions include ASCAT microwave soil moisture observations. This raises an important point that the operational NWP assimilation and forecast systems are undergoing constant development. This makes it imperative that the characteristics of these models in predicting the WAM are kept up to date, so forecasters interpreting numerical products can use them more effectively. This is an area where increased collaboration between African forecasters and the operational modelling centres would be of great benefit.

#### *10.2.1.1.2 Diurnal Cycle of Convection*

Current numerical models of the atmosphere with parametrised moist processes (convection, large-scale condensation, etc.) have well-known deficiencies in predicting West African rainfall, particularly the diurnal cycle over land (e.g. Bechtold *et al.*, 2004). As an example, Figure 10.9b shows some of the errors in the seasonally averaged diurnal cycle of West African (5–2°N, 10°W–10°E) precipitation in the MetUM from Marsham *et al.* (2013a). For a 12 km regional model with parametrised convection we see a clear systematic bias in the diurnal cycle over land, with precipitation peaking at local solar maximum (1200 LT) and very small values overnight. In contrast, the observational estimates from TRMM (and CMORPH) show a peak in precipitation later in the day (1800 LT) and have significant precipitation overnight. So even at relatively high horizontal resolutions of 12 km, a model with parametrised convection has significant diurnal cycle systematic biases. The MetUM was also run with explicit convection (allowing the convection to occur on the grid scale and be handled by the model's numerics) at 12 km and 4 km. Clearly, 12 km is still too coarse to truly represent individual convective plumes and 4 km can still only be considered 'convection permitting' rather than truly 'convection resolving'. However, both of these explicit convection runs do a much better job at capturing the diurnal cycle, although the precipitation maximum at 1800 LT is too intense.

Similar studies with cloud-resolving model simulations (Chaboureau *et al.*, 2004) have identified the important physical ingredients needed to capture an accurate diurnal cycle in convective parametrisations. These include a period of shallow convection in the morning that begins once the convective inhibition (CIN) is lower than the vertical (turbulent) kinetic energy, a well-defined transition from shallow to deep convection in the afternoon following significant moistening of the PBL by detrainment from shallow convection. A convective triggering that links CIN and vertical kinetic energy is also thought to be important. The goal is to use the information from the explicit convection or cloud-resolving simulations and AMMA observations to help inform the parametrisation developments to improve the diurnal cycle over land. Stratton and Stirling (2012) have shown some improvement to the diurnal cycle over land through changes to make convective entrainment[5] a function of height of the LCL, which reduces the depth of convection early in the day. More recently, the work of Bechtold *et al.* (2014) shows improved diurnal cycle from changing the closure in the ECMWF cumulus convection involving a convective adjustment timescale for the free troposphere, and a coupling coefficient between the free troposphere and the boundary layer based on different timescales over land and ocean. These examples show that progress in convective parametrisation is being made from knowledge gained from fine-resolution cloud-resolving models and detailed process-based observations.

The 2012 TIGGE forecasts show a range of behaviour in terms of the diurnal cycle over the WAM region (Figure 10.9a). MetUM shows the behaviour already discussed with too intense rainfall at 1200 LT and too little overnight, although the daily mean (shown in the legend) is close to the TRMM value (the aforementioned Stratton and Stirling (2012) model changes are not included). ECMWF also shows a similar phasing of the diurnal cycle but with more precipitation at night and more realistic amounts during the day (Bechtold *et al.* (2014) convection changes not included). NCEP also has peak rainfall at 1200 UTC but has much more precipitation overnight than either MetUM or ECMWF. JMA and CMC both have good phase in their diurnal cycles, but too little precipitation overall (2.7 mm day$^{-1}$ and 2.5 mm day$^{-1}$ respectively compared with 3.7 mm day$^{-1}$ for TRMM). Clearly, the details of how the PBL, convection and cloud/precipitation are parametrised in these models is important for capturing the details of the diurnal cycle. What is not clear is how important having a correct diurnal cycle is for predicting precipitation at other timescales of a day to a week. Evidence from Marsham *et al.* (2013a) suggests that incorrectly modelling the diurnal cycle interacts with the north–south pressure gradients, which in turn changes the northward moisture flux out of the Sahel (Birch *et al.*, 2014).

#### *10.2.1.1.3 Precipitation Variability at Daily and Weekly Timescales*

The variability in daily mean precipitation over the Sahel is considerable, with typically around 20 precipitation events a month exceeding 5 mm day$^{-1}$, interspersed with relatively dry periods (Figure 10.10a and AEW example in Figure 10.19). As discussed in Section 10.1.2.3, this variability is associated with westward-travelling AEWs and organised MCSs evolving on the AEJ. As the monsoon progresses through June, July and August these precipitation maxima become more intense. Even at 12–36 h the TIGGE control forecasts typically underestimate the amplitude of these maxima (Figure 10.10a and b), and the phasing of the maxima and minima is not well predicted, as shown by the weak time-series correlations (e.g. 0.1 to 0.3) between daily averaged TRMM and TIGGE forecasts (see legend in Figure 10.10a). NCEP stands out as having a more reasonable amplitude for the maxima, but still little day-to-day predictability.

Janicot *et al.* (2011) have discussed the precipitation variability across West Africa at a number of intraseasonal timescales (see also Section 7.1.3). They showed that, as well as precipitation variability on daily timescales associated with westward-travelling weather systems (AEW), the precipitation is also modulated by lower frequency modes of variability; two on a 15-day timescale (quasi bi-weekly zonal dipole mode and the Sahel mode) and one with a periodicity around 40 days (Africa MJO mode). If we apply a 5-day smoothing to the 12–36 h Sahel precipitation (Figure 10.10b) we begin to see that the TIGGE models are able to capture the lower frequency swings in precipitation through the season with higher correlations.[6] The models tend to systematically underestimate the amount of rainfall in the Sahel at this 12–36 h range, with the exception of NCEP (Figure 10.10b). Looking at the longer forecast ranges of 96–120 h (Figure 10.10c) and 192–216 h (Figure 10.10d) we see more clearly the drift of each model toward its own preferred climate. NCEP tends to overestimate the precipitation, although it shows reasonable prediction of some of the low-frequency swings in precipitation. This increased precipitation in NCEP can be seen in the seasonally averaged precipitation (Figure 10.8) as higher rain rates much further north than TRMM or the other models and a rapid decrease across a small latitude range from these higher values to dry conditions over the Sahara. CMC shows a considerable underestimation of Sahel precipitation and not much predictability. The other three models maintain more realistic levels of precipitation, and in this regard they have better precipitation distributions than the 12–36 h forecasts, although the predictability is diminished. This suggests a considerable initial spin-up in these models from the initial analysis time.

Considering the other climatic zones outlined in Chapter 1, the Guinea Coast region (5–9°N, 10°W–10°E; Figure 10.11a) has the models overestimating the 10-day smoothed precipitation at 12–36 h, particularly during August and September. The Soudanian region (9–12°N) shows a more marked seasonal cycle (Figure 10.11b). The models again are able to capture some of the subseasonal swings in precipitation at this early forecast range, but again there is considerable scatter amongst the various

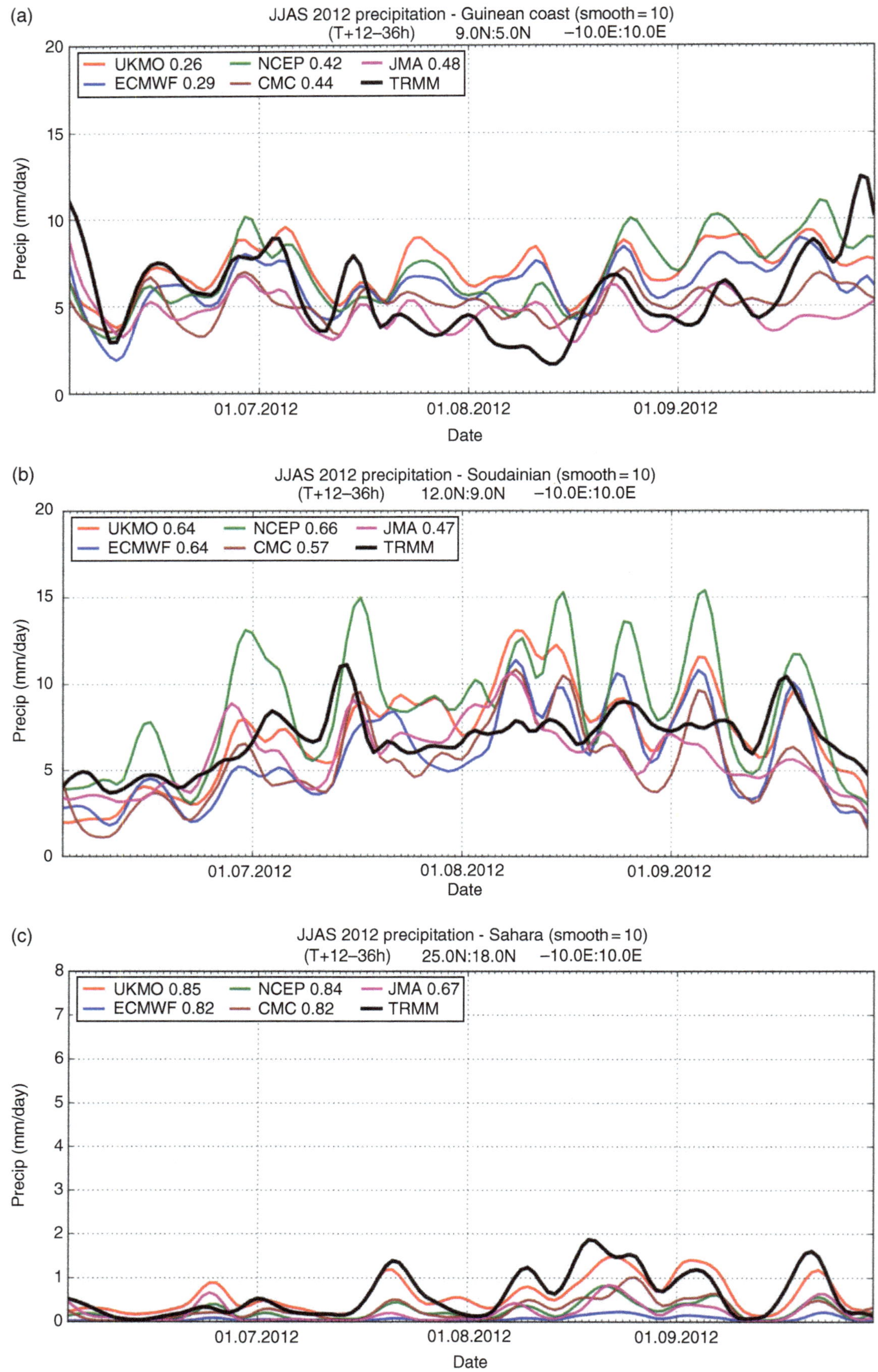


**Figure 10.11** Evolution of 10-day smoothed daily mean precipitation during the June–September 2012 WAM (10°W–10°E) season for the (a) Guinean Coast region (5–9°N), (b) Soudanian region (9–12°N) and (c) Sahara (18–25°N).

predictions as well as a clear hierarchy in terms of overall model precipitation bias, with NCEP overestimating and CMC and JMA having the lowest precipitation and the other models falling in between. Finally, for the small amounts of precipitation falling in the (southern) Sahara (Figure 10.11c) the Met Office provides the best estimate of amounts and the highest time-series correlation, consistent with the seasonal mean precipitation (Figure 10.8).

##### 10.2.1.1.4 Case Study: Heavy Rainfall Event over Mali on 7–8 August 2012

From the daily and 5-day averaged precipitation time series (Figure 10.10a and b) there is a period of intense precipitation observed in TRMM around 7–8 August 2012 within the Sahel region for which the NWP models show good predictability out to a few days ahead. The spatial distribution of daily accumulated precipitation for 1200 UTC on 7–8 August 2012 from TRMM shows a large organised area of rain covering southern Mali (Figure 10.12). The forecast day 2–3 accumulated precipitation from the TIGGE control forecasts initialised on 5 August 2012, 1200 UTC, is also shown. Most of the models do a good job of capturing the position of this event. This is just an example to show that on occasions with strong dynamical forcing (associated here with a trough and active AEW in this case – see Figure 10.20) the NWP models are capable of capturing these extremes. This case is discussed further in the context of ensemble predictions and probabilistic forecasts in Section 10.2.2.2.

##### 10.2.1.1.5 Understanding of The Water Cycle From Amma 2006: Impact Studies on Numerical Weather Prediction Forecasts and Analyses

The AMMA campaign in 2006 (Redelsperger *et al.*, 2006; Lebel *et al.*, 2010; Janicot *et al.*, 2008) provided a marked increase in the density of ground-based observations. During June–September 2006, some 7000 soundings were made, and six AMMA ground-based Global Positioning System (GPS) stations were operating during the SOP, allowing two north–south transects for the study of the water cycle (Bock *et al.*, 2008). These observations have been used both to evaluate physical processes for the water cycle within the WAM and also to potentially improve initial conditions for the operational NWP analyses and forecasts. It was found that there were significant dry biases in the radiosonde observations of relative humidity made with the Vaisala RS80-A taken during 2006 when compared with other independent data, such as measurements of precipitable water vapour from GPS (Bock *et al.*, 2007). These dry biases were also apparent in NWP analyses from ECMWF and Meteo-France, and corrections to the sonde measurements were developed (Agusti-Panareda *et al.*, 2009; Nuret *et al.*, 2008).

Applying these bias corrections in humidity and assimilating the 'corrected' sonde measurements had a marked impact on the NWP analyses of humidity and precipitation and AEJ structure in both ECMWF (Agusti-Panareda *et al.*, 2010) and Meteo-France (Faccani *et al.*, 2009) global NWP systems. However, owing to large systematic errors in the NWP models themselves, Agusti-Panareda *et al.* (2010) reported that the benefit of this additional (bias-corrected) sonde data does not last beyond the first 1–2 days of NWP forecasts. Faccani *et al.* (2009) noted that the impact of the improved AMMA analyses does have a beneficial remote impact downstream over the Europe region on a timescale of 2–3 days.

A comprehensive study by Meynadier *et al.* (2010a,b) used a hybrid data set, including satellite observations, land-surface models and NWP analyses, to study the West African water cycle and evaluate model analyses and reanalyses. Some key biases in the water cycles of the model products were highlighted (Figure 10.13). First, it is shown that the rain band is too far south in the models. In reality the Sahel is a net sink of water for the atmosphere during the wet season (i.e. $P > E$ ), but biases of 1–2 mm day$^{-1}$ in the $P$ and $E$ rates mean that the Sahel is a moisture source in the models (i.e. $P < E$ ). Second, the observed hybrid dataset suggests that MFC (Moisture Flux Convergence) occurs in the Sahel during the wet season but the models produce only very weak MFC, or even MFD (Moisture Flux Divergence), in this region (Figure 10.13). The unrealistic MFD, coupled with a deep layer of dry air advected from the north at mid-levels, is thought to block the development of deep convection and the northward propagation of the monsoon rain band in the models. Birch *et al.* (2014) argued that these errors are strongly influenced by the erroneous diurnal cycle of deep convection in the models and the difficulty of triggering model convection in the northern Sahel, where there can be strong CIN.

#### 10.2.1.2 Near-surface Temperatures and Humidities

The meridional temperature and moisture gradients are important factors in the dynamics of the WAM. In addition, forecasters are interested in the ability of NWP models to capture both extreme dry conditions (drought) and their associated extremes in near-surface temperatures, both for agricultural and health impacts on local communities. The seasonally averaged 2 m temperatures and dew point temperatures in TIGGE 1200 UTC analyses (Figure 10.14) show the strong gradients north–south in both temperature and moisture, but even amongst the analyses there are significant differences. For example, the Met Office analysis has the warmest 2 m temperatures over the Sahara (>42°C) and JMA the weakest (38–40°C). The dew point temperatures show the Met Office, ECMWF and NCEP analyses have similar levels of dryness over the Sahara (Td < 4), but NCEP and Met Office are moister than ECMWF further south in the Guinea and Soudanian regions. Both CMC and JMA are moister over the Sahara than the other analyses (see Td), consistent with the cooler 2 m temperatures in these two models.

The day-9 ($T$ + 216 h) drift in 2 m temperatures and dew points are also shown in Figure 10.14. The Met

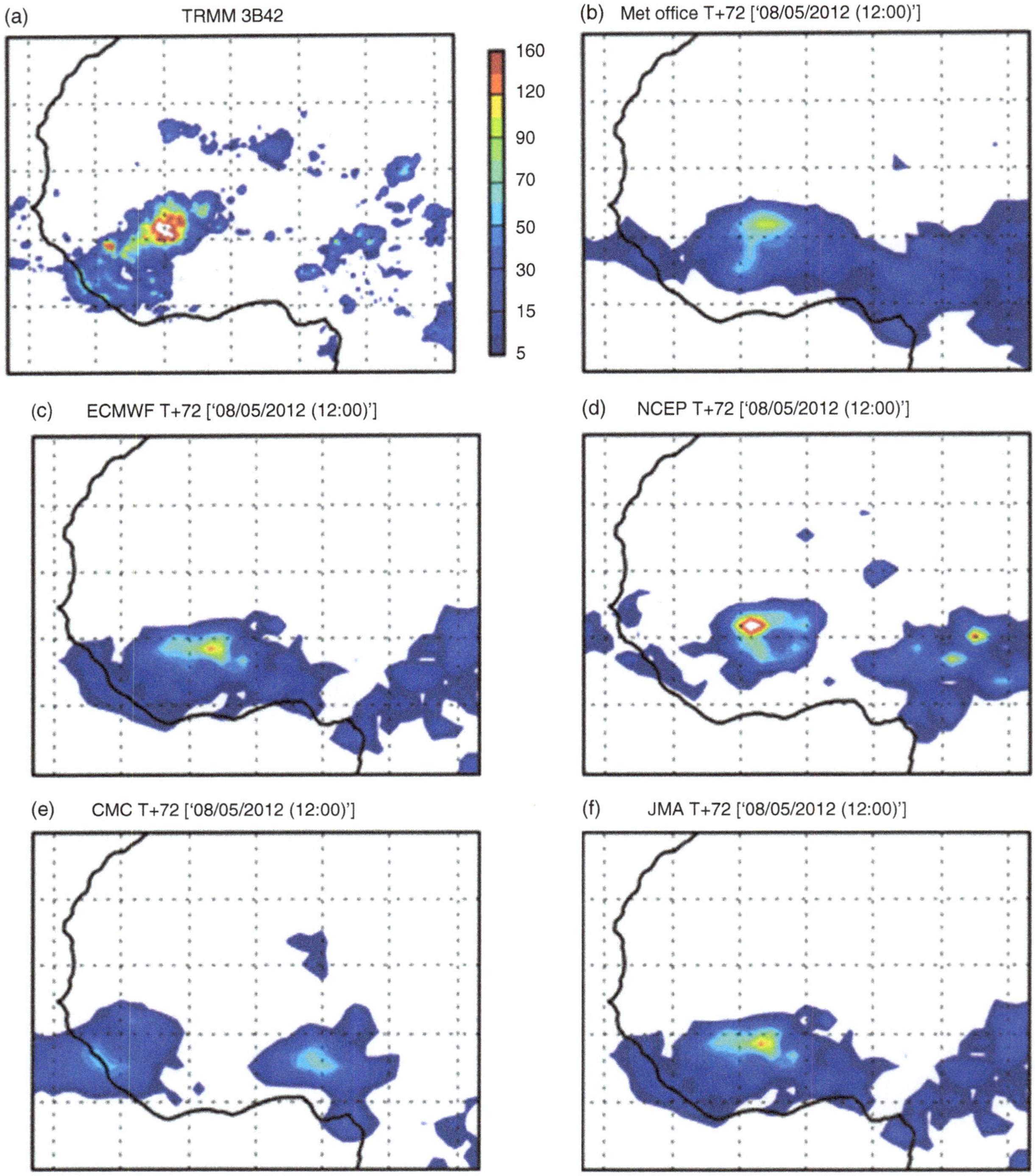


**Figure 10.12** Daily accumulated precipitation (mm $day^{-1}$). Case study for 7–8 August 2012 showing TRMM and TIGGE control ($T + 48$–72 h) forecast precipitation accumulation for forecasts initialised at 1200 UTC on 5 August 2012.

Office, ECMWF and JMA models show similar biases with cooling and moistening across the Sahel. This is consistent with a spin-up in precipitation in these models from initial analysis time that wets the soils in these regions and leads to increased (evaporative) cooling and moistening of the near-surface layers of the model atmosphere. Over the Sahara the Met Office forecasts continue to warm slightly and dry out, whereas NCEP and ECMWF show a cooling in Saharan temperatures. CMC has a marked warming of up to 3°C over the Sahara and a moistening. It is worth stating that some of the differences over the Sahara between the analyses (e.g. Met Office–CMC) are as large as the drift of any one model from its own analysis. This reflects the fact that there are very few observations here to constrain the model analyses and was also one of the motivating factors for studying this region in more detail in the recent Fennec field campaign (Marsham *et al.* 2013b).

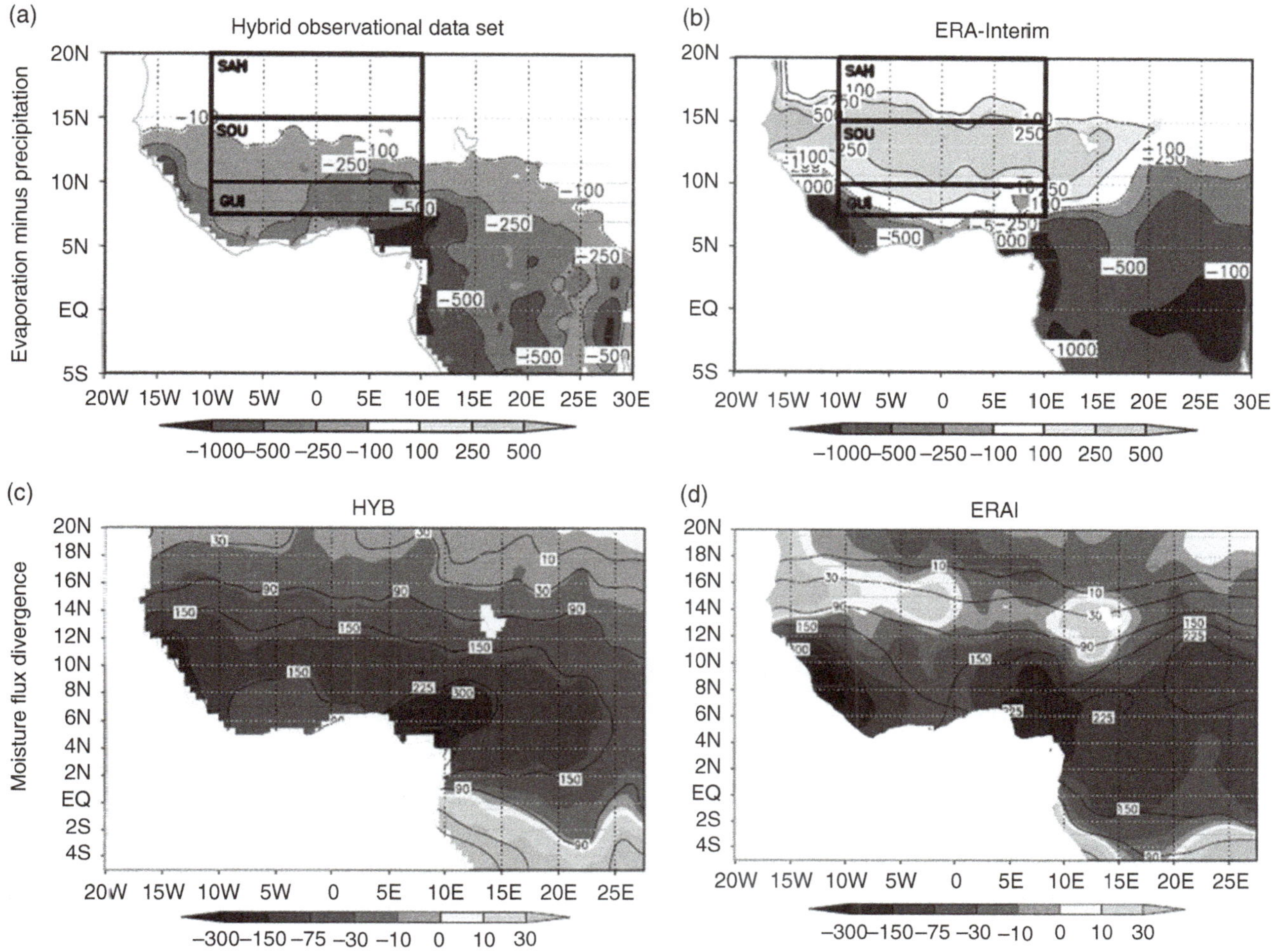


**Figure 10.13** Estimates for evaporation $E$ minus precipitation $P$ (top) and moisture flux divergence (MFD – bottom) during August (2002–2007) from the hybrid observational dataset (left) of Meynadier *et al.* (2010a) and from ERA-interim (ERA-I) analysis right. In contrast to the hybrid observations, the ERA-I reanalysis has, incorrectly, $E - P > 0$ at 10–15°N due to the precipitation band being too far south in the reanalysis (not shown). In this same latitude band ERA-I also shows some regions of MFD where the hybrid observational dataset only has moisture flux convergence (MFC) over West Africa during the monsoon season. *Source:* Meynadier *et al.* (2010b). ©John Wiley and Sons. Reproduced with permission..

Recent research comparing observations taken during Fennec with the MetUM has suggested that a large part of the MetUM warm biases in the Saharan PBL can be traced to the model's inability to adequately capture the cold pool outflows associated with deep organised convection (Garcia-Carreras *et al.*, 2013). These cold pool outflows can propagate large distances and act to ventilate and cool the Saharan PBL. This and other research suggests that parametrising these non-local effects in our convection schemes may be important (Grandpeix and Lafore, 2010).

#### 10.2.1.3 Low-level Monsoon Flow and Saharan Heat Low

There is considerable uncertainty in the depth and structure of the Saharan heat low in the TIGGE analyses (Figure 10.15). Consistent with its warmer 2 m temperatures, the Met Office analysis has lower pressure and the model continues to drift toward lower pressure in the day 5 forecasts, although the 2 m temperatures themselves only warm slightly in the forecasts. CMC also shows a drift toward lower pressures over the Sahara that are consistent with a larger warming seen in its 2 m temperatures. In contrast, JMA, NCEP and ECMWF show a tendency toward higher pressures over the northern Sahara in their day 5 forecasts. These errors in the pressure gradients across West Africa are important in forcing errors in the low-level monsoon flow, which itself is important in the mean transports of moisture into the Sahel and interactions with the synoptic weather systems (AEW, etc.). A more detailed discussion of Saharan errors and the role of missing processes in the Met Office global model can be found in Garcia-Carreras *et al.* (2013). The errors in the low-level 10 m wind fields reflect these MSLP (Mean Sea Level Pressure) biases with convergent flow toward the negative MSLP biases. For the Met Office this means an eastward flow error into the Sahara from the west coast of Africa, whereas for JMA we see the reverse, with a westward flow error across West Africa.

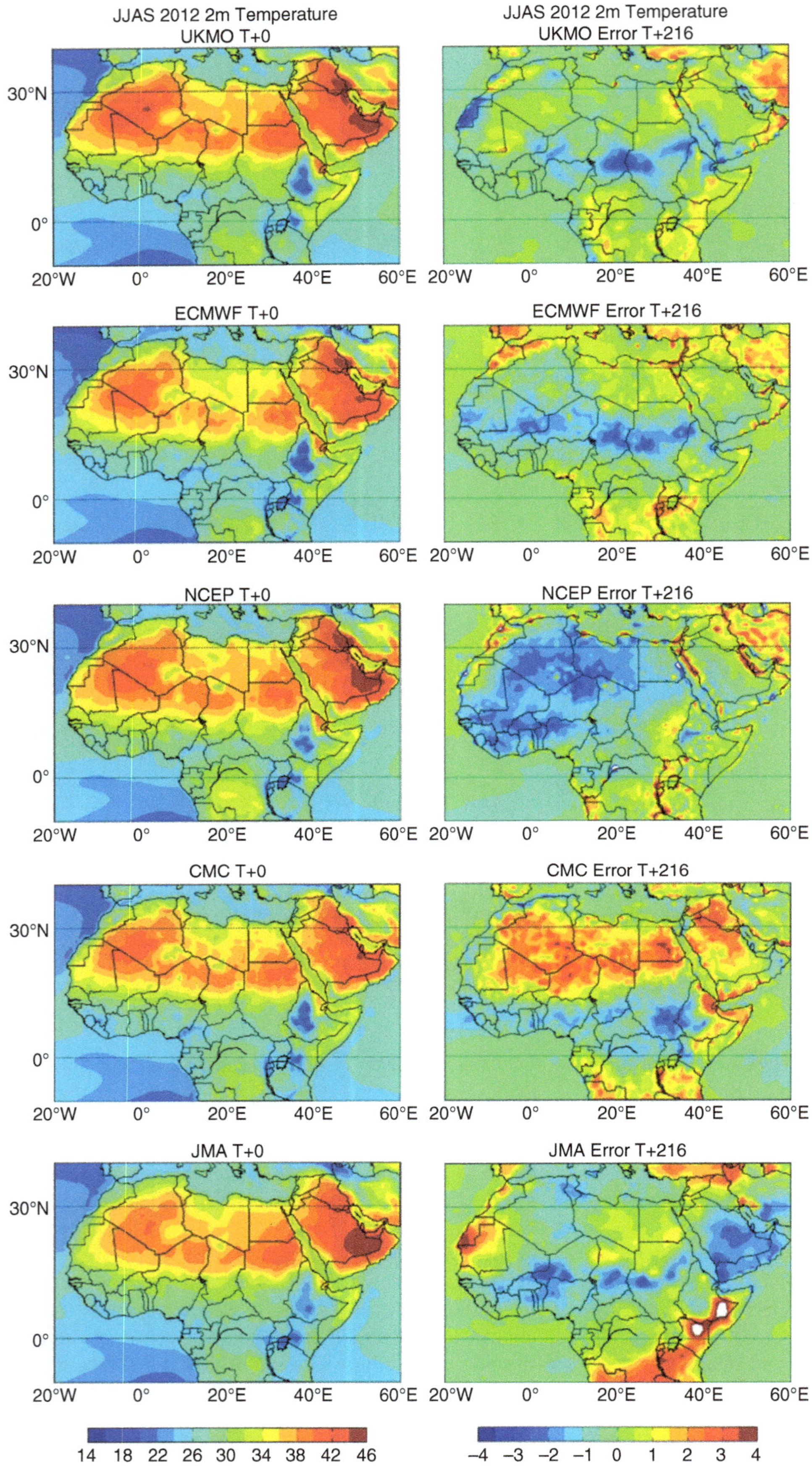


**Figure 10.14** Seasonally averaged near-surface (2 m) temperatures (column 1) and dew point temperatures (column 3) from TIGGE analyses (see Table 10.1 for details of TIGGE systems) for June–September 2012. The $T + 216$ h (day 9) mean model errors/drifts (forecast minus analysis) are also shown for 2 m temperature (column 2) and 2 m dew point temperature (column 4).

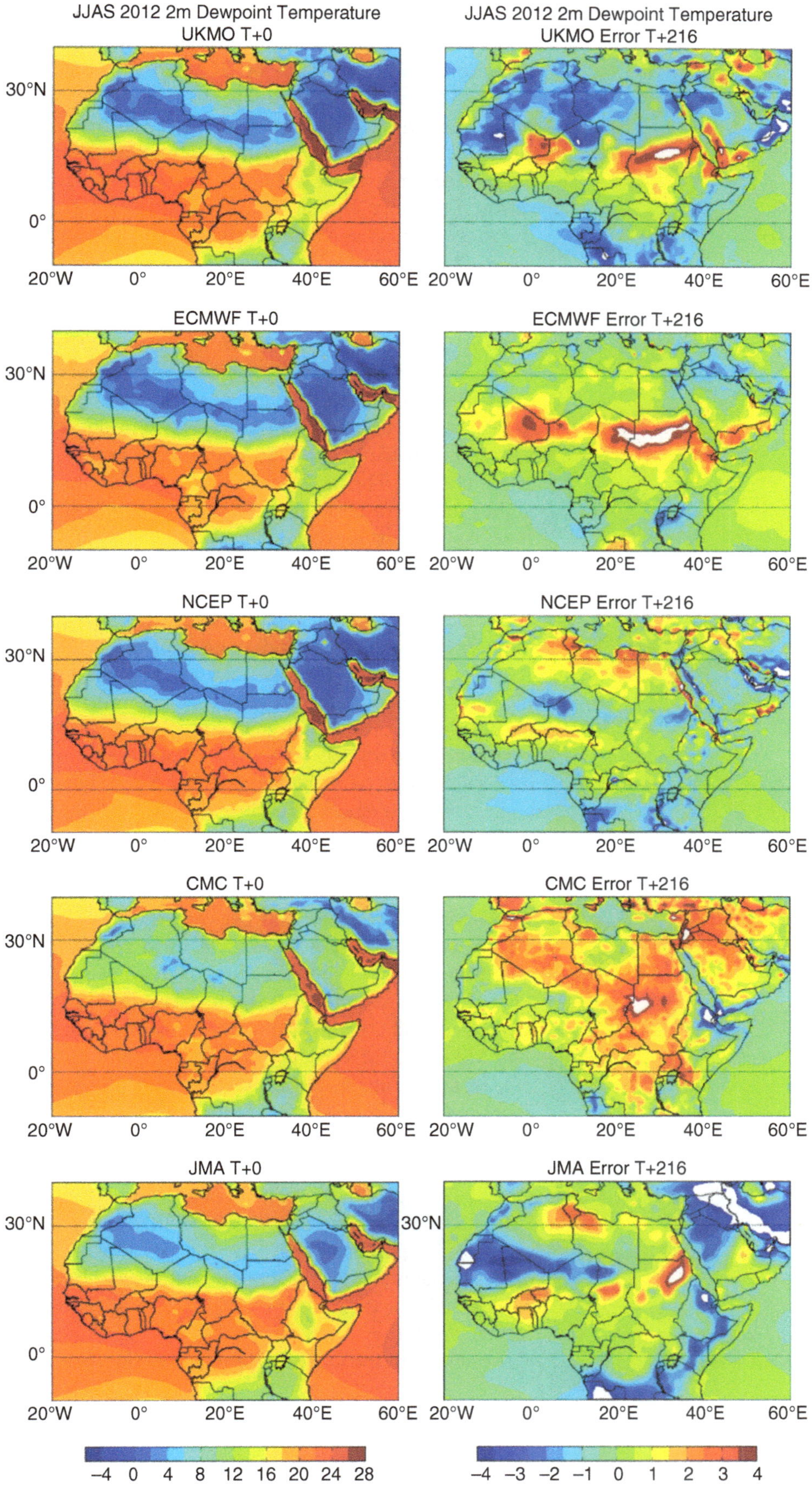


Figure 10.14 (Cont'd)

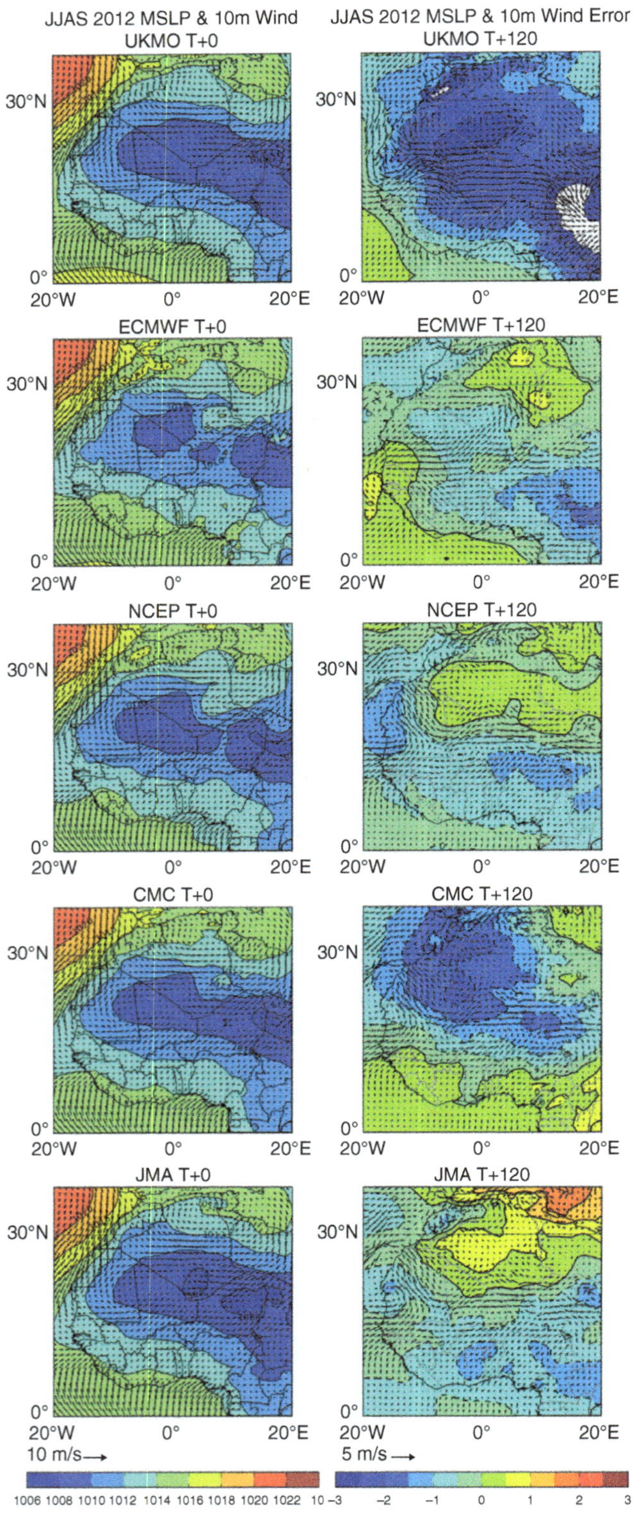


**Figure 10.15** Low level monsoon flow (10 m wind vectors) and Saharan heat low (mean sea-level pressure (MSLP)) for TIGGE analyses (left) and control forecast errors at $T+120$ h (right).

#### 10.2.1.4 African Easterly Jet

The AEJ is an important component of the WAM (Chapter 2). The JET2000 campaign (Thorncroft *et al.*, 2003) was designed to provide detailed observations of the AEJ structure (using dropsondes from research aircraft) and to investigate the role of additional observations in improving analyses and forecasts from the operational NWP systems at ECMWF and the Met Office. The north–south temperature gradient determines the latitude for deep dry convection (maximum over the Sahara), and moisture gradients determine where deep moist convection occurs (peaking further south). The moist and dry convection largely determine the free troposphere temperature gradients and are key to the dynamics of the jet, which is in thermal balance with this temperature gradient (Thorncroft and Blackburn, 1999). The low-level PBL potential temperature ($\theta$) gradient from the aircraft measurements shows the increase in $\theta$ from south to north, with the moisture decreasing to the north; the effect of these opposing gradients is characterised by a maximum in the equivalent potential temperature ($\theta_e$) peaking at around 13°N (Figure 10.16). The analysis from ECMWF at this time tended to underestimate the temperatures (too cool) and the 5-day forecast is even cooler, and also underestimates the temperature gradient. For $\theta_e$, the model analysis has a peak too far north at 16°N and the 5-day forecast is even worse. These errors act to weaken the north–south temperature gradients, and consistent with this the AEJ was too weak in forecasts and analyses. Adding the JET2000 dropsondes into the data assimilation did improve the $\theta_e$ gradient dramatically (Figure 10.16), and a more detailed discussion of the analysis sensitivity tests in Tompkins *et al.* (2005) found improved dynamical and thermodynamical structure of the jet. This result emphasises the importance of the PBL state to the wider monsoon dynamics, including the AEJ.

The AEJ in TIGGE analyses for June–September 2012 shows reasonable agreement between all of the centres (Figure 10.17). All have an easterly jet located between 40°E and 20°W with wind speeds increasing gradually toward the west with a peak jet core of 10–12 m s$^{-1}$ located around 0°E and also further west over the coast. The AEJ in the JMA analysis is weaker than in the other models and stronger in the Met Office. This is perhaps consistent with a weaker low level temperature gradient in JMA and a stronger one in the Met Office (Figure 10.14). At day 5 most of the TIGGE forecasts show some weakening of the AEJ core from greater than 10 m s$^{-1}$ to 8 m s$^{-1}$. The exception is CMC, which maintains a stronger jet, perhaps consistent with its warming bias over the Sahara. Met Office day 5 forecasts have the weakest AEJ in the 20–30°E longitude range, although further west they maintain jet core strength in good agreement with the analyses from the various centres.

#### 10.2.1.5 African Easterly Waves

Accurate prediction of AEWs (Section 2.1.3) is required for good NWP forecasts of precipitation on the timescales of days to a week over West Africa. Most NWP models still struggle to capture the delicate thermodynamic and

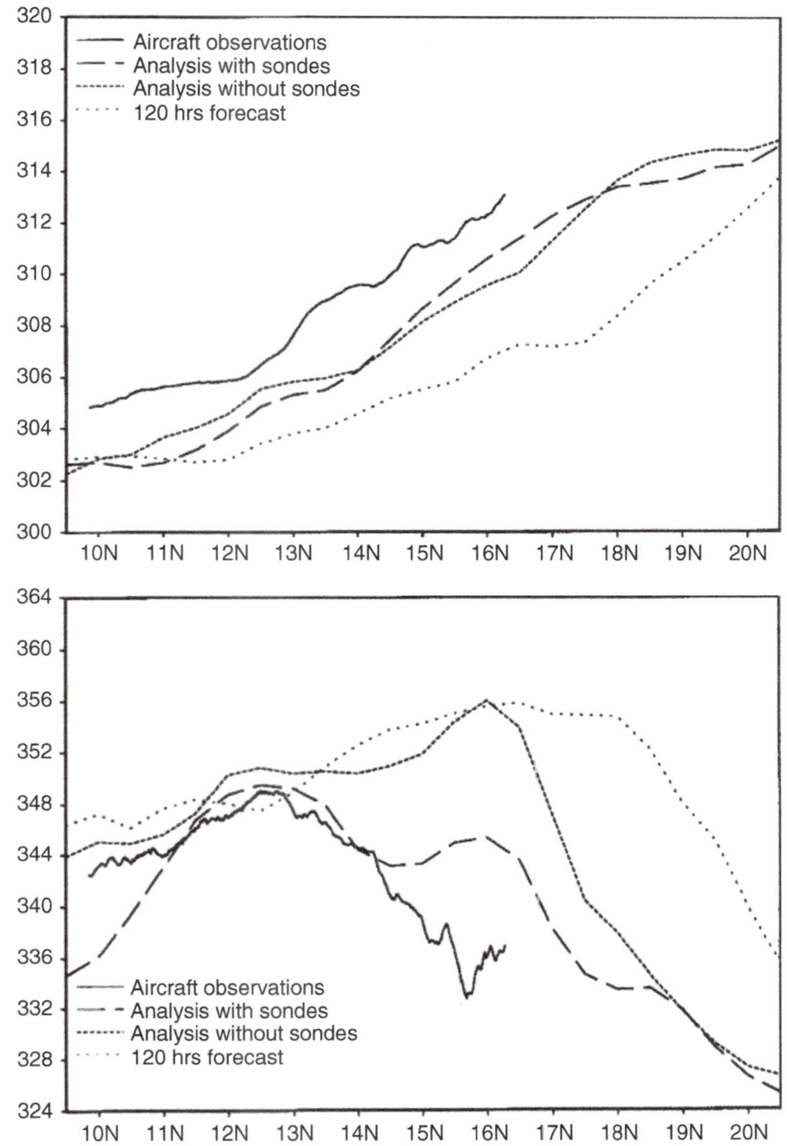


**Figure 10.16** Meridional profiles of (top) potential temperature $\theta$ and (bottom) equivalent potential temperature $\theta_e$ based on aircraft observations on 28 August at a height of approximately 875 hPa (solid), between 1317 and 1538 UTC, ECMWF analysis for 1200 UTC 28 August with the dropsondes (long dashed), ECMWF analysis for 1200 UTC 28 August without dropsondes (small dashed), and the 5-day forecast from ECMWF for 1200 UTC 28 August (dotted). *Source:* Thorncroft *et al.* (2003). © American Meteorological Society. Used with permission.

dynamical balances that are responsible for the initiation, growth and decay of these waves. However, on occasions with strong dynamical forcing the models seem to do a better job. An example is shown from the Meso-NH model with 32 km grid resolution from the AMMA 2006 period (Figure 10.18) in a study by Söhne *et al.* (2008). The skill of day 1 cloud predictions (measured with the Heidke skill score (HSS)) seems to be higher when there is strong AEW activity. The authors suggest that increased predictability and skill for MCS might be achieved during these periods of enhanced wave activity if improvements could be made in the analysis of atmospheric flows (temperature and moisture gradients, AEW, AEJ, etc.). A detailed study of the structure of one of the AEWs seen in the Söhne *et al.* study from 25 to 29 July was given by Bain *et al.* (2011) and was well predicted by various models 5 days in advance.

Finally, we present an example of the AEWs observed and modelled during August 2011 from the Met Office global NWP analyses and forecasts (Figure 10.19) taken from Bain *et al.* (2013). The AEWs are clear in the analyses of curvature vorticity (following Berry *et al.* (2007)), and at least two formed into Atlantic tropical cyclones. The forecasts out to $T + 144$ h show that propagating AEWs are still present, but as the forecast range progresses the waves weaken considerably over land and fewer are initiated in the east. The rainfall is reasonably well organised by the AEW in the analysis, although the poorly modelled diurnal cycle over land is clearly visible. At later forecast ranges this organisation weakens, suggesting the convection-wave coupling is not well captured. Clearly, much more research is required to diagnose these model systematic biases. Bain *et al.* (2013) reported some improvement in AEW prediction in climate simulations through an increase in resolution and through increasing the entrainment in the convective parametrisation. This latter change made convection more intense and less frequent, and this stronger

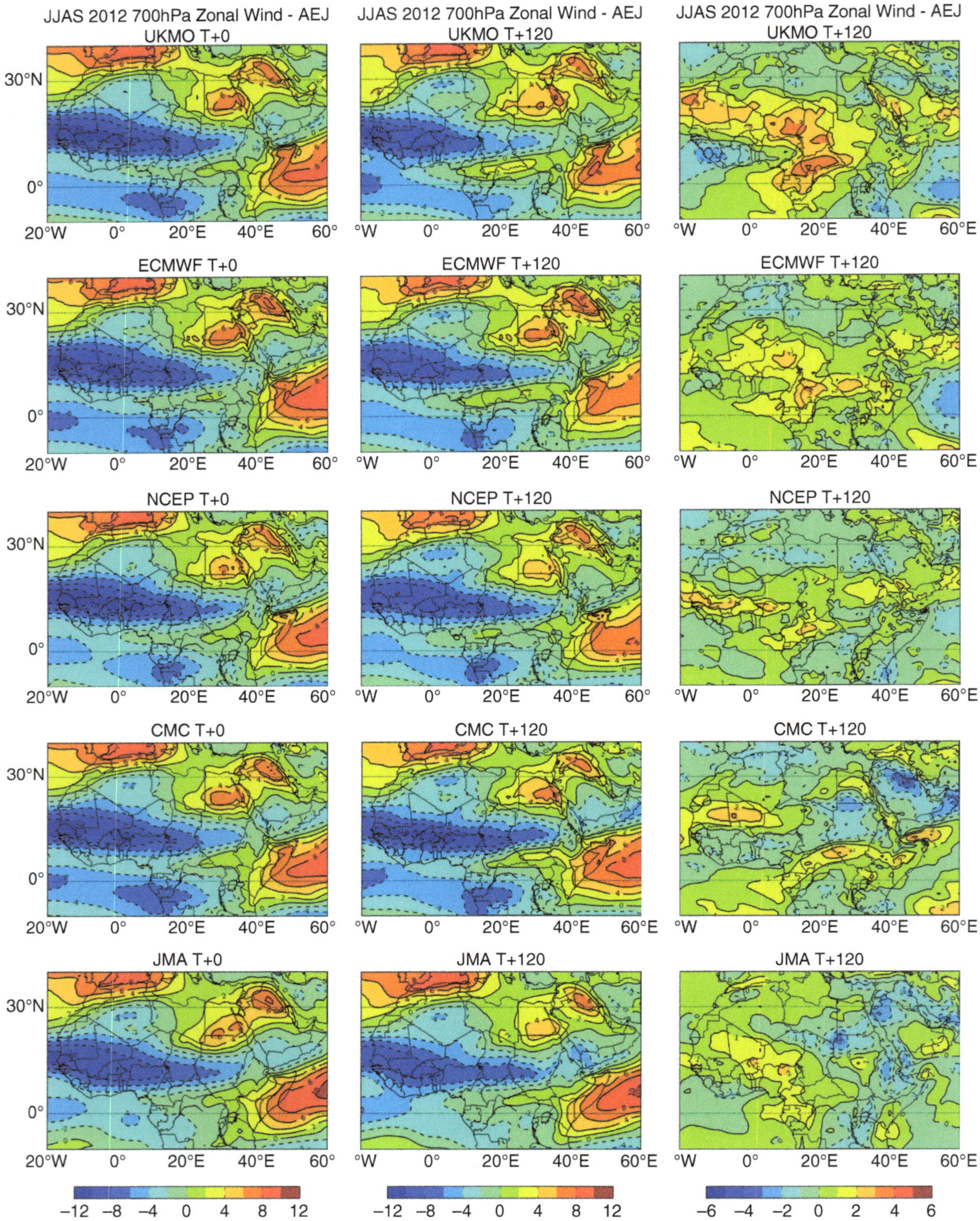


**Figure 10.17** AEJ in TIGGE analyses (left) and $T + 120$ h forecasts (middle) shown by 700 hPa zonal winds. The $T + 120$ h mean errors in the AEJ are also shown (right).

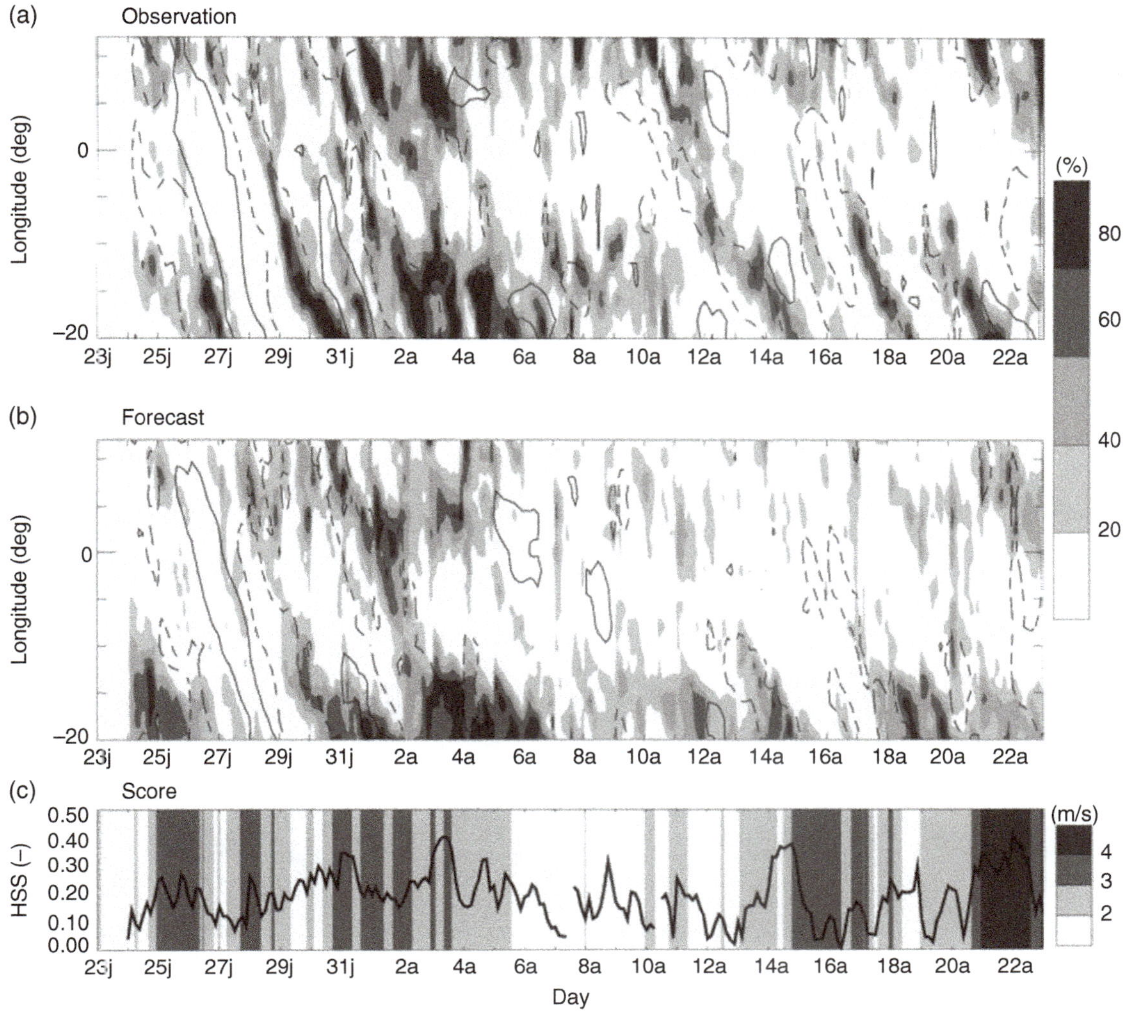


**Figure 10.18** (a, b) Time–longitude diagram of wind and cloud fraction averaged between 10° and 15°N from 23 July to 22 August 2006: MSG cloud fraction (shading, %) and 700 hPa meridional wind contours at $-3$ m s$^{-1}$ (dashed) and at 3 m s$^{-1}$ (solid). Fields are taken from (a) MSG observations and ECMWF analyses and (b) Méso-NH forecasts at $D + 1$. (c) The time evolution of the HSS calculated point by point (line) and absolute meridional wind intensity from the analyses (shading, m s$^{-1}$) averaged over 10–15°N. *Source:* Söhne *et al.* (2008). © American Meteorological Society. Used with permission.

convection suppressed smaller systems and allowed larger convective systems to grow, reinforcing AEW strength.

### 10.2.2 Predicting Severe Weather and Modelling Uncertainty

The two main climate and weather hazards in Africa are at opposite ends of the spectrum. On the one hand, prolonged periods of drought or anomalous breaks in the rainy season can lead to crop failure and famine. On the other hand, too much rainfall can lead to problems with devastating flooding events (e.g. Mozambique in 2000, West and Central Africa in 2007), causing pests such as locusts, increased incidence of cholera and malaria or, again, crop failure if the wrong type of plant is sown. At the shorter immediate weather timescales the main hazards are primarily associated with squall lines and large active storm systems. Downdrafts spread out when they hit the surface, causing a radial-spreading cold pool outflow. Such cold density currents are often associated with dust fronts, particularly at the beginning of the rainy season, which can spread over large distances (see Section 5.1.2 and Figure 5.6). In addition, lightning and flash flooding associated with extreme rainfall are also a real threat to life and livestock.

Predicting storm events is extremely challenging for NWP systems. The inherent character of convection is extremely chaotic, often initiating and propagating due to very small perturbations in the atmosphere or anomalies in the land surface.

There are two main ways to mitigate against these modelling problems. The first is to attempt some modelling of the chaos by running multiple versions of the same forecast with slightly different initial conditions (ensemble techniques). The second is to run a high-resolution,

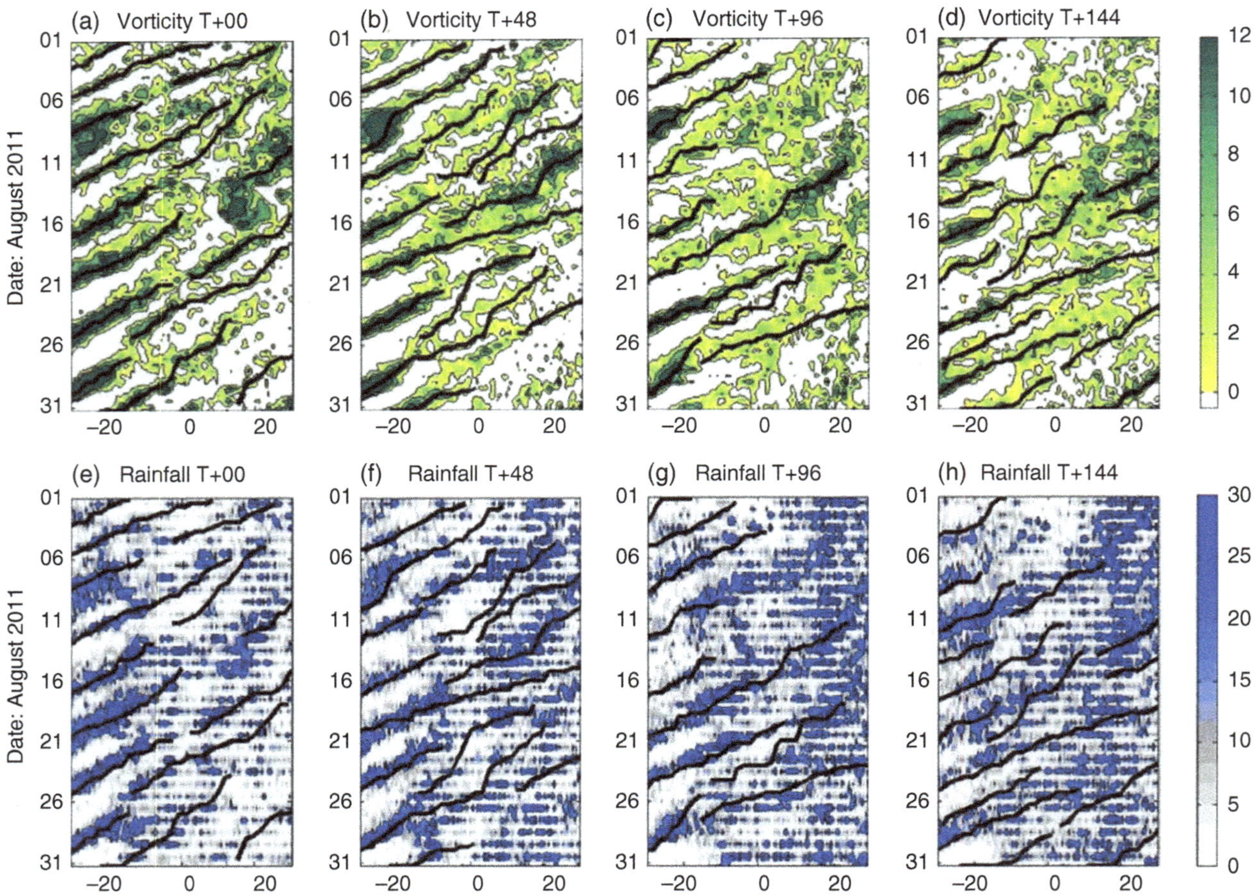


**Figure 10.19** Example of NWP capability in Met Office operational deterministic global forecasts from August 2011. The top panels show curvature vorticity and the bottom precipitation from the analysis and forecasts. The solid black lines are the tracks of the AEWs using the objective tracking algorithm discussed in Bain *et al.* (2013). *Source:* Bain *et al.* (2013). Reproduced with permission of the Royal Meteorological Society.

convective-permitting (4 to 1 km) model in an attempt to adequately resolve the convection processes rather than relying on the physical parametrisations of coarser resolution models. A third way currently being explored in the UK and elsewhere is to combine the convection-permitting model with an ensemble technique to give a high-resolution 'convection-permitting' probabilistic forecast.

The use of NWP for hazard prediction is still in its infancy. Despite the challenges models face, there is great potential for their use to be expanded.

#### 10.2.2.1 Ensemble Forecasts from Deterministic Predictions

Over the years, West African forecasters have based their forecast decisions purely on subjective methods after laborious analysis of the data field. Such methods include persistence, trend, climatology, statistical and analogue. Results were encouraging, especially in the areas of nowcasting and short-range (1–2 days) forecasting.

Currently, all forecast offices in West Africa use a variety of model outputs for their forecasts. We note here that model sensitivity varies from products of one centre to another and on a day-to-day basis, when certain conditions (e.g. pressure, convection, precipitable water) are considered. As we have seen from the examples with the TIGGE control forecasts earlier, there is considerable variation in the forecasts and even in the analyses produced from each centre; and although they often have common deficiencies, they also show characteristics and biases that are specific to a given model. The use of a single deterministic forecast imposes a strong limitation on the ability and capability of the forecaster. Increasingly, the African forecasters overcome this challenge by comparing and combining model outputs from the operational deterministic models (e.g. Diop-Kane *et al.*, 2008). This simple technique is a method of 'ensemble forecasting' and is based on two scenarios:

- Comparison of model outputs from different model centres with the same initialisation time (a 'poor man's' multi-model ensemble).
- Comparison of model outputs from the same model centre but with different initialisation times (a so-called lagged average forecast ensemble).

A practical implementation of this process might proceed as follows:

- Take a field variable such as the wind vector.
- Consider the initialisation time of 0000Z.
- View/print model charts of precipitation or wind vectors from different model centres (Meteo France, ECMWF, Met Office, NCEP, etc.) for 0000Z (for example, see Figure 10.12).
- Compare them and see how closely the wind vectors agree with each other.
- Suppose we now view/print out wind vector charts for initialisation times of 0000Z, 0600Z, 1200Z and so on from a particular model centre. On comparing, we shall be able to note changes (if any) on the model outputs with respect to time.

Apart from precipitation or wind vectors, other variables (e.g. temperature, divergence, pressure) can be considered for this process. The big advantage is that, given clear uncertainties in both the initial analyses and the differing characteristics of each modelling system, intermodel comparison provides a better platform for the forecaster to make informed decisions, and forecast skill is enhanced.

This same process has been used in a more objective technique to construct a multi-model 'super ensemble', first proposed by Krishnamurti *et al.* (1999) using a weighted ensemble of several operational global models, where the weights are determined from the past performance of the various models. Mutemi *et al.* (2007) have used this super-ensemble technique for NWP prediction over a variety of regions of Africa. They showed that the super ensemble mean skill for precipitation was superior to any of the individual component models, although the skill declined rapidly beyond day 3.

#### 10.2.2.2 World Meteorological Organization Severe Weather Forecast Demonstration Project

More recently, with the recognition that the prediction of weather is an inherently probabilistic process, we have seen the development of EPSs at many NWP centres. This is particularly relevant for prediction of extreme or hazardous weather, where EPSs may be able to detect a small predictable signal for extreme or severe weather and give some indication of the risk. However, it has also been recognised that the capabilities of these prediction systems are not being adequately communicated to the developing world (Webster, 2013) for a number of complex reasons, including poor investment, failing infrastructure, and lack of training and guidance in the use of these systems and their products. The WMO SWFDP is an attempt to close this gap and bring current NWP/EPS capability to countries where these facilities are not currently available.

The goals are to provide forecast data to national meteorological and hydrological services (NMHSs) to improve ability to predict severe weather at longer lead times and provide risk-based advice and warnings to disaster management and civil protection authorities and to feed back to operational NWP centres on performance and potential improvements to the service.

The first SWFDP region was in southern Africa during 2006–2007 with a focus on heavy precipitation and strong winds. It involved NMHSs in Botswana, Mozambique, Madagascar, Tanzania and Zimbabwe (now extended to 16 southern African countries), with coordination from regional African centres in Pretoria and La Réunion using NWP/EPS products from ECMWF, NCEP and the UK Met Office. This has included annual training and workshops for forecasters.

As an example, we show some output from the TIGGE EPS systems for the 7–8 August case study discussed earlier (Figure 10.12). This example (Figure 10.20) was generated from the TIGGE/SWFDP ensemble extreme weather warnings site.[7] Figure 10.20 shows the forecast probability (50%, 70%, 90%) of extreme precipitation, defined in this example as greater than the 90th percentile,[8] for each centre's EPS (Met Office, ECMWF, NCEP and JMA) at 0–1 day into the forecast. The probabilities are determined as the number of ensemble members that exceed the 90th percentile. Also shown is the MCGE (Multi-Centre Grand Ensemble) comprising information from all ensemble members from all centres. The observed extreme precipitation is also shown, and for West Africa we see two locations of extreme precipitation (registering as >95th observed percentile): one over southern Mali and a second further north and east over the Niger–Algeria border. These are associated with a trough/AEW system shown by an MSLP minimum in the observations (and forecasts). The MCGE shows successful predictions of these extremes, with over 70% of ensemble members agreeing there will be extreme precipitation (>90th percentile) over southern Mali and between 30 and 50% agreeing to extreme precipitation over Niger–Algeria. Individual centre's probabilistic predictions show good agreement for the southern precipitation event, all with greater than 70% of their ensemble members suggesting extreme precipitation. For the northern event the Met Office and NCEP EPS both have the majority (>70%) of their ensemble members agreeing on an extreme event, whereas ECMWF and JMA are less certain, with fewer members (10–30%) predicting an extreme. At later forecasts ranges of 3–4 days, from the ensembles initialised on 4 August and also verifying on 7–8 August, we still see good predictability with 50–70% probability for the precipitation extreme over southern Mali, but less predictability for the event over Niger–Algeria (Figure 10.21).

Perhaps of even more interest than this predictable case are those with more marginal predictability, where

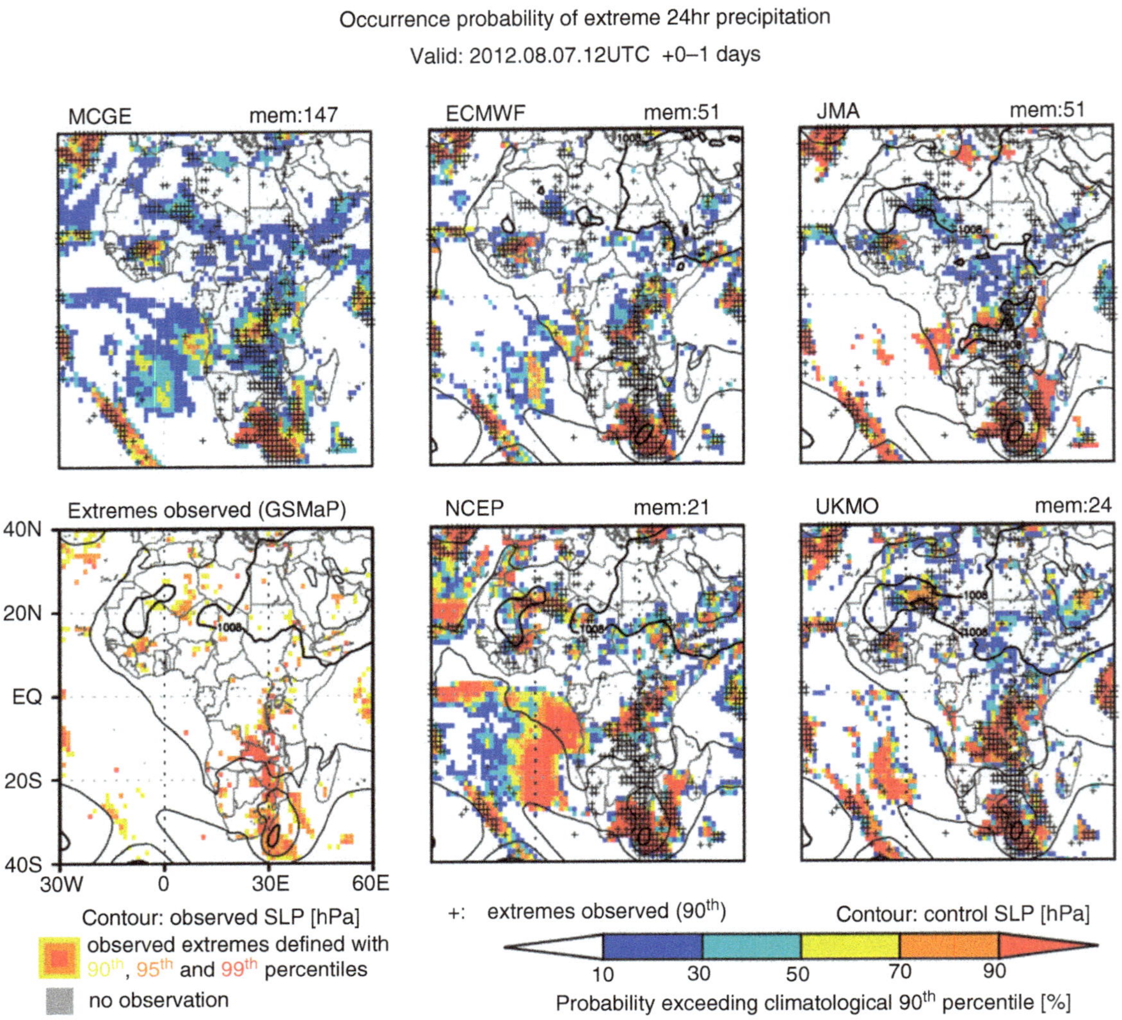


**Figure 10.20** Ensemble- based prediction of the risk of extreme precipitation. Example from TIGGE web site of ensemble-based predictions. This shows the ensemble prediction probabilities from the TIGGE models and the multi-centre grand ensemble (MCGE) at 0–1 day for prediction of extreme precipitation (exceeding the 90th percentile). This is for the forecasts valid at 1200 UTC 7–8 August 2012, and corresponds to the case study shown in Figure 10.12 for the deterministic control forecasts. The observed extremes at 90th, 95th and 99th percentiles are shown from the GSMaP. These are also overlaid on each forecast with '+'. *Source:* Reproduced with the kind permission of Dr Mio Matsueda.

perhaps the operational deterministic forecasts fail to capture the extreme event but a few of the perturbed ensemble members may give a small, but useful, probability of extreme weather. It is important in assessing these ensemble predictions that they produce reliable statistics so that, for example, they do not have a high false alarm rate for such extreme events (e.g. predict many more extremes than are observed). This is an area where forecasters and model developers can (and should) work more closely together to evaluate these characteristics of the individual ensemble systems.

#### 10.2.2.3 Convective (Kilometre)-scale Modelling

Alongside the EPSs, a practical application of high-resolution (1–4 km 'convection-permitting') modelling is to provide additional guidance of severe weather through 'dynamical downscaling'. Here, we provide a few examples of how convective-scale models have been used in an NWP framework.

##### *10.2.2.3.1 Convective-Scale Amma Case Studies With Arome and Cosmo Models*

An NWP case study by Beucher *et al.* (2014) for an active convection period during AMMA 2006 contrasted the performance of the ARPEGE global model (50 km and 25 km resolutions) with the AROME convective-scale model (Seity *et al.*, 2011) run at 5 km grid spacing. The 5 km AROME model was initialised from the global ARPEGE 4D-Var initial conditions. AROME showed a much improved representation of all precipitation regimes over diurnal to synoptic scales ranging from light rain to two active MCSs connected to an AEW

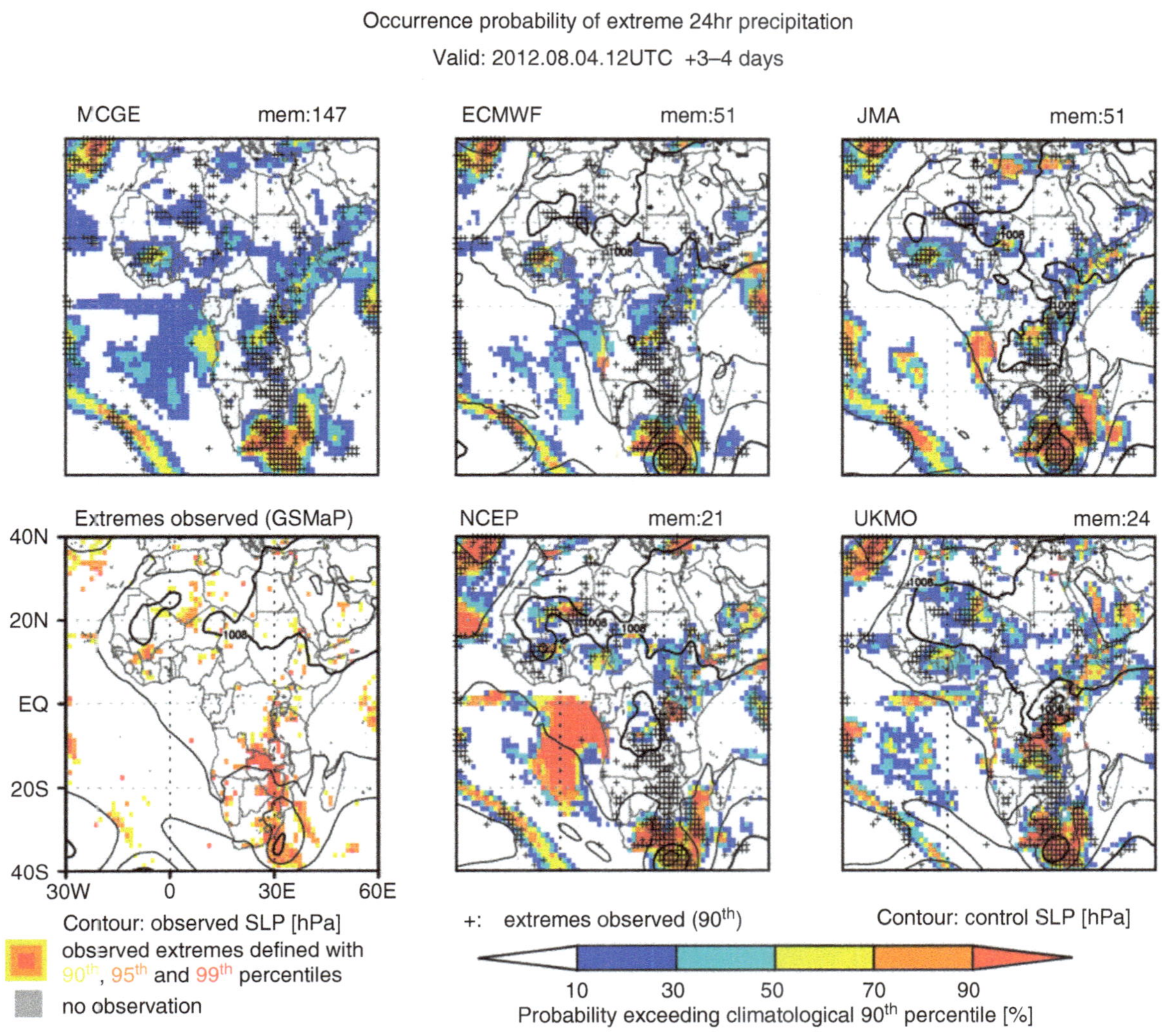


**Figure 10.21** As Figure 10.20, but for days 3–4 forecasts valid at 1200 UTC 7 August 2012 and initialised on the 4 August at 1200 UTC. *Source:* reproduced with the kind permission of Dr Mio Matsueda.

(Bain *et al.*, 2011) and to a monsoon burst. However, the rainfall intensity was typically overestimated by 25% in AROME. In contrast, the ARPEGE global model, with parametrised convection, showed too much light rain and underestimated the more extreme precipitation events, although its daily mean precipitation was closer to the TRMM observations. Another interesting component of this study was the sensitivity of ARPEGE and AROME forecasts to the assimilation of additional surface-sensitive microwave observations from AMSU-B (following Karbou *et al.*, 2010a,b). While this additional observational data improved the ARPEGE analysis, it had a small positive impact on the ARPEGE forecasts. When used to force the AROME model, these improved initial conditions gave a much larger benefit with improved total column water (TCW) and quantified precipitation forecasts. A similar AMMA case study with the COSMO model run at 2.2 km (Schwendike and Jones, 2010) looked at the role of convective systems in the development of an AEW that later evolved into Hurricane Helene. This study used the convective-scale COSMO model to study the complex scale interactions between convection in evolving MCSs and a synoptic-scale AEW. It provides a very nice example of using the convective-scale models to study and understand the mesoscale dynamics of these extreme weather events.

##### 10.2.2.3.2 Case Study: Severe Storms over Lake Victoria with the Met Office 4 Km Unified Model

This second example is over East Africa but demonstrates the potential utility of convective-scale modelling for a specific end-user impact over Lake Victoria. In 2011, the Department of Meteorology in Uganda initiated a mobile weather alert system aimed at warning local fishermen of severe weather events over Lake Victoria. The lake is the largest tropical freshwater body and is notoriously

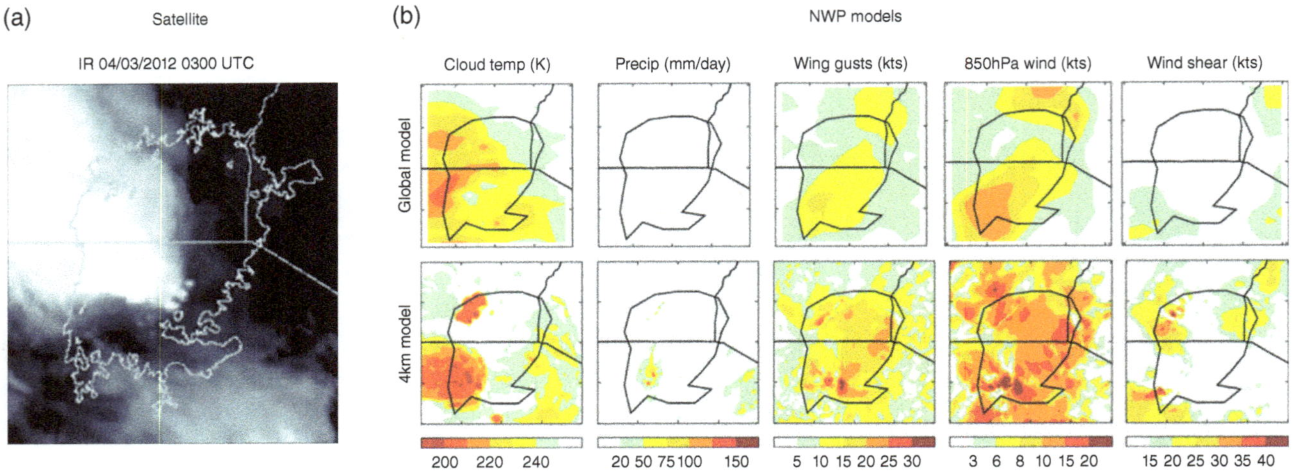


**Figure 10.22** Case study of Lake Victoria storm from 4 March 2012. (a) Satellite IR imagery. (b) Met Office NWP operational model performance for global and 4 km models. *Source:* Chamberlain *et al.* (2014). Reproduced with permission of the Royal Meteorological Society.

dangerous, with several hundred deaths a year attributed to its frequent and extreme (thunder) storm events.

The model weather alert system was set up to provide information on the likelihood of storms. The Ugandan forecasters used numerical weather forecasts combined with satellite and local knowledge to issue weather alerts directly to users by SMS text message. The weather alerts were colour coded according to likelihood of event and potential impact.

As part of the pilot study, a high-resolution 4 km version of the MetUM was set up. The model was limited area and used the operational MetUM global for boundary conditions. The Met Office global model was run four times a day, initialising at 0000, 0600, 1200 and 1800 UTC. From the 1800 UTC global forecast, a 3-h forecast was run to create the initial conditions used for the boundaries of the limited-area 4 km model. The 4 km model was then run out to 36 h and effectively acted as a 'downscaler' of the global model forecast, with no data assimilation within the domain. Hence, although more processes were able to be resolved, this LAM did not contain any more observational information to drive it than the global model.

A validation study was carried out to assess the performance of the 4 km model in comparison with the global operational model (Chamberlain *et al.*, 2014). The authors found that the 4 km model was able to capture higher wind speeds and more extreme rainfall rates. Figure 10.22 shows a case study from 5 March 2012. On this date a severe storm occurred in the southeast of Lake Victoria, and local news reports attributed the tragic deaths of two Tanzanian fishermen to the event. The satellite image in Figure 10.22a shows a mature storm anvil covering a large area of the lake. The storm initiated at 0000 UTC (0300 LT) and was at a maximum between 0300 and 0600 UTC (0600 and 0900 LT). The model fields shown in Figure 10.22b are 12-h forecasts valid at 0600 UTC. Both the global and the high-resolution models capture some storm signal in several of the fields – wind gusts are reasonably high, wind at 850 hPa is high and cloud top temperatures are low. However, in all of the fields, the 4 km model is showing a greater range of values, and has a stronger indication that a storm event may occur.

Using both models gives stronger confidence in the forecast – high-resolution models are not always adding any new initial information (i.e. there are no additional observations driving them) to the forecast but do develop alternative solutions for convective processes and assist decision-makers as an additional tool.

Performance comparison of global with high-resolution models is perhaps unfair in some regards owing to the inherent differences in detail of output. Caution must be exercised when interpreting high-resolution model information – the high-resolution models will have better mesoscale representation but will still suffer from inaccurate initial conditions and misrepresentation of local-scale and microscale processes.

#### *10.2.2.3.3 Summary of Convective-scale Modelling*

- High-resolution (kilometre-scale) models can add detail to model forecasts and have improved representation of mesoscale meteorological processes (e.g. diurnal cycle of convection over land, organised MCSs).
- High-resolution models combined with detailed observations (e.g. AMMA) are used in research to inform scientists of the potential processes that are missing from parametrisations in the coarser scale models.
- High-resolution models can be used as 'dynamical downscalers' in conjunction with global models to give confidence to forecast decision-making.

## Acknowledgements

This work has drawn heavily on the NWP research carried out at numerous operational NWP centres, and particularly on the research coming from the AMMA campaign in West Africa in 2006 involving many scientists across multiple institutions in Africa, Europe and the USA.

## Appendix: Acronyms

| | |
|---|---|
| ASCAT | Advanced Scatterometer |
| CAPE | convective available potential energy |
| CIN | convective inhibition |
| CMC | Canadian Meteorological Centre |
| ECMWF | European Centre for Medium-Range Weather Forecasts |
| EPS | ensemble prediction system |
| GATE | GARP Atlantic Tropical Experiment |
| GFS | Global Forecast System |
| GPS | Global Positioning System |
| GWD | gravity wave drag |
| IASI | Infrared Atmospheric Sounding Interferometer |
| JMA | Japan Meteorological Agency |
| LAM | limited-area model |
| LCL | lifting condensation level |
| MCGE | Multi-Centre Grand Ensemble |
| MetUM | Met Office Unified Model |
| MFC | moisture flux convergence |
| MFD | moisture flux divergence |
| MJO | Madden–Julian oscillation |
| MOS | model output statistics |
| MSLP | mean sea-level pressure |
| NCEP | National Centers for Environmental Prediction |
| NMHSs | national meteorological and hydrological services |
| NMS | National Met Services |
| NWP | Numerical Weather Prediction |
| PBL | planetary boundary layer |
| RTTOV | Radiative Transfer for TOVS |
| SMAP | NASA's Soil Moisture Active Passive mission, January 2015 |
| SOP | Special Observing Period |
| SST | sea surface temperature |
| SWFDP | WMO Severe Weather Forecast Demonstration Project |
| TCW | total column water |
| TIGGE | THORPEX Interactive Grand Global Ensemble |
| TIROS | Television and Infrared Observational Satellite |
| TRMM | Tropical Rainfall Measuring Mission |
| WAM | West African monsoon |
| WMO | World Meteorological Organization |

## Notes

1 Pilot balloons are cheaper than radiosondes but only measure wind profiles.
2 http://nwpsaf.eu/site/software/rttov/.
3 EKF is a nonlinear version of Kalman filtering.
4 A procedure referred to as Reynolds averaging after the fluid dynamicist Osborne Reynolds (1842–1912).
5 The process by which convective clouds draw air from their environment into the cloud.
6 Some of this increased correlation is simply the ability of the models to capture the background seasonal cycle/trend, but there is still also some skill in representing the peaks and troughs in weekly precipitation.
7 At http://tparc.mri-jma.go.jp/TIGGE/tigge_SWFDP.html. The TIGGE Museum web site has further resources at http://tparc.mri-jma.go.jp/TIGGE/.
8 The model 90th percentile is defined separately for each centre and measured from each model's own climatological distribution estimated from past TIGGE forecasts; this effectively bias-corrects the EPS forecasts for known deficiencies in each model's own climate.

## References

Agusti-Panareda A, Vasiljevic D, Beljaars A, *et al.* 2009. Radiosonde humidity bias correction over the West African region for the special AMMA reanalysis at ECMWF. *Q. J. R. Meteorol. Soc.* **135**: 595–617.

Agustí-Panareda A, Beljaars ACM, Cardinali C, *et al.* 2010. Impacts of assimilating AMMA soundings on ECMWF analyses and forecasts. *Weather Forecast.* **25**: 1142–1160.

Arakawa A. 2000a. A personal perspective on the early years of general circulation modelling at UCLA. In: *General Circulation Model Development – Past, Present and Future*, DA Randall (ed.). International Geophysics Series, vol. 70. Academic Press; chapter 1.

Arakawa A. 2000b. Future development of general circulation models. In: *General Circulation Model Development – Past, Present and Future*, DA Randall (ed.). International Geophysics Series, vol. 70. Academic Press; chapter 23.

Arakawa A, Schubert W. 1974. Interaction of a cumulus ensemble with the large-scale environment. Part I. *J. Atmos. Sci.* **31**: 674–701.

Bain CL, Parker DJ, Dixon N, *et al.* 2011. Anatomy of an observed African easterly wave in July 2006. *Q. J. R. Meteorol. Soc.* **137**: 923–933. doi: 10.1002/qj.812.

Bain CL, Williams KD, Milton SF, Heming JT. 2014. Objective tracking of African easterly waves in Met Office models. *Q. J. R. Meteorol. Soc.* **140**: 47–57. doi: 10.1002/qj.2110.

Baldauf M, Seifert A, Förstner J, *et al.* 2011. Operational convective-scale numerical weather prediction with the COSMO model: description and sensitivities. *Mon. Weather Rev.* **139**: 3887–3905. doi: http://dx.doi.org/10.1175/MWR-D-10-05013.1.

Bechtold P, Bazile E, Guichard F, *et al.* 2001. A mass-flux convection scheme for regional and global models. *Q. J. R. Meteorol. Soc.* **127**: 869–886.

Bechtold P, Chaboureau J-P, Beljaars A, et al. 2004. The simulation of the diurnal cycle of convective precipitation over land in a global model. *Q. J. R. Meteorol. Soc.* **130**: 3119–3137. doi: 10.1256/qj.03.103.

Bechtold P, Semane N, Lopez P, *et al.* 2014. Representing equilibrium and non-equilibrium convection in large-scale models. *J. Atmos. Sci.* **71**(2): 734–753. doi: 10.1175/JAS-D-13-0163.1.

Berry G, Thorncroft C, Hewson T. 2007. African easterly waves during 2004 – analysis using objective techniques. *Mon. Weather Rev.* **135**: 1251–1267. doi: http://dx.doi.org/10.1175/MWR3343.1.

Betts AK, Miller M. 1986. A new convective adjustment scheme. Part II: single-column tests using GATE wave, BOMEX, ATEX, and arctic-air mass data sets. *Q. J. R. Meteorol. Soc.* **112**: 693–709.

Beucher F, Lafore J-F, Karbou F, Roca R. 2014. High-resolution prediction of a major convective period over West Africa. *Q. J. R. Meteorol. Soc.* **140**: 1409–1425. doi: 10.1002/qj.2225.

Birch CE, Parker DJ, Marsham JH, *et al.* 2014. A seamless assessment of the role of convection in the water cycle of the West African monsoon. *J. Geophys. Res. Atmos.* **119**: 2890–2912. doi: 10.1002/2013JD020887.

Bock O, Bouin M-N, Walpersdorf A, *et al.* 2007. Comparison of ground-based GPS precipitable water vapour to independent observations and numerical weather prediction model reanalyses over Africa. *Q. J. R. Meteorol. Soc.* **133**: 2011–2027.

Bock O, Bouin MN, Doerflinger E, *et al.* 2008. The West African monsoon observed with ground-based GPS receivers during AMMA. *J. Geophys. Res.* **113**: D21105. doi: 10.1029/2008JD010327.

Bougeault P. 1985. A simple parameterization of the large-scale effects of cumulus convection. *Mon. Weather Rev.* **113**: 2108–2121.

Bougeault P, Toth Z, Bishop C, *et al.* 2010. The THORPEX Interactive Grand Global Ensemble (TIGGE). *Bull. Am. Meteorol. Soc.* **91**: 1059–1072.

Bouttier F, Courtier P. 1999. Data assimilation concepts and methods. ECMWF training material. http://www.ecmwf.int/en/learning/education-material/lecture-notes.

Chaboureau J-P, Guichard F, Redelsperger J-L, Lafore J-P. 2004. The role of stability and moisture in the diurnal cycle of convection over land. *Q. J. R. Meteorol. Soc.* **130**: 3105–3117.

Chamberlain JM, Bain CL, Boyd DFA, *et al.* 2014. Forecasting storms over Lake Victoria using a high resolution model. *Meteorol. Appl.* **21**: 419–430. doi: 10.1002/met.1403.

Charney JG. 1947. The dynamics of longwaves in a baroclinic westerly current. *J. Meteorol.* **4**: 135–163.

Charney JG, Fjortoft R, von Neumann J. 1950. Numerical integration of the barotropic vorticity equation. *Tellus* **2**: 237–254.

Côté J, Gravel S, Méthot A, *et al.* 1998a. The operational CMC-MRB global environmental multiscale (GEM) model. Part I: design considerations and formulation. *Mon. Weather Rev.* **126**(6): 1373–1395.

Côté J, Desmarais JG, Gravel S, *et al.* 1998b. The operational CMC-MRB global environmental multiscale (GEM) model. Part II: results. *Mon. Weather Rev.* **126**(6): 1397–1418.

Courtier P, Geleyn J-F. 1988. A global numerical weather prediction model with variable resolution: application to the shallow-water equations. *Q. J. R. Meteorol. Soc.* **114**: 1321–1346.

Daley R. 1991. *Atmospheric Data Analysis*. Cambridge University Press: Cambridge.

Davies T, Cullen MJP, Malcolm AJ, *et al.* 2005. A new dynamical core for the Met Office's global and regional modelling of the atmosphere. *Q. J. R. Meteorol. Soc.* **131**: 1759–1782. doi: 10.1256/qj.04.101.

Dharssi I, Bovis KJ, Macpherson B, Jones CP. 2011. Operational assimilation of ASCAT surface soil wetness at the Met Office. *Hydrol. Earth Syst. Sci. Disc.* **8**(2): 4313–4354.

Diop-Kane M, Mumba Z, Abebe A, *et al.* 2008. The use of ECMWF products at ACMAD; their performance in forecasting severe weather in Africa. In: *ECMWF Proceedings of 11th Workshop on Meteorological Operational Systems*, 12–16 November 2007, UK; pp. 100–103.

Donner LJ. 1993. A cumulus parameterization including mass fluxes, vertical momentum dynamics, and mesoscale effects. *J. Atmos. Sci.* **50**: 889–906.

Durran DR. 1999. *Numerical Methods for Wave Equations in Geophysical Fluid Dynamics*. Springer.

Eady ET. 1949. Long waves and cyclone waves. *Tellus* **1**: 33–52.

Epstein E. 1969: Stochastic dynamic prediction. *Tellus* **21**: 739–759.

Faccani C, Rabier F, Fourrié N, *et al.* 2009. The impacts of AMMA radiosonde data on the French global assimilation and forecast system. *Weather Forecast.* **24**: 1268–1286.

Fink AH, Agustí-Panareda A, Parker DJ, *et al.* 2011. Operational meteorology in West Africa: observational networks, weather analysis and forecasting. *Atmos. Sci. Lett.* **12**: 135–141.

Garcia-Carreras L, Marsham JH, Bain DJ, *et al.* 2013. The impact of convective cold pool outflows on model biases in the Sahara. *Geophys. Res. Lett.* **40**: 1647–1652.

Grandpeix J, Lafore J-P. 2010. A density current parametrization coupled with Emanuel's convection scheme. *Part I – the models. J. Atmos. Sci.* **67**: 881–897.

Gregory D, Rowntree PR. 1990. A mass flux convection scheme with representation of cloud ensemble characteristics and stability-dependent closure. *Mon. Weather Rev.* **118**: 1483–1506. doi: 10.1175/1520-0493 (1990)118<1483:AMFCSW>2.0.CO;2.

Haltiner GJ, Williams RT. 1980. *Numerical Prediction and Dynamic Meteorology*, 2nd edition. John Wiley & Sons.

Holtslag AAM, Boville BA. 1993. Local versus non-local boundary layer diffusion in a global climate model. *J. Climate* **6**: 1825–1842.

Huffman GJ, Adler RF, Bolvin DT, *et al.* 2007. The TRMM multi-satellite precipitation analysis: quasi-global, multi-year, combined-sensor precipitation estimates at fine scale. *J. Hydrometeorol.* **8**: 38–55.

Janicot S, Thorncroft CD, Ali A, *et al.* 2008. Large-scale overview of the summer monsoon over West Africa during the AMMA field experiment in 2006. *Ann Geophys.* **26**: 2569–2595.

Janicot S, Caniaux G, Chauvin F, *et al.* 2011. Intraseasonal variability of the West African monsoon. *Atmos. Sci. Lett.* **12**: 58–66. doi: 10.1002/asl.280.

Kain JS, Fritsch JM. 1990. A one-dimensional entraining/detraining plume model and its application in convective parameterizations. *J. Atmos. Sci.* **47**: 2784–2802.

Kalnay E. 2003. *Atmospheric Modelling, Data Assimilation, and Predictability*. Cambridge University Press: Cambridge.

Karbou F, Gérard E, Rabier F. 2010a. Global 4DVar assimilation and forecast experiments using AMSU observations over land. Part I: impact of various land surface emissivity parameterizations. *Weather Forecast.* **25**: 5–19. doi: 10.1175/2009WAF2222243.1.

Karbou F, Rabier F, Lafore J-P, *et al.* 2010b. Global 4DVar assimilation and forecast experiments using AMSU observations over land. Part II: impact of assimilating surface sensitive channels on the African monsoon during AMMA. *Weather Forecast.* **25**: 20–36. doi: 10.1175/2009WAF2222244.1.

Kim Y-J, Eckermann SD, Chun H-Y. 2003. An overview of the past, present and future of gravity-wave drag parametrization for numerical climate and weather prediction models. *Atmos. Ocean* **41**: 65–98.

Krishnamurti TN, Kishtawal CM, LaRow TE, *et al.* 1999. Improved weather and seasonal climate forecasts from multimodel superensemble. *Science* **285**(5433): 1548–1550.

Kuettner JP. 1974. General description and central program of GATE. *Bull. Am. Meteorol. Soc.* **55**: 712–719.

Kuo HL. 1965. On formation and intensification of tropical cyclones through latent heat release by cumulus convection. *J. Atmos. Sci.* **22**: 40–63.

Lebel T, Parker DJ, Flamant C, *et al.* 2010. The AMMA field campaigns: multiscale and multidisciplinary observations in the West African region. *Q. J. R. Meteorol. Soc.* **136**: 8–33. doi: 10.1002/qj.486.

Leith CE. 1974. Theoretical skill of Monte Carlo forecasts. *Mon. Weather Rev.* **102**: 409–418.

Lewis JM. 2005. Roots of ensemble forecasting. *Mon. Weather Rev.* **133**: 1865–1885. doi:http://dx.doi.org/10.1175/MWR2949.1.

Lorenc AC. 1986. Analysis methods for numerical weather prediction. *Q. J. R. Meteorol. Soc.* **112**(474): 1177–1194.

Lorenz EN. 1963. Deterministic nonperiodic flow. *J. Atmos. Sci.* **20**(2): 130–141.

Lynch P. 2011. From Richardson to early numerical weather prediction. In: *The Development of Atmospheric General Circulation Models: Complexity, Synthesis and Computation*, L Donner, W Schubert, R Somerville (eds). Cambridge University Press.

Mahfouf J-F. 1991. Analysis of soil moisture from near-surface parameters: a feasibility study. *J. Appl. Meteorol.* **30**: 1534–1547.

Manabe S, Smagorinsky J, Strickler RF. 1965. Simulated climatology of a general circulation model with a hydrologic cycle. *Mon. Weather Rev.* **93**: 769–798. doi: http://dx.doi.org/10.1175/1520-0493(1965)093<0769:SCOAGC>2.3.CO;2.

Marsham JH, Dixon N, Garcia-Carreras L, *et al.* 2013a. The role of moist convection in the West African monsoon system: insights from continental-scale convection-permitting simulations. *Geophys. Res. Lett.* **40**: 1843–1849. doi: 10.1002/grl.50347.

Marsham JH, Brooks BJ, Hobby M, *et al.* 2013b. Meteorology and dust in the central Sahara: observations from Fennec supersite-1 during the June 2011 intensive observation period, *J. Geophys. Res. D Atmos.* **118**: 4069–4089. doi: 10.1002/jgrd.50211.

Mason PJ. 1986. On the parametrization of orographic drag. In: *Seminar/Workshop on Observation, Theory and Modelling of Orographic Effects*, 15–20 September, ECMWF, Shinfield Park, Reading, UK, vol. 1; pp. 167–194.

Meynadier R, Bock O, Guichard F, *et al.* 2010a. West African monsoon water cycle: 1. A hybrid water budget data set. *J. Geophys. Res.* **115**(D19): D19106.

Meynadier R, Bock O, Gervois S, *et al.* 2010b. West African monsoon water cycle: 2. Assessment of numerical weather prediction water budgets. *J. Geophys. Res.* **115**(D19): D19107.

Miller MA, Slingo A. 2007. The ARM mobile facility and its first international deployment: measuring radiative flux divergence in West Africa. *Bull. Am. Meteorol. Soc.* **88**: 1229–1244.

Mutemi JN, Ogallo LA, Krishnamurti TN, *et al.* 2007. Multimodel based superensemble forecasts for short and medium range NWP over various regions of Africa. *Meteorol. Atmos. Phys.* **95**: 87–113.

Nebeker F. 1995. *Calculating the Weather: Meteorology in the 20th Century*. Academic Press.

Nuret M, Lafore J-P, Bock O, *et al.* 2008. Correction of humidity bias for Vaisala RS80-A sondes during the AMMA 2006 observing period. *J. Atmos. Oceanic Technol.* **25**(11): 2152–2158.

Parker DJ, Fink A, Janicot S, *et al.* 2008. The AMMA radiosonde program and its implications for the future of atmospheric monitoring over Africa. *Bull. Am. Meteor. Soc.* **89**: 1015–1027. doi: http://dx.doi.org/10.1175/2008BAMS2436.1.

Peixoto J-P, Oort A. 1992. *Physics of Climate*. American Institute of Physics: New York.

Redelsperger J-L, Thorncroft CD, Diedhiou A, *et al.* 2006. African Monsoon multi-disciplinary analysis: an international research project and field campaign. *Bull. Am. Meteorol. Soc.* **87**: 1739–1746.

Richardson LF. 1922. *Weather Prediction by Numerical Process*. Cambridge University Press.

Rossby CG. 1940. Planetary flow patterns in the atmosphere. *Q. J. R. Meteorol. Soc.* **66**: 68–87.

Schwendike J, Jones SC. 2010. Convection in an African easterly wave over West Africa and the eastern Atlantic: a model case study of Helene (2006). *Q. J. R. Meteorol. Soc.* **136**: 364–396. doi: 10.1002/qj.566.

Seity Y, Brousseau P, Malardel S, *et al.* 2011. The AROME–France convective-scale operational model. *Mon. Weather Rev.* **139**: 976–991. doi: http://dx.doi.org/10.1175/2010MWR3425.1.

Shutts G. 2005. A kinetic energy backscatter algorithm for use in ensemble prediction systems. *Q. J. R. Meteorol. Soc.* **131**: 3079–3102. doi: 10.1256/qj.04.106.

Söhne N, Chaboureau J-P, Guichard F. 2008. Verification of cloud cover forecast with satellite observation over West Africa. *Mon. Weather Rev.* **136**: 4421–4434.

Stephens GL, Vane DG, Boain RJ, *et al.* 2002. The CloudSat mission and the A-train: a new dimension of space-based observations of clouds and precipitation. *Bull. Am. Meteorol. Soc.* **83**: 1771–1790. doi: 10.1175/BAMS-83-12-1771.

Staniforth A, Côté J. 1991. Semi-Lagrangian integration schemes for atmospheric models – a review. *Mon. Weather Rev.* **119**: 2206–2223. doi: http://dx.doi.org 10.1175/1520-0493(1991)119<2206:SLISFA>2.0.CO;2.

Stratton R, Stirling A. 2012. Improving the diurnal cycle of convection in GCMs. *Q. J. R. Meteorol. Soc.* **138**: 1121–1134.

Thorncroft CD, Blackburn M. 1999. Maintenance of the African easterly jet. *Q. J. R. Meteorol. Soc.* **125**: 763–786.

Thorncroft CD, Parker DJ, Burton RR, *et al.* 2003. The JET2000 project – aircraft observations of the African easterly jet and African easterly waves. *Bull. Am. Meteorol. Soc.* **84**: 337–351. doi: 10.1175/BAMS-84-3-33.

Tiedtke M. 1989. A comprehensive mass flux scheme for cumulus parameterization in large-scale models. *Mon. Weather Rev.* **117**: 1779–1800.

Tiedtke M. 1993. Representation of clouds in large-scale models. *Mon. Weather Rev.* **121**: 3040–3061.

Tompkins AM, Diongue A, Parker DJ, Thorncroft C. 2005. The African easterly jet in the ECMWF integrated forecast system: 4D-Var analysis. *Q. J. R. Meteorol. Soc.* **131**(611): 2861–2885.

Wagner W, Lemoine G, Rott H. 1999. A method for estimating soil moisture from ERS scatterometer and soil data. *Remote Sens. Environ.* **70**(2): 191–207.

Wilson DR, Bushell AC, Kerr-Munslow AM, *et al.* 2008. PC2: a prognostic cloud fraction and condensation scheme. I: scheme description. *Q. J. R. Meteorol. Soc.* **134**: 2093–2107. doi: 10.1002/qj.333.

Webster PJ. 2013. Meteorology: improve weather forecasts for the developing world. *Nature* **493**(7430): 17–19.

# 11

# West African Synthetic Analysis and Forecast: WASA/F

**Lead authors:** *Jean Philippe Lafore*[a] *and Nicolas Chapelon*[b]

**Contributing authors:** *Florent Beucher*[a], *Mariane Diop Kane*[c], *Alexandre Gaymard*[b], *Abdou Kasimou*[d], *Serge Lepape*[b], *Zilore Mumba*[e], *Bernard Orji*[f], *David Osika*[g], *Douglas J. Parker*[h], *Emmanuel Poan*[a,i], *Léon Guy Razafindrakoto*[d] *and Jean Christophe Vincendon*[b]

**Reviewers:** *ACMAD and the West African forecasting community (feedback coordinated by Abdou Kassimou, ACMAD, Niamey, Niger)*

[a] *CNRM-UMR3589, Météo-France and CNRS, Toulouse, France*
[b] *CISMF, Météo-France, Toulouse, France*
[c] *Agence Nationale de l'Aviation Civile et de la Météorologie du Sénégal (ANACIM), Dakar, Sénégal*
[d] *African Centre of Meteorological Applications for Development (ACMAD), Niamey, Niger*
[e] *University of Zambia, Lusaka, Zambia*
[f] *Nigerian Meteorological Agency (NIMET), Lagos, Nigeria*
[g] *Ghana Meteorological Agency, Accra, Ghana*
[h] *University of Leeds, Leeds, UK*
[i] *Université du Québec à Montréal, Montreal, Canada*

## 11.1 Introduction

The task of a forecaster involves the analysis of numerous observations and numerical weather prediction (NWP) products (including both analysis and forecast products) before deciding on the weather forecast for a given location and range. This is a very complex process involving both objective and subjective criteria and where the experience of the forecaster can play an important role in the skill of the final forecast. The difficulty is even greater for tropical regions, where, in contrast to mid-latitudes, the atmospheric flow is weakly balanced, resulting in weaker predictability, and the NWP skill is weak, especially for convective events. A further challenge is the speed with which the process needs to be performed.

To help forecasters in West Africa with this difficult task, a conceptual framework was developed and tested in 2006, during the Specific Observation Period of the African Monsoon and Multidisciplinary Analyses (AMMA) project. The framework adopts an object-oriented approach to forecasting based on available conceptual models, as presented in previous chapters. Similar methods have already been developed for the mid-latitudes, such as the ANASYG (Graphic SYnoptic ANAlysis)/PRESYG (Graphic Synoptic Prediction of Météo-France) (Santurette and Joly, 2002) or the SigMet charts of the UK Met Office. This forecasting approach consists of drawing single synthetic maps that summarise all the key features of the West African monsoon (WAM), analysed or forecast at a given time. To produce the final forecast of weather events, a combined analysis of the selected main features of the WAM (following the rules presented and illustrated in this guide) is conducted. These synthetic maps are named West African synthetic analysis (WASA) and West African synthetic forecast (WASF). This method has been used at the African Center of Meteorological Applications for Development (ACMAD) since 2006 and provides a unique and common framework within which forecasters in different West African countries can discuss the current situation. Due to this operational practice, and to major advances achieved recently with the AMMA program, the method has been updated and improved since 2006, in particular to be more repeatable, and to be based on quantitative measures.

The following 11 features are considered as important and will figure on the WASA/F (Figures 11.1 and 11.2) maps using the symbols listed in Table 11.1, in order to capture the main characteristics of the situation and to forecast the occurrence of deep convection, dust and the weather over West Africa.

1) The intertropical discontinuity (ITD).
2) The heat low (HL).
3) The subtropical jet (STJ).
4) Features associated with mid-latitudes.

*Meteorology of Tropical West Africa: The Forecasters' Handbook*, First Edition. Edited by Douglas J. Parker and Mariane Diop-Kane.

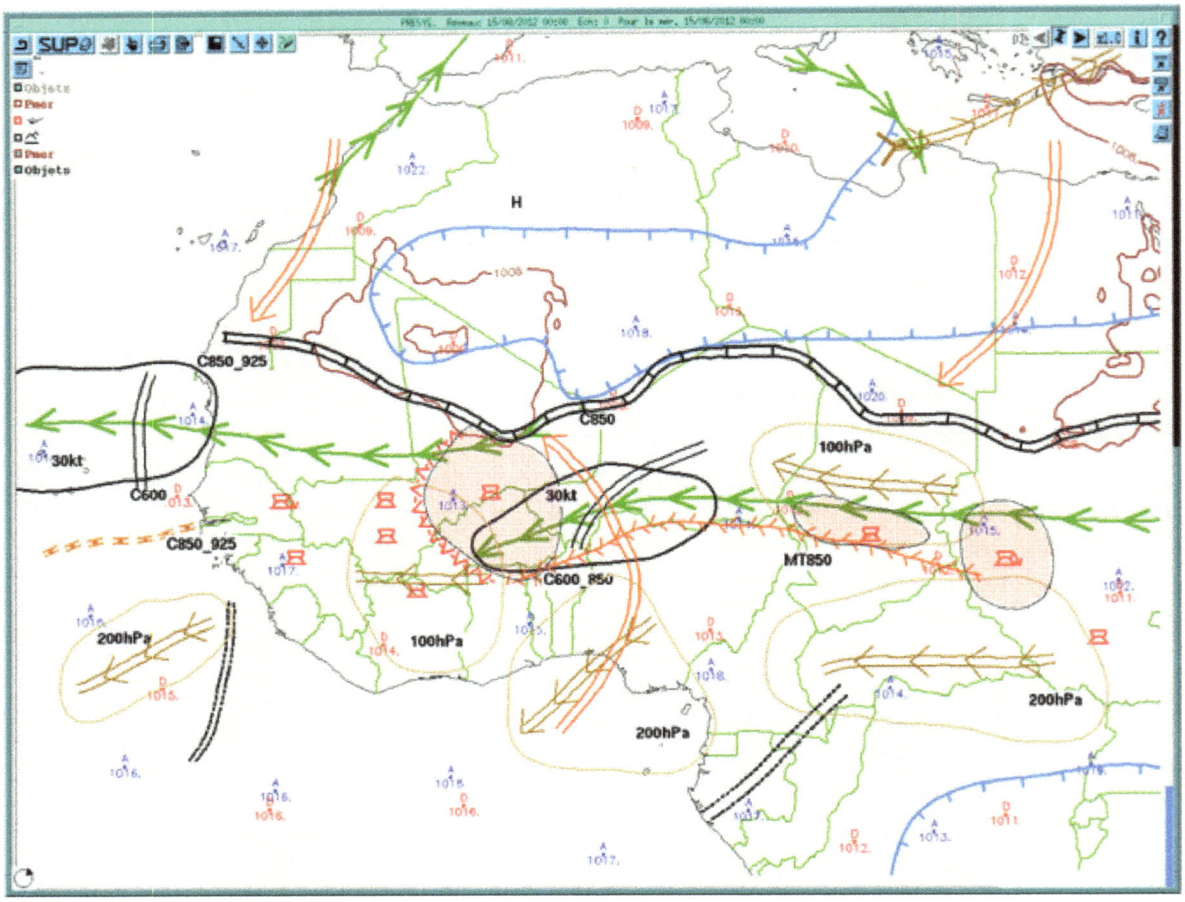


**Figure 11.1** Illustration of the different key features figuring on a WASA/F, for 15 August 2012 at 0000 UTC.

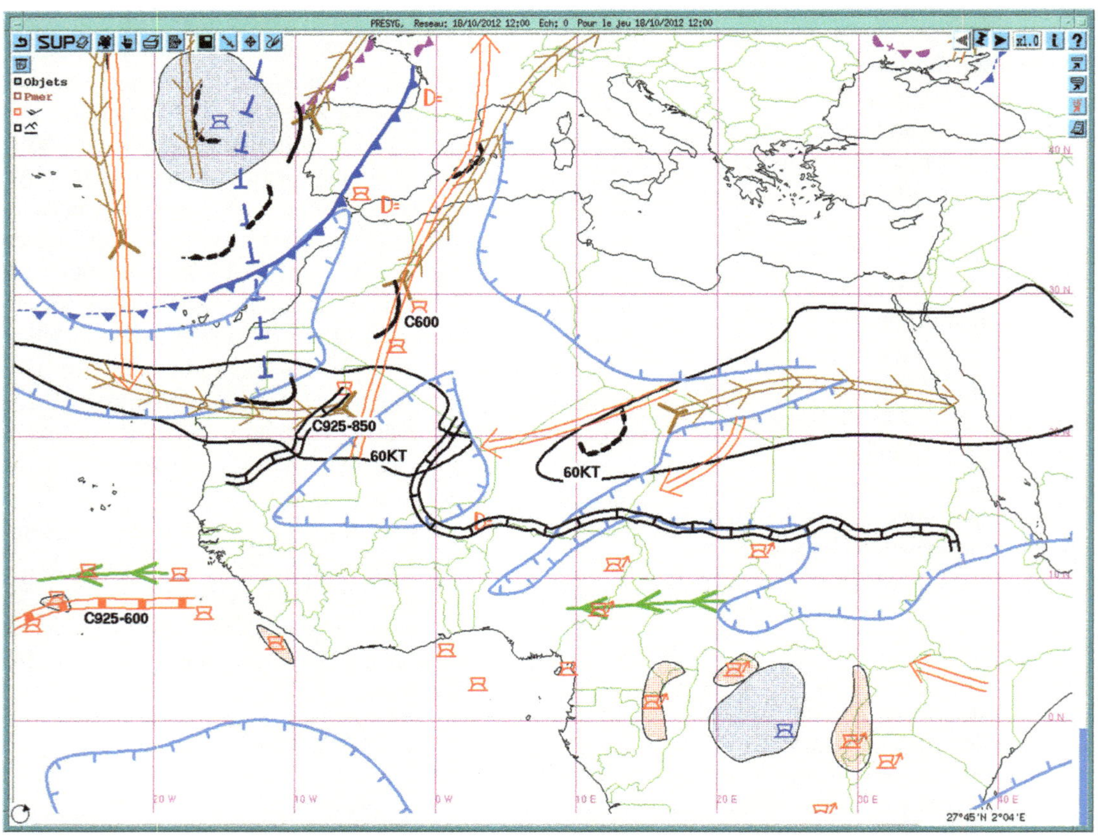


**Figure 11.2** Illustration of the different key features figuring on a WASA/F, for a dust situation associated with an interaction with mid-latitudes; here for 19 October 2012 at 1200 UTC. The ITD is more normally a continuous line except at the coast, where we can have a mixed zone (particularly at the beginning and end of the rainy season) with two lines that will meet and continue as a single line inland; see Section 11.2.

**Table 11.1** List of symbols figuring on the WASA/F maps. Caution: some symbols may evolve or depend on the weather analysis and forecasting system used; here, they are illustrated for the 'Synergie' system[a] developed by Météo-France and Meteo France International.

| | |
|---|---|
|   | **ITD** (intertropical discontinuity) |
| 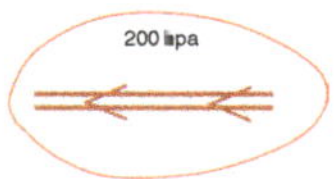 | **TEJ** (tropical easterly jet) axis<br>the core must be outlined with an indication of the considered level 200 or 100 hPa |
|  | **STJ** (subtropical jet) or **PJ** (polar jet) axis |
| 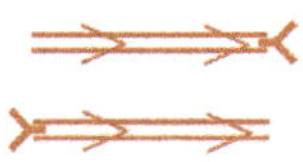 | Exit/entrance of a jet |
|   | **UTT** (upper tropospheric trough) |
|   | **Potential vorticity (PV) anomaly** active/not active |
|  | Cold and warm surface front |
|   | **Mid-level dry intrusion** boundary |
| 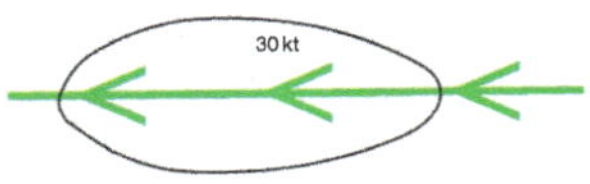 | **AEJ** (African easterly jet) axis<br>The core must be outlined with an indication of the threshold (25 or 30 kt) |
|  | **AEW (African easterly wave) trough**<br>**AEW ridge** |
| **C850** | **Cyclonic vortex** at a specific level – here 850 hPa (or layer) |
| **A and D** | **Surface high (Anticyclone 'A') and low (Depression 'D') pressure centres (alternatively, in anglophone usage, 'H' and 'L')** |
|  | Northerly **dry burst**<br>Southerly **monsoon burst** |
|  | **MT** (monsoon trough) at 850 hPa |
|  | **ITCZ** (intertropical convergence zone) over ocean<br>active/not active for convection |
|  | **Dust or sand** symbols (see Table 5.1) |
|  | **Convection** and growing or decaying stages<br>NB: organised convective system can be outlined and shaded in red |

[a] http://www.mfi.fr/en/page/meteorological-information-systems/synergie-forecasting.php.

5) Mid-level dry air.
6) The Tropical Easterly Jet (TEJ).
7) The African Easterly Jet (AEJ).
8) African easterly waves (AEWs) and cyclonic vortices.
9) The African monsoon layer and monsoon trough (MT).
10) Dust monitoring and forecast is extensively treated in Section 5.2 and is therefore not elaborated here. Nevertheless, dust symbols listed in Table 5.1 when relevant, must figure on the WASA/F maps, as illustrated by Figure 11.2.
11) Convective activity with the distinction between three cases:
    - suppressed convection areas;
    - unorganised isolated convective cells;
    - mesoscale convective systems (MCSs) and squall lines.

## 11.2 The Intertropical Discontinuity/Intertropical Front, or Intertropical Boundary in Ghana

The intertropical discontinuity (ITD)/intertropical front (ITF), or intertropical boundary (ITB) in Ghana, is represented by the following symbol: 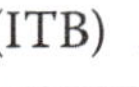

### 11.2.1 Main Characteristics

The ITD is a sharp gradient close to the surface, which divides the south to southwest flow of monsoon air to the south of the ITD, with high thermodynamic energy and thus high wet bulb potential temperature $\theta_w'$, from the northerly flow of dry air with low dew point temperature ($T_d$ or low $\theta_w'$) to the north of the ITD. On average, it has almost a zonal orientation, and a strong diurnal cycle in its northward penetration.

Over the ocean, the surface convergence line occurs between southeasterlies and northeasterlies and coincides with the intertropical convergence zone (ITCZ). Over West Africa, however (i.e. over land), the ITD is shifted about 5° to the north of the maximum rainy zone. Also, over land we prefer to avoid the use of the term 'ITCZ' and instead refer to the MT (monsoon trough), which is defined as the location of convergence and vorticity at 850 hPa, and is where vigorous tropical disturbances actually occur. On satellite images, the MT is clearly identified by large convective clouds or cluster of clouds.

### 11.2.2 Drawing Rules

As illustrated by Figure 11.3, the three criteria needed to draw the ITD, classified by decreasing degree of importance, are:

- *Convergence line at the surface (10 m or 950 hPa) between southerly and northerly winds.* This criterion is easier to use during the night (0000 and 0600 UTC).
- *Strong gradient of humidity at the surface (synoptic observation (SYNOP) at 2 m or analysis).* You can use $T_d$ (limit near 15°C), $\theta_w'$ (with 0.5°C spacing) or relative humidity (limit near 40%).
- *Surface pressure.* In general, the line of pressure minimum (the Saharan HL) will, in the summer months, lie quite close to (and to the north of) the ITD. In general, these lines do not exactly coincide, but consideration of the pressure pattern can be useful near to the western coast of Africa where the first two criteria may be difficult to use, as illustrated by Figure 11.4.

The drawing based on model-analysed fields must be checked and corrected by the use of surface observations (SYNOP and Meteorological Aviation Report) when available. Also, convective cells detected from satellite images can guide ITD drawing, following the observation that convection is rare north of the ITD.

All these criteria do not necessarily converge on the same drawing, so the forecaster will have to use their best judgement, according more importance to one of the criteria listed earlier. Figure 11.3 illustrates the drawing of the ITD and of the HL (Section 11.3) at two times of the day to account for the strong diurnal cycle of these features. It is worth noticing that most of the time the ITD is slightly south of the surface pressure minimum corresponding to the HL. The ITD is drawn over Africa from the Atlantic coast to Sudan, where it disappears.

Over the Atlantic Ocean, the ITCZ is drawn by the coincidence of three features, namely the convergence of the trades, the maximum of energy ($\theta_w' > 21$ °C ) at 950 hPa, and the location of convection, either isolated or organised (MCSs). There is a continuity with the convective systems going out of Africa, but not with the ITD. Therefore, there is no need to link the ITD over land with the ITCZ over the ocean.

### 11.2.3 Specific Cases

In winter, when the ITD lies to the south of West Africa, it can be difficult to define it accurately. For instance, in the case of Harmattan dry intrusions from the north the ITD can even be pushed southward over the sea, in to the Gulf of Guinea. In other cases of penetration over Africa of mid-latitude depressions, these events can alter the HL structure and perturb the ITD. Illustrations of such configurations are given in the online case studies of the Handbook project at www.cnrm.meteo.fr/waf_handbook_casestudies.

## 11.3 The Heat Low or Thermal Depression

### 11.3.1 Main Characteristics

A look at the MSL pressure field at large scale (Figure 11.5) allows monitoring of the location, intensity and evolution of the main anticyclonic and cyclonic centres governing the circulation over Africa; that is, Azores, St Helena and Mascarene highs (in blue), West African and Arabic thermal lows (in red) and northern mid-latitude pressure centres. However, the penetration and withdrawal of the monsoon over the continent is largely driven by the West African HL to be represented on the WASA/F maps. In general, the thermal depressions are limited to the lowest levels of the atmosphere, but over West Africa the HL has a larger vertical extent, sometimes up to 600 hPa, and a large magnitude. Also, the location and extension of the HL appear to be strongly forced by mid-latitude circulations and perturbations, as detailed in Sections 2.1.4 and 2.1.5. The large diurnal cycle of the HL is such that:

- The HL reaches its maximum (minimum of pressure) in late afternoon as a result of the intense diurnal heating.
- Nevertheless, the wind response is not instantaneous, so that the maximum of wind occurs at the end of the night.

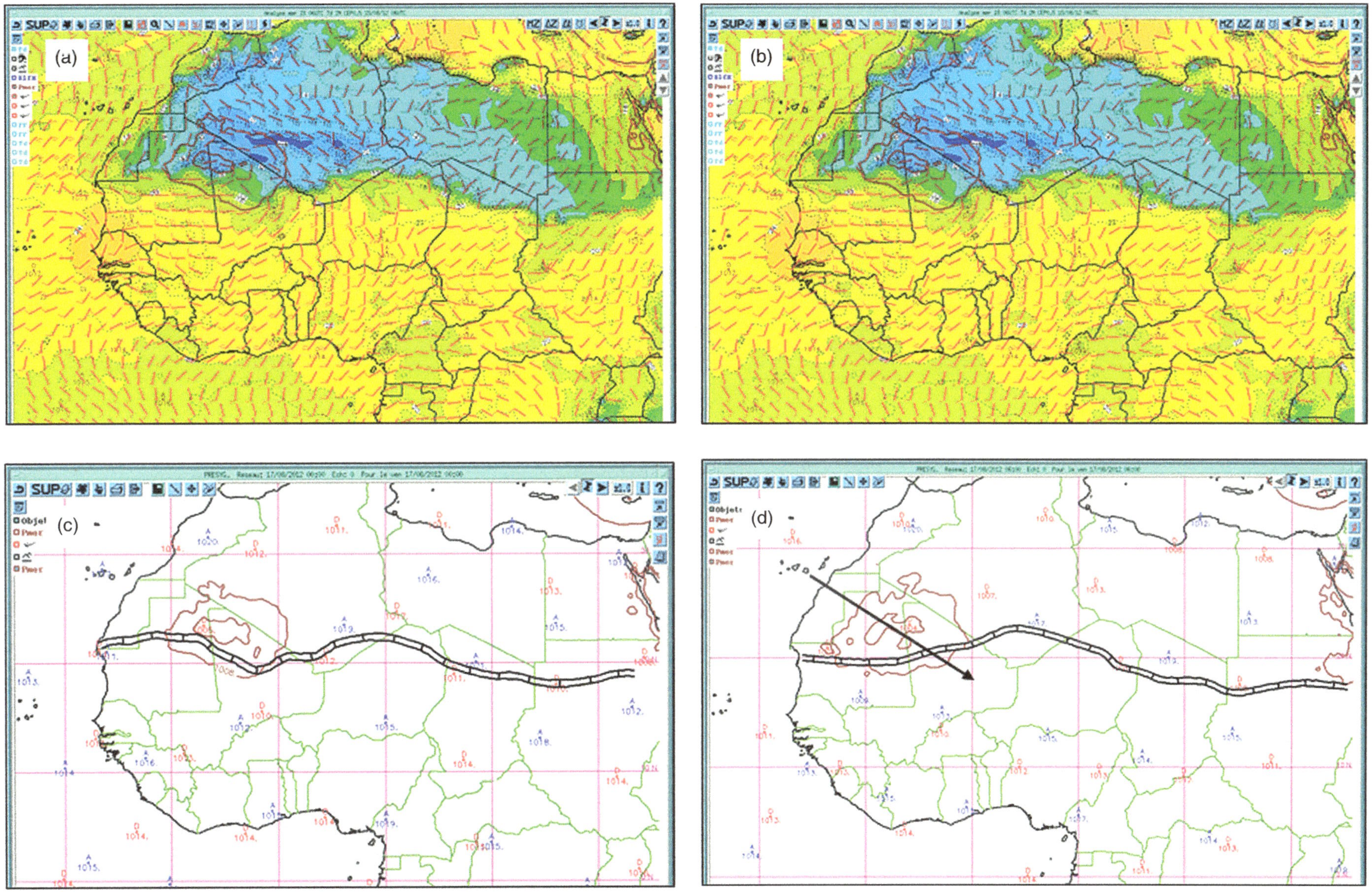


**Figure 11.3** Illustration of plotting the ITD from the ARPEGE (operational forecasting global model of Meteo-France) operational analysis. (a, b) Maps of $T_d$ at 2 m (colour), horizontal wind vectors at 925 hPa and mean sea level (MSL) pressure field (2 hPa isoline interval), to outline the HL (threshold of 1008 hPa and 1006 hPa at 0600 UTC (a) and 1800 UTC (b)) for 15 August 2012. (c, d) Corresponding WASA for the ITD and HL features. The arrow in (d) indicates the location of the vertical cross-section displayed by Figure 11.6.

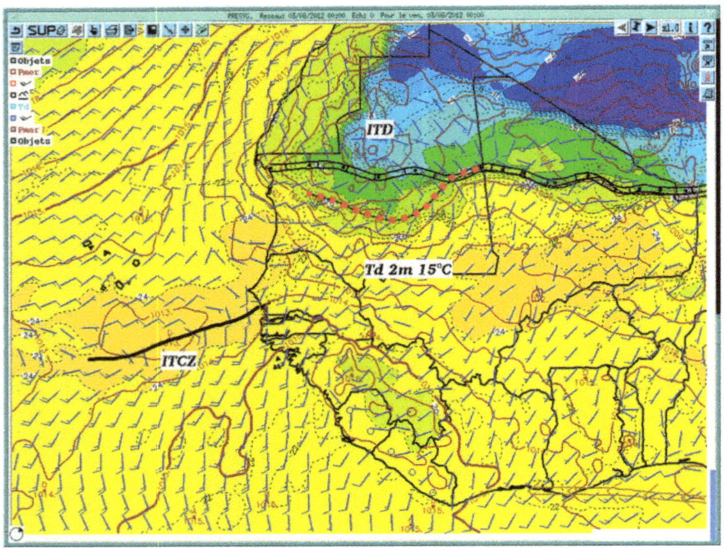


**Figure 11.4** *Illustration of ITD drawing near the coast, using the pressure and wind criteria, where the $T_d$ value (15°C) and maximum gradient (red dotted line) are not coincident. Note the discontinuity between the ITD and the ITCZ over ocean. Fields are provided by the* ARPEGE operational analysis on 3rd August 2012.

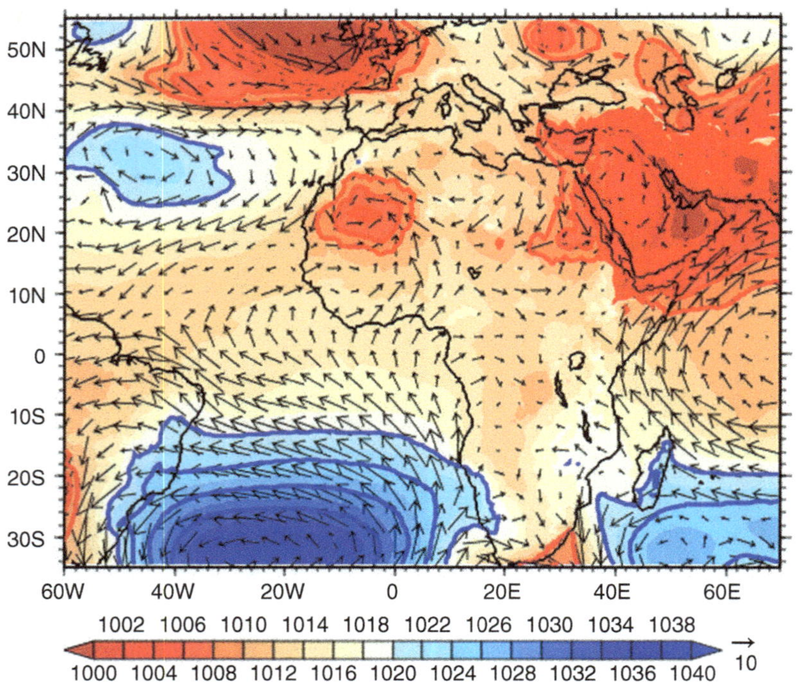


**Figure 11.5** Daily-average of MSL pressure field and wind at 925 hPa for 15 August 2012, as available on the Monitoring and forecasting of Intra Seasonal Variability over Africa (MISVA) site (http://isv.sedoo.fr/index) for the European Centre for Medium-Range Weather Forecasts (ECMWF) operational analysis. Red and blue areas correspond to low- and high-pressure centres respectively.

- The ITD in general fits in the position of the HL pressure minimum, but there is an extension of the HL to the north in the afternoon.

The maximum of surface convergence at the ITD occurs between 0000 and 0600 UTC; it is weaker during the day.

### 11.3.2 Drawing Rules

The HL is outlined by pressure lower than a given threshold. But we need to take into account its strong diurnal cycle, with the minimum of pressure obtained around 1800 UTC and the weakest depression around 1200 UTC. We propose drawing surface pressure every 2 hPa with the following criteria, as illustrated by Figure 11.3:

- at 1200 UTC lower than 1010 hPa;
- at 1800 UTC lower than 1006 hPa;
- at 0000 UTC lower than 1008 hPa;
- at 0600 UTC lower than 1008 hPa.

If the HL is weak and not visible with these thresholds, the location and intensity of the minimum of pressure has

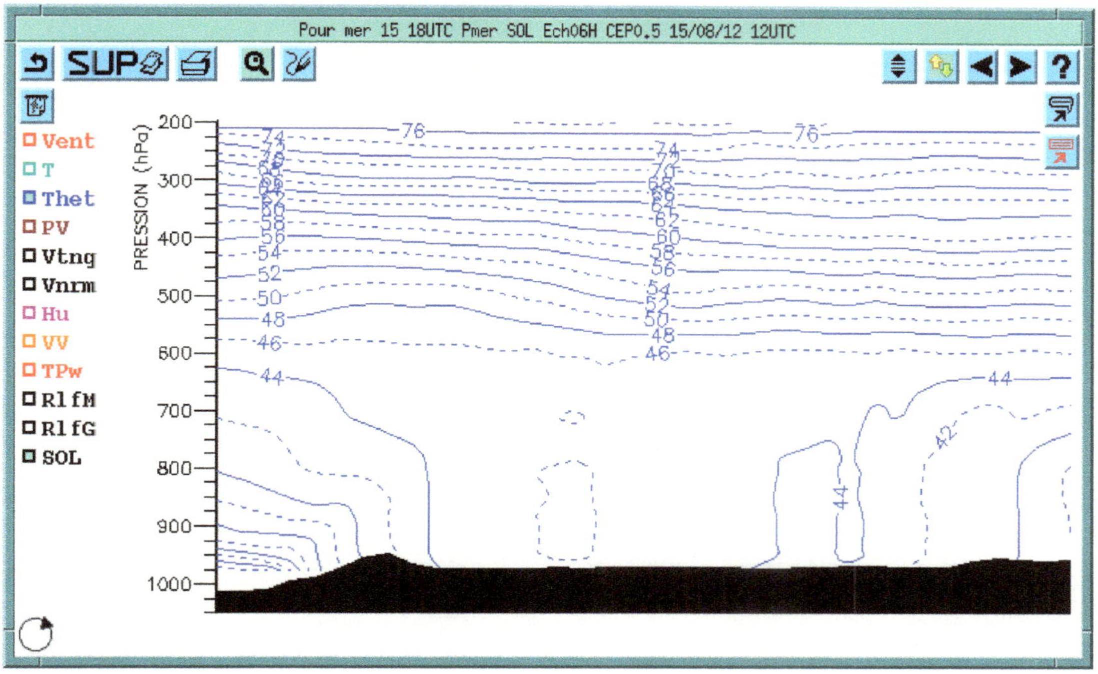


Figure 11.6 Vertical cross-section of potential temperature $\theta$, of the HL for 15 August 2012 at 1800 UTC, along the arrow drawn on Figure 11.3d from [28°N; 15°W] to [18°N; 1°E]. The warm boundary-layer air associated with the HL extends vertically up to around 600 hPa in this case.

to be mentioned anyway. Most of the time the HL is located in the north of Mali and Mauritania, but there are also HL modulations travelling from east to west along the ITD and associated with the propagation of AEWs.

To investigate the structure of the HL, two complementary diagnostics are useful:

1) Vertical cross-sections in $\theta$ through the HL (or vertical sounding) can help, especially at 1800 UTC, as the dry convective activity has produced a homogeneous layer between the surface and 700–600 hPa ($\theta \approx 44-45$°C), as illustrated by Figure 11.6.
2) Potential temperature maps at 850 hPa (if available) mark the trace of the HL (>38°C), as provided by the MISVA site[1] (Figure 11.7). Associated temperature and wind anomalies explain the location and extension of the HL and its evolution respectively. For instance, in the case illustrated in Figure 11.7, owing to a strong southerly wind anomaly over western Europe the HL is expected to shift northward and intensify from 15 to 17 August 2012.

## 11.4 The Subtropical Jet

Symbol: 

At upper levels, the subtropical jet (STJ) core marks the separation between the deep tropical troposphere (~16 km deep) and the less deep mid-latitude troposphere (~10 km). The STJ is associated with an intense potential vorticity (PV) band. To identified the STJ, the use of wind vectors on PV iso-surfaces is recommended (0.7 or 2 PVU (potential vorticity unit)), as illustrated by Figure 11.8. If PV surfaces are not available, wind maps around 200 hPa could be used.

The STJ is identified and outlined as areas with wind intensity larger than a threshold (Figure 11.8c). The difficulty comes from the fact that the threshold depends on the season. We propose to use 45 kt during the monsoon (e.g. May to September), since the STJ is less active during this period, and 60 kt during the winter. The threshold value must be indicated on the WASA/F map. The STJ axis is also drawn with the jet symbol core following the streamlines and along its maximum. The STJ climatological position is over the Maghreb during the monsoon, but the STJ can be over the Sahel during the transition periods of June and September and is permanent during winter.

The role of the STJ in West African weather is not well understood, but it is recommended that the forecaster monitors its position. The STJ is connected to the Libyan High in winter, and is involved in the interaction between the tropics and mid-latitudes, in relation to mid-latitude troughs. Chapter 2 provides more detailed explanations and a conceptual model of these interactions (Figure 2.20). The zones of acceleration and slowing down of the STJ, named entrance and

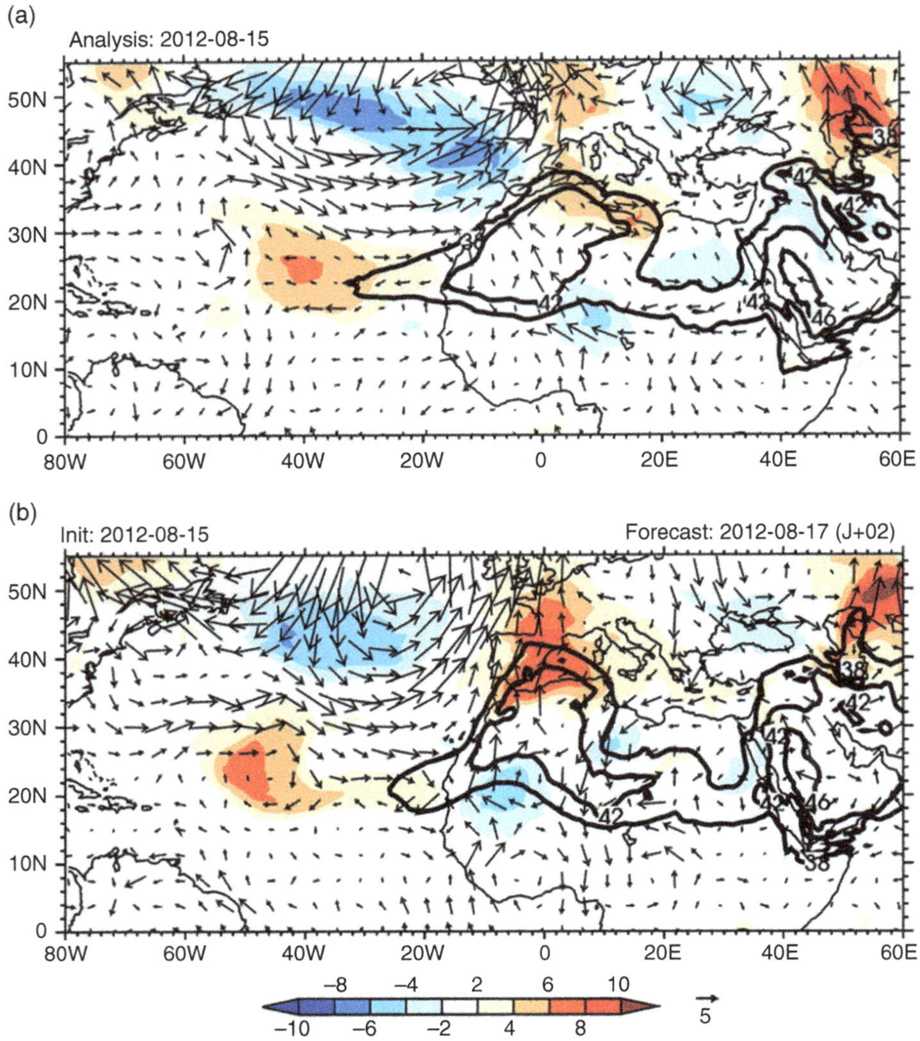


**Figure 11.7** Daily mean maps of 850 hPa potential temperature $\theta$ (K – colours) and wind anomalies (m s$^{-1}$ – arrows) for (a) the ECMWF analysis of 15 August 2012 and (b) the 2 days range forecast valid on 17 August 2012. Contours indicate the raw 850 hPa $\theta$ (°C) with one contour every 4°C, from 38°C to outline the HL area. All anomalies are determined against a climatological daily seasonal cycle, computed with ERA-Interim. Six-hourly ECMWF data were averaged to obtain daily data. Maps available from the MISVA site.
*Source*: Lafore *et al.* (2012). Reproduced with permission of Météo-France.

exit zones, can force ageostrophic circulation. A consequence of this is that the lifting effect occurs at the right of the jet entrance and the left of the jet exit; the opposite rule applies for the subsidence, which occurs on the left of the entrance and right of the exit. These effects are apparent on the vertical velocity field in the mid-troposphere (e.g. 700 or 600 hPa), where the vertical velocity can force convection through ascent or inhibit convection in regions of subsidence. In such cases the forecaster can use the symbols (see Table 11.1) that identify entrance or exit zones.

## 11.5 Features Associated with Mid-latitudes

Symbols:

As developed in Section 2.1.4, features associated with mid-latitudes can force weather over North and West Africa. They are routinely monitored and drawn by all mid-latitude forecasting offices, and as standard products from the major NWP centres. For consistency with the forecasters' work in mid-latitudes, we apply the same drawing rules for WASA/F. Mid-latitude features to be drawn are as follows, and as displayed in Figure 11.8b:

- The core of the polar jet (PJ) detected on the PV surface (0.7 or 2 PVU), with indication of entrance and exit zones (as for the STJ core).
- Surface cold and warm fronts with classical symbols.
- Stratospheric foliations are drawn as detected with the water vapour satellite images (black areas) coupled with a drop of the geopotential high of the PV surface (Figure 11.8a). If they are coupled with precipitation bands, they are considered active and drawn as heavy black lines. If not, they are drawn as dashed heavy lines (Table 11.1).
- From experience in the tropics, it appears useful to add the UTT (line of blue 'T's) at 200 hPa, detected from the vorticity and streamline fields. Indeed, often only the UTT reaches the tropics, whereas the PJ stays in the mid-latitudes.

Figure 11.8 provides an illustration of the drawing of those mid-latitude features that can interact with the STJ and impact weather over Africa more frequently in winter or spring. In the case-study of tropical–extratropical interaction (see CS03 at www.cnrm.meteo.fr/waf_handbook_casestudies) the UTT southward extension crossing the STJ favours a tropical plume and the breaking of the STJ, dust storms (Chapter 5, Figure 5.6) and cold events.

## 11.6 Mid-level Dry Air

Symbol:

### 11.6.1 Main Characteristics

Dry air can have major and contradictory impacts on convection, either as an inhibiting factor, especially during its initiation stage, or as an organising factor through rain evaporation feeding density currents (see Sections 11.12, 3.1.2.2 and 3.2.2). This contradictory role is of primary importance for the understanding of the WAM, as often mid-level dry air prevails over the Sahel and Sahara. The origin of this dry air and its interaction with convection are still an issue for research (see Section 2.1.5). A typical trajectory comes from the anticyclonic southern flank of the PJ in its exit zone over western Europe. Air parcels typically sink below the subtropical jet, before continuing their path over the Sahara toward the Sahel.

### 11.6.2 Drawing Rules

Dry air boundaries are drawn on WASA/F maps (hatched blue lines) through zones where the horizontal gradient of $T_d$ (around −36°C), or of relative humidity (less than 15%), or of $\theta'_w$ (less than 19°C) is maximum (Figure 11.9). Dry air is generally at 500 hPa north of the ITD, but over the monsoon layer the altitude may vary from 500 to 700 hPa. Vertical cross-sections can help to define the limit and the associated level, which is on rare occasions at 400 or 800 hPa. A look at 600 hPa wind will show if this dry air is being advected to the south or southwest. Figure 11.8b also provides some examples of drawing mid-level dry air boundaries on the WASA.

In some cases, during the monsoon season particularly, tropical dry air over the Gulf of Guinea penetrates over the West African continent, south of the maximum rainy zone. Such dry air areas must also be outlined, just as for those coming from mid-latitudes. However, their degree of dryness is less extreme, reaching a minimum of about 15% relative humidity for air masses of tropical origin, to be compared with 5% for those from the extratropics.

## 11.7 The Tropical Easterly Jet

Symbol: 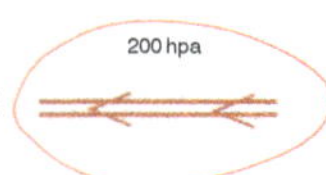


### 11.7.1 Main Characteristics

The TEJ is an easterly upper level jet found during the African monsoon south of the ITD and north of the Equator. It corresponds to the upper branch of the Walker circulation, initiated by the Indian monsoon. Its July–August–September climatology exhibits two cores above 35 kt, at 100 and 200 hPa, around 15°N and 8°N respectively. The appearance of the later TEJ core along the coast is synchronous with the installation of the little dry season over the Gulf of Guinea and the rainy season in the Sahel. Its intensification above 45 kt corresponds to the suppression of deep convection in the Gulf of Guinea.

The relation of the TEJ with the convective activity is not straightforward; however, the upper level core acceleration seems closely related to the strengthening of the Indian monsoon, propagating westward over Africa. The second core at 200 hPa results from an acceleration of the easterly wind south of big MCSs. This is due to the anticyclonic divergent circulation generated at upper levels by the sustained convective activity in the ITCZ.

Although at the present time we have no good physical understanding nor clear observational evidence of the role played by the TEJ, we recommend its systematic drawing

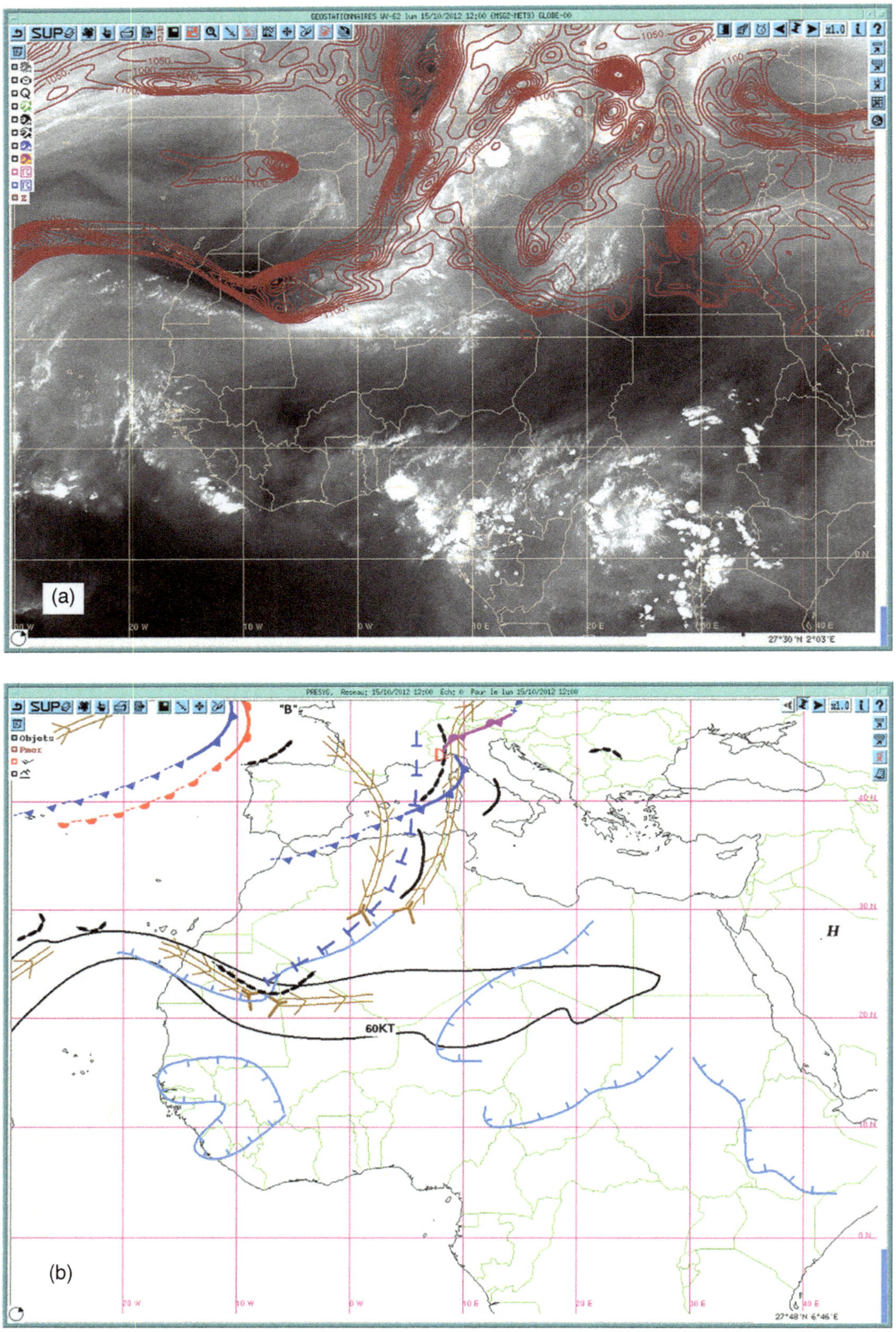


Figure 11.8 (a) Water vapour Meteosat satellite image and geopotential (red isolines: 50 dmgp below 1100 dmgp) at 0.7 PVU. (b) Corresponding WASA for the STJ, mid-latitude and dry intrusion features for 15 October 2012 at 1200 UTC. Fields are provided by ARPEGE operational analysis.

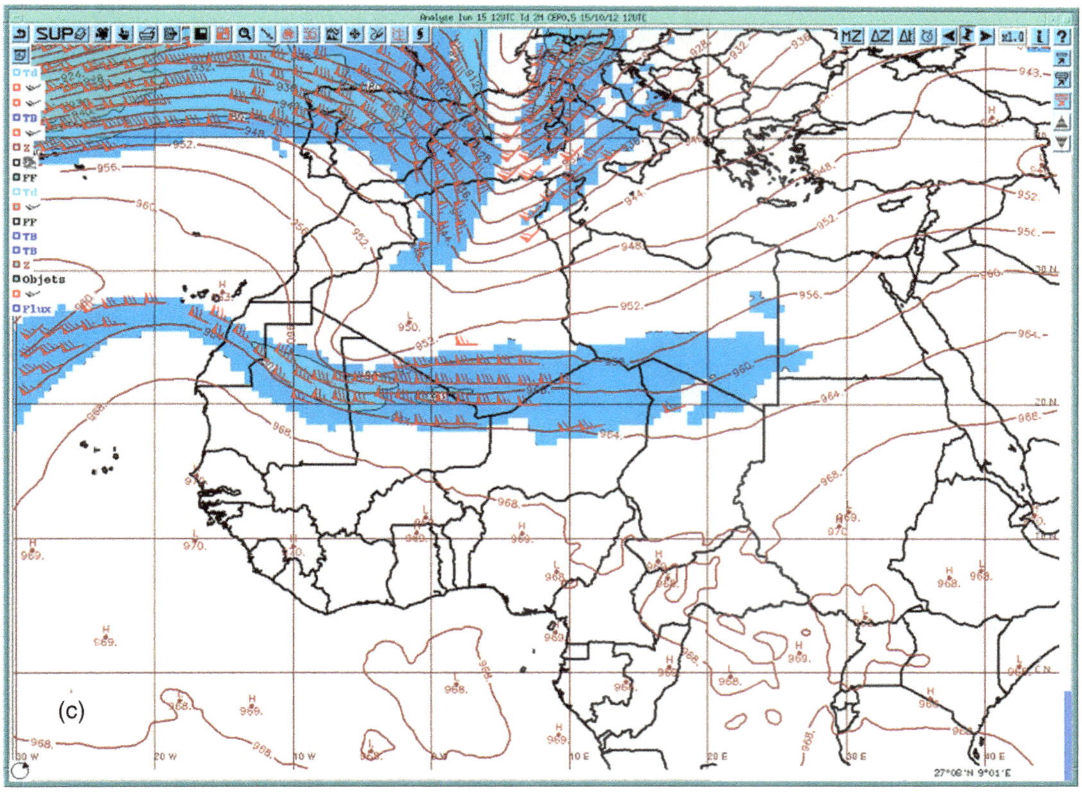


Figure 11.8 (Cont'd) (c) Geopotential height (dmgp) at 300 hPa and wind vectors at 0.7 PVU (blue shaded areas for wind stronger than 60 kt).

on WASA/F to characterise the synoptic situation and to tackle its behaviour in relation to the synoptic and convective activity. For instance, the initiation of some convective periods over East Africa (propagating later over West Africa) seems to follow the intensification of TEJ due to active periods of the Indian monsoon.

### 11.7.2 Drawing Rules

To localise TEJ cores, an analysis of the horizontal wind at different levels (200 to 100 hPa, Figure 11.10a) is recommended, or, if possible at a PV level (1 PVU). North–south, vertical cross-sections may provide a better view of the structure of the jet (Figure 11.10c and d). Cores are outlined for a threshold between 40 and 50 kt and its axis of maximum wind is drawn along the streamlines with the jet symbol. It is possible to find several TEJ cores, so the corresponding pressure level should be indicated, as in Figure 11.10b.

## 11.8 The African Easterly Jet

Symbol: 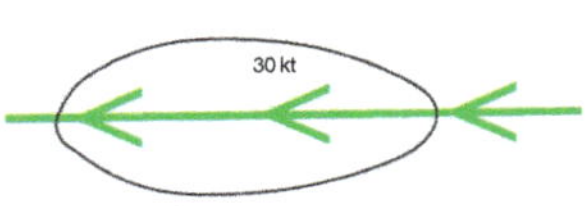


### 11.8.1 Main Characteristics

The AEJ is an easterly jet found in the 500–700 hPa layer with a climatological maximum around 600 hPa, and core speeds of about 15 m $s^{-1}$, or around 30 kt. Its summer mean position is about 16°N with a maximum core over northern Sénégal and southern Mauritania. Its latitudinal variations are large (8–20°N), limited to the north by the latitude of the ITD (typically between 20 and 25°N), and it can in some cases shift southward to the coast during monsoon breaks.

The AEJ axis corresponds to the zonal band of strong baroclinicity separating moist and colder Sudanese regions from dry and warm Saharan zones. The AEJ is a key feature to monitor, together with its oscillations (i.e. AEWs, developed in Section 11.9).

### 11.8.2 Drawing Rules

- Similarly to TEJ drawing, the AEJ cores areas are outlined with heavy black lines. The threshold usually chosen is between 20 and 30 kt (10–15 m $s^{-1}$). The choice of threshold is evaluated by the forecaster, who indicates the threshold value in the outlined area. The cores are drawn using wind vectors and intensity in the 600–700 hPa layer (Figure 11.11a and b).

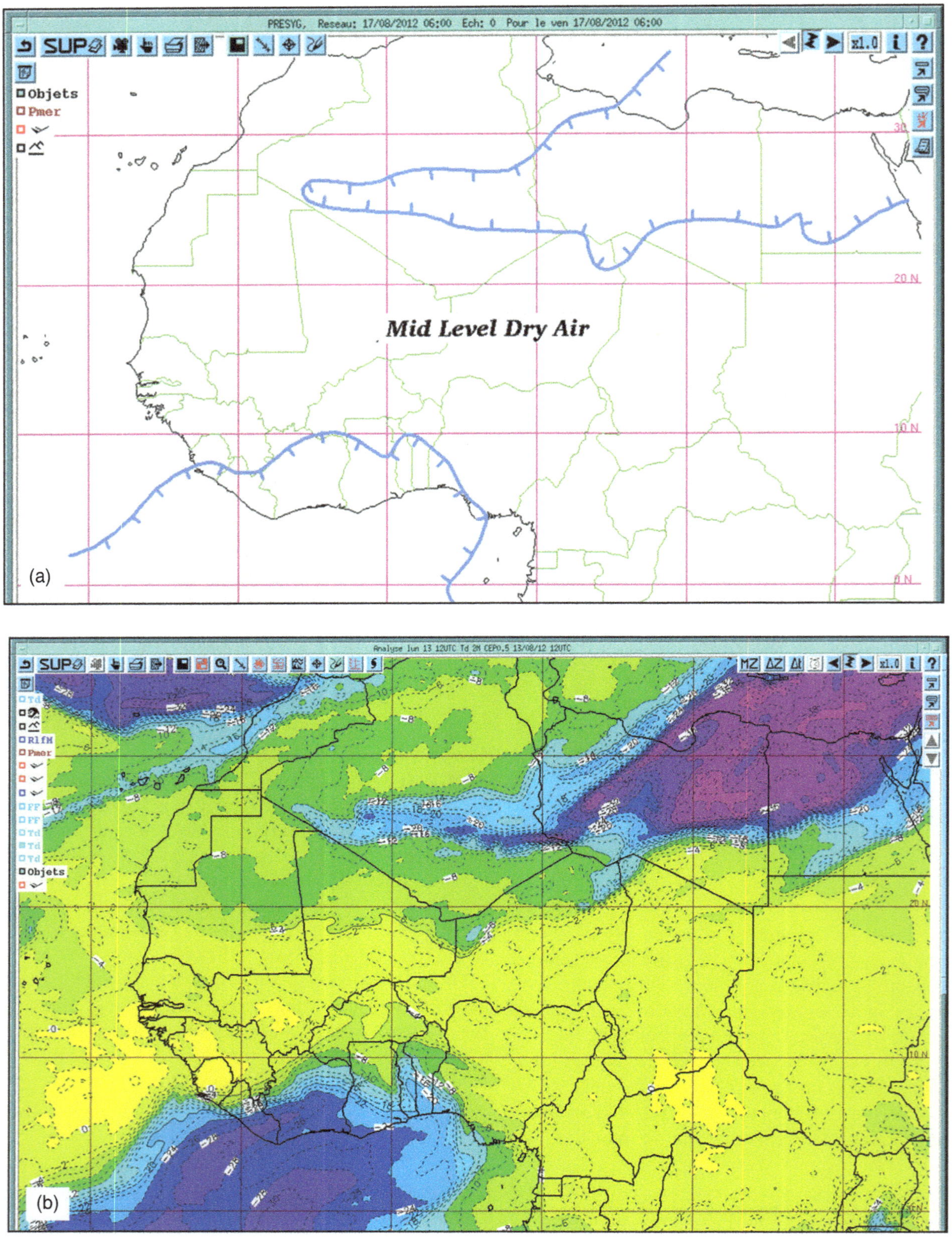


Figure 11.9 Drawing of mid-level dry air (a) using the $T_d$ map at 600 hPa (b) provided by the ARPEGE operational analysis for 13 August 2012 at 1200 UTC.

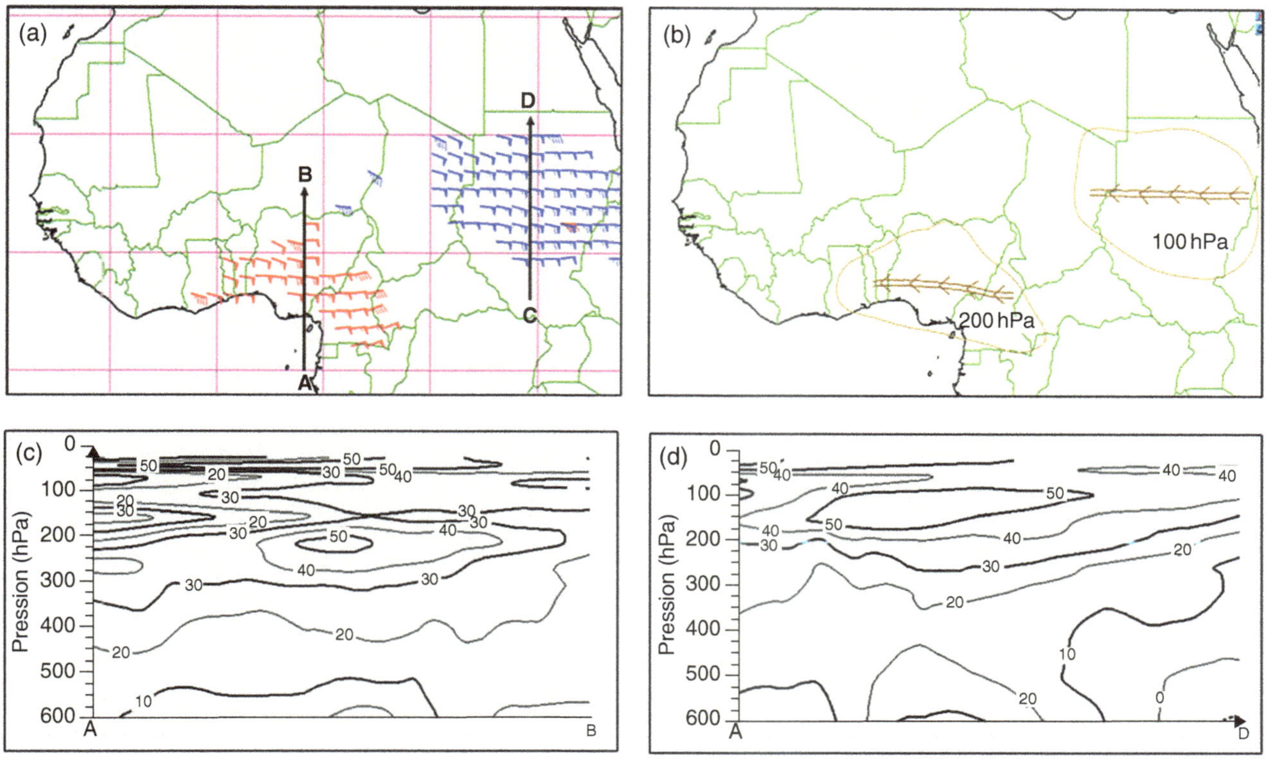


**Figure 11.10** Illustration of TEJ drawing for 13 August 2012 at 1200 UTC. (a) Wind vector (kt) at 200 hPa (threshold 45 kt) in red, and at 100 hPa (threshold 45 kt) in blue. (b) Resulting WASA for the TEJ drawing. Vertical south–north cross-sections of the zonal wind intensity along the segments (c) AB and (d) CD respectively, indicated in (a). Fields are provided by ARPEGE operational analysis.

- The threshold chosen to draw the AEJ depends on the NWP model, and for an analysis at the 700 hPa level the threshold is lower. The threshold may range from approximately 10 m $s^{-1}$ for the UK model to 13 m $s^{-1}$ for the French ARPEGE model at 600 hPa. In order to maintain the AEJ coherence in time and longitude, we must avoid thresholds that are too strong.
- The axis of maximum AEJ is drawn too, over the whole of Africa (green lines with arrows). A weaker threshold between 15 and 20 kt is proposed to plot wind vector and intensity (Figure 11.11b) used to draw this axis. It can be interrupted if too weak, and if possible it is drawn along streamlines. When a double maximum appears for a given longitude, another axis can be drawn. Such a configuration is a typical signature of an active situation ahead of an AEW trough, with possible further breaking (see Figures 2.30 and 2.31 in Section 2.2.2.2).
- North of the ITD, other mid-level jets can be found, but they are not AEJs; they are connected with the passage of mid-latitude Rossby waves penetrating over North Africa. Although these jets north of the ITD are not AEJs, they can interact with and force the AEJ, and modulate the AEW activity. So we recommend the drawing of such jets north of the ITD on the WASA/F (Figure 11.11c). In Section 11.9, on AEWs, we propose a diagnostic to monitor this forcing mechanism by mid-latitude waves.

The Guinea Coast in the south and the ITD in the north are natural limits for the AEJ.

## 11.9 African Easterly Waves and Cyclonic Vortices

Symbols: 

### 11.9.1 Main Characteristics

AEWs are synoptic perturbations occurring during the summer monsoon (May to October) over West Africa. They typically appear over the Eastern Sahel (15–30°E; 10–15°N). Their maximum amplitude in terms of meridional wind reaches about 5 m $s^{-1}$ around the level of the AEJ, or around 650 hPa. Below this level, the axis typically has a pronounced eastward tilt, while above the AEJ the axis typically has a westward tilt; however, significant variability in this tilt can occur from case to case. Strong convergence occurs ahead of the trough in the lowest 150 hPa, with divergence above this, around 600 hPa.

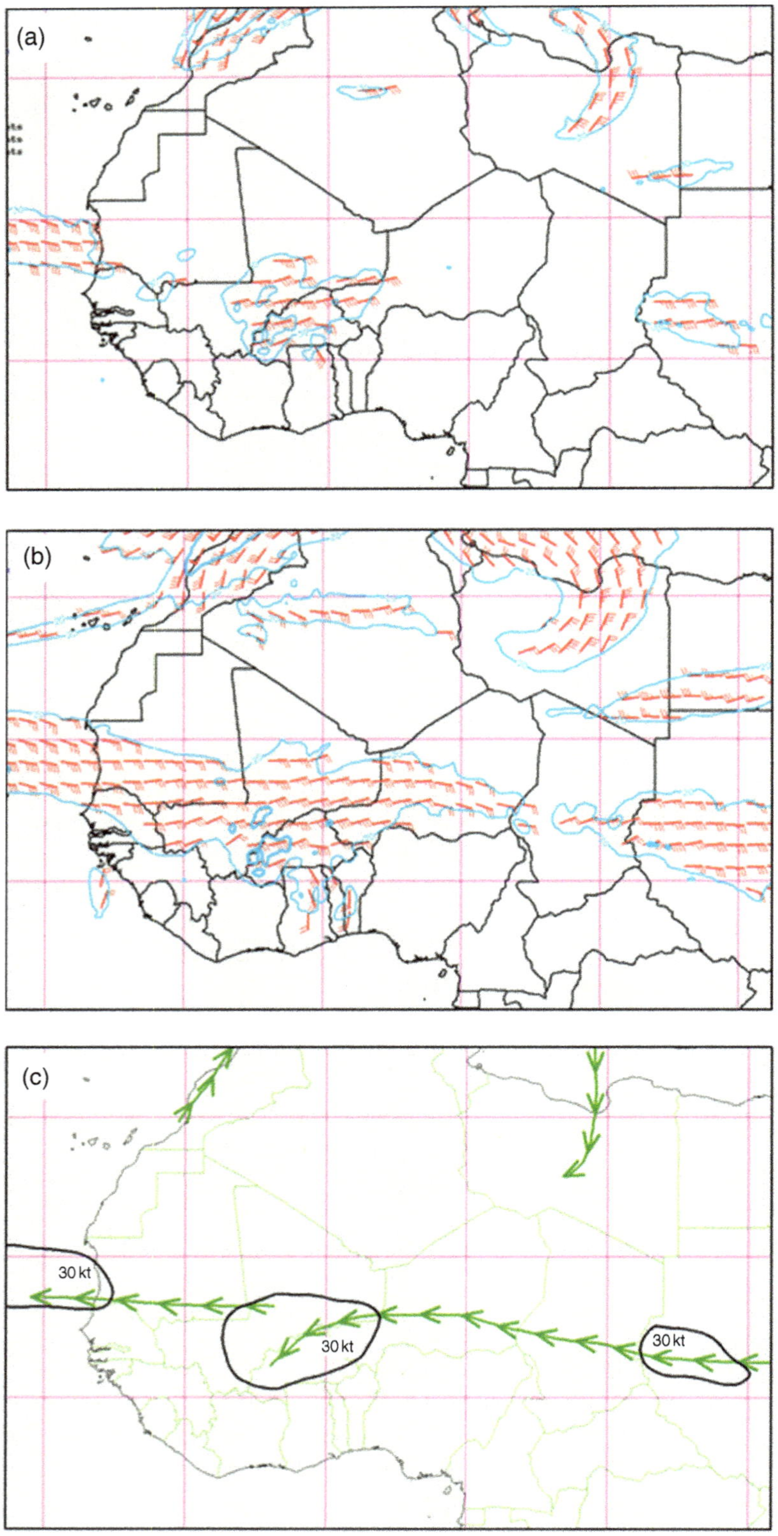


**Figure 11.11** Illustration of AEJ drawing for 15 August 2012 at 1200 UTC using the 'Synergie' forecasting system and the ARPEGE operational analysis. (a) Wind vectors at 600 hPa, for intensity thresholds of 30 kt and (b) 20 kt. (c) The resulting WASA for the AEJ axis and cores, and for mid-level wind cores north of the ITD.

AEWs have an averaged wavelength of about 2500 km and they propagate westward at a speed of about 6° to 7° per day, resulting in a typical period of oscillation of 3–5 days. They strengthen as they propagate westward and influence a great depth of the atmosphere. In fact, we observe a large variability of the AEW characteristics and structure from case to case, in part due to their strong coupling with convection and the surface.

As a consequence of the variations from case to case, and the variability and intensity of convection embedded in the waves, the drawing of AEWs is one of the most difficult exercises for the African forecaster, as it is not easy to avoid a subjective judgement. A necessary criterion to have in mind is the temporal continuity of this structure. Indeed, we can often follow a wave over several days as it crosses Africa from Chad to the Atlantic. To guide the forecaster in this difficult task, one must keep in mind conceptual models presented in detail in Chapter 2 (Section 2.1.3), enriched by one's own experience of reality in real time.

### 11.9.2 Drawing Rules

Two primary AEW features are drawn on WASA/F maps to characterise the wave: troughs and ridges (continuous and dashed heavy black double lines respectively). The trough axis is in general oriented north–south and corresponds to the line of '0' meridional wind at 600 or 700 hPa, or equivalently the northernmost limit of the curved streamlines. It is easier to find the wave on streamlines and even more clearly on the field of vorticity (threshold at $6 \times 10^{-5}\,s^{-1}$) as illustrated by Figure 11.12a and b respectively. If the vorticity signature has a longitudinal extension, it is often related to the horizontal shear on the southern flank of the AEJ core, but otherwise it may be a trough of the wave.

To avoid a subjective drawing of the trough (heavy black double line), it is recommended to analyse the vertical structure of the system in one or both of two ways:

1) To understand the nature and intensity of the AEW structure, forecasters will normally analyse the vertical structures of the AEW on horizontal maps, layer after layer. In particular, it is advised to check the consistency of monsoon inflow at the surface, cyclonic circulation at 850 hPa and AEW trough at 700 or 600 hPa. Typical structures to be analysed are shown in Section 2.1.3.3, and schematically in Figure 2.10.
2) Where possible, it is advised to inspect zonal vertical cross-sections of the meridional wind field through the AEW trough, first evidenced by wind and streamlines at mid-levels (Figure 11.12c–e). In this case, the trough axis (heavy black '0' isoline) slopes rearward (toward the east with height) as in conceptual baroclinic models. Several longitudinal vertical cross-sections are necessary to draw the full trough axis, noting in particular that convection can perturb the trough structure by destroying its slope, as illustrated by the cross-section E–F. This method also gives a good indication on the AEW intensity, providing values of northerly and southerly wind on each side of the trough axis.

Based on the precipitable water (PW) and the 'barotropic' analysis, Figure 11.13a–c proposes three complementary diagnostics available on the MISVA site, aimed at helping analyse AEWs.

1) Figure 11.13a allows detection of the moist and dry phases of the AEW, corresponding to PW anomalies (PW*) and the associated low-level wind anomalies. In the present case, we have two moist events at 10°E and 16°W and a dry event in between (2°W). In agreement with the conceptual model (Section 2.1.3.3; Figure 2.10), north of the AEJ the wet and dry events occur behind (to the east of) troughs and ridges respectively. Also, the ITD latitudinal position (see the WASA map in Figure 11.13d) reflects those events, with the ITD pushed northward around the longitudes where there is a wet phase of PW* in the AEW.
2) The 'barotropic' analysis consists of looking at the mean circulation in a layer below the AEJ level. This approach naturally filters the smallest and shortest structures, helping the detection of the major synoptic features with a better temporal continuity. It results in two complementary diagnostics illustrated by Figure 11.13b and c. Figure 11.13b represents the mean meridional wind in the 600–925 hPa layer in colour and the monsoon mean horizontal circulation at low levels (925–850 hPa layer) indicated by wind vectors. It is an efficient diagnostic to monitor deep monsoon surges (blue positive areas) or northerly bursts above a shallow monsoon layer (red negative areas). These monsoon surges or northerly bursts are almost in phase with moist and dry phases of the AEW respectively, in agreement with the conceptual model (Section 2.1.3.3; Figure 2.10). This diagnostic also allows the monitoring of mid-latitude Rossby waves travelling eastward and affecting African weather, and monsoon surges coming from the Gulf of Guinea and travelling westward.
3) Figure 11.13c represents streamlines (blue) of the mean flow in the 950–600 hPa layer with its intensity (isolines, above $7.5\,m\,s^{-1}$) and vorticity (colour). Combined with the meridional wind intensity (colour on Figure 11.13b), it helps the drawing of the axis of mean troughs and ridges (red solid and dashed lines respectively). The trough at 600 hPa drawn on the WASA (Figure 11.13d) is found a few degrees to the west of the highest PW*, in agreement with the AEW composite.

The combination of these three diagnostics is also powerful in the early detection of the initiation of AEWs to the east. This is the case over Sudan, as revealed by the closed mean circulation in Figure 11.13c. It is confirmed by a dipole of southerly and northerly winds (Figure 11.13b) coupled with a PW* (Figure 11.13a), suggesting a growing stage. Also, a strong zonal dipole in the Sahel band of dry/moist anomalies associated with a dipole of northerly/southerly low-level wind anomalies (as between north Niger and north Mali – Figure 11.13b) is an indicator of intense growing of the

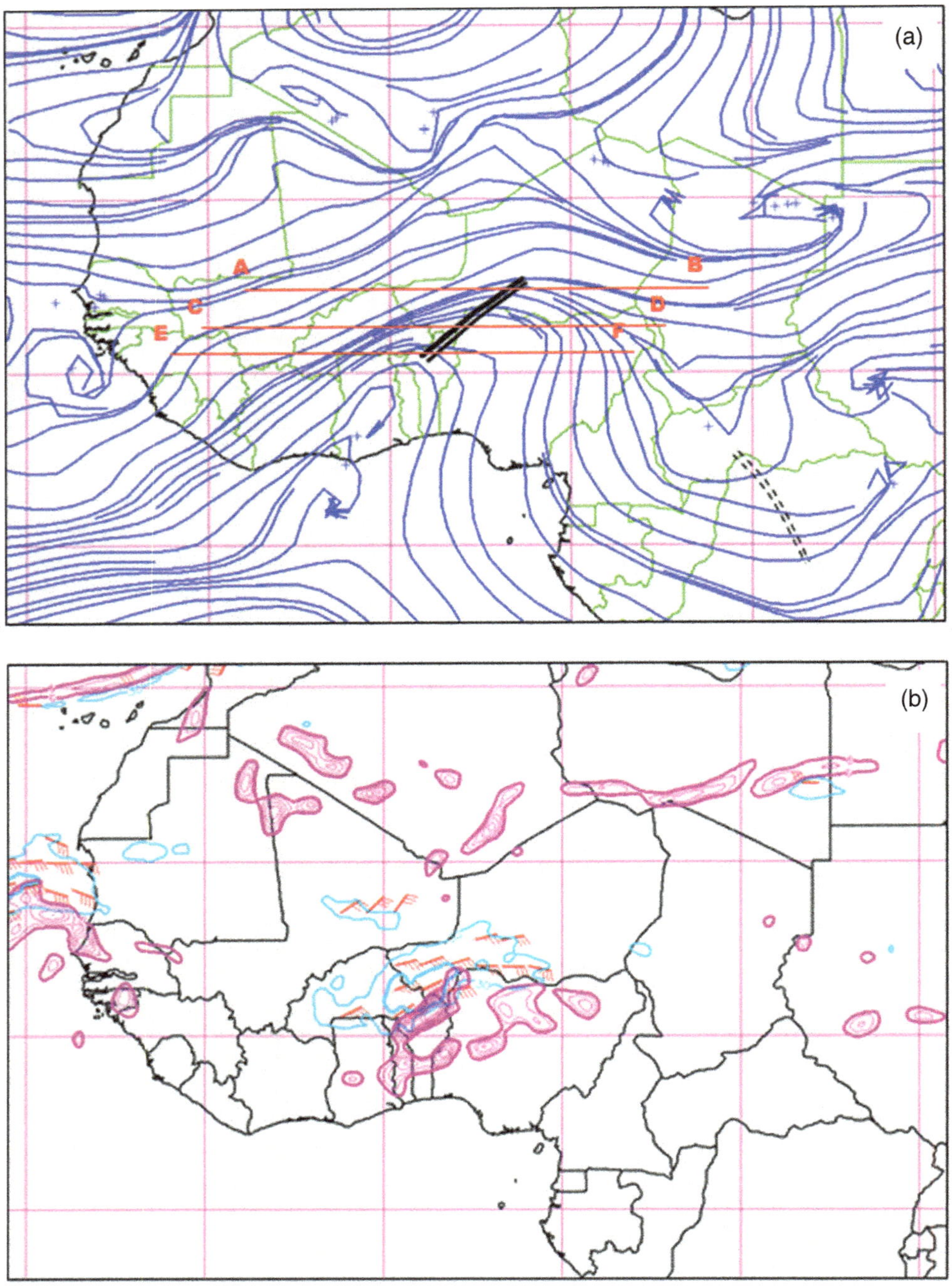


Figure 11.12 Illustration of AEW trough drawing (heavy black double line) for 15 August 2012 at 0000 UTC using the 'Synergie' forecasting system and the ARPEGE operational analysis. (a) Streamlines (blue) and (b) vorticity field (pink isolines above $6 \times 10^{-5}$ $s^{-1}$) at 600 hPa with superposition of wind vectors (red) above 30 kt to outline the AEJ core (light blue intensity isolines) at 600 hPa.

AEW (Section 2.1.3). It should also be recognised that the use of PW* accentuates the northern structure of AEWs, as the anomalies are strongest where the mean gradients in PW are most pronounced.

Ridge drawing (dashed heavy double black line) is less precise than trough drawing, as it is a wider area of anticyclonic circulation. The same method, using layers and/or vertical cross-sections, as for troughs, can be used. The combination of mean streamlines (Figure 11.13c) and of the mean meridional wind in the 600–925 hPa layer (Figure 11.13b) can be used efficiently. For instance, the dashed red lines in Figure 11.13 allow the identification of an intense ridge and mean anticyclonic circulation over the Atlantic and southern Sénégal corresponding to an area of suppressed convection (see Figure 2.29d).

In addition, when intense, the vorticity field at the AEJ level (Figure 11.13c) allows detection of active

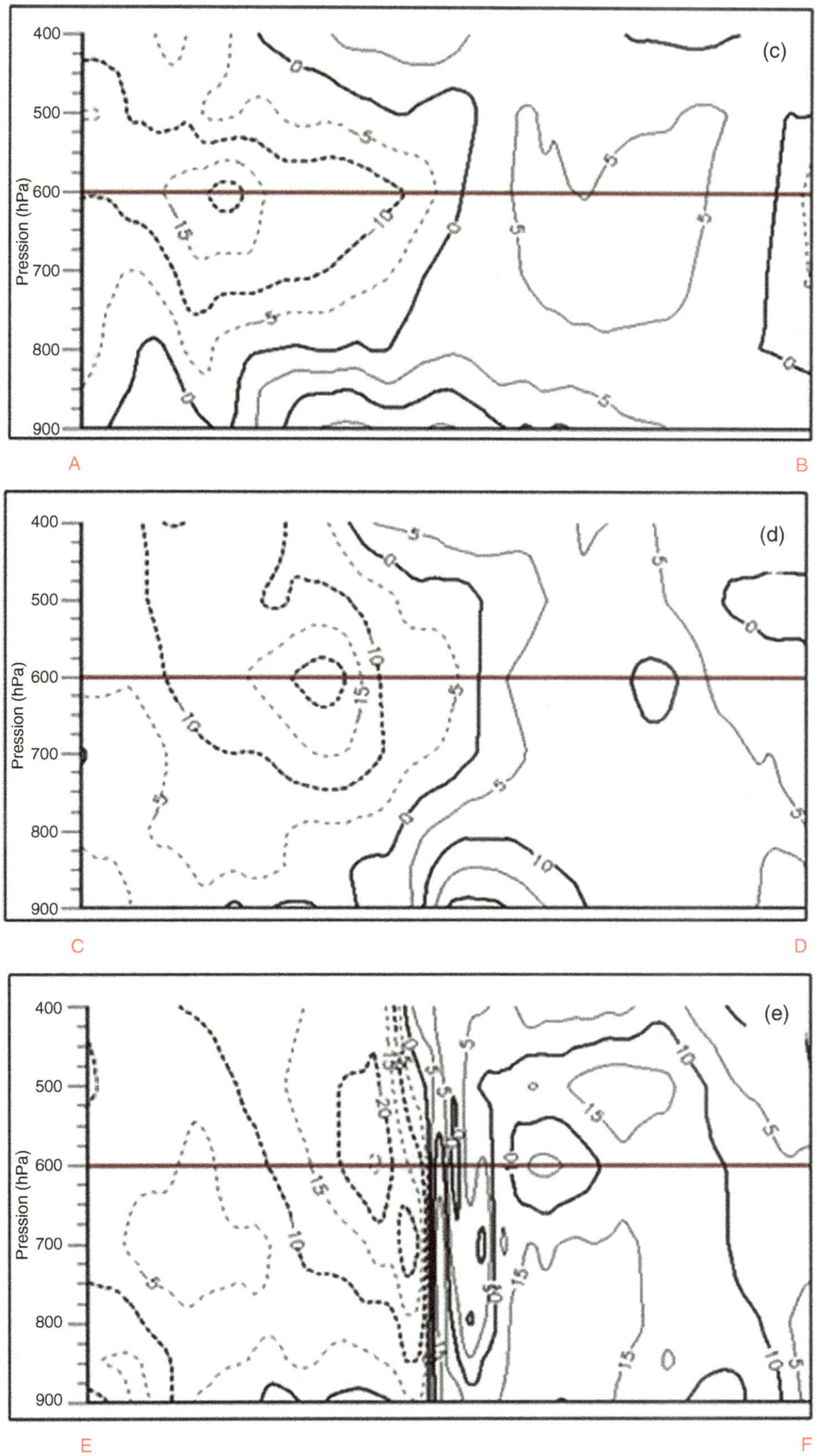


**Figure 11.12 (Cont'd)** zonal vertical cross-sections of meridional wind (kt) at three latitudes: (c) AB, (d) CD and (e) EF transects, indicated in red in (a). The solid dashed double lines in (a) correspond to the ridge.

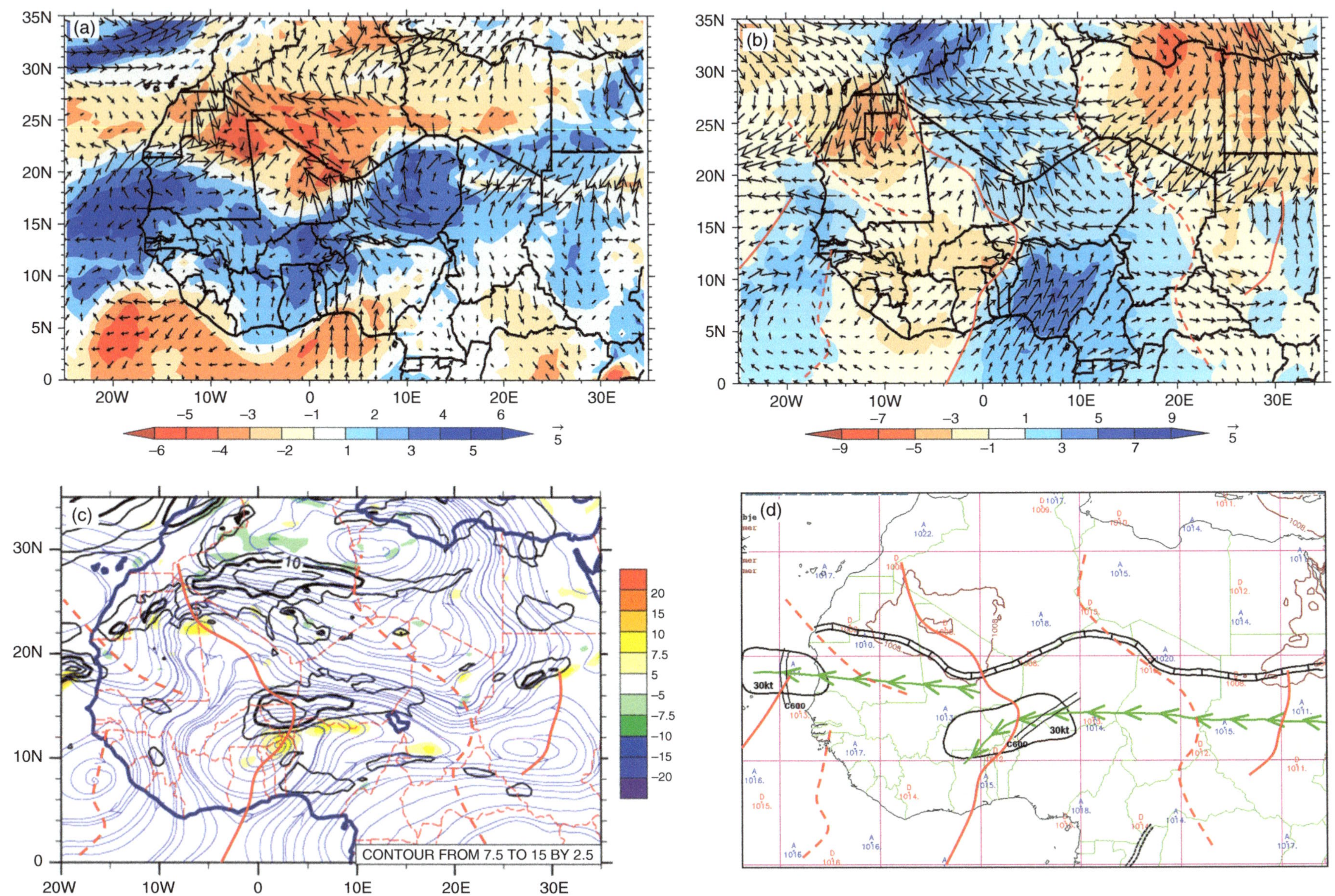


**Figure 11.13** Complementary diagnostics relating to the PW perspective on AEW structure and to the 'barotropic analysis', available on the MISVA site for 15 August 2012 at 0000 UTC: (a) PW* (mm – colour) and 925 hPa horizontal wind anomalies (m $s^{-1}$); (b) mean meridional wind in the 925–600 hPa layer (m $s^{-1}$ – colour) and mean horizontal wind vector in the 925–850 hPa layer; (c) streamlines (blue) for the mean flow in the 950–600 hPa layer, with mean horizontal wind intensity (black isolines above 7.5 m $s^{-1}$) and vorticity ($10^{-5}$ $s^{-1}$ – colour); (d) corresponding WASA for the AEJ and AEWs. The locations of the trough and ridge mean axis, as detected by the mean circulation in the 950–600 hPa layer (c), are superposed to (b), (c) and (d) with solid and dashed red lines respectively. Fields are provided by the ECMWF operational analysis.

cyclonic centres that are indicated on WASA/F maps with a $\mathbf{C}_{600}$ symbol and a subscript (in hectopascals) for the level.

Chapter 2 provides some illustrations of the use of these diagnostics (see Section 2.2.2).

## 11.10 The African Monsoon Layer and the Monsoon Trough

Symbols: 

### 11.10.1 Main Characteristics

Drawing the ITD, marking the northward penetration of the monsoon, is not enough to characterise the monsoon flux, which is the major moisture conveyor feeding the convective activity. We also need a description of its circulation, depth and energy directly related to the amount of water available in this layer. Thus, variables such as $\theta'_w$ or relative humidity associated with the horizontal wind are crucial for the forecasters to analyse this layer in detail. A major feature within the monsoon layer is the MT. Contrary to its oceanic counterpart, called the ITCZ, the MT is a convergence line of southwesterlies and northeasterlies, which is not vertical, because the surface convergence at the ITD is shifted about 5° to the north. It results in a line of vorticity around the 850 hPa level located in general on the southern flank of the AEJ. As the mean horizontal wind is weak at 850 hPa, streamlines typically exhibit closed circulations wrapping around the cyclonic centres located along the MT.

The MT is a key feature to monitor, as it is a favourable area for convective activity (triggering and organising), even more so when the associated MT vorticity centres are in phase with those located just above and associated with the AEW passage.

### 11.10.2 Drawing Rules

To characterise the monsoon, two features are drawn on WASA/F maps: the MT (red convergence line symbol shown in Table 11.1) and the core of the monsoon bursts and northerly bursts with double red arrows (Figure 11.14b).

- The MT is diagnosed at the 850 hPa level, as a line of maximum vorticity (Figure 11.14a), lying to the south of the ITD, and usually just south of the AEJ core. This line will connect closed cyclonic vortices at this level. For the most intense vorticies (threshold $6 \times 10^{-5}\ \mathrm{s}^{-1}$), the centre of the circulation will be marked by a **C**. The vorticity map at 850 hPa also exhibits cyclonic structures near or north of the ITD (Figure 11.14a), related to the vertical extension of the thermal depression (e.g. panel 4 of Figure 2.10). Satellite infrared (IR) images can help in some cases to localise the MT, in particular in the afternoon, at the onset or growing stages of convection, as illustrated by Figure 11.15; in this case the monsoon burst clearly pushes the MT toward the ITD between 0 and 10°E.
- Over the ocean the ITCZ symbol (Table 11.1) is used instead of the MT symbol, to localise the convective area coinciding with the low-level convergence and a band of vorticity at about 850 hPa.
- Monsoon bursts are also drawn on the WASA/F maps with the help of the mean meridional wind in the 600–925 hPa layer (Figure 11.13b) following the axis of maximum southerly wind (blue area). Inversely, northerly bursts are also drawn with the help of the same diagnostics and corresponding to red areas.

To help the forecaster in the analysis of the monsoon layer, two complementary diagnostics are proposed (Figure 11.16) based on the PW field. The PW field (Figure 11.16a) accounts for the water available (i.e. latent energy) in the monsoon layer and above it, but it is reduced over mountainous areas. To overcome this limitation, we propose a diagnostic of an equivalent monsoon depth (MD):

$$\mathrm{MD} = \frac{z_s q(T_{850}) + \mathrm{PW}}{q_s(T_{850})} = z_s \mathrm{RH}_{850} + \frac{\mathrm{PW}}{q_s(T_{850})}$$

where $z_s$ is the surface altitude and $q_s(T_{850})$ is the saturated mixing ratio of water vapour at 850 hPa, so that $\mathrm{RH}_{850}$ is the relative humidity at that level. This level is chosen to avoid a diurnal cycle of the MD, which is measured relative to the sea level. The advantage of this empirical formulation is that over warm areas, such as within the HL, the equivalent MD decreases drastically. To avoid an artificial decrease of MD over mountains due to the decrease of PW, PW is corrected by adding a virtual part in the layer between the MSL and the surface (inside the mountain) by assuming an extrapolated mean value of $q(T_{850})$ down to the MSL (the first term in the MD equation), to compute its contribution to the MD.

As defined here, the MD is a thermodynamic measure of the depth of humid air in the monsoon, and represents a smoother field (Figure 11.16b) than dynamic measures of the monsoon depth, based on the wind fields.[2] Because MD is relatively smooth, it may help to identify monsoon penetrations and retreats, and the MT and ITD location. The addition of the wind shear vector in the 600–950 hPa layer also provides useful information to forecast convective organisation (see Section 11.12).

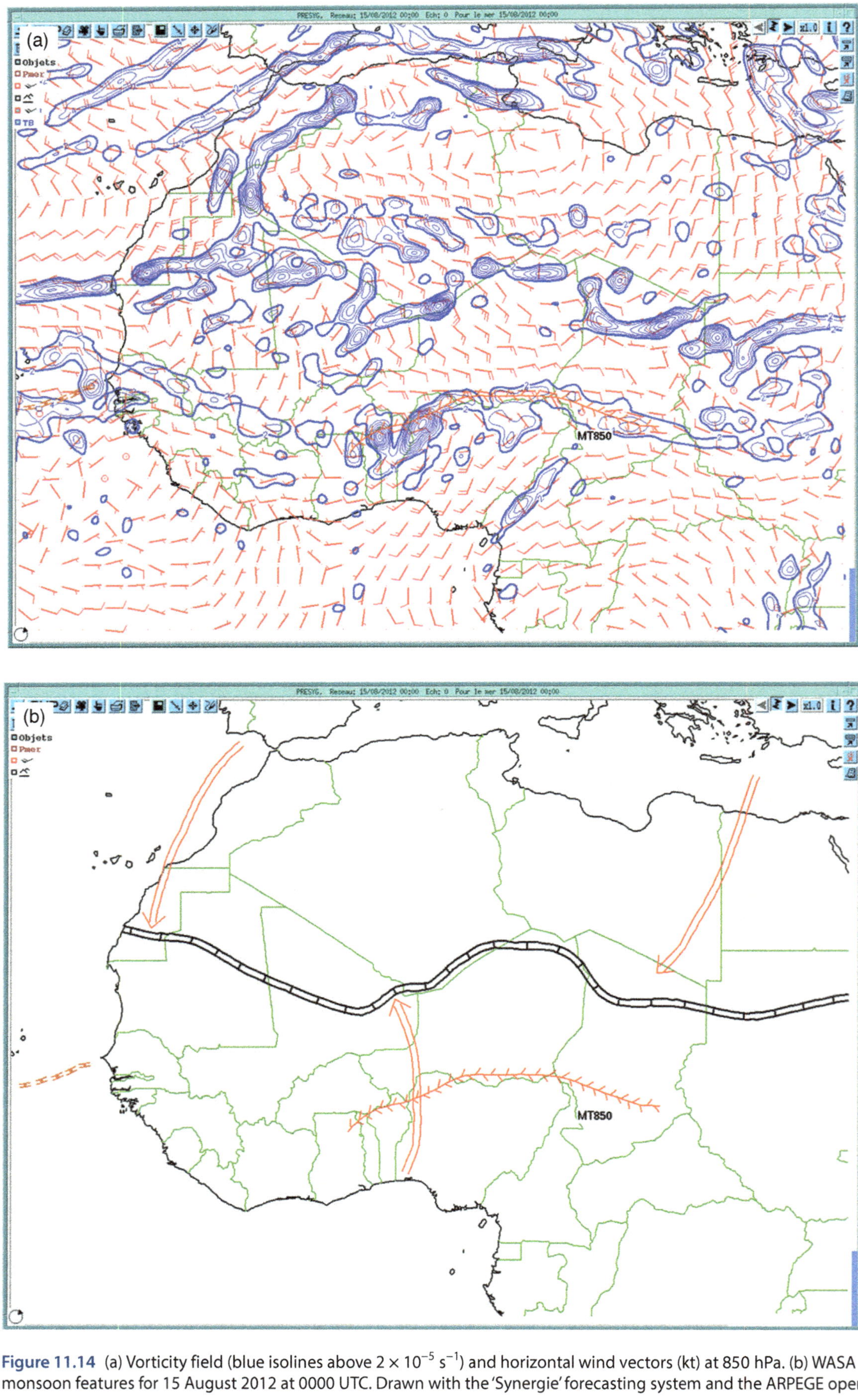


Figure 11.14 (a) Vorticity field (blue isolines above $2 \times 10^{-5}$ $s^{-1}$) and horizontal wind vectors (kt) at 850 hPa. (b) WASA map concerning monsoon features for 15 August 2012 at 0000 UTC. Drawn with the 'Synergie' forecasting system and the ARPEGE operational analysis.

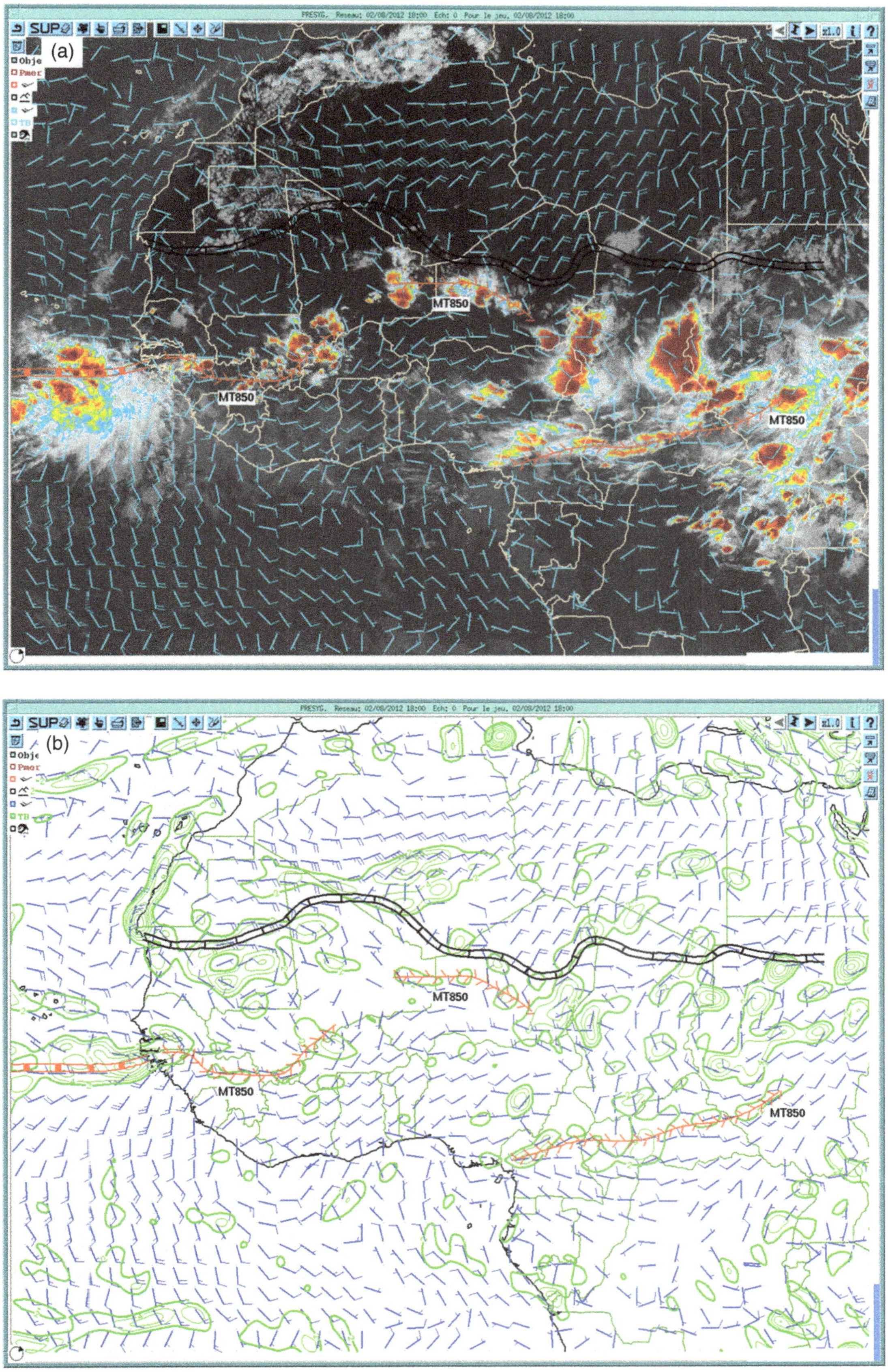


Figure 11.15 Illustration of MT drawing: (a) IR Meteosat image from EUMETSAT on 2 August 2012 at 1800 UTC, with superposition of the wind vector (kt – blue) at 850 hPa and the drawing of the ITD and MT. (b) As (a) with the IR image replaced by the relative vorticity (green $>6 \times 10^{-5}\ s^{-1}$) at 850 hPa. In some cases troughs are better identified with streamlines. Fields are provided by the ARPEGE operational analysis.

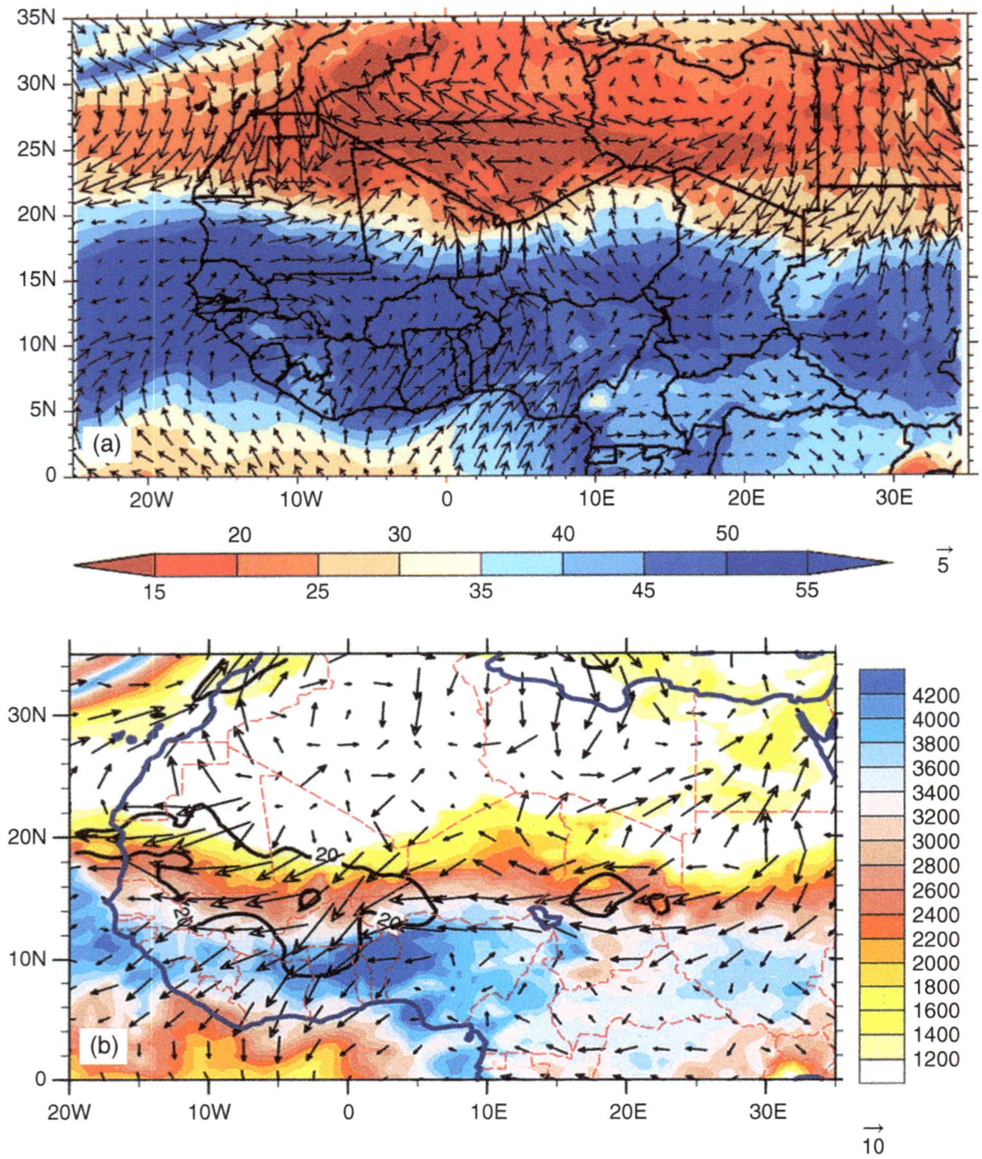


Figure 11.16 (a) PW field superposed on the 925 hPa wind vector (m s$^{-1}$) and (b) monsoon equivalent depth MD (metres) with the superposition of the wind shear vector (m s$^{-1}$) in the 600–950 hPa layer for the 15 August 2012 at 00UTC. The heavy isoline in (b) outlines areas with shear above 20 m s$^{-1}$. Data are provided by the ARPEGE operational analysis.

Comparing the WASA drawing (Figure 11.14b) with the aforementioned diagnostics (Figure 11.13a,b; Figure 11.16) shows strong consistency.

- The MT coincides with areas where the MD is at its deepest. This is also true for the ITCZ over the Atlantic.
- The northward penetration of the ITD over northeastern Niger is clearly seen in the MD, and this wet event is preceded to the west by a monsoon burst.
- The reinforcement and rotation to the north of the shear on the western side of the wet event corresponds to a reinforcement of the baroclinicity, and of the HL. It is also a signature of the intensification of the AEJ.

## 11.11 Dust or Sand

Symbols: 

Dust monitoring and forecasting is extensively treated in Section 5.2. The corresponding dust areas and appropriate symbols must figure on the WASA/F maps.

## 11.12 Convection

Symbols: 

Analysing and forecasting the areas of convection, including the type and intensity, is one of the main objectives of the forecaster during the WAM period.

### 11.12.1 Analysis of the Convective Activity

Concerning the analysis (WASA), the forecaster must outline the most intense convection area as estimated from the IR imagery (Figure 11.17a) and use MCS tracking (see Chapter 6) to have an estimation of the speed of the MCS (use the last three or four images at a 1 h frequency to estimate the speed). The drawing must be precise and adjusted with the latest IR image. The MCS will appear as a zone enclosed with the convective symbol inside (e.g. in the region of Burkina and eastern Mali, in Figure 11.1), while for zones with isolated convective cells only the symbol has to be drawn (e.g. over Sénégal, Guinea and western Mali in Figure 11.1).

Following this satellite analysis, the forecaster has to look at the convection in models and verify its conformity with the location of convection derived from the satellite. This is not such an easy task as there is no simple quantification of the convective activity, but the forecaster can use the following criteria to detect the strongest and long-lived MCS in the model:

- The accumulated rain over a short period (3 or 6 h) shows a maximum. The rain may have low values for an MCS close to the ITD because evaporation of rain is so important in the north of the monsoon.
- There is ascending vertical velocity in the mid-troposphere (600–700 hPa) and a secondary maximum around 400–300 hPa close to this maximum. The diabatic heating coming from convection produces this vertical velocity.
- There is relative humidity (>80%) at 600–700 hPa.
- For some of the most clear cases, there is divergence at 200 hPa and convergence at 700 hPa nearby.
- This exercise is easier at 6 h as we have only a few MCSs at the end of the night. At 18 h the diurnal convection is at its maximum and we cannot make a clear separation of isolated convection from the MCSs.

### 11.12.2 Forecast Rules and Required Diagnostics for Convection

For the forecast (WASF) of convection, the forecaster must combine different information. It is the most difficult task for forecasters, as models have weak skill in predicting convection over tropical regions. Indeed, present parameterisations of convection do not treat correctly all the different convective processes and stages (triggering, inhibition, propagation, evaporation) that are key features over West Africa. But the model skill is better for some other larger scale structures, as listed before (notably the ITD, HL, jets, dry air intrusions and AEWs, and especially when there are strong mid-latitude influences). For each case, where one of these structures has an influence on convection, the predictability of convection grows and can be the same as the predictability of the structure.

Forecasting of convection can be separated into two tasks:

1) Prediction of locations at which convection will initiate (usually in the afternoon over land), and regions which will be free of convection (Section 3.2.3).
2) Prediction of the form of convection that will result, once it has started; for instance, whether the convection will be isolated and short-lived, or organised, and the likely movement of any organised convection (Section 3.2.4).

Once storms have formed, then their subsequent behaviour should be the subject of nowcasting (see Chapter 6).

In each of these tasks, there are several bodies of information and evidence that a forecaster can make use of:

1) Synoptic information and observations of storm environment (e.g. profile stability measures – see Section 3.2.3.1.2 (**1**)). These data include observations such as the radiosonde profile, pilot balloon profiles, and evidence from NWP models (to be used with caution).
2) History of recent storms, which influence the atmospheric state and the land surface and may provide triggers for new storms.

Table 11.2 summarises the forecast rules and required diagnostics for convection, to guide the forecaster's task. The forecaster's experience will be crucial to succeed in this difficult task, on the basis of the available information from observations and NWP products, which need to be provided on time and with the adequate diagnostics. In order to train new forecasters and to better handle methods and diagnostics proposed in this book, a set of case studies has been prepared and made available at www.cnrm.meteo.fr/waf_handbook_casestudies/ as an e-learning component of the handbook project. The vocation of this e-learning component will be updated regularly, to include new case studies and improvements in terms of new diagnostics, tools and interpretation.

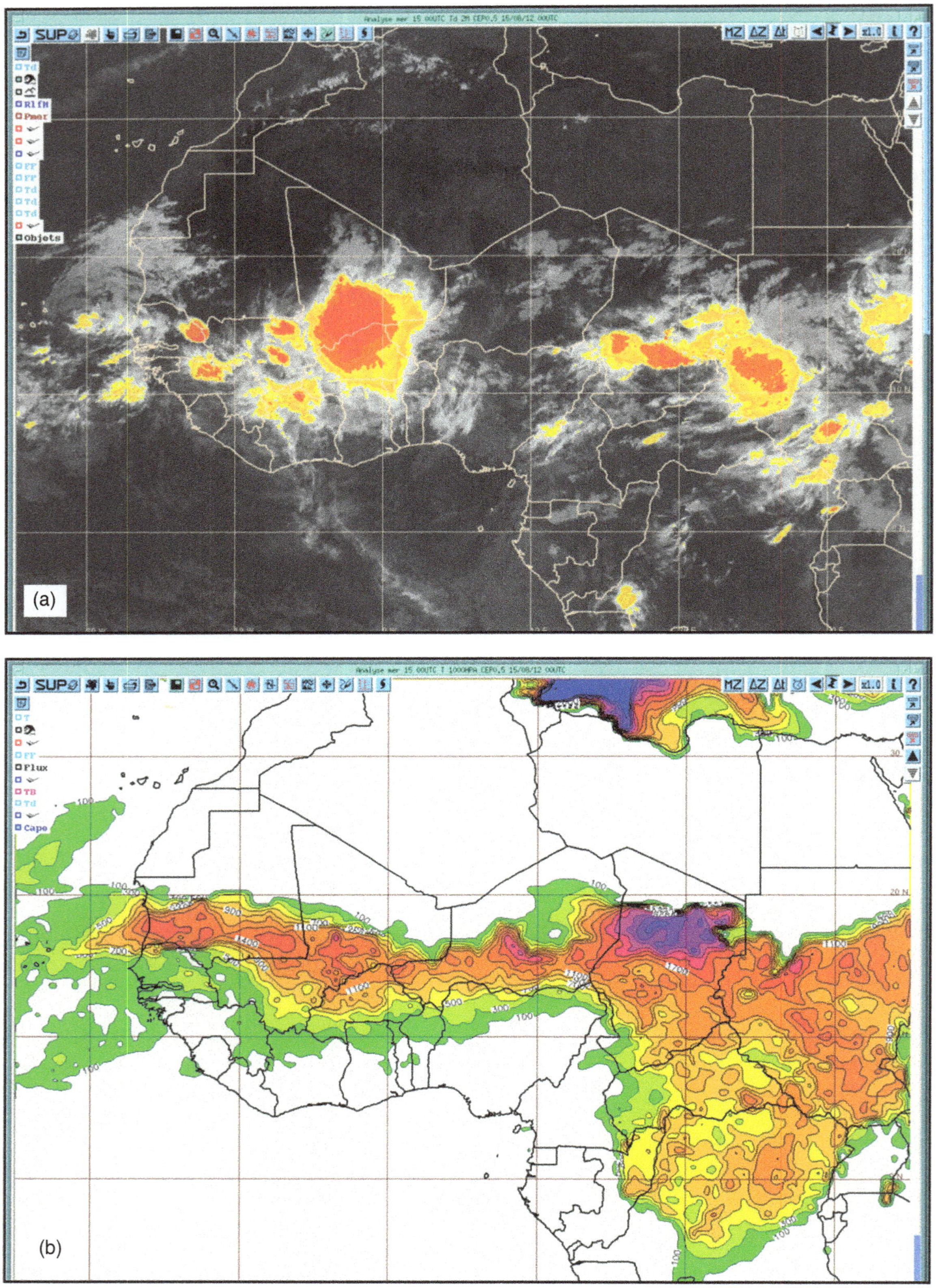


**Figure 11.17** (a) IR Meteosat image (coloured areas for $T_b$ less than −65°C) and (b) CAPE field (J kg$^{-1}$) for the 15 August 2012 at 0000 UTC from the ARPEGE operational analysis.

Table 11.2 Forecast fields and structures.

| Parameter/ phenomenon | Application in forecasting | Other uses, and references | Further comments |
|---|---|---|---|
| **0. Past activity for last 3 days** | | | |
| Persistence | Detect consistent, slow-moving or stationary synoptic situations for which convective patterns may be repeated from day to day. | | |
| Tracking of coherent convective structures | Useful to extrapolate a possible scenario based on the propagation of the coherent structures and on the evolution of their environment.<br>Note that a tracked feature may become inactive for as much as 24 h, before reactivating along the same extrapolated path. | Useful also to evaluate the NWP analysis/forecast.<br>Section 2.1.3.6<br>Section 6.1.2.5 | Important to focus on upstream regions such as eastern Sahel (Chad, Sudan, Ethiopia), Gulf of Guinea, etc. |
| Last day's convective activity | 'Wake effects' resulting from convective storms may lead to a greater chance of convective initiation the next day. Such wake features include vortices (850 hPa), moist patches at the surface, cold pools resulting in northward monsoon pushes, convergence lines, patches of cloud cover impacting incoming solar energy, and various other processes.<br>In particular, gravity waves may propagate long horizontal distances from a storm, and initiate new systems hundreds of kilometres away, several hours after the parent system. | Section 3.1.4.1<br>Section 4.1.5 | |
| **1. Large-scale forcing** | | | |
| Large-scale context | Ongoing activity of convection in the region may be modulated by subseasonal variability: i.e. HL, Guinean, Sahelian modes of variability, mid-latitudes and equatorial waves (Rossby, Kelvin, Madden–Julian oscillation, etc.). Also sea surface temperature anomaly patterns must be monitored. | See Chapters 2 and 7 for synoptic and subseasonal variability respectively. | |
| 1. ITD | No convection is expected north of the ITD, except if strongly forced from mid-latitudes or by orography.<br>Possible short-lived systems close to the ITD. | See Section 11.2 | There are strong meridional ITD oscillations due to the diurnal cycle and to the passage of AEWs (3–5 days period). |
| 2. HL | The induced, low-level convergence may force convection if enough humidity has been injected into the HL region. But the key feature will be the strength of the subsidence capping the HL corresponding to secondary circulation of the Hadley cell that prevents deep convective development. | Look at the vertical thermodynamic profiles.<br>See Section 11.3 | Not explicitly used in the convection forecast but key feature of AEWs. |
| 3. STJ | May force convection (right entrance, left exit of jet) or inhibit it (left entrance, right exit of jet). | See Section 11.4 | Seasonality: stronger in spring and autumn |
| 4. Midlatitude forcing | UTTs: ascent ahead (to the east) of an UTT may force convection. Upper-level dry subsidence occurs behind the UTT (to the west).<br>Crucial for dry-season thunderstorms. Enhance convection during wet season. | UTT favours southerly moist advection on its eastern side.<br>See Section 11.5 | Exhibits significant seasonality and depends on the PJ activity. |
| 5. Dry intrusions | Weaken updraughts and therefore suppress onset of convection.<br>Reinforce downdraughts and therefore increase organisation of convection. | More convective activity may be found on the leading edge of a dry layer (diagnosed ~500 hPa in practice).<br>See Section 11.6 | |

*(Continued)*

Table 11.2 (Continued)

| Parameter/ phenomenon | Application in forecasting | Other uses, and references | Further comments |
|---|---|---|---|
| 6. TEJ | The role of the TEJ is uncertain, but it has to be there for the rainy season to start. | As quite close to the Equator (weak Coriolis parameter) the STJ rules cannot be applied (e.g. we cannot assume ascent at the right entrance and left exit).<br>See Section 11.7 | There is a lack of confidence in the relationship between TEJ and convection; the extent to which the TEJ is the cause or consequence of convection is not well understood (TEJ reinforcement occurs south of the convection). |
| 7. AEJ | Fast-moving MCSs occur within the AEJ core.<br>Slower MCSs occur to the south of the AEJ core.<br>Determines the storm type (storm organisation and movement) as a result of the shear (700 hPa to surface). | The AEJ marks the transition between moist tropical profiles to the south, and Saharan-influenced profiles, with greater CIN and drier mid-levels to the north.<br>See Section 11.8 | Part of the AEJ's horizontal momentum is delivered to the storms, especially their downdraughts. Conversely, the AEJ is accelerated by convection, so the AEJ is partly a convection signature. |
| 8. AEW | Strongly modulates convection:<br>Long-lived MCSs form in or just ahead of trough close to AEJ axis and move to the southwest.<br>To the south of the AEJ on the AEW trough, slower moving systems are formed.<br>Behind the trough, less-organised convection may form in high PW air. | | |
| | Suppression of convection occurs typically in and ahead of the ridge, especially where there is northerly flow above the monsoon layer. Exceptions can occur (e.g. Figure 3.20) of the formation of a line of storms along the MT within the ridge of the wave. | See Sections 11.9 and 2.1.3.6 | |
| 9. MT | Key for triggering storms when associated with moisture. | South of the ITD | |
| | Serves as a conveyor belt for travelling vortices, which may initiate convection at a later time. | See Section 11.10 | |
| **2. Local forcing** | | | |
| Orography | Dominates the geographic and climatological distribution of noon-to-afternoon triggering. | See Section 3.1.4.1.1 | Must be kept in mind always. |
| Diurnal cycle | Peak of triggering at 1500 local solar time (LST).<br>Minimum convective activity in the morning. | See Sections 3.1.4.2 | 1500 LST is a key time in the convective-forecasting day, when much of the triggering occurs, and forecasting makes a transition to nowcasting. |
| Sea and land breezes | Strongly determine the diurnal cycle of convection along coastal regions, and modulate its intensity. Effects are very important within around 50 km of the coast, both onshore and offshore. Showers along the coast in late morning typically move inland in the afternoon. | See Section 4.1.4.2 | |
| Soil moisture | Over the Sahel, soil moisture patches from a previous storm provide favourable conditions for triggering for a 2-day period after a rain event. | See Sections 3.1.4.1.3 and 4.1.5.2 | |

Table 11.2 (Continued)

| Parameter/ phenomenon | Application in forecasting | Other uses, and references | Further comments |
|---|---|---|---|
| Surface properties (rocks, vegetation, etc.) | Hot spots favourable to triggering may be detected in climatologies. | See Sections 3.1.4.1.3, 4.1.4 and 4.1.5 | Knowledge of the local terrain and climatology is essential. |
| Convergence lines | Convergence lines provide lines of afternoon convective triggering. | Difficult to detect owing to the lack of observation, but in some situations (e.g. topographically forced) NWP products may provide good indications.<br>See Sections 3.1.4.1.4, 4.1.4 and 6.1.2.6 | Some convergence lines may be climatologically repeated; for instance, convergence lines downstream of coastal headlands or downstream of hills. |
| **3. Vertical stability** | | | |
| Sounding analysis: all parameters | Before convection: look at instabilities not only from the surface, as convection can start from mid-levels. | Provides information on cloud layers to complete observations. | Important step with observed soundings. |
| CAPE: not very informative alone! | Necessary, favourable, but not sufficient.<br>Almost always positive in the monsoon season.<br>Note that convection decreases CAPE.<br>Useful during onset and cessation of the rainy season as an index for convective potential. | Strong diurnal cycle → max. in the afternoon explaining the max. of triggering.<br>For instance, in Figure 11.17 maximum CAPE is not collocated with the maximum convection.<br>See Section 3.1.2.1 | *Caution 1:* there are different ways to compute CAPE, which differ between the NWP products.<br>*Caution 2:* CAPE in the NWP model data will be sensitive to the convection occurring in that model. |
| CIN | Tends to suppress the development and maintenance of convection. | Delays the convection onset → allows the storage of more energy in the subcloud layer, favouring a build-up of CAPE and PW, leading to more intense storms later in the day, or the following days.<br>See Section 3.1.2.1 | CIN is in general large over Africa especially over the Sahel (~100 J/kg) due to the prevailing mean subsidence associated with dry air conditions. |
| Low level $\theta_w'$ up to 850 hPa (summer) | Similar characteristics to CAPE. | See Section 3.1.2.1 | Commonly used by forecasters. |
| PW | A PW maximum is favourable and necessary (~36 mm), but not sufficient to get convective precipitation.<br>However, intense convection can occur with weaker PW, exhibiting elevated cloud bases (north of the ITD) but local and weak precipitation at ground.<br>Probability of accumulated precipitation increases with PW. | Over Sahel PW ≥36 mm is needed.<br>See Section 3.1.2.1 | The equivalent MD (see Section 11.10) is a diagnostic directly derived from PW. It allows us to monitor the MD and penetration. |

(*Continued*)

Table 11.2 (Continued)

| Parameter/ phenomenon | Application in forecasting | Other uses, and references | Further comments |
|---|---|---|---|
| **4. Humidity profile** | | | |
| Cloud base height | Elevated cloud base favours triggering in the afternoon if there is a well-mixed, deep PBL.<br>Favours intense downdraughts, precipitation evaporation and thus intense cold pools → helps to organise and maintain convection. | Generally anti-correlated with PW (over Africa).<br>Severe winds possible (wind gusts, intense updraughts, in-cloud turbulence).<br>See Section 3.1.2.2 | Not available as a standard product (up to now). |
| Mid-level dryness (700–500 hPa) as measured by $\theta'_{w\,min}$, RH or saturation deficit. | Causes inhibition of updraughts (due to entrainment).<br>Favours intense downdraughts, precipitation evaporation and thus intense cold pools and therefore helps to organise and maintain convection. | $\theta'_{w\,min}$ provides the minimum possible temperature in the cold pools (a lower bound: in practice the cold pool is commonly 10°C warmer than this value).<br>When compared with $\theta$ in the PBL, $\theta'_{w\,min}$ provides cold pool maximum possible temperature drop, and therefore a measure of gust severity.<br>See Section 3.1.2.2 | Downdraught characteristics are highly variable, with some being significantly warmer than the value given by $\theta'_{w\,min}$ |
| Downward CAPE | Favours intense downdraughts, precipitation evaporation and thus intense cold pools → helps to organise and maintain convection. | Similar information as mid-level dryness parameters | Not available (up to now). |
| **5. Wind profile** | | | |
| Shear (surface to 700 hPa) | Favours fast-moving MCSs in the region of the AEJ.<br>Direction of shear vector is crucial and indicates the propagation direction for the whole MCS. | See Section 3.1.2.3 | MISVA provides a useful shear diagnostic combined with MD allowing to detect areas favourable for squall lines.<br>See Figure 11.16b. |
| Mean wind in convective layer. | Provides the speed and direction of individual convective cells. | See Section 3.1.2.3 | Convective layer: cloud base to ~400 hPa. |
| Low-level wind (in the monsoon layer) | Strengthening of the monsoon advection ahead of convection can reinforce the convection. | | |
| Upper-level wind (~200 hPa) | | Determines the direction of anvil flow. | Useful for satellite interpretation. |
| Vertical velocity Velocity potential at 200 hPa | Areas of ascending motion (700 and 400 hPa) are favourable to trigger and intensify convection, whereas subsiding conditions suppress convection.<br>*Caution:* this variable, especially on small scales, is dominated by the convection within the model. In using vertical velocity to forecast convection, you are mostly just following the model convection. | There is more confidence at large scales, when looking at the divergent flow at upper level.<br>A useful parameter is the velocity potential at 200 hPa allowing monitoring and forecasting of large-scale convective areas. | *Caution:* the accuracy of this variable depends directly on the skill of model convective schemes that still have poor skill in the tropics and especially over Africa. |

## Appendix: Acronyms

| | |
|---|---|
| AEJ | African easterly jet |
| AEW | African easterly waves |
| AMMA | African Monsoon and Multidisciplinary Analyses |
| ARPEGE | Operational forecasting global model of Météo-France |
| CAPE | convective available potential energy |
| CIN | convective inhibition energy |
| ECMWF | European Centre for Medium-Range Weather Forecasts |
| HL | heat low |
| IR | infrared |
| ITCZ | intertropical convergence zone |
| ITD | intertropical discontinuity |
| LST | local solar time |
| MCS | mesoscale convective system |
| MD | monsoon depth |
| MISVA | Monitoring and forecasting of Intra Seasonal Variability over Africa |
| MSL | mean sea level |
| MT | monsoon trough |
| NWP | numerical weather prediction |
| PJ | polar jet |
| PV | potential vorticity |
| PVU | potential vorticity unit |
| PW | precipitable water |
| STJ | subtropical jet |
| SYNOP | synoptic observation |
| UTT | upper tropospheric trough |
| WAM | West African monsoon |
| WASA | West African synthetic analysis |
| WASF | West African synthetic forecast |

## Notes

1 http://isv.sedoo.fr/index.

2 From inspection of soundings, MD corresponds approximately to the upper limit of the pseudoadiabatic layer, and the lower limit of the Saharan air layer over the Saharan air layer (i.e. ITD).

## References

Lafore J-P, Asencio N, Bouniol D, *et al.* 2012. Évolution de notre compréhension du système de mousson ouest-africain. *La Météorologie* (No. Spécial AMMA – Octobre): 11–16. http://http://documents.irevues.inist.fr/handle/2042/48127 (accessed 1 August 2016). doi: 10.4267/2042/48127.

Santurette P, Joly A. 2002. ANASYG/PRESYG, Météo-France's new graphical summary of the synoptic situation. *Meteorol. Appl.* **9**: 129–154. doi: 10.1017/S1350482702002013

# Index



*Meteorology of Tropical West Africa: The Forecasters' Handbook*, First Edition. Edited by Douglas J. Parker and Mariane Diop-Kane.

## b

## d

## e

## f



## g

## h

## m

### y



### z

Printed in the USA/Agawam, MA
July 21, 2026

907992.046